NATURAL PHENOMENA

SPORTS APPLICATIONS

OTHER APPLICATIONS

"I highly prefer **Inquiry Into Physics** *over all the other leading texts available at this level."*
—Antonio Pagnamenta,
University of Illinois,
Chicago

"This is my favorite book for this course."
—Jim Hamm,
Big Bend
Community College

INQUIRY INTO PHYSICS
FIFTH EDITION

Vern J. Ostdiek
Donald J. Bord

■ Interact

■ Inspire

■ Inquire

Balancing facts and techniques with mystery and majesty, **Inquiry Into Physics, Fifth Edition**, is the ideal text for students who are encountering physics for the first time. Authors Vern J. Ostdiek and Donald J. Bord emphasize the many ways in which physics connects to our everyday lives—the people, the technology, and the discoveries that have shaped our past and will continue to impact our future. By integrating simple but essential mathematics, the authors continually demonstrate how the conceptual side of physics is interwoven with the quantitative side.

This fifth edition is infused with timely new content while expanding on the authors' successful inquiry approach. In every chapter, readers are encouraged to try things, to discover relationships, and to look for answers in the world around them. This approach, combined with an engaging narrative and an abundance of applied examples, ensures that **Inquiry Into Physics** will reveal for students the everyday practicality—and beauty—of physics.

Your special preview begins on the next page

THOMSON
™
BROOKS/COLE

Encourage exploration

New

Explore It Yourself activities encourage students to try hands-on experiments. Formerly titled ***Do-It-Yourself***, these boxes now involve more exploration and often appear before the associated topics to pique reader interest in the upcoming material. Students are encouraged to draw conclusions based on their observations before proceeding to the text's discussion of the principles of physics involved.

Enhanced

Self-contained ***Physics Potpourri*** sections enrich students' understanding of physics and its applications through the exploration of topics drawn from astronomy, the history of science, biophysics, and other areas. All ***Physics Potpourri*** boxes are updated in this edition, and five are completely new, including "Putting Sound to Work," "Hooke-d!," "Friction: A Sticky Subject," "Fresnel, Pharos, and Physics," and "The Camera Obscura: A Room with a View." Most now conclude with suggested Web resources. Students can use a ***Physics Potpourri*** to begin a short research project or writing assignment.

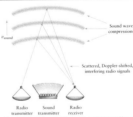

The text's outstanding art program further highlights the everyday applicability of physics. Colorful photographs and illustrations allow students to visualize even the most abstract concepts.

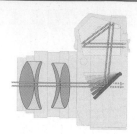

● **Figure 9.17** A single-lens reflex (SLR) camera. Light enters through the lens, strikes the mirror, travels to the prism and then to the photographer's eye. When the shutter is released, the mirror pops up, allowing light to pass directly to the film. The mirror then drops down.

"One-Way Mirror"

A "one-way mirror" is made by partially coating glass so that it reflects some of the light and allows the rest to pass through. This is called a *half-silvered mirror* (● Figure 9.18). When used as a window or wall between two rooms, it will function as a one-way mirror if one of the rooms is brightly lit and the other is dim. It will appear to be an ordinary mirror to anyone in the bright room, but it will appear to be a window to anyone in the dim room. This is because, in the bright room, the light reflected off the half-silvered mirror is much more intense than the light that passes through from the other room. In the dim room, the transmitted light from the bright room dominates (see ● Figure 9.19).

A person in the dim room can see what is happening in the bright room without being seen by anyone in the bright room (see ● Figure 9.20). This device is often used in interview and interrogation rooms and as a means of observing customers in stores and gambling casinos. Note that if a bright light is turned on in the dimmer room, the one-way effect is destroyed. Ordinary window glass is a crude one-way mirror because it does reflect some of the light that strikes it. At night, one can see into a brightly lit room through a window, but anyone in the room has difficulty seeing out.

● **Figure 9.18** Light striking a half-silvered mirror. Part of the light is reflected, and part passes through.

● **Figure 9.19** A half-silvered mirror used as a "one-way mirror" between two rooms. In the room on the left, a person sees mostly reflected light (a mirror). In the room on the right, a person sees mostly transmitted light (a window).

New

This edition contains a new introductory chapter that covers the application of physics, the scientific method, and measurement. The chapter, entitled "Prologue: Getting Started," answers basic questions such as: *What is physics?, How does one do physics?*, and *How does a beginner learn physics?* With this new chapter, students are able to immediately see why the study of physics is a useful—and stimulating—endeavor.

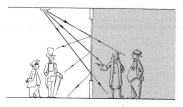

Why Learn Physics?

The answer is easy for those majoring in physics, engineering, or other sciences: Physics will provide them with important tools for their academic and professional lives. The technology that our modern society relies on comes from applying the discoveries of physics and other sciences. From designing safe, efficient passenger jets to producing sophisticated, inexpensive computers, engineers apply physics every day.

Landing astronauts on the Moon and returning them safely to Earth, one of the greatest feats of the twentieth century, is a good example of physics applied on many levels (see ● Figure P.1). The machines involved—from powerful rockets to on-board computers—were designed, developed, and tested by people who knew a lot about physics. The planning of the orbits and the timing of the rocket firings to change orbits involved scientists with a keen understanding of basic physics, like gravity and the "laws of motion." Often, the payoff is not so tangible or immediate as a successful Moon landing. Behind great technological advances are years or even decades of basic research into the properties of matter.

For instance, take a simple portable CD player (see ● Figure P.2). A laser reads data off a spinning disc, integrated circuit chips inside "translate" the digital data into electrical signals, tiny magnets in the headphones help convert those signals into sound, and a liquid crystal display provides information for the user. If you could send this marvelous device back in time to when your grandparents were children, it would astound the greatest physicists and electrical engineers of the day. But even at that time, scientists were studying the properties of semiconductors (the raw material for lasers and integrated circuit chips) and liquid crystals.

For you and others like you, taking perhaps just one physics course in your life, the usefulness of physics is probably not a big reason for studying it. We will see that with even one course, you can use physics to determine, for example, how large a raft has to be to support you or whether using a toaster and a hair dryer at the same time will trip a circuit breaker. But you are not going to make a living with your understanding of physics, nor will you be using it (knowingly) every day. So why should *you* study physics? There are both aesthetic and practical reasons for learning physics. Seeing the order that exists in nature and that it follows a relatively small number of "rules" can be fascinating—similar to learning the inspirations behind a musician's or an artist's work. Learning how common devices operate gives you a better understanding of how to use them and may reduce any frustration you have with them. An elementary knowledge of physics also helps you make more informed decisions regarding important issues facing you, your

● **Figure P.1** It took a lot of physics to get astronaut Neil Armstrong to the Moon.

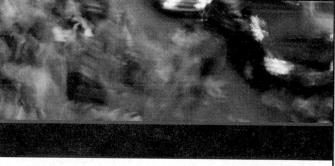

Make connections

New

Chapter openers include **new** flowchart diagrams explaining how chapter concepts integrate with concepts elsewhere in the book. These flow-charts guide students in their concept review. The chapter openers are also more concise in this edition, and most are new or completely revised to include timely new topics, such as SmartPaper, metal detectors, plasma TVs, and PET scans.

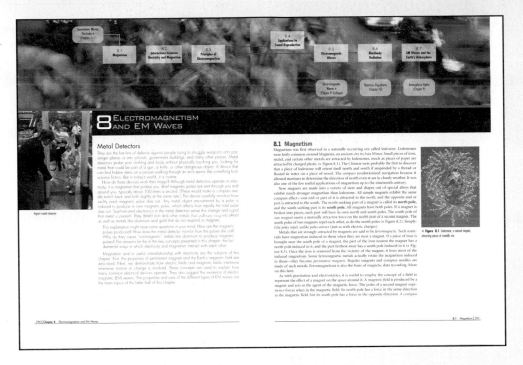

Concept Maps visually summarize abstract concepts, helping students link key concepts, processes, and systems. Between one and three of these maps appear in each chapter.

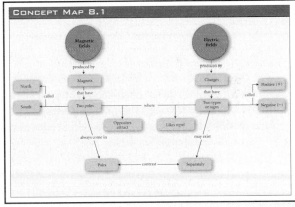

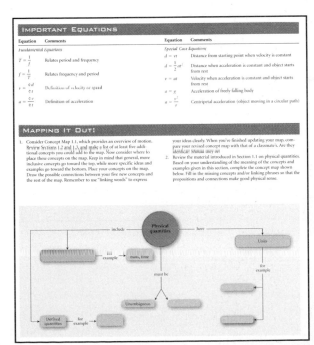

Mapping It Out! exercises help students use concept mapping techniques to improve their understanding of important concepts and relationships introduced in each chapter. These exercises ask students to fill in concepts on a pre-existing map skeleton or to create their own maps.

Preview

We finish this chapter with a brief look at how our method of analyzing motion evolved. The first great strides toward the development of exact sciences were made by the ancient Greeks. Pythagoras, in the sixth century B.C., showed that numbers, often regarded merely as mental abstractions, are related to the natural world and even to human perception. His most famous discoveries were in geometry and acoustics. Other ancients, before the year 200 B.C., accomplished accurate measurements of the diameter of the Earth, the Moon, and the Moon's orbit.

Aristotle

The philosopher Aristotle (384–322 B.C.; ● Figure 1.31) is regarded as the first person to attempt physics. In fact, he gave *physics* its name. Born during the height of ancient Greek civilization, Aristotle became one of the central figures in the explosion of intellectual development that took place in that era. Starting at the age of seventeen, Aristotle was a student of the great philosopher Plato for 20 years. He became the tutor of Alexander the Great and later spent years studying the cultures, flora, and fauna of the exotic lands conquered by Alexander. In 335 B.C., Aristotle opened a school at the Lyceum, near Athens, where he taught until his death. Most of what we know about Aristotle's thoughts and teachings are based on the lectures he gave at the Lyceum.

Aristotle was a master of virtually all of the academic disciplines that existed at the time and actually invented some of them. Psychology is one example. His ideas about physics were influenced a great deal by the methodology that he used in logic, biology, and other areas. Aristotle took the important step of realizing that a number of complicating factors in the physical world often mask hidden order. His analysis of motion, the first mechanics, rested on

● **Figure 1.31** *Aristotle teaching Alexander the Great* (painting by J. L. G. Ferris)

the distinction between *natural motion*, like that of a falling body, and *unnatural motion*, like that of a cart being pulled down a road.

According to Aristotle, heavy objects fall because they are seeking their natural place. In his model, the speed of a falling body is constant and depends on its weight and the medium through which it falls. Heavy objects fall faster than lighter ones. Also, a rock drops faster through air than it does through water (● Figure 1.32). This analysis is only partly correct: It applies only after an object has been falling for a period of time and friction has become important. Speed first increases at a fixed rate—9.8 m/s each second, as we have seen. A marble and a rock dropped at the same time will build up speed together and hit the ground at the same time (● Figure 1.33). Only after a rock has fallen a greater distance will air resistance cause its speed to level off at a constant value. This final speed, called the *terminal speed*, depends as much on the shape and size of the object as it does on the weight. These same factors affect the terminal speeds of bodies falling through water, but such speeds are much slower (as Aristotle pointed out) and are reached much more quickly. We will take a closer look at this in Section 2.6.

Aristotle's description of the motion of falling bodies fits only after their motions are dominated by air resistance. In a similar way, his analysis of "unnatural motion" indirectly includes friction as the dominating influence. A rock's "natural" tendency is to remain at rest on the ground or to fall toward the Earth's center if dropped. Making the rock move horizontally by pushing it is unnatural and requires some external agent or force. Aristotle's view that a force is always required to maintain horizontal motion fits well whenever there is a great deal of friction but not so well in cases like those of a marble thrown horizontally or of a smooth, heavy ball

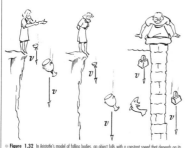

● **Figure 1.32** In Aristotle's model of falling bodies, an object falls with a constant speed that depends on its weight and the medium through which it falls. (Left) A heavy object falls faster than a light object. (Right) An object falls faster through air than through water.

---

rolling over a flat, hard surface. In these situations, motion can continue for quite some time with no force acting to maintain it. Because of Aristotle's overwhelming reputation as a scholar, zealous supporters of his ideas allowed his theories to dominate physics, almost without question, for about 2,000 years.

Galileo

Galileo Galilei (1564–1642; ● Figure 1.34) lived during one of the most fruitful periods of human civilization—the Renaissance. His birth came in the same year as the death of Michelangelo and the birth of Shakespeare. Galileo made many important discoveries in mechanics and astronomy, some of which were disputed bitterly because they contradicted accepted views passed down from Aristotle's time. Galileo's strong support of the heliocentric (Sun-centered) model of the solar system and other factors led to his being placed on trial by leaders of the Inquisition. But this was the age when superstition was giving way to rational thought, when evidence and logical proof were beginning to win out over blind acceptance of doctrines. Though he was censured and forced to publicly recant his support of the heliocentric theory, Galileo's work gained wide acceptance. He is considered one of the founders of modern physical science.

Galileo was one of the first to rely on observation and experimentation. In 1583, he noticed that a lamp hanging from the ceiling of the cathedral at Pisa would swing back and forth with a constant period, even though the length of the arc of its motion decreased. His discovery, that the frequency of a pendulum depends only on its length, is the basis for the pendulum clock.

Another important contribution by Galileo was his insistence that scientific terms,

● **Figure 1.34** *Galileo Galilei*

statements, and analyses be logically consistent. He sought to establish a "scientific method" and believed that, as a first step, the language of science should be unambiguous. To Galileo, mathematics was necessary to help accomplish this task.

Although his astronomical discoveries brought Galileo most of his fame (and controversy), his conclusions about motion are of interest here. Like Aristotle, Galileo recognized that the simple rules that govern motion can be hidden by other phenomena. He realized that the two things that have the largest effect on objects moving on the Earth are gravity and friction. More importantly, he reasoned that friction is often complicated and unpredictable, and that it is best to first focus on systems in which friction doesn't dominate completely.

The physical system that Galileo used to analyze uniform motion and uniformly accelerated motion was a smooth, heavy ball rolling on a smooth, straight surface that can be tilted. If the surface is tilted (an inclined plane), the ball's speed will increase as it rolls down. The opposite occurs when the ball is initially rolling up the plane—its speed decreases. But what if the surface is level? Logically, the ball's speed should neither increase nor decrease but should stay the same (● Figure 1.35). So Galileo reasoned correctly that, if there is no friction, an object moving on a level surface will proceed with constant speed.

Galileo discovered the law of falling bodies, also using a ball rolling on an inclined plane (see ● Figure 1.36). He realized that the speed of a dropped object increases as it falls. The biggest problem he faced was measuring the speed. The only clocks available at

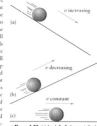

● **Figure 1.35** (a) As a ball rolls down an inclined plane, its speed increases. (b) As a ball rolls up an inclined plane, its speed decreases. (c) When a ball rolls on a level surface, it should therefore have a constant speed.

● **Figure 1.33** Strobe photograph of two falling bodies. Even though one is much heavier than the other, they have the same acceleration.

Each chapter ends with a fascinating look at the human side of physics. These vignettes, called the ***Physics Family Album*** (formerly ***Historical Notes***), trace the lives and work of ground-breaking individuals who discovered and developed the physics concepts presented in the chapter. From Aristotle to Murray Gell-Mann, these sections engage readers in the challenges and achievements of physicists throughout history.

New

InfoTrac® College Edition keywords are now included at the end of most chapters, giving students the opportunity to research and explore chapter topics online. Each new copy of the text is accompanied by four months of free access to this powerful online library. See page 8 for more information.

For additional readings, explore InfoTrac® College Edition, your online library. Go to http://www.infotrac-college.com/wadsworth and use the passcode that came on the card with your book. Try these search terms: blimp, fullerene, carbon nanotubes, aneroid barometer, superfluid, Bernoulli's principle.

Develop understanding

Expanded

Additional *Learning Checks* have been added for consistency, clarity, and a focus on primary concepts. These short, in-chapter self-quizzes occur at the ends of most sections and are written to help students assess their basic comprehension. Approximately 100 *Learning Check* questions are new in this edition.

LEARNING CHECK

1. As a skier gains speed while gliding down a slope, _____ energy is being converted into _____ energy.
2. (True or False.) Work done inside an isolated system can increase the total energy in the system.
3. (Choose the *incorrect* statement.) The total kinetic energy plus potential energy of a body
   a) can be negative.
   b) always remains constant if the body is falling freely.
   c) always remains constant if friction is acting.
   d) can remain constant even if the body's speed is decreasing.
4. Diver A jumps off a platform and is going 5 m/s when entering the water. To be going 10 m/s when entering the water, diver B would have to jump off a platform that is _____ times as high as the platform A used.

ANSWERS: 1. potential, kinetic 2. False 3. (c) 4. four

Enhanced

Questions check the reader's basic knowledge of the material as well as the ability to extend that understanding to new or hypothetical situations. New in the fifth edition— an icon distinguishes between review-type *Questions* and more inquiry-based, conceptual ones. Nearly 100 new *Questions* have been added, most of them conceptual.

Problems, a number of which are new to this edition, are simple mathematical exercises based on realistic applications of the physics developed in each chapter.

QUESTIONS

(▶ Indicates a review question, which means it requires only a basic understanding of the material to answer. The others involve integrating or extending the concepts presented thus far.)

1. Three bar magnets are placed near each other along a line, end to end, on a table. The net magnetic force on the middle one is zero, and its north pole is to the left. Make two sketches showing the possible arrangements of the poles of the other magnets.
2. ▶ Sketch the shape of the magnetic field around a bar magnet.
3. ▶ What happens to a ferromagnetic material when it is placed in a magnetic field?
4. ▶ What causes magnetic declination? Is there a place where the magnetic declination is 180° (a compass points south)? If so, approximately where?
5. ▶ Describe the three basic interactions between electricity and magnetism.
6. ▶ Explain what *superconducting electromagnets* are. What advantages do they have over conventional electromagnets? What disadvantages do they have?
7. ▶ Name five different *basic* devices that use at least one of the electromagnetic interactions.
8. ▶ In many cases, the effect of an electromagnetic interaction is perpendicular to its cause. Describe two different examples that illustrate this.
9. To test whether a material is a superconductor, a scientist decides to make a ring out of the material and then to see whether a current will flow around in the ring with no steady energy input.
   a) Explain how a magnet could be used to initiate the current.
   b) At some later time, how could the scientist check to see whether the current is still flowing in the ring but without touching the ring?
10. ▶ A coil of wire has a large alternating current flowing in it. A piece of aluminum or copper placed near the coil becomes warm even if it does not touch the coil. Explain why.
11. In the particular accelerator shown in Figure 8.16, what are the possible directions of the magnetic field that keeps the particles traveling around the circular path?
12. ▶ What is the "motor-generator duality"? Explain how it is used.
13. Explain why two wires, each with a current flowing in it, exert forces on each other even when they are not touching each other.
14. ▶ Explain why a transformer doesn't work with DC.
15. What would a speaker do at the instant a low-voltage battery is connected to it?
16. If a magnetic tape or floppy disk is placed near a strong magnet, the information on it is erased. Why is that? What happens?
17. ▶ What is *analog-to-digital conversion*, and how is it used in sound reproduction?
18. ▶ Explain how electromagnetic waves are a natural outcome of the principles of electromagnetism.
19. An electromagnetic wave travels in a region of space occupied only by a free electron. Describe the resulting motion of the electron.
20. ▶ List the main types of electromagnetic waves in order of increasing frequency. Give at least one useful application for each type of wave.
21. ▶ Alternating current with a frequency of 1 million Hz flows in a wire. What in particular could be detected traveling outward from the wire?
22. ▶ What are the main uses of microwaves? Explain how each process works.
23. A liquid compound is not heated by microwaves the way water is. What can you conclude about the nature of the compound's molecules?
24. Aircraft equipped with powerful radar units are forbidden from turning them on when on the ground near people. Explain why this is so.
25. ▶ Which type of EM wave does your body emit most strongly?
26. ▶ What is different about our perceptions of the different frequencies within the visible light band of the EM spectrum?
27. A heat lamp is designed to keep food and other things warm. Would it also make a good tanning lamp?
28. ▶ How are x rays produced?
29. ▶ Why are x rays more strongly ab[...] and other tissues?
30. ▶ What is *blackbody radiation*? Ho[...] blackbody change as its temperatur[...]
31. A light bulb manufacturer makes b[...] atures," meaning that the spectrum[...] blackbody with that temperature. W[...] appearance of the light from bulbs [...] K and 4,000 K?
32. Explain how infrared light can be u[...] mals in the dark. Can you think of [...] not work?
33. ▶ What is different about a star tha[...] one that appears bluish?
34. ▶ What effect does the ozone layer [...] Sun? What is currently threatening [...]
35. ▶ Describe the greenhouse effect th[...] atmosphere.
36. ▶ How does the ionosphere affect [...] munications?

PROBLEMS

1. A radio is designed to operate on a 9-V battery or on household current. It contains a transformer that reduces 120 V AC to 9 V AC. What is the ratio of the number of turns in the output coil to the number of turns in the input coil?
2. The generator at a power plant produces AC at 24,000 V. A transformer steps this up to 345,000 V for transmission over power lines. If there are 2,000 turns of wire in the input coil of the transformer, how many turns must there be in the output coil?
3. Compute the wavelength of the carrier wave of your favorite radio station.
4. What is the wavelength of the 76-Hz ELF wave (used to communicate with submarines)?
5. Compute the frequency of an EM wave with a wavelength of 1 in. (0.0254 m).
6. The wavelength of an electromagne[...]
   a) What is the frequency of the wa[...]
   b) What type of EM wave is it?
7. Determine the range of wavelength[...]
8. A piece of iron is heated with a tor[...] How much more energy does it emit as blackbody radiation at 900 K than it does at room temperature, 300 K?
9. The filament of a light bulb goes from a temperature of about 300 K up to about 3,000 K when it is turned on. How many times more radiant energy does it emit when it is on than when it is off?
10. What is the wavelength of the peak of the blackbody radiation curve for the human body ($T = 310$ K)? What type of EM wave is this?

Problems | 331

CHALLENGES

1. The Earth's magnetic field lines are not parallel to its surface except in certain places: They actually "dip" downward at some angle to the ground. (An ordinary compass does not show this.)
   a) At what places on Earth would the "magnetic dip" be the greatest?
   b) Where on the Earth would the dip be zero?
2. A solenoid connected to a 60-Hz AC source will produce an oscillating magnetic field, as we have seen. If a permanent magnet is inserted into the solenoid, it will oscillate, but *not* with the same frequency than an unmagnetized piece of iron would. Why? (This is why the "hum" or "buzz" from electrical devices is sometimes 60-Hz sound and sometimes 120-Hz sound.)
3. ▶ The "right-hand rule" is a way to determine the direction of the magnetic field produced by moving charges. Imagine wrapping your right hand around the path of the charges so that the positive charges (or the current) flow from the "little finger" side of your fist to the thumb side (= Figure 8.55). Then your fingers circle the path in the same direction as the magnetic field lines. Use this rule to verify the directions of the magnetic fields shown in Figures 8.8 and 8.9. How would you use the rule to find the direction of the magnetic field lines around a moving negative charge?

▶ **Figure 8.55**

4. Sketch a diagram of an atom with one electron orbiting the nucleus. Use the right-hand rule (see Challenge 3) to determine the direction of the magnetic field produced by the electron.
5. If a coil of wire is connected to a very sensitive ammeter and then waved about in the air, a current will be induced in it even if there are no magnets around. Why?
6. Even though the output voltage of a transformer can be much larger than the input voltage, the power output is nearly the same as the power input. (There is some energy loss due to ohmic heating.) Use this to determine the relationship between the input and output currents and the number of turns in the input and output coils.
7. The highest frequency sound that can be recorded by a tape recorder depends on the *size* of the gap in the recording head. Why? Would a wider or a narrower gap be capable of recording higher frequencies?
8. Describe what happens to an electric charge as an electromagnetic wave passes through the region around it. Explain why the charge will produce another electromagnetic wave. (This process occurs in the ionosphere.)
9. Would an electromagnetic pump be able to move a liquid superconductor? Why or why not?
10. The nuclei of carbon atoms that are found in nature come in two main "varieties" called *isotopes* (more on this in Chapter 11). The carbon-14 nuclei have a greater mass than the carbon-12 nuclei. The two types of atoms can be separated from each other by ionizing them, accelerating them all to some uniform velocity, and then passing them between the poles of a magnet. Why does this separate the two isotopes? Where does each type go? What would happen if all the atoms did not have the same speed?

Challenges provide more advanced questions and problems that test the student's mastery of the material at a deeper level. Many of these challenge problems can be used as starting points for class discussions.

Also New

End-of-chapter *Summaries* are now structured in an easy-to-read bulleted list, helping students review the major points of each chapter quickly and thoroughly.

Enhance your course

Classroom Preparation/Lecture and Presentation Tools

Instructor's Manual with Test Bank

0-534-49332-7

This manual contains outlines, chapter overviews, learning objectives, teaching suggestions, lecture hints, and an exciting new addition, **InfoTrac® College Edition** activities. The manual also contains an updated **Test Bank** of approximately 120–140 problems and answers (multiple-choice, matching, essays, etc.) per chapter—20 percent of the questions are new to this edition.

CNN® Today Video: Physics

0-534-37089-6

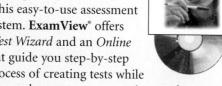

Launch your lectures with footage found on **CNN Today Videos**. Just as the text shows students how physics touches their everyday lives, these videos also demonstrate the relevance of course topics to day-to-day life. Organized by topics covered in a typical course, this video is divided into short segments—perfect for introducing key concepts. High-interest clips are followed by questions designed to spark class discussion.

Book-Specific Instructor's Web Site

http://physics.brookscole.com/ostdiek

The **Brooks/Cole Physics Resource Center** and our password-protected **Book-Specific Instructor's Web Site** give you access to Web links, lecture outlines, a downloadable *Instructor's Manual,* Microsoft® *PowerPoint*® slides, an online image bank with electronic text images for lecture use, and more.

Accessible 24 hours a day, the **Book-Specific Instructor's Web Site** ensures that substantial resources for your course are never more than a mouse click away!

Transparency Acetates

0-534-49334-3

This set of 200 full-color transparency acetates is newly updated to contain many graphs and graphics from the fifth edition.

Assessment and Course Management

ExamView® Computerized Testing

0-534-49331-9

Create, deliver, and customize tests and study guides (both print and online) in minutes with this easy-to-use assessment and tutorial system. **ExamView®** offers both a *Quick Test Wizard* and an *Online Test Wizard* that guide you step-by-step through the process of creating tests while allowing you to see the test you are creating on the screen exactly as it will print or display online. You can build tests of up to 250 questions using up to 12 question types. Using **ExamView's** complete word-processing capabilities, you can enter an unlimited number of new questions or edit existing questions.

WebTutor™ ToolBox on Blackboard and WebCT

Packaged with the text:
Blackboard: 0-534-20920-3 • WebCT: 0-534-20929-7

Available **FREE** with this text, **WebTutor ToolBox** is preloaded with content and available via PIN code. **WebTutor ToolBox** pairs all the content of the text's **Book Companion Web Site** with all the sophisticated course management functionality of a WebCT or Blackboard product. Instructors can assign materials (including online quizzes) and have the results flow automatically to their gradebook. **WebTutor ToolBox** is ready to use as soon as you log on—or you can customize its preloaded content by uploading images and other resources, adding Web links, or creating your own practice materials. Students have access only to student resources, while instructors have the option to access password-protected instructor resources.

Enrich student learning
Content Mastery: Study Aids, Quizzing, and Research

The Brooks/Cole Physics Resource Center and Book Companion Web Site
http://physics.brookscole.com/ostdiek

Our activity-packed and student-focused Web site includes a bounty of learning aids for your students. Available 24 hours a day, this site gives students instant access to valuable resources that will increase their mastery of course material. At the site, students will find a wide variety of chapter-by-chapter resources to help them succeed, including: *Tutorial Quizzes*, an *Online Glossary*, *Flashcards*, links to *Online Tutorial Programs* and *Key Concept Demonstrations*, **InfoTrac® College Edition** periodical research key terms and exercises, *Suggested Readings*, and *Chapter/Topic Updates*.

InfoTrac® College Edition

When you adopt **Inquiry Into Physics, Fifth Edition**, you and your students receive four months of free anytime, anywhere access to the reliable resources found on **InfoTrac College Edition**. This fully searchable online library offers 22 years' worth of full-text articles from almost 5,000 scholarly journals and popular publications such as: *American Scientist, Discover, Physics Review, Science, Scientist, Scientific American,* and thousands more. This incredible depth and breadth of material—available 24 hours a day from any computer with Internet access—makes conducting research so easy, your students will want to use it to enhance their work in every course. Plus, students now also gain instant access to critical thinking and paper writing tools through *InfoWrite*. **NEW—InfoTrac College Edition** search terms are now added to the end of most chapters in the text.

Essential Study Skills for Science Students

0-534-37595-2

by Daniel D. Chiras

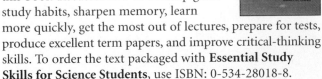

Written specifically for science students, this book discusses how to develop good study habits, sharpen memory, learn more quickly, get the most out of lectures, prepare for tests, produce excellent term papers, and improve critical-thinking skills. To order the text packaged with **Essential Study Skills for Science Students**, use ISBN: 0-534-28018-8.

The World of Physics: Mysteries, Magic, and Myth

0-03-031944-7

by John W. Jewett

Intended for non-science majors taking a liberal arts physics course and for anyone intrigued by the physical world, the mysteries, magic, and myth of the physics of everyday life are revealed in this engaging resource. The book follows the subjects of more traditional physics books, but with a truly enlightening presentation. To order the text packaged with **The World of Physics**, use ISBN: 0-534-28027-7.

A Beginner's Guide to Scientific Method, Third Edition

0-534-58450-0

by Stephen Carey

This concise book provides an introduction to the scientific method of inquiry. This book not only presents a methodical approach to the proper conduct of science but also contains comprehensive coverage of pseudoscience and fallacies. The text assists students in using the scientific method to design and assess experiments. To order the text packaged with **A Beginner's Guide to Scientific Method** at significant savings, use ISBN: 0-534-28009-9.

The Physics ToolBox: A Survival Guide for Introductory Physics

0-03-034652-5

by Kirsten A. Hubbard and Debora M. Katz

The **Physics ToolBox** is a "paperback mentor" specifically created to supplement other materials in the introductory physics course. It is designed to help students get the most out of the course by providing study skills advice, math help, tips on dealing with instructors and classmates, and—best of all—a fresh look at the information in a typical introductory physics course. To order the text packaged with **The Physics ToolBox** at significant savings, use ISBN: 0-534-27822-1.

To my good friends Ruthie, Jeannette, and John W.

Vern Ostdiek

❧

To my teachers Charles G. Lee and Ethel F. Laine, who set me on an intellectual quest that has taken me far beyond my wildest childhood dreams.

Don Bord

Fifth Edition

INQUIRY INTO PHYSICS

Vern J. Ostdiek
Benedictine College

Donald J. Bord
University of Michigan–Dearborn

THOMSON
BROOKS/COLE

Australia • Canada • Mexico • Singapore • Spain
United Kingdom • United States

THOMSON

™

BROOKS/COLE

Physics Acquisitions Editor: *Chris Hall*
Development Editor: *Rebecca Heider*
Editorial Assistant: *Seth Dobrin, Jessica Howard*
Marketing Manager: *Kelley McAllister, Erik Evans*
Marketing Assistant: *Leyla Jowza*
Advertising Project Manager: *Stacey Purviance*
Project Manager, Editorial Production: *Teri Hyde*
Print/Media Buyer: *Rebecca Cross*
Permissions Editor: *Sarah Harkrader*
Production Service: *Lachina Publishing Services*

Text Designer: *Patrick Devine Design*
Copy Editor: *Donald Gecewicz*
Illustrator: *Lachina Publishing Services, Greg Gambino*
Photo Researcher: *Lynne Marie Sanders*
Cover Designer: *Patrick Devine Design*
Cover Image: © *Royalty-Free/CORBIS*
Cover Printer: *The Lehigh Press, Inc.*
Compositor: *Lachina Publishing Services*
Printer: *Quebecor World/Dubuque*

For more information about our products, contact us at:

Thomson Learning Academic Resource Center
1-800-423-0563
For permission to use material from this text, contact us by:
Phone: 1-800-730-2214 **Fax:** 1-800-730-2215
Web: http://www.thomsonrights.com

Library of Congress Control Number: 2003115202

Student Edition: ISBN 0-534-49168-5

Instructor's Edition: ISBN 0-534-49330-0

Brooks/Cole   Thomson Learning
10 Davis Drive
Belmont, CA 94002
USA

Asia
Thomson Learning
60 Albert Street, #15-01
Albert Complex
Singapore 189969

Australia
Nelson Thomson Learning
102 Dodds Street
South Melbourne, Victoria 3205
Australia

Canada
Nelson Thomson Learning
1120 Birchmount Road
Toronto, Ontario M1K 5G4
Canada

Europe/Middle East/Africa
Thomson Learning
Berkshire House
168-173 High Holborn
London WC1V 7AA
United Kingdom

Latin America
Thomson Learning
Seneca, 53
Colonia Polanco
11560 Mexico D.F.
Mexico

Spain
Paraninfo Thomson Learning
Calle/Magallanes, 25
28015 Madrid, Spain

BRIEF CONTENTS

CONTENTS

Welcome to the new edition of *Inquiry Into Physics*. This book is for beginners, students getting their first experience with a course in physics. It is a journey that is likely to fascinate you, because physics connects to us in a variety of ways. It touches our everyday lives through technology like computers and processes like collisions. It satisfies our intellectual curiosity by revealing nature's secrets, even about something far removed from our everyday experience, like the Big Bang. Physics discoveries, like nuclear fission and fusion, have shaped the political landscape of the world and are likely to continue to do so. A look back at the people who made important discoveries connects us to our past, while pointing us toward our future. These aspects of physics are blended together throughout this text. The tools we use to present them are basic: The written word, visual connections via hundreds of photographs, diagrams, and graphs, and simple mathematics for showing how the conceptual side of physics is interwoven with the quantitative side.

In this edition, more than in previous ones, we are emphasizing the inquiry approach: We ask readers to try things, to discover relationships, to look for answers in the world around them, not just in books. In a sense, we are guides on a tour into the realm of physics. We hope you don't just sit on the tour bus the whole time, looking at the sights and listening to our spiel. We urge you to get out of the bus once in a while. Walk around in this world, touch things, pick them up, try stuff yourself. Ignite and focus that curiosity we are all born with. Learning is more fun when you are out there and engaged, and it's more interesting.

What's New?

This new edition includes a number of changes.

- A new chapter, **Prologue: Getting Started,** has been added. It attempts to answer questions like what is physics, how does one do physics, and how does a beginner learn physics.
- The inquiry-based approach has been reemphasized throughout.
- **Table-of-Contents Diagrams** at the beginning of the chapters show how the sections in the chapters are interconnected with each other and with topics elsewhere in the book.
- **Explore It Yourself** activities have been expanded and completely revamped. They are now more exploratory and often appear before the associated topics to motivate the reader about the upcoming material. "Answers," where relevant, appear in an appendix.
- The **Learning Checks** (short, in-chapter quizzes) have been expanded and now occur at the ends of most sections. Nearly 100 of the questions are new.

- The **chapter openers** are shorter, and most are new or completely revised. They focus on topics of current interest and include questions the reader can answer by studying the material in the chapter.
- Five of the **Physics Potpourris** (special-topics essays) are new, and the rest have been updated. Most now conclude with suggested Web resources.
- **Suggested Readings** have been shortened and emphasize recent and/or the most relevant sources of information in print.
- **InfoTrac® College Edition** search terms have been added to the end of most chapters to help students use InfoTrac® for further research into relevant topics.
- The end-of-chapter **Questions** have been "flagged" to distinguish review-type ones from more inquiry-based, conceptual ones. Close to 100 new questions have been added, most of them of the latter type.
- Chapter **Summaries** were redone as bulleted lists to help students review the material easily.
- New, revised and updated material is integrated throughout the book on topics such as kaleidoscopes, metal detectors, plasma TVs, and much more.

What's the Plan?

The basic organization of the text remains unchanged, with topics sequenced in traditional order. Each of the numbered chapters is built around the following features:

Table of Contents Diagrams. Similar to flowcharts, they show how the sections in the chapters are interconnected with each other and with topics elsewhere in the book.

Chapter Openers. These motivate the reader by showing how a topic of current interest involves the ideas presented in the chapter.

Examples. Worked problems illustrating the roles of physics and simple mathematics in real-life situations are included as models for problem solving.

Explore It Yourself activities. These experiments and exercises give students the chance to see and do physics without using specialized equipment or highly sophisticated techniques. Most are placed before the relevant discussion to motivate the reader about the upcoming concepts. "Answers," where relevant, appear in an appendix.

Learning Checks. Simple self-quizzes, designed to test a reader's basic comprehension of the material, appear at the ends of most sections.

Physics Family Album. The final section in each chapter presents a look at the human side of physics by describing the lives and work of some of those who discovered and developed concepts presented in the chapter.

Physics Potpourris. These self-contained explorations of selected topics drawn from astronomy, the history of science, biophysics, and other areas are designed to deepen and enrich the student's understanding of physics and its applications. Here *potpourri* means a miscellaneous collection, but two metaphors based on its other meanings might also be appropriate: An aromatic blend that arouses a sense of mystery and intrigue, or a fine stew consisting of a mixture of hearty ingredients—comfort food—and tangy spices personally selected by the chefs.

Concept Maps. Based on principles developed from educational research, these are visual displays of the relationships between important concepts.

Summary. This is a brief, bulleted review of the major points in the chapter.

Important Equations. An annotated list of the key equations in the chapter is presented for quick reference.

Mapping It Out! These exercises, some intended for group collaboration, are included to help students use concept-mapping techniques to improve their understanding of important concepts and relationships introduced in each chapter.

Questions. These queries check students' basic understanding of the material and their ability to extend that understanding to new or hypothetical situations. An icon is used to distinguish the two types.

Problems. These are simple mathematical exercises based on realistic applications of the physics developed in each chapter.

Challenges. More advanced questions and problems that test a reader's mastery of the material at a deeper level are included for the benefit of highly motivated students. Many of these can be used as starting points for class discussions.

Suggested Readings. At the end of each chapter are a set of suggested readings that may serve as resource material for further investigation (for term papers, special projects, and so on) of many of the subjects and applications introduced. A list of suggested key terms to be used with InfoTrac® is also included.

In the appendixes and back endsheets you will find: **Table of Conversion Factors and Other Information** (including tables of metric prefixes, physical constants, and other often-used data); **Periodic Table of the Elements; Winners of the Nobel Prize in Physics; Math Review;** and **Answers** to the odd-numbered **Problems** and **Challenges** and to the **Explore It Yourself** activities (where appropriate). There is also a list of **Selected Applications** of physics concepts and principles in the front endsheets.

There is more than enough material in the text for a typical one-semester course. However, about 30 of the more specialized sections can be omitted with minimal effect on later material. Specifically, these include the **Physics Family Albums** and sections 2.9, 4.6, 4.7, 5.2, 5.5, 5.7, 6.4, 6.5, 6.6, 8.4, 8.7, 9.5, 9.7, 10.7, 10.8, and 12.5. The table-of-contents diagrams at the beginning of each chapter show more detail about the interdependence of sections.

What Else Is There?

A comprehensive teaching and learning package accompanies this book.

Brooks/Cole Physics Resource Center is Brooks/Cole's Web site for physics, which contains a home page for *Inquiry Into Physics,* fifth edition, at http://physics .brookscole.com/ostdiek. Students can access practice quizzes for every chapter and hyperlinks that relate to each chapter's contents.

InfoTrac® College Edition is an online library available FREE with each copy of *Inquiry Into Physics.* (Due to license restrictions, *InfoTrac College Edition* is available free with the purchase of a new book only to college students in North America.) It gives students access to full-length articles—not simply abstracts—from more than 700 scholarly and popular periodicals, updated daily and dating back as much as four years. Student subscribers receive a personalized account ID that gives them four months of unlimited Internet access—at any hour of the day.

Instructors can find information about the following in the Instructor's edition or at www.brookscole.com: **Instructor's Manual with Test Bank, Color Transparencies, ExamView®,** and **CNN Physics Videos.**

There is also an authors' web site, www.iip5.com, where you will find: annotated links to Web sites and references to recent articles relevant to material in the text, both grouped by chapter; updates to Web addresses given in the text; other features that may evolve with time; and a way to contact us with your suggestions or questions. We hope that it will be useful to both instructors and students.

Who's Responsible?

The creation of this new edition involved the collaboration of many talented professionals. Our biggest thank you goes to Rebecca Heider at Brooks/Cole, who oversaw the revision from start to finish and helped us in countless ways. Others at Brooks/Cole who contributed in various capacities: Chris Hall provided guidance early in the process; Sarah Harkrader handled permissions; Jessica Howard worked on the ancillary package; Teri Hyde was the project manager; Kelley McAllister was the marketing manager; and Sam Subity did the B/C web site. Jeff Lachina, Mandy Hetrick, and others at Lachina Publishing Services produced the final product. Also contributing to the project were Christine Davis and Lynne Sanders. Thanks to all of you.

We wish to express our gratitude to the instructors who contributed to the development of the fifth edition with their comments, corrections, and suggestions. Farhang Amiri, Weber State University, and Richard Rothschild, University of California, San Diego, gave in-depth reviews of certain chapters. Jo Wadehra of Wayne State University did an accuracy review of the entire text. John McClelland at the University of Richmond volunteered useful suggestions. The following individuals participated in a survey about their liberal arts physics course: Howard L. Brooks. DePauw University; Jeffrey Christafferson, Ferris State University; Claire Dewberry, Florida Community College at Jacksonville; David R. Dinsmore, Liberty University; F. Paul Esposito, University of Cincinnati; Johnny Evans, Lee University; Carlos E. Figueroa, Cabrillo College; Robert B. Hallock, University of Massachusetts, Amherst; Jim Hamm, Big Bend Community College; J. F. Hasbun, State University of West Georgia; Kyle Hathcox, Union University; Don Haywood, Front Range Community College; Sunil Labroo, State University of New York–Oneonta; Rizwan Mahmood, Slippery Rock University; Pete Markowitz, Florida International University; Xavier K. Maruyama, Naval Postgraduate School; Devon B. Mason, Albright College; Linda McDonald, North Park University; Antonio Pagnamenta, University of Illinois at Chicago; Ali Piran, Stephen F. Austin State University; John K. Pribram, Bates College; Constantin Rasinariu, Columbia College Chicago; Brian A. Raue, Florida International University; Dan P. Smith, Taylor University; Walther Spjeldvik, Weber State University.

Lastly, we would like to thank those instructors and students who have used *Inquiry Into Physics* over the years. Working on this book allows us to combine our love of physics with our love of teaching. The fact that so many newcomers to the world of physics have used the fruit of our labor as their guide makes the process even more gratifying. Thanks.

Vern Ostdiek
Donald Bord

PROLOGUE: GETTING STARTED

"In a century that will be remembered foremost for its science and technology—in particular for our ability to understand and then harness the forces of the atom and universe—one person clearly stands out as both the greatest mind and paramount icon of our age: The kindly, absent-minded professor whose wild halo of hair, piercing eyes, engaging humanity, and extraordinary brilliance made his face a symbol and his name a synonym for genius, Albert Einstein." (*Time* magazine.)

© Time Life Pictures

In 1999, *Time* magazine, known for naming its annual "person of the year," set out to choose its "person of the century." A daunting task it was, considering the events of the turbulent twentieth century and the inevitable criticism that would come from those favoring someone not named. Would it be a world leader who shaped significant spans of the century—for good or bad—such as Franklin D. Roosevelt (a runner-up for person of the century) or Joseph Stalin (1939 and 1942 person of the year)? Perhaps a military figure like Dwight D. Eisenhower (1944 person of the year) or a spiritual leader like Pope John Paul II (1994 person of the year)? Would it be a champion of peace or justice such as Mahatma Gandhi (the other runner-up) or Martin Luther King, Jr. (1963 person of the year)? In the end, the selection was someone with enormous name recognition, but whose work most people would confess ignorance about, the physicist Albert Einstein.

Einstein was chosen because he symbolized the great strides made during the 1900s in deciphering and harnessing fundamental aspects of the material universe. But his style, his manner, and his allure had something to do with it as well. At times, he dominated physics with spectacular results, in a way that is reminiscent of certain "athletes of the 20th century" (Ali, Gretzky, Jordan, Montana, Navratilova, Ruth . . .). Like them, only his last name is needed to identify him. In 1905, while working as a civil servant far removed from the great centers of physics research, Einstein had three scientific papers—about three different subjects—published in a German physics journal. They were so extraordinary that any one would likely have led to his receiving the Nobel Prize in physics—then, as now, the highest award in the field.

Had *Time* magazine been in business during previous centuries, the editors might well have honored other physicists in the same way, perhaps Galileo or Newton for the seventeenth century or Maxwell for the nineteenth. Such is the high regard that Western civilization has for the field of physics and those who excel in it. Partly, it is the impact their discoveries often have on our lives, by way of technological gadgets or civilization-threatening weaponry. But often it is the intellectual resonance we have with the revolutionary insights they give us about the universe (it is the Earth that moves around the Sun, it is matter being converted into energy that makes the Sun shine . . .).

Welcome to the world of physics. You are embarking on an introduction to a field that continues to fascinate people in all walks of life. Tell a friend or family member that you are now studying physics. That will quite likely impress them in a way that most other subjects would not. Whether it should do so is an interesting question. We hope that when you finish this endeavor, you will answer yes.

Why Learn Physics?

The answer is easy for those majoring in physics, engineering, or other sciences: Physics will provide them with important tools for their academic and professional lives. The technology that our modern society relies on comes from applying the discoveries of physics and other sciences. From designing safe, efficient passenger jets to producing sophisticated, inexpensive computers, engineers apply physics every day.

Landing astronauts on the Moon and returning them safely to Earth, one of the greatest feats of the twentieth century, is a good example of physics applied on many levels (see ● Figure P.1). The machines involved—from powerful rockets to on-board computers— were designed, developed, and tested by people who knew a lot about physics. The planning of the orbits and the timing of the rocket firings to change orbits involved scientists with a keen understanding of basic physics, like gravity and the "laws of motion." Often, the pay-off is not so tangible or immediate as a successful Moon landing. Behind great technological advances are years or even decades of basic research into the properties of matter.

For instance, take a simple portable CD player (see ● Figure P.2). A laser reads data off a spinning disc, integrated circuit chips inside "translate" the digital data into electrical signals, tiny magnets in the headphones help convert those signals into sound, and a liquid crystal display provides information for the user. If you could send this marvelous device back in time to when your grandparents were children, it would astound the greatest physicists and electrical engineers of the day. But even at that time, scientists were studying the properties of semiconductors (the raw material for lasers and integrated circuit chips) and liquid crystals.

For you and others like you, taking perhaps just one physics course in your life, the usefulness of physics is probably not a big reason for studying it. We will see that with even one course, you can use physics to determine, for example, how large a raft has to be to support you or whether using a toaster and a hair dryer at the same time will trip a circuit breaker. But you are not going to make a living with your understanding of physics, nor will you be using it (knowingly) every day. So why should *you* study physics? There are both aesthetic and practical reasons for learning physics. Seeing the order that exists in nature and that it follows a relatively small number of "rules" can be fascinating—similar to learning the inspirations behind a musician's or an artist's work. Learning how common devices operate gives you a better understanding of how to use them and may reduce any frustration you have with them. An elementary knowledge of physics also helps you make more informed decisions regarding important issues facing you, your

● **Figure P.1** It took a lot of physics to get astronaut Neil Armstrong to the Moon.

● **Figure P.2** Portable CD player: one product of decades of research in physics.

community, your nation, and the world. As you progress through this book, keep track of events through the news media of your choice. You may be surprised at just how often physics is in the news—directly or indirectly.

If you start this excursion into the world of physics with a sense of curiosity and a thirst for knowledge, you won't be disappointed. And you will have the two most important characteristics needed to make the endeavor both easy and successful. Learning how sunlight and raindrops combine to make a rainbow will deepen your appreciation of its beauty. Knowing about centripetal force will help you understand why ice or gravel on a curved road is dangerous. Learning the basics of nuclear physics will help you understand the danger of radon gas and the promise of nuclear fusion. Knowing the principles behind stereo speakers, aircraft altimeters, refrigerators, lasers, microwave ovens, guitars, and Polaroid sunglasses will give you a better appreciation of what they do.

Throughout this book, we encourage the reader to be inquisitive. Just memorizing definitions and facts doesn't lead you to a real understanding of a subject, any more than memorizing a manual on playing soccer means you can jump into a game and do well. You have to practice, try things, think of situations and how events would evolve, and so on. So it is with physics. Being able to recite Newton's third law of motion is good, but understanding what it means and how it works in the real world is what's really important.

Often, we pose questions or ask the reader to try something, so that when you realize what the answer or outcome is you will have truly learned something. You might think of it as "learning by inquiry." The Explore It Yourself activities are particularly designed for this purpose. Many of the questions at the ends of the chapters are inquiry-based. Once you get used to this method of learning, you will find that you will learn the material faster and more deeply than before.

What Is Physics?

Because physics is one of the basic sciences, it is important to have an idea of just what science is. Science is the *process* of seeking and applying knowledge about our universe. Science also refers to the *body of knowledge* about the universe that has been amassed by humankind. Pursuing knowledge for its own sake is pure or basic science; developing ways to use this knowledge is applied science. Astronomy is mostly a pure science, while engineering fields are applied science. The material we present in this book is a combination of fundamental concepts that we believe are important to know for their own sake and of examples of the many ways that these concepts are applied in the world around us.

There are other ways to classify the different areas of science besides pure and applied. There are physical sciences (physics and geology are just two examples), life sciences (biology and medicine), and social sciences (psychology and sociology). As with most such schemes, there are overlaps: The subfield of biophysics is a good example.

Physics is not as easy to define as some areas of science like biology, the study of living organisms. If you ask a dozen physicists to define the term, you are not likely to get two answers exactly alike. One suitable definition is that physics is the study of the fundamental structures and interactions in the physical universe. In this book, you will find much about the structures of things like atoms and nuclei, along with close looks at how things interact by way of gravity, electricity, magnetism, and so on. Within physics, there is a wide range of divisions. ● Table P.1 lists some of the common areas based on one measure of research activity. There is a lot of overlap between the divisions, and some of them are clearly allied with other sciences like biology and chemistry.

The field of physics is divided differently when the basics are being taught to beginners. The topics presented to students in their first exposure to physics are usually ordered according to their historical development (study of motion first, elementary particles and cosmology last). This ordering also approximates the ranking of areas by our everyday experience with them. We've all watched people in motion and things collide, but few people encounter the idea of quarks before taking a physics class—even though we and all of the stuff we deal with are mainly composed of quarks.

The vast majority of students who take an introductory course in physics are not majoring in it. Most of those who do earn a degree in physics find employment in busi-

Area	Topics of Investigation
● Table P.1 Some commonly identified divisions of physics, ranked roughly by number of doctorates earned each year. (Based on information from the American Institute of Physics.)	
1. Condensed Matter	structures and properties of solids and liquids
2. Astronomy and Astrophysics	stars, galaxies, evolution of the universe
3. Particles and Fields	fundamental particles and fields, high-energy accelerators
4. Optics and Photonics	light, laser technology
5. Atomic and Molecular	atoms, molecules
6. Nuclear Physics	nuclei, nuclear matter, quarks and gluons
7. Plasma and Fusion Physics	plasmas, fusion research
8. Materials Science	applications of condensed-matter physics
9. Biophysics	physics of biological phenomena
10. Atmospheric and Space Physics	meteorology, Sun, planets

ness, industry, government, or education. Data from late 1990s indicate that the majority of individuals with bachelor's and master's degrees are employed in the first two areas, whereas those with doctorates were mainly found in the last two. In addition to the expected occupations like researcher and teacher, people with physics degrees also have job titles like engineer, manager, computer scientist, and technician. Often, physicists are hired not so much for their knowledge of physics but for their experience with problem solving and advanced technology.

How Is Physics Done?

So how does one "do" physics, or science in general? How did humankind come by this mountain of scientific knowledge that has been amassed over the ages? A blueprint exists for scientific investigation that makes an interesting starting point for answering these questions. It is at best an oversimplification of how scientists operate. Perhaps we should regard it as a game plan that is quickly modified when the action starts. It is called the **scientific method.** One version of it goes something like this: Careful *observation* of a phenomenon induces an investigator to question its cause. A *hypothesis* is formed that purports to explain the observation. The scientist devises an *experiment* that will test this hypothesis, hoping to show that it is correct—at least in one case—or that it is incorrect. The outcome of the experiment often raises more questions that lead to a *modification* of the hypothesis and further experimentation. Eventually, an accepted hypothesis that has been verified by different experiments can be elevated to a *theory* or a *law*. The term that is used—*theory* or *law*—is not particularly important in physics: Physicists hold Newton's second *law* of motion and Einstein's special *theory* of relativity in roughly the same regard in terms of their validity and their importance.

One nice thing about the scientific method is that it is a logical procedure that is practiced by nearly everyone from time to time (see ● Figure P.3). Let's say that you get into your car and find that it won't start (observation). You speculate that maybe the battery is dead (hypothesis). To see if this is true, you turn on the radio or the lights to see if they work (experiment). If they don't, you may look for someone to give you a jump start. If they do, you probably guess that something else, such as the starter, is causing the problem. Clearly, a good mechanic must be proficient at this way of investigating things. A doctor making a medical diagnosis uses similar procedures.

Does the outline of the scientific method appear in some "how-to" manual for scientific discovery? Are scientists required to take an oath to follow it faithfully every day at work? Of course not. But the individual elements of the method are essential tools of the scientist. They are useful to students as well. In the Explore It Yourself activities found throughout this book, we ask you to try things (experiment) and then draw conclusions based on the outcomes. Understanding how nature works based on what you do and observe is a great way to learn.

● **Figure P.3** The basics of the scientific method, with an example.

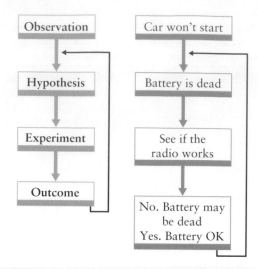

One of the architects of the scientific method was Galileo Galilei (1564–1642). He thought a great deal about how science should be done and applied his ideas to his study of motion. Galileo believed that science had to have a strong logical basis that included precise definitions of terms and a mathematical structure with which to express relationships. He introduced the use of controlled experiments, which he used with great success in his studies of how objects fall. By ingenious experimental design, he overcame the limitations of the crude timing devices that existed and measured the acceleration of falling bodies. Galileo was a shrewd observer of natural phenomena, from the swinging of a pendulum to the moons orbiting Jupiter. By drawing logical conclusions from what he saw, he demonstrated that rules could be used to predict and explain natural phenomena that had long seemed mysterious or magical. We will take a closer look at Galileo's work and life in Chapter 1.

The scientific method is important, particularly in the day-to-day process of filling in the details about a phenomenon being studied. But it is not the whole story. Even a brief look at the history of physics reveals that there is no simple recipe that scientists have followed to lead them to breakthroughs. Some great discoveries were made by traditional physicists working in labs, proceeding in a "scientific method" kind of way. Usually, it's not that simple. Other factors often come into play, such as accidents (Galvani discovered electric currents while performing biology experiments) or luck (Becquerel stumbled upon nuclear radiation because of a string of cloudy days in Paris). Sometimes "thought experiments" were required because the technology of the time didn't allow "real" experiments. Newton predicted artificial satellites, and Einstein unlocked relativity this way. Often, it was hobbyists, not professional scientists, who made discoveries (the statesman Benjamin Franklin and schoolteacher Georg Simon Ohm). Sometimes, it is scientists correctly interpreting the results of others who failed to "connect the dots" themselves (Lise Meitner and her nephew Otto Frisch identified nuclear fission that way). One of the best ways to learn about the nature of scientific discovery is to study the past. Throughout this book, the Physics Family Album sections, along with a few of the Physics Potpourris, give you some idea of the variety of ways that discoveries have been made and glimpses of the personalities of the discoverers.

How Does One Learn Physics?

The goal of learning physics and any other science is to gain a better understanding of the universe and the things in it. We generally consider only a small segment of the universe at one time, so that its structure and the interactions within it are manageable. We call this a *system*. Some examples of systems that we will talk about are the nucleus of an atom, the atom itself, a collection of atoms inside a laser, air circulating in a room, a rock moving near the Earth's surface, and the Earth with satellites in orbit about it (see ● Figure P.4).

The kinds of things a person might want to know about a system include: (1) its structure or configuration; (2) what is going on in it and why; and (3) what will happen in it in

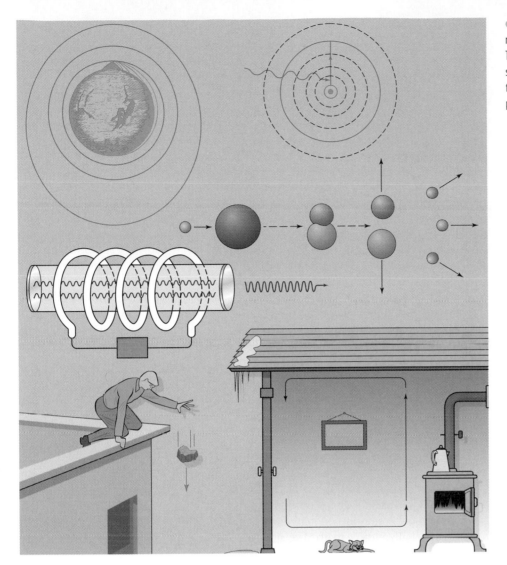

● **Figure P.4** Combination of six figures representing different systems we will be examining. The scale of the different parts varies greatly, from smaller than can be seen with a microscope to thousands of miles. See if you can connect each part to the system described in the text.

the future. The first step is easy—identifying the objects in the system. Protons, electrons, chromium atoms, heated air, a rock, and the Moon are some of the things in the systems mentioned above. Often, the items in a physical system are already familiar to us. A host of other intangible things in a system must also be identified and labeled before real physics can begin. We must define things like the *speed* of the rock, the *density* of the air, the *energy* of the atoms in the laser tube, and the *angular momentum* of a satellite to understand what is going on in a system and where it is headed. We will call these things and others like them **physical quantities.** Most will be unfamiliar to you unless you have studied physics before. Together with the named objects, they form what can be called the *vocabulary of physics*. There are hundreds of physical quantities in regular use in the various fields of physics, but for our purposes in this book, we will need only a fraction of these.

Physics seeks to discover the basic ways in which things interact. Laws and principles express relationships that exist between physical quantities. For example, the law of fluid pressure expresses how the pressure at some location in a fluid depends on the weight of the fluid above. This law can be used to find the water pressure on a submerged submarine or the air pressure on a person's chest. These "rules" are used to understand the interactions in a system and to predict how the system will change with time in the future. The laws and principles themselves were formed after repeated, careful observations of countless systems by scientists throughout history. They withstood the test of time and repeated experimentation before being elevated to this status. You might regard physics as the continued search for, and the application of, basic rules that govern the interactions in the universe.

The process of learning physics has two main thrusts: You need to develop an understanding of the different physical quantities used in each area (establish a vocabulary), and you must grasp the significance of the laws and principles that express the relationships among these physical quantities. Let us caution you again: Memorizing the definitions and laws is only a first step—that alone won't do it. See, do, think, interact, visualize. Get involved in the physical world. That's how you learn physics. This book includes dozens of Explore It Yourself activities and worked-out examples based on real-world situations to help you in this process.

Another tool we use to help you visualize the relationships in physics is the *concept map*. Concept maps were developed in the 1960s and are used in a wide range of fields in a variety of ways. A concept map presents an overview of issues, examples, concepts, and skills in the form of a set of interconnected *propositions*. Two or more concepts joined by linking words or phrases make a proposition. The meaning of any particular concept is the sum of all the links that contain the concept. To "read" a concept map, start at the top with the most general concepts, and work your way down to the more specific items and examples at the bottom. Concept Map P.1 is one example. It is used to show some of the connections between the general concept, science, and one branch of physics, biophysics. This particular concept map could easily be expanded by, for example, showing all of the social sciences or all of the branches of physics.

In this book, each chapter contains concept maps designed as summaries to help you organize the ideas, facts, and applications of physics. You should understand that there are many possible maps that could be constructed from a given set of concepts. The maps drawn in this book represent one way of organizing and understanding a particular set of concepts.

You will have opportunities at the end of each chapter to develop lists of important concepts and to construct from them your own concept maps. Most people find it easier to understand relationships if they are displayed visually. You should find that the process of completing a concept map yourself gives you deeper insights into the ideas that are involved.

One of the main reasons physics has been so successful is that it harnesses the power of mathematics in useful ways. Many of the most important relationships involving physical quantities are best expressed mathematically. Predictions about the future conditions in a system usually involve math. The successes of Einstein, Newton, and others largely

CONCEPT MAP P.1

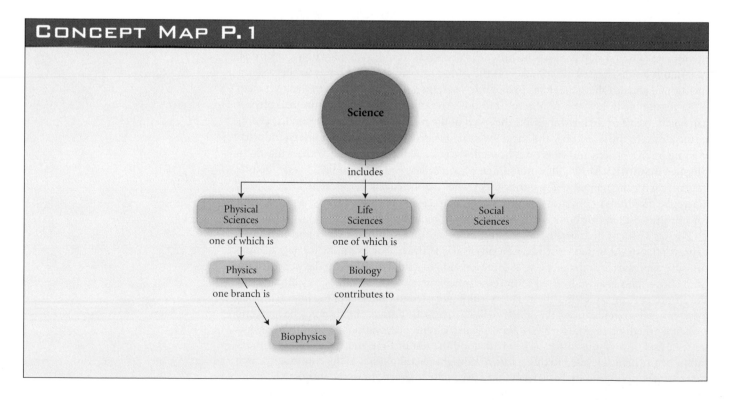

came about because they used mathematics to predict or explain things that no one could before (moving clocks appear to run slow, tides are caused by the Moon's gravity). An essential part of learning physics is developing an understanding of, and an appreciation for, this powerful side of physics.

The good news for beginners is that the simplest of mathematics—what most of us learned before age 16 or so—is all that is needed for this purpose. So hand in hand with the conceptual side of physics, we give you a taste of the mathematical side, through worked-out examples and end-of-chapter problems. Over the years, we have found that even the most math-wary students often become very comfortable with this aspect of the material. An added benefit of your excursion into the world of physics is that you are likely to emerge with a better feel for the usefulness of simple mathematics.

Physical Quantities and Measurement

To be useful in physics, physical quantities must satisfy some conditions. A physical quantity must be *unambiguous*, its meaning clear and universally accepted. Understanding the meaning of a term involves more than just memorizing the words in its definition. Words can only describe something. To understand a concept, you must go beyond words. For example, the simple definition "speed equals distance divided by time" does not really convey the conceptual understanding of speed that we have. Many physical quantities (speed, pressure, power, density, and others) can be defined by an equation. Mathematical statements tend to be more precise than ones in words, making the meanings of these terms clearer.

Observation yields *qualitative* information about a system. Measurement yields *quantitative* information, which is central in any science that strives for exactness. Consequently, physical quantities must be measurable, directly or indirectly. One must be able to assign a numerical value that represents the amount of a quantity that is present. It is easy to visualize a measurement of distance, area, or even speed, but other quantities, like pressure, voltage, or power, are a bit more abstract. Each of these can still be measured in prescribed ways. They would be useless if this were not so.

The basic act of measuring is one of comparison. To measure the height of a person, for instance, one would compare the distance from the floor to the top of the person's head with some chosen standard length, such as a foot or a meter (● Figure P.5). The height of the person is the number of units 1 foot or 1 meter long (including fractions) that have to be put together to equal that distance. The **unit of measure** is the standard used in the measurement—the foot or meter in this case. A complete measurement of a physical quantity, then, consists of a number and a unit of measure. For example, a person's height might be expressed as

$$\text{height} = 5.75 \text{ feet} \qquad \text{or} \qquad h = 5.75 \text{ ft}$$

Here h represents the quantity (height) and ft the unit of measure (feet). The same height in meters is

$$h = 1.75 \text{ m}$$

So, when we introduce a physical quantity into our physics vocabulary—another "tool," so to speak—we must specify more than just a verbal definition. We should also give a mathematical definition (if possible), relate it to other familiar physical quantities, and include the appropriate units of measure.

In the world today, there are two common systems of measure. The United States uses the **English system,** and the rest of the world, for the most part, uses the **metric system.** An attempt has been made in the United States to switch completely to the metric system, but so far it has not succeeded. The metric system has been used by scientists for quite some time, and we will use it a great deal in this book. It is a convenient system to use because the different units for each physical quantity are related by powers of 10. For example, a kilometer equals 1,000 meters, and a millimeter equals 0.001 meter. The prefix itself designates the power of 10. *Kilo-* means 1,000, *centi-* means 0.01 or $\frac{1}{100}$, and *milli-* means 0.001 or $\frac{1}{1000}$. A *kilometer*, then, is 1,000 meters. ● Table P.2 illustrates the common

● **Figure P.5** Measurement is an act of comparison. A person's height is measured by comparison with the length of a chosen standard. In this case, height equals five 1-foot lengths plus a segment 0.75 feet long. The same person's height is also equal to 1 meter plus 0.75 meter.

1 foot

1 meter

metric prefixes. You may not know what an *ampere* is, but you should see immediately that a *milliampere* is one thousandth of an ampere.

The Physics Potpourri *The Metric System:* "For All People, for All Time" gives a brief look at how the metric system came about. About two dozen of these features appear throughout the book. They are intended to give you a deeper, richer view of selected topics in the history and applications of physics.

Having to use two systems of units is like living near the border between two countries and having to deal with two systems of currency. Most people who grew up in the United States have a better feel for the size of English-system units like feet, miles per hour, and pounds than for metric-system units like meters, kilometers per hour, and newtons. Often, the examples in this book will use units from both systems, so that you can compare them and develop a sense of the sizes of the metric units. A table relating the units in the two systems is included in the back inside cover. Fortunately we won't have to deal with two systems of units after we reach electricity (Chapter 7).

A prologue is an introductory development. This prologue is an introduction to the field of physics, our approach to teaching it, and how to get started learning it. The groundwork has been laid, and we are now ready to proceed.

● **Table P.2** Common Metric Prefixes and Their Equivalents	
1 *centi*meter = 0.01 meters	1 meter = 100 centimeters
1 *milli*meter = 0.001 meters	1 meter = 1,000 millimeters
1 *kilo*meter = 1,000 meters	1 meter = 0.001 kilometers
EXAMPLES	
189 centimeters = 1.89 meters	72.39 meters = 7,239 centimeters
25 millimeters = 0.025 meters	0.24 meters = 240 millimeters
7.68 kilometers = 7,680 meters	23.4 meters = 0.0234 kilometers
(There is a more complete list of metric prefixes on the inside back cover.)	

The Metric System: "For All People, for All Time."

The French Revolution, beginning with the storming of the Bastille on 14 July 1789, gave birth not only to a new republic but also to a new system of weights and measures. Eighteenth-century France's system of weights and measures had fallen into a chaotic state, with unit names that were confusing or superfluous and standards that differed from one region to another. Seizing upon the opportunity presented to them by the political and social turmoil accompanying the revolution, scientists and merchants, under the leadership of Charles-Maurice de Talleyrand, presented a plan to the French National Assembly in 1790 to unify the system. The plan proposed two changes: (1) the establishment of a decimal system of measurement and (2) the adoption of a "natural" scale of length. Neither of these two notions was new to scholars of this period. The first had been discussed as early as 1585 by Simon Stevin, a hydraulic engineer in Holland, in a pamphlet called *La Disme* (i.e., *The Tenth Part*). The second notion was introduced in 1670 by Abbé Gabriel Mouton, who proposed that a standard of length be defined in terms of a fraction of the length of the meridian arc extending from the North Pole to the equator.

The plan was finally adopted into law on 7 April 1795. The new legislation defined the meter as the measure of length equal to 1 ten-millionth of the meridian arc passing through Paris from the North Pole to the equator and the gram as the mass of pure water contained in a cube one-hundredth of a meter (a *centi*meter) on a side at the temperature of melting ice. It also made this system obligatory in France.

The tasks of actually determining the sizes of these newly defined units were assigned to Jean-Baptiste Delambre and Pierre Méchain, who were to survey the length of the meridian arc through Paris, and to Louis Lefèvre-Gineau and Giovanni Fabbroni, who were to determine the absolute weight of water. As it turned out, these measurements were not made without difficulty and, sometimes, danger. For example, during the period between 1792 and 1798 Delambre and Méchain made measurements along the meridian between Dunkirk, France, and Barcelona, Spain, amid the riot and turmoil still present in many parts of Europe. They

● **Figure P.7** Many countries have issued stamps commemorating the metric system. These two, from Australia, highlight conversions between the metric and English units for length and volume.

were frequently arrested as spies, often had their equipment confiscated, and were generally harassed at every turn. Finally in 1798, with the job done and the length of the centimeter accurately known, Lefèvre-Gineau and Fabbroni set about their work. They, too, encountered difficulties, largely having to do with reaching and maintaining the required measurement temperature, but they completed their task in only one year.

Beginning in 1798, an international committee with representatives from nine nations undertook to carry out the calculations required to produce the standards needed to define and extend the new system of weights and measures. It submitted its report to the French legislature for ratification on 27 June 1799, and the bill passed on 10 December 1799. That document is the first official text in which the metric system is mentioned. According to this law, the definitive standards of length and mass to be used in commercial and scientific interactions throughout France were "the meter and kilogram of platinum deposited with the legislative body" (see ● Figure P.6).

Since then, the definitions of the standards of length and mass have undergone several revisions, and other units of measure have been incorporated. Nevertheless, the basic tenets of the metric system—simplicity and convenience stemming from its use of a decimal system of measure, and uniformity and reproducibility deriving from its reliance on a set of standards—survive. Since its adoption in Europe, first in France, then in Holland (1816) and Greece (1836), the use of the metric system has proliferated so that no nation is without knowledge of it (● Figure P.7). The metric system has become, in the motto adopted by its founders, a system "for all people, for all time."

The United States is one of only a few countries that have not officially adopted the metric system for their manufacturing and commercial activities, despite its arguable merits. What are your opinions and views on this matter? What do you believe would be the benefits of converting our economy to the metric system? What do you see as the downsides to such a transformation? Check out **InfoTrac® College Edition** (key words "metric system") or other on-line resources for further information on this subject and to see what the experts have to say on this issue.

Courtesy National Institute of Standards and Technology

● **Figure P.6** The international 1 kg standard of mass, a platinum–iridium cylinder.

© Vern Ostdiek

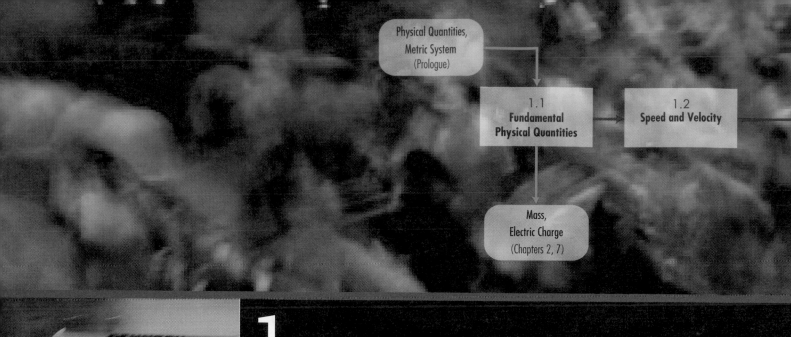

Physical Quantities,
Metric System
(Prologue)

1.1
**Fundamental
Physical Quantities**

1.2
Speed and Velocity

Mass,
Electric Charge
(Chapters 2, 7)

1 THE STUDY OF MOTION

Drag racer at the starting line.

Drag Racing

Acceleration, pure and simple. That's the point of drag racing. Increase the speed of the dragster as quickly as possible so it will travel the one-quarter of a mile (about 400 meters) faster than your opponent. The elite machines have engines that are much more powerful than those on the largest airplanes in World War II (about 5,000 horsepower). They go from 0 to over 300 mph in less than 5 seconds. At the end of their brief run, another critical feat must be accomplished: They have to be brought to a stop safely. How do you slow a vehicle that is going much faster than a passenger jet when it lands? You deploy a parachute.

Imagine driving such a machine: Five seconds of forward acceleration pressing you to the back of your seat, followed by many more seconds of acceleration in the opposite direction—backward—that strains the straps holding you in. And for much of that time you are going hundreds of miles per hour. The faint of heart need not apply for this job.

Much of this chapter is an introduction to the study of motion. One of the key concepts in this area of physics is acceleration, which is used to describe changes in motion. Drag racing is one of the most dramatic examples of human beings deliberately accelerating themselves. The attendance at these events suggests that many people are thrilled to watch it happen.

Two types of acceleration are involved in drag racing, speeding up and slowing down. Is there another type of acceleration? Can a car accelerate without changing its speed? Read on to find out.

In this first chapter, we build on the groundwork for the study of physics that was begun in the Prologue. Most of this chapter is an introduction to the branch of physics called *mechanics*—the study of motion and its causes. The main topic of this chapter is motion and how it is described using the concepts of speed, velocity, and acceleration. Both examples and graphs are used to illustrate and define these three concepts. An important example treated in detail is the motion of a body undergoing free fall. We conclude with a brief account of the work of two men who made important contributions to the development of mechanics.

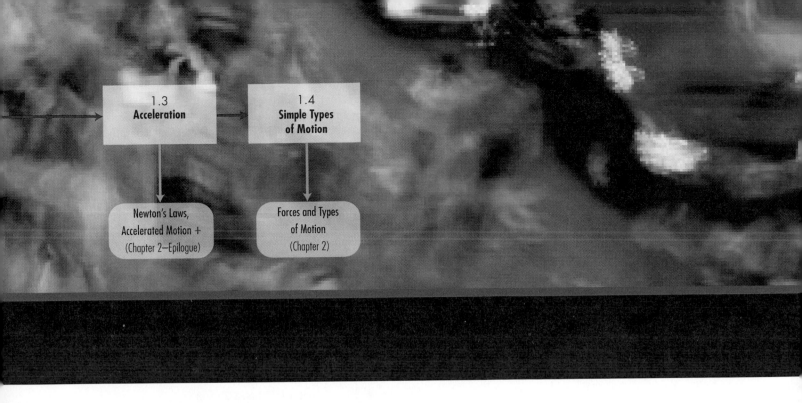

1.1 Fundamental Physical Quantities

The next phase of our journey into physics involves a closer look at motion—what it is, how one quantifies it, what the simplest kinds of motion are, and so on. Fortunately for beginners, most of the concepts and terms are familiar because motion is a part of our everyday world. We have a sense of how fast 60 miles per hour (or 97 kilometers per hour) is and how to compute speed if we know the distance traveled and the time elapsed. The unit of measure pretty much tells us how to do that: "Miles per hour" means divide the distance in miles by the time in hours. But there are some important subtleties that must be examined, and our basic idea of acceleration needs to be expanded. The ways that we will express relationships throughout the text, oftentimes involving things not so familiar as motion, will be presented. This is also a perfect time to take a closer look at those "physical quantities" introduced in the Prologue.

In physics, there are three basic aspects of the material universe that we must describe and quantify in various ways: *space, time,* and *matter.* All physical quantities used in this textbook involve measurements (or combinations of measurements) of space, time, and the properties of matter. The units of measure of all of these quantities can be traced back to the units of measure of distance, time, and two properties of matter called *mass* and *charge.* We will not deal with charge until Chapter 7.

Distance, time, and mass, known as the **fundamental physical quantities,** are such basic concepts that it is difficult to define them, particularly time. **Distance represents a** measure of space in one dimension. Length, width, and height are examples of distance measurements. The following table lists the common distance units of measure, in both the metric system and the English system, and their abbreviations. (You might want to review the basics of the metric system presented in the Prologue.) This same format will be used for all physical quantities that have several common units.

Physical Quantity	Metric Units	English Units
Distance d (or l, w, h)	meter (m)	foot (ft)
	millimeter (mm)	inch (in.)
	kilometer (km)	mile (mi)
	centimeter (cm)	

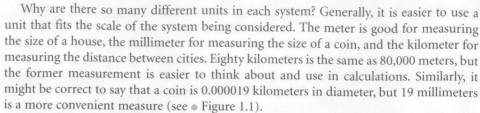

19 mm

0.000019 km

● **Figure 1.1** The two measurements represent the same distance, but the one in millimeters is more convenient to use and visualize.

Why are there so many different units in each system? Generally, it is easier to use a unit that fits the scale of the system being considered. The meter is good for measuring the size of a house, the millimeter for measuring the size of a coin, and the kilometer for measuring the distance between cities. Eighty kilometers is the same as 80,000 meters, but the former measurement is easier to think about and use in calculations. Similarly, it might be correct to say that a coin is 0.000019 kilometers in diameter, but 19 millimeters is a more convenient measure (see ● Figure 1.1).

The sizes of all of the distance units, including the English units, are defined in relation to the meter. In this book, we will use meters most often in our examples. You should try to get used to distance measurements expressed in meters and have an idea, for example, of how long 25 and 0.2 meters are. ● Table 1.1 shows some representative distances expressed in metric and in English units.*

It is a rather simple matter to convert a distance expressed in, say, meters to a distance expressed in another unit. For example, you might solve a problem and find that the answer is, "The sailboat travels 23 meters in 10 seconds." Just how far is 23 meters? Think of any numerical measurement as a number multiplied by a unit of measure. In this case, 23 meters equals 23 *times* 1 meter. Then the 1 meter can be replaced by the corresponding number of feet—a conversion factor found in the back inside cover.

$$23 \text{ meters} = 23 \times 1 \text{ meter}$$

But 1 meter = 3.28 feet; therefore:

$$23 \text{ meters} = 23 \times 1 \text{ meter} = 23 \times 3.28 \text{ feet}$$

$$23 \text{ meters} = 75.4 \text{ feet}$$

Two other physical quantities that are closely related to distance are *area* and *volume*. Area commonly refers to the size of a surface, such as the floor in a room or the outer skin of a basketball. The concept of area can apply to surfaces that are not flat and to "empty," two-dimensional spaces such as holes and open windows (see ● Figure 1.2). Area is a much more general idea than "length times width," an equation you may have learned that applies only to rectangles. The area of something is just the number of squares 1 inch by 1 inch (or 1 meter by 1 meter, or 1 mile by 1 mile, etc.) that would have to be added or placed together to cover it.

In a similar manner, the volume of a solid is the number of cubes 1 inch or 1 centimeter on a side needed to fill the space it occupies. For common geometric shapes such as a rectangular box or a sphere, there are simple equations to compute the volume. We use volume measures nearly every time we buy food in a supermarket.

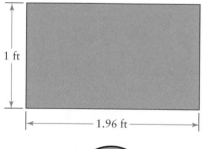

1 ft

1.96 ft

0.79 ft

● **Figure 1.2** Every surface, whether it is flat or curved, has an area. The area of this rectangle and the area of the surface of the basketball are equal.

Physical Quantity	Metric Units	English Units
Area (*A*)	square meter (m²)	square foot (ft²)
	square centimeter (cm²)	square inch (in.²)
	square kilometer (km²)	square mile (mi²)
	hectare	acre

● **Table 1.1** Some Representative Sizes and Distances		
Size/Distance	**Metric**	**English**
Size of a nucleus	1×10^{-14} m	4×10^{-13} in.
Size of an atom	1×10^{-10} m	4×10^{-9} in.
Size of a red blood cell	8×10^{-6} m	3×10^{-4} in.
Typical height of a person	1.75 m	5.75 ft
Tallest building	452 m	1,483 ft
Diameter of Earth	1.27×10^7 m	7,920 miles
Earth-Sun distance	1.5×10^{11} m	9.3×10^7 miles
Size of our galaxy	9×10^{20} m	6×10^{17} miles

*We will use scientific notation occasionally. See Appendix B for a review.

Physical Quantity	Metric Units	English Units
Volume (V)	cubic meter (m^3)	cubic foot (ft^3)
	cubic centimeter (cm^3 or cc)	cubic inch ($in.^3$)
	liter (L)	quart, pint, cup
	milliliter (mL)	teaspoon, tablespoon

Area and volume are examples of physical quantities that are based on other physical quantities—in this case, just distance. Their units of measure are called *derived units* because they are derived from more basic units (1 square meter = 1 meter × 1 meter = 1 meter^2). Until we reach Chapter 7, all of the physical quantities will have units that are derived from units of distance, time, mass, or a combination of these.

The measure of **time** is based on periodic phenomena—processes that repeat over and over at a regular rate. The rotation of the Earth was originally used to establish the universal unit of time, the *second*. The time it takes for one rotation was set equal to 86,400 seconds (24 × 60 × 60). Both the metric system and the English system use the same units for time.

Physical Quantity	Metric Units	English Units
Time (t)	second (s)	second (s)
	minute (min)	minute (min)
	hour (h)	hour (h)

Explore It Yourself 1.1

For this, you need a pendulum—something dangling on the end of a string about 1 to 2 feet (or $\frac{1}{2}$ meter) long. A shoe hanging from its shoelace works, or keys tied to a thread. You also need a timer—a watch that displays seconds is OK, or search for a "web stopwatch" to use on-line.

1. Set the pendulum swinging back and forth, and time how long it takes for it to do 10 full cycles (left to right then back to left is one cycle). Divide that time by ten to get the *time for one cycle*. This should be around 1 to 2 seconds.
2. Now set it swinging again, and this time count how many full cycles happen in 10 seconds. (You can estimate fractions, like 7.5 cycles.) Divide this number by 10 to get the *number of cycles that occur per second*.
3. See if you can figure out the approximate mathematical relationship between the two numbers.
4. Make the string twice as long or half as long and repeat. What is different with these new data?

Clocks measure time by using some process that repeats. Many mechanical clocks use a swinging pendulum. The time it takes to swing back and forth is always the same. This is used to control the speed of a mechanism that turns the hands on the clock face (see Figure 1.3). Mechanical wristwatches and stopwatches use an oscillating balance wheel for the same purpose. Quartz electric clocks and digital watches use regular vibrations of an electrically stimulated crystal made of quartz.

The first step in designing a clock is to determine exactly how much time it takes for one cycle of the oscillation. If it were 2 seconds for a pendulum, for example, the clock would then be designed so that the second hand would rotate once during 30 oscillations. The **period** of oscillation in that case is 2 seconds.

● **Figure 1.3** The regular swinging of the pendulum—by its repetitive motion—controls the speed of this clock.

DEFINITION

Period The time for one complete cycle of a process that repeats. It is abbreviated T, and the units are seconds, minutes, and so forth.

Time Out!

While the notion of time may be hard to define, the process of timekeeping is straightforward, requiring only a system for accurately counting the cycles of regularly occurring events. The operative word here is *regularly*: To keep time or to measure a time interval, we need something that cycles or swings or oscillates at a constant rate. Up until 1956, the fundamental clock used to tell time was the Earth-Sun system, and the fundamental unit of time, the second, was defined as 1/86,400 of a *mean solar day*. A mean solar day is the average interval between successive crossings of your local meridian by the Sun. Unfortunately, the rate of rotation of the Earth is not *strictly* constant, and a mean solar day varies in length over long time spans because of a gradual slowdown in the Earth's rotation rate brought about by friction between its oceans and its crust. In addition, short-term variations in the rate of spin of the Earth, some of which are seasonal, alter the length of a mean solar day.

For these reasons, in the late 1950s, a still more uniform cycle within the Earth-Sun system was sought, and the second was redefined as 1/31,556,925.9747 of the length of the year beginning in January 1900. This *ephemeris second* is based on the motion of the Earth around the Sun, which is governed (as we shall see in Chapter 2) by Newton's laws of motion. As a result, its evaluation requires astronomical observation and calculation and is not easily obtained with high accuracy except after many years of careful work. Thus, although ephemeris time appears to be extremely uniform, it suffers from not being able to be found quickly and accurately.

In 1967, a new *atomic second* was defined. In this case, 1 second equals the interval of time containing 9,192,631,770 oscillations of light waves given off by isolated cesium atoms (see Chapter 10 for a discussion of atoms and their light-emitting properties). With modern cesium atomic clocks, it is possible to establish the length of a second in less than a minute, with an accuracy of a few billionths of a second. Even higher precision and accuracy may soon be achieved with so-called trapped ion clocks, which use the elements mercury, cesium, or ytterbium and are being developed at the Jet Propulsion Laboratory in Pasadena, California, the National Institute of Standards and Technology (NIST) in Boulder, Colorado, and elsewhere.

Time measurement is so accurate that the length of the standard meter is set by how long it takes light to travel that distance—3.33564095 billionths of a second.

The introduction of atomic time leaves us with a problem. Our daily lives are generally governed by day-night cycles—that is, by the Earth-Sun clock, not by phenomena associated with cesium atoms. As the Earth continues to spin down, the length of a mean solar day (and hence the mean solar second) will continue to grow larger relative to the atomic second. As time goes on, Earth-Sun clocks will gradually fall further and further behind atomic clocks. This is clearly not a desirable situation.

In the early 1970s, the French Bureau International de l'Heure, the world's official timekeepers, introduced Coordinated Universal Time (UTC). Under this system, the length or duration of the second is dictated by atomic time, but the time commonly reported by time services is required to remain within 0.9 seconds of mean solar time. In this system, a *leap second* is added or subtracted as needed to compensate for changes in the rate of the Earth's rotation relative to

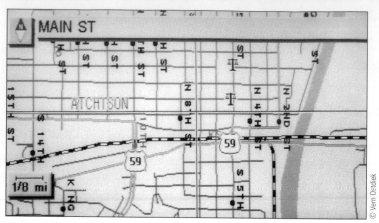

● **Figure 1.4** Map display on a sport-utility vehicle's in-dash GPS navigation system.

atomic time. Recent criticism of the leap second has led to calls to abandon it, but defenders argue that doing so will disrupt satellite communications and cost millions in computer upgrades.

If all this discussion seems a bit complicated to you, that's understandable. The topic of time and its determination is one with which physicists and philosophers have struggled for centuries. The point to be emphasized is that as more sophisticated experiments have been devised to probe nature's secrets, the need for ever more reliable and stable clocks has grown. The development of atomic clocks is the latest attempt to meet the requirements of scientists and engineers in this regard, and their use has now spilled over into our everyday lives. The adoption of UTC is an example. Another is the *Global Positioning System* (GPS). This system consists of an array of 24 satellites, five to eight of which are always visible from any point on the globe. Each satellite circles the Earth every 12 hours and carries up to four cesium and rubidium atomic clocks to generate the accurate time signals used in making precise position measurements. Small handheld GPS receivers costing under $200 are available that provide horizontal and vertical positional accuracies to better than 150 m and timing accuracies of the order of 3.5×10^{-7}s or less. Originally developed for military missions, the civilian use of GPS today extends to commercial ship and aircraft navigation; nationwide truck and freight-car tracking; "precision" farming, land surveying and geological studies; and worldwide, digital communication networks (including some that control banking functions at local ATMs). GPS receivers have even been integrated with data bases of maps and street directories and installed in autos to assist drivers in reaching their destinations (● Figure 1.4). GPS technology will soon be routinely available as part of personal cellular phone service. But regardless of how it is measured or used by humankind, one fact about time remains true: *Tempus fugit*, "time flies."

One of your authors (VO) uses GPS technology routinely to monitor his location while hang-gliding. What personal use do you make or have you made of this innovation? Even if you do not employ GPS data directly in your daily life, the odds are that you benefit from it in countless ways in terms of the goods and services you regularly purchase and enjoy. To learn just how reliant our society has become on this system, check out the World Wide Web using any of the common search engines or try **InfoTrac® College Edition** (key word "GPS"). The range of application of the GPS is truly astounding!

There is another way to look at this. The clock designer must determine how many cycles must take place before 1 second (or 1 minute or 1 hour) elapses. In our example, one-half of a cycle takes place during each second. This is called the **frequency** of the oscillation.

DEFINITION

Frequency The number of cycles of a periodic process that occur per unit time. It is abbreviated f.

The standard unit of frequency is the hertz (Hz), which equals 1 cycle per second.

$$1 \text{ Hz} = 1/s = 1 \text{ s}^{-1}$$

The frequency of AM radio stations is expressed in kilohertz (kHz) and those of FM stations in megahertz (MHz). *Mega-* is the metric prefix signifying 1 million. So, 91.5 MHz equals 91,500,000 Hz. Another metric prefix, *giga-*, is also commonly used with hertz. For example, the "speed" of a computer's processor—actually, the frequency of the electrical signal it uses—might be 2 gigahertz, which is 2 billion (2,000,000,000) hertz.

The relationship between the period of a cyclic phenomenon and its frequency is simple: The period equals 1 divided by the frequency, and vice versa.

$$\text{period} = \frac{1}{\text{frequency}}$$

$$T = \frac{1}{f}$$

Also:

$$f = \frac{1}{T}$$

Example 1.1

A mechanical stopwatch uses a balance wheel that rotates back and forth 10 times in 2 seconds. What is the frequency of the balance wheel?

$$\text{frequency} = \text{number of cycles per time}$$

$$f = \frac{10 \text{ cycles}}{2 \text{ s}}$$

$$= 5 \text{ Hz}$$

What is the period of the balance wheel?

$$\text{period} = \text{time for one cycle}$$

$$= 1 \text{ divided by the frequency}$$

$$T = \frac{1}{5 \text{ Hz}}$$

$$= 0.2 \text{ s}$$

The balance wheel oscillates 300 times each minute.

The third basic physical quantity is mass. The **mass** of an object is basically a measure of how much matter is contained in it. (This statement illustrates the sort of circular definition that arises when one tries to define fundamental concepts.) We know intuitively that a large body, such as a locomotive, has a large mass because it is composed of a great deal of material. Mass is also a measure of what we sometimes refer to in everyday speech as *inertia*. The larger the mass of an object, the greater its inertia and the more difficult it is to speed up or slow down.

Physical Quantity	Metric Units	English Units
Mass *m*	kilogram (kg) gram (g)	slug

Mass is not in common use in the English system; note the unfamiliar unit, the slug. Weight, a quantity that is related to, but is *not* the same as mass, is used instead. We can contrast the two ideas: When you try to lift a shopping cart, you are experiencing its weight. When you try to speed it up or slow it down, you are experiencing its mass (see ● Figure 1.5). We will take another look at mass and weight in Chapter 2.

● **Figure 1.5** (a) When lifting something, you must overcome its weight. (b) When changing its speed, you experience its mass.

(a) (b)

LEARNING CHECK

Simple quizzes like this are included at the end of most sections. They are not designed to be a thorough examination of the material. If you have trouble answering these questions, it probably means you should go back and reread the section. Answers are given at the end of each Learning Check.

1. Three of the fundamental physical quantities in physics are
   a) distance, time, and weight
   b) distance, time, and mass
   c) distance, time, and speed
   d) distance, area, and volume
2. (True or False.) When the period of a pendulum is less than 1 second, the frequency is always greater than 1 hertz.
3. One kilogram is the same as _____ grams.

ANSWERS: 1. (b) 2. True 3. 1,000

1.2 Speed and Velocity

A key concept to use when quantifying motion is **speed.**

Speed Rate of movement. Rate of change of distance from a reference point. The distance something travels divided by the time elapsed.

Physical Quantity	Metric Units	English Units
Speed (*v*)	meter per second (m/s) kilometer per hour (km/h)	foot per second (ft/s) mile per hour (mph)

A couple of aspects of speed are worth highlighting. First, speed is *relative*. A person running on the deck of a ship cruising at 20 mph might have a speed of 8 mph relative to the ship, but the speed relative to the water or a pier would be 28 mph (if headed toward

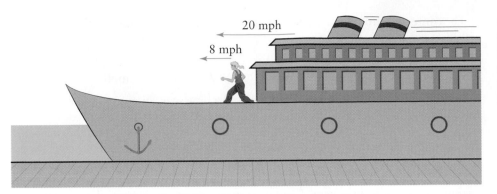

● **Figure 1.6** Speed is relative. The speed of a person running on a ship is 8 mph relative to the deck. If the ship's speed is 20 mph, the person's speed relative to the pier is either 28 mph (if headed toward the front of the ship) or 12 mph (if headed toward the rear).

the front of the ship). If we use the ship as the reference point, the speed is 8 mph. With the water or a pier as the reference point, the speed is 28 mph (● Figure 1.6). If you are traveling at 55 mph on a highway and a car passes you going 60 mph, its speed is 5 mph relative to your car. Most of the time speed is measured relative to the surface of the Earth; this will be the case in this book, unless stated otherwise.

Second, it is important to distinguish between *average speed* and *instantaneous speed*. An object's average speed is the total distance it travels during some period of time divided by the time that elapses:

$$\text{average speed} = \frac{\text{total distance}}{\text{total elapsed time}}$$

If a 1,500-mile airline flight lasts 3 hours, the airplane's average speed is 500 mph. Of course, the airplane's speed changes during the 3 hours, so its speed is not 500 mph at each instant. (You can average 75% on four exams without actually getting 75% on each one.) Similarly, a sprinter who runs the 100-meter dash in 10 seconds has an average speed of 10 m/s but is not traveling with that speed at each moment. This gives rise to the concept of instantaneous speed—the speed that an object has at an instant in time. A car's speedometer actually gives instantaneous speed. When it shows 55 mph, it means that, if the car traveled with exactly that speed for 1 hour, it would go 55 miles. How can one determine the instantaneous speed of something that is not equipped with a speedometer? Instantaneous speed can't be measured exactly using the basic meaning of speed (distance traveled divided by the time it takes) because an "instant" implies that *zero* time elapses. But we can get a good estimate of an object's instantaneous speed by timing how long it takes to travel a very short distance:

$$\text{instantaneous speed} = \frac{\text{very short distance}}{\text{very short time}}$$

For example, given sophisticated equipment, we might measure how long it takes a car to travel 1 meter. If that time is found to be 0.05 seconds (not an "instant" but a very short time), a good estimate of the car's instantaneous speed is

$$\text{instantaneous speed} = \frac{1 \text{ m}}{0.05 \text{ s}} = 20 \text{ m/s} = 44.8 \text{ mph}$$

In drag racing (see p. 12), the maximum instantaneous speed of a dragster is estimated by timing how long it takes to travel the final 60 feet (about 18 meters) of the quarter-mile race. This is typically less than 0.15 seconds.

Global positioning system (GPS) receivers used by pilots, drivers, hikers, and others employ radio signals from satellites to determine location. They can compute the device's approximate instantaneous speed by computing how far it travels in a short time and then dividing by that time. One common GPS receiver updates its position every second, so it can estimate instantaneous speed using 1 second as its "short time." A typical bicycle speedometer uses a sensor on the front wheel. The device can time how long it takes the wheel to make one rotation and estimate the instantaneous speed by using the circumference of the wheel (typically around 2.1 meters) as the "short distance."

The conversion from meters per second to miles per hour is done the same way it was with distance in Section 1.1. From the Table of Conversion Factors on the back inside cover, 1 m/s = 2.24 mph; so

$$20 \text{ m/s} = 20 \times 1 \text{ m/s} = 20 \times 2.24 \text{ mph} = 44.8 \text{ mph}$$

Actually, this is the car's average speed during the time span. In normal situations, a car's speed will increase or decrease by at most 1 mph during a twentieth of a second, so this is a pretty good estimate. Our answer is within ±1 mph of the true value. During a collision, a car's speed can change a great deal during 0.05 seconds, so one would have to somehow use a much shorter time interval to calculate instantaneous speed in this case. An automaker might use high-speed videotapes of a crash test to measure much shorter periods and corresponding distances. In general, "very short time" means an interval during which the object's speed won't change by an amount greater than the desired error limit of the estimate. The concept of instantaneous speed is what is important here, more so than the technical details of how it is measured in different situations.

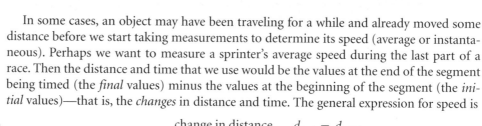

Explore It Yourself 1.2

Look up the record times for Olympic racing events in running, swimming, ice skating, or bicycling. (These can be found in a book of world records or on-line.) For several events, calculate the average speed, and notice how it is lower for longer races. Why is this so? For comparison, compute the average speeds for the same distance in two different events, such as 1,500 meters running and swimming. Why are they so different?

In some cases, an object may have been traveling for a while and already moved some distance before we start taking measurements to determine its speed (average or instantaneous). Perhaps we want to measure a sprinter's average speed during the last part of a race. Then the distance and time that we use would be the values at the end of the segment being timed (the *final* values) minus the values at the beginning of the segment (the *initial* values)—that is, the *changes* in distance and time. The general expression for speed is

$$\text{speed} = \frac{\text{change in distance}}{\text{change in time}} = \frac{d_{\text{final}} - d_{\text{initial}}}{t_{\text{final}} - t_{\text{initial}}}$$

$$v = \frac{\Delta d}{\Delta t}$$

(The symbol Δ is the Greek letter delta and is used to represent a "change in" the physical quantity.) This equation can represent both average and instantaneous speed. When Δt is the total elapsed time for a trip, v is the average speed. When Δt is a very short time, then v is the instantaneous speed.

● **Figure 1.7** Florence Griffith Joyner, who set world records in the 100 meters and 200 meters.

© Paul Merdity/Tony Stone Worldwide

Example 1.2

An analysis of a videotape of Olympic gold-medal winner Florence Griffith Joyner (1959–1998; ● Figure 1.7) running a 100-meter dash might yield the data shown in the table in the margin. Compute her average speed for the race and estimate her peak instantaneous speed. For the entire race, this would be as follows:

$$\text{average speed} = v = \frac{\Delta d}{\Delta t} = \frac{d_{\text{final}} - d_{\text{initial}}}{t_{\text{final}} - t_{\text{initial}}}$$

$$v = \frac{100 \text{ m} - 0 \text{ m}}{10.50 \text{ s} - 0 \text{ s}} = \frac{100 \text{ m}}{10.50 \text{ s}}$$

$$= 9.52 \text{ m/s} \ (= 21.3 \text{ mph})$$

d (meters)	t (seconds)
0	0
60	6.85
70	7.76
80	8.67
90	9.58
100	10.50

The equally spaced times between 60 and 90 meters indicate that her speed was constant. Therefore, we can use any segment from this part of the race to compute her instantaneous speed. For the segment between 80 and 90 meters, this would be:

$$v = \frac{\Delta d}{\Delta t} = \frac{d_{\text{final}} - d_{\text{initial}}}{t_{\text{final}} - t_{\text{initial}}}$$

$$= \frac{90\,\text{m} - 80\,\text{m}}{9.58\,\text{s} - 8.67\,\text{s}} = \frac{10\,\text{m}}{0.91\,\text{s}}$$

$$= 11.0\,\text{m/s} \; (= 24.6\,\text{mph})$$

The speed of a car being driven around in a city changes quite often. The instantaneous speed may vary from 0 mph (at stoplights) to 45 mph. The average speed for the trip is the total distance divided by the total time, maybe 20 mph.

When the speed of an object is constant, the average speed and the instantaneous speed are the same. In this case, we can express the relationship between the distance traveled and the time that has elapsed as follows:

$$d = vt \quad \text{(when speed is constant)}$$

This is an example of what is called a *proportionality:* We say that d is proportional to t (abbreviated $d \propto t$). If the time is doubled, the distance is doubled. The constant speed v is called the *constant of proportionality.* We will encounter many examples in which one physical quantity is proportional to another.

Explore It Yourself 1.3

When lightning strikes (see ● Figure 1.8), the flash of light reaches us in a fraction of a second, but the sound (thunder) is delayed. Why is that? The information in Table 1.2 should help you answer that. The time delay between the flash and the sound can be used to estimate how far away a lightning strike is. The sound travels about 340 meters in 1 second or 1 mile in 5 seconds. So what is the simple rule that relates the distance to the lightning to the number of seconds between the light and the sound? (Note: Follow the safety guidelines whenever there is a potential for lightning: Go into a building or a vehicle with a metal roof, and stay away from windows, plumbing, telephones with cords, and so on.)

Most drivers are accustomed to speed measured in miles per hour or kilometers per hour. This is most convenient when talking about travel times for distances greater than a few miles. Often, it is more enlightening to use feet per second. For example, a car going 65 mph travels 130 miles in 2 hours (using the preceding equation). But for a potential accident, it is relevant to consider how far the car will travel in seconds. Since 65 mph equals 95.6 feet per second, if a driver takes 2 seconds to decide how to avoid an accident, the car will have traveled 191 feet—more than 10 car lengths.

Try to develop a feel for the different speed units, particularly meters per second. You might keep in mind this comparison:

$$65\,\text{mph} = 95.6\,\text{ft/s} = 29.1\,\text{m/s}$$

Here is one last example of doing conversions. From the Table of Conversion Factors, 1 mph = 0.447 m/s:

$$65\,\text{mph} = 65 \times 1\,\text{mph} = 65 \times 0.447\,\text{m/s} = 29.1\,\text{m/s}$$

Apparently, the universe was created with an absolute speed limit—the speed of light in empty space. This speed is represented by the letter c. Nothing has ever been observed traveling faster than c. The value of c and some other speeds are included in ● Table 1.2.

● **Figure 1.8** A distant lightning strike provides a good demonstration of the difference between the speed of sound and the speed of light.

An important aspect of motion is *direction*. We will see that changing the direction of motion of a moving body is equivalent to changing the speed. **Velocity** is a physical quantity that incorporates both ideas.

Velocity Speed in a particular direction (same units as speed).

The speed of a ship might be 10 m/s, while its velocity might be 10 m/s east. Whenever a moving body changes direction, such as a car going around a curve or someone walking around a corner, the velocity changes even if the speed does not. Being told that an airplane flies for 2 hours from a certain place and averages 100 mph does not tell you enough to determine where it is. Knowing that it travels in a straight line due north would allow you to pinpoint its location.

A speedometer alone gives the instantaneous speed of a vehicle. A speedometer used in conjunction with a compass would give velocity: speed and the direction of motion. In

● **Table 1.2** Some Speeds of Interest

Description	Metric	English
Speed of light, *c* (in vacuum)	3×10^8 m/s	186,000 miles/second
Speed of sound (in air, room temperature)	344 m/s	771 mph
Highest instantaneous speeds:		
Running (cheetah)	28 m/s	63 mph
Swimming (sailfish)	30.4 m/s	68 mph
Flying—level (merganser)	36 m/s	80 mph
Flying—dive (peregrine falcon)	97 m/s	217 mph
Humans (approximate):		
Swimming	2.5 m/s	5.6 mph
Running	12 m/s	27 mph
Ice skating	14 m/s	31 mph

a car traveling along a winding road, the movement of the compass needle indicates that the velocity is changing even if the speed is not.

One of the standard displays on simple GPS (global positioning system) receivers includes both the speed and the direction of motion, in other words, the velocity (● Figure 1.9). Whenever either the speed or the heading (direction) is changing, the velocity is changing.

Velocity is an example of a physical quantity called a **vector.** Vectors have both a numerical size (magnitude) and a direction associated with them. Quantities that do not have a direction are called **scalars.** Speed by itself is a scalar. Only when the direction of motion is included do we have the vector velocity. Similarly, we can define the vector *displacement* as distance in a specific direction. For the airplane referred to earlier, the *distance* it travels in 2 hours is 200 miles. Its actual location can be determined only from its *displacement*—200 miles due north, for example. The basic equation for speed, $v = \Delta d/\Delta t$, is also the equation for velocity (that's why v is used) with d representing a vector displacement.

We can classify most physical quantities as scalars or vectors. Time, mass, and volume are all scalars because there is no direction associated with them.

Vectors are represented by arrows in drawings, the length of the arrow being proportional to the size of the vector (see ● Figure 1.10). If a car is traveling twice as fast as a pedestrian, then the arrow representing the car's velocity is twice as long as that representing the pedestrian's velocity.

When an object can move forward or backward along a line, its velocity is positive when it is going in one direction and negative when it is going in the other direction. When going forward in a car, you might give the velocity as +10 m/s (positive). If the car stops and then goes backward, the velocity is negative, −5 m/s, for example. The velocity of a person on a swing is positive, then negative, then positive, and so on. Which direction is associated with positive velocity is somewhat arbitrary. When dribbling a basketball, you could say its velocity is positive when it is moving downward and negative when it is moving upward, or vice versa. Speed does not become negative when the velocity does. The negative sign is associated with the direction of motion.

When dealing with a system in which the direction of motion can change, even if forward and backward are the only possible directions, it is better to use velocity than speed because the + or − sign indicates the direction of motion. In a situation in which the direction of motion does not change, like a falling object, the terms *speed* and *velocity* are often used interchangeably. However, in the remainder of this text, *velocity* will be used in all situations in which the direction of motion can be important. In cases where an object's direction of motion doesn't change, we will use the direction of the object's initial motion as the positive direction (unless stated otherwise).

Vector Addition

Sometimes a moving body has two velocities at the same time. The runner on the deck of the ship in Figure 1.6 has a velocity relative to the ship and a velocity because the ship itself is moving. A bird flying on a windy day has a velocity relative to the air and a velocity because the air carrying the bird is moving relative to the ground. The velocity of the runner relative to the water or that of the bird relative to the ground is found by adding the two velocities together to give the *net,* or *resultant, velocity.* Let's consider how two velocities (or two vectors of any kind) are combined in *vector addition.*

When adding two velocities, you represent each as an arrow with its length proportional to the magnitude of the velocity—the speed. For the runner on the ship, the arrow representing the ship's velocity is $2\frac{1}{2}$ times as long as the arrow representing the runner's velocity because the two speeds are 20 mph and 8 mph ($8 \times 2\frac{1}{2} = 20$). Each arrow can be moved around for convenience, provided its length and its direction are not altered. Any such change would make it a different vector. The procedure for adding two vectors is as follows.

> Two vectors are added by representing them as arrows and then positioning one arrow so its tip is at the tail of the other. An arrow drawn from the tail of the first

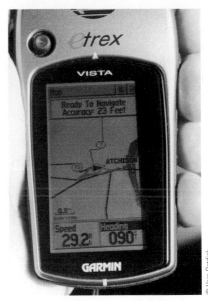

● **Figure 1.9** The information at the bottom of this GPS display gives the velocity—speed and direction. ("Heading" is based on north = 0°, east = 90°, south = 180°, and west = 270°.) In this case the unit's velocity is 29.2 mph due east.

● **Figure 1.10** Vector quantities can be represented by arrows. Arrow length indicates the vector's size, and arrow direction shows the vector's direction. In both figures, velocity is represented with an arrow. The car's velocity (and speed) is much greater than the pedestrian's, so the arrow representing the car's velocity is longer.

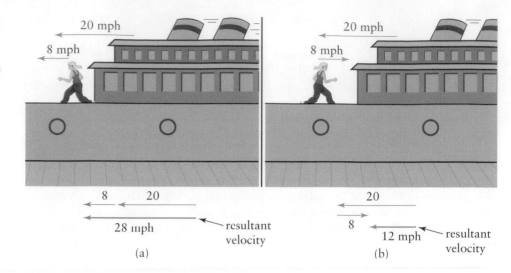

● **Figure 1.11** (a) The resultant velocity of a runner on the deck of a ship is found by adding the runner's velocity and the ship's velocity. The result is 28 mph forward. (b) Using the same procedure when the runner is headed toward the rear of the ship, the resultant velocity is 12 mph forward.

arrow to the tip of the second is the arrow representing the resultant vector—the sum of the two vectors.

● Figure 1.11 shows this for the runner on the deck of the ship. In Figure 1.11a, the runner is running forward, so the two arrows are parallel. When the arrows are positioned "tip to tail," the resultant velocity vector is parallel to the others, and its magnitude—the speed—is 28 mph (8 mph + 20 mph). In Figure 1.11b, the runner is running toward the rear of the ship, so the arrows are in opposite directions. The resultant velocity is parallel to the ship's velocity, and its magnitude is 12 mph (20 mph − 8 mph).

Vector addition is done the same way when the two vectors are not along the same line. ● Figure 1.12a shows a bird with velocity 8 m/s north in the air while the air itself has velocity 6 m/s east. The bird's velocity observed by someone on the ground, Figure 12b, is the sum of these two velocities. This we determine by placing the two arrows representing the velocities tip to tail as before and drawing an arrow from the tail of the first to the tip of the second (Figure 1.12c and d). The direction of the resultant velocity is toward the northeast. Watch for this when you see a bird flying on a windy day: Often the direction the bird is moving is not the same as the direction its body is pointed.

What about the magnitude of the resultant velocity? It is not simply 8 + 6 or 8 − 6, because the two velocities are not parallel. With the numbers chosen for this example, the magnitude of the resultant velocity—the bird's speed—is 10 m/s. If you draw the two original arrows with correct relative lengths and then measure the length of the resultant

● **Figure 1.12** The velocity of a bird relative to the ground (b) is the vector sum of its velocity relative to the air and the velocity of the air (wind). (c) and (d) show that the vectors can be added two different ways, but the resultant is the same vector.

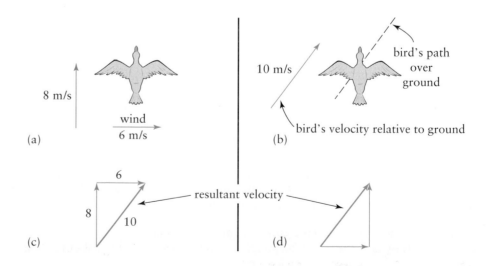

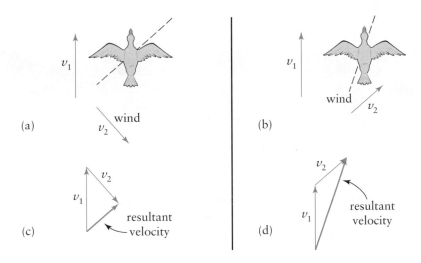

(a)

(b)

(c)

resultant velocity

(d)

resultant velocity

arrow, it will be $\frac{5}{4}$ times the length of the arrow representing the 8 m/s vector. Then 8 m/s times $\frac{5}{4}$ equals 10 m/s.*

Vector addition is performed in the same manner, no matter what the directions of the vectors. ● Figure 1.13 shows two other examples of a bird flying with different wind directions. The magnitudes of the resultants are best determined by measuring the lengths of the arrows. There are many other situations in which a body's velocity is the sum of two (or more) velocities (for example, a swimmer or boat crossing a river). Displacement vectors are added in the same fashion. If you walk 10 meters south, then 10 meters west, your net displacement is 14.1 meters southwest.

The process of vector addition can be "turned around." Any vector can be thought of as the sum of two other vectors, called *components* of the vector. When we observe the bird's single velocity in Figure 1.12b, we would likely realize that the bird has two velocities that have been added. Even when a moving body only has one "true" velocity, it may be convenient to think of it as two velocities that have been added together. For example, a soccer player running southeast across a field can be thought of as going south with one velocity and east with another velocity at the same time (● Figure 1.14). A car going down a long hill has one velocity component that is horizontal and another that is vertical (downward).

---

* The Pythagorean theorem can be used to calculate the magnitude of the resultant vector. The arrows in Figure 1.12c form a right triangle. For any right triangle:

$$c^2 = a^2 + b^2$$

In this case, $a = 8$ and $b = 6$. Therefore: $c^2 = a^2 + b^2 = 8^2 + 6^2$

$$= 64 + 36 = 100$$

$$c = \sqrt{100} = 10$$

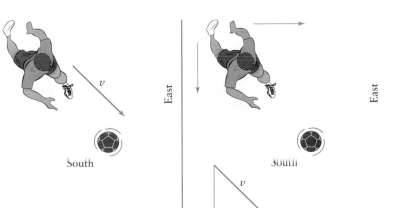

South

East

South

East

● **Figure 1.14** A soccer player running southeast can be thought of as having a velocity toward the east and a velocity toward the south at the same time. When these two velocities, called *components*, are added together, the resultant is the original velocity.

Learning Check

1. For objects moving with constant speed,
   a) the instantaneous speed is always zero.
   b) the average speed is always zero.
   c) the instantaneous speed and average speed are always equal.
   d) none of the above.
2. The highest speed that can be observed in nature is
   a) the speed of light.
   b) the speed of sound.
   c) 1,000,000 m/s.
   d) there is no limit.
3. A physical quantity that has a direction associated with it is called a _______________.
4. (True or False.) As a car is traveling down a road, it is possible for its speed to be changing while its velocity is constant.
5. (Choose the *incorrect* statement.) When two velocities are added, the resultant velocity
   a) can be parallel to one or both of the original velocities.
   b) cannot have zero magnitude.
   c) can have a greater magnitude than either of the original velocities.
   d) can be perpendicular to one of the original velocities.

ANSWERS: 1. (c) 2. (a) 3. vector 4. False 5. (b)

1.3 Acceleration

The physical world around us is filled with motion. But think about this for a moment: Cars, bicycles, pedestrians, airplanes, trains, and other vehicles all change their speed or direction often. They start, stop, speed up, slow down, and make turns. The velocity of the wind usually changes from moment to moment. Even the Earth as it moves around the Sun is constantly changing its direction of motion and its speed, though not by much. The main thrust of Chapter 2 is to show how the change in velocity of an object is related to the force acting on it. For these reasons, a very important concept in physics is **acceleration.**

DEFINITION

Acceleration Rate of change of velocity. The change in velocity divided by the time elapsed.

$$a = \frac{\Delta v}{\Delta t}$$

Physical Quantity	Metric Units	English Units
Acceleration (a)	meter per second^2 (m/s^2)	foot per second^2 (ft/s^2) mph per second (mph/s)

Whenever something is speeding up or slowing down, it is undergoing acceleration. As you travel in a car, anytime the speedometer's reading is changing, the car is accelerating. Acceleration is a vector quantity, which means it has both magnitude and direction. Note that the relationship between acceleration and velocity is the same as the relationship between velocity and displacement. Acceleration indicates how rapidly velocity is changing, and velocity indicates how rapidly displacement is changing.

Example 1.3 A car accelerates from 20 to 25 m/s in 4 seconds as it passes a truck (● Figure 1.15). What is its acceleration?

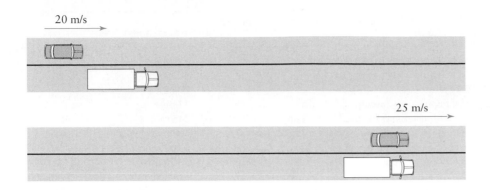

Since the direction of motion is constant, the change in velocity is just the change in speed—the later speed minus the earlier speed.

$$a = \frac{\Delta v}{\Delta t} = \frac{\text{final speed} - \text{initial speed}}{\Delta t}$$

$$= \frac{25 \text{ m/s} - 20 \text{ m/s}}{4 \text{ s}}$$

$$= \frac{5 \text{ m/s}}{4 \text{s}}$$

$$= 1.25 \text{ m/s}^2$$

This means that the car's speed increases 1.25 m/s during each second.

When something is slowing down (its speed is decreasing), it is undergoing acceleration. In everyday speech, the word *acceleration* is usually applied when speed is increasing, *deceleration* is used when speed is decreasing, and a change in direction of motion is not referred to as an acceleration. In physics, one word, *acceleration,* describes all three cases because each represents a change in velocity.

Example 1.4

After a race, a runner takes 5 seconds to come to a stop from a speed of 9 m/s. The acceleration is as follows:

$$a = \frac{\Delta v}{\Delta t} = \frac{0 \text{ m/s} - 9 \text{ m/s}}{5 \text{ s}}$$

$$= \frac{-9 \text{ m/s}}{5 \text{ s}}$$

$$= -1.8 \text{ m/s}^2$$

The minus sign means the acceleration and the velocity (which is positive) are in opposite directions.

Perhaps the most important example of accelerated motion is that of an object falling freely near the Earth's surface (see ● Figure 1. 16). By an object falling freely, we mean that only the force of gravity is acting on it. We can ignore things like air resistance. A rock falling a few meters would satisfy this condition, but a feather would not.

Freely falling bodies move with a constant downward acceleration. The magnitude of this acceleration is represented by the letter g.

$$g = 9.8 \text{ m/s}^2 \qquad \text{(acceleration due to gravity)}$$

$$32 \text{ ft/s}^2 \qquad 22 \text{ mph/s}$$

The downward velocity of a falling rock increases 22 mph each second that it falls. The letter g is often used as a unit of measure of acceleration. An acceleration of 19.6 m/s² equals 2 g. (Some representative accelerations are given in ● Table 1.3.)

● **Figure 1.16** Freely falling Sara has constant acceleration.

© Ver Ostdiek

• **Table 1.3** Some Accelerations of Interest

Description	Acceleration	
Freely falling body (on the Moon)	1.6 m/s²	0.16 g
Freely falling body (on the Earth)	9.8 m/s²	1 g
Space shuttle (maximum)	29 m/s²	3 g
Drag racing—car (average for ¼ mile)	32 m/s²	3.3 g
Highest (sustained) survived by human	245 m/s²	25 g
Clothes—spin cycle in a typical washing machine	400 m/s²	41 g
Tread of a typical car tire at 65 mph	2,800 m/s²	285 g
Click beetle jumping	3,920 m/s²	400 g
Bullet in a high-powered rifle	2,000,000 m/s²	200,000 g
Projectile in an electromagnetic "rail gun"	1.96×10^9 m/s²	2×10^8 g

Explore It Yourself 1.4

It's easy to compute the acceleration of a car (in g's) as it speeds up or slows down. Simply time how long (in seconds) it takes for the speed to change by 22 mph. Do this when you are a passenger and can see the speedometer. For example, it may take 4 seconds for a car to speed up from 20 mph to 42 mph. How do you find the acceleration in g's using these numbers? (Hint: If it took 1 second for the speed to increase by 22 mph, the acceleration would be 22 mph/s = 1 g.)

Another way to think of acceleration is in terms of the changes in the arrows drawn to represent velocity. When a car accelerates from one speed to a higher speed, the *length* of the arrow used to represent the velocity *increases* (see • Figure 1.17). If we place the two arrows side by side, the change in velocity can be represented by a third arrow, Δv, drawn from the tip of the first arrow to the tip of the second. The later velocity equals the initial velocity plus the change in velocity. The original arrow plus the arrow representing Δv equals the later arrow. The acceleration vector is this change in velocity divided by the time. Note that the acceleration vector points forward. Increasing the velocity requires a forward acceleration.

Centripetal Acceleration

The concept of acceleration includes changes in *direction* of motion as well as changes in speed. A car going around a curve and a billiard ball bouncing off a cushion are accelerated, even if their speeds do not change.

We can once again use arrows to indicate the change in velocity. • Figure 1.18 shows a car at two different times as it goes around a curve. If we place the two arrows representing the car's velocity next to each other, we see that the direction of the arrow has changed. The change in velocity, Δv, is represented by the arrow drawn from the tip of v_1 to the tip of v_2. In other words, the original arrow plus the arrow representing Δv equals the later arrow, just like in straight-line acceleration (Figure 1.17).

Note the direction of the change in velocity Δv: It is directed toward the center of the curve. Because the acceleration is in the same direction as Δv, it also is directed toward

• **Figure 1.17** As a car accelerates in a straight line, the length of the arrow representing its velocity increases. The arrow marked Δv represents the change in velocity from v_1 to v_2 and indicates the direction of the acceleration—forward.

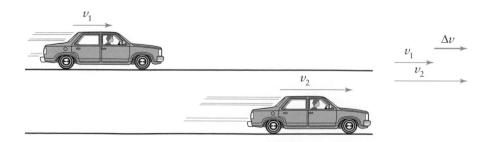

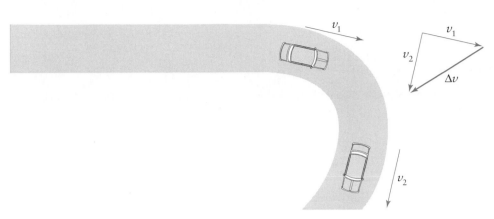

the center. For this reason, the acceleration of an object moving in a circular path is called **centripetal acceleration** (for "center-seeking"). The centripetal acceleration is always perpendicular to the object's velocity—directed either to its right or to its left.

So we know the direction of the acceleration of a body moving in a circular path, but what about its magnitude? The faster the body is moving, the more rapidly the direction of its velocity is changing. Consequently, the magnitude of the centripetal acceleration depends on the speed v. It also depends on the radius r of the curve (● Figure 1.19). A larger radius means the path is not as sharply curved, so the velocity changes more slowly and the acceleration is smaller. The actual equation for the size of the acceleration is

$$a = \frac{v^2}{r} \quad \text{(centripetal acceleration)}$$

We can also describe the relationship between a, v, and r by saying that the *acceleration is proportional to the square of the speed*:

$$a \propto v^2$$

and the *acceleration is inversely proportional* to the *radius r*:

$$a \propto \frac{1}{r}$$

This means that when the speed is doubled, the acceleration becomes four times as large. If the radius is doubled, the acceleration becomes one-half as large. We will encounter these two relationships again.

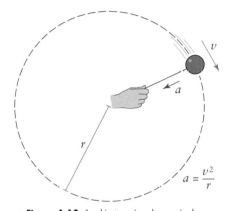

● **Figure 1.19** An object moving along a circular path is accelerated because its direction of motion, and therefore its velocity, are changing. Its centripetal acceleration equals the square of its speed, divided by the radius of its path. Doubling its speed would quadruple its acceleration. Doubling the radius of its path would halve the acceleration.

Let's estimate the acceleration of a car as it goes around a curve. The radius of a segment of a typical cloverleaf is 20 meters, and a car might take the curve with a constant speed of 10 m/s (about 22 mph; see ● Figure 1.20).

Because the motion is circular,

$$a = \frac{v^2}{r} = \frac{(10 \text{ m/s})^2}{20 \text{ m}}$$

$$= \frac{100 \text{ m}^2/\text{s}^2}{20 \text{ m}} = 5 \text{ m/s}^2$$

If the car could go 20 m/s and stay on the road (it could not), its acceleration would be four times as large—20 m/s² or about 2 g.*

Example 1.5

---

* The speed, heading, and time displayed on a GPS receiver can be used to compute the centripetal acceleration of a vehicle going around a curve. The equation is different because the radius is not used. If v = speed in mph, Δhdg = change in heading in degrees, and Δt = time elapsed in seconds, then the acceleration in mph/s is:

$$a = v \times \frac{\Delta hdg}{\Delta t} \times 0.0175$$

(The "0.0175" converts degrees into proper units.) For example, if the heading changes from 30° to 90° in 10 s, and the car is going 65 mph, then $\Delta hdg = 60°$ and $a = 6.8$ mph/s $= 0.31$ g. (This all works without a GPS receiver if the change in heading is known—for example, a 90° turn.)

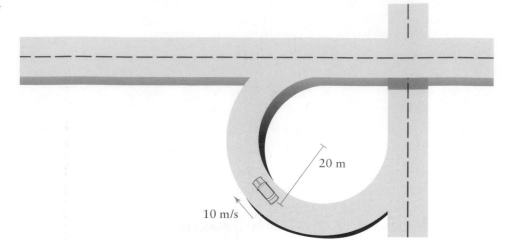

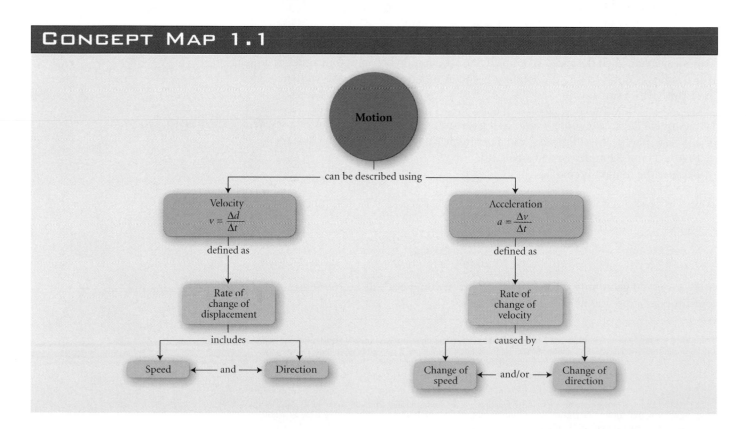

Many people have difficulty accepting the idea of centripetal acceleration, or they don't see how changing the direction of motion of a body is the same kind of thing as changing its speed. But some common experiences show that it is. Let's say that you are riding in a bus and that a book is resting on the slick seat next to you. The book will slide over the seat in response to the bus's acceleration. As the bus speeds up, the book slides backward. As the bus slows down, the book slides forward. In both cases, the reaction of the book is to move in the *direction opposite* the bus's acceleration. What happens when the bus goes around a curve? The book slides toward the *outside* of the curve, showing that the bus is accelerating toward the *inside* of the curve—a centripetal acceleration.

The cornering ability of a car is often measured by the maximum centripetal acceleration it can have when it rounds a curve. Automotive magazines often give the "cornering acceleration" or "lateral acceleration" of a car they are evaluating. A typical value for a sports car is 0.85 *g*, which is equal to 8.33 m/s².

Concept Map 1.1 summarizes the concepts of velocity and acceleration.

CONCEPT MAP 1.1

Motion

— can be described using —

Velocity
$$v = \frac{\Delta d}{\Delta t}$$

defined as

Rate of change of displacement

— includes —

Speed ← and → **Direction**

Acceleration
$$a = \frac{\Delta v}{\Delta t}$$

defined as

Rate of change of velocity

— caused by —

Change of speed ← and/or → **Change of direction**

1. When an object's velocity is changing, we say that the object is _____ .
2. A freely falling body experiences
   a) constant velocity.
   b) centripetal acceleration.
   c) constant acceleration.
   d) steadily increasing acceleration.
3. (True or False.) The acceleration of a slow-moving object can be greater than the acceleration of a fast-moving object.

4. When something moves in a circle with constant speed,
   a) its acceleration is perpendicular to its velocity.
   b) its acceleration is zero.
   c) its velocity is constant.
   d) its acceleration is parallel to its velocity.

ANSWERS: 1. accelerating 2. (c) 3. True 4. (a)

1.4 Simple Types of Motion

Now let's take a look at some simple types of motion. In each example, we'll consider a single body moving in a particular way. Our goal is to show how distance and speed depend on time. Take note of the different ways that the relationships can be shown or expressed.

Zero Velocity

The simplest situation is one in which *no motion* occurs: a single body sitting at a fixed position in space. We characterize this system by saying that the distance of the object, from whatever fixed reference point we choose, is constant. Nothing is changing. This means that the object's velocity and acceleration are both zero.

Constant Velocity

The next simplest case is *uniform motion*. Here the body moves with a constant velocity—a constant speed in a fixed direction. An automobile traveling on a straight, flat highway at a constant speed is a good example. A hockey puck sliding over smooth ice almost fits into this category, because it slows down only slightly due to friction. Note that the acceleration equals zero. That much should be obvious. The interesting relationship here is between distance and time. We can express that relationship in four different ways: with words, mathematics, tables, or graphs.

Let's take the example of a runner traveling at a steady pace of 7 m/s. If you are standing on the side of the road, how does the distance from you to the runner change with time, after the runner has gone past you? In words, the distance increases 7 meters each second: Two seconds after the runner passes by, the distance would be 14 meters (see • Figure 1.21). Stated mathematically, the distance in meters equals the time multiplied by the velocity, 7 m/s. The time is measured in seconds, starting just as the runner passes you. The shorthand way of writing this is the mathematical equation

$$d = 7t \quad (d \text{ in meters, } t \text{ in seconds})$$

This is an example of the general equation that we saw earlier with $v = 7$ m/s:

$$d = vt$$

The same information can also be put in a table of values of time and distance (see the margin table). These values all satisfy the equation $d = 7t$. The margin table shows the values of distance (d) at certain times (t). You could make the table longer or shorter by using different time increments.

Time (s)	Distance (m)
0	0
1	7
2	14
3	21
4	28

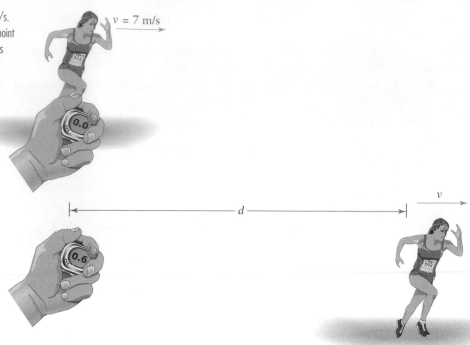

● **Figure 1.21** The velocity of a runner is 7 m/s. This means that the runner's distance from a fixed point (you standing behind the runner) increases 7 meters each second.

$v = 7$ m/s

0.0

0.6

d

v

The fourth way to show the relationship between distance and time is to graph the values in the table (see ● Figure 1.22a). The usual practice is to graph distance versus time, which means distance on the vertical axis. Note that the data points lie on a straight line. Remember this simple rule: When the speed is constant, the graph of distance versus time is a straight line. In general, when one quantity is proportional to another, the graph of the two quantities is a straight line.

An important feature of this graph is its *slope.* The slope of a graph is a measure of its steepness. In particular, the slope is equal to the *rise* between two points on the line divided by the *run* between the points. This is illustrated in Figure 1.22b. The rise is a distance, Δd, and the run is a time interval, Δt. So the slope equals Δd divided by Δt, which is also the object's velocity. The slope of a distance-versus-time graph equals the velocity.

The graph for a faster-moving body, a racehorse for instance, would be steeper—it would have a larger slope. The graph of d versus t for a slower object (a person walking) would have a smaller slope (see ● Figure 1.23). When an object is standing still (when it

● **Figure 1.22** (a) Graph of distance versus time when velocity is a constant 7 m/s. (b) Same graph with the slope indicated.

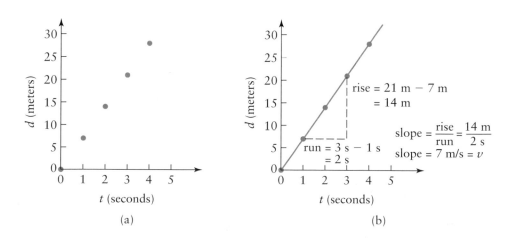

rise = 21 m − 7 m
= 14 m

$$\text{slope} = \frac{\text{rise}}{\text{run}} = \frac{14 \text{ m}}{2 \text{ s}}$$

slope = 7 m/s = v

run = 3 s − 1 s
= 2 s

(a)

(b)

has no motion), the graph of *d* versus *t* is a flat line parallel with the horizontal axis. The slope is zero because the velocity is zero.

Even when the velocity is not constant, the slope of a *d* versus *t* graph is still equal to the velocity. In this case, the graph is not a straight line because, as the slope changes (a result of the changing velocity), the graph curves or bends. The graph in ● Figure 1.24 represents the motion of a car that starts from a stop sign, drives down a street, and then stops and backs into a parking place. When the car is stopped, the graph is flat. The distance is not changing and the velocity is zero. When the car is backing up, the graph is slanting downward. The distance is decreasing, and the velocity is negative.

Constant Acceleration

The next simple motion example is *constant acceleration in a straight line*. This means that the object's velocity is changing at a fixed rate. A freely falling body is the best example. A ball rolling down a straight, inclined plane is another. Often, cars, runners, bicycles, trains, and aircraft have nearly constant acceleration when they are speeding up or slowing down.

Let's use free fall as our example. Assume that a heavy rock is dropped from the top of a building and that we can measure the instantaneous velocity of the rock and the distance that it has fallen at any time we choose (see ● Figure 1.25). The rock falls with an acceleration equal to *g*. This is 9.8 m/s², which is the same as 22 mph per second. First, consider how the rock's velocity changes. The velocity increases 9.8 m/s, or 22 mph, each second. This means that the rock's velocity equals the time (in seconds) multiplied by 9.8 (for m/s) or 22 (for mph). Mathematically,

$$v = 9.8t \quad (v \text{ in m/s}, t \text{ in seconds})$$

$$v = 22t \quad (v \text{ in mph}, t \text{ in seconds})$$

The general form of the equation that applies to any object that starts from rest and has a constant acceleration *a* is

$$v = at \quad (\text{when acceleration is constant})$$

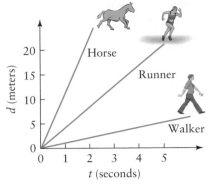

● **Figure 1.23** The slope of a distance-versus-time graph equals the velocity. The graph for a body with a higher velocity (the horse) has a larger slope—that is, a steeper graph. When the velocity is very low, the graph is nearly a horizontal line.

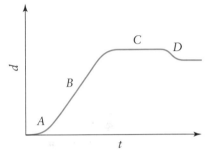

● **Figure 1.24** Graph of distance versus time for a car with varying velocity. At point *A* on the graph, the slope starts to increase as the car accelerates. At *B*, its velocity is constant. The slope decreases to zero at point *C* when the car is stopped. At *D*, the car is backing up, so its velocity is negative.

● **Figure 1.25** A rock falls freely after it is dropped from the top of a building. The distance *d* is measured from the top of the building. The rock's velocity increases at a steady rate—9.8 m/s each second.

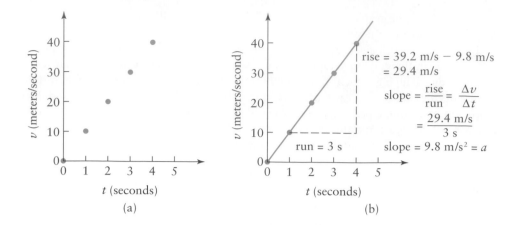

rise = 39.2 m/s − 9.8 m/s
    = 29.4 m/s

$$\text{slope} = \frac{\text{rise}}{\text{run}} = \frac{\Delta v}{\Delta t}$$

$$= \frac{29.4 \text{ m/s}}{3 \text{ s}}$$

run = 3 s   slope = 9.8 m/s² = a

(a)

(b)

So in constant acceleration, the velocity is proportional to the time. The proportionality constant is the acceleration a. A table of values for this example is shown below.

Time	Velocity	
(s)	(m/s)	(mph)
0	0	0
1	9.8	22
2	19.6	44
3	29.4	66
4	39.2	88

The graph of velocity versus time is a straight line (see ● Figure 1.26). The slope of a graph of velocity versus time equals the acceleration. This is because the rise is a change in velocity, Δv, and the run is a change in time Δt; so

$$\text{slope} = \frac{\Delta v}{\Delta t} = a$$

The corresponding graph for a body with a smaller acceleration, say, a ball rolling down a ramp, would have a smaller slope.

The similarity between the graphs in Figure 1.26 and the graphs in Figure 1.22 for uniform motion is obvious. But keep in mind that the graph of *distance* versus time is a straight line for uniform motion. Here it is the graph of *velocity* versus time.

What is the relationship between distance and time when the acceleration is constant? It is a bit more complicated, as expected. ● Figure 1.27 shows that the distance a falling body travels during each successive time interval grows larger as it falls. Since the velocity is continually changing, the distance equals the average velocity times the time. What is the average velocity? The object starts with velocity equal to zero, and after accelerating for a time t, its velocity is at. Its average velocity is:

$$\text{average velocity} = \frac{0 + at}{2} = \frac{1}{2}at$$

(If you take two quizzes and get 0 on the first and 8 on the second, your average grade is 4—one-half of 0 plus 8.)

The distance traveled is this average velocity times the time:

$$d = \text{average velocity} \times t = \frac{1}{2}at \times t$$

$$d = \frac{1}{2}at^2 \quad \text{(when acceleration is constant)}$$

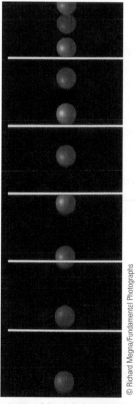

● **Figure 1.27** A falling ball photographed with a flashing strobe light. Each image shows where the ball was at the instant when the light flashed. The images are close together near the top because the ball is moving more slowly at first. As it falls, it picks up speed and thus travels farther between flashes.

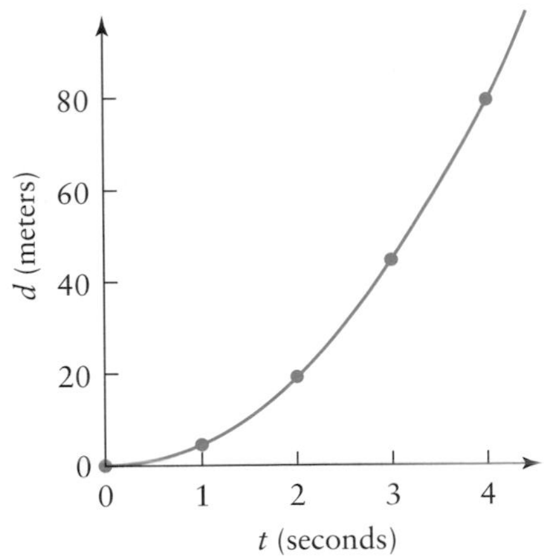

● **Figure 1.28** Graph of distance versus time for a freely falling body. The slope increases, indicating that the velocity increases.

In the case of a falling body, the acceleration is 9.8 m/s²; therefore,

$$d = \frac{1}{2}at^2 = \frac{1}{2} \times 9.8 \times t^2$$

$$= 4.9t^2 \qquad (d \text{ in meters, } t \text{ in seconds})$$

So in the case of constant acceleration, the distance is proportional to the square of the time. The constant of proportionality is one-half the acceleration.*

A table of distance values for a falling body is shown in the margin. This distance increases rapidly. The graph of distance versus time curves upward (see ● Figure 1.28). This is because the velocity of the body is increasing with time, and the slope of this graph equals the velocity. ● Table 1.4 and Concept Map 1.2 summarize these three simple types of motion.

Rarely does the acceleration of an object stay constant for long. As a falling body picks up velocity, air resistance causes its acceleration to decrease (more on this in Section 2.6). When a car is accelerated from a stop, its acceleration usually decreases, particularly when the transmission is shifted into a higher gear. ● Figure 1.29 shows the velocity of a car as it accelerates from 0 to 80 mph. Note that acceleration decreases (the slope gets smaller). During the short time the transmission is shifted, the acceleration is zero.

Graphs may be the best way to show the relationships between physical quantities. Mathematics is the most precise way, and it is not limited to two (or sometimes three)

Time (s)	Distance (m)
0	0
1	4.9
2	19.6
3	44.1
4	78.4

---

* For an object that is already moving with a velocity $v_{initial}$ and then undergoes constant acceleration, the average velocity after a time t is

$$v_{average} = \frac{v_{initial} + (v_{initial} + at)}{2} = v_{initial} + \frac{1}{2}at$$

Therefore,

$$d = v_{initial}t + \frac{1}{2}at^2$$

● **Table 1.4** Summary of Examples of Motion

Type of Motion	Behavior of Physical Quantities	Equations
Stationary object	Distance constant	$d = $ constant
	Speed zero	$v = 0$
	Acceleration zero	$a = 0$
Uniform motion†	Distance proportional to time	$d = vt$
	Velocity constant	$v = $ constant
	Acceleration zero	$a = 0$
Uniform acceleration† (from rest)	Distance proportional to time squared	$d = \frac{1}{2}at^2$
	Velocity proportional to time	$v = at$
	Acceleration constant	$a = $ constant
† Distance measured from object's initial location.		

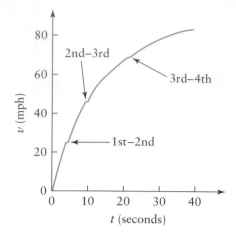

Figure 1.29 Graph of a car's velocity versus time as the car is accelerated. The notches in the curve occur at transmission shifts when the engine is momentarily disconnected from the drivetrain. Between the shifts, the acceleration (slope) decreases as the velocity increases.

quantities, as are graphs. However, math is inherently abstract and also somewhat like a language.

In this book, we will "translate" most of the mathematics we use into statements and also express many of the relationships graphically. But we will also retain some of the "original" language, mathematics, and show that even an understanding of high school math allows you to solve physics problems. The application of physics in our society by engineers and others requires mathematics. The Golden Gate Bridge could not have been designed and built through the use of words only.

When you see a graph, the first thing you should do is take careful note of which quantities are plotted. You should notice that some of the graphs for the examples of motion are straight lines. When the graph shows *distance* versus time, a straight line means the *velocity* is constant. But when it shows *velocity* versus time, a straight line means that the *acceleration* is constant. Even though the shapes of the two graphs are similar, because the quantities plotted are different for each graph, they represent very different situations. To a business executive, a rising graph showing profits will inspire joy, and a rising graph showing expenses will cause concern.

The most important thing to look for in the shape of a graph is trends. Is the slope always positive or always negative—is the graph continually going up or going down? If the slope changes, consider what that signifies. If it is a graph of velocity versus time, for example, choose a particular time, and note whether the velocity is increasing, decreasing, or remaining the same.

Let's look at an application of the kinds of graphs we've been using. During a karate demonstration, a concrete block is broken by a person's fist. ● Figure 1.30a shows the dis-

Figure 1.30 (a) Graph of the position of a fist versus time during a karate blow. Contact is made at about 6 milliseconds. (b) Graph of the velocity of the fist versus time. At contact, the fist accelerates rapidly. [From S. R. Wilke, R. E. McNair, M. S. Feld. "The Physics of Karate." *American Journal of Physics* 51 (September 1983): 783–790.]

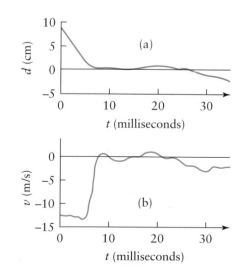

Simple types of motion

include

None — so

- Distance — given by — $d = $ constant
- Speed — given by — $v = 0$
- Acceleration — given by — $a = 0$

Uniform — means

No change in speed ← and → No change in direction — so

- Distance — given by — $d = vt$
- Speed — given by — $v = $ constant
- Acceleration — given by — $a = 0$

Constant linear acceleration — means

Constant change in speed ← and → No change in direction — so

- Distance — given by — $d = (\frac{1}{2}) at^2$
- Speed — given by — $v = at$
- Acceleration — given by — $a = $ constant

tance of the hand above the block versus time, measured using high-speed photography. The fist travels downward until it contacts the block, at about 6 milliseconds. This causes a large acceleration as the fist is brought to a sudden stop. Figure 1.30b shows the velocity of the fist, just the slope of the distance versus time graph at each time. Contact with the concrete is indicated by the steep part of the graph as the velocity goes to zero. If we take the slope of this segment of the velocity graph, we find that the acceleration of the fist at that moment is about 3,500 m/s², or 360 g (ouch!). What happened at about 25 milliseconds?

LEARNING CHECK

1. What is different about the graph of distance versus time for a fast-moving object compared to that of a slow-moving object?
2. (True or False.) A rock dropped from the top of a tall building travels the same distance during each 1-second interval that it falls.
3. If the graph of a car's velocity versus time curves upward, we can conclude its acceleration is
   _____.

4. Graphs of distance versus time and velocity versus time are useful for analyzing
   a) the motion of a falling body.
   b) the performance of a car.
   c) laboratory measurements of an accelerating object.
   d) all of the above.

ANSWERS: 1. larger slope **2.** False **3.** increasing **4.** (d)

We finish this chapter with a brief look at how our method of analyzing motion evolved. The first great strides toward the development of exact sciences were made by the ancient Greeks. Pythagoras, in the sixth century B.C., showed that numbers, often regarded merely as mental abstractions, are related to the natural world and even to human perception. His most famous discoveries were in geometry and acoustics. Other ancients, before the year 200 B.C., accomplished accurate measurements of the diameter of the Earth, the Moon, and the Moon's orbit.

Aristotle

The philosopher Aristotle (384–322 B.C.; ● Figure 1.31) is regarded as the first person to attempt physics. In fact, he gave *physics* its name. Born during the height of ancient Greek civilization, Aristotle became one of the central figures in the explosion of intellectual development that took place in that era. Starting at the age of seventeen, Aristotle was a student of the great philosopher Plato for 20 years. He became the tutor of Alexander the Great and later spent years studying the cultures, flora, and fauna of the exotic lands conquered by Alexander. In 335 B.C., Aristotle opened a school at the Lyceum, near Athens, where he taught until his death. Most of what we know about Aristotle's thoughts and teachings are based on the lectures he gave at the Lyceum.

Aristotle was a master of virtually all of the academic disciplines that existed at the time and actually invented some of them. Psychology is one example. His ideas about physics were influenced a great deal by the methodology that he used in logic, biology, and other areas. Aristotle took the important step of realizing that a number of complicating factors in the physical world often mask hidden order. His analysis of motion, the first mechanics, rested on

● **Figure 1.31** *Aristotle teaching Alexander the Great* (painting by J. L. G. Ferris).

the distinction between *natural motion*, like that of a falling body, and *unnatural motion*, like that of a cart being pulled down a road.

According to Aristotle, heavy objects fall because they are seeking their natural place. In his model, the speed of a falling body is constant and depends on its weight and the medium through which it falls. Heavy objects fall faster than lighter ones. Also, a rock drops faster through air than it does through water (● Figure 1.32). This analysis is only partly correct: It applies only after an object has been falling for a period of time and friction has become important. Speed first increases at a fixed rate—9.8 m/s each second, as we have seen. A marble and a rock dropped at the same time will build up speed together and hit the ground at the same time (● Figure 1.33). Only after a rock has fallen a much greater distance will air resistance cause its speed to level off at some constant value. This final speed, called the *terminal speed*, depends as much on the shape and size of the object as it does on the weight. The same factors affect the terminal speeds of bodies falling through water, but such speeds are much slower (as Aristotle pointed out) and are reached much more quickly. We will take a closer look at this in Section 2.6.

Aristotle's description of the motion of falling bodies fits only after their motions are dominated by air resistance. In a similar way, his analysis of "unnatural motion" indirectly includes friction as the dominating influence. A rock's "natural" tendency is to remain at rest on the ground or to fall toward the Earth's center if dropped. Making the rock move horizontally by pushing it is unnatural and requires some external agent or force. Aristotle's view that a force is always required to maintain horizontal motion fits well whenever there is a great deal of friction but not so well in cases like those of a rock thrown horizontally or of a smooth, heavy ball

● **Figure 1.32** In Aristotle's model of falling bodies, an object falls with a constant speed that depends on its weight and the medium through which it falls. (Left) A heavy object falls faster than a light object. (Right) An object falls faster through air than through water.

rolling over a flat, hard surface. In these situations, motion can continue for quite some time with no force acting to maintain it. Because of Aristotle's overwhelming reputation as a scholar, zealous supporters of his ideas allowed his theories to dominate physics, almost without question, for about 2,000 years.

Galileo

Galileo Galilei (1564–1642; ● Figure 1.34) lived during one of the most fruitful periods of human civilization—the Renaissance. His birth came in the same year as the death of Michelangelo and the birth of Shakespeare. Galileo made many important discoveries in mechanics and astronomy, some of which were disputed bitterly because they contradicted accepted views passed down from Aristotle's time. Galileo's strong support of the heliocentric (Sun-centered) model of the solar system and other factors led to his being placed on trial by leaders of the Inquisition. But this was the age when superstition was giving way to rational thought, when evidence and logical proof were beginning to win out over blind acceptance of doctrines. Though he was censured and forced to publicly recant his support of the heliocentric theory, Galileo's work gained wide acceptance. He is considered one of the founders of modern physical science.

Galileo was one of the first to rely on observation and experimentation. In 1583, he noticed that a lamp hanging from the ceiling of the cathedral at Pisa would swing back and forth with a constant period, even though the length of the arc of its motion decreased. His discovery, that the frequency of a pendulum depends only on its length, is the basis for the pendulum clock.

Another important contribution by Galileo was his insistence that scientific terms, statements, and analyses be logically consistent. He sought to establish a "scientific method" and believed that, as a first step, the language of science should be unambiguous. To Galileo, mathematics was necessary to help accomplish this task.

Although his astronomical discoveries brought Galileo most of his fame (and controversy), his conclusions about motion are of interest here. Like Aristotle, Galileo recognized that the simple rules that govern motion can be hidden by other phenomena. He realized that the two things that have the largest effect on objects moving on the Earth are gravity and friction. More importantly, he reasoned that friction is often complicated and unpredictable, and that it is best to first focus on systems in which friction doesn't dominate completely.

The physical system that Galileo used to analyze uniform motion and uniformly accelerated motion was a smooth, heavy ball rolling on a smooth, straight surface that can be tilted. If the surface is tilted (an inclined plane), the ball's speed will increase as it rolls down. The opposite occurs when the ball is initially rolling up the plane—its speed decreases. But what if the surface is level? Logically, the ball's speed should neither increase nor decrease but should stay the same (● Figure 1.35). So Galileo reasoned correctly that, if there is no friction, an object moving on a level surface will proceed with constant speed.

Galileo discovered the law of falling bodies, also using a ball rolling on an inclined plane (see ● Figure 1.36). He realized that the speed of a dropped object increases as it falls. The biggest problem he faced was measuring the speed. The only clocks available at

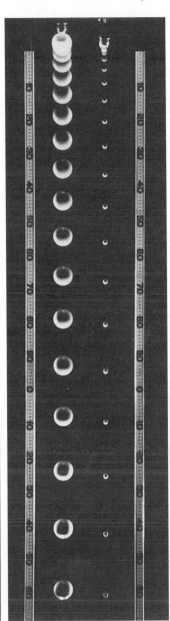

● **Figure 1.33** Strobe photograph of two falling bodies. Even though one is much heavier than the other, they have the same acceleration.

● **Figure 1.34** Galileo Galilei.

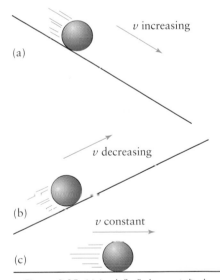

● **Figure 1.35** (a) As a ball rolls down an inclined plane, its speed increases. (b) As a ball rolls up an inclined plane, its speed decreases. (c) When a ball rolls on a level surface, it should therefore have a constant speed.

© Art Resource

● **Figure 1.36** Galileo using an inclined plane to experiment with acceleration.

vacuum, devoid of air resistance and other forms of friction. Only by understanding a simplified model of reality can one hope to comprehend its complexities and subtleties.

Aristotle and other ancient Greeks initiated the science of physics. Galileo corrected Aristotle's mechanics and established the importance of mathematics and experimentation in physics. But the greatest name in the development of mechanics is that of Isaac Newton, who was born on Christmas Day in the year of Galileo's death.

that time were water clocks, based on how long it took water to drain out of (or to fill) a container. Galileo realized that a ball rolling down an inclined plane is also accelerated because of gravity. He reasoned that tilting the surface more and more steeply would make the ball's motion become closer to that of free fall (see ● Figure 1.37). Using the inclined plane and his crude clocks, Galileo discovered the correct relationship between distance and time for uniformly accelerated motion.

Legend has it that Galileo once dropped two different-sized objects from the top of the Leaning Tower of Pisa to show that they would hit the ground at nearly the same time. Whether he actually did this or not is unknown. His conclusion about the motion of falling bodies is correct whenever air resistance is negligible. This was illustrated dramatically by astronaut David R. Scott while performing experiments on the Moon. He simultaneously dropped a hammer and a feather. Both fell at the same rate and hit the lunar surface at the same time because there is no air on the Moon to slow the feather.

Galileo used an approach that is now standard procedure in physics: If you wish to understand a real physical process, first consider an idealized system in which complicating factors (like friction) are absent. He promoted the idea of imagining how bodies would move in a perfect

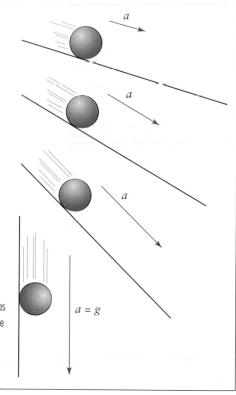

● **Figure 1.37** A ball is accelerated as it rolls down an inclined plane. Making the plane steeper increases the ball's acceleration. When the plane is vertical, the acceleration equals g.

SUMMARY

- **Distance, time,** and **mass** are the three **fundamental physical quantities** used in mechanics, the branch of physics dealing with the motion of objects.

- **Speed, velocity,** and **acceleration** are derived quantities (based on distance and time) that are used to specify how an object moves.

- **Acceleration,** the rate of change of velocity, is zero for bodies that are moving with a constant speed in a straight line.

- On the Earth, freely falling objects have a constant downward acceleration of 9.8 m/s^2 or 32 ft/s^2.

- Velocity and acceleration are **vectors,** meaning that they have direction as well as magnitude.

- A body moving along a circular path undergoes **centripetal acceleration** because the direction of its velocity is changing.

- Aristotle was the first to attempt a scientific analysis of motion. His conclusions about "natural motion" and "unnatural motion," while flawed, were still on the right track.

- Galileo made several important discoveries in physics and correctly described the motion of freely falling bodies. He was one of the first to use logical observation, experimentation, and mathematical analysis in his work, all important components of modern scientific methodology.

IMPORTANT EQUATIONS

Equation	Comments
Fundamental Equations	
$T = \dfrac{1}{f}$	Relates period and frequency
$f = \dfrac{1}{T}$	Relates frequency and period
$v = \dfrac{\Delta d}{\Delta t}$	Definition of velocity or speed
$a = \dfrac{\Delta v}{\Delta t}$	Definition of acceleration

Equation	Comments
Special-Case Equations	
$d = vt$	Distance from starting point when velocity is constant
$d = \dfrac{1}{2}at^2$	Distance when acceleration is constant and object starts from rest
$v = at$	Velocity when acceleration is constant and object starts from rest
$a = g$	Acceleration of freely falling body
$a = \dfrac{v^2}{r}$	Centripetal acceleration (object moving in a circular path)

MAPPING IT OUT!

1. Consider Concept Map 1.1, which provides an overview of motion. Review Sections 1.2 and 1.3, and make a list of at least five additional concepts you could add to the map. Now consider where to place these concepts on the map. Keep in mind that general, more inclusive concepts go toward the top, while more specific ideas and examples go toward the bottom. Place your concepts on the map. Draw the possible connections between your five new concepts and the rest of the map. Remember to use "linking words" to express your ideas clearly. When you've finished updating your map, compare your revised concept map with that of a classmate's. Are they identical? Should they be?

2. Review the material introduced in Section 1.1 on physical quantities. Based on your understanding of the meaning of the concepts and examples given in this section, complete the concept map shown below. Fill in the missing concepts and/or linking phrases so that the propositions and connections make good physical sense.

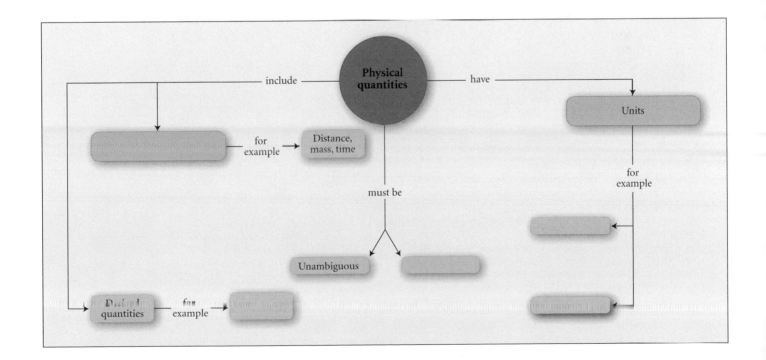

QUESTIONS

(▶ Indicates a review question, which means it requires only a basic understanding of the material to answer. The others involve integrating or extending the concepts presented thus far.)

1. Two rugs have exactly the same shape, but one is twice as long as the other. Does that mean its area is also twice as large? Explain.
2. ▶ Give a couple of examples of derived units of measure.
3. A pendulum clock is taken to a repair shop. Its pendulum is replaced by a shorter one that oscillates with a smaller period than the original. What effect, if any, does this have on how the clock runs?
4. ▶ What are the "basic" or "fundamental" physical quantities? Why are they called that?
5. Many countries that formerly used the English system of measure have converted to the metric system. Why is the metric system simpler to use, once you are familiar with it?
6. To prepare an aircraft carrier to launch airplanes, it is turned "into the wind" (so it is sailing against the wind). Why does this aid in launching the aircraft? (The aircraft take off in the direction the carrier is going.)
7. Scenes in films or television programs sometimes show people jumping off moving trains and having unpleasant encounters with the ground. If someone is on a moving flat-bed train car and wishes to jump off, how could the person use the concept of relative speed to make a safer dismount?
8. ▶ List the physical quantities identified in this chapter. From which of the fundamental physical quantities is each derived? Which of them are vectors, and which are scalars?
9. ▶ What is the distinction between speed and velocity? Describe a situation in which an object's speed is constant but its velocity is not.
10. ▶ What is "vector addition," and how is it done?
11. Can the resultant of two velocities have zero magnitude? If so, give an example.
12. A swimmer heads for the opposite bank of a river. Make a sketch showing the swimmer's two velocities and the resultant velocity.
13. A basketball player shoots a free throw. Make a sketch showing the basketball's velocity just after the ball leaves the player's hands. Draw in two components of this velocity, one horizontal and one vertical. Repeat the sketch for the instant just before the ball reaches the basket. What is different?

14. ▶ What is the relationship between velocity and acceleration?
15. ▶ How does the velocity of a freely falling body change with time? How does the distance it has fallen change? How about the acceleration?
16. One suggested way to improve the chances of survival if caught in an elevator falling out of control is to jump upward just before impact. Explain why doing so would not help much.
17. ▶ What is centripetal acceleration? What is the direction of the centripetal acceleration of a car going around a curve?
18. During 200-meter and 400-meter races, runners must stay in lanes as they go around a curved part of the track. If runners in two different lanes have exactly the same speed, will they also have exactly the same centripetal acceleration as they go around a curve? Explain.
19. An insect is able to cling to the side of a car's tire when the car is going 5 mph. How much harder is it for the insect to hold on when the speed is 10 mph?
20. As a car goes around a curve, the driver increases its speed. This means the car has two accelerations. What are the directions of these two accelerations?
21. The following are speeds and headings displayed on a GPS receiver. (Heading gives the direction of motion based on: north = 0°, east = 90°, south = 180°, etc.) In each case, indicate whether the receiver was accelerating during the time between the displays, and if it was, describe in what way the receiver was accelerating.
    a) Initially: 60 mph, 70°. 5 seconds later: 50 mph, 70°.
    b) Initially: 50 mph, 70°. 5 seconds later: 70 mph, 70°.
    c) Initially: 60 mph, 70°. 5 seconds later: 60 mph, 90°.
22. In Figure 1.19, arrows show the directions of the velocity and the acceleration of a ball moving in a circle. Make a similar sketch showing these directions for a car (a) speeding up from a stop sign and (b) slowing down as it approaches a stop sign.
23. ▶ If a ball is thrown straight up into the air, what is its acceleration as it moves upward? What is its acceleration when it reaches its highest point and is stopped at an instant?
24. ▶ What does the slope of a distance-versus-time graph represent physically?
25. Sketch a graph of velocity versus time for the motion illustrated in Figure 1.24. Indicate what the car's acceleration is at different times.
26. ▶ To what extent was Aristotle's model of falling bodies correct? How was it wrong?

PROBLEMS

1. A yacht is 20 m long. Express this length in feet.
2. Express your height (a) in meters and (b) in centimeters.
3. A convenient time unit for short time intervals is the *millisecond*. Express 0.0452 s in milliseconds.
4. One mile is equal to 1,609 m. Express this distance in kilometers and in centimeters.
5. A hypnotist's watch hanging from a chain swings back and forth every 0.8 s. What is the frequency of its oscillation?
6. The quartz crystal used in an electric watch vibrates with frequency 32,768 Hz. What is the period of the crystal's motion?
7. A passenger jet flies from one airport to another 1,200 miles away in 2.5 h. Find its average speed.
8. In 2003, the world record in the 200-m dash was 19.32 s. What was the average speed in m/s? In mph?

9. In Figure 1.13, assume that $v_1 = 8$ m/s and $v_2 = 6$ m/s. Use a ruler to estimate the magnitudes of the resultant velocities in (c) and (d).
10. On a day when the wind is blowing toward the south at 3 m/s, a runner jogs west at 4 m/s. What is the velocity (speed and direction) of the air relative to the runner?
11. How far does a car going 25 m/s travel in 5 s? How far would a jet going 250 m/s travel in 5 s?
12. A long-distance runner has an average speed of 4 m/s during a race. How far does the runner travel in 20 min?
13. Draw an accurate graph showing the distance versus time for the car in Problem 11. What is the slope?
14. The graph in ● Figure 1.38 shows the distance versus time for an elevator as it moves up and down in a building. Compute the elevator's velocity at the times marked a, b, and c.

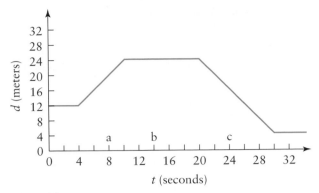

Figure 1.38

15. A Ferrari GTO can go from 0 to 100 mph (44.7 m/s) in 11 s.
    a) What is its average acceleration?
    b) The same car can come to a complete stop from 40 m/s in about 5 s. What is its average acceleration?
16. As a baseball is being thrown, it goes from 0 to 40 m/s in 0.15 s.
    a) What is the acceleration of the baseball?
    b) What is the acceleration in *g*'s?
17. A child attaches a rubber ball to a string and whirls it around in a circle overhead. If the string is 0.5 m long, and the ball's speed is 10 m/s, what is the ball's centripetal acceleration?

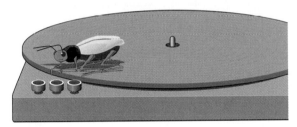

Figure 1.39

18. An insect sits on the edge of a spinning record that has a radius of 0.15 m. The insect's speed is about 0.5 m/s when the record is turning at $33\frac{1}{3}$ rpm. What is the insect's acceleration?
19. A runner is going 10 m/s around a curved section of track that has a radius of 35 m. What is the runner's acceleration?
20. During a NASCAR race a car goes 50 m/s around a curved section of track that has a radius of 250 m. What is the car's acceleration?
21. A rocket accelerates from rest at a rate of 60 m/s².
    a) What is its speed after it accelerates for 40 s?
    b) How long does it take to reach a speed of 7,500 m/s?
22. A train, initially stationary, has a constant acceleration of 0.5 m/s².
    a) What is its speed after 15 s?
    b) What would be the total time it would take to reach a speed of 25 m/s?
23. a) Draw an accurate graph of the speed versus time for the train in Problem 22.
    b) Draw an accurate graph of the distance versus time for the train in Problem 22.

24. Draw an accurate graph of the velocity versus time for the elevator in Problem 14.
25. A skydiver jumps out of a helicopter and falls freely for 3 s before opening the parachute.
    a) What is the skydiver's downward velocity when the parachute opens?
    b) How far below the helicopter is the skydiver when the parachute opens?
26. A rock is dropped off the side of a bridge and hits the water below 2 s later.
    a) What was the rock's velocity when it hit the water?
    b) What was the rock's average velocity as it fell?
    c) What is the height of the bridge above the water?
27. A roller coaster starts at the top of a straight track that is inclined 30° with the horizontal. This causes it to accelerate at a rate of 4.9 m/s² (1/2 g).
    a) What is the roller coaster's speed after 3 s?
    b) How far does it travel during that time?

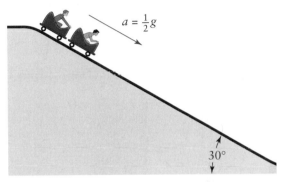

Figure 1.40

28. During takeoff, an airplane goes from 0 to 50 m/s in 8 s.
    a) What is its acceleration?
    b) How fast is it going after 5 s?
    c) How far has it traveled by the time it reaches 50 m/s?
29. The graph in ● Figure 1.41 shows the velocity versus time for a bullet as it is fired from a gun, travels a short distance, and enters a block of wood. Compute the acceleration at the times marked a, b, and c.

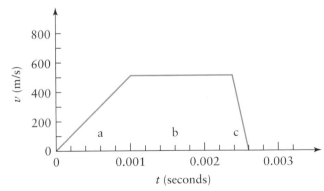

Figure 1.41

30. A bungee jumper falls for 1.3 s before the bungee cord begins to stretch. Until the jumper has bounced back up to this level, the bungee causes the jumper to have an average acceleration upward of 4 m/s².
    a) How fast is the jumper going when the bungee cord begins to stretch?
    b) How far below the diving platform is the jumper at that moment?

c) How long after the bungee cord begins to stretch does the jumper reach the low point of the drop?
d) How far below the diving platform is the jumper at the instant the speed is zero?

31. A drag-racing car goes from 0 to 300 mph in 5 s. What is its average acceleration in *g*'s?

CHALLENGES

1. Communication satellites are stationed 22,000 miles above the Earth's surface. How long does it take a radio signal traveling at the speed of light to travel from the Earth to the satellite and back?
2. Most cars can decrease their speed much more rapidly than they can increase it (see Problem 15). What factors contribute to cause this?
3. The Moon's mass is 7.35×10^{22} kg, and it moves in a nearly circular orbit with radius 3.84×10^8 m. The period of its motion is 27.3 days. Use this information to determine the Moon's (a) speed and (b) acceleration.
4. A sports car is advertised to have a maximum cornering acceleration of 0.85 *g*.
    a) What is the maximum speed that the car can go around a curve with a 100-m radius?
    b) What is its maximum speed for a 50-m radius curve?
    c) If wet pavement reduces its maximum cornering acceleration to 0.6 *g*, what do the answers to (a) and (b) become?
5. A spacecraft lands on a newly discovered planet orbiting the star Antares. To measure the acceleration due to gravity on the planet, an astronaut drops a rock from a height of 2 m. A precision timer indicates that it takes the rock 0.71 s to fall to the ground. What is the acceleration due to gravity on this planet?
6. When an object is thrown straight *upward*, gravity causes it to decelerate at a rate of 1 *g*—its speed decreases 9.8 m/s or 22 mph each second.

a) Explain how we can use the equations, tables, and graphs for a freely falling body in this case.
b) A baseball is thrown vertically up at a speed of 39.2 m/s (88 mph). How much time elapses before it reaches its highest point, and how high above the ground does it get?

7. For an object starting at rest with a constant acceleration, derive the equation that relates its speed to the distance it has traveled. In other words, eliminate time in the two equations relating *d* to *t* and *v* to *t*. Test the equation by showing that an object reaches a final speed of 9.8 m/s if it is dropped from a height of 4.9 m.

8. A race car starts from rest on a circular track with radius 100 m and begins to increase its speed by 5 m/s each second. At what point in time is the car's vector acceleration directed 45° away from straight ahead? What is the magnitude of the resultant acceleration at that moment?

9. Of the three kinds of acceleration a car undergoes during normal driving (speeding up, slowing down, centripetal acceleration), which typically has the largest magnitude? The smallest? (This will depend on driving habits of the driver, the performance of the car, road conditions, and so on.)

SUGGESTED READINGS

Bergquist, James C., Steven R. Jefferts, and David J. Wineland. "Time Measurement at the Millennium." *Physics Today* 54, no. 3 (March 2001): 37–42. "The latest clocks use a single ion to measure time with an anticipated precision of one part in (10 to the 18th)."

Brancazio, Peter J. *Sport Science.* New York: Simon and Schuster, 1984. A good source for applications of physics to sports, with material relevant to Chapters 1 through 4.

*The Guinness Book of Records****.* (Latest edition), Guinness Publishing, Ltd. A good source for a rich variety of record times, speeds, weights, and so on.

Hawking, Stephen W. *A Brief History of Time.* New York: Bantam Books, 1988. Best-selling book by a famous physicist on the "role of time in physics." (Only one equation included.)

Herring, Thomas A. "The Global Positioning System." *Scientific American* 274 no. 2 (February 1996): 44–50. Describes the GPS system and many of the diverse ways in which it is used.

Kleppner, Daniel. "On the Matter of the Meter." *Physics Today* 54, no. 3 (March 2001): 11–12. A look at the different ways that the "meter" has been defined over the years.

Segré, Emilio. *From Falling Bodies to Radio Waves.* New York: W. H. Freeman and Co., 1984. A history of physics up to a century ago, by a Nobel laureate.

Sobel, Dava. *Galileo's Daughter.* New York: Walker & Company, 1999. A biography of Galileo focusing on his daughter Sister Maria Celeste and letters she wrote to him.

"Special Issue: A Matter of Time." *Scientific American* 287, no. 3 (September 2002): 36–93. Ten articles on time, under headings that include physics, philosophy, and biology.

Wilke, S. R., R. E. McNair, and M. S. Feld, "The Physics of Karate." *American Journal of Physics* 51, no. 9 (September 1983): 783–790. Ronald E. McNair was one of the seven astronauts killed in the disaster of the space shuttle *Challenger* in 1986. For an obituary of this remarkable man, see *Physics Today* 39(4) (April 1986): 72–73. Part IV of the article by Wilke et al. (page 786) explains how the graphs in Figure 1.30 were produced.

For additional readings, explore InfoTrac® College Edition, your online library. Go to http://www.infotrac-college.com/wadsworth and use the passcode that came on the card with your book. Try these search terms: GPS, Galileo Galilei, Aristotle, metric system, atomic clock, Ronald McNair.

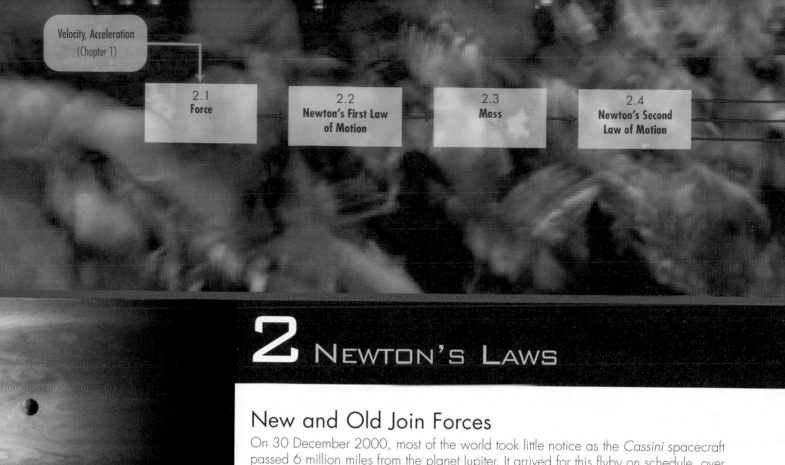

Velocity, Acceleration
(Chapter 1)

| 2.1 Force | 2.2 Newton's First Law of Motion | 2.3 Mass | 2.4 Newton's Second Law of Motion |

2 NEWTON'S LAWS

Courtesy of NASA

The moon Io appears to float above the cloudtops of Jupiter in this image captured by NASA's *Cassini* spacecraft on the dawn of the new millennium, January 1, 2001, two days after *Cassini's* closest approach to Jupiter.

New and Old Join Forces

On 30 December 2000, most of the world took little notice as the *Cassini* spacecraft passed 6 million miles from the planet Jupiter. It arrived for this flyby on schedule, over three years after it was launched, having traveled more than a billion miles. *Cassini* was on its way to Saturn for a detailed study of that planet and its moons. This was another routine event in the 40 or so years of exploring celestial bodies with space probes. We've gotten used to seeing amazing images they send back and to hearing of new discoveries about things like planetary rings and exotic moons.

The devices on *Cassini* (rocket engines, scientific instruments, computers, radio equipment, and so on) are products of twentieth-century technology. But the principles behind interplanetary travel are rooted in laws established by Isaac Newton over three hundred years ago. His third law of motion explains how a vehicle can accelerate itself in empty space where there is nothing to push against. His second law makes it possible to figure out how long to fire a rocket engine to change a spacecraft's velocity by a desired amount. And his law of universal gravitation is the tool for taking into account the effect of the Earth, the Sun, and other planets on a spacecraft's path. As if that weren't enough, Newton also invented calculus, the branch of mathematics essential for performing the latter two tasks. The validity and usefulness of these tools are reconfirmed every time a spacecraft arrives on target, on time, at a planet far away.

So while *Cassini's* dazzling images and wealth of scientific data represent another triumph of modern electronics, it is basic mechanics first spelled out in the seventeenth century that mission planners use to get it to Jupiter and beyond. After you finish this chapter, you will have some understanding of how this is done.

What is the difference between a rocket accelerating a craft in space and a car accelerating on a highway? How do mission controllers figure out how long to fire a rocket engine? Is the "gravity" affecting a spacecraft the same as the "gravity" making a rock fall? Read on to find the answers to these questions and more.

In this chapter we present Newton's three laws of motion and his law of universal gravitation. These laws form the basis of mechanics. Force, a central concept in physics, is introduced, and several common examples are given. Each law is used to extend your understanding of motion, of how forces affect motion, and of gravity. The cause of an object's acceleration, the principle behind rockets, and the nature of orbits are some of the applications of these laws that are described. The chapter concludes with a close look at the life of Sir Isaac Newton.

2.5 The International System of Units (SI)	2.6 Examples: Different Forces, Different Motions	Simple Types of Motion (Chapter 1)		2.8 The Law of Universal Gravitation	2.9 Tides
		2.7 Newton's Third Law of Motion			
Numerous Physical Quantities (Chapters 3–8, 10)	Oscillation, Waves (Chapters 3, 6, 8, 9)	Momentum, Energy Pressure + (Chapter 3–Epilogue)		Fundamental Forces (Chapters 7, 8, 12, Epilogue)	

2.1 Force

Sir Isaac Newton (1642–1727) was an English scholar who made many fundamental discoveries in both physics and mathematics. Newton is often regarded as the father of modern physical science, and for this reason, it is difficult to overestimate his importance in the development of today's civilization. The dominant scientific, industrial, and technological advancements of the last two centuries were triggered in part by Newton's work.

In formulating his mechanics, Newton began with Galileo's ideas about motion and then sought systematic rules that govern motion and, more importantly, *changes in motion.* The key concept in Newtonian mechanics is **force.**

DEFINITION

Force A push or a pull acting on a body. Force usually causes some distortion of the body, a change in its velocity, or both. Force is a vector.

Physical Quantity	Metric Units	English Units
Force (F)	newton (N)	pound (lb)
	dyne	ounce (oz)
	metric ton	ton

The conversion factors between the primary units of force in the two systems are

$$1 \text{ N} = 0.225 \text{ lb} \qquad 1 \text{ lb} = 4.45 \text{ N}$$

This means that a force of 150 pounds is equal to one of 668 newtons.

The distortion caused by a force is often obvious, such as the compression of a sofa cushion when you sit on it. Sometimes it cannot be observed without help. For example, high-speed photography reveals that a golf ball is flattened as a club hits it (● Figure 2.1). Spring scales, like the produce scales in supermarkets, use distortion to measure forces. The greater the force acting to stretch or compress a spring, the greater the distortion (● Figure 2.2).

Force is a bit difficult to define and is sometimes regarded as a fundamental quantity like time and distance. The English-system units should indicate to you that the idea of force is quite common. The words *push, shove, lift, pull,* and *yank* are some everyday synonyms that we use to describe force. The concepts of force and energy are probably the two

● **Figure 2.1** Force on the golf ball distorts it and accelerates it.

● **Figure 2.2** (a) A force of 5 newtons stretches the spring 9 centimeters. (b) Doubling the force on the spring doubles the distance it is stretched. The scale itself has a spring inside and makes use of this principle.

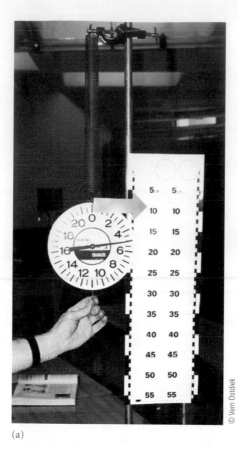

(a)

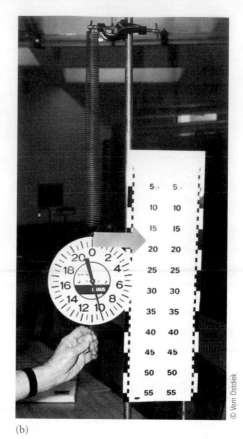

(b)

most ubiquitous and useful in all of physics. Newton's laws of motion are simple, direct statements about forces in general and their relationship to motion.

Let's consider some examples of force to show how versatile the concept is. We exert forces on objects in dozens of everyday situations, such as in pushing or pulling a door open, lifting a box, pulling down a window shade, and throwing a ball (● Figure 2.3). Many machines we use are designed to exert forces. Cranes, hoists, jacks, and vises are all examples of such machines. Automobiles and other propelled vehicles function by causing forces to act on them. If this last concept seems a bit odd to you, come back to it after you have read about Newton's third law of motion in Section 2.7.

The most common force in our lives is **weight.** Most of us measure the weight of our bodies regularly. Many things that we buy, such as flour, cat food, and nails, are sold by weight.

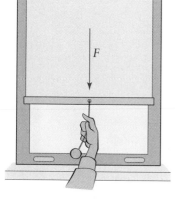

DEFINITION

Weight The force of gravity acting on a body. Abbreviated *W.*

● **Figure 2.3** We exert forces every day.

The direction of the force of gravity is what determines the directions up and down. We are so accustomed to living with this force that pulls everything toward the Earth's surface that we often forget that it is a force (● Figure 2.4). A child's simple observation that things naturally "fall down" leads one to Aristotle's idea of motion toward a natural place. But Newton made the important observation that objects are pulled toward the Earth by a force in much the same way that a sled is pulled by a person. The *law of universal gravitation*, the topic of Section 2.8, generalizes the concept of gravitational force.

The weight of an object depends on two things: the amount of matter comprising the object (its mass) and whatever is outside the object that is causing the gravitational force. The second point means that a body's weight depends on where it is. The weight of an object on the Moon is about one-sixth the weight it would have on Earth because the

● **Figure 2.4** Weight (*W*) is the downward force of gravity. This force acts on objects whether they are stationary, moving horizontally, or moving vertically. (Arrows are not to scale.)

gravitational pull of the Moon is less than the Earth's. Even on the Earth, the weight of an object varies slightly with location. A 190-pound person would weigh about 1 pound less at the equator than at the North or South Pole.

Another important force is **friction.**

Friction A force of resistance to relative motion between two bodies or substances in physical contact.

Friction is at work when a chair slides across a floor, when brakes keep a car from rolling down a hill, when air resistance slows down a baseball, and when a boat glides over the water. Notice that friction occurs at the surfaces or boundaries of the materials involved. We can distinguish two types of friction, static and kinetic. When there is no relative motion between two objects, the friction that acts is **static friction.** If you try to push a refrigerator across a floor, you must first overcome the force of static friction between it and the floor. A person who is walking or running relies on the static friction between his or her shoes and the ground. Even though the person's body is moving relative to the ground, there is no relative motion between the shoes and the ground when they are in contact. It is difficult to walk on ice because the force of static friction is reduced. Even a moving car relies on the static friction between its tire surfaces and the pavement. As long as the tires are not skidding or spinning, there is no relative motion between them and the pavement where they are in contact with one another.

Explore It Yourself 2.1

1. Place a coin, an eraser, a small book, or some other object on a large hardcover book or notebook, and gradually tilt the book until the object slides off. Note how steep the book has to be before the object starts sliding. Do this with several different objects. Is the steepness of the book when the different objects begin to slide the same?
2. Now repeat this with something that can roll, like a ball or a pen. What is different? What can you conclude about the difference between the force of sliding friction and that of rolling friction?

A block of wood resting on an inclined plane is a simple system that relies on static friction (● Figure 2.5). If the plane is horizontal, there is no friction. When the plane is tilted a small amount, friction acts to keep the block from sliding down. Here the force of friction opposes the component of the force of gravity—weight—which acts parallel to the plane. The steeper the plane is, the greater the force of static friction must be to keep the block from sliding down. When the angle of the plane reaches a certain value, the block will begin to move down. The component of the weight parallel to the plane will

Refer to "vector addition" in Section 1.2.

Figure 2.5 Static friction opposes gravity, which acts to pull the block down the incline. As the plane is tilted, both forces (F_g and F_f) increase until the maximum possible force of static friction is reached (c). Further tilting makes F_g larger than F_f, and the block accelerates down the incline (d).

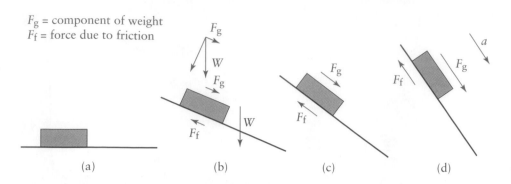

F_g = component of weight
F_f = force due to friction

(a)   (b)   (c)   (d)

have exceeded the maximum force of static friction. Notice that the force of static friction between two surfaces can have any value between zero and some maximum.

Kinetic friction acts when there is relative motion between two substances in contact, such as an aircraft when moving through the air, a fish swimming underwater, or a tire skidding on pavement. In Figure 2.5d, kinetic friction acts on the block of wood when it is sliding down the inclined plane. The force of kinetic friction that acts between two solids is usually less than the maximum static friction that can act. This is why a car can be stopped more quickly when its tires are not skidding.

Explore It Yourself 2.2

1. As in Explore It Yourself 2.1, place something on a large hardcover book or notebook, and tilt it until the object barely slides off. As it slides, which type of friction is acting on it?
2. Decrease the steepness of the book slightly, and put the object back on it. It should stay put. Which type of friction is acting on it now?
3. If you have set up things just right, when you give the object a slight push it will continue to slide on its own, at least for a bit. How does this relate to the difference between the force of static friction and that of kinetic friction?

The effects of kinetic friction are often undesirable. A car that is traveling on a flat road at a constant speed consumes fuel mainly because it must act against the forces of kinetic friction—air resistance acting on the car's exterior and friction between various moving parts in the axles, transmission, and engine. This also applies to aircraft and ships. Brakes represent a useful application of kinetic friction. On most bicycles, the brakes consist of pads that rub against the wheel rims (• Figure 2.6). When the brakes are applied, the force of kinetic friction between the pads and the rim slows the bicycle.

Often it is difficult to include the effects of friction in mathematical models of physical systems. Consequently, we will find it helpful to consider situations in which frictional forces are small enough to be ignored. It is easier to analyze the motion of a falling rock than that of a falling feather. We need to understand these simple systems before we can incorporate the complexities of friction. For a closer look at the scientific study of friction, see the Physics Potpourri, "Friction: A Sticky Subject," on page 52.

Figure 2.6 Bicycle brakes use kinetic friction as the pads rub against the rim of the wheel.

LEARNING CHECK

1. (True or False.) Every force that acts on anything has a specific direction associated with it.
2. Which of the following is *not* a unit of force?
   a) newton   b) kilogram   c) ounce
   d) ton   e) All are units of force.
3. The force of gravity acting on an object is called
   _____.

4. The air exerts a force on a person standing in a strong wind. This type of force is one example of _____ friction.

ANSWERS: 1. True 2. (b) 3. weight 4. kinetic

2.2 Newton's First Law of Motion

As we have just seen, many forces act in everyday situations. Newton's laws make no reference to specific types of forces. They are general statements that apply whether a force is caused by gravity, friction, or a direct push or pull.

> **Newton's First Law of Motion** An object will remain at rest or in motion with constant velocity unless acted on by a net external force.

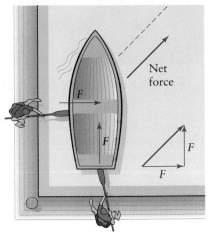

● **Figure 2.8** Two sailors push on a boat in different directions. The net force is the vector sum of the two applied forces.

Refer to "vector addition" in Section 1.2.

The last phrase needs some explanation. An *external* force is one that is caused by some outside agent. Weight is an external force because it is caused by something outside an object (for example, the Earth). If your car stalls, you cannot move it by sitting in the driver's seat and pushing on the windshield. This would be an internal force. You must get out of the car and push from the outside. The *net* force is the vector sum of the forces acting on the body. If one person pushes forward on your car while another pushes backward with equal effort, the net force is zero. If two forces act in the same direction, the net force is the sum of the two (● Figure 2.7). If two forces act in different directions, the net force is found by adding the two vectors together (● Figure 2.8).

At first glance, Newton's first law does not seem to be profound. Obviously, an object will remain stationary unless a net force causes it to move. But the law also states that anything that is already moving will not speed up, slow down, or change direction unless a net force acts on it. This means that the states of no motion and of uniform motion are equivalent as far as forces are concerned. Aristotle's flawed concept of motion (pp. 38–40) implies that a force is required to maintain it. Newton's first law implies that a force is required only to *change* the state of motion. This may seem to run counter to your intuition, but that is because you rarely see a moving object with no forces acting on it. A car traveling at a constant velocity has zero net force acting on it: The various forces (including air resistance and gravity) cancel each other.

As you throw a ball, your hand exerts a net force that is in the same direction as the ball's velocity. The ball's speed increases because the force acts in the same direction as its

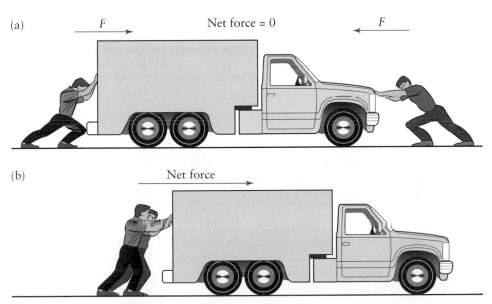

● **Figure 2.7** (a) Equal forces in opposite directions produce a net force equal to zero. (b) Forces in the same direction add together.

Friction: A Sticky Subject

"Can't live with it and can't live without it." This old adage can easily be used to describe friction. On the one hand, friction can be a nuisance. Friction and the general wear produced by it on machine parts sliding past one another have enormous economic impact and affect our national security and quality of life. It has been estimated that the equivalent of between 2 and 6 percent of the U.S. gross domestic product could be saved each year by improved attention to reducing friction and its associated wear and energy losses. Adopting even the lowest estimates, this amounts to a staggering $120 billion annually. On the other hand, without friction it would be impossible to walk or drive a car, to play a violin, or to skydive. In some cases, it's beneficial to maximize friction instead of minimize it, as in the friction between the tires of a car and the road during braking.

Tribology (from the Greek *tribein*, to slide or rub) is a branch of science dedicated to the study of friction and wear and how to control them by lubrication. The use of friction by human beings dates to neolithic times when fires were started by rubbing sticks together or by generating sparks by striking flint. The value of lubricants was recognized more than 4000 years ago by Sumerian and Egyptian engineers who used oil and mud to ease the transportation of large stones on sledges at construction sites (● Figure 2.9). The first systematic study of friction is found in the work of Leonardo da Vinci more than 500 years ago. (Among other things, Leonardo appears to have been the first to experiment with objects sliding on inclined planes as shown in Figure 2.5.) However, many aspects of friction are still not well understood today, including its origins in the electromagnetic forces that dominate at the submicroscopic level.

Because Leonardo's experiments lay undiscovered in his notebooks until the 1960s, first priority in describing the nature of the frictional interaction between two solid surfaces sliding against one another is usually given to the French physicist Guillaume Amontons. In 1699, Amontons noted that the force of friction between two surfaces in relative motion is directly proportional to the force pressing the surfaces together. The constant of proportionality is called the *coefficient of friction*. Amontons also reported that the friction force was independent of the apparent area of contact: A small block experiences the same frictional force as a large one made of the same material as long as the force pressing each of them against a third surface is the same.

Additional aspects of the frictional interaction were discovered more than fifty years later by Charles Augustin Coulomb, better known for his work on electrostatics (see Section 7.2), in a comprehensive experimental study based on Newtonian mechanics. While confirming the earlier results of Amontons, Coulomb found in addition that the

● **Figure 2.9** Portion of a relief from the tomb of Ti, a royal hairdresser and later steward in Egypt's 5th dynasty. Dating from around 2400 B.C., the image shows one of the first documented uses of lubrication to reduce friction. In particular, a worker is seen pouring an unknown liquid ahead of and beneath the sledge on which Ti's statue is being dragged to the tomb.

friction force is generally independent of the speed with which the two objects slide past one another, at least as long as the relative velocity is not too high. Coulomb also studied the differences between static and kinetic friction. He was the first to begin to model the frictional interaction by imagining that the surfaces were covered with elastic fibers that became intermeshed, thus impeding the smooth slippage of one object past the other.*

---

* The conditions described apply only for sliding friction between solids. For "viscous friction" involving solids moving through fluids like air or water, experiments reveal that the frictional force is directly proportional to the relative speed of the object through the fluid.

motion. When you catch the ball, the force is opposite to the direction of the ball's velocity. Here the force slows down the ball.

Another important point implied by the first law is that a force is required to change the direction of motion of an object. Velocity includes direction. Constant velocity implies constant direction as well as constant speed. To make a moving object change its direction of motion, a net external force must act on it. To deflect a moving soccer ball, a player must exert a sideways force on it with the foot or head (● Figure 2.11). Without such a force, the ball will not change direction.

Refer to "acceleration" in Section 1.3.

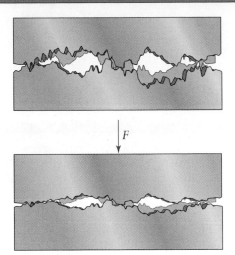

● **Figure 2.10** A highly simplified schematic of our modern conception of interacting surfaces. Here, the contact points (or *asperities*) are the locations where friction occurs between two rough surfaces sliding past one another (upper). If the force that squeezes the two surfaces together increases, then so too does the total area of microscopic contact. That increase, and not merely the degree of surface roughness, governs the size of the resulting frictional force.

Although these "laws" of sliding friction have been known for centuries, attempts to explain them have only relatively recently begun to show promise. The beginnings of a proper explanation of why the friction force is independent of the apparent area of contact was not presented until around 1940. At this time, Frank Bowden and David Tabor suggested that when two surfaces touch one another, the actual microscopic area of contact, determined when *asperities*—atomic pits or mountains—on one surface meet those of the other, is far smaller than the apparent macroscopic area (● Figure 2.10). At each of these microscopic contact points, a local bond is formed that welds the two surfaces together. According to this so-called *adhesion model* of friction, to initiate sliding motion between two objects, a sufficiently large force must be applied to sever the existing microscopic bonds (overcome the static frictional force) between them. Once in relative motion, a force (albeit of lesser magnitude) must still be applied to keep the surfaces sliding past one another because new contact regions are continuously being formed as others are being broken.

Modern studies of friction at the atomic level using atomic-force microscopes (close relatives to the scanning tunneling microscopes discussed in Chapter 10) have inaugurated the era of *nanotribology* and have shown that the adhesion model is not completely correct. In particular, investigations carried out in the 1990s reveal that friction does not correlate with the strength of the adhesive bond but rather with the so-called *adhesive irreversibility*, that is, how easily two surfaces become stuck relative to becoming unstuck. The processes of sticking and unsticking are not, it turns out, symmetric. It's their difference that determines the frictional force. Unfortunately, precisely characterizing this difference in terms of fundamental physics has proved elusive. Current treatments of friction including sound-wave mechanics (see Chapter 6) are helping to model the energy losses that accompany this process.

What now seems clear is that Amontons's observations that the friction force is proportional to the force pressing the surfaces together can be understood from an atomic perspective: The harder you squeeze two surfaces together, the greater the true area of microscopic contact (asperity to asperity) and the greater the frictional force. Even with the deeper insight into friction that modern surface studies have afforded, however, we still do not have a comprehensive physical model for friction that allows us to connect directly our knowledge of the microscopic contact point forces and the macroscopic "laws" of friction. Although prospects for further elucidating the nature of frictional interactions, the role of lubrication in reducing such interactions, and wear appear excellent, it still seems fair to say that friction remains a sticky business.

You can get some idea of the frictional forces that exist between different surfaces under various circumstances by comparing the coefficients of friction measured by scientists and engineers over the years. For example, the static friction coefficient for rubber on dry concrete is about 1.0, while the value for rubber on wet concrete is 30% lower, only 0.7. Little wonder, then, why it's harder to start and stop your car on wet pavement than on dry. Similarly, the static friction coefficient for shoes on wood is 0.9, while that for shoes on ice is 0.1—results not too surprising to anyone who's taken a fall on an icy sidewalk in winter. Go on-line and search out some other values of the coefficients of friction for other systems. Do the tabulated values confirm your personal experience? Be sure to note the differences between the coefficients of static and kinetic friction. Can you explain these differences in terms of our model of friction? Try to secure some measurements for biological systems, if possible. Joint wear and the ameliorating effects of fluid-film lubrication are currently among the hottest topics in the field of biomechanics. Good luck!

Explore It Yourself 2.3

For this you need a sock, a piece of thread about an arm's length or more long, a reasonably sharp knife (be careful!), and a place where a flying, rolled-up sock isn't going to hurt anything. Roll up the sock and tie it to one end of the thread. Grasp the other end and whirl the sock in a horizontal circle above your head. With your free hand, or with the aid of an assistant, quickly (but carefully!) move the knife into the path of the thread so that the knife cuts the thread near its middle at the moment the sock is passing directly in front of you. Stow the knife, and then find the sock. In which direction did it go after the force on it was removed (the string was cut)? Not sure? Tie the two pieces of thread together and do it again.

The implication of Newton's first law is that a net external force must act on an object to speed it up, slow it down, or change its direction of motion. Because each of these is an acceleration of the object, the first law tells us that a force is required to produce any acceleration. The centripetal acceleration inherent in circular motion requires a force; it is called the **centripetal force.** Like centripetal acceleration, the centripetal force on an object is directed toward the center of the object's path. As a car goes around a curve, the centripetal force that acts on it is a sideways frictional force between the tires and the road. Without this friction, as is nearly the case on ice, there is no centripetal force, and the car will go in a straight line—off the curve.

Imagine a rubber ball tied to a string and whirled around in a horizontal circle overhead. The ball "wants" to move in a straight line (by the first law) but is prevented from doing so by the centripetal force acting through the string. If the string breaks, the path of the ball will be a straight line in the direction the ball was moving at the instant the string broke (● Figure 2.12). You may be tempted to think that the ball would move in a direct line away from the center of its previous circular motion. But that is not the case. If you ever see a hammer thrower, a discus thrower, or someone using a sling (like the one David used against Goliath), notice the path of the projectile after it is released. It moves along a line tangent to its original circular path.

There are other examples of this effect. Children (or even physics professors) riding on a spinning merry-go-round must hold on because their bodies "want" to travel in a straight line, not in a circle. Similarly, clothes are partially dried during the spin cycle in an automatic washer because the water droplets tend to travel in straight lines and move through the holes in the side of the tub. The clothes are "pulled away" from the water as they spin. The same effect could be used to create an "artificial gravity" on a space station by making it spin (● Figure 2.13). (The classic movie *2001: A Space Odyssey* showed this well.) You may have tried out a "spinning room" at an amusement park; it uses the same principle.

● **Figure 2.12** Centripetal force must act on any object to keep it moving on a circular course. If the force is removed, the object will move at the same speed in a straight line.

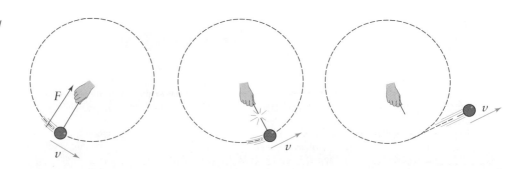

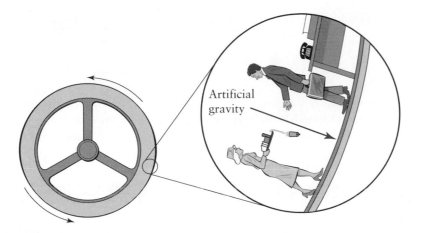

● **Figure 2.13** Model of a rotating space station. The circular paths of the occupants would make them experience artificial gravity.

Artificial gravity

LEARNING CHECK

1. For an object to move with constant velocity,
   a) the net force on it must be zero.
   b) a constant force must act on it.
   c) it must move in a circle.
   d) there must be no friction acting on it.

2. (True or False.) It is possible for the net force on an object to be zero even if several forces are acting on it at the same time.

3. The force that must act on a body moving along a circular path is called the _____.

ANSWERS: 1. (a) 2. True 3. centripetal force

2.3 Mass

Newton's second law of motion states the exact relationship between the net force acting on an object and its acceleration. But before stating the second law, we will take another look at mass. Imagine the effect of a small net force acting on a car (see ● Figure 2.14). The resulting acceleration would be quite small. The same net force would cause a much larger acceleration if it acted on a shopping cart.

The property of matter that makes it resist acceleration is often referred to as *inertia*. A car is more difficult to accelerate than a shopping cart because it has more inertia. The concept of inertia is embodied in the physical quantity *mass*.

> **DEFINITION**
>
> **Mass** A measure of an object's resistance to acceleration. A measure of the quantity of matter in an object.

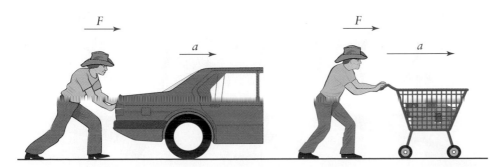

F a F a

● **Figure 2.14** The effect of a net force on an object depends on the object's mass. If the mass is very large, the acceleration of the object will be small. Conversely, if the mass is small, the acceleration will be large.

Physical Quantity	Metric Units	English Units
Mass (m)	kilogram (kg) gram (g)	slug

The concept of mass is not in everyday use in the English system. For example, flour is purchased by weight (pounds), not by mass (slugs). In the metric system, the situation is the reverse. Mass is commonly used instead of weight. Flour is purchased by the kilogram, not by the newton.*

The actual mass of a given object depends on its size (volume) and its composition. A rock has more mass than a pebble because of its size; it has more mass than an otherwise identical piece of Styrofoam because of its composition. The rock is harder to accelerate in the sense that a larger force is required. Fundamentally, the mass of a body is determined mainly by the total numbers of subatomic particles that comprise it (protons, neutrons, and electrons).

Mass is *not* the same as weight. Mass and weight are related to each other in that the weight of an object is proportional to its mass. But weight is a force that is an effect of gravity. Mass is an intrinsic property of matter that does not depend on any external phenomenon. One major cause of confusion is that users of both the English system and the metric system often incorrectly lump together the two concepts in everyday use. Weight is often incorrectly used in place of mass in countries using the English system, and mass is often incorrectly used in place of weight in countries using the metric system. It usually doesn't matter since they are proportional to each other, but if sometime in the future people routinely travel to the Moon or to other places where the acceleration due to gravity is different, then it will.

If a hammer has a mass of 1 kilogram, that will not change if it is taken into space aboard a spacecraft and then to the Moon's surface. Its mass is 1 kilogram wherever it is. The hammer's weight, however, varies with the location, because weight depends on gravity (see ● Figure 2.15). Its weight on Earth is 9.8 newtons, its weight would appear to be zero in a spacecraft traveling to the Moon, and its weight would be only 1.6 newtons on the Moon's surface. Though "weightless" in the spacecraft, the hammer is not massless and would still resist acceleration as it does on Earth.

Another way of approaching mass and weight is to think of them as two different characteristics of matter. We might call these aspects *inertial* and *gravitational*. Mass is a measure of the inertial property of matter—how difficult it is to change its velocity. Weight illustrates the gravitational aspect of matter: any object experiences a pull by the Earth, the Moon, or any other body near it.

---

* To help you get a feel for the size of the kilogram, we offer the following "mixed" conversion:

1 kilogram *weighs* 2.2 pounds (on Earth)

One kilogram *does not equal* 2.2 pounds! On Earth, anything with that mass has a *weight* of 9.8 newtons—which happens to equal 2.2 pounds.

● **Figure 2.15** The hammer's weight depends on where it is, but its mass is always the same—1 kilogram.

On the Earth

In space

On the Moon

2.4 Newton's Second Law of Motion

We are now ready to state Newton's second law of motion. This law is our most important tool for applying mechanics in the real world.

> **LAWS**
>
> **Newton's Second Law of Motion** An object is accelerated whenever a net external force acts on it. The net force equals the object's mass times its acceleration.*
>
> $$F = ma$$

This law expresses the exact relationship between force and acceleration. For a given body, a larger force will cause a proportionally larger acceleration (see ● Figure 2.16). Force and acceleration both are vectors. The direction of the acceleration of an object is the same as the direction of the net force acting on it.

An airplane with a mass of 2,000 kilograms is observed to be accelerating at a rate of 4 m/s². What is the net force acting on it?

$$F = ma = 2,000 \text{ kg} \times 4 \text{ m/s}^2$$
$$= 8,000 \text{ N}$$

Example 2.1

---

* The unit of measure of force, the newton, is established by this law. One newton is the force required to give a 1-kilogram mass an acceleration of 1 m/s². In other words:

$$1 \text{ newton} = 1 \text{ kilogram–meter/second}^2$$

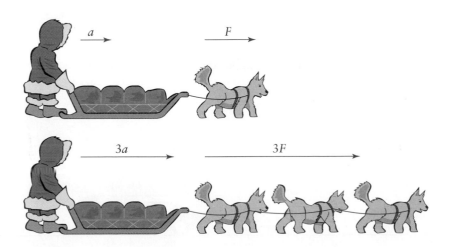

● **Figure 2.16** The acceleration of an object is proportional to the net force. Tripling the force on an object triples its acceleration.

Another way of stating Newton's second law is that an object's acceleration is equal to the net force acting on the object divided by its mass.

$$a = \frac{F}{m}$$

This means that a large force acting on a large mass can result in the same acceleration as a small force acting on a small mass. A force of 8,000 newtons acting on a 2,000-kilogram airplane causes the same acceleration as a force of 4 newtons acting on a 1-kilogram toy.

The following example illustrates what can be done with the mechanics that we have learned so far.

Example 2.2   An automobile manufacturer decides to build a car that can accelerate uniformly from 0 to 60 mph in 10 s (see ● Figure 2.17). In metric units, this is from 0 to 27 m/s. The car's mass is to be about 1,000 kilograms. What is the force required?

First, we must determine the acceleration and then use Newton's second law to find the force. As we did in Chapter 1:

$$a = \frac{\Delta v}{\Delta t} = \frac{27 \text{ m/s} - 0 \text{ m/s}}{10 \text{ s}}$$

$$= 2.7 \text{ m/s}^2$$

So the force needed to cause this acceleration is

$$F = ma = 1{,}000 \text{ kg} \times 2.7 \text{ m/s}^2$$

$$= 2{,}700 \text{ N}$$

(This is equal to $2{,}700 \times 0.225$ lb $= 607.5$ lb.) In the next chapter, we will use this information to determine the size (power output) of the car's engine.

Newton's second law establishes the relationship between mass and weight. Recall that any freely falling body has an acceleration equal to g (9.8 m/s²). By the second law, the size of the gravitational force needed to cause this acceleration is

$$F = ma = mg$$

We call this force the object's weight. So in this case the expression "force is equal to mass times acceleration" translates to "weight is equal to mass times acceleration due to gravity."

$$F = ma \rightarrow W = mg$$

On Earth, where the acceleration due to gravity is 9.8 m/s², the weight of an object is

$$W = m \times 9.8 \qquad \text{(on Earth, } m \text{ in kg, } W \text{ in N)}$$

The weight of a 2-kilogram brick is

$$W = 2 \text{ kg} \times 9.8 \text{ m/s}^2 = 19.6 \text{ N}$$

On the Moon, the acceleration due to gravity is 1.6 m/s². Therefore,

$$W = m \times 1.6 \qquad \text{(on the Moon, } m \text{ in kg, } W \text{ in N)}$$

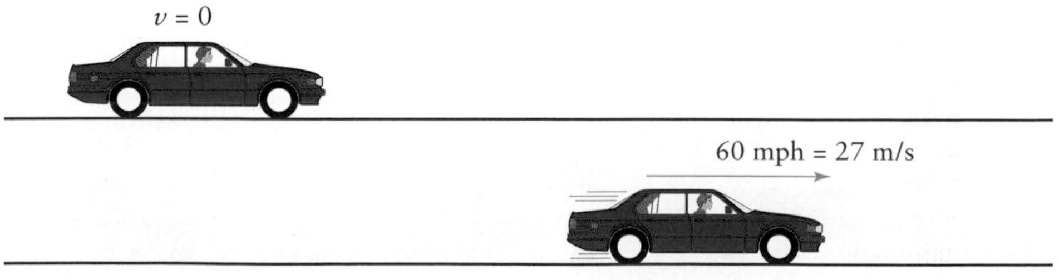

● **Figure 2.17** Car accelerating from rest to 60 mph or 27 m/s.

$v = 0$

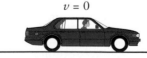

60 mph = 27 m/s

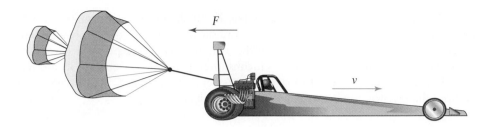

● **Figure 2.18** The dragster slows down because the net force on it (*F*), caused mainly by the parachute, is opposite its velocity (*v*).

Explore It Yourself 2.4

For this you need the same "sock on a string" used in Explore It Yourself 2.3, but no knife. Whirl the sock in circles overhead at a low speed (less than one complete circle per second). Then do it with a higher speed. (If the thread is weak, it may break.) What do you feel in the string when the sock's speed is high compared to when its speed is low? What does this tell you about the relationship between the speed of an object moving in a circle and the size of the centripetal force that acts on it?

Forces that cause deceleration or centripetal acceleration also follow the second law. To slow down a moving object, a force must act in the direction opposite the object's velocity (see ● Figure 2.18). A force acting sideways on a moving body causes a centripetal acceleration. The object moves along a circular path as long as the net force remains perpendicular to the velocity. The size of the centripetal force required is

Refer to "centripetal acceleration" in Section 1.3.

$$F = ma$$

and since

$$a = \frac{v^2}{r}$$

$$F = \frac{mv^2}{r} \qquad \text{(centripetal force)}$$

In Example 1.5, we computed the centripetal acceleration of a car going 10 m/s around a curve with a radius of 20 meters. If the car's mass is 1,000 kilograms, what is the centripetal force that acts on it?

Example 2.3

$$F = \frac{mv^2}{r} = \frac{1,000 \text{ kg} \times (10 \text{ m/s})^2}{20 \text{ m}}$$

$$= 5,000 \text{ N} = 1,120 \text{ lb}$$

Since we had already computed the acceleration (*a* = 5.0 m/s²), we could have used *F* = *ma* directly.

$$F = ma = 1,000 \text{ kg} \times 5.0 \text{ m/s}^2$$

$$= 5,000 \text{ N}$$

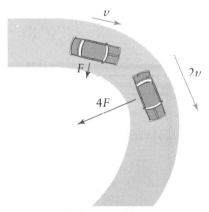

The centripetal force acting on a car going around a flat curve is supplied by the friction between the tires and the road. Note that the faster the car goes, the greater the force required to keep it moving in a circle. If two identical cars go around the same curve, and one is going two times as fast as the other, the faster car needs four times the centripetal force (see ● Figure 2.19). The required centripetal force is inversely proportional to the radius of the curve. A tighter curve (one with a smaller radius) requires a larger force or a smaller speed. If a car goes around a curve too fast, the force of friction will be too small to maintain the centripetal force, and the car will go off the outside of the curve.

● **Figure 2.19** The centripetal force necessary to keep a car on a curved road is supplied by static friction between the tires and the road. The size of the force is proportional to the square of the speed. One car going twice as fast as the other would need four times the centripetal force.

This form of Newton's second law, $F = ma$, is not the original version, but it is the most useful one for our purposes. It applies only when the mass of the object doesn't change. This is almost always the case, but there are important exceptions. For example, the mass of a rocket decreases rapidly as it consumes its fuel. Consequently, its acceleration will increase even if the net force on the rocket doesn't change. Mathematics more advanced than we will use is needed to deal with such cases.

LEARNING CHECK

1. When a given net force acts on a body, the body's acceleration depends on its _____.
2. The weight of an object equals its mass times _____.
3. Identical cars A and B go around the same curve, with A traveling two times as fast as B. The centripetal force on A

a) is two times as large as the centripetal force on B.
b) is four times as large as the centripetal force on B.
c) is one-half as large as the centripetal force on B.
d) is the same as that on B, because they are on the same curve and have the same mass.

ANSWERS: 1. mass 2. acceleration due to gravity 3. (b)

2.5 The International System of Units (SI)

The metric unit of force—the newton—is the force required to cause a 1-kilogram mass to accelerate 1 m/s². So when a mass in kilograms is multiplied by an acceleration in meters per second squared, the result is a force in newtons. If grams or centimeters per second squared (cm/s²) were used instead, the force unit would not be newtons.

At this point, you may be getting confused by the different units of measure, particularly when different physical quantities are combined in an equation. To help alleviate this problem, a separate system of units within the metric system was established. It is called the **international system,** or **SI,** after the French phrase *Système International d'Unités.* This system associates only one unit of measure with each physical quantity (● Table 2.1). Each unit is chosen so that the system is internally consistent, that is, when two or more physical quantities are combined in an equation, the result will be in SI units if the original physical quantities are in SI units. For example, the SI units of distance and time are the meter and the second, respectively. Consequently, the SI unit of speed is the meter per second. Table 2.1 gives the SI units of the physical quantities that we have used so far.

You may notice that, for the most part, we have been using SI units in our examples. From now on, we will use SI units when working in the metric system.

2.6 Examples: Different Forces, Different Motions

In Chapter 1, we considered three simple states of motion, zero velocity, constant velocity, and constant acceleration. Now, let's see how force is involved in these cases and then look at some other types of motion. Here the particular cause of the force is not impor-

● **Table 2.1** SI Units (Partial List)	
Physical Quantity	**SI Unit**
Distance (d)	meter (m)
Area (A)	square meter (m²)
Volume (V)	cubic meter (m³)
Time (t)	second (s)
Frequency (f)	hertz (Hz)
Speed and velocity (v)	meter per second (m/s)
Acceleration (a)	meter per second squared (m/s²)
Force (F) and weight (W)	newton (N)
Mass (m)	kilogram (kg)

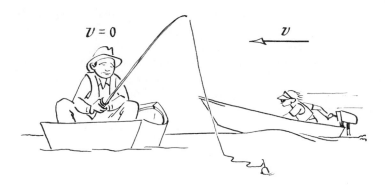

● **Figure 2.20** Whether the boats are moving with constant velocity or sitting at rest, the net force on both boats is zero.

tant. A constant net force acting on a moving object has the same effect whether it is due to gravity, friction, or someone's pushing the object. What interests us now is the relationship between the force—both its magnitude and direction—and the motion. Our concerns are the direction of the force compared to the object's velocity, whether the force is constant, and the way in which the force varies if it is not constant.

The acceleration of a stationary object or of one moving with uniform motion (constant velocity) is zero. By the second law, this means that the net force is also zero (see ● Figure 2.20). It is that simple. When the net force on a body is zero, we say that it is in *equilibrium,* whether it is stationary or moving with constant velocity.

In uniform acceleration, the second law implies that a constant force acts to cause the constant acceleration. A steady net force acting in a fixed direction will make an object move with a constant acceleration. A freely falling body is a good example: The force (weight) is constant and always acts in a downward direction.

A net force that acts opposite to the direction of motion of an object will cause it to slow down. If the force continues to act, the object will come to a stop at an instant and then accelerate in the direction of the force, opposite its original direction of motion. This is what happens when you throw a ball straight up into the air (see ● Figure 2.21). It is accelerating downward the whole time. While on the way up, its speed decreases until it reaches its highest point. There it has zero speed at an instant and then it falls with increasing speed. Its acceleration is always g, even at the instant that it is stopped.

What happens when an object is thrown upward at an angle to the Earth's surface? This is one example of a classic mechanics problem—projectile motion. The motion is a composite of horizontal and vertical motions. The path that a projectile takes has a characteristic arc shape that is nicely illustrated by the stream of water from a water fountain (see ● Figure 2.22). This shape, an important one in mathematics, is called a *parabola*.

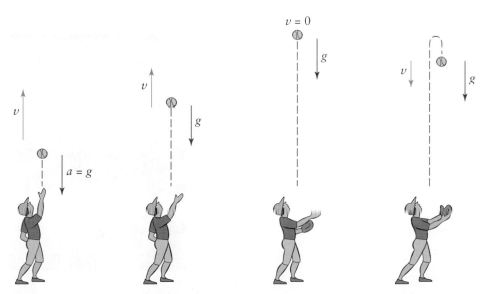

● **Figure 2.21** When a ball is thrown straight up, its acceleration is 1 g downward. This causes it to slow down on its way upward, stop at an instant, and then speed up on its way downward.

The key to understanding projectile motion is to realize that the vertical force of gravity has no effect on the horizontal motion. An object initially moving horizontally has the same downward acceleration g as an object that is simply dropped. So, ignoring air resistance, a projectile moves horizontally with constant speed and vertically with constant downward acceleration (● Figure 2.23). As an object moves along its path, the vertical component of its velocity decreases to zero (at the highest point of the arc) and then increases downward. The horizontal component of its velocity stays constant. Over level ground, a projectile will travel farthest if it starts at an angle of 45° to the ground. When the force of air resistance is large enough to affect the motion, such as when a softball is thrown very far, the maximum range occurs at smaller angles.

© Vern Ostdiek

● **Figure 2.22** The stream of water from a water fountain shows the shape of the path of projectiles—a parabola.

Explore It Yourself 2.5

Investigate projectile motion with a garden hose over level ground on a day with no wind. Shoot the water up at an angle to the ground, and note how the distance to where it hits the ground varies with the angle of the nozzle. What angle gives the maximum range? For any range shorter than this there are two different angles that will give the same range. Demonstrate this.

Simple Harmonic Motion

Here is a simple situation with a force that is not constant. Imagine an object, say, a block, attached to the end of a spring (● Figure 2.24). When it is not moving, the block is in equilibrium, and the net force on it is zero. If you lift the block up a bit and then release it, it will experience a net force downward, toward its original (rest) position. The reverse occurs if you pull it down a bit below its rest position and release it. Then the net force on it will be up, again toward its original position. Such a force is called a *restoring force* because it acts to restore the system to the original configuration. In this case, the net force is proportional to the distance from the rest position. The farther the block is displaced, the greater the force acting to move it back. (Figure 2.2 also shows this.)

What kind of motion does this force cause? If an object is pulled down and then released, the upward force will cause it to accelerate. It will pick up speed as it moves upward, but the force and acceleration will decrease as it nears the rest position. When it reaches that point, the force is zero, and it stops accelerating but continues its upward motion (Newton's first law). Once it has moved past the rest position, the force is down-

● **Figure 2.23** (a) The path of a projectile is shown with its velocity vector at selected moments. At the very top its velocity is horizontal. (b) The same projectile with the vertical and horizontal components of the velocity shown separately. The horizontal velocity component stays constant, while the vertical component decreases, and then increases downward, because of the constant downward acceleration due to gravity. (c) A strobe view of three "projectiles" being juggled.

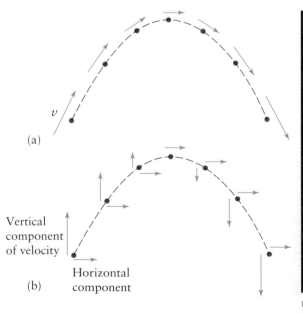

v

(a)

Vertical component of velocity

Horizontal component

(b)

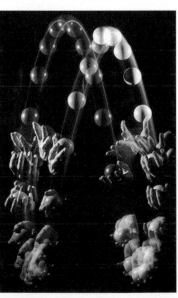

© Richard Megna/Fundamental Photographs

(c)

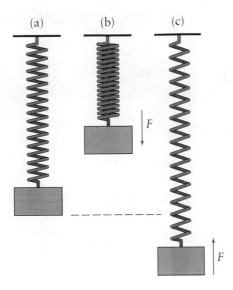

F

F

● **Figure 2.24** The net force that acts on a block hanging from a spring depends on the displacement of the block from its rest (equilibrium) position (a). If raised and then released, the block experiences a net force downward (b). If the block is pulled downward and then released (c), the net force is upward.

ward. The object will slow down, stop at an instant, and then gain speed downward. This process is repeated over and over: The object oscillates up and down.

This type of motion, which is very important in physics, is called **simple harmonic motion.** It occurs in many other systems—a pendulum swinging through a small angle, a cork bobbing up and down in water, a car with very bad shock absorbers, and air molecules vibrating with the sound from a tuning fork. In fact, simple harmonic motion is involved in all kinds of waves, as we shall see.

The graphs of distance versus time and velocity versus time show the characteristic oscillation (● Figure 2.25). Both graphs have what is known as a *sinusoidal* shape. This shape arises often in physics, and we will see it again when we talk about waves (Chapter 6).

Simple harmonic motion is cyclical motion with a constant frequency. The frequency of oscillation depends only on the mass of the object and the strength of the spring.* The same mass on a stronger spring will have a higher frequency. For a given spring, a larger

---

* For the mathematically inclined—the strength of a spring is found by measuring the force F needed to stretch it a distance d. Then one computes:

$$k = \frac{F}{d}$$

where k is called the *spring constant*. A strong (stiff) spring has a large k, and a weak spring has a small k. A mass m attached to the spring would oscillate with a frequency given by:

$$f = \frac{1}{2\pi} \times \sqrt{\frac{k}{m}} = 0.159 \times \sqrt{\frac{k}{m}}$$

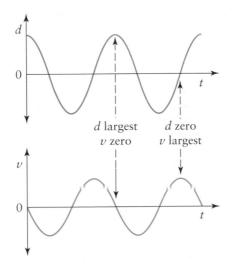

d largest
v zero

d zero
v largest

● **Figure 2.25** Graphs of d versus t and v versus t for simple harmonic motion. Notice that the velocity is zero when the distance is largest, and vice versa. At the high and low points of the motion, the mass is not moving at an instant. As it passes through the rest position, it has its highest speed.

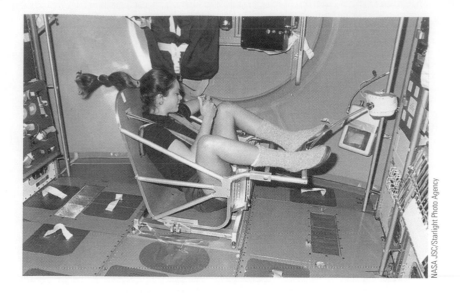

mass will have a smaller (lower) frequency. This principle is employed by an inertial balance, a device that uses simple harmonic motion to measure the mass of an object. To use it, one measures the frequency of oscillation of the object and then compares this frequency to those of known masses listed in a table or plotted on a graph. In an orbiting spacecraft, this sort of scale must be used because of the weightless environment (see ● Figure 2.26).

Falling Body with Air Resistance

The force of air resistance that acts on things moving through the air, like a thrown baseball or a falling skydiver, is one example of kinetic friction. This force is in the opposite direction of the object's velocity and will cause the object to slow down if no other force opposes it. The faster an object goes, the larger the force of air resistance. There is no simple equation for the size of the force of air resistance. For things like baseballs, bicyclists, cars, and aircraft, the force of air resistance is approximately proportional to the square of the object's speed relative to the air. For example, on a day with no wind the force of air resistance on a car going 60 mph is about four times as large as when it is going 30 mph.

Explore It Yourself 2.6

You need a shoe, and two identical pieces of waste paper—like two pages from a discarded magazine. Wad up one piece very tightly, to the approximate size of a golf ball. Wad up the other more loosely, to about the size of a grapefruit.

1. Hold the two wadded up pieces of paper above your head and drop them at the same time. Do they reach the floor at the same time? Does the weight of a body alone determine the effect of air resistance on it as it falls?
2. Similarly, drop the shoe and the loosely wadded piece of paper. Do they reach the floor at the same time? What does this show about the effect of air resistance on falling bodies?

Without air resistance the constant force on a falling body (its weight) gives it a constant acceleration g. Its speed increases steadily until it hits the ground. But as a body falls through the air, the force of air resistance grows as the speed increases and eventually affects the motion (● Figure 2.27). This increasing force acts opposite to the downward force of gravity, so the net force decreases. This continues as the body gains speed until the force of air resistance acting upward equals the weight acting downward. At this point,

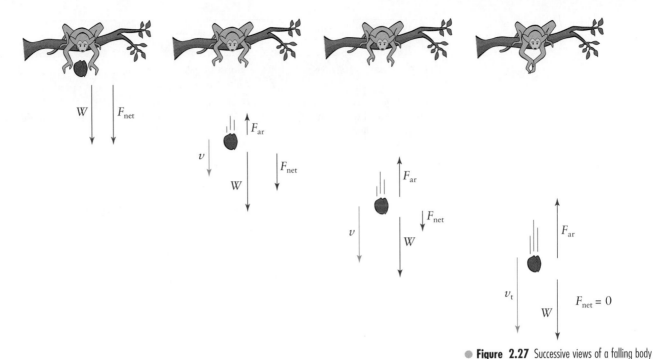

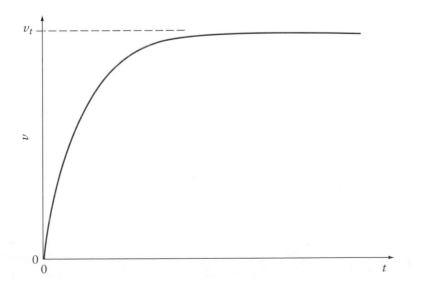

Figure 2.27 Successive views of a falling body affected by air resistance (ar). The upward force of air resistance increases as the object's speed increases. When this force is large enough to offset the downward weight (W), the net force (F_{net}) is zero (far right). The object's speed is constant and is called the terminal speed (v_t).

the net force is zero, so the speed stays constant from then on. This speed is called the *terminal speed* of the body. ● Figure 2.28 shows a graph of speed versus time for this motion.

Rocks and other dense objects have large terminal speeds and may fall for many seconds before air resistance affects their motion appreciably. Feathers, dandelion seeds, and balloons take less than a second to reach their terminal speeds, which are quite low. If a skydiver jumps out of a hovering helicopter or balloon, he or she will fall for about 2 or 3 s before the force of air resistance starts to have a major effect. The terminal speed depends on the skydiver's size and orientation when falling, but it is typically around 120 mph (about 54 m/s).* Then, when the parachute opens, the increased air resistance slows the skydiver to a much lower terminal speed—maybe 10 mph.

---

* Terminal speed also depends on how thin the air is. In 1960, Captain Joseph Kittinger jumped from a balloon at 100,000 feet high and may have fallen faster than the speed of sound! As he descended into denser air he slowed down—even before he opened his parachute.

Figure 2.28 Graph of speed versus time for a falling body with air resistance. At first the speed increases rapidly as in free fall (compare to Fig. 1.26b). But the increasing force of air resistance gradually reduces the acceleration to zero, at which point the speed is constant—the terminal speed v_t.

● Table 2.2 Summary of Examples of Forces

Nature of Net Force	Description of Motion
Zero net force	Constant velocity: stationary or motion in straight line with constant speed
Constant net force	Constant acceleration
Force parallel to velocity	Motion in a straight line with increasing speed
Force opposite to velocity	Motion in a straight line with decreasing speed
Force perpendicular to velocity	Motion in a circle: radius depends on speed and force
Restoring force proportional to displacement	Simple harmonic motion (oscillation)
Net force decreases as speed increases	Acceleration decreases: velocity reaches a constant value

In the spring-and-block example in Figure 2.24, the net force depends on the object's distance from its equilibrium position. When air resistance affects the motion of a falling body, the net force depends on the body's speed. These examples illustrate how interconnected the different physical quantities are in real physical systems. Things can get quite complicated: Imagine a spring and mass suspended under water. Here the net force would depend on both the block's position and its speed. See if you can describe how the block would move. ● Table 2.2 summarizes the types of forces that we have considered thus far.

In all of the examples in this section, the position of the object can be predicted for any instant of time in the future through the use of Newton's laws of motion and appropriate mathematical techniques—provided we have accurate information about the initial position and velocity of the object and the forces that affect its motion. This ability is the main reason Newton's laws are so important. It allows us to send satellites to planets billions of miles away and to predict what a roller coaster will do before it is built. But discoveries in the last century show that we can't always predict the future configurations of systems. In Chapter 10, we describe how the science of quantum mechanics, the essential tool for dealing with systems on the scale of atoms or smaller, tells us that such arbitrarily accurate predictions cannot be made because it is impossible to measure precisely both the position and velocity of particles such as electrons. The study of chaos has revealed that even in some relatively simple systems there is inherent randomness: The future configuration at some instant in time cannot be predicted no matter how accurately we know the forces, initial positions, and initial velocities. However, the mechanics based on Newton's laws remains one of the most valuable tools for applying physics in the world around us.

LEARNING CHECK

1. (True or False.) The speed of a projectile is constant as it travels even though its velocity is not constant.
2. The motion of a child on a swing moving back and forth is an example of _____.
3. A napkin is dropped and falls to the floor. As it falls,
   a) the force of air resistance acting on it increases.
   b) its speed increases.
   c) the net force on it decreases.
   d) all of the above.

4. In which of the following situations is the net force constant?
   a) simple harmonic motion
   b) freely falling body
   c) a dropped object affected by air resistance
   d) all of the above

ANSWERS: 1. False   2. simple harmonic motion   3. (d)   4. (b)

2.7 Newton's Third Law of Motion

Newton's third law of motion is a statement about the nature of forces in general. It is a simple law that adds an important perspective to the understanding of forces.

Newton's Third Law of Motion Forces always come in pairs: When one object exerts a force on a second object, the second exerts an equal and opposite force on the first.

Explore It Yourself 2.7

Stand facing a wall. Push on the wall so hard that you have to take a step backward to keep from falling over. What was the direction of the force you exerted? What was the direction in which your body accelerated? Is there a contradiction here?

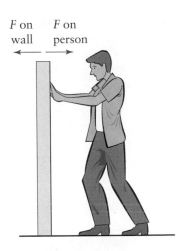

F on wall F on person

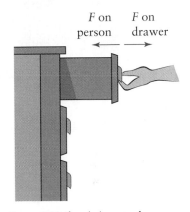

F on person F on drawer

● **Figure 2.29** If one body exerts a force on a second body, the second exerts an equal and opposite force on the first.

If object A causes a force on object B, object B exerts an equal force in the opposite direction on A.

$$F_{B \text{ on } A} = -F_{A \text{ on } B}$$

If you push against a wall with your hand, the wall exerts an equal force back on your hand (see ● Figure 2.29). A book resting on a table exerts a downward force on the table equal to the weight of the book. The table exerts an upward force on the book also equal to the book's weight. The Earth pulls down on you with a force that is called your weight. Consequently, you exert an equal but upward force on the Earth. When you dive off a diving board, the Earth's gravitational force accelerates you downward. At the same time, your equal and opposite pull upward on the Earth accelerates it toward you. But the Earth's mass is about one hundred thousand billion billion (10^{23}) times your mass, so its acceleration is negligible.

The third law gives a new insight into what actually happens in many physical systems. If you are on roller skates and push hard against a wall, you accelerate backward (see ● Figure 2.30). Think about this for a moment: A *forward* force by your hands makes you accelerate in the *opposite direction*. In reality, the equal and opposite force of the wall on your hands causes the acceleration. If you stand in the middle of the floor, you can't use your hands to push yourself because you need to push against something. Similarly, when a car speeds up, the engine causes the tires to push backward on the road. It is the road's

● **Figure 2.30** When a roller skater pushes against a wall, the wall's equal and opposite force on the skater causes the skater's acceleration backwards.

a

F

F

v

Figure 2.31 A small cart is fitted with a spring-loaded plunger and trigger. The cart is accelerated only if the plunger can exert a force on something else (b and c); then the equal and opposite force on the cart causes the acceleration.

(a) (b) (c)

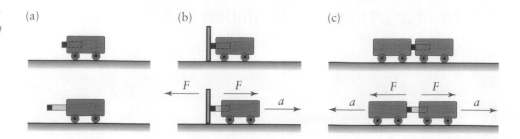

Courtesy of NASA

Figure 2.32 Rockets use Newton's third law.

equal and opposite force on the tires that causes the car to accelerate forward. The same thing happens when the brakes slow the vehicle. The recoil of a gun is caused by the third law: The large force accelerating the bullet produces an equal and opposite force on the gun—the "kick."

Figure 2.31 illustrates Newton's third law. A small cart is fitted with a spring-loaded plunger that can be pushed into the cart and retained. When the spring is released, the plunger pushes out. If there is nothing for the plunger to push against (Figure 2.31a), the cart does not move afterwards. The plunger can exert no force, so there is no equal and opposite force to accelerate the cart. If the cart is next to a wall when the spring is released (Figure 2.31b), the force on the wall results in an opposite force on the cart, and the cart accelerates away from the wall. If a second cart is next to the plunger (Figure 2.31c), both carts feel the same force in opposite directions, and they accelerate away from each other. In Chapter 3, we will show that the ratio of the speeds of the carts depends on the ratio of their masses. If one has twice the mass of the other, it will have half the speed.

Rockets and jet aircraft are propelled by ejecting combustion gases at a high speed (Figure 2.32). The engine exerts a force that ejects the gases, and the gases exert an equal and opposite force on the rocket or jet. Unlike cars, they do not need to push against anything to be propelled.

Birds, airplanes, and gliders can fly because of the upward force exerted on their wings as they move through the air (Figure 2.33a). As the air flows under and over the wing, it is deflected (forced) downward. This results in an equal and opposite upward force on the wing, called *lift* (see Figure 2.33b). Propellers on airplanes, helicopters, boats, and ships employ the same basic idea. They pull or push the air or water in one direction, resulting in a force on the propeller in the opposite direction. Propellers are useless in a vacuum.

Explore It Yourself 2.8

When riding in a car on a highway, put your hand a short distance out of the window, palm down, and use it as a wing to illustrate lift. The vertical force is the lift; the force acting backward is known as *drag*. Vary the angle of your hand, and notice how the lift varies. At a certain large angle, the lift actually goes to zero: This is known as a *stall*.

Whenever an object is accelerating, it exerts an equal and opposite force on whatever is accelerating it. You have probably noticed this when riding in a car, bus, or airplane as it accelerates: The seat exerts a forward force on you that causes you to accelerate. Your body pushes back on the seat with an equal and opposite force. It seems that there is some force "pulling" you back against the seat. This is not a real force, just a reaction of your mass to acceleration. The same effect is observed when an object is undergoing a centripetal acceleration. As a car or bus goes around a curve, you seem to be "pulled" to the side. Again, it is just a reaction to a net force causing an acceleration.

(a)

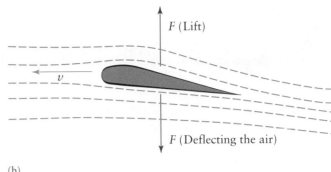

F (Lift)

v

F (Deflecting the air)

(b)

© Ruth Krusemark

● **Figure 2.33** (a) Author Vern Ostdiek having fun with Newton's third law. (b) A wing on a flying aircraft deflects the air downward. This downward force on the air causes an equal and opposite upward force on the wing.

Like Newton's first law, the third law is mainly conceptual rather than mathematical. It emphasizes that forces arise only during interactions between two or more things. By thinking in terms of pairs of forces, we can more easily distinguish the causes and effects in such interactions.

Concept Map 2.1 summarizes the relationship between force and Newton's laws of motion.

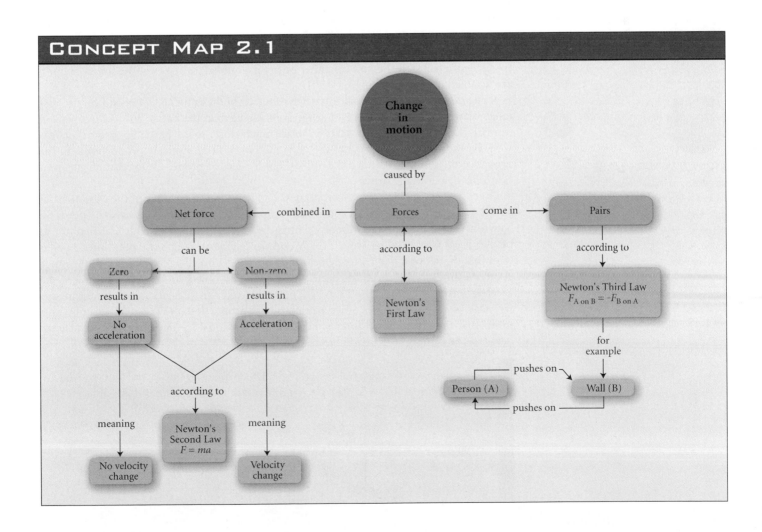

CONCEPT MAP 2.1

Change in motion

caused by

Net force ← combined in ← Forces → come in → Pairs

can be

according to

according to

Zero ⟷ Non-zero

Newton's First Law

Newton's Third Law
$F_{A \text{ on } B} = -F_{B \text{ on } A}$

results in

results in

No acceleration

Acceleration

for example

pushes on

Person (A)

Wall (B)

pushes on

according to

meaning

meaning

No velocity change

Newton's Second Law
$F = ma$

Velocity change

1. (True or False.) At this moment you are exerting a force upward on the Earth.
2. In the process of throwing a ball forward
   a) the ball exerts a rearward force on your hand.
   b) the ball exerts a forward force on your hand.
   c) your hand exerts a rearward force on the ball.
   d) the only force present is your force on the ball.

3. At one moment in a football game, player A exerts a force of 200 N to the east on player B. The size of the force that B exerts on A is _____, and it is directed to the _____.
4. (True or False.) When you jump upward, the floor exerts a force on you that makes you accelerate.

ANSWERS: 1. True 2. (a) 3. 200 N, west 4. True

2.8 The Law of Universal Gravitation

Newton's fourth major contribution to the study of mechanics is not a law of motion but a law relating to gravity. Newton made an important intellectual leap: He realized that the force that pulls objects toward the Earth's surface also holds the Moon in its orbit. Moreover, he claimed that every object exerts an attractive force on every other object. This concept is called *universal gravitation:* Gravity acts everywhere and on all things. The force that the Earth exerts on objects near its surface—weight—is just one example of universal gravitation.

What determines the size of the gravitational force that acts between two bodies? Newton used his deep understanding of mathematics and mechanics along with information about the orbits of the Moon and the planets to reason what the force law must be. The third law of motion states that when two bodies exert forces on one another, the forces are equal in size. Since the Earth's gravitational force depends on an object's mass, Newton reasoned, the force on each object in a pair is proportional to each object's mass (see ● Figure 2.34). If the mass of either object is doubled, the sizes of the forces on both are doubled.

The size of the gravitational force between two bodies must also depend on the distance between them. The Sun is much more massive than the Earth, but its force on you is much less than the Earth's, because you are much closer to the Earth. For geometric reasons, Newton felt that the size of the gravitational force is inversely proportional to the square of the distance between the bodies. But he needed proof—proof that he got from examining the Moon's orbit.

Long before Newton's time, astronomers had measured the radius of the Moon's orbit and, knowing this and the period of its motion, had determined its orbital speed. So Newton could calculate the Moon's (centripetal) acceleration, v^2/r, which turned out to be about $g/3{,}600$. In other words, the Moon's acceleration is about 3,600 times smaller than that of an object falling freely near the Earth. Newton also realized that the Moon is about 60 times farther from the Earth's center than an object on the Earth's surface. The acceleration of the Moon is 60 squared, or 3,600, times smaller because it is 60 times as far away (see ● Figure 2.35). Since acceleration is proportional to force, Newton had his proof that the gravitational force is inversely proportional to the square of the distance between the centers of the two objects.

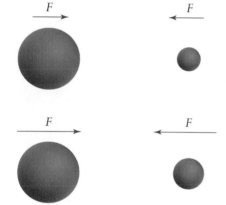

● **Figure 2.34** The equal and opposite gravitational forces that act on any pair of objects depend on the masses of both objects. If either object is replaced by another with a larger mass, the two forces are proportionally larger.

LAWS

Newton's Law of Universal Gravitation Every object exerts a gravitational pull on every other object. The force is proportional to the masses of both objects and inversely proportional to the square of the distance between their centers.

$$F \propto \frac{m_1 m_2}{d^2}$$

where m_1 and m_2 are the masses of the two objects, and d is the distance between their centers.

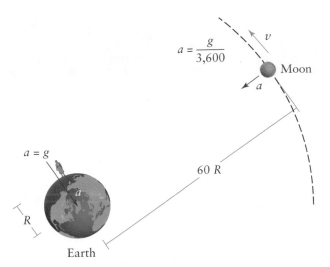

$$a = \frac{g}{3,600}$$

Moon

$a = g$

$60\ R$

R

Earth

● **Figure 2.35** The gravitational force that the Earth exerts on other objects is inversely proportional to the square of the distance from the Earth's center to the object's center. That is why the Moon's centripetal acceleration is much less than g. (Not drawn to scale.)

This force acts between the Earth and the Moon, the Earth and you, two rocks in a desert—any pair of objects. Your weight depends on the distance between you and the Earth's center. If you were twice as far from the Earth's center (8,000 miles instead of 4,000 miles), you would weigh one-fourth as much (see ● Figure 2.36). If you were three times as far, you would weigh one-ninth as much.

In 1798, the English physicist Henry Cavendish performed precise measurements of the actual gravitational forces acting between masses. Cavendish used a delicate torsion balance (● Figure 2.37). Two masses were balanced on a beam that was suspended from a thin wire attached to the beam's midpoint. Two large masses were placed next to the smaller ones on the beam. The gravitational forces on the smaller masses were strong enough to rotate the beam slightly and twist the wire. Cavendish used the amount of twist to measure the size of the gravitational force. His results showed that the force between two 1-kilogram masses 1 meter apart would be 6.67×10^{-11} newtons. With this result, Newton's law of universal gravitation can be expressed as an equation:

$$F = \frac{6.67 \times 10^{-11}\ m_1 m_2}{d^2} \qquad \text{(SI units)}$$

The constant of proportionality in the equation is called the gravitational constant G.

$$G = 6.67 \times 10^{-11}\ \text{N-m}^2/\text{kg}^2$$

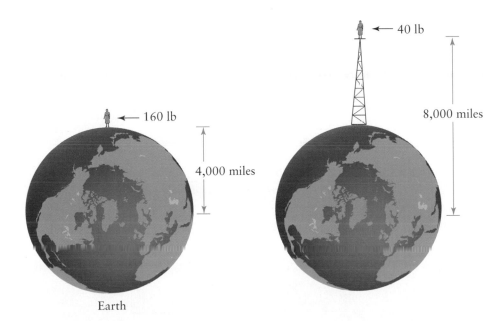

160 lb

40 lb

4,000 miles

8,000 miles

Earth

● **Figure 2.36** On a tower 4,000 miles high, you would be twice as far from the Earth's center as you are when standing on its surface. On the tower, you would weigh one-fourth as much. (Person not drawn to scale.)

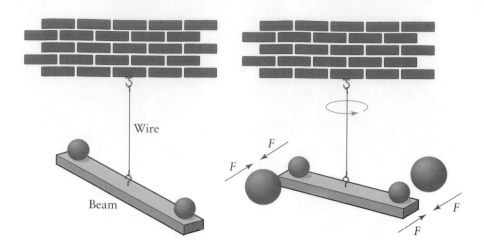

● **Figure 2.37** Simplified sketch of the torsion balance used by Cavendish to measure the gravitational force between masses. The two large objects exert forces on the two small ones, causing the suspending wire to twist.

Since G is such a small number, the gravitational force between objects is usually quite small. For example, the force between two persons, each with mass equal to 70 kilograms, when they are 1 m apart is only 0.00000033 newtons, or 0.0000012 ounces. The Earth's gravitational force on us (and our force on the Earth) is so large because the Earth's mass is huge.

In fact, we can use the law of universal gravitation to compute the mass of the entire Earth. First compute the gravitational force exerted on a body of mass m by the Earth—the body's weight—using the equation $W = mg$. This force can also be calculated using:

$$F = \frac{GmM}{R^2}$$

Here M represents the mass of the Earth, and R is the radius of the Earth, the distance the body is from the Earth's center. The value of R, first measured by the Greeks over 2,000 years ago, is about 6.4×10^6 meters. Since these two forces are equal to each other:

$$W = F$$

$$mg = \frac{GmM}{R^2}$$

Canceling the m:

$$g = \frac{GM}{R^2}$$

The values of g, G, and R can be inserted and the resulting equation solved for M, the mass of the Earth. The result: $M = 6 \times 10^{24}$ kg.

The acceleration due to gravity on the Moon is not the same as it is on the Earth. Likewise, each of the other planets has its own "g." This variation exists simply because the masses and the radii are all different. The values of the acceleration due to gravity have been computed for the Moon, the Sun, and the planets, using the preceding equation and each one's mass M and radius R (see ● Table 2.3 for some of these).

● **Table 2.3** Acceleration due to Gravity in Our Solar System	
Location	**Acceleration due to Gravity**
Earth	1.0 g
Sun	27.9 g
Moon	0.16 g
Mercury	0.38 g
Venus	0.88 g
Mars	0.39 g
Jupiter	2.65 g

Newton used his law of universal gravitation to explain a variety of phenomena that had been mysteries before. These included the cause of tides, the motion of comets, and the theoretical basis for orbital motion.

Orbits

Newton used an elegant "thought experiment" to illustrate that orbital motion about the Earth is actually an extension of projectile motion. Imagine that a cannon is placed at the top of a very high mountain and that it can shoot a cannonball horizontally at any desired speed (see ● Figure 2.38). If the cannonball just rolls out of the barrel, it will fall in a straight line to the Earth. Given some small initial speed, its trajectory to Earth will be a parabola (path D in Figure 2.38). But if its speed is increased more, the ball travels farther and farther around the Earth before it hits the ground (paths E, F, and G). If it were possible to shoot a cannonball with a high enough speed, it would travel in a full circle around the Earth and hit the cannon in the rear. It would be in orbit about the Earth, just as the Moon is. An object in orbit is continually "falling" toward the Earth. On an even higher mountain, one could place the cannonball in an orbit with a larger radius.

Realistically, this could not be done because of the force of air resistance. To orbit the Earth, an object must be above (outside) the Earth's atmosphere. But the idea is a handy way of showing the connection between a terrestrial phenomenon—projectile motion—and the motion of the Moon about the Earth. It also shows what Newton predicted—that we can put satellites into orbit.

It is quite easy to estimate how fast something has to move to stay in orbit. When an object is moving in a circle around the Earth, the centripetal force on it is the Earth's gravitational pull. For a satellite orbiting just above the Earth's surface, the gravitational force is approximately equal to the satellite's weight when it is on the Earth's surface—mg. Also, the radius of the satellite's orbit is approximately equal to the radius of the Earth, R (see

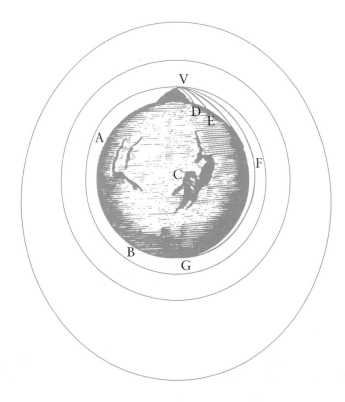

● **Figure 2.38** A re-creation of the original drawing of Newton's cannon "thought experiment." The cannon is placed on a very high mountaintop (V). The path of the cannonball depends on its initial speed.

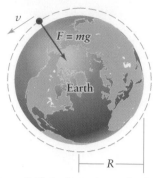

Figure 2.39 Satellite in a low, circular orbit about the Earth. The gravitational (centripetal) force on the satellite is about equal to the satellite's weight when on the Earth, *mg,* and the radius of the orbit is about equal to the Earth's radius, *R.*

Figure 2.39). The required speed is found by equating the centripetal force to the gravitational force. The centripetal force is mv^2/R, where R is the radius of the Earth. So

$$\frac{mv^2}{R} = mg$$

$$v^2 = gR = 9.8 \text{ m/s}^2 \times (6.4 \times 10^6 \text{ m})$$

$$= 63{,}000{,}000 \text{ m}^2/\text{s}^2$$

$$v = 7{,}900 \text{ m/s}$$

Newton's mathematical analysis of orbital motion is not restricted to Earth orbits. It also applies to the motions of the planets around the Sun and to the moons in orbit about the other planets. His results allowed astronomers to calculate the orbits of celestial objects with higher accuracy and with fewer observational data than before. He showed that comets, such as Halley's Comet, are moving in orbits around the Sun. The orbits of most of them are flattened ellipses with the Sun near one end, at a point called the *focus* of the ellipse (Figure 2.40). This is why Halley's Comet is only near enough to the Earth to be seen every 76 years. It spends most of its time far from the Sun and the Earth.

The orbits of all of the planets, Earth included, are actually ellipses, although they are much closer to being circular than the orbit of Halley's Comet. The Sun is at one focus of these orbits. Note in Figure 2.40 that the elliptical nature of Pluto's orbit causes it to spend part of the time inside the orbit of Neptune.

Explore It Yourself 2.9

You can draw the correct shape of an ellipse using two tacks or pins, some string or thread, a ruler, a piece of paper, and a surface into which you can push the tacks (a bulletin board works).

1. Place the tacks into the paper 10 centimeters apart. Wrap the string around the tacks, and tie it to form a tight loop around them. Move the tacks a few centimeters closer together. Insert a pen into the loop, move it outward until the string is tight, and draw the complete path around the tacks while holding the pen against the string (Figure 2.41). The resulting figure is an ellipse with each tack at one focus of the ellipse.
2. Place the tacks 4 centimeters apart. The ellipse you draw has the shape of Pluto's orbit. (The Sun would be at one of the pins.) Note that it is hard to distinguish it from a circle. How should you place the pins to draw a true circle?
3. Place the tacks 8.6 centimeters apart to draw the shape of the orbit of Nereid, a moon of Neptune that has a highly elliptical orbit.

Gravitational Field

Gravitation is an example of action at a distance. Objects exert forces on each other, even though they may be far apart and there is no matter between them to transmit the forces. The other forces we have talked about involve direct contact between things. Gravitation does not.

How is force possible without contact? One way to get better insight into this situation is to use the concept of a *field*. Imagine an object situated in space. The matter in the object causes an effect or a disturbance in the space around it. We call this a *gravitational field*. This field extends out in all directions but becomes weaker at greater distances from the object. In this model, the field itself causes a force to act on any other object. It plays the role of an invisible agent for the gravitational force. Whenever a second body is in the field of the first, it experiences a gravitational force. But the field is present even when there is no other object around to experience its effect.

We might call this a "force field" because it causes forces on other bodies. But do not imagine it to be the kind of "invisible wall" one sees in science-fiction movies. One way to represent the shape of the gravitational field around an object is by drawing arrows at

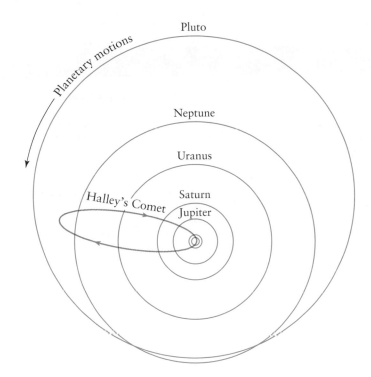

Planetary motions

Pluto

Neptune

Uranus

Saturn

Halley's Comet Jupiter

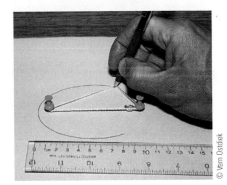

Figure 2.40 Orbit of Halley's Comet around the Sun. The comet is seen from Earth only when it is in that part of its orbit near the Sun. The Earth's orbit is the smallest circle.

Figure 2.41

© Vern Ostdiek

different points in space. They show the magnitude and the direction of the force that would act on anything placed at each point (Figure 2.42a). These arrows are long near the object and short farther away because the gravitational force decreases as the distance increases. Another way to represent the gravitational field is to connect the arrows, making "field lines." The direction of the field line at any point in space again shows the direction of the force that would act on an object placed there. The strength of the gravitational field is represented by the spacing of the lines: The lines are farther apart where the field is weaker (Figure 2.42b).

All of the forces in nature can be traced to four fundamental forces. The gravitational force is one of them. The others are the electromagnetic force, responsible for electric and magnetic effects; the strong nuclear force; and the weak nuclear force. These will be discussed in later chapters. But what about friction and other forces involving direct contact? Most of these can be traced back to the electromagnetic force. This force determines the sizes and shapes of atoms and molecules. For example, a stretched spring pulls back because of the electrical forces between the atoms in it.

Figure 2.42 Two ways of showing the gravitational field around an object: (a) with arrows and (b) with field lines.

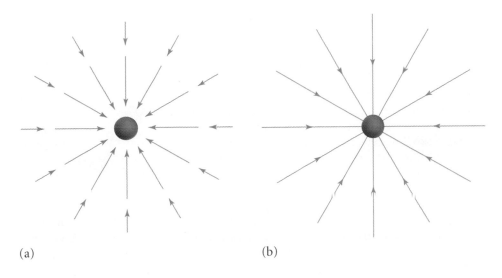

(a) (b)

Hooke-d!

The year 2003 marks the three-hundredth anniversary of the death of Robert Hooke (1635–1703). Described as "England's Leonardo," Hooke was a highly accomplished scientist who contributed significantly to fields as diverse as astronomy and anatomy, geology and geometry, acoustics and aerodynamics, not to mention physiology and physics. Little wonder, then, that Hooke has also been called "Europe's last Renaissance man."

Born on the Isle of Wight, Robert Hooke displayed early talents as a draftsman, practical mechanic, and inventor. By the end of his career, Hooke could lay claim to the invention of the first efficient vacuum pump, the universal joint now found in most motor vehicles, the filar micrometer used in astronomy to measure small angular separations, a pedometer, the compound microscope, a marine barometer, and a portable camera (cameras are discussed in Chapter 9). After completing his master's degree, he was employed first as an assistant to chemist Robert Boyle (see Physics Family Album in Chapter 4), and then, in 1662, as Curator of Experiments for the Royal Society of London. Three years later, Hooke was appointed professor of geometry at Gresham College and a fellow of the Royal Society. In this way, Hooke became the first salaried research scientist in Great Britain.

Although he had built a solid reputation as a capable and clever innovator in the field of scientific instrumentation, Hooke's renown was secured with the publication in 1665 of a work entitled *Micrographia*. In it, Hooke recounts a series of experiments carried out between 1661 and 1664 involving the microscopic examination of things as diverse as ice crystals, blue flies, cork cells, silk fabric, and sewing needles (● Figure 2.43). Besides providing painstakingly detailed descriptions of the structures of these items (as well as finely engraved images that he created himself), Hooke sought to understand each system in terms of broader, overarching, if not universal, principles. Thus, his examination of ice crystals led him to speculations about fundamental atomic structure; his characterization of the cellular structure of wood prompted a series of investigations of the role of air in combustion; and his study of the anatomy of fly wings

and the phenomena of "buzzing" produced a treatise on aerodynamics and acoustics.

The impact of *Micrographia* on both the scientific community and the literate public was immediate and unprecedented. Hooke was acknowledged as a scientist of genius and became an instant celebrity. There was even a play, *The Virtuoso*, based on Hooke's work (and, to some extent, his life). *Micrographia* has been called one of the formative books of its age and cited as recommended reading for all undergraduate science students because of the model of controlled experimentation and careful observation it presents. But if this work is thought to have been so important, why, one wonders, is it so little known or appreciated today? The answer lies in part with the later conflict that arose between Newton and Hooke and the subsequent overshadowing of Hooke by Newton, largely at the hands of Newton's followers. But before commenting further on this issue, it is

Courtesy of University of Pennsylvania

● **Figure 2.43** An engraving of Hook's compound microscope, including a specially designed specimen illuminator.

LEARNING CHECK

1. (True or False.) If the distance between a spacecraft and the Earth is doubled, the gravitational force on the spacecraft will be one half as large.
2. (Choose the *incorrect* statement.) The gravitational force exerted on a satellite by the Earth
   a) is directed toward the center of the Earth.
   b) depends on the satellite's speed.
   c) depends on the satellite's mass.
   d) depends on the Earth's mass.

3. (True or False.) Gravity supplies the centripetal force needed to keep a planet in a circular orbit.
4. Every object creates a _____ in the space around it.

ANSWERS: 1. False 2. (b) 3. True 4. gravitational field

beneficial to consider some of the other contributions to the scientific enterprise that Hooke made. Hooke possessed a unique capacity to integrate information from a rich variety of sources and to distill from them the common, unifying principles. One such principle that he frequently invoked to explain a variety of systems displaying periodic and/or vibratory motion was elasticity. The culmination of his efforts in this area was his work "Of Springs" (1678). It was here that Hooke articulated his famous law ("Ut pondus sic tensia"—"the [suspended] weight is equal to the [spring] tension"). His conclusion was the result of a widely diverse set of experiments on such things as beating fly wings, pendulum watches, and vibrating violin strings.

Robert Hooke also had an abiding interest in astronomy and gravitation. Indeed, over half of the papers he delivered to the Royal Society dealt with matters astronomical. Hooke investigated the rotation of Jupiter and Mars and crater formation on the Moon. (Crater Hooke on the Moon honors his pioneering investigations in lunar and planetary geology.) In a lengthy and careful study of the Pleiades, a young cluster of stars in Taurus, Hooke demonstrated that the light-gathering power (the ability to see faint objects) and the resolution (the ability to see fine detail) of a telescope both increase with the size of the aperture (opening) of the instrument. He also made detailed observations of the comets of 1664 and 1677 and concluded from them that cometary nuclei are solid, that the tails of comets develop by erosion of material from the head of the comet, and that some of a comet's light originates from the comet itself instead of coming solely from reflected sunlight. Modern research has confirmed all of Hooke's speculations about telescope optics and the nature of comets.

However, it was in the field of gravitation that Hooke's most controversial work was carried out, work that ultimately led to his conflict with Newton. In a paper entitled, "Attempt to Prove the Motion of the Earth," presented in 1674, Hooke states his conclusions about gravity, based presumably on his comet studies and his experiments with falling bodies released from towers at St. Paul's Cathedral and Westminster Abbey. Hooke argued that the gravitational force exists between all bodies and acts towards their centers, that all bodies tend to continue their motion in a straight line but can be deviated into orbits by the action of the gravitational force, and that the gravitational force is greater when the bodies are closer together. By 1680, he had quantified the last point, writing to Newton that the gravitational force "is always in duplicate proportion to the distance from the center reciprocal." As you have learned in this chapter, Hooke's intuitions about gravity were generally correct, although unlike Newton, he did not possess the mathematical virtuosity to prove his assertions.

The problem arose when, in 1687, Newton published his now famous *Principia Mathematica* detailing his study of mechanics and refused to acknowledge Hooke's insights and contributions regarding the nature of the inverse square law of gravitation. A bitter feud between Hooke and Newton ensued that lasted until Hooke's death in 1703. Among the sad results of this nearly 15-year conflict was the transformation of Robert Hooke from a generous, convivial, and highly social man into a reclusive, melancholy, mistrustful curmudgeon whose reputation was irreparably damaged by the rancorous disagreement. In ill health for several years before the end, Hooke died in a pitiful condition, dirty, ragged, alone, and for most of posterity, largely forgotten.

No portrait or contemporary engraving of Robert Hooke survives today. Similarly, there remains only one example of the elegant townhouses and public buildings that Hooke designed, some in collaboration with Sir Christopher Wren, in the wake of the Great Fire that destroyed much of London in 1666. And yet, as the foregoing glimpse into his life testifies, there are ample reasons to honor and celebrate Hooke's achievements for their eclecticism as well as for their foresight. Here is a true Renaissance man, a man deserving of as much recognition in our own time as he enjoyed for most of his life in his own.

For more information on Robert Hooke, two good places to start are Stephen Inwood's recent biography entitled "The Man Who Knew Too Much" (Macmillan: 2003) and the BBC's website (http://www.bbc.co.uk/history/discovery/revolutions/hooke_robert_beavon_03.shtml/), featuring articles and interviews produced in conjunction with their three-part television series on Robert Hooke aired in the spring of 2003.

2.9 Tides

For the most part, tides are the result of gravitational forces exerted on the Earth by the Moon. To understand this, consider ● Figure 2.44. Points A, B, and C all lie on a line through the center of the Earth and the center of the Moon. According to Newton's law of universal gravitation, Earth material at A experiences a stronger attractive force toward the Moon than does material at C because it is closer to the Moon. Similarly, material at C experiences a greater attractive force toward the Moon than does material at B. Thus, the material at A is pulled away from material at C, while the material at C is in turn pulled away from the material at B. The net effect is to separate these three points. Thus the shape of the Earth is elongated or stretched by the gravitational pull of the Moon along a line connecting the two bodies.

This view is one that might be seen by an observer outside the Earth-Moon system looking in. One might ask what an observer at point C sees. Such an observer would

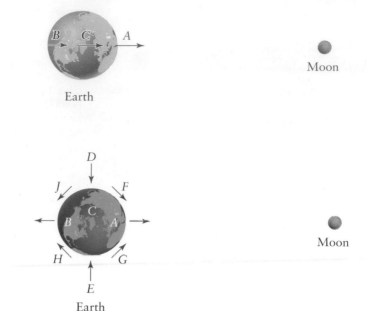

Figure 2.44 Gravitational forces exerted by the Moon on material at three different locations on Earth. (Not drawn to scale.)

B C A

Earth

Moon

Figure 2.45 Gravitational forces exerted by the Moon on material at different locations on Earth, relative to point *C*, the Earth's center. (Not drawn to scale.)

D

J   F

B   C   A

H   G

E

Earth

Moon

notice that point *A* is pulled toward the Moon and away from *C*. The observer would also see that point *B*, relative to *C*, seems to be pushed away from *C*, in the direction away from the Moon.

We can redraw our earlier figure now indicating how an Earthbound observer at *C* views the forces exerted by the Moon at points *A* and *B* (● Figure 2.45). Notice that the same stretching or elongation of the Earth along the direction of the Moon occurs as before, but now the symmetry of the situation as seen from *C* is manifested.

But what has this to do with tides? Suppose we were to perform this same kind of analysis for points *D, E, F, G, H,* and *J*. We would find that, relative to *C*, forces along the directions of the arrows shown in Figure 2.45 would exist because of the gravitational presence of the Moon. Now imagine the Earth to be covered with an initially uniform depth of water. How would this fluid move in response to these forces? Again, applying the laws of mechanics, we are led to the following, perhaps somewhat startling, results. (1) Water at points *D* and *E* would weigh slightly more than water elsewhere because, in addition to the Earth's own gravitational pull toward *C*, there is a small component of force toward *C* produced by the Moon. (2) Conversely, water at points *A* and *B* would weigh slightly less than water elsewhere because the Moon exerts small forces in directions opposite to the Earth's own inward gravitational attraction toward *C*. These forces work to counteract gravity and, hence, to reduce the weight of the water. (3) Water at points *F, G, H,* and *J* would experience a force parallel to the Earth's surface and would begin to flow in the directions of the arrows at each point, much as water flows down any

Figure 2.46 (a) The "tidal bulges" in the oceans caused by the gravitational pull of the Moon. (b) As the Earth rotates, places on its surface move from low tide to high tide, back to low tide, and so on. (Not drawn to scale.)

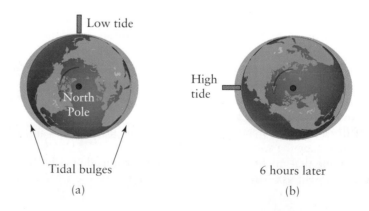

Low tide

North Pole

Tidal bulges

(a)

High tide

6 hours later

(b)

(a)

(b)

● **Figure 2.47** High tide (a) and low tide (b).

inclined surface in response to gravity. This flow causes the water to pile up at opposite sides of the Earth along the line joining the Earth to the Moon. These "piles" of water are called the tidal bulges, and as the Earth rotates, points on its surface move into and then out of the bulges, resulting in the high and low tides observed at intervals of roughly 6 hours (● Figure 2.46 and Figure 2.47).

Many complications to this simple picture cause the behavior of real tides to deviate somewhat from that predicted by this model. In particular, we have ignored such things as tidal effects caused by the Sun, the influence of the Earth's rotation on the motion of the tidal bulges, and the effects on the local heights of tides caused by the Earth's rugged, uneven surface. The first of these complications leads to such phenomena as the lower-than-average *neap tides,* which occur when the Moon is in either the first- or third-quarter phase (and hence at right angles to the Sun in the sky), and the higher-than-average *spring tides,* which happen when the Moon is in either the new or the full phase (and hence in line with the Sun in the sky). Despite having ignored these problems, we are still able to understand the basic characteristics of tides in terms of this simple Newtonian model. So while the tides may not wait for us, their behavior is sufficiently predictable that we can at least be aware of the time of their arrivals and departures.

LEARNING CHECK

1. (True or False.) Tides are caused by the fact that the Earth's center and the different parts of its oceans are not all the same distance from the Moon.

2. Before going to the beach, you try to estimate what the tide is (high, low, or in between) by noting where in the sky the moon is. It is low tide

a) when the moon is near its highest point in the sky.
b) when the moon is near the horizon.
c) whenever the moon is not visible.
d) none of the above.

ANSWERS: 1. True **2.** (b)

Isaac Newton was born in rural Woolsthorpe, England on Christmas Day in 1642 (● Figure 2.48). His father died shortly before Newton's birth, and when his mother remarried three years later, he was entrusted to the care of his maternal grandparents. His interest in science was kindled while he attended a boarding school, and he entered Trinity College at Cambridge in 1661. There he studied mathematics and physics until the university was closed in the summer of 1665 because of the Great Plague. Newton returned to Woolsthorpe and during the next two years made his great intellectual discoveries in mechanics—the laws of motion and gravitation. As if this were not enough for someone in his early twenties, he also made important contributions to mathematical analysis, independently invented calculus, and explained how prisms produce the spectrum of colors from sunlight. Any of these accomplishments alone would have ensured him a place in the history of science and mathematics. The fact that they were performed in only two years by an "amateur" in isolation from the great intellectual centers of the world makes Newton's feat one of the greatest in the history of human thought.

● **Figure 2.48** Isaac Newton (1642–1727) experimenting with light.

Modern physical science and mathematics would be impossible without calculus. (Newton invented this entirely new branch of mathematics at about the same time as did the German philosopher and mathematician Baron Gottfried Wilhelm von Leibniz.) Calculus extended the reach of mathematics to the realm of continuously changing physical processes. With his calculus and his laws of motion, Newton was able to analyze physical systems that have nonconstant forces and, consequently, varying accelerations. A mass on a spring and the resulting simple harmonic motion are a good example (see Section 2.6). Many of the equations in this book were originally derived using calculus. After Newton, mechanics was no longer limited to the ideal systems of Galileo. Perhaps the best indication of the importance of calculus is that virtually all students in mathematics or physical science take courses in calculus at the beginning of their college education.

After Cambridge reopened in 1667, Newton returned to pursue his studies, but he did not publish his discoveries. He recognized their importance but was apparently reluctant to reveal them before they were completely developed. One of the details that he eventually proved, using his calculus, was that the gravitational force between spherical bodies depends on the distance between their centers. Depending on how you look at it, this may or may not seem logical, but Newton wanted proof. If this were not true, Newton's law of universal gravitation would not tie together the Earth's force on the Moon and the Earth's force on objects on its surface. Newton's instructors were aware of his accomplishments, and one of his mathematics professors turned over his position to Newton in 1669.

During his 20-year tenure as a professor of mathematics at Cambridge, Newton concentrated more on the study of light than on mechanics. He invented the reflecting telescope, giving astronomers their most valuable tool (● Figure 2.49). Nearly all of the great telescopes in use today, including

● **Figure 2.49** Newton's reflecting telescope, which he presented to the Royal Society in London.

SUMMARY

- **Force** is the most important physical quantity in Newtonian mechanics. Forces are all around us; **weight** and **friction** play dominant roles in our physical environment.

- The three **laws of motion** are direct, universal statements about forces in general and how they affect motion.

- **Newton's first law of motion** states that an object's velocity is constant unless a net force acts on it.

- Newton's **second law of motion** states that the net force on an object equals its mass times its acceleration. This law is the key to the application of mechanics. By knowing the size of the force acting on a the Hubble Space Telescope, are reflectors. Newton's work with lenses and the prismatic analysis of light were important contributions to optics. He did publish some of these discoveries, but they were met with contentious criticism. As a result, Newton resolved to withhold his other works. His major critic argued that he had made some interesting findings about how light behaved but hadn't dealt with the fundamental question of what exactly light is. Newton realized that one must first know as much as possible about the properties and behavior of light before one can hope to explain what it is.

The great astronomer Edmond Halley, after whom Halley's Comet is named, recognized the importance of Newton's mechanics to physics in general and to astronomy in particular. Halley succeeded in convincing Newton to publish a treatise on mechanics. After 18 months of prodigious concentration and labor, Newton completed his greatest work, the *Philosphiae Naturalis Principia Mathematica* (*Mathematical Principles of Natural Philosophy;* ● Figure 2.50). Published in 1687, the *Principia* was soon recognized as one of the greatest books ever written. In it, Newton presented his basic principles of mechanics and gravitation and used them to explain a number of physical phenomena that had puzzled scientists for centuries. Most of the material in this chapter is based on the *Principia.*

In 1696, Newton left Cambridge and the academic world and became Warden of the Mint in London. Three years later, he was appointed Master of the Mint and served in this capacity until his death. Newton put England on the gold standard and oversaw the introduction of milling on the edges of coins to discourage people from shaving off the precious metals. His duties included sending convicted counterfeiters to the gallows. In 1704, he published another important book, *Opticks,* which summarized his work with light and color that had spanned several decades. In 1705, he was knighted by Queen Anne. Sir Isaac Newton was buried with full honors at Westminster Abbey in 1727.

Newton was not simply a scientist and mathematician. He spent much of his "spare" time throughout his life analyzing biblical text. He was a religious man and considered his great discoveries in physical science evidence of the hand of God. Newton also invested considerable time and effort in the field of alchemy. He was even more secretive about his findings in this area, and

almost nothing was published.

Throughout much of his adult life Newton was embroiled in a series of bitter disputes with other scientists and mathematicians. (See Physics Potpourri "Hooke-d!" on p. 76.) A major one was a 40-year battle with Leibniz concerning who first invented calculus. Accusations of plagiarism were made by both sides. Newton's conduct in these conflicts was often characterized by vindictiveness and heavy-handedness. He had a sense of his place in history and would not allow anyone to take partial credit for his discoveries. Newton possessed many talents and virtues, but magnanimity toward professional rivals was not one of them.

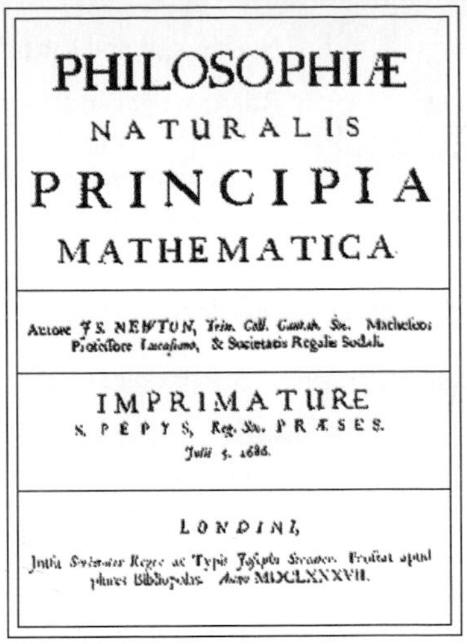

It is a bit difficult to fully appreciate Newton's work at this point in history because we are on the inside looking out, so to speak. The very manner in which we "do" science is based, to a large extent, on Newton's discoveries. He showed that nature has special built-in rules (laws) that govern diverse phenomena. Individual types of motion need not be considered in isolation; they are all subject to three basic laws. He fully united the terrestrial world and the cosmos by showing that these rules apply in both realms. Newton revealed the immense usefulness and power gained by incorporating advanced mathematics into physics. He broke from the abstract, speculative approach to science and opened a new way with a combination of ingenious experimentation and theoretical analysis. One can argue that Newton's intellectual accomplishments are unsurpassed in their influence on history.

PHILOSOPHIÆ
NATURALIS
PRINCIPIA
MATHEMATICA

Autore *JS. NEWTON,* Trin. Coll. Cantab. Soc. Mathefeos Profeffore Lucafiano, & Societatis Regalis Sodali.

IMPRIMATURE
S. PEPYS, Reg. Soc. PRÆSES.
Julii 5. 1686.

LONDINI,
Juffu Societatis Regiæ ac Typis Jofephi Streater. Proftat apud plures Bibliopolas. Anno MDCLXXXVII.

© CORBIS

● **Figure 2.50** Title page of Newton's *Principia Mathematica.*

body, one can determine its acceleration. Then its future velocity and position can be predicted using the concepts in Chapter 1.

■ **Newton's third law of motion** states that whenever one body exerts a force on a second one, the second exerts an equal and opposite force on the first. This deceptively simple statement gives us new insight into the nature of forces.

■ **The law of universal gravitation** states that equal and opposite forces of attraction act between every pair of objects. The size of each force is proportional to the masses of the objects and inversely proportional to the square of the distance between their centers.

■ The nature of orbits and the causes of tides are just two of the many phenomena that can be explained using the law of universal gravitation.

■ **Isaac Newton** is the founder of modern physical science. His laws of motion, law of gravitation, and calculus form a nearly complete system for solving problems in mechanics. He also made important discoveries in optics. Newton's combination of logical experimentation and mathematical analysis shaped the way science has been done ever since.

Equation	Comments	Equation	Comments
Fundamental Equations		*Special-Case Equations*	
$F = ma$	Newton's second law of motion	$F = \dfrac{mv^2}{r}$	Centripetal force (object moving in a circular path)
$W = mg$	Relates weight and mass		
$F = \dfrac{Gm_1m_2}{d^2}$	Law of universal gravitation		

MAPPING IT OUT!

1. Reread Section 2.8 on the law of universal gravitation and make a list of concepts and examples that might serve as a basis for developing a concept map summarizing the material in this section. After creating your list, reorder the items, ranking them from most general to least general (that is, most specific).

2. In this chapter, you've encountered a large number of concepts related to forces and motion. Organizing a concept map might help clarify the meanings of many of these concepts for you. As a start, you examine a concept map (shown below) pertaining to the concept of "net force" created by a student who had previously taken this course. This student had some misconceptions about this topic, however, so there are some blatant errors in the concept map. Locate and correct as many of these errors as you can. [*Hint:* Carefully inspect the linking words used to connect the concepts and consider the meanings of the "propositions" they make.]

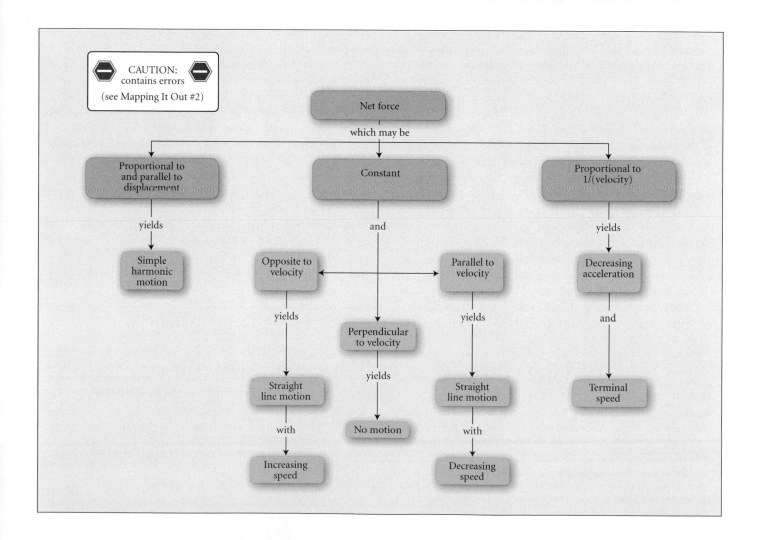

(▶ *Indicates a review question, which means it requires only a basic under-standing of the material to answer. The others involve integrating or extending the concepts presented thus far.)*

1. ▶ What is force? Identify several of the forces that are acting on or around you.

2. ▶ What is weight? Under what circumstances might something be weightless?

3. A person places a book on the roof of a car and drives off without remembering to remove it. As the book and the car move down a street at a steady speed, there are two horizontal forces acting on the book. What are they?

4. ▶ What are the two types of friction? Can both types act on the same object at the same time?

5. A person places a hand on a closed book resting on a table, and then presses downward while pushing outward. Either the book slides across the table or the hand slides across the book. What determines which of these happens? Which type(s) of friction is (are) involved?

6. ▶ What do we mean by "external" force? In light of Newton's third law of motion, why can't an internal force cause a net force?

7. At one moment in a football game, player A exerts a force to the east on player B. At the same time, a teammate of A exerts the same-sized force to the south on player B. In what direction is B likely to go because of these forces?

8. ▶ How does an object move when it is subject to a steady centripetal force? How does it move if that force suddenly disappears?

9. A person is riding on a train while watching the display on a GPS unit (refer back to Figure 1.9). The person notices that both the "speed" and the "heading" readings are not changing. What can the person conclude about the net force acting on the train car?

10. ▶ Discuss the distinction between mass and weight.

11. ▶ Two astronauts in an orbiting space station "play catch" (throw a ball back and forth to each other). Compared to playing catch on Earth, what effect, if any, does the "weightless" environment have on the process of accelerating (throwing and catching) the ball?

12. An amusement park ride called the Detonator lifts 12 people verti-cally and then plunges them "back to Earth faster than free fall." What is the direction of the net force that the mechanism must exert on the passengers? Describe what the riders feel as they descend.

13. Single-engine airplanes usually have their propeller at the front. Boats and ships usually have their propeller(s) at the rear. From the perspective of Newton's second law of motion, is this significant?

14. As a rocket ascends, its acceleration increases even though the net force on it stays constant. Why?

15. ▶ What is the international system of units (SI)?

16. An archer aims an arrow exactly horizontal over a flat field and shoots it. At the same instant, the archer's watchband breaks and the watch falls to the ground. Does the watch hit the ground before, at the same time as, or after the arrow hits the ground? Defend your answer.

17. ▶ Describe the variation of the net force on, and the acceleration of, a mass on a spring as it executes simple harmonic motion.

18. ▶ Explain how the change in the force of air resistance on a falling body causes it to eventually reach a terminal speed.

19. The terminal speed of a Ping-Pong ball is about 20 mph. From the top of a building a Ping-Pong ball is thrown *downward* with an ini-tial speed of 50 mph. Describe what happens to the ball's speed as it moves downward from the moment it is thrown to the moment when it hits the ground.

20. At least two forces are acting on you right now. What are these forces? On what is the equal and opposite force to each of these act-ing?

21. As any car travels on a flat stretch of highway with a constant speed, the road exerts both a vertical force and a horizontal force on the car. What are the nature of these forces and what are their direc-tions? (There are two possibilities for each.)

22. How is Newton's third law of motion involved when you jump straight upward?

23. Jane and John are both on roller skates and are facing each other. First Jane pushes John with her hands and they move apart. Later they get together, and John pushes Jane equally hard with his hands and they move apart. Do they move any differently in the two cases? Why or why not?

24. What would be the gravitational force on you if you were at a point in space a distance R (the Earth's radius) above the Earth's surface?

25. ▶ Describe how the magnitude of the gravitational force between objects was first measured.

26. If suddenly the value of G, the gravitational constant, increased to a billion times its actual value, what sorts of things would happen?

27. The first "Lunar Olympics" is to be held on the Moon inside a huge dome. Of the usual Olympic events—track and field, swimming, gymnastics, and so on—which would be drastically affected by the Moon's gravity? In which events would Earth-based records be bro-ken? In which events would the performances be no better—or per-haps worse—than on the Earth?

28. ▶ In the broadest terms, what causes tides?

29. The Sun exerts much larger forces on the Earth and its oceans than the Moon does, yet the tides caused by the Sun are much lower than those caused by the Moon. Why?

30. We have studied four different laws authored by Isaac Newton. For each of the following, indicate which law is best for the task described.
    a) Calculating the net force on a car as it slows down.
    b) Calculating the force exerted on a satellite by the Earth.
    c) Showing the mathematical relationship between mass and weight.
    d) Explaining the direction that a rubber stopper takes after the string that was keeping it moving in a circle overhead is cut.
    e) Explaining why a gun recoils when it is fired.
    f) Explaining why a wing on an airplane is lifted upward as it moves through the air.

31. ▶ What mathematical "tool" did Isaac Newton develop that has been essential to advanced physics since then?

1. Express your weight in newtons. From this determine your mass in kilograms.

2. A child weighs 300 N. What is the child's mass?

3. Suppose an airline allows a maximum of 30 kg for each suitcase a passenger brings along.
    a) What is the weight in newtons of a 30-kg suitcase?
    b) What is the weight in pounds?

4. The mass of a certain elephant is 1,130 kg.
    a) Find the elephant's weight in newtons.
    b) Find its weight in pounds.

5. A 400-kg sailboat accelerates at a rate of 1.5 m/s^2. What is the net force acting on it?

6. A motorcycle and rider have a total mass equal to 300 kg. The rider applies the brakes, causing the motorcycle to accelerate at a rate of -5 m/s^2. What is the net force on the motorcycle?

7. As a 2-kg ball rolls down a ramp, the net force on it is 10 N. What is the acceleration?

8. In an experiment performed in a space station, a force of 60 N causes an object to have an acceleration equal to 4 m/s^2. What is the object's mass?

9. The engines in a supertanker carrying crude oil produce a net force of 20,000,000 N on the ship. If the resulting acceleration is 0.1 m/s², what is the ship's mass?

10. A person stands on a scale inside an elevator at rest (● Figure 2.51). The scale reads 800 N.
    a) What is the person's mass?
    b) The elevator accelerates upward momentarily at the rate of 2 m/s². What does the scale read then?
    c) The elevator then moves with a steady speed of 5 m/s. What does the scale read?

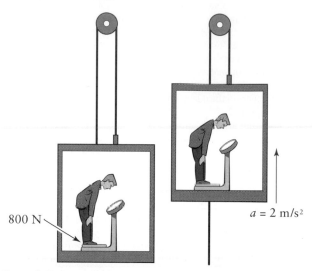

800 N

$a = 2 \text{ m/s}^2$

● **Figure 2.51**

11. A jet aircraft with a mass of 4,500 kg has an engine that exerts a force (thrust) equal to 60,000 N.
    a) What is the jet's acceleration when it takes off?
    b) What is the jet's speed after it accelerates for 8 s?
    c) How far does the jet travel during the 8 s?

12. At the end of Section 1.4, we mentioned that the maximum acceleration of a fist during a particular karate blow was measured to be about 3,500 m/s² (Figure 1.30).
    a) If the mass of the fist was approximately 0.7 kg, what was the maximum force?
    b) What was the maximum force on the concrete block?

13. A sprinter with a mass of 80 kg accelerates from 0 m/s to 9 m/s in 3 s.
    a) What is the runner's acceleration?
    b) What is the net force on the runner?
    c) How far does the sprinter go during the 3 s?

14. As a baseball is being caught, its speed goes from 30 to 0 m/s in about 0.005 s. Its mass is 0.145 kg.
    a) What is the baseball's acceleration in m/s² and in g's?
    b) What is the size of the force acting on it?

15. On aircraft carriers, catapults are used to accelerate jet aircraft to flight speeds in a short distance. One such catapult takes a 18,000-kg jet from 0 to 70 m/s in 2.5 s.
    a) What is the acceleration of the jet (in m/s² and g's)?
    b) How far does the jet travel while it is accelerating?
    c) How large is the force that the catapult must exert on the jet?

16. At the end of an amusement park ride, it is desirable to bring a gondola to a stop without having the acceleration exceed 2 g. If the total mass of the gondola and its occupants is 2,000 kg, what is the maximum allowed braking force?

17. An airplane is built to withstand a maximum acceleration of 6 g. If its mass is 1,200 kg, what size force would cause this acceleration?

18. Under certain conditions, the human body can safely withstand an acceleration of 10 g.
    a) What net force would have to act on someone with mass of 50 kg to cause this acceleration?
    b) Find the weight of such a person in pounds, then convert the answer to (a) to pounds.

19. A race car rounds a curve at 60 m/s. The radius of the curve is 400 m, and the car's mass is 600 kg.
    a) What is the car's (centripetal) acceleration? What is it in g's?
    b) What is the centripetal force acting on the car?

20. A hang glider and its pilot have a total mass equal to 120 kg. While executing a 360° turn, the glider moves in a circle with an 8-m radius. The glider's speed is 10 m/s.
    a) What is the net force on the hang glider?
    b) What is the acceleration?

21. A 0.1-kg ball is attached to a string and whirled around in a circle overhead. The string breaks if the force on it exceeds 60 N. What is the maximum speed the ball can have when the radius of the circle is 1 m?

22. On a highway curve with radius 50 m, the maximum force of static friction (centripetal force) that can act on a 1,000-kg car going around the curve is 8,000 N. What speed limit should be posted for the curve so that cars can negotiate it safely?

23. A centripetal force of 200 N acts on a 1,000-kg satellite moving with a speed of 5,000 m/s in a circular orbit around a planet. What is the radius of its orbit?

24. As a spacecraft approaches a planet, the rocket engines on it are fired (turned on) to slow it down so it will go into orbit around the planet. The spacecraft's mass is 2,000 kg and the thrust (force) of the rocket engines is 400 N. If its speed must be decreased by 1,000 m/s, how long must the engines be fired? (Ignore the change in the mass as the fuel is burned.)

25. A space probe is launched from the Earth, headed for deep space. At a distance of 10,000 miles from the Earth's center, the gravitational force on it is 600 lb. What is the size of the force when it is at each of the following distances from the Earth's center?
    a) 20,000 miles
    b) 30,000 miles
    c) 100,000 miles

CHALLENGES

1. The force on a baseball as it is being hit with a bat can be over 8,000 lb. No human can push on a bat with that much force. What is happening in this instance?

2. Why does banking a curve on a highway allow a vehicle to go faster around it?

3. As a horse and wagon are accelerating from rest, the horse exerts a force of 400 N on the wagon (● Figure 2.52). Illustrating Newton's third law, the wagon exerts an equal and opposite force of 400 N. Since the two forces are in opposite directions, why don't they cancel each other and produce zero acceleration?

4. Perform the calculation of the force acting between two 70-kg people standing 1 m apart as given on p. 72.

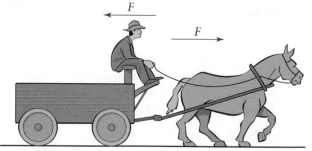

F

F

● **Figure 2.52**

5. Perhaps you've noticed that the rockets used to put satellites and spacecraft into orbit are usually launched from pads near the equator. Why is this so? Is the fact that rockets are usually launched to the east also important? Why?

6. The acceleration of a freely falling body is not exactly the same everywhere on Earth. For example, in the Galapagos Islands at the equator, the acceleration of a freely falling body is 9.780 m/s², while at the latitude of Oslo, Norway, it is 9.831 m/s². Why does the acceleration differ?

7. A 200-kg communications satellite is placed into a circular orbit around the Earth with a radius of 4.23×10^7 m (26,300 miles) (● Figure 2.53).
   a) Find the gravitational force on the satellite. (There is some useful information in Section 2.8.)
   b) Use the equation for centripetal force to compute the speed of the satellite.
   c) Show that the period of the satellite—the time it takes to complete one orbit—is 1 day. (The distance it travels during one orbit is 2π, or 6.28, times the radius.) This is a *geosynchronous orbit:* The satellite stays above a fixed point on the Earth's equator.

8. Complete the calculation of the mass of the Earth as outlined in Section 2.8.

9. During our discussion of the motion of a falling body, we said that the gravitational force acting on it—its weight—is constant. But the

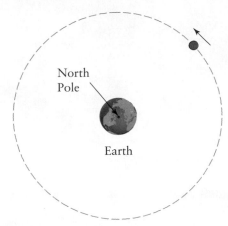

● **Figure 2.53**

law of universal gravitation tells us that the gravitational force on a body increases as it gets closer to the Earth's center. Is there a contradiction here? Explain.

SUGGESTED READINGS

Bell, Robin. "Gravity Gradiometry." *Scientific American* 278, no. 6 (June 1998): 74–79. A discussion of how small variations in the Earth's gravitational field are used by geologists and others.

Gehrels, Tom. "Collisions with Comets and Asteroids." *Scientific American* 274, no. 3 (March 1996): 74–80. An overview of origins, detection, and possible deflection of celestial objects that might hit the Earth.

Krim, Jacqueline. "Friction at the Atomic Scale." *Scientific American* 275, no. 4 (October 1996): 74–80. Describes some of the causes of sliding friction and how they are investigated.

Murphy, Kenneth L., and Evelyn T. Patterson. "Mars Global Surveyor Aerobraking." *Physics Teacher* 36, no. 3 (March 1998): 154–156. Describes how the spacecraft carrying the Mars *Pathfinder* used "air" resistance in the upper atmosphere of Mars to change its orbit.

Musser, George, and Mark Alpert. "How to Go to Mars." *Scientific American* 282, no. 3 (March 2000): 44–51. Includes descriptions of eight current and proposed propulsion systems.

O'Neill, Gerard K. *The High Frontier.* New York: Morrow, 1976. A physicist's description of how space colonies could be built. They would be designed so that rotation produced an artificial gravity inside.

Sawicki, Mikolaj. "Myths about Gravity and Tides." *Physics Teacher* 37, no. 7 (October 1999): 438–441. A more detailed look at tides.

Schwarzschild, Bertram. "Beam Balance Helps to Settle Down Measurement of the Gravitational Constant." *Physics Today* 55, no. 11 (November 2002): 19–21. Describes recent techniques for measuring *G* "by far, the most imprecisely known constant" on a standard list of fundamental constants.

Sir Isaac Newton's Mathematical Principles of Natural Philosophy and His System of the World. Translated by A. Motte (1729); edited by F. Cajori. Berkeley, CA: University of California Press, 1934. English translation of Newton's *Principia.*

White, Michael. *Isaac Newton: The Last Sorcerer.* Reading, MA: Addison–Wesley, 1997. A biography of Newton that emphasizes his work in alchemy. It includes much about his personal history and provides a historical context for his work.

For additional readings, explore InfoTrac® College Edition, your online library. Go to http://www.infotrac-college.com/wadsworth and use the passcode that came on the card with your book. Try these search terms: Isaac Newton, Cassini, tribology, chaos, comets, tides, Robert Hooke.

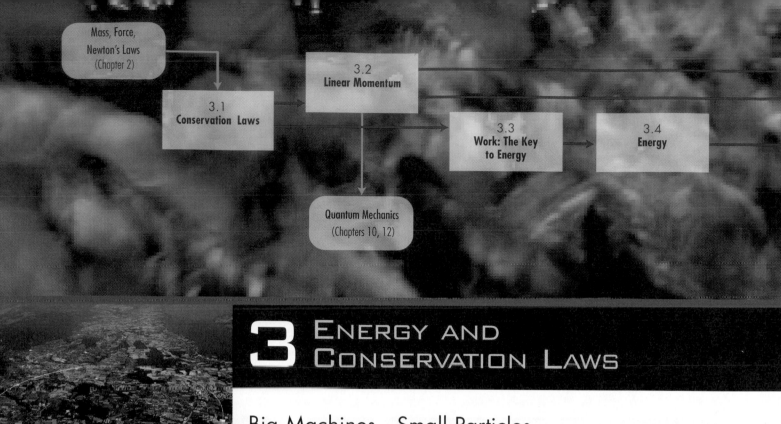

Mass, Force,
Newton's Laws
(Chapter 2)

3.1
Conservation Laws

3.2
Linear Momentum

3.3
Work: The Key
to Energy

3.4
Energy

Quantum Mechanics
(Chapters 10, 12)

Aerial photograph taken near Geneva, Switzerland, with the paths of CERN particle accelerators marked.

3 ENERGY AND CONSERVATION LAWS

Big Machines—Small Particles

The largest scientific instruments ever built are used to study the tiniest things known to exist. The main accelerator ring at the European Organization for Nuclear Research (CERN) is 27 kilometers (16.8 miles) in circumference. Tevatron, near Chicago, measures 6.4 kilometers (4 miles) in circumference (shown later in Figure 8.16). With these instruments and their smaller cousins, scientists study the fundamental particles that form the basis of all the matter that surrounds us, as well as esoteric particles that existed shortly after the birth of the universe in the Big Bang. These particles are created by accelerating electrons, protons, and other particles to nearly the speed of light, then making them collide with other particles. The objects of investigation are the colliding particles and the "debris" from the collisions.

How do colliders enable scientists to study the properties of particles a billion times smaller than atoms, the smallest things ever imaged with the most sophisticated of microscopes? The key is the use of what are called *conservation laws*. By knowing what is going on before two particles collide and then studying what the particles produced by the collision do afterward, it is possible to infer what was going on during the collision and to extract information about the nature of the forces that were acting. These laws allow physicists to unlock the mysteries of elementary particles.

The application of conservation laws for linear momentum, energy, and angular momentum is not restricted to the realm of subatomic particles. These laws apply to atoms, baseballs, trucks, planets, galaxies—everything—and to *all* processes, not just collisions. They give us useful insight into situations in our everyday experiences.

How are these laws used to reconstruct a fender bender or to learn more about the forces in a nucleus? Read on to find out.

In this chapter, we introduce the use of conservation laws in the study of how objects move. The laws of conservation of linear momentum, energy, and angular momentum are shown to be simple but powerful tools for analyzing processes—such as collisions—that we could not handle using only the concepts from Chapter 2. Work and power, two important physical quantities related to energy, are illustrated as well. The concept of energy is one of the most important in physics, and its usefulness extends well beyond the area of mechanics.

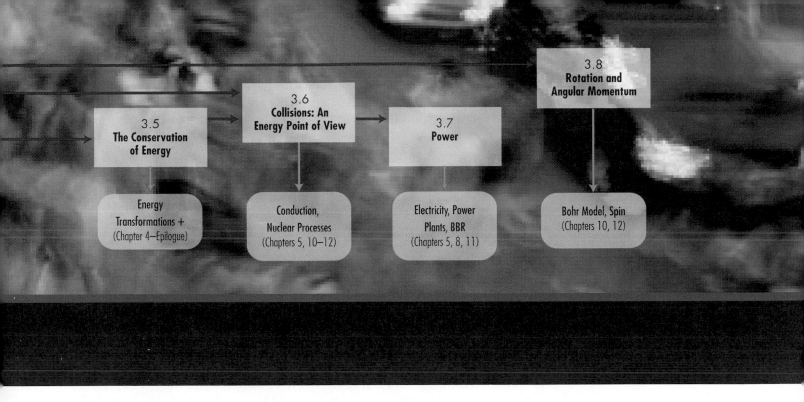

| 3.5 The Conservation of Energy | 3.6 Collisions: An Energy Point of View | 3.7 Power | 3.8 Rotation and Angular Momentum |

Energy Transformations + (Chapter 4–Epilogue)

Conduction, Nuclear Processes (Chapters 5, 10–12)

Electricity, Power Plants, BBR (Chapters 5, 8, 11)

Bohr Model, Spin (Chapters 10, 12)

3.1 Conservation Laws

Newton's laws of motion, in particular the second law, govern the instantaneous behavior of a system. They relate the forces that are acting at any instant in time to the resulting changes in motion. **Conservation laws** involve a different approach to mechanics, more of a "before-and-after" look at systems. A conservation law states that the total amount of a certain physical quantity present in a system stays constant (is conserved). For example, we might state the following conservation law.

> **PRINCIPLES**
>
> The total mass in an isolated system is constant.*

An "isolated system" in this case means that no matter enters or leaves the system. The law states that the total mass of all the objects in a system doesn't change regardless of the kinds of interactions that go on in it.

A simple example of this rather obvious conservation law is given by aerial refueling. One aircraft, called a tanker, pumps fuel into a second aircraft while both are in flight (● Figure 3.1). If we ignore the small amount of fuel that both aircraft consume during the refueling, this can be regarded as an isolated system. The law of conservation of mass tells us that the total mass of both aircraft remains constant. So if the receiving aircraft gains 2,000 kilograms of fuel, we automatically know that the tanker loses 2,000 kilograms of fuel. If one tanker refuels several aircraft in a formation, we include all of the aircraft in the system. The total mass of fuel dispensed by the tanker equals the total mass of fuel gained by the other aircraft. This fact may be useful. Let's say that the fuel gauge on one of the receiving aircraft is faulty. To determine how much fuel that aircraft was given, the crews use the conservation of mass: The total mass of fuel unloaded from the tanker minus the total mass of fuel given to the other aircraft in the formation equals the mass of fuel given to the aircraft in question.

● **Figure 3.1** Aerial refueling represents a simple example of the conservation of mass.

---

* As we shall see in Chapter 11, Einstein showed that energy must be included in this law because energy can be converted into matter, and vice versa. We will not need this refinement until then, however.

The preceding example illustrates how we can use a conservation law. Without knowing the details about an interaction (for example, the actual rate at which fuel is transferred between the aircraft), we can still extract quantitative information by simply comparing the total amount of mass before and after. Three more conservation laws are presented in this chapter. Although they are a bit less intuitive and a bit more complicated to use than the law of conservation of mass, they are applied in the same way.

3.2 Linear Momentum

The conservation law for **linear momentum** follows directly from Newton's laws of motion, and we will consider it first.

> **DEFINITION**
>
> **Linear Momentum** The mass of an object times its velocity. Linear momentum is a vector.
>
> $$\text{linear momentum} = mv$$

Linear momentum is often referred to simply as *momentum*. It is a vector quantity (since velocity is a vector), and its SI unit of measure is the kilogram-meter/second (kg m/s).

Linear momentum incorporates both mass and motion. Anything that is stationary has zero momentum. The faster a body moves, the larger its momentum. A heavy object moving with a certain velocity has more momentum than a light object moving with the same velocity (● Figure 3.2). For example, the momentum of a bicycle and rider with a total mass of 80 kilograms and a speed of 10 m/s is

$$mv = 80 \text{ kg} \times 10 \text{ m/s} = 800 \text{ kg-m/s}$$

The linear momentum of a 1,200-kilogram car with the same speed is 12,000 kg-m/s. The bicycle and rider would have to be going 150 m/s (not likely) to have this same momentum.

Newton originally stated his second law of motion using linear momentum. In particular, we can restate this law as follows.

> **LAWS**
>
> **Newton's Second Law of Motion (alternate form)** The net external force acting on an object equals the rate of change of its linear momentum.
>
> $$\text{force} = \frac{\text{change in momentum}}{\text{change in time}}$$
>
> $$F = \frac{\Delta(mv)}{\Delta t}$$

● **Figure 3.2** Linear momentum depends on the mass and the velocity. If a car and a bicycle have the same velocity, the car has a larger momentum because it has a larger mass.

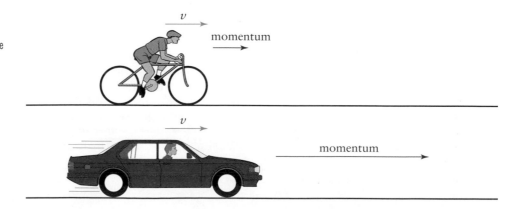

To change an object's linear momentum, a net force must act on it. The larger the force, the faster the momentum will change. If the mass of the object stays constant, which is true in most cases, this equation is equivalent to the first form of Newton's second law (Section 2.4) because

$$\frac{\Delta(mv)}{\Delta t} = m\frac{\Delta v}{\Delta t} = ma \qquad \text{(if mass is constant)}$$

Either form of the second law could be used for cars, airplanes, baseballs, and so on. For rockets and similar things with changing mass, only the alternate form should be used.

We can get yet another useful form of the second law by multiplying both sides by Δt to obtain

$$\Delta(mv) = F\,\Delta t$$

The quantity on the right side is called the *impulse.* The same change in momentum can result from a small force acting for a long time or a large force acting for a short time. When you throw a tennis ball, a small force acts on it for a relatively long period of time (as your hand moves through the air). When the ball is served at the same speed with a racquet, a large force acts for a short period of time.

This equation is useful for analyzing what goes on during impacts in sports that use balls and clubs. When a tennis racquet hits a tennis ball, a large force, F, acts for a short time, Δt. The result is a change in momentum of the ball, $\Delta(mv)$. One reason to have good follow-through on a shot is to prolong the time of contact, Δt, between the ball and the racquet. This leads to a greater change in momentum, so the ball will leave the racquet with a higher speed.

Example 3.1

Let's estimate the average force on a tennis ball as it is served (● Figure 3.3). The ball's mass is 0.06 kilograms, and it leaves the racquet with a speed of, say, 40 m/s (90 mph). High-speed photographs indicate that the contact time is about 5 milliseconds (0.005 s).

Since the ball starts with zero speed, its change in momentum is

$$\Delta(mv) = \text{momentum afterwards} = 0.06 \text{ kg} \times 40 \text{ m/s}$$

$$= 2.4 \text{ kg-m/s}$$

So the average force is

$$F = \frac{\Delta(mv)}{\Delta t} = \frac{2.4 \text{ kg-m/s}}{0.005 \text{ s}}$$

$$= 480 \text{ N} = 108 \text{ lb}$$

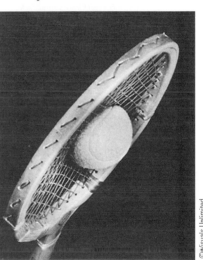

● **Figure 3.3** The tennis racquet exerts a large force on the tennis ball for a short time.

Our main application of the idea of linear momentum is based on the following conservation law.

LAWS

Law of Conservation of Linear Momentum The total linear momentum of an isolated system is constant.

For this law, an isolated system means that there are *no outside forces* causing changes in the linear momenta of the objects inside the system. The momentum of an object can change only because of interaction with other objects in the system. For example, once the cue ball on a pool table has been shot (given some momentum), the pool table and the balls form an isolated system. If the cue ball collides with another ball initially at rest, its momentum is changed (decreased) (● See Figure 3.4). This change occurs because of an interaction with another object in the system. The momentum of the ball with which

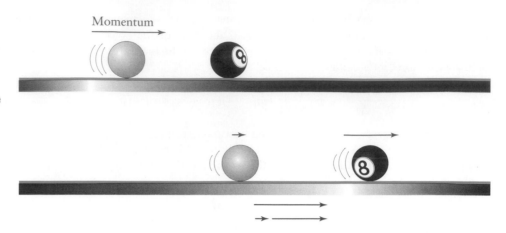

Momentum

it collides is increased, although the total linear momentum of the system remains the same. If someone put a hand on the table and stopped the cue ball, the system would no longer be isolated, and we couldn't assume that the total linear momentum of the system was constant.

The most important use of the law of conservation of linear momentum is in the analysis of collisions. Two billiard balls colliding, a traffic accident, and two skaters running into each other are some familiar examples of collisions. We will limit our examples to collisions involving only two objects that are moving in one dimension (along a line).

Explore It Yourself 3.1

For this one, you and a friend have to be on skates (ice, roller, or in-line), and you have to be comfortable colliding with each other and moving backward (slowly) on the skates. Face each other with you stationary, and each with hands in front ready to clasp the other person's hands. Have your friend skate into you (not too fast), and then clasp each other's hands as you move backward. Do the two of you move as fast as your friend was initially moving? If one of you is lighter than the other, switch places and do it again. What is different?

During any collision, the objects exert equal and opposite forces on each other that cause them to accelerate (in opposite directions). These forces are usually quite large and are often due to direct contact between the bodies, as is the case with billiard balls and automobiles. "Action-at-a-distance" forces between objects that don't actually come into contact, like the gravitational pull between a spacecraft and a planet, can also be involved in collisions. The following statement is the key to applying the law of conservation of linear momentum to a collision:

LAWS

The total linear momentum of the objects in the system before the collision is the same as the total linear momentum after the collision.

$$\text{total } mv \text{ before} = \text{total } mv \text{ after}$$

Example 3.2   We can use the law of conservation of linear momentum to analyze a simple automobile collision. A 1,000-kilogram automobile (car 1) runs into the rear of a stopped car (car 2) that has a mass of 1,500 kilograms. Immediately after the collision, the cars are hooked together, and their speed is estimated to have been 4 m/s (● Figure 3.5). What was the speed of car 1 just before the collision?

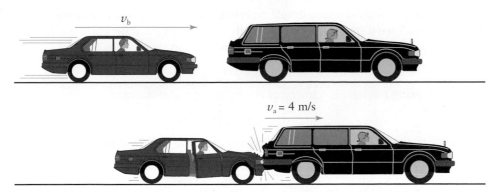

● **Figure 3.5** A 1,000-kilogram car collides with a 1,500-kilogram car that is stationary. Afterward, the two cars are hooked together and move with a speed of 4 m/s. The conservation of linear momentum allows us to determine that the speed of the first car was 10 m/s.

$$v_a = 4 \text{ m/s}$$

Our conservation law tells us that the total linear momentum of the system will be constant:

$$\text{total } (mv)_{\text{before}} = \text{total } (mv)_{\text{after}}$$

Before the collision, only car 1 is moving. The total linear momentum before the collision is

$$(mv)_{\text{before}} = m_1 \times v_{\text{before}} = (1{,}000 \text{ kg}) \times v_{\text{before}}$$

After the collision, both cars are moving as one body. The total momentum after the collision is

$$(mv)_{\text{after}} = (m_1 + m_2) \times v_{\text{after}} = (1{,}000 \text{ kg} + 1{,}500 \text{ kg}) \times 4 \text{ m/s}$$

$$= 2{,}500 \text{ kg} \times 4 \text{ m/s} = 10{,}000 \text{ kg-m/s}$$

These two linear momenta are equal. Therefore,

$$1{,}000 \text{ kg} \times v_{\text{before}} = 10{,}000 \text{ kg-m/s}$$

$$v_{\text{before}} = \frac{10{,}000 \text{ kg-m/s}}{1{,}000 \text{ kg}}$$

$$= 10 \text{ m/s}$$

This type of analysis is routinely used to reconstruct traffic accidents. For example, it can be used to determine whether a vehicle was exceeding the speed limit just before a collision. In this problem the car was going 10 m/s, or about 22 mph. If the accident happened in a 15-mph speed zone, the driver of car 1 would have been speeding.

The speed of a bullet or a thrown object can be measured similarly. A bullet is fired into, and becomes embedded in, a block of wood hanging from a string (● Figure 3.6). If one measures the masses of the bullet and the wood block and the speed of the block immediately afterward, the initial speed of the bullet can be found by using the conservation of linear momentum. You can measure the speed of your "fast ball" by throwing a lump of sticky clay at a block and proceeding in the same way. The key in both cases is measuring the speed of the block of wood after the collision. One can do this most easily by using the law of conservation of energy. We will show how in Section 3.5.

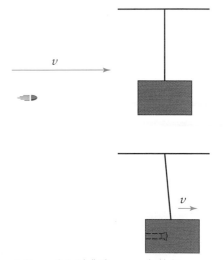

● **Figure 3.6** A bullet becomes embedded in a block of wood. If the speed of the block and the masses of the block and the bullet are measured, the initial speed of the bullet can be computed using the conservation of linear momentum.

Explore It Yourself 3.2

Another skating exercise, as in Explore It Yourself 3.1. Stand facing each other, then push each other so you both go backwards. Are you and your friend going the same speed after the push? If not, which one is going faster?

In Section 2.7 we described a simple experiment in which we used two carts (Figure 2.31c). One cart has a spring-loaded plunger that pushes on the other cart, causing both carts to be accelerated. By the law of conservation of linear momentum,

$$(mv)_{before} = (mv)_{after}$$

Before the spring is released, neither cart is moving, so the momentum is zero:

$$(mv)_{before} = 0$$

Therefore, the total linear momentum afterward is zero. But since both carts are moving, this momentum equals:

$$(mv)_{after} = 0 = (mv)_1 + (mv)_2$$

So:

$$(mv)_1 = -(mv)_2$$

The momentum of one of the carts is negative. This has to be the case, because they are moving in opposite directions. (Remember: linear momentum is a vector.) Since $(mv)_1 = m_1 \times v_1$:

$$m_1 \times v_1 = -m_2 \times v_2$$

$$v_1 = -\frac{m_2}{m_1} \times v_2$$

$$\frac{v_1}{v_2} = -\frac{m_2}{m_1}$$

So we have derived the statement made in Section 2.7: The ratio of the speeds of the two carts is the inverse of the ratio of their masses. If the mass of cart 2 is 3 kilograms and the mass of cart 1 is 1 kilogram, the ratio is 3 to 1. Cart 1 moves away with three times the speed of cart 2 (● Figure 3.7). Note that this gives us only the ratio of the speeds, not the actual speed of each cart, which would depend on the strength of the spring. With a weak spring, the speeds might be 1 m/s and 3 m/s. With a stronger spring, the speeds might be 2.5 m/s and 7.5 m/s.

We can use linear momentum conservation to get a different view of some of the situations described in Section 2.7. There we used forces and Newton's third law of motion. When a gun is fired, the bullet acquires momentum in one direction, and the gun gains equal momentum in the opposite direction—the "kick" of the gun against the shoulder or hand. If an ice skater throws an object forward, he or she will move backward with equal momentum (● Figure 3.8). Rockets and jets give momenta to the ejected exhaust gases. They in turn gain momentum in the forward direction. For a rocket with no external forces on it, the increase in momentum during each second will depend on how much gas is ejected—which equals the mass of fuel burned—and on the speed of the gas (● Figure 3.9).

● **Figure 3.7** The mass of the cart on the right, cart 2, is three times the mass of the cart on the left, cart 1. After the spring is released, the speed of the lighter cart is three times the speed of the more massive cart.

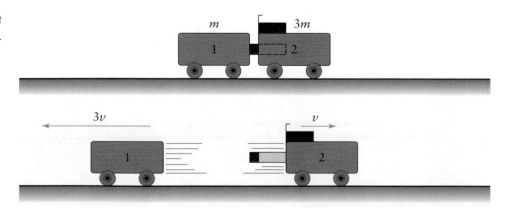

Momentum

Why is linear momentum conserved? We can use Newton's second and third laws of motion to answer this question. When two objects exert forces on each other by colliding or via a spring plunger, the forces are equal and opposite. They push on each other with the same size force but in opposite directions. By the second law (alternate form), these equal forces cause the momenta of both objects to change at the same rate. As long as the objects are interacting, *they change one another's momentum by the same amount but in opposite directions.* The momentum gained (or lost) by one object is exactly offset by the momentum lost (or gained) by the other. The total linear momentum is not changed. In Example 3.2 the 1,000-kilogram car is slowed from 10 m/s before the collision to 4 m/s after the collision (Figure 3.5). Its momentum is decreased by 6,000 kg-m/s (1,000 kilograms $\times$ 10 m/s $-$ 1,000 kilograms $\times$ 4 m/s). The 1,500-kilogram car goes from 0 m/s before to 4 m/s after. So its momentum is increased by 6,000 kg-m/s (1,500 kilograms $\times$ 4 m/s).

These examples illustrate the usefulness of conservation laws. The approach is different from that used with Newton's second law in Chapter 2, where it was necessary to know the size of the force that acts on the object at each moment to determine its velocity. With the law of conservation of linear momentum, we do not have to know the details of the interactions—how large the forces are and how long they act. All we need is some information about the system before and after the interaction. Note that in Example 3.2 we used information from after the collision to determine the speed of the car *before* the collision. In the example with the carts, we determined the ratio of the speeds *after* the interaction.

v

● **Figure 3.9** As a rocket gives momentum to the exhaust gases, it gains momentum in the opposite direction.

LEARNING CHECK

1. (True or False.) A conservation law can be used without knowing all of the details about what is occurring inside a system.
2. The linear momentum of a truck will always be greater than that of a bus if
   a) the truck's mass is larger than the bus's but its speed is the same.
   b) the truck's speed is larger than the bus's but its mass is the same.
   c) both its mass and its speed are larger than the bus's.
   d) Any of the above.
3. In a collision, the total _____ is the same before and after.
4. During a collision of two football players, the total linear momentum of the system cannot be zero.

ANSWERS: 1. True   2. (d)   3. linear momentum   4. False

3.3 Work: The Key to Energy

The law of conservation of energy is arguably the most important of the conservation laws. Not only is it useful for solving problems, it is also a powerful theoretical statement that can be used to understand widely diverse phenomena and to show what hypothetical processes are or are not possible. As we mentioned earlier, the concept of **energy** is one of the most important in physics. This is because energy takes many forms and is involved in all physical processes. One could say that every interaction in our universe involves a transfer of energy or a transformation of energy from one form to another.

We can compare the concept of energy to that of financial assets, which can take the form of cash, real estate, material goods, or investments, among other things. The study of economics is in part a study of these forms of financial assets and how they are transferred and transformed. Much of physics deals with the forms of energy and the transformations that occur during interactions.

When first encountered, the concept of energy is a bit difficult to understand because there is no simple way to define it. As an aid, we will first introduce **work,** a physical quantity that is quite basic and that gives us a good foundation for understanding energy.

The idea of work in physics arises naturally when one considers *simple machines* like the lever and the inclined plane when used in situations with negligible friction. Let's say that you use a lever to raise a heavy rock (● Figure 3.10). If you place the fulcrum close to the rock, you find that a small, downward force on your end results in a larger, upward force on the rock. However, the distance your end moves as you push it down is correspondingly larger than the distance the rock is raised. By measuring the forces and distances, we find that dividing the larger force by the smaller force gives the same number (or ratio) as dividing the larger distance by the smaller distance. In particular:

$$\frac{F \text{ on left end}}{F \text{ on right end}} = \frac{d \text{ right end moves}}{d \text{ left end moves}}$$

We can multiply both sides by *F on right end* and *d left end moves* and get the following result:

$$(F \text{ on left}) \times (d \text{ left moves}) = (F \text{ on right}) \times (d \text{ right moves})$$

$$F_{\text{left}} \, d_{\text{left}} = F_{\text{right}} \, d_{\text{right}}$$

In other words, even though the two forces and the two distances are different, the quantity *force times distance* has the same value for both ends of the lever. We might say that raising the rock is a fixed task. One can perform the task by lifting the rock directly or by using a lever. In the former case, the force is large, equal to the rock's weight, but the distance moved is small. When using the lever, the force is smaller, but the distance is larger. The quantity *Fd* is the same, regardless of which way the task is accomplished.

● **Figure 3.10** When a lever is used to raise a rock, a small force on the right end results in a larger force on the left end. But the right end moves a greater distance than the left end. The force multiplied by the distance moved *is the same* for both ends.

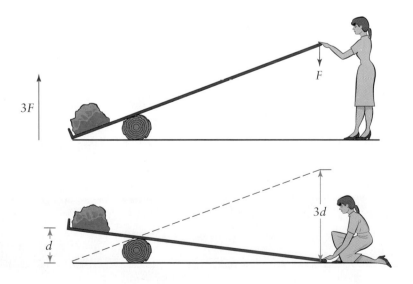

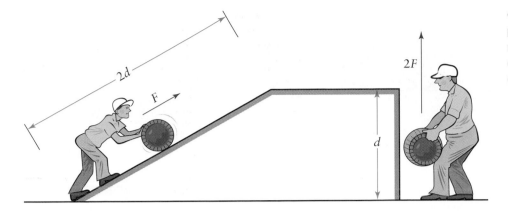

We reach the same conclusion when considering an inclined plane. Let's say that a barrel must be placed on a loading dock (● Figure 3.11). Lifting the barrel directly requires a large force acting through a small distance, the height of the dock. If the barrel is rolled up a ramp, a smaller force is needed, but the barrel must be moved a greater distance. Again, the product of the force and the distance moved is the same for the two methods.

$$(F \text{ lifting}) \times \text{height} = (F \text{ rolling}) \times (\text{ramp length})$$

$$F_{\text{lifting}} \, d_{\text{lifting}} = F_{\text{rolling}} \, d_{\text{rolling}}$$

The quantity *force times distance* is obviously a useful way of measuring the "size" of a task. It is called *work*.

DEFINITION

Work The force that acts times the distance moved in the direction of the force:

$$\text{work} = Fd$$

Physical Quantity	Metric Units	English Units
Work	joule (J) [SI unit] erg calorie (cal) kilowatt-hour (kWh)	foot-pound (ft-lb) British thermal unit (Btu)

Since force is a vector, work equals the distance moved times the component of the force parallel to the motion. Work itself is not a vector. There is no direction associated with work.

Whenever an object moves and there is a force acting on the object in the same or opposite direction that it moves, work is done. When the force and the motion are in the same direction, the work is positive. When they are in opposite directions, the work is negative.

Example 3.3

Because of friction, a constant force of 100 newtons is needed to slide a box across a room (● Figure 3.12). If the box moves 3 meters, how much work is done?

$$\text{work} = Fd$$

$$= 100 \text{ N} \times 3 \text{ m}$$

$$= 300 \text{ N-m}$$

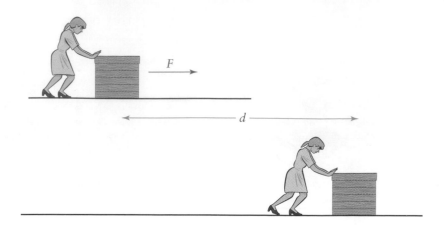

Figure 3.12 Pushing on a box with a force of 100 newtons causes it to slide over the floor. If the box moves 3 meters, you do 300 joules of work.

The unit in the answer is the newton-meter (N-m). This is called the *joule*.

$$1 \text{ joule} = 1 \text{ newton-meter} = 1 \text{ newton} \times 1 \text{ meter}$$

$$1 \text{ J} = 1 \text{ N-m}$$

The joule is a derived unit of measure, and it is the SI unit of work and energy. Just remember that if you use only SI units for force, distance, and so on, your unit for work and energy will always be the joule (J).

Example 3.4   Let's say that the barrel in Figure 3.11 has a mass of 30 kilograms and that the height of the dock is 1.2 meters. How much work would you do when lifting the barrel?

$$\text{work} = Fd$$

The force is just the weight of the barrel, *mg*.

$$F = W = mg = 30 \text{ kg} \times 9.8 \text{ m/s}^2 = 294 \text{ N}$$

Hence:

$$\text{work} = Fd = Wd$$

$$= 294 \text{ N} \times 1.2 \text{ m}$$

$$= 353 \text{ J}$$

The work you do when rolling the barrel up the ramp would be the same. The force would be smaller, but the distance would be larger.

Figure 3.13 When you carry a box across a room, the force on the box is perpendicular to the direction the box moves. No work is done on the box by the force holding it up.

Explore It Yourself 3.3

You will have to be in a multistory building with stairs. Walk around on a completely level surface (perhaps back and forth in a hallway) for about 20 seconds. Wait a minute or two, and then climb the stairs upward for about 20 seconds. Do you feel any different? How is work involved in the two cases?

It is important to note that work is not done if the force is perpendicular to the motion. When you simply carry a box across a room, your force on the box is vertical, whereas the motion of the box is horizontal. Hence, you do no work on the box (● Figure 3.13).

Uniform circular motion is another situation in which a force acts on a moving body but no work is done. Recall that a centripetal force must act on anything to keep it moving along a circular path (● Figure 3.14). This force is always toward the center of the circle and perpendicular to the object's velocity at each instant. Therefore, the force does not do work on the object.

Work can be done in a circular motion by a force that is not a centripetal force. When you turn the crank on a pencil sharpener or a fishing reel, for instance, you exert a force on the handle that is in the same direction as the handle's motion. Hence, you do work on the handle.

Work is done on an object when it is accelerated in a straight line. The following example shows how the amount of work can be computed. (In the next section, we will see that there is an easier way to do this.)

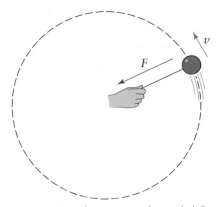

● **Figure 3.14** The string exerts a force on the ball that is perpendicular to its motion. No work is done by this force.

Example 3.5

In Example 2.2, we used Newton's second law to compute the force needed to accelerate a 1,000-kilogram car from 0 to 27 m/s in 10 seconds. Our answer was $F = 2,700$ newtons. How much work is done?

$$\text{work} = Fd$$

To find the distance that the car travels, we use the fact that the car accelerates at 2.7 m/s^2 for 10 seconds. Using the equation from Section 1.4:

$$d = \frac{1}{2}at^2$$

$$= \frac{1}{2} \times 2.7 \text{ m/s}^2 \times (10 \text{ s})^2$$

$$= 1.35 \text{ m/s}^2 \times 100 \text{ s}^2 = 135 \text{ m}$$

So the work done is

$$\text{work} = Fd$$

$$= 2,700 \text{ N} \times 135 \text{ m}$$

$$= 364,500 \text{ J}$$

We have seen that work is done (a) when a force acts to move something against the force of gravity (Figures 3.10 and 3.11), (b) when a force acts to move something against friction (Figure 3.12), and (c) when a force accelerates an object. There are many other possibilities. When a force distorts something, work is done. For example, to compress or stretch a spring, a force must act on it. This force acts through a distance in the same direction as the force, so work is done.

Work is also done when a force causes something to slow down. When you catch a ball, your hand exerts a force on the ball. As the ball slows down, it pushes your hand back with an equal and opposite force (see ● Figure 3.15). In this case, the ball does work on your hand. Your hand does negative work on the ball. The amount of work that the ball does on your hand is equal to the amount of work that was originally done on the ball to accelerate it (ignoring the effect of air resistance). If you allow your hand to move back as you catch the ball, the force of the ball on you will be less than if you try to keep your hand stationary. The work that the ball will do on your hand is the same either way. Since work = force × distance, the force on your hand will be smaller if the distance that the ball moves while you catch it is larger.

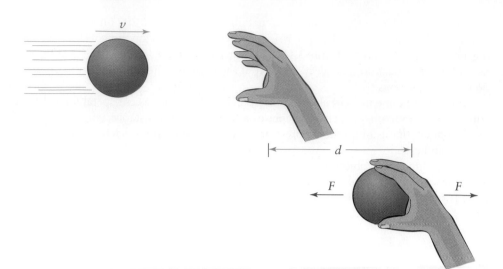

Figure 3.15 As you catch a ball, your hand exerts a force on the ball. By Newton's third law, the ball exerts an equal and opposite force on your hand. This force does work on you as you slow down the ball. The work done is equal to the work that was done to accelerate the ball in the first place.

One last example: When an object falls freely, the force of gravity does work on it. As in the previous example with the car, this work goes to accelerate the object. In particular, if a body falls a distance d, the work done on it by the force of gravity is

$$\text{work} = Fd$$

But

$$F = W = mg$$

So:

$$\text{work} = Wd = mgd$$

The work that the force of gravity does on an object as it falls is equal to the work that was done to lift the object the same distance (● Figure 3.16). When something is lifted, we say that work is done *against* the force of gravity. The movement is in the opposite direction of this force. When something falls, work is done *by* the force of gravity. The movement is in the same direction as the force.

In summary, work is done by a force whenever the point of application moves in the direction of the force. Work is done against the force whenever the point of application moves opposite the direction of the force. Forces always come in pairs that are equal and opposite (Newton's third law of motion). Consequently, when work is done by one force, this work is being done against the other force in the pair.

Figure 3.16 The work done lifting an object equals the work done by gravity as the object falls.

3.4 Energy

In Section 3.3, we saw that work can often be "recovered." The work that is done when a ball is thrown is equal to the work done by the ball when it is caught. The work done when lifting an object against the force of gravity is equal to the work done by the force of gravity when the object falls. When work is done on something, it gains **energy.** This energy can then be used to do work. A thrown ball is given energy. This energy is given up to do work when the ball is caught.

> **DEFINITION**
>
> **Energy** The measure of a system's capacity to do work. That which is transferred when work is done. Abbreviated E.

The units of energy are the same as the units of work. Like work, energy is a scalar.

The more work done on something, the more energy it gains, and the more work it can do in return. We might say that energy is "stored work." To be able to do work, an object must have energy. When you throw a ball, you are transferring some energy from you to the ball. The ball can then do work. In Figures 3.13 and 3.14, no energy is transferred because no work is done.

There are many forms of energy corresponding to the many ways in which work can be done. Some of the more common forms of energy are chemical, electrical, nuclear, and gravitational, as well as the energy associated with heat, sound, light, and other forms of radiation. Anything possessing any of these forms of energy is capable of doing work (see ● Figure 3.17).

In mechanics, there are two main forms of energy, which we can classify under the single heading of *mechanical energy.* Anything that has energy because of its motion or because of its position or configuration has mechanical energy. We refer to the former as **kinetic energy** and to the latter as **potential energy.**

● **Figure 3.17** The crane can do work because it has energy—chemical energy in its fuel.

> **DEFINITION**
>
> **Kinetic Energy** Energy due to motion. Energy that an object has because it is moving. Abbreviated KE.

Anything that is moving has kinetic energy. The simplest example is an object moving in a straight line. The amount of kinetic energy an object has depends on its mass and speed. In particular:

$$KE = \frac{1}{2}mv^2 \quad \text{(kinetic energy)}$$

The kinetic energy that an object has is equal to the work done when accelerating the object from rest.* So another way to determine the amount of work done when accelerating an object is to compute its kinetic energy. This also shows that the work done depends only on the object's final speed and not on how rapidly or slowly it was accelerated.

Example 3.6

In Example 3.5, we computed the work that is done on a 1,000-kilogram car as it accelerates from 0 to 27 m/s. The car's kinetic energy when it is traveling 27 m/s is

$$KE = \frac{1}{2}mv^2 = \frac{1}{2} \times 1,000 \text{ kg} \times (27 \text{ m/s})^2$$

$$= 500 \text{ kg} \times 729 \text{ m}^2/\text{s}^2$$

$$- 364,500 \text{ J}$$

This is the same as the work done as the car accelerates. The car can do 364,500 joules of work because of its motion.

● **Figure 3.18** A spinning dancer has kinetic energy, even though she stays in one place.

The kinetic energy of a moving body is proportional to the square of its speed. If one car is going twice as fast as a second, identical car, the faster one has four times the kinetic energy. It takes four times as much work to stop the faster car. Note that the kinetic energy of an object can never be negative. (Why? Because m is always positive and v^2 is positive even when v is negative.)

Since speed is relative, kinetic energy is also relative. A runner on a moving ship has KE relative to the ship and a different KE relative to something at rest in the water.

Another way that an object can have kinetic energy is by rotating. To make something spin, work must be done on it. A dancer or skater performing a pirouette has kinetic energy (see ● Figure 3.18). A spinning top has kinetic energy, as do the Earth, Moon, Sun, and other astronomical objects as they spin about their axes. The amount of kinetic energy that a spinning object has depends on its mass, its rotation rate, and the way its mass is distributed. Some devices and toys use rotational kinetic energy as a way to "store" energy. ● Figure 3.19 shows a toy car and a shaver that operate this way.

DEFINITION

Potential Energy Energy due to an object's position or orientation. Energy that a system has because of its configuration. Abbreviated *PE*.

---

* For the mathematically inclined—in the case of constant acceleration, we can compute the work done to accelerate something from rest.

$$\text{work} = Fd$$

Now we use what we learned in Chapters 1 and 2. The force needed is $F = ma$, and the distance traveled is $d = \frac{1}{2}at^2$. So:

$$\text{work} = Fd = ma \times \frac{1}{2}at^2 = \frac{1}{2}m \times a^2 \times t^2 = \frac{1}{2}m \times (at)^2$$

But at equals the speed v that the body has. So:

$$\text{work} = \frac{1}{2}mv^2 = KE$$

(a)

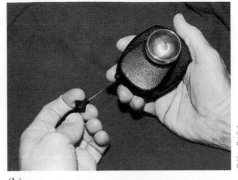

(b)

● **Figure 3.19** This transparent toy car (left) and shaver (right) use rotational kinetic energy stored in spinning flywheels. The car's flywheel is the disc with the spiral yellow, orange, and red pattern.

The amount of potential energy that a system acquires is equal to the work done to put it in that configuration. When an object is lifted, it is given potential energy. It can use this energy to do work. For example, when the weights on a cuckoo clock or any other gravity-powered clock are raised, they are given potential energy (● Figure 3.20). As they slowly fall, they do work in operating the clock.

In Section 3.3, we computed the work done when an object is lifted. This is equal to the potential energy that it is given. Since the work is done against the force of gravity, it is called **gravitational potential energy.**

$$PE = \text{work done} = \text{weight} \times \text{height} = Wd = mgd$$

$$PE = Wd = mgd \qquad \text{(gravitational potential energy)}$$

Gravitational potential energy is the most common type of potential energy, and it is often referred to simply as *potential energy.* Potential energy is a relative quantity because height can be measured relative to different levels.

● **Figure 3.20** The source of energy to run this clock is the potential energy of the weights.

A 3-kilogram brick is lifted to a height of 0.5 meters above a table (● Figure 3.21). Its potential energy relative to the table is

$$PE = mgd$$

$$PE = 3 \text{ kg} \times 9.8 \text{ m/s}^2 \times 0.5 \text{ m}$$

$$= 14.7 \text{ J} \quad \text{(relative to the table)}$$

But the tabletop itself may be 1 meter above the floor. The brick's height above the floor is 1.5 meters, and so its potential energy relative to the floor is

$$PE = mgd$$

$$= 3 \text{ kg} \times 9.8 \text{ m/s}^2 \times 1.5 \text{ m}$$

$$= 44.1 \text{ J} \quad \text{(relative to the floor)}$$

Example 3.7

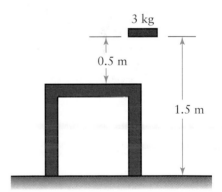

● **Figure 3.21** The potential energy (*PE*) of an object depends on how its height (*d*) is measured. The brick has 14.7 joules of potential energy relative to the table and 44.1 joules of potential energy relative to the floor. In both cases, the potential energy equals the work done to raise the brick from that level.

A person sitting in a chair has gravitational potential energy relative to the floor, to the basement of the building, and to the level of the oceans. Usually, some convenient reference level is chosen for determining potential energies. In a room it is logical to use the floor as the reference level for measuring heights and, consequently, potential energies.

An object's potential energy is negative when it is *below* the chosen reference level. Often the reference level is chosen so that negative potential energies signify that an object cannot "leave." For example, it is logical to measure potential energy relative to the ground level on flat ground outdoors. Anything on the ground has zero potential energy, and anything above the ground has positive potential energy. If there is a hole in the

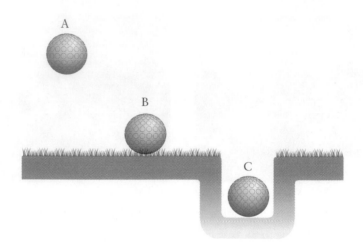

● **Figure 3.22** The potential energy of golf ball A is positive relative to the ground. The potential energy of B is zero, and that of C is negative because it is below ground level. Balls A and B can move horizontally while C is restricted to the hole.

● **Figure 3.23** Olympic archer Joanne Edens. While pulling the bowstring back, she does work bending the bow. This gives the bow elastic potential energy.

ground, any object in the hole will have negative potential energy (● Figure 3.22). Any object that has zero or positive potential energy can move about horizontally if it has any kinetic energy. Objects with negative potential energy are confined to the hole. They must be given enough energy to get out of the hole before they can move horizontally.

Springs and rubber bands can possess another type of potential energy, **elastic potential energy.** Work must be done on a spring to stretch or compress it. This gives the spring potential energy (see ● Figure 3.23). This "stored energy" can then be used to do work. The actual amount of potential energy a spring has depends on two things: how much it was stretched or compressed and how strong it is. In Figure 3.7, the combined kinetic energies of the carts after the spring is released equal the original elastic potential energy of the spring. A stronger spring would possess more potential energy and would give the carts more *KE,* so they would go faster.

Many devices use elastic potential energy. Toy dart guns have a spring inside that is compressed by the shooter. When the trigger is pulled, the spring is released and does work on the dart to accelerate it. The potential energy of the spring is converted to kinetic energy of the dart. The bow and arrow operate this same way, with the bow acting as a spring. Windup devices such as clocks, toys, and music boxes use energy stored in springs to operate the mechanism. Usually, the spring is in a spiral shape, but the principle is the same. Rubber bands provide lightweight energy storage in some toy airplanes.

Another form of energy important in mechanical systems is **internal energy.** Internal energy, heat, and temperature are discussed formally in Chapter 5. Basically, the internal energy of a substance is the total energy of all the atoms and molecules in the substance. To raise something's temperature, to melt a solid, or to boil a liquid all require increasing the internal energy of the substance. Internal energy decreases when a substance's temperature decreases, when a liquid freezes, or when a gas condenses.

Internal energy is involved whenever there is kinetic friction. In Figure 3.12, the work done on the box as it is pushed across the floor is converted into internal energy because of the friction between the box and the floor. The result is that the temperatures of the floor and the box are raised (although not by much). As a car or a bicycle brakes to a stop, its kinetic energy is converted into internal energy in the brakes. Automobile disc brakes can become red hot under extreme braking. Meteors (shooting stars) are a spectacular example of kinetic energy being converted into internal energy (● Figure 3.24). When they enter the atmosphere at very high speed, air resistance heats them enough to glow and melt. Gravitational potential energy can also be converted into internal energy. A box slowly sliding down a ramp has its gravitational potential energy converted into internal energy by friction. If you climb a rope and then slide down it, you can burn your hands severely as some of your potential energy is converted into internal energy because of the friction between your hands and the rope.

Internal energy can also be produced by internal friction when something is distorted. The work you do when stretching a rubber band, pulling taffy, or crushing an aluminum

© Reuters/CORBIS

● **Figure 3.24** A meteoroid leaves a glowing trail in the night sky. Air resistance causes its kinetic energy to be converted into internal energy.

can generates internal energy. When you drop something and it doesn't bounce, like a book, most of the energy the object had is converted into internal energy on impact.

Internal energy arising from friction, unlike kinetic energy and potential energy, usually cannot be recovered. Work done to lift a box and give it potential energy can be recovered as work or some other form of energy. Work done to slide a box across a floor becomes internal energy that is "lost" (made unavailable). In every mechanical process, some energy is converted into internal energy. It has been estimated that in the United States the annual financial losses associated with overcoming friction are several hundred billion dollars. (See "Friction: A Sticky Subject" on p. 52.)

In summary, work always results in a transfer of energy from one thing to another, in a transformation of energy from one form to another, or both. In our earlier analogy in which we compared energy to financial assets, work plays the role of a transaction such as buying, selling, earning, or trading. These transactions can be used to increase or decrease the net worth of an individual or to convert one form of asset into another. Work done *on* a system increases its energy. Work done *by* the system decreases the energy of the system. Work done *within* the system results in one form of energy being changed into another.

LEARNING CHECK

1. To be able to do work, a system must have
   _____.

2. Identical cars A and B are traveling down a highway, with A going two times as fast as B. The kinetic energy of A is
   a) 2 times as large as the kinetic energy of B.
   b) 4 times as large as the kinetic energy of B.
   c) one-half as large as the kinetic energy of B.
   d) the same as that of B, because they have the same mass.

3. (True or False.) When you lift an object, its gravitational potential energy decreases.

4. As a baseball player slides into home, the player's kinetic energy is converted into _____ energy.

ANSWERS: 1. energy **2.** (b) **3.** False **4.** internal

3.5 The Conservation of Energy

In the preceding section, we described several situations in which one form of energy was converted into another. These included a dart gun (potential energy in a spring converted to kinetic energy of the dart), a car braking (kinetic energy converted into internal energy), and a box sliding down an inclined plane (gravitational potential energy becoming internal energy). There are countless situations involving many other forms of energy.

Many devices in common use are simply energy converters. Some examples are listed in ● Table 3.1. Some of these involve more than one conversion. In a hydroelectric dam (see ● Figure 3.25), the potential energy of the water behind the dam is converted into kinetic energy of the water. The moving water then hits a turbine (propeller), which gives it kinetic energy of rotation. The rotating turbine turns a generator, which converts the kinetic energy into electrical energy.

Internal energy is an intermediate form of energy in both the car engine and the nuclear power plant. In a car engine, the chemical energy in fuel is first converted into internal energy as the fuel burns (explodes). The heated gases expand and push against pistons (or rotors in rotary engines). These in turn make a crankshaft rotate. In a nuclear power plant, nuclear energy is converted into internal energy in the reactor core. This internal energy is used to boil water into steam. The steam is used to turn a turbine, which turns a generator that produces electrical energy.

Internal energy is also a "wasted" by-product in all of the devices in Table 3.1. Any device that has moving parts has some kinetic friction. Some of the input energy is converted into internal energy by this friction. The generator, electric motor, car engine, and the electrical power plants all have unavoidable friction (● Figure 3.26). In some of the

● Table 3.1 Some Energy Converters

Device	Energy Conversion
Light bulb	Electrical energy to radiant energy
Car engine	Chemical energy (in fuel) to kinetic energy
Battery	Chemical energy to electrical energy
Elevator	Electrical energy to gravitational potential energy
Generator	Kinetic energy to electrical energy
Electric motor	Electrical energy to kinetic energy
Solar cell	Radiant energy (light) to electrical energy
Flute	Kinetic energy (of air) to acoustic energy (of sound wave)
Nuclear power plant	Nuclear energy to electrical energy
Hydroelectric dam	Gravitational potential energy (of water) to electrical energy

● **Figure 3.25** A hydroelectric power station uses several energy conversions.

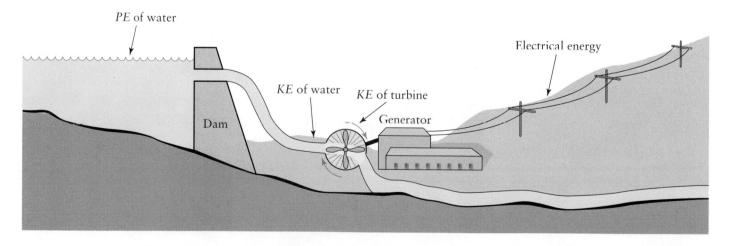

devices, internal energy is produced because of the nature of the process. Over 95% of the electrical energy used by an incandescent light bulb is converted to internal energy, not usable light. Over 60% of the available energy in coal-fired and nuclear power plants goes to unused internal energy. We will investigate this further in later chapters.

Even though there are many different forms of energy and countless devices that involve energy conversions, the following law always holds.

> **LAWS**
>
> **Law of Conservation of Energy** Energy cannot be created or destroyed, only converted from one form to another. The total energy in an isolated system is constant.

For a system to be isolated, energy cannot leave or enter it. For a mechanical system, work cannot be done on the system by an outside force, nor can the system do work on anything outside of it.

This law means that energy is a commodity that cannot be produced from nothing or disappear into nothing. If work is being done or a form of energy is "appearing," then energy is being used or converted somewhere. Unlike money, which you can counterfeit or burn, you cannot manufacture or eliminate energy.

The law of conservation of energy is both a practical tool and a theoretical tool. It can be used to solve problems, notably in mechanics, and it is a necessary condition that proposed theoretical models must satisfy. As an example of the latter, a theoretical astrophysicist may develop a model that explains how stars convert nuclear energy into heat and radiation. A first test of the validity of the model is whether or not energy is conserved.

We now illustrate the practical usefulness of the law by considering some mechanical systems. In each case, we assume that friction is negligible, so we do not have to take into account any conversion of potential energy or kinetic energy into internal energy.

The basic approach in using the law of conservation of energy is the same as that used with the corresponding law for linear momentum. If there is a conversion of energy in the system from one form to another, the total energy before the conversion equals the total energy after the conversion.

total energy before = total energy after

A good example that we have encountered before (in Sections 1.4 and 2.6) is the motion of a freely falling body. If an object is raised to a height d, it has gravitational potential energy. If it is released, the object will fall and convert its potential energy into kinetic energy. It is a continuous process. As it falls, its potential energy decreases because its height decreases, while its kinetic energy increases because its speed increases. If it is falling freely, there is no air resistance, and the only two forms of energy are kinetic and gravitational potential. Energy conservation then means that the sum of the kinetic energy and the potential energy of the object is always the same (see ● Figure 3.27).

$$E = KE + PE = \text{constant}$$

We can use this fact to show how the object's speed just before it hits the floor depends on the height d. We do this by looking at the object's energy just as it is dropped and then just before it hits. At the instant it is released, its kinetic energy is zero, because its speed is zero, and its potential energy is mgd.

$$E = KE + PE = 0 + PE$$
$$= PE = mgd \quad \text{(just released)}$$

When it reaches the floor, at the instant before impact, its potential energy is zero. So:

$$E = KE + PE = KE + 0$$
$$= KE = \frac{1}{2}mv^2 \quad \text{(just before impact)}$$

● **Figure 3.26** This wind generator converts kinetic energy of the wind into electrical energy plus some internal energy because of friction.

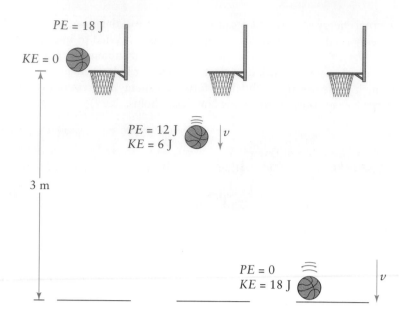

● **Figure 3.27** A basketball rolls off the rim and falls to the floor. Initially, it has potential energy only. As it falls, its potential energy decreases as its kinetic energy increases. Just before it hits the floor, it has kinetic energy only. At each point as it falls, its total energy, kinetic plus potential, is the same, 18 joules.

These two quantities are equal since the amount of the object's energy has not changed, only the form. The potential energy the object had when it was released equals the kinetic energy it has just before impact.

$$\frac{1}{2}mv^2 = mgd$$

Dividing both sides by m and multiplying by 2, we get:

$$v^2 = 2gd$$

$$v = \sqrt{2gd} \qquad \text{(speed after falling distance } d)^\star$$

Example 3.8 In 2003, a man went over Horseshoe Falls, part of Niagara Falls, and survived. He is thought to be the first person to do so without the aid of any safety devices. The height of the falls is about 50 meters. Estimate the speed of the man when he hit the water at the bottom of the falls.

Assuming that the air resistance is too small to affect the motion appreciably, we can use the preceding result:

$$v = \sqrt{2gd} = \sqrt{2 \times 9.8 \text{ m/s}^2 \times 50 \text{ m}}$$

$$= \sqrt{980 \text{ m}^2/\text{s}^2}$$

$$= 31.3 \text{ m/s} \qquad \text{(about 70 mph)}$$

The speed of a freely falling body does not depend on its mass. It only depends on how far it has fallen *(d)* and the acceleration due to gravity. ● Table 3.2 shows the speed of an object after it has fallen various distances. In Section 1.4 we illustrated the relationships between speed and time and between distance and time. This completes the picture.

---

* This is the answer to Challenge 7 in Chapter 1.

Table 3.2 Speed Versus Distance for a Freely Falling Body			
A. SI Units		**B. English Units**	
Distance *d* (m)	Speed *v* (m/s)	Distance *d* (ft)	Speed *v* (mph)
0	0	0	0
1	4.4	1	5.5
2	6.3	2	7.7
3	7.7	3	9.5
4	8.9	4	11
5	9.9	5	12
10	14	10	17
20	20	20	24
100	44	100	55

Note: Parts A and B are independent. The distances and speeds in A and B are not equivalent.

We have the reverse situation when an object is thrown or projected straight up. It starts with kinetic energy, rises until all of that has been converted into potential energy, and then falls (Figure 2.21). The conservation of energy tells us that the kinetic energy that it begins with equals its potential energy at the highest point. This results in the same equation relating the initial speed *v* to the maximum height reached, *d*:

$$v^2 = 2gd$$

which we can rewrite as:

$$d = \frac{v^2}{2g}$$

To compute how high something will go when thrown straight upward, insert its initial speed for *v* in this equation. We can also use Table 3.2 "backward": A ball thrown upward at 12 mph will reach a height of 5 feet.

Explore It Yourself 3.4

Battery race. You will need two cylindrical batteries (AAs or AAAs work well), a ramp for the batteries to roll on (a piece of cardboard or a thin hardcover book), and a book or some other kind of backstop.

1. Place the backstop in front of the ramp a distance from the ramp's lower end equal to its length (see Figure 3.28). Release a battery from the top of the ramp and then from places lower down the ramp. Is it traveling the same speed at the bottom each time? Why?
2. Hold one battery at the top of the ramp and the other one one-fourth of the way up from the bottom of the ramp. Release them at the same time. Which one wins the race to the backstop? Repeat with the lower battery starting at different places on the ramp. What happens in these races?

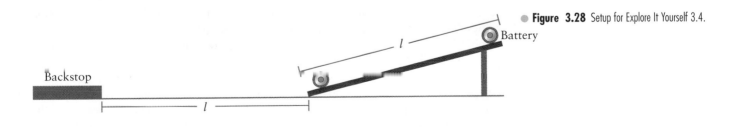

● **Figure 3.28** Setup for Explore It Yourself 3.4.

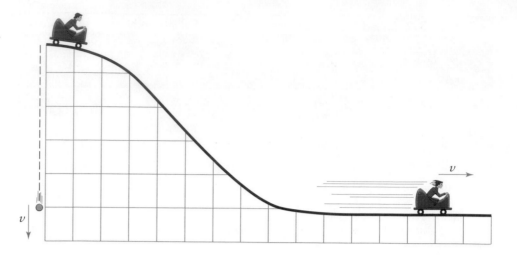

● **Figure 3.29** As a roller coaster travels down a hill, its potential energy is converted into kinetic energy. If there is no friction, its kinetic energy at the bottom equals its potential energy at the top. Its speed at the bottom is the same as that of an object dropped from the same height.

Now let's consider a similar problem. A roller coaster starts from rest at a height d above the ground. It rolls without friction or air resistance down a hill (see ● Figure 3.29). What is its speed when it reaches the bottom?

Again, the only forms of energy are gravitational potential energy and kinetic energy, since we assume there is no friction. The total energy, which is kinetic energy plus potential energy, is constant. The kinetic energy of the roller coaster at the bottom of the hill must equal its potential energy at the top.

$$KE \text{ (at bottom)} = PE \text{ (at top)}$$

$$\frac{1}{2}mv^2 = mgd$$

This is the same result obtained for a freely falling body. The speed at the bottom is consequently given by the same equation:

$$v = \sqrt{2gd}$$

It would have been impossible to solve this problem using only the tools from Chapter 2. To use Newton's second law, $F = ma$, one needs to know the net force that acts at every instant. The net force pulling the roller coaster along its path varies as the slope of the hill changes, so it is a very complicated problem. The principle of energy conservation allows us to easily solve a problem that we could not have solved before. In the process, we also come up with the following general result: The law of conservation of energy tells us that for an object affected by gravity but not friction, the speed that it has at a distance (d) below its starting point is given by the preceding equation regardless of the path it takes. The speed of a roller coaster that rolls down a hill is the same as that of an object that falls vertically the same height. The roller coaster does take more time to build up that speed (its acceleration is smaller), and the falling body therefore reaches the ground sooner.

The motion of a pendulum involves the continuous conversion of gravitational potential energy into kinetic energy and back again. Let's say that a child on a swing is pulled back (and up; see ● Figure 3.30). The child then has gravitational potential energy because of his or her height above the rest position of the swing. When released, the child swings downward, and the potential energy is converted into kinetic energy. At the lowest point in the arc, the child has only kinetic energy, which equals the original potential energy. The child then swings upward and converts the kinetic energy back into potential energy. This continues until the swing stops at a point nearly level with the starting point. This process is repeated over and over as the child swings. Air resistance takes away some of the kinetic energy. If the child is pushed each time, the work done puts energy into the system and counteracts the effect of the air resistance. Without air resistance or any other

● **Figure 3.30** As a child swings back and forth, gravitational potential energy is continually converted into kinetic energy and back again. The potential energy at the highest (turning) points equals the kinetic energy at the lowest point.

friction, the child would not have to be pushed each time and would continue swinging indefinitely.

The maximum height that a pendulum reaches (at the turning points) depends on its total energy. The more energy a pendulum has, the higher the turning points. In Section 3.2 we described a way to measure the speed of a bullet or a thrown object (Figure 3.6). The law of conservation of linear momentum is used to relate the speed of the bullet before the collision to the speed of the block (and bullet) afterward. If the wood block is hanging from a string, the kinetic energy it gets from the impact causes it to swing up like a pendulum. The more energy it gets, the higher it will swing. We can determine the speed of the block after impact by measuring how high the block swings. The potential energy of the block (and bullet) at the high point of the swing equals the kinetic energy of the block (and bullet) right after impact. This results in the same equation relating the speed at the low point to the height reached.

$$v = \sqrt{2gd}$$

In Section 2.6, we discussed the motion of an object hanging from a spring (Figure 2.24). The motion also consists of a continual conversion of potential energy into kinetic energy and back again.

● Figure 3.31 shows someone putting a ball at a miniature golf course. Since the ball rests in a small valley below ground level, its potential energy is negative relative to the level ground. When the ball is not moving, its total energy is negative because its kinetic energy is zero and its potential energy is negative. The golfer gives the ball kinetic energy by hitting it with a club. A weak putt gives it enough energy to roll back and forth but not to "escape" from the hole (Figure 3.31a). The ball's total energy is larger but still negative. In Figure 3.31b, the golfer gives the ball more energy by hitting it harder. Because its total energy is still negative, the ball again oscillates back and forth, although reaching a higher point on each side before turning around. In Figure 3.31c, the golfer hits the ball just hard enough for the ball to roll out of the valley and stop once it is out. The ball is given just enough kinetic energy to make its total energy equal to zero, and it "escapes" from the little valley. If the ball were hit even harder, it would escape and have excess kinetic energy—it would continue to roll on the level ground.

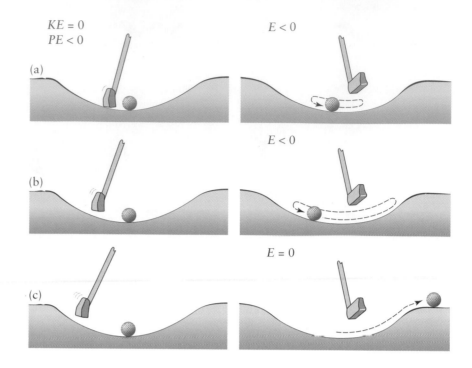

Figure 3.31 A golf ball at rest in the small valley has negative potential energy. Hitting the golf ball gives it kinetic energy, but it oscillates inside the valley if its total energy is negative, (a) and (b). If the golf ball is given enough kinetic energy to make its total energy zero, it rolls out of the valley and stops (c).

$KE = 0$
$PE < 0$
(a)

$E < 0$

$E < 0$
(b)

$E < 0$

$E = 0$
(c)

$E = 0$

This is the principle behind rocking a car when it is stuck. If a tire is in a hole, it is best to make the car oscillate back and forth. By giving it some energy during each cycle, by pushing or by using the engine, one can often give the car enough energy to leave the hole.

There are many analogous systems in physics in which an object is bound unless its total energy is equal to or greater than some value. A satellite in orbit around the Earth is an important example. The satellite's motion from one side of the Earth to the other and back is similar to the motion of the golf ball in the valley. If it is given enough energy, the satellite will escape from the Earth and move away, much like the golf ball. The minimum speed that will give a satellite enough energy to leave the Earth is called the *escape velocity*. Its value is approximately 11,200 m/s or 25,000 mph.

When water boils, the individual water molecules are given sufficient energy to break free from the liquid (Chapter 5). Sparks and lightning occur only after electrons are given enough energy to break free from their atoms (Chapter 7). The transition of a system from a bound state to a free state is quite common in physics and is not limited to mechanical systems.

LEARNING CHECK

1. As a skier gains speed while gliding down a slope, _____ energy is being converted into _____ energy.
2. (True or False.) Work done inside an isolated system can increase the total energy in the system.
3. (Choose the *incorrect* statement.) The total kinetic energy plus potential energy of a body
   a) can be negative.
   b) always remains constant if the body is falling freely.

c) always remains constant if friction is acting.
d) can remain constant even if the body's speed is decreasing.

4. Diver A jumps off a platform and is going 5 m/s when entering the water. To be going 10 m/s when entering the water, diver B would have to jump off a platform that is _____ times as high as the platform A used.

ANSWERS: 1. potential, kinetic 2. False 3. (c) 4. four

3.6 Collisions: An Energy Point of View

Earlier in this chapter we pointed out that the main "tool" for studying all collisions is the law of conservation of linear momentum (Section 3.2). In this section, we look at collisions from an energy standpoint. In some collisions, the only form of energy involved, before and after, is kinetic energy. In other collisions, forms of energy like potential energy and internal energy play a role. Collisions can be classified as follows.

An Elastic Collision is one in which the total kinetic energy of the colliding bodies after the collision *equals* the total kinetic energy before the collision.

An Inelastic Collision is one in which the total kinetic energy of the colliding bodies after the collision *is not equal to* the total kinetic energy before. The total kinetic energy after can be greater than, or less than, the total kinetic energy before.

In an elastic collision, kinetic energy is conserved. The total energy is *always* conserved in both types of collisions, but in elastic collisions no energy conversions take place that make the total kinetic energy after different from the total kinetic energy before.

● Figure 3.32 illustrates examples of these two types of collisions. Two equal-mass carts traveling with the same speed but in opposite directions collide. In both collisions, the total linear momentum before the collision equals the total after. (This total is equal to zero. Why?) In Figure 3.32a, the carts bounce apart because of a spring attached to one of them. After the collision, each cart has the same speed it had before, but it is going in the opposite direction. Consequently, the total kinetic energy of the two carts is the same after the collision as it was before. This is an elastic collision.

Figure 3.32b is an example of an inelastic collision. This time the two carts stick together (because of putty on one of them) and stop. The total kinetic energy after the collision is zero in this case. The automobile collision analyzed in Example 3.2 (Figure 3.5) is also an inelastic collision. To see this, use the information to calculate the total kinetic energy before and after the collision.

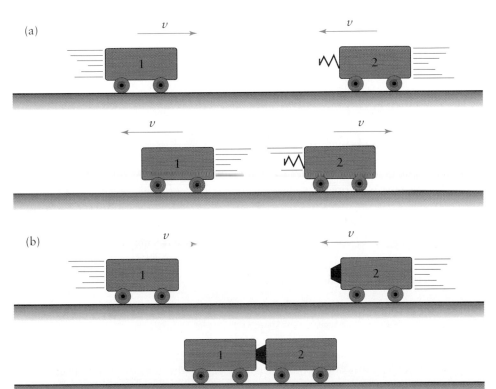

● **Figure 3.32** (a) Two carts with the same mass *m* and speed *v* collide head on and bounce apart. The total kinetic energy of the carts is the same before and after the collision. (b) This time the two carts stick together after the collision. In this case, the kinetic energy after the collision is zero.

Energy Conservation, Consumption, and Crisis

We hear a lot about the need to conserve resources, particularly energy, in the news these days. In the light of our discussion in this chapter, one might wonder why people are concerned about conservation of energy. After all, energy *must* be conserved according to fundamental physical principles, mustn't it? What's all the fuss about something that happens naturally? The paradox here derives from the ways physicists and the general public use and interpret the phrase "energy conservation." It is an example of the kind of confusion that can arise when the same words are used with different meanings by different groups.

As we have seen in Section 3.5, conservation of energy as *a physical law* refers to the fact that energy can neither be created nor destroyed in any interaction but merely changed in form. Put another way, in any system the total amount of energy remains constant. However, conservation of energy *as an economic, environmental, or social principle* involves reducing our reliance on certain types of energy sources like coal, oil, and natural gas and becoming more efficient in their use in cases where switching to other sources is deemed unacceptable and/or unfeasible. For the general public, then, conservation of energy means husbanding precious "nonrenewable" natural resources and, overall, using less energy to accomplish the myriad tasks we carry out daily in our society.

Among the elements that have prompted society's attempts to use less energy is the way energy consumption, especially in the United States, has increased over the last 150 years. ● Figure 3.33 displays the annual amount of energy consumed in the United States since 1850 from sources like wood, coal, oil, and natural gas, as well as from hydroelectric, geothermal, and nuclear power generating plants. Note that the graph is *not* a straight line. From one year to the next, the amount of energy consumed does not increase by a fixed amount as would be the case if the relationship between energy use and time were linear. Instead, energy consumption grows increasingly rapidly with each passing year.

If one analyzes this behavior carefully, it may be demonstrated that the *change* in energy consumption during any given interval of time depends directly on the amount of energy being used at the start of the time interval. Symbolically, using the notation of Chapter 1, we may write

$$\Delta E/\Delta t \propto E$$

where E is the energy consumption, Δ means "change in," and $\propto$ is the mathematical symbol for "proportional to." A quantity that increases (or decreases) in this manner is said to exhibit *exponential* growth (or decay). Evidently, energy consumption in the United States has grown exponentially since 1850, and in the judgment of many experts the origins of the energy crisis are to be found in the exponential growth of energy consumption.

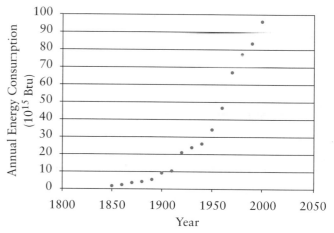

● **Figure 3.33** Annual energy consumption in the United States since 1850. The total consumption from all sources rose exponentially until about 1980, with an annual growth rate of 4.3%.

Example 3.9   Recall the automobile collision analyzed in Example 3.2 (Figure 3.5). Compare the amounts of kinetic energy in the system before and after the collision.

The kinetic energy before the collision was

$$KE_{\text{before}} = \frac{1}{2} \times 1{,}000 \text{ kg} \times (10 \text{ m/s})^2$$

$$= 50{,}000 \text{ J}$$

The kinetic energy after the collision was

$$KE_{\text{after}} = \frac{1}{2} \times 2{,}500 \text{ kg} \times (4 \text{ m/s})^2$$

$$= 20{,}000 \text{ J}$$

So 30,000 joules (60%) of the kinetic energy before the collision was converted into other forms of energy.

The problem, as these investigators see it, lies in the so-called *doubling time* for energy use. It is possible to show that the doubling time is related to the percent increase of the quantity in a given interval of time (a year, a week, a day, a second, and so on). Specifically, if we let t_D represent the doubling time, then

$$t_D \cong 70/PC$$

where PC is the percent change over some unit of time. The units of time for t_D will be those used to express PC. (The "$\cong$" in the formula means "approximately equal to.") For example, if a certain quantity grew steadily at a rate of 10% per year, it would double in size in 70/10 or 7 years.

The data shown in Figure 3.33 yield a growth rate in energy consumption in the United States of a little over 4% a year, at least for many years preceding 1980. At this rate, the doubling time for energy use in this country would be about 16 years. This means that if such growth were sustained, every 16 years the amount of energy consumed in the United States would double in size, requiring a similar doubling of the production capabilities of all the energy industries in the nation. Similarly, at this rate, in only 80 years (five doubling times), the U.S. energy consumption (and production) would increase by a factor of $2 \times 2 \times 2 \times 2 \times 2 = 2^5$ or 32. Exponential quantities rise (or fall) extremely quickly, even for what might be considered modest growth (or decay) rates.

Pushing this example just a bit further, we can begin to see why scientists, engineers, economists, government officials, and others have called for profound changes in the way our society uses energy to avoid a severe energy crisis: At 4% annual growth, in just 320 years the amount of energy required by our citizenry would be over 1 million times what it is today! This is far beyond even the most optimistic estimates of our capacity to find and/or perfect new energy sources.

Clearly, for exponentially increasing quantities, as time goes on, their values quickly become enormous, eventually approaching infinity. For this reason, it is impossible for any real quantity to continue to grow exponentially for a long period of time, much less forever. The ability of any system to sustain such growth is rapidly outstripped, and the growth is halted. Since the 1970s, the public's appreciation of the severity of the energy problem has deepened considerably, as has its commitment to adopting measures to reduce the growth in energy consumption. In part because of energy conservation measures implemented during the past 20 years, particularly the production of more fuel-efficient cars, total energy consumption in the nation has slowed considerably.

Among the potential sources of domestic fuel that have caught the attention of the media, if not the scientific community, is hydrogen. When combined with oxygen in a chemical reaction, hydrogen produces energy much as the combustion of gasoline vapor and air does in a typical auto engine. The advantages of hydrogen combustion (which is used to power the launches of many spacecraft, including the Space Shuttles) are that it generates only water as a by-product (which is better for the environment than the gases typically released by most gasoline engines) and that it relies on reactants that are potentially very plentiful—hydrogen and oxygen. Both elemental gases may be obtained by electrolyzing seawater (of which the world's oceans hold vast supplies).

The catch is that it takes energy to carry out the separation of water into its constituent elements, just as it takes energy to distill gasoline and other lighter petroleum products from crude oil. From an economic point of view, we have yet to reach the stage where the benefits of using hydrogen as a fuel outweigh the costs of producing, storing, and dispensing it.

Spend some time on the World Wide Web investigating the economic and scientific pros and cons of using hydrogen fuel cells as a major energy source. You may find significant differences of opinion among the various interested parties concerning the promise of this fuel source. But even if it becomes possible in the future to tap hydrogen as a major source of energy to power our civilization, we will still have to curb our appetite for energy to avoid the consequences of exponential growth in consumption. The supply of hydrogen may be large, but it is not infinite! We must continue to use our resources wisely.

In these two examples of inelastic collisions, part or all of the original kinetic energy of the colliding bodies is converted into other forms of energy, mostly internal energy, but also some sound (the "crash" that we would hear). In Figure 3.32b, *all* of the kinetic energy is converted into other forms of energy.

In some collisions, the total kinetic energy after the collision is greater than the total kinetic energy before the collision. If our old friend the cart, with its plunger pushed in ("loaded"), is struck by a second cart, the plunger will be released by the shock (● Figure 3.34). The plunger's potential energy is transferred to both carts as kinetic energy. Therefore, the total kinetic energy after the collision is greater than the total kinetic energy before the collision. It is an *inelastic* collision. The stored energy is released by the collision.

The use of collisions is an invaluable tool in studying the structure and properties of atoms and nuclei. Much of the information in Chapters 10, 11, and 12 was gleaned from the careful analysis of countless collisions. Linear accelerators, cyclotrons, betatrons, and other devices produce high-speed collisions between atoms, nuclei, and subatomic particles. The collisions are recorded and analyzed using the law of conservation of linear

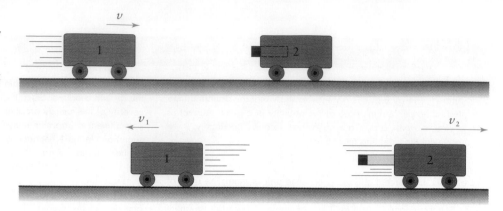

Figure 3.34 Cart 2 has energy stored in its spring-loaded plunger. When this cart is struck by cart 1, this potential energy is converted into kinetic energy, which is then shared by both carts. The total kinetic energy after the collision is greater than the total kinetic energy before the collision.

momentum and other principles. If a collision is inelastic, the amount of kinetic energy "lost" or "gained" in the collision is useful for determining the properties of the colliding particles. (See "Big Machines—Small Particles," p. 86.)

Collisions are also responsible for other phenomena such as gas pressure and the conduction of heat. Air molecules colliding with the inner surface of a balloon keep the balloon inflated. When you touch a piece of ice, the molecules in your finger collide with, and lose energy to, the molecules of the ice. This lowers the temperature of your finger.

Elastic collisions involving the force of gravity (and no physical contact) are used in space exploration. Called the *slingshot effect* or *gravity assist,* the technique involves having a spacecraft overtake a planet and pass it on its side away from the Earth. The spacecraft gains kinetic energy from the planet, the way the eight ball gains kinetic energy in the collision depicted in Figure 3.4. The planet loses kinetic energy, but the planet is so huge that its decrease in speed is imperceptible—like the effect of a volleyball bouncing off the front of a moving locomotive. Space probes sent to the outer part of the solar system (beyond Jupiter) have relied on gravity assists. The *Voyager 2* spacecraft used gravity assists from Jupiter, Saturn, and Uranus on its journey to Neptune and beyond. (Its speed was increased by 10 miles per second by the Jupiter gravity assist.)

Two recent missions exploited gravity assists from the inner planets, thereby allowing heavy spacecraft to be started on their journeys using relatively small rockets. On the way to its 1995 arrival at Jupiter, the *Galileo* spacecraft used two gravity assists from Earth and one from Venus. The 6-ton *Cassini* spacecraft used two gravity assists from Venus, one from Earth, and one from Jupiter to reach Saturn in 2004 (Figure 3.35).

Figure 3.35 The path of the *Cassini* spacecraft on its journey to Saturn. It gains speed as it passes each planet.

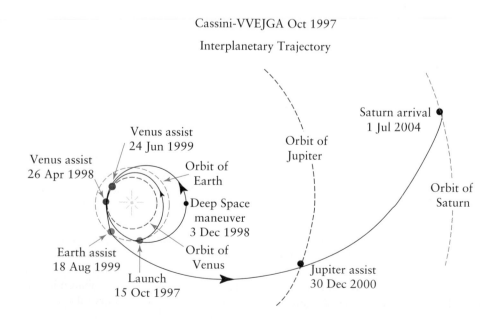

3.7 Power

We have seen many examples of work being done and energy being transformed into other forms. The amount of time involved in these processes has not entered into the discussion until now. Let's say that a ton of bricks needs to be loaded from the ground onto a truck (see ● Figure 3.36). This might be done in one of two ways. First, a person could lift the bricks one at a time and place them on the truck. This might take the person an hour. Second, a forklift could be used to load the bricks all at once. This might take only 10 seconds. In both cases, the same amount of work is done. The force on each brick (its weight) times the distance it is moved (the height of the truck bed) is the same whether the bricks are loaded one at a time or all at once. The work done, *Fd*, is the same in both cases, but the **power** is different.

DEFINITION

Power The rate of doing work. The rate at which energy is transferred or transformed. Work done divided by the time. Energy transferred divided by the time.

$$P = \frac{\text{work}}{t} \qquad P = \frac{E}{t}$$

● **Figure 3.36** A ton of bricks is loaded onto a truck in two different ways. The work done is the same, but since the forklift does the job much faster, the power is much greater.

Physical Quantity	Metric Units	English Units
Power (P)	watt (W)	foot-pound/second (ft-lb/s) horsepower (hp)

In SI units, the weight of a ton of bricks is 8,900 newtons. If the height of the truck bed is 1.2 meters, then the work done is

$$\text{work} = Fd = 8,900 \text{ N} \times 1.2 \text{ m} = 10,680 \text{ J}$$

If the forklift does this work in 10 seconds, the power is

$$P = \frac{\text{work}}{t} = \frac{10,680 \text{ J}}{10 \text{ s}}$$

$$= 1,068 \text{ J/s} = 1,068 \text{ W}$$

The unit joule per second (J/s) is defined to be the watt (W), the SI unit of power.

$$1 \text{ watt} = \frac{1 \text{ joule}}{1 \text{ second}}$$

$$1 \text{ W} = 1 \text{ J/s}$$

The watt should be familiar to you because it is commonly used to measure the power consumption of electrical devices. A 60-watt light bulb uses electrical energy at the rate of 60 joules each second. A 1,600-watt hair dryer uses 1,600 joules of energy each second (see ● Figure 3.37).

Horsepower is the most commonly used unit of power in the English system of units. Automobile engines, lawn mowers, and many other motorized devices are rated in horsepower. The basic power unit, foot-pound per second, is the unit of work, foot-pound, divided by the unit of time, second. The conversion factors are

$$1 \text{ hp} = 550 \text{ ft-lb/s} = 746 \text{ W}$$

A device that could raise 550 pounds a distance of 1 foot in 1 second would output 1 horsepower. Raising 110 pounds a distance of 5 feet in 1 second would also require 1 horsepower.

The relationship between power and work (or energy) is the same as that between speed and distance. In each case, the former is the rate of change of the latter.

$$P = \frac{\text{work}}{\text{time}} \longleftrightarrow v = \frac{\text{distance}}{\text{time}}$$

A runner and a bicyclist can both travel a distance of 10 miles, but the latter can do it much faster because a bicycle is capable of much higher speeds. A person and a forklift can both do 10,680 joules of work raising the bricks, but the forklift can do it much faster because it has more power.

● **Figure 3.37** The number, 1600, indicates how much power this hair dryer requires.

© Vern Ostdiek

Example 3.10

In Examples 2.2 and 3.5, we computed the acceleration, force, and work for a 1,000-kilogram car that goes from 0 to 27 m/s in 10 seconds. We can now determine the required power output of the engine. The work, 364,500 joules, is done in 10 seconds. Hence the power is

$$P = \frac{\text{work}}{t} = \frac{364,500 \text{ J}}{10 \text{ s}}$$

$$= 36,450 \text{ W} = 48.9 \text{ hp}$$

The car's kinetic energy when going 27 m/s is also 364,500 joules (see Example 3.6). It takes the engine 10 seconds to give the car this much kinetic energy, so we get the same result using energy divided by time.

● **Figure 3.38** The *Gossamer Albatross*, here flying low over the Atlantic Ocean, became the first human-powered aircraft to cross the English Channel (12 June 1979). The pilot turned the propeller using a bicycle-type mechanism. A steady power output of about 250 watts was required for the flight, which lasted 2 hours and 49 minutes.

Given enough food or fuel, there is usually no limit to how much work a device or a person can do. But there is a limit on how fast the work can be done. The power is limited. In other words, only so much work can be done each second. The power output can be anything from zero (no work) to some maximum. For example, a 100-horsepower automobile engine can put out from 0 to 100 horsepower. When accelerating as fast as possible, the engine is putting out its maximum power. While cruising down a flat highway, the engine may be putting out only 10–20 horsepower, enough to counteract the effects of air resistance and other frictional forces.

The human body has a maximum power output that varies greatly from person to person. In the act of jumping, an outstanding athlete can develop more than 8,000 watts, but only for a fraction of a second. The same person would have a maximum of less than 800 watts if the power level had to be maintained for an hour. The average person can produce 800 watts or more for a few seconds and perhaps 100–200 watts for an hour or more (see ● Figure 3.38). When running, the body uses energy to overcome friction and air resistance. In short races, the best runners can maintain a speed of 10 m/s for about 20 seconds. In longer races, the speeds are lower because the power level has to be maintained longer. For a race lasting about 30 minutes, the best average speed is about 6 m/s.

Explore It Yourself 3.5

Compute your own power output when walking or running up a flight of stairs (● Figure 3.39). First you need to compute the work you do by measuring the vertical height of the stairs and then multiplying this number by your weight. (You may want to use SI units—meters and newtons.) The power is this work divided by the time it takes to climb the stairs.

If you go hiking or biking up a long steep hill, you can estimate your steady power output by performing the same calculation. The vertical distance can be found using a topographic map or altimeter. (Do not forget to include the weight of your backpack and/or bicycle.)

● **Figure 3.39** To measure your power output when going up a flight of stairs, multiply the height of the stairs by your weight, and divide this by the time it takes you to climb the stairs.

1. (True or False.) When more people ride upward in an elevator, it requires more power.

2. Identical cars A and B are being driven up the same steep hill. If the power output of A is larger than that of B, what must be happening?

ANSWERS: 1. True 2. A is going faster.

3.8 Rotation and Angular Momentum

Our final conservation law applies to rotational motion. A spinning ice skater and a satellite moving in a circular path around the Earth are examples. You might say that this law is the rotational analogue or counterpart of the law of conservation of linear momentum.

> **LAWS**
>
> **Law of Conservation of Angular Momentum** The total angular momentum of an isolated system is constant.

For a system to be isolated so that the law of conservation of angular momentum applies, the only net external force that can act on the object must be directed toward or away from the center of the object's motion. The centripetal force required to keep an object moving in a circle fits this condition. A force that acts in any direction other than toward or away from the center of motion produces what is called *torque* (from the Latin word for "twist"). When a spacecraft fires its rocket engines to reenter the atmosphere, the force on it acts opposite to its direction of motion and therefore produces a torque that decreases its angular momentum. Torque is the rotational analogue of force: A net external force changes an object's linear momentum, and a net external torque changes an object's angular momentum.

But what exactly is **angular momentum**? We first introduce angular momentum for the simple case of a body moving in a circle, and then extend it to motion along noncircular paths. Imagine a small object moving along a circular path, like a satellite in orbit around the Earth. In this case, the angular momentum equals the product of the object's mass, its speed, and the radius of its path.

$$\text{angular momentum} = mvr \qquad \text{(circular path)}$$

Notice that the angular momentum of the object is also equal to its linear momentum (mv) multiplied by the radius of its circular path.

To illustrate the law of conservation of angular momentum, imagine a ball circling overhead on the end of a string that passes through a tube (● Figure 3.40a). (We assume there is no friction or air resistance.) The faster the ball goes, the greater its orbital angular momentum. Using a longer string for the same speed would also make its angular momentum larger. Imagine suddenly shortening the string by pulling downward on the end with your free hand, letting the string slide through the tube. This makes the ball move in a circle with a smaller radius but with a higher speed (Figure 3.40b). Since the force exerted on the ball is directed toward the center of its motion, angular momentum is conserved—it has the same value before and after the change. Because the radius of the ball's path is now smaller, its speed must be higher. If you let the string out so the radius is larger, the ball will slow down, keeping the angular momentum constant.

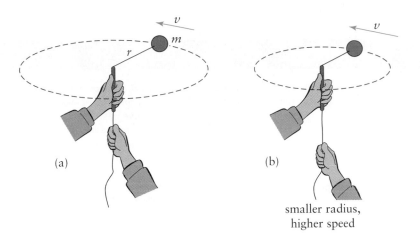

(a)

(b)

smaller radius,
higher speed

In the process of pulling the string downward, you do work. This work goes to increase the kinetic energy of the ball.

With caution, we can use this definition of angular momentum—*mvr*—for an object moving in a path other than a circle. ● Figure 3.41 shows the elliptical orbit of a satellite moving around the Earth. At points *A* and *B*, the satellite's velocity will be perpendicular to a line from it to the Earth's center. At these points the satellite's path is like a short segment of a circle. Consequently, its angular momentum is *mvr*. At point *B*, *r* is smaller than at point *A*. Because the angular momentum is the same at *A* as at *B*, the satellite's speed is greater at *B*. For example, if the satellite is 13,000 kilometers (about 8,000 miles, twice the Earth's radius) from the Earth's center at point *B* and 26,000 kilometers from the Earth's center at point *A*, its speed at *B* will be two times its speed at *A*. The actual values for the speeds are about 6,400 m/s (about 14,000 mph) when it is closest to the Earth and about 3,200 m/s when it is farthest away.

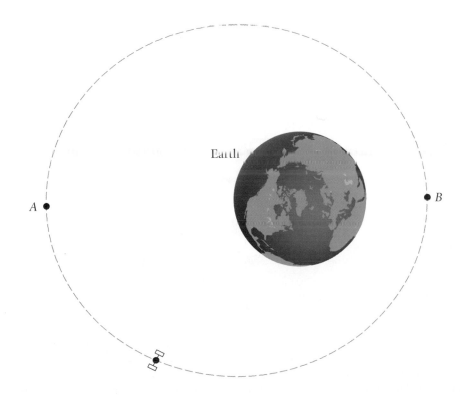

● **Figure 3.41** A satellite in orbit around the Earth is twice as far from the Earth's center at point *A* as it is at point *B*. Conservation of angular momentum then tells us that its speed at *A* is one-half its speed at *B*.

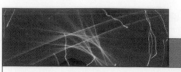
Starquakes: A Glitch *in Time*

Pulsars are astronomical objects that emit a regular sequence of pulses of radio energy. The pulse period (the interval of time between one pulse and the next) is usually quite short (a second or less for most pulsars), and it is generally accepted that pulsars are rapidly rotating *neutron stars.*

A neutron star is a remnant of a star that may have ended its life in a brilliant flash called a "supernova explosion." Though small (about 20 kilometers across), neutron stars are very dense: A teaspoon of neutron star stuff could easily weigh a billion tons.

The most probable mechanism by which pulsars produce the radio energy we receive involves the very strong magnetic fields that these objects are also known to possess. Charged particles (mostly electrons) near the surface of the neutron star experience a force due to the magnetic field (see Section 8.2) and are accelerated toward the magnetic poles. As they do so, they begin to radiate energy. Specifically, radio waves are beamed out into space from what might be termed "radio hot spots" near the magnetic poles of the neutron star. As these hot spots are rotated across our line of sight by the spinning neutron star, we receive regularly spaced, short bursts of radio radiation. What we observe then is an astronomical "lighthouse effect" (● Figure 3.42).

When pulsars were first discovered in 1967, their most prominent feature was the precision with which the pulses were spaced. However, after continued observation, scientists found that these "pulsar clocks" were slowing down. Their periods were gradually increasing. We now think that the neutron star associated with a pulsar is losing energy through interactions with its environment and rotating more and more slowly with time. Astronomers expected a continual decay in the rotation rates of these compact stars and a corresponding lengthening of the periods of pulsars. The analogy with a toy top slowly spinning down because of frictional losses to the table on which it rotates is perhaps not a bad one to have in mind here.

But nature is full of surprises. In 1969, a pulsar in the constellation of Vela suddenly sped up—that is, its period abruptly decreased!

Afterward, it once again resumed its steady decline in spin at the same rate as before. This was not an isolated event. At least 12 similar speed-ups in the Vela pulsar have been observed. They have also been seen in 9 other pulsars, including the one embedded in the Crab Nebula supernova remnant (● Figure 3.43). These brief events in which a pulsar increases its spin by a small amount are called *glitches.* Their explanation has its root in the law of conservation of angular momentum.

Current models for the structure of neutron stars suggest that they have solid, crystalline crusts up to a few kilometers thick. Beneath the

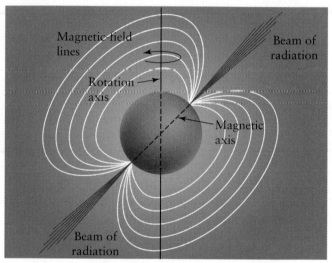

● **Figure 3.42** A simplified "lighthouse" model of a pulsar. Radio "hot spots" associated with the strong magnetic field of the neutron star emit copious quantities of radio radiation. As these hot spots are carried across our line of sight by the spin of the neutron star, we receive sharp pulses of radio energy.

Explore It Yourself 3.6

For this, you need a chair that can spin in circles easily. (Caution: Don't try this if you are prone to dizziness.) Move it to a place at least 5 feet from walls, furniture, and other objects. Sit in it with both of your arms straight out from your sides; then use your feet to make the chair spin. Lift your feet, and then pull your arms tight against your body. What happens? Extend your arms out to the side again. What happens? (You can enhance the effect by holding a weight in each hand.)

An object spinning about an axis, like a top or an ice skater doing a pirouette, has angular momentum. We can think of each part of the object as moving in a circle and having angular momentum. For example, a spinning skater's hands, arms, shoulders, and other body parts are all moving in circles. The combined angular momentum of the parts of a spinning body remains constant if no torque acts on it. The rate of spinning can be increased or decreased by repositioning parts of the object. For example, the spinning ice skater can start with arms extended outward. When the arms are pulled in closer to the

● **Figure 3.43** The Crab Nebula—debris of a supernova explosion that was observed and recorded by the Chinese in the year 1054.

crust, and to some extent permeating it, is a sea of superfluid neutrons (see the Potpourri on p. 156 for more on superfluidity). During the steady slowdown in the spin of the neutron star crust, there is a lag in the response of the interior fluid so that it is always spinning somewhat faster than the crust. (The dynamics are similar to that of an uncooked egg which, when set spinning, slows down much more slowly than a hardcooked one due to differential rotation of the liquid white and yoke inside. Try it! This is a neat way to distinguish cooked eggs from uncooked ones.)

The first proposed cause of pulsar spin-ups, and one which may explain small glitches like those observed in the Crab pulsar, was "starquakes." Most spinning objects, including the Earth and the Sun, are oblate in shape, that is, they have a larger equatorial radius than polar radius. In other words, they are flattened at the poles and bulged out along the equator. As a spinning body slows down, the oblateness decreases, and the object becomes more nearly spherical in shape. In a rapidly rotating neutron star, the crust is sufficiently rigid that it cannot respond to the necessary reduction in oblateness in a smooth and continuous manner, so internal stresses build up until the crust cracks and the "bulges" abruptly move slightly inward. To the extent that the neutron star may be considered an isolated body, conservation of angular momentum must apply to it, yielding an increase in the star's rotation rate as a result. This effect is analogous to an ice skater pulling in her arms during a pirouette and spinning faster.

Major glitch events like those seen in the Vela pulsar may require an additional, more complicated explanation, in which the inner neutron superfluid plays a significant role in the spin-up, but again all this occurs within the framework of conservation of angular momentum.

Our astronomical application of the principle of conservation of angular momentum focuses on a star that has reached the end of its luminous lifetime. But angular momentum is also important at the beginning of a star's life. For example, if a cloud of interstellar gas and dust is spinning too rapidly, the process of collapse that concentrates the mass to stellar sizes can be hindered to the point where no stars are produced. Even if rotational effects are small enough so that star formation can start through gravitational contraction of the interstellar cloud, conservation of angular momentum guarantees that as the collapse proceeds the rate of spin of the star-forming material will increase. How the system deals with this increasing spin determines, in part, whether it ends as a single star with planets or a double (or multiple) star. The study of star formation and role of angular momentum in establishing the properties of young stellar systems is one of the "hot" areas in astronomical research these days. Check out the World Wide Web for information on this topic using keys words like "star formation," "protostars," "T Tauri stars," or "cocoon nebulae." You might even try "extrasolar planets" if you're inclined to learn the latest about the discovery and characteristics of planetary systems outside our own. Happy hunting!

body, the skater spins faster. Each part of the skater's arms has a certain amount of angular momentum as it moves in a circle with some radius. Pulling the arms in decreases the radius, which makes the angular momentum of that part of the arm decrease. Consequently the rest of the skater's body spins faster, so that the total angular momentum remains constant. The reverse happens if the arms are moved out—the skater slows down.

LEARNING CHECK

1. The centripetal force acting on a body moving in a circle changes its linear momentum, but it does not change its _____ momentum.

2. (True or False.) If both the linear momentum and the kinetic energy of a satellite in orbit about the Earth increase, then its angular momentum must also increase.

ANSWERS: 1. angular 2. False

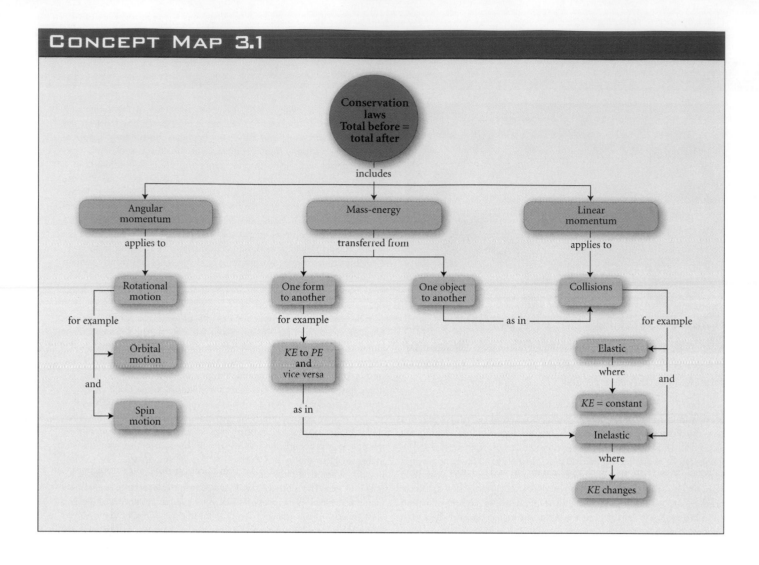

Concept Map 3.1 illustrates the three conservation laws for linear momentum, energy, and angular momentum.

Most of the principles discussed in this chapter were developed piecemeal by a host of physicists and mathematicians in the seventeenth, eighteenth, and nineteenth centuries. They are extensions of Newton's mechanics, not fundamentally new formulations. As mentioned near the end of Section 3.2, the law of conservation of linear momentum has its roots in Newton's laws of motion. The ideas of work and potential energy arise naturally from Newton's second law of motion when the force on an object depends only on its position.

Before Newton's *Principia*, the importance of momentum and kinetic energy had been discovered experimentally. Galileo used the product of weight and velocity as momentum, a quantity that is nearly the same as our linear momentum, since weight and mass are proportional. In 1665, the Dutch physicist Christian Huygens reported that the quantity mv^2 was conserved during the collision of two hard balls. (Huygens later acquired fame for his work in optics. His explanation of the fundamental nature of light was contrary to Newton's but eventually was proven to be correct.) The actual use of the word *energy*, the identification of kinetic energy as being equal to one-half the mass times the speed squared, and the statement of the law of conservation of energy came about around the middle of the nineteenth century.

The simple machines discussed in Section 3.3 were first analyzed by the famous Greek mathematician and scientist Archimedes (287?–212 B.C.). A citizen of Syracuse, Sicily, Archimedes (Figure 3.44) explained how a force could be amplified using a lever, a combination of pulleys, or other simple machines. The ancient historian Plutarch described how he demonstrated the usefulness of pulleys by moving a large, dry-docked ship by himself. This so impressed King Hiero of Syracuse that he persuaded Archimedes to construct machines to defend the city. These proved to be quite useful when the Romans attacked Syracuse by land and sea. Plutarch includes this account in his *Life of Marcellus:*

● **Figure 3.44** Archimedes.

> Archimedes began to ply his engines, and shot against the land forces of the assailants all sorts of missiles and immense masses of stones, which came down with incredible din and speed: nothing whatever could ward off their weight, but they knocked down in heaps those who stood in their way, and threw their ranks into confusion. At the same time huge beams were suddenly projected over the ships from the walls, which sank some of them with great weights plunging down from on high: others were seized at the prow by iron claws, or beaks like the beaks of cranes, drawn straight up into the air, and then plunged stern foremost into the depths, or were turned round and round by means of enginery within the city, and dashed upon the steep cliffs that jutted out beneath the wall of the city, with great destruction of the fighting men on board, who perished in the wrecks.

Archimedes' machines forced Marcellus, the Roman general, to abandon the direct assault in favor of a long siege. Two years later Syracuse fell, and Archimedes was killed by the conquerors.

Archimedes' most famous discovery was the principle of buoyancy that is named after him (Section 4.5). We should also note that he was one of the first to engage in two pursuits that have been central to physics since that time. First, as an engineer, he used his discoveries to construct useful devices. Second, he was a scientific consultant for the military. To this day, the military is a primary source of funding for research in physics and other sciences. Many great discoveries in physics and engineering, some quite terrifying, have come from efforts to devise offensive and defensive weapons of war.

From the two forms of mechanical energy introduced in this chapter, the concept of energy has been expanded to encompass many other diverse forms. A major part of the growth of science and technology in the last two centuries has been the discovery and application of new forms of energy. One hundred years ago, Albert Einstein showed that the concept of energy even includes matter: Matter can be converted into energy, and vice versa (see Chapter 11). The energy shortage experienced by the United States in the 1970s and the resulting increase in energy costs quickly illustrated how dependent our society is on devices that require energy sources. Energy is now a necessity of contemporary society, almost equal in importance to food, labor, and raw materials.

SUMMARY

- **Conservation laws** are powerful tools for analyzing physical systems, particularly those in mechanics. Their main advantage is that it is not necessary to know the details of what is going on in the system at each instant.

- The use of conservation laws is based on a "before-and-after" approach: The total amount of the conserved physical quantity *before* an interaction is equal to the total amount *after* the interaction.

- **Linear momentum, energy,** and **angular momentum** are physical quantities that are defined and used mainly because they are conserved in isolated systems.

- The main application of the **law of conservation of linear momentum** is to collisions. The total linear momentum before a collision equals the total linear momentum after the collision, if the system is isolated. This applies to all collisions, both elastic and inelastic.

- **Work** is done whenever a force acts through a distance in the same direction as the force.

- To be able to do work, a device or a person must have **energy.** The act of doing work involves the transfer of energy from one thing to another, the transformation of energy from one form to another, or both.

- In mechanics, the main forms of energy are **kinetic energy, potential energy,** and **internal energy.** There are many other forms of energy corresponding to different sources of work. Any form of energy can be used to do work if a suitable conversion device is available.

- The **law of conservation of energy** tells us that in all such conversions, the total amount of energy remains constant.

- **Power** is the rate of doing work or using energy. It is the measure of how fast energy is transferred or transformed.

- **Angular momentum** is a conserved quantity in rotational motion.

IMPORTANT EQUATIONS

Equation	Comments	Equation	Comments
Fundamental Equations		*Special-Case Equations*	
linear momentum $= mv$	Definition of linear momentum	$v = \sqrt{2gd}$	Speed after falling a distance d
$F = \dfrac{\Delta(mv)}{\Delta t}$	Alternate form of Newton's second law	$d = \dfrac{v^2}{2g}$	Height reached given initial speed v
work $= Fd$	Definition of work	angular momentum $= mvr$	Angular momentum (circular motion)
$KE = \dfrac{1}{2}mv^2$	Kinetic energy		
$PE = Wd = mgd$	Gravitational potential energy		
$P = \dfrac{\text{work}}{t}; \qquad P = \dfrac{E}{t}$	Definition of power		

MAPPING IT OUT!

1. Reexamine Section 3.3 on work. Make a list of at least *ten* key concepts and applications relating to work; write each of the concepts or examples on small Post-It® notes. Rank the items on your list from most inclusive to least inclusive, and organize your notes on a large piece of paper or poster board in a manner that reflects your ranking. Leave enough space between concepts to permit the insertion of appropriate linking words and phrases. Now, on separate Post-It®'s, perhaps of a different color, write down the required linking words to form meaningful propositions with your concepts; place these notes at the correct locations on the chart paper or poster board. You have just constructed a concept map for work! Examine the map carefully. Do all the propositions make sense? Are the most general concepts located near the top of the map? Are major concepts or connections missing? After you have finished refining your concept map, compare it with one that a classmate has produced. What similarities do they share? What differences are there between them? How might the maps be connected to ones designed to promote understanding of the concept of energy?

2. Repeat Exercise 1 above for Section 3.2 and relate your concept map directly to Concept Map 3.1 on p. 122 in the text.

QUESTIONS

(▶ Indicates a review question, which means it requires only a basic understanding of the material to answer. The others involve integrating or extending the concepts presented thus far.)

1. ▶ What is a conservation law? What is the basic approach taken when using a conservation law?
2. ▶ Why is the alternate form of Newton's second law of motion given in this chapter the more general form?
3. Could the linear momentum of a turtle be greater than the linear momentum of a horse? Explain why or why not.
4. An astronaut working with many tools some distance away from a spacecraft is stranded when the "maneuvering unit" malfunctions. How can the astronaut return to the spacecraft by sacrificing some of the tools?
5. ▶ For what type of interaction between bodies is the law of conservation of linear momentum most useful?
6. Describe several things you have done today that involved doing work. Are you doing work right now?

7. If we know that a force of 5 N acts on an object while it moves 2 meters, can we calculate how much work was done with no other information? Explain.

8. During a head-on collision of two automobiles, the occupants are decelerated rapidly. Use the idea of work to explain why an air bag that quickly inflates in front of an occupant reduces the likelihood of injury.

9. When climbing a flight of stairs, do you do work on the stairs? Do the stairs do work on you? What happens?

10. People and machines around us do work all the time. But is it possible for things like magnets and the Earth to do work? Explain.

11. How can you give energy to a basketball? Is there more than one way?

12. ▶ Identify as many different forms of energy as you can that are around you at this moment.

13. When you throw a ball, the work you do equals the kinetic energy the ball gains. If you do *twice* as much work when throwing the ball, does it go *twice as fast*? Explain.

14. ▶ An object has kinetic energy, but it stays in one place. What must it be doing?

15. ▶ How can the gravitational potential energy of something be negative?

16. ▶ What is elastic potential energy? Is there anything around you now that possesses elastic potential energy?

17. Identify the energy conversions taking place in each of the following situations. Name all of the relevant forms of energy that are involved.
    a) A camper rubbing two sticks together to start a fire.
    b) An arrow shot straight upward, from the moment the bowstring is released by the archer to the moment when the arrow reaches its highest point.
    c) A nail being pounded into a board, from the moment a carpenter starts to swing a hammer to the moment when the nail has been driven some distance into the wood by the blow.
    d) A meteoroid entering the Earth's atmosphere.

18. Solar-powered spotlights have batteries that are charged by solar cells during the day and then operate lights at night. Describe the energy conversions in this entire process, starting with the Sun's nuclear energy and ending with the light from the spotlight being absorbed by the surroundings. Name all of the forms of energy that are involved.

19. Truck drivers approaching a steep hill that they must climb often increase their speed. What good does this do, if any?

20. If you hold a ball at eye level and drop it, it will bounce back, but not to its original height. Identify the energy conversions that take place during the process, and explain why the ball does not reach its original level.

21. A ball is thrown straight upward on the Moon. Is the maximum height it reaches less than, equal to, or greater than the maximum height reached by a ball thrown upward on the Earth with the same initial speed (no air resistance in both cases)? Explain.

22. ▶ Describe the distinction between elastic and inelastic collisions. Give an example of each.

23. Many sports involve collisions, between things—like balls and rackets—and between people—as in football or hockey. Characterize the various sports collisions as elastic or inelastic.

24. Carts A and B stick together whenever they collide. The mass of A is twice the mass of B. How could you roll the carts toward each other in such a way that they would be stopped after the collision? (Assume there is no friction.)

25. Is it possible for one object to gain mechanical energy from another without touching it? Explain.

26. Two cranes are lifting identical steel beams at the same time. One crane is putting out twice as much power as the other. Assuming friction is negligible, what can you conclude is happening to explain this difference?

27. A person runs up several flights of stairs and is exhausted at the top. Later the same person walks up the same stairs and does not feel as tired. Why is this? Ignoring air resistance, does it take more work or energy to run up the stairs than to walk up?

28. ▶ How can a satellite's speed decrease without its angular momentum changing?

29. ▶ Why do divers executing midair somersaults pull their legs in against their bodies?

30. It is possible for a body to be both spinning and moving in a circle in such a way that its total angular momentum is zero. Describe how this can be.

PROBLEMS

1. A sprinter with a mass of 65 kg reaches a speed of 10 m/s during a race. Find the sprinter's linear momentum.

2. Which has the larger linear momentum: a 2,000-kg houseboat going 5 m/s or a 600-kg speedboat going 20 m/s?

3. In Section 2.4, we computed the force needed to accelerate a 1,000-kg car from 0 to 27 m/s in 10 s. Compute the force using the alternate form of Newton's second law. The change in momentum is the car's momentum when going 27 m/s minus its momentum when going 0 m/s.

4. A runner with a mass of 80 kg accelerates from 0 to 9 m/s in 3 s. Find the net force on the runner using the alternate form of Newton's second law.

5. A pitcher throws a 0.5-kg ball of clay at a 6-kg block of wood. The clay sticks to the wood on impact, and their joint velocity afterward is 3 m/s. What was the original speed of the clay?

6. A 3,000-kg truck runs into the rear of a 1,000-kg car that was stationary. The truck and car are locked together after the collision and move with speed 9 m/s. What was the speed of the truck before the collision?

7. A 50-kg boy on roller skates moves with a speed of 5 m/s. He runs into a 40-kg girl on skates. Assuming they cling together after the collision, what is their speed?

8. Two persons on ice skates stand face to face and then push each other away (● Figure 3.45). Their masses are 60 and 90 kg. Find the ratio of their speeds immediately afterward. Which person has the higher speed?

● **Figure 3.45**

9. A loaded gun is dropped on a frozen lake. The gun fires, with the bullet going horizontally in one direction and the gun sliding on the ice in the other direction. The bullet's mass is 0.02 kg, and its speed is 300 m/s. If the gun's mass is 1.2 kg, what is its speed?

10. A running back with a mass of 80 kg and a speed of 8 m/s collides with, and is held by, a 120-kg defensive tackle going in the opposite direction. How fast must the tackle be going before the collision for their speed afterward to be zero?

11. A motorist runs out of gas on a level road 200 m from a gas station. The driver pushes the 1,200-kg car to the gas station. If a 150-N force is required to keep the car moving, how much work does the driver do?

12. In Figure 3.10, the rock weighs 100 lb and is lifted 1 ft by the lever.
    a) How much work is done?
    b) The other end of the lever is pushed down 3 ft while lifting the rock. What force had to act on that end?

13. A weight lifter raises a 100-kg barbell to a height of 2.2 m. What is the barbell's potential energy?

14. A microwave antenna with a mass of 80 kg sits atop a tower that is 50 m tall. What is the antenna's potential energy?

15. A windsurfer and sailboard have a combined mass of 90 kg. What is their kinetic energy when they are going 10 m/s?

16. While in orbit about the Earth, a space shuttle's mass is 80,000 kg and its speed is 7,900 m/s. What is its kinetic energy?

17. The kinetic energy of a motorcycle and rider is 60,000 J. If their total mass is 300 kg, what is their speed?

18. In compressing the spring in a toy dart gun, 0.5 J of work is done. When the gun is fired, the spring gives its potential energy to a dart with a mass of 0.02 kg.
    a) What is the dart's kinetic energy as it leaves the gun?
    b) What is the dart's speed?

19. A worker at the top of a 629-m-tall television transmitting tower in North Dakota accidentally drops a heavy tool. If air resistance is negligible, how fast is the tool going just before it hits the ground?

20. A student drops a water balloon out of a dorm window 12 m above the ground. What is its speed when it hits the ground?

21. A child on a swing has a speed of 7.7 m/s at the low point of the arc (●Figure 3.46). How high will the swing be at the high point?

22. The cliff divers at Acapulco, Mexico, jump off a cliff 26.7 m above the ocean. Ignoring air resistance, how fast are the divers going when they hit the water?

23. In the Drop Tower at Bremen, Germany, the 300-kg test chamber falls freely from a height of 110 m.
    a) What is the chamber's potential energy at the top of the tower?
    b) How fast is it going when it reaches the bottom of the tower? (You may want to convert your answer to mph for comparison to highway speeds.)

24. The fastest that a human has run is about 12 m/s.
    a) If a pole vaulter could run this fast and convert all of his or her kinetic energy into gravitational potential energy, how high would he or she go?
    b) Compare this height with the world record in the pole vault.

25. A bicycle and rider going 10 m/s approach a hill. Their total mass is 80 kg.
    a) What is their kinetic energy?
    b) If the rider coasts up the hill without pedaling, at what point will the bicycle come to a stop?

26. In January 2003, an 18-year-old student gained a bit of fame for surviving—with only minor injuries—a remarkable traffic accident. The vehicle he was driving was "clipped" by another one, left the road, and rolled several times. He was thrown upward from the vehicle (he wasn't wearing a seat belt) and ended up dangling from an overhead telephone cable and a ground wire about 8 meters above the ground. Rescuers got him down after 20 minutes. It is estimated that he reached a maximum height of about 10 meters.
    a) How fast was the driver's body going when he was thrown from the vehicle?
    b) If he had not landed in the wires, how fast would he have been going when he hit the ground?

27. The ceiling of an arena is 20 m above the floor. What is the minimum speed that a thrown ball would have to have to reach the ceiling?

28. Compute how much kinetic energy was "lost" in the collision in Problem 6.

29. Compute how much kinetic energy was "lost" in the collision in Problem 7.

30. A 1,000-W motor powers a hoist used to lift cars at a service station.
    a) How much time would it take to raise a 1,500-kg car 2 m?
    b) If it is replaced with a 2,000-W motor, how long would it take?

31. How long does it take a worker producing 200 W of power to do 10,000 J of work?

32. An elevator is able to raise 1,000 kg to a height of 40 m in 15 s.
    a) How much work does the elevator do?
    b) What is the elevator's power output?

33. A professor's little car can climb a hill in 10 s. The top of the hill is 30 m higher than the bottom, and the car's mass is 1,000 kg. What is the power output of the car?

34. In the annual Empire State Building race, contestants run up 1,575 steps to a height of 1,050 ft. In 1983, the winner of this race was Al Waquie, with a time of 11 min and 36 s. Mr. Waquie weighed 108 lb.
    a) How much work was done?
    b) What was the average power output (in ft-lb/s and in hp)?

35. A bicyclist rides to the top of Mt. Nebo, Arkansas, in 23 min. The vertical height that the bicyclist climbs is 360 m, and the total mass of bicycle and rider is 85 kg. What was the bicyclist's power output?

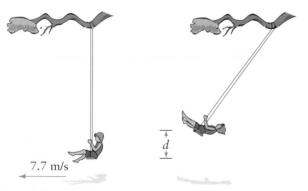

7.7 m/s

●Figure 3.46

1. Rank the following three collisions in terms of the extent of damage that the car would experience. Explain your reasons for ranking the collisions as you did.
   a) A car going 10 m/s striking an identical car that was stationary.
   b) A car going 10 m/s running into an immovable concrete wall.
   c) A head-on collision between identical cars, both going 10 m/s.

2. The system depicted in Figure 3.7 can be used to measure the mass of a body. Specify what quantities would have to be known and what would have to be measured to do this. Suggest how these measurements might be performed.

3. A bullet with a mass of 0.01 kg is fired horizontally into a block of wood hanging on a string. The bullet sticks in the wood and causes it to swing upward to a height of 0.1 m. If the mass of the wood block is 2 kg, what was the initial speed of the bullet?

4. In a head-on, inelastic collision, a 4,000-kg truck going 10 m/s east strikes a 1,000-kg car going 20 m/s west.
   a) What is the speed and direction of the wreckage?
   b) How much kinetic energy was lost in the collision?

5. A person on a swing moves so that the chain is horizontal at the turning points (● Figure 3.47). Show that the centripetal acceleration of the person at the low point of the arc is exactly 2 *g*, regardless of the length of the chain. This means that the force on the chains at the low point is equal to *three times* the person's weight.

(*Hint:* the vertical distance between the turning point and the low point equals the length of the chain.)

6. Assume that as a car brakes to a stop it undergoes a constant acceleration (deceleration). Explain why the stopping distance becomes *four times* as large if the initial speed is *doubled*.

7. The "shot" used in the shot-put event is a metal ball with a mass of 7.3 kg. When thrown in Olympic competition, it is accelerated to a speed of about 14 m/s. As an approximation, let's say that the athlete exerts a constant force on the shot while throwing it and that it moves a distance of 3 m while accelerating.
   a) What is the shot's kinetic energy?
   b) Compute the force that acts on the shot.
   c) It takes about 0.5 s to accelerate the shot. Compute the power required. Convert your answer to horsepower.

8. At the point in its orbit when it is closest to the Sun, Halley's Comet travels 54,500 m/s (● Figure 3.48). When it is at its most distant point, the separation between it and the Sun is about 60 times that when it is at its closest point. What is the speed of the comet at the distant point?

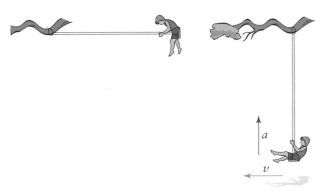

● **Figure 3.47**

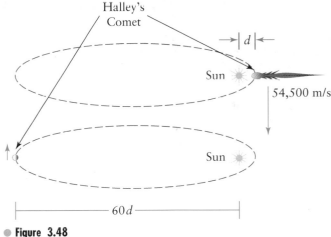

● **Figure 3.48**

SUGGESTED READINGS

Adair, Robert. "The Physics of Baseball." *Physics Today* 48, no. 5 (May 1995): 26–31. Included are discussions of energy and power in connection with throwing and hitting baseballs.

Bartlett, Albert A. "Forgotten Fundamentals of the Energy Crisis." *American Journal of Physics* 46, no. 9 (September 1978): 74–79. A classic paper on exponential growth and the consumption of fossil fuels.

Bartlett, Albert A., and Charles W. Hord. "The Slingshot Effect: Explanations and Analogies." *The Physics Teacher* 23, no. 8 (November 1985): 466–473.

Carlson, Shawn. "The Lure of Icarus." *Scientific American* 277, no. 4 (October 1997): 117–119. A look at the past and future of human-powered flight.

Drela, Mark, and John S. Langford. "Human-Powered Flight." *Scientific American* 253, no. 5 (November 1985): 144–151. Along with drawings and specifications of different aircraft, it includes a discussion of human power output.

Langford, John. "Triumph of *Daedulus*." *National Geographic* 174, no. 2 (August 1988): 191–199. Describes the 72-mile flight of pedal-powered *Daedulus* from Crete to Santorin on 23 April 1988.

Lederman, Leon M. "The Tevatron." *Scientific American* 264, no. 3 (March 1991): 48–55. A close look at the huge proton-antiproton collider near Chicago, written by a 1988 Nobel laureate.

Myers, Stephen, and Emilio Picasso. "The LEP Collider." *Scientific American* 263, no. 1 (July 1990): 54–61. Describes the Large Electron-Positron Collider near Geneva, Switzerland, which was shut down in 2000 and is being rebuilt as the Large Hadron Collider (LHC).

"Preventing the Next Oil Crunch." *Scientific American* 278, no. 3 (March 1998): 77–95. Four articles about the world's oil supplies and advanced techniques for liquid-fuel extraction.

Rosen, Harold A., and Deborah R. Castleman. "Flywheels in Hybrid Vehicles." *Scientific American* 277, no. 4 (October 1997): 75–77. Describes a car that would use an engine and a flywheel for energy storage.

For additional readings, explore InfoTrac® College Edition, your online library. Go to http://www.infotrac-college.com/wadsworth and use the passcode that came on the card with your book. Try these search terms: particle accelerator, Tevatron, escape velocity, exponential growth, gravity assist, pulsar, Daedulus.

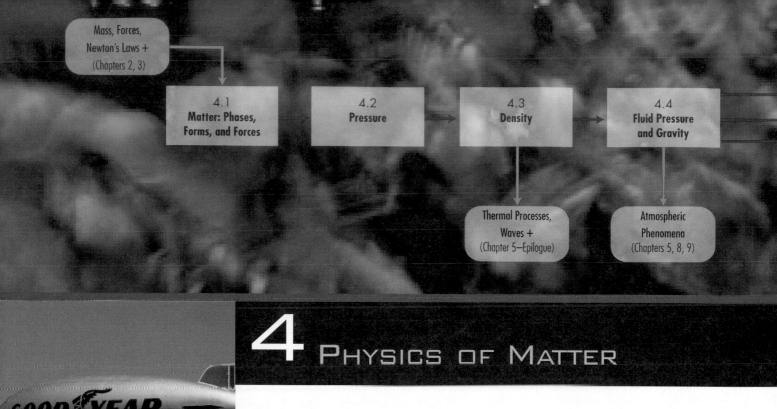

```
Mass, Forces,
Newton's Laws +
(Chapters 2, 3)
```

```
4.1
Matter: Phases,
Forms, and Forces
```

```
4.2
Pressure
```

```
4.3
Density
```

```
4.4
Fluid Pressure
and Gravity
```

```
Thermal Processes,
Waves +
(Chapter 5—Epilogue)
```

```
Atmospheric
Phenomena
(Chapters 5, 8, 9)
```

4 PHYSICS OF MATTER

Airships

You've probably seen them on TV so often you hardly notice them anymore. Blimps, floating lazily above major sporting events, functioning as a combination camera platform and giant billboard in the sky. But such "lighter-than-air" craft have a long and colorful history. Humans first took to the air in controlled flight aboard balloons more than a century before the Wright brothers took off. The era of powered human flight began before the Civil War when an airship was equipped with a steam engine. The first regular airline service across the Atlantic Ocean featured giant rigid airships called zeppelins, each carrying dozens of passengers in luxurious accommodations. Over a hundred zeppelins were built, some being used to bomb London and other European cities during World War I (that's right, I, not II). In the 1930s the U.S. Navy operated two flying aircraft carriers—huge, 200-ton airships that could launch and retrieve small airplanes. (Both were destroyed in storms.) And one of the first—and still one of the most spectacular—disasters caught on film was the flaming destruction of the *Hindenburg* zeppelin in 1937.

So how can a metal aircraft weighing so many tons float in the air like a party balloon? Why are modern airships spared the fate of the *Hindenburg*? Answers to these and many more related questions are found in considerations of the properties of matter, density, pressure, and buoyancy—all topics of this chapter.

In this chapter, we look into the physics of extended matter, particularly liquids and gases. The first section is an overview of the types, properties, and submicroscopic compositions of matter. Then the crucial physical quantities, *pressure* and *density*, are presented. The remainder of the chapter covers several laws and principles that explain a variety of phenomena involving fluids. The ideas presented in this chapter—particularly the law of fluid pressure, Archimedes' principle, and the concept of density—make it easy to understand how things like airships and altimeters work.

4.1 Matter: Phases, Forms, and Forces

The subject of this section is matter—anything that has mass and occupies space. The Earth, the water we drink, the air we breathe, our bodies, everything we touch is composed of matter. Obviously, matter exists in different forms that have different physical

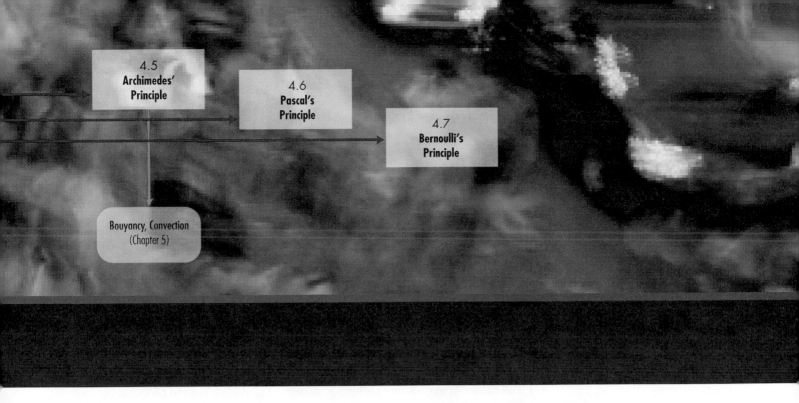

properties. We can classify matter into four categories: **solid, liquid, gas,** and **plasma.** These are called the four **phases,** or *states,* of matter. (In physics, *gas* does *not* refer to gasoline, and *plasma* does *not* refer to the liquid part of blood.) Briefly, we can distinguish the four phases as follows.

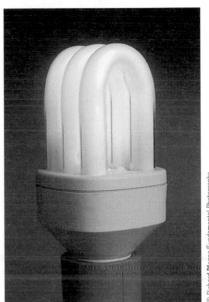

© Richard Megna/Fundamental Photographs

DEFINITION

Solids Rigid; retain their shape unless distorted by a force. Examples: rock, wood, plastic, iron.

Liquids Flow readily; conform to the shape of a container; have a well-defined boundary (surface); have higher densities than gases. Examples: water, beverages, gasoline, blood.

Gases Flow readily; conform to the shape of a container; do not have a well-defined surface; can be compressed (squeezed into a smaller volume) readily. Examples: air, carbon dioxide, helium.

Plasmas Have the properties of gases but also conduct electricity; interact strongly with magnetic fields; commonly exist at higher temperatures. Examples: gases in operating fluorescent, neon, and vapor lights (see ● Figure 4.1); matter in the Sun and stars.

● **Figure 4.1** Fluorescent lights use glowing plasmas.

Nearly all of the matter in our everyday experience appears as solid, liquid, or gas. Traditionally, these have been referred to as the three states of matter, with plasmas being a special "fourth" state of matter. Although plasmas are rare on Earth, most of the visible matter in the universe is in the form of plasmas in stars. In the last 50 years, the study of plasmas has grown to be one of the major subfields of physics because of the interest in nuclear fusion (the topic of Section 11.7). Nuclear fusion is the source of energy for stars, the Sun included. One of the main goals of plasma physics is to artificially produce star-like plasmas in which fusion can occur.

Many substances do not fit easily into one of these slots. Granulated sugar and salt flow readily and take the shape of a container. But they are considered to be solids because a single granule of each does fit the description of a solid and because both can be crystallized into larger, solid chunks. Tar and molasses do not flow readily, particularly when they are cold, and so they act somewhat like solids. But given time they do flow and take the shape of their container; they are considered liquids.

Many substances are composites of matter in two different phases. Styrofoam behaves like a solid but is composed mostly of a gas trapped in millions of tiny, rigidly connected bubbles. The water in many rivers carries along tiny, solid particles that will settle if the water is allowed to stand. The mists that fill a shower stall and comprise fog and low clouds consist of millions of small, round droplets of water mixed in with the air. Apples and potatoes are solid but contain a great deal of liquid.

Another factor that complicates our neat classification of matter is that the phase of a given substance can change with temperature and pressure. Water is a good example. Normally a liquid, water becomes a solid (ice) when cooled below 0°C (32°F; ● Figure 4.2). Under normal pressure, water becomes a gas (steam) when its temperature is raised above 100°C (212°F). Even at room temperature, water can be made to boil if the air pressure is reduced. Propane, carbon dioxide, and many other gases can be forced into the liquid phase at room temperature by increasing the pressure. Most refrigerators and air conditioners depend on the pressure-induced liquification of a gaseous refrigerant. (More on this in Section 5.7.)

To simplify our discussion in this chapter, we will consider only matter in one "pure" phase: Solid like a rock, liquid like pure water, and gaseous like air. When we say that a substance exists in a particular phase, we mean its phase at normal room temperature and pressure (unless stated otherwise).

The phases of matter refer to the macroscopic (external) form and properties of matter. These in turn are determined by the microscopic (internal) composition of matter. Around 2,500 years ago, some Greek philosophers theorized that all matter as we see and experience it is composed of tiny, indivisible pieces. It turns out that this is true up to a point: Diamond, water, and oxygen are all composed of extremely tiny "building blocks" or units that are the smallest entities that retain the identity of the substance. (But, as we shall see, the building blocks themselves are composed of still smaller particles.) Water is a liquid, and diamond is a solid because of the properties of the small units that comprise each. We can reclassify all substances by the nature of their intrinsic composition. We will begin with the simplest class of matter and proceed in order of increasing complexity.

The chemical **elements** represent the simplest and purest forms of everyday matter. At this time, scientists have identified 114 different elements and have agreed upon names for 110 of them. Some of the common substances used in our society are elements. These include hydrogen, helium, carbon, nitrogen, oxygen, neon, gold, iron, mercury, and aluminum. Each element is composed of incredibly small particles called **atoms.** There are 114 different atoms, one for each of the different known elements. Only about 90 of the elements exist naturally on Earth. The others are artificially produced in laboratories. The majority of the elements are quite rare and have names familiar only to chemists and other scientists.

The atom is not an indivisible particle: It has its own internal structure. Every atom has a very dense, compact core called the **nucleus,** which is surrounded by one or more particles called **electrons.** The nucleus itself is composed of two kinds of particles, **protons** and **neutrons*** (● Figure 4.3). The protons and electrons have equal but opposite electric charges and attract each other. The electrons are much lighter than protons and neutrons, and they move in orbits with the attraction of the protons supplying the centripetal force.

Every atom associated with a particular element has the same unique number of protons. For example, atoms that have 2 protons are atoms of helium, those with 8 protons are atoms of oxygen, those with 79 protons are atoms of gold, and so on. The *atomic number* of an element is the number of protons that are in each atom of the element. The atomic number of helium is 2 because every atom of helium has 2 protons in it. For oxygen it is 8, for gold it is 79, and so on. Each element is also given an abbreviation called its *chemical symbol*. ● Table 4.1 contains the chemical symbols and atomic numbers of

● **Figure 4.2** The phase of any substance depends on its temperature.

© Vern Ostdiek

---

* Protons and neutrons are themselves composed of smaller particles called *quarks.* More on this in Chapter 12.

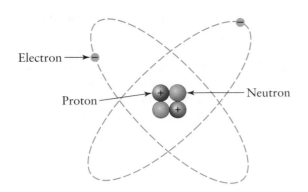

Electron →

Proton →

← Neutron

● **Figure 4.3** Simplified diagram of an atom. An atom consists of electrons in orbit around a compact nucleus, which in turn contains protons and neutrons. All atoms of a particular element have the same number of protons. This number is the element's atomic number. (Not drawn to scale.)

some familiar elements. It also includes the phase of each element at room temperature and normal pressure. The periodic table of the elements on the back inside cover shows all of the known elements and some of their properties. (We will take a closer look at the structure of atoms in Chapters 7 and 10.)

Chemical **compounds** are the next simplest form of everyday matter. There are millions of different compounds, including many common substances such as water, salt, sugar, and alcohol. Compounds are similar to elements in that each compound also has a unique building block called a **molecule.** Every molecule of a particular compound consists of the same unique combination of two or more atoms held together by electrical forces. For example, each molecule of water consists of two atoms of hydrogen attached to one atom of oxygen (● Figure 4.4). Similarly, one atom of carbon attached to one atom of oxygen forms a molecule of carbon monoxide. Each compound can be represented by a "formula"—a shorthand notation showing both the kinds and the numbers of atoms in each of the compound's molecules. You are probably familiar with the formula for water, H_2O. Some others are NaCl (table salt), CO_2 (carbon dioxide), CO (carbon monoxide),

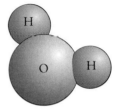

● **Figure 4.4** Water molecules consist of two hydrogen atoms attached to an oxygen atom. This structure is common to all molecules of water, ice, and steam.

● **Table 4.1** Some Common Chemical Elements			
Element	**Symbol**	**Atomic Number**	**Phase**
Hydrogen	H	1	Gas
Helium	He	2	Gas
Carbon	C	6	Solid
Nitrogen	N	7	Gas
Oxygen	O	8	Gas
Neon	Ne	10	Gas
Sodium	Na	11	Solid
Aluminum	Al	13	Solid
Silicon	Si	14	Solid
Chlorine	Cl	17	Gas
Calcium	Ca	20	Solid
Iron	Fe	26	Solid
Cobalt	Co	27	Solid
Nickel	Ni	28	Solid
Copper	Cu	29	Solid
Zinc	Zn	30	Solid
Silver	Ag	47	Solid
Barium	Ba	56	Solid
Gold	Au	79	Solid
Mercury	Hg	80	Liquid
Lead	Pb	82	Solid
Uranium	U	92	Solid

$C_{12}H_{22}O_{11}$ (sugar), and C_2H_5OH (ethyl alcohol). The study of how atoms combine to form molecules is a major part of chemistry.

Some elements in the gas phase are also composed of molecules. The oxygen in the air we breathe is an element, but most of the oxygen atoms are paired up to form O_2 molecules. Ozone (O_3) is a rarer form of oxygen that is present in air pollution and also in the ozone layer, centered about 15 miles above the Earth's surface. Each ozone molecule is composed of three oxygen atoms. Nitrogen gas also exists in the form of molecules, N_2. Helium and neon are both gaseous elements whose atoms do not form molecules; they remain separate.

Many substances, such as air, stone, and seawater, are composed of two or more different compounds or elements that are physically mixed together. These are classified as **mixtures** and **solutions.** Air consists of dozens of different gases that are mixed together. ● Table 4.2 shows the composition of clean, dry air. The air we breathe also contains water vapor and pollutants (like carbon monoxide). The amount of such gases present in the air varies considerably from place to place and from day to day. The air in the Sahara Desert does not have quite the same composition as the air in Los Angeles.

A mixture of two elements is not the same as a compound. If you mix hydrogen and oxygen, you simply have a gaseous mixture. The individual hydrogen atoms and oxygen molecules remain separate. If you ignite the mixture, the hydrogen and oxygen atoms will combine to form water molecules. Energy is released explosively in the form of heat (fire) because the atoms have less total energy when they are bound to each other.

The example of hydrogen, oxygen, and water also illustrates that the properties of a compound (water) are usually quite different from the properties of the constituent elements (hydrogen and oxygen). Sodium (Na) is a solid that reacts violently with water. Chlorine (Cl) is a gas that is used to kill bacteria in drinking water. If either element were ingested alone, it could be fatal. But when sodium and chlorine are combined chemically, the result is table salt (NaCl), a necessary part of our diet (● Figure 4.5).

The basic unit of life is the cell, and each cell is composed of many different compounds. Many of these contain billions of atoms in each molecule, the DNA molecule being an important example. The main constituents of such "organic" molecules are hydrogen, carbon, oxygen, and nitrogen. Many other elements are present in smaller amounts. For example, your bones and teeth contain calcium.

So atoms are basic to compounds as well as to elements. It is difficult to imagine how small atoms are and just how many there are in common objects. Atoms are about 1 ten-millionth of a millimeter in diameter. A ball 1 inch in diameter is about midway between the size of an atom and the size of the Earth (● Figure 4.6). In other words, if a 1-inch ball were expanded to the size of the Earth, each atom in the ball would expand to about 1 inch in diameter.

● **Table 4.2** Composition of Clean, Dry Air

Gas	Percent Composition (by volume)
Nitrogen (N_2)	78.1
Oxygen (O_2)	20.9
Argon (Ar)	0.93
Carbon dioxide (CO_2)	0.03
Other gases (Ne, He, and so on)	0.04

● **Figure 4.5** Separately, chlorine and sodium are poisonous. When chemically combined, they form table salt.

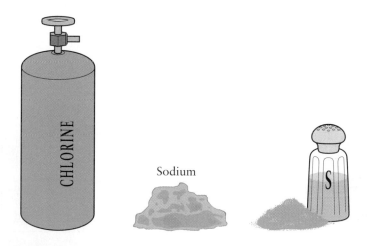

● **Figure 4.6** Think about how many golf balls it would take to fill up a hollow ball the size of the entire Earth. That's about how many atoms there are in each golf ball.

Since atoms are so small, huge numbers of them are present in anything large enough to be seen. There are about 100 thousand billion billion (one followed by 23 zeros) atoms in each of your fingernails. Even the smallest particles that can be seen with the naked eye contain far more atoms than there are people on Earth.

Atoms are nearly indestructible—only nuclear reactions affect them—and they are continually recycled. The Earth and everything on it are believed to be composed of the debris of stars that exploded billions of years ago. Chemical processes such as fire, decay, and growth result in atoms being combined with or dissociated from other atoms. A single carbon atom in your earlobe may once have been part of a dinosaur, a redwood tree, a rose, or Leonardo da Vinci—or all four.

Behavior of Atoms and Molecules

The constituent particles of matter, atoms and molecules, exert electrical forces on each other. ("Static cling" is an example of an electrical force.) The nature of these forces determines the properties of the substance. The forces between atoms in an element depend on the configuration of the electrons in each atom. In a compound, the size and shape of the molecules, as well as the forces between the molecules, affect its observed form and properties. We can relate the three common phases of matter to the interparticle forces as follows.

DEFINITION

Solids Attractive forces between particles are very strong; the atoms or molecules are rigidly bound to their neighbors and can only vibrate.

Liquids The particles are bound together, though not rigidly; each atom or molecule can move about relative to the others but is always in contact with other atoms or molecules.

Gases Attractive forces between particles are too weak to bind them together; atoms or molecules move about freely with high speed and are widely separated; particles are in contact only when they collide.

A standard model for representing the interparticle forces in a solid consists of each atom connected to each neighbor by a spring. The atoms are free to oscillate like a mass hanging from a spring. (The vibration of the atoms or molecules is related to the temperature, as we shall see in Chapter 5.) Often atoms or molecules in a solid form a regular geometric pattern called a *crystal* (● Figure 4.7). Table salt is a crystalline compound in which the sodium and chlorine atoms alternate with each other. In solids that do not have a regular crystal structure, called *amorphous solids,* the atoms or molecules are "piled together" in a random fashion. Glass is a good example of such a solid.

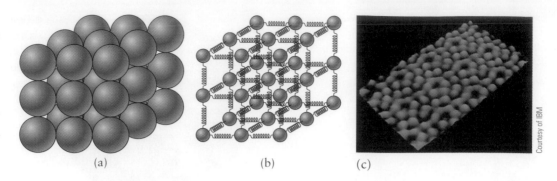

Figure 4.7 (a) The atoms or molecules in a crystalline solid are arranged in a regular three-dimensional pattern, similar to the way rooms are arranged in a large building. The atoms or molecules exert forces on each other. (b) A crystal behaves much like an array of particles that are connected to each other by springs. (c) Image showing the geometric arrangement of silicon atoms, produced with a scanning tunneling microscope (STM) at the IBM Thomas J. Watson Research Center.

(a)   (b)   (c)

Courtesy of IBM

Carbon is very interesting because it is an element with two common crystalline forms (graphite and diamond) that possess very different properties, plus a large number of recently discovered molecular forms. In diamond, each carbon atom is strongly bonded to each of its four nearest neighbors resulting in a crystalline solid that is the hardest known natural material (● Figure 4.8a). In graphite, the primary ingredient in the mis-named "lead" in pencils, the carbon atoms form sheets, with each atom strongly bonded to its three nearest neighbors in the same layer to form a mesh of hexagons (Figure 4.8b). These sheets can easily be forced to slide relative to each other making graphite an excellent "dry" lubricant. Carbon atoms can also bond together to form large molecules, the

Figure 4.8 Three forms of solid carbon. Each small sphere represents a carbon atom, and the connecting rods represent the force holding pairs of carbon atoms together: (a) diamond; (b) graphite; and (c) C_{60} crystal. Copyright 1993, Henry Hill, Jr.

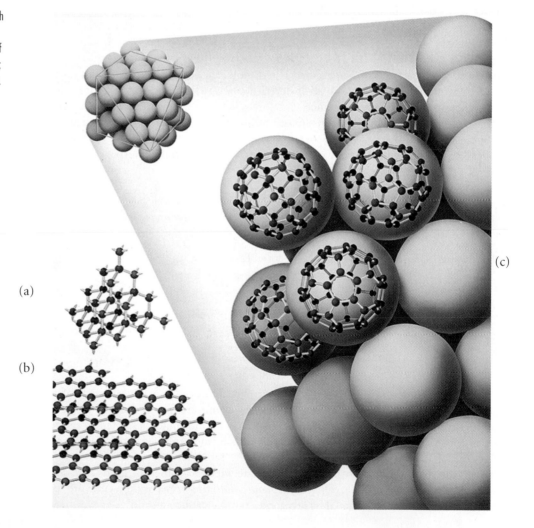

(a)

(b)

(c)

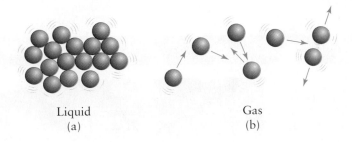

Liquid
(a)

Gas
(b)

● **Figure 4.9** (a) The atoms or molecules in a liquid remain in contact with each other, but they are free to move about. (b) In a gas, the atoms or molecules are not bound to each other. They move with high speed and interact only when they collide.

most famous being C_{60}, named buckminsterfullerene ("buckyball" for short) after the famous engineer and philosopher R. Buckminster Fuller. The molecule reminded its discoverers of geodesic domes invented by Fuller. The 60 atoms in each buckminsterfullerene molecule arrange themselves into a soccer-ball shape consisting of 12 pentagons and 20 hexagons. The C_{60} molecules can in turn bond together to form a crystal (Figure 4.8c). Dozens of larger and smaller hollow carbon molecules can also form, as well as "carbon nanotubes" consisting of graphitelike sheets of carbon atoms rolled up into tubes. The C_{60} and other molecular carbon forms may have useful applications in microelectronics and medicine.

In a liquid, the forces between the particles are not strong enough to bind them together rigidly. The atoms or molecules are free to move around as well as vibrate (● Figure 4.9a). This is similar to a collection of ball-shaped magnets clinging to each other. The forces between the particles are responsible for surface tension and the spherical shape of small drops.

Many compounds have an interesting intermediate phase between solid and liquid called the *liquid crystal* phase. The molecules have some mobility, as in a liquid, but they are arranged regularly, as in a solid. Liquid crystal displays (LCDs) in calculators, flat video displays, and watches use electrically induced movement of the molecules to change the optical properties of the liquid crystal. (More on this in Section 9.1.)

In a gas, the atoms and molecules are widely separated and move around independently, except when they collide (see Figure 4.9b). Their speeds are surprisingly high: The oxygen molecules you are now breathing have an average speed of over 1,000 mph. As each molecule collides randomly with other molecules, its speed is sometimes increased and other times decreased.

Often, the surface of a solid or a liquid forms a boundary for a gas. The surface of water in a glass, a lake, or an ocean forms a lower boundary for the air above. The inside of the walls of a tire are a boundary for the air inside (● Figure 4.10). The high-speed atoms or molecules in the gas exert a force on the surface as their random motions cause them to collide with it. This is the basis for gas pressure, as mentioned in Section 3.6. In other words, the weight of a car is supported by the collisions of air molecules with the inner walls of the tires.

At normal temperatures and pressures, the average distance between the centers of the atoms or molecules in a gas is about 10 times that in a solid or liquid (● Figure 4.13). There is a great deal of empty space in a gas, and that is why gases can be compressed easily.

Let's summarize the information in this section. The 114 different types of atoms (the elements) form the basis of matter as we experience it. Molecules are composed of two or

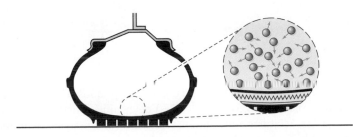

● **Figure 4.10** The air molecules inside a tire collide with the walls and exert forces on it. If the molecules weren't moving, the tire would be flat.

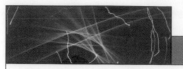
What's in a Name?

What's in a name? Well, for the elements, names contain clues to their chemical and physical properties, to their places and means of discovery, to the origins of their chemical symbols, and to the scientists who figured prominently in their isolation. In short, a great deal can be learned about the chemical elements from a study of their names.

Consider the element chlorine (Cl). The word *chlorine* derives from the Greek word *chloros* meaning "pale green"—an apt description for this greenish-yellow gas employed as a weapon during World War I to take lives (the infamous "mustard gas") and now used as a disinfectant and water-purifying agent to help save lives. The element iridium (Ir) takes its name from the Greek word for rainbow, *iris*. This is also a rather poetic description of an element whose crystalline salts are brilliantly colored, spanning the entire rainbow of hues from deep blue to dark red. Or take zirconium (Zr). Its name comes from the Persian word *zargun* meaning "goldlike." This, too, seems an appropriate way to characterize the element responsible for the luster and sheen of the mineral and gemstone that we now call *zircon* but that was well known to the ancients and described in biblical writings.

Many elements obtain their names from their places of discovery. Probably the best examples are the elements erbium (Er), terbium (Tb), ytterbium (Yb), and yttrium (Y), all of which take their names from the village of Ytterby in Sweden near Stockholm (● Figure 4.11). Ytterby is the site of a quarry that yielded many unusual minerals containing these and other uncommon elements. Further examples of elements named after places first associated with their discovery are californium (Cf), named for the state and university in which it was first produced by a team of Berkeley scientists in 1950, and strontium (Sr), whose name comes from the mineral stronianite found outside the village of Strontian in Scotland, where the element was first discovered in 1790. A final example of this type is helium (He), meaning "the Sun"; it serves to remind us that the element was first detected in 1868 in the outer atmosphere of the Sun during a solar eclipse!

The symbols of most of the chemical elements bear a direct correspondence to their names, usually consisting of the first letter and, at most, one additional letter from the name. However, for many of the

● **Figure 4.11** Dr. "Skip" Simmons (center) and geology colleagues from the University of New Orleans outside the village of Ytterby, Sweden, where neighboring mines have yielded up some of the most exotic elements in the periodic table.

Courtesy William B. Simmons

common elements that have been known for centuries, this is not the case. Classic examples include iron (Fe), gold (Au), mercury (Hg), and copper (Cu). For such elements, the symbols often originate from their Latin equivalents and date from a time hundreds of years ago when Latin was the language of scholarship the world over. Iron is an Anglo-Saxon word whose Latin equivalent is *ferrum,* hence the symbol Fe. Similarly, gold derives from an Anglo-Saxon word of the same spelling, but its symbol, Au, is taken from the Latin word *aurum* meaning "shining dawn." Mercury, named after the fleet-footed Roman god, takes its symbol, Hg, from the Latin word *hydrargyrum,* which translates to "liquid silver"—a perfect description of this shiny, liquid metal. Lastly, the symbol for copper, Cu, comes from the Latin *cuprum;* this word is a shortened corruption of the phrase *aes Cyprium,* meaning "metal from Cyprus." The Romans obtained virtually all of their supply of copper from this Mediterranean island. Other instances of this type of symbol derivation include silver (Ag from the Latin *argentum*) and lead (Pb from the Latin *plumbum*—compare our word "plumber").

● **Figure 4.13** Gases can be compressed because the atoms and/or molecules are widely separated. Increasing the force (pressure) on the boundaries of a gas squeezes the particles closer together.

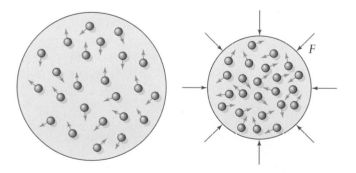

more atoms "stuck" together to form the basic unit of compounds. The nature of the forces between the building blocks (atoms or molecules) determines whether the substance is a solid, a liquid, or a gas. Mixtures and solutions consist of different elements and/or compounds mixed together. Concept Map 4.1 summarizes this section in another way.

A number of the element names derive from the names of scientists who have contributed to the discovery or the isolation of the elements. Specific cases include mendelevium (Md), named after Dmitri Mendeleev, a Russian chemist who developed the periodic table of the elements; curium (Cm), named for Pierre and Marie Curie, pioneers in the discovery and characterization of radioactive elements; and gadolinium (Gd), after Finnish chemist Johan Gadolin, the discoverer of yttrium.

As mentioned in the text, 114 elements have been identified. Traditionally, the individual or group responsible for first isolating a new element earns the right to propose its name. At the time of this writing, names have been established for the first 110 elements, those of elements 104–110 having only been recently agreed upon by the Committee on Inorganic Chemistry Nomenclature of the International Union of Pure and Applied Chemistry. The new names follow the same general patterns observed in elements with atomic numbers less than 103. For example, element 104 is named rutherfordium (Rf), after Ernest Rutherford, a New Zealand physicist who won the Nobel Prize in chemistry in 1907 for establishing that radioactivity causes a transmutation of an element (see Chapter 11 for more on this topic). Element 105, dubnium (Db), is named for the city of Dubna, where Russian investigators first identified this species. Seaborgium (Sg) is the name adopted for element 106; this choice honors U.S. nuclear chemist Glenn T. Seaborg, who first synthesized plutonium and nine other transuranium elements. Names for elements 107 through 109 are bohrium (Bh), after Niels Bohr, a Danish physicist instrumental in founding quantum mechanics (see Chapter 10); hassium (Hs), from the Latin name for the state of Hesse in Germany, where the Laboratory for Heavy Ion Research (GSI), which first produced elements 107–112, is located ( Figure 4.12); and meitnerium (Mt), for Lise Meitner, a German scientist who contributed significantly to our understanding of nuclear fission and beta decay (see Figure 11.36). Recently, element 110 was named darmstadtium (Ds), honoring the site of the GSI. Names for elements 111, 112, 114, and 116 have yet to be agreed upon.

● **Figure 4.12** GSI facility in Darmstadt, Germany.

In summary, there can be a lot in a name, especially if it is the name of a chemical element. Perhaps this short excursion into the lexicology of the elements will make you want to know more about this subject or will at least make it a little easier for you to remember the symbols of the more common elements that make up our environment. For those interested in pursuing this topic, a good place to start is the most recent edition of *The Handbook of Chemistry and Physics*, Section B, "The Elements and Inorganic Compounds." The World Wide Web also abounds in sites that provide additional information on the discovery, properties, and uses of the chemical elements, as well as their organization in the periodic table. Many Web pages provide interesting and amusing exercises to help you learn the element names and symbols, like this fill-in activity: "The poor fellow crashed his (Ag) _____ (Ne) _____ into a wall. Luckily, a good (Sm) _____ came along and tried to (He) _____, but, alas, it was too late. All the hapless rescuer could do was (Ba) _____ in a (K) _____ the hill nearby." BeSTe OF LuCK In YOURe SeArCH.

LEARNING CHECK

1. The three common phases of matter on Earth are _____ , _____ , and _____ .
2. (True or False.) The number of protons in the nucleus of an atom determines which element it is.
3. The molecule is the basic unit of all
   a) elements.
   b) compounds.
   c) solutions.
   d) mixtures.

4. Graphite, diamond, and buckminsterfullerene are different forms of the element _____ .
5. (True or False.) Air can be compressed easily because there is a great deal of empty space between air molecules.

ANSWERS: 1. solid, liquid, gas 2. True 3. (b) 4. carbon 5. True

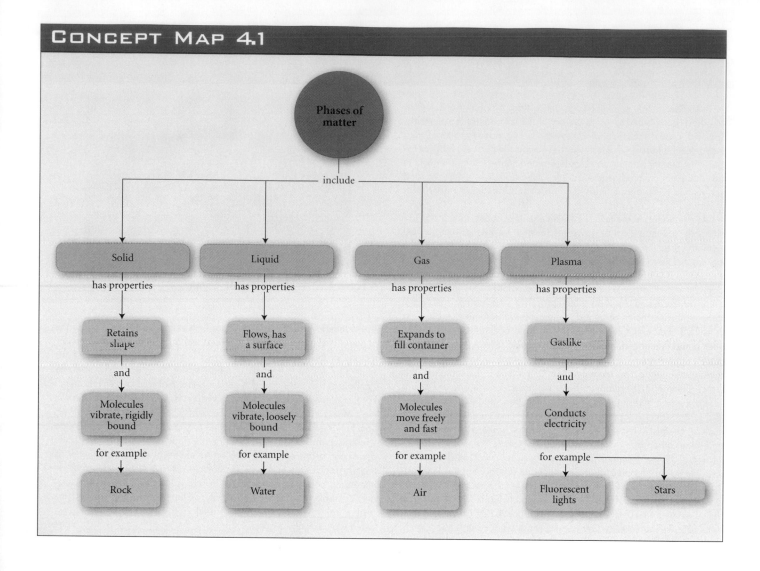

4.2 Pressure

Now that we have established some of the basic properties of solids, liquids, and gases, it is time to consider how the mechanics we've presented in the preceding chapters relates to extended matter. We have already seen in Chapters 1, 2, and 3 that extended objects can often be treated as particles (for example, a rock as it falls, satellites and even planets as they move in orbits, trucks during collisions, and so on). Furthermore, the law of conservation of angular momentum (Section 3.8) allows us to examine interesting types of rotational motions of solid bodies. But what about gases and liquids? The fluids in and around us—blood in our veins and arteries, the atmosphere, streams and oceans—are in constant motion. Can we extend our study of mechanics to fluid motion? The answer is a qualified yes.

Newton's laws and the conservation laws can be applied to fluids using appropriate extensions of physical quantities like mass and force, but the mathematics is much more complicated and well beyond the level of this text. (Some of the most powerful computers in the world are used exclusively to solve problems involving fluids, such as the motions of the atmosphere.) So we will limit our study of fluid motion to one (restricted) conservation law (Section 4.7). However, there are several phenomena related to fluids at rest that are important in our daily lives and that can be dealt with using simple mathematics. These phenomena are the principal topics of most of the rest of this chapter. We first introduce the physical quantities **pressure** and **density,** which are extensions of force and mass, respectively. These are two essential quantities for studying fluids both at rest and in motion.

Forces that are exerted by gases, liquids, and solids normally are spread over a surface. The force that the floor exerts on you when you are standing is distributed over the bottoms of your feet. A boat floating on water has an upward force spread over its lower surface (● Figure 4.14). The air around a floating balloon exerts forces on the balloon's surface. In situations like these, the physical quantity pressure is quite useful.

● **Figure 4.14** The water exerts an upward force that is spread over the submerged surface of the boat.

> **DEFINITION**
>
> **Pressure** The force per unit area for a force acting perpendicular to a surface. The perpendicular component of a force acting on a surface divided by the area of the surface.
>
> $$p = \frac{F}{A}$$

Pressure is a scalar: There is no direction associated with it.

Note: p is the abbreviation for pressure, and P is the abbreviation for power.

Physical Quantity	Metric Units	English Units
Pressure (p)	pascal (Pa) (1 Pa = 1 N/m²)	pound per square foot (lb/ft²) pound per square inch (psi)
	millimeters of mercury (mm Hg)	inches of mercury (in. Hg)

The standard pressure units consist of a force unit divided by an area unit. The SI unit of pressure is the pascal (Pa), which equals 1 newton per square meter. The English unit for pressure (psi) may be the most familiar to you. Air pressure in tires is commonly measured in psi. For comparison:

$$1 \text{ psi} = 6{,}890 \text{ Pa}$$

This number is so large because a force of 1 pound on each square inch would produce a huge force on 1 square meter—a much larger area. The two pressure units involving mercury will be explained in Section 4.4. Another common unit of pressure is the atmosphere (atm). It equals the average air pressure at sea level. (Later, we will see that the air pressure is lower at higher altitudes.) It is related to the other units as follows:

$$1 \text{ atm} = 1.01 \times 10^5 \text{ Pa} = 14.7 \text{ psi}$$

This unit is like the g used as a unit of acceleration. We are all subject to the pressure of the air around us and to the acceleration due to gravity, so it is natural to compare other pressures and accelerations to them. The old practice of using the length of a person's foot as a unit of distance is another example of taking a unit of measure provided by nature. ● Table 4.3 lists some representative pressures.

● Table 4.3 Some Pressures of Interest

Description	p (Pa)	p (psi)	p (atm)
Lowest laboratory pressure	7×10^{-11}	1×10^{-14}	7×10^{-16}
Atmospheric pressure at an altitude of 100 km	0.06	9×10^{-6}	6×10^{-7}
Lowest recorded sea level atmospheric pressure	0.87×10^5	12.6	0.86
Average sea level atmospheric pressure	1.01×10^5	14.7	1
Highest recorded sea level atmospheric pressure	1.08×10^5	15.7	1.07
Inside a tire (typical)	3.1×10^5	45	3.1
4,000 m underwater (*Titanic*'s remains)	4×10^7	5,800	390
Center of the Earth	1.7×10^{11}	2.5×10^7	1.7×10^6
Highest sustained laboratory pressure	3×10^{11}	4.4×10^7	3×10^6
Center of the Sun	1.3×10^{14}	1.9×10^{10}	1.3×10^9

Example 4.1

A 160-pound person stands on the floor. The area of each shoe that is in contact with the floor is 20 square inches. What is the pressure on the floor? Assuming the person's weight is shared equally between the two shoes, the force of one shoe is 80 pounds. So:

$$p = \frac{F}{A} = \frac{80 \text{ lb}}{20 \text{ in.}^2}$$

$$= 4 \text{ psi} \qquad \text{(standing on both feet)}$$

By Newton's third law of motion, the floor exerts an equal and opposite force on the shoes. Thus the pressure of the floor on the shoes is also 4 psi.

If the person stands on one foot instead, that shoe has all of the weight and the pressure is

$$p = \frac{160 \text{ lb}}{20 \text{ in.}^2}$$

$$= 8 \text{ psi} \qquad \text{(standing on one foot)}$$

What if the person put on high-heeled shoes and balanced on the heel of one shoe (Figure 4.15)? The bottom of the heel might measure 0.5 inches by 0.5 inches. The area is then 0.25 square inches. So the pressure in this case would be

$$p = \frac{160 \text{ lb}}{0.25 \text{ in.}^2}$$

$$= 640 \text{ psi} \qquad \text{(balanced on a narrow heel)}$$

● **Figure 4.15** The pressure on the floor is much higher with narrow heels.

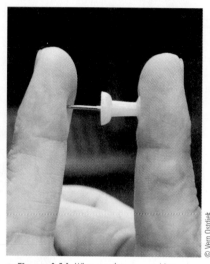

● **Figure 4.16** When a tack is squeezed between two fingers, the force on each finger is the same.

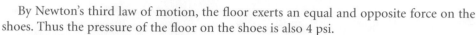

Explore It Yourself 4.1

Lightly squeeze a push pin or thumbtack between two fingers, as shown in ● Figure 4.16. The force is the same on each finger. Do you feel the same thing in each finger? How is pressure involved?

Example 4.1 and Explore It Yourself 4.1 show that the same force causes much higher pressure when it acts over a smaller area. We can think of pressure as a measure of how "concentrated" a force is.

There are many situations in which a liquid or a gas is under pressure and exerts a force on the walls of its container. The relationship between force and pressure can be used to determine the force on a particular area of the walls. Since pressure equals force divided by area, the pressure times the area equals the force. In other words, the total force on a surface equals the force on each square inch (the pressure) times the total number of square inches (the area).

$$F = pA$$

Example 4.2

In the late 1980s, there were several spectacular (and tragic) aircraft mishaps involving rapid loss of air pressure in the passenger cabins (● Figure 4.17). (The causes included failure of a cargo door, outer skin rupture due to corrosion or cracks or both, and small bombs.) Because the cabins are pressurized, there are large outward forces acting on windows, doors, and the aircraft skin.* Let's estimate the sizes of these forces.

---

* One may ask why aircraft do not fly at low altitudes so that cabins would not need to be pressurized. There are several reasons why jets fly at 10,000 meters or higher. (1) If something goes wrong and an aircraft starts losing altitude, the pilot has more time to recover when the plane is very high to begin with. More than once an aircraft has been saved after diving over 7,000 meters. (2) Because the air is thinner at high altitudes, the force of air resistance on an aircraft is smaller. Consequently, it can travel faster and use less fuel than when flying at low altitudes. (3) The air is usually much "smoother" at high altitudes. There is much less turbulence, so passengers are more comfortable.

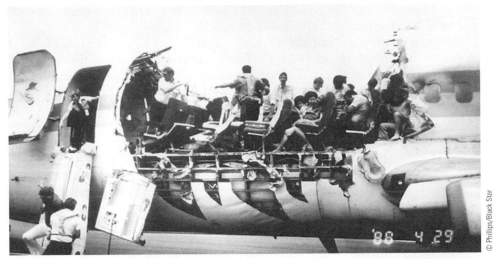

● **Figure 4.17** A large section of the aluminum skin on this older, heavily used Boeing 737 jet ripped away while the aircraft was flying at 7,300 meters (24,000 feet) above sea level (April 1988). One person was killed, and dozens were injured, but the aircraft landed safely.

The pressure inside a passenger jet cruising at high altitude (about 7,500 meters or 25,000 feet) is about 6 psi (0.41 atmospheres) greater than the pressure outside. What is the outward force on a window measuring 1 foot by 1 foot and on a door measuring 1 meter by 2 meters? The area of the window is

$$A = 1 \text{ ft} \times 1 \text{ ft} = 12 \text{ in.} \times 12 \text{ in.}$$

$$= 144 \text{ in.}^2$$

(The area has to be in square inches because the pressure is in pounds per square *inch.*)

The force on the window is

$$F = pA = 6 \text{ psi} \times 144 \text{ in.}^2$$

$$= 864 \text{ lb}$$

A rather small pressure causes a large force on the window. For the door, we use SI units:

$$p = 6 \text{ psi} = 6 \times 1 \text{ psi} = 6 \times 6{,}890 \text{ Pa}$$

$$= 41{,}340 \text{ Pa}$$

$$A = 1 \text{ m} \times 2 \text{ m} = 2 \text{ m}^2$$

Consequently, the force on the door is

$$F = pA = 41{,}340 \text{ Pa} \times 2 \text{m}^2$$

$$= 82{,}680 \text{ N}$$

The force on the door is greater than the weight of 10 small automobiles, or 100 people.

Pressure is a relative quantity. When you test the air pressure in a tire, you are comparing the pressure of the air inside the tire with the pressure outside the tire. When a tire is flat, there is still air inside it, but the pressure inside is the same as the atmospheric pressure outside.

For example, let us say that the atmospheric pressure is 14 psi. A tire is tested, and the air pressure gauge shows 30 psi. This means that the air pressure inside the tire is 30 psi higher than the air pressure outside the tire. So the actual pressure on the inner walls of the tire is 30 + 14 = 44 psi. This is sometimes referred to as the *absolute pressure.* The pressure relative to the outside air (30 psi in this case) is then called the *gauge pressure* (● Figure 4.18).

We can look at this another way. The pressure inside the tire, 44 psi, causes an outward force of 44 pounds on each square inch of the tire wall. The air pressure outside the tire

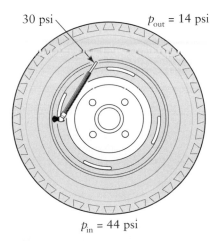

30 psi p_{out} = 14 psi

p_{in} = 44 psi

● **Figure 4.18** An absolute pressure inside of 44 psi produces a gauge pressure of 30 psi.

causes an inward force of 14 pounds on each square inch. The net force due to air pressure on each square inch of the tire is then 30 pounds.

The gauge pressure of the air in a tire changes if the outside air pressure changes. What happens if the car is driven into a large chamber and the air pressure in the chamber is increased from 14 psi to 44 psi? The gauge pressure in the tires will then be zero, and the tires will go flat even though no air has been removed from them. When the air pressure is reduced to 14 psi, the tires will expand again to their normal shape.

In the previous example of the pressurized aircraft, the 6 psi is the gauge pressure. The absolute pressure inside the cabin might be 11 psi, and the air pressure outside might be 5 psi. The atmospheric pressure at 26,000 feet is about 5 psi.

● **Figure 4.19** The outcome of Explore It Yourself 4.2.

© Vern Ostdiek

Explore It Yourself 4.2

Put an ounce or so of water into an empty aluminum can. Heat the can until the water is boiling vigorously and you can clearly see vapor coming out of the opening. Let the water boil for about a minute. With a gloved hand, quickly but carefully turn the (hot!) can upside down over a pan or sink with cold water in it and plunge the top of it an inch or so into the water. What happens? (See ● Figure 4.19.) How is the atmosphere involved in this?

The standard pen-shaped tire-pressure tester nicely illustrates some of the physics that we have considered so far. It consists of a hollow tube (cylinder) fitted with a piston that can slide back and forth in the cylinder (● Figure 4.20). Air from the tire enters the cylinder at the left end and pushes on the piston. The right end allows air from the outside to push on the other side of the piston. If the pressure inside the tire is greater than the outside air pressure, there is a net force to the right on the piston. A spring placed behind the piston is compressed by this net force. The greater the net force on the piston, the greater the compression of the spring. A calibrated shaft extends from the right side of the piston and out of the right end. When the piston is pushed to the right, the shaft protrudes from the right end the same distance that the spring is compressed. The length of shaft showing indicates the gauge pressure in the tire, since that is what causes the force on the piston.

We conclude this section with one last important note about pressure. Since gases are compressible, the volume of a gas can be changed (Figure 4.13). Whenever the volume of a fixed amount of gas is changed, the pressure in the gas changes also. Increasing the volume of a gas reduces the pressure. Decreasing the volume increases the pressure. To understand why this is the case, recall from the previous section that the collisions of the

● **Figure 4.20** In a common tire-pressure tester, the higher pressure of the air in the tire pushes the piston to the right and compresses the spring. The higher the pressure, the greater the force and the greater the compression of the spring.

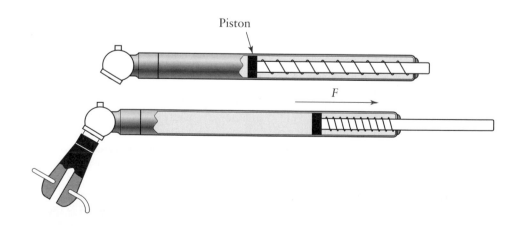

Piston

F

atoms and molecules in a gas with a surface cause gas pressure. If the volume is decreased, the particles are squeezed together so there are more of them near each square inch of the boundaries. More collisions occur each second, which means more force on each square inch and higher pressure. The opposite happens when the volume is increased.

The temperature of a gas influences the speeds of the atoms or molecules. Consequently, the pressure is also affected by the gas temperature. When the temperature of a given quantity of gas is kept constant, the pressure p is related to the volume V as follows:

$$pV = \text{a constant} \qquad \text{(gas at fixed temperature)}$$

This means that the *volume of a gas is inversely proportional to the pressure*. If the pressure is doubled, the volume is halved. In Chapter 5, we will see what effect temperature has on pressure and volume.

LEARNING CHECK

1. If the same-sized force is made to act over a smaller area,
   a) the pressure is decreased.
   b) the pressure is not changed.
   c) the pressure is increased.
   d) the result depends on the shape of the area.
2. (True or False.) A sealed box contains air under pressure. The outward forces on the walls will be largest on those walls with the smallest area.

3. The pressure difference between the inside and the outside of a tire is the _____ pressure.
4. (True or False.) If the mass and temperature of the gas in a container stay constant while the volume is decreased, the pressure will also decrease.

ANSWERS: 1. (c)   2. False   3. gauge   4. False

4.3 Density

Pressure is an extension of the idea of force. Similarly, **mass density** is an extension of the concept of mass. Just as pressure is a measure of the concentration of force, density is a measure of the concentration of mass.

Mass Density The mass per unit volume of a substance. The mass of a quantity of a substance divided by the volume it occupies.

$$D = \frac{m}{V}$$

Note: D is the abbreviation for mass density, and d is the abbreviation for distance.

Physical Quantity	Metric Units	English Units
Mass density (D)	kilogram per cubic meter (kg/m^3) gram per cubic centimeter (g/cm^3)	slug per cubic foot

To find the mass density of a substance, one measures the mass of a part or "sample" of it and divides by the volume of that part or sample. It doesn't matter how much is used: The greater the volume, the greater the mass.

Example 4.3

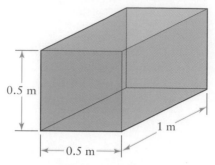

● **Figure 4.21** The mass of the water in this aquarium is 250 kilograms. Using this and the volume of the aquarium, we can compute the mass density of water.

The dimensions of a rectangular aquarium are 0.5 meters by 1 meter by 0.5 meters. The mass of the aquarium is 250 kilograms larger when it is full of water than when it is empty (● Figure 4.21). What is the density of the water? First, the volume of the water is

$$V = l \times w \times h$$
$$= 1 \text{ m} \times 0.5 \text{ m} \times 0.5 \text{ m}$$
$$= 0.25 \text{ m}^3$$

So:

$$D = \frac{m}{V} = \frac{250 \text{ kg}}{0.25 \text{ m}^3}$$
$$= 1,000 \text{ kg/m}^3 \qquad \text{(mass density of water)}$$

If a tank with twice the volume were used, the mass of the water in it would be twice as great, and the density would be the same. The mass density of any amount of pure water is 1,000 kg/m³.

When the same tank is filled with gasoline, the mass of the gasoline is found to be 170 kilograms. Therefore, the density of gasoline is

$$D = \frac{170 \text{ kg}}{0.25 \text{ m}^3}$$
$$= 680 \text{ kg/m}^3$$

Except for small changes caused by variations in temperature or pressure, the mass density of any pure solid or liquid is constant. It is an identifying trait of that substance. The mass densities of pure gases, on the other hand, vary greatly with changes in temperature or pressure. We have seen that doubling the pressure on a gas halves the volume. That would double the mass density as well. In Chapter 5, we will describe how changing the temperature of a gas also alters its volume and, therefore, its density. By convention, the tabulated densities of gases are given at the standard temperature and pressure (STP): 0°C temperature and 1 atmosphere pressure.

The mass density of each element or compound is fixed. Water, lead, mercury, salt, oxygen, gold, and so on, all have unique mass densities that have been measured and catalogued. The density of a mixture containing two or more substances depends on the density and percentage of each component. Most metals in common use are alloys, consisting of two or more metallic elements and other elements like carbon. For example, 14-karat gold is only about 58% gold. However, the mass density of a mixture with a particular composition is constant.

● Table 4.4 lists the mass densities (column 3) of several common substances. The values for mixtures can vary and so are merely representative. The mass densities of the gases are for standard conditions.

Having a list of the densities of common substances is quite useful for three reasons. First, one can use mass density to help *identify* a substance. For example, one can determine whether or not a gold ring is solid gold by measuring its density and comparing it to the known density of pure gold. Second, density measurement is used routinely to determine how much of a particular substance is present in a mixture. The coolant in an automobile radiator is usually a mixture of water and antifreeze. These two liquids have different densities (Table 4.4), so the density of a mixture depends on the ratio of the amount of water to the amount of antifreeze. The higher the density, the greater the antifreeze content and the lower the freezing temperature of the coolant (● Figure 4.22). By simply measuring the coolant density, one can determine the coolant's freezing temperature. If you have donated blood to a blood bank, part of the screening included checking to see whether the hemoglobin content of your blood was high enough. This is done by determining whether the blood's density is greater than an accepted minimum value.

● **Figure 4.22** Ruth determines the freezing point of the car's coolant by measuring the coolant's density. This indicates the relative amount of antifreeze in the mixture.

© Vern Ostdiek

Table 4.4 Densities of Some Common Substances

Substance	Type*	Mass Density, D (kg/m³)	Weight Density, D_W (lb/ft³)	Specific Gravity
Solids				
Juniper wood	m	560	35	0.56
Ice	c	917	57.2	0.917
Ebony wood	m	1,200	75	1.2
Concrete	m	2,500	156	2.5
Aluminum	e	2,700	168	2.7
Diamond	e	3,400	210	3.4
Iron	e	7,860	490	7.86
Brass	m	8,500	530	8.5
Nickel	e	8,900	555	8.9
Copper	e	8,930	557	8.93
Silver	e	10,500	655	10.5
Lead	e	11,340	708	11.34
Uranium	e	19,000	1,190	19
Gold	e	19,300	1,200	19.3
Liquids				
Gasoline	m	680	42	0.68
Ethyl alcohol	c	791	49	0.791
Water (pure)	c	1,000	62.4	1.00
Seawater	m	1,030	64.3	1.03
Antifreeze	m	1,100	67	1.1
Sulfuric acid	c	1,830	114	1.83
Mercury	e	13,600	849	13.6
Gases (at 0°C and 1 atm)				
Hydrogen	e	0.09	0.0056	0.00009
Helium	e	0.18	0.011	0.00018
Air	m	1.29	0.08	0.00129
Carbon dioxide	c	1.98	0.12	0.00198
Radon	e	10	0.627	0.010

*Note: "e" stands for element, "c" for compound, and "m" for mixture.

Third, one can *calculate the mass* of something if one knows what its volume is. The mass of a substance equals the volume that it occupies times its mass density.

$$m = V \times D$$

Example 4.4

The mass of water needed to fill a swimming pool can be computed by measuring the volume of the pool. Let's say a pool is going to be built that will be 10 meters wide, 20 meters long, and 3 meters deep. How much water will it hold? The volume of the pool will be

$$V = l \times w \times h = 20 \text{ m} \times 10 \text{ m} \times 3 \text{ m} = 600 \text{ m}^3$$

$$m = V \times D = 600 \text{ m}^3 \times 1,000 \text{ kg/m}^3$$

$$= 600,000 \text{ kg}$$

That is a lot of water (about 3,000 bathtubs full).

In some cases, it is practical to use another type of density called **weight density.** It is commonly used in the English system of units because weight is in more common use than mass.

DEFINITION

Weight Density The weight per unit volume of a substance. The weight of a quantity of a substance divided by the volume it occupies.

$$D_W = \frac{W}{V}$$

Physical Quantity	Metric Units	English Units
Weight density D_W	newton per cubic meter (N/m³)	pound per cubic foot (lb/ft³) pound per cubic inch (lb/in.³)

The weight density of a substance is equal to the mass density times the acceleration due to gravity. This is because the weight of a substance is just g times its mass.

$$W = m \times g \rightarrow D_W = D \times g$$

Column 4 in Table 4.4 shows representative weight densities in English units. These can be used in the same ways that mass densities are used. One interesting exercise is to compute the weight of the air in a room. (Often people do not realize that air and other gases *do* have weight.)

Example 4.5 A college dormitory room measures 12 feet wide by 16 feet long by 8 feet high. What is the weight of air in it under normal conditions (● Figure 4.23)? The weight is related to the volume by the equation:

$$W = D_W \times V$$

But

$$V = l \times w \times h - 16 \text{ ft} \times 12 \text{ ft} \times 8 \text{ ft}$$

$$= 1{,}536 \text{ ft}^3$$

So the weight is

$$W = D_W \times V = 0.08 \text{ lb/ft}^3 \times 1{,}536 \text{ ft}^3$$

$$= 123 \text{ lb}$$

This is for sea-level pressure and 0°C temperature. At normal temperature, 20°C, the weight would be 115 pounds.

Explore It Yourself 4.3

The next time you go grocery shopping, rank several different packaged foods and/or ingredients by density (different breakfast cereals, flour, and marshmallows are possibilities). Before you go, devise a method that you can use. Is the type of container important? Can you always determine if one substance has a higher density than another without using measuring devices?

● **Figure 4.23** Air does have weight.

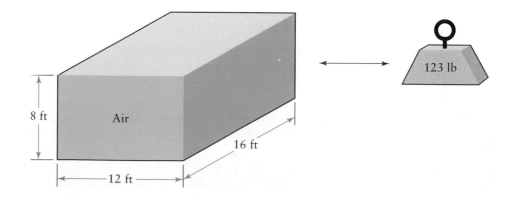

Traditionally, the term *density* has referred to mass density for those using the metric system and to weight density for those using the English system. When comparing the densities of different substances, yet another quantity is often used, the **specific gravity.** The specific gravity of a substance is the ratio of its density to the density of water. Column 5 of Table 4.4 shows the specific gravities of the substances. Diamond is 3.4 times as dense as water, so its specific gravity is 3.4. This means that a certain volume of diamond will have 3.4 times the mass and 3.4 times the weight of an equal volume of water.

Water is used as the reference for specific gravities simply because it is such an important and yet common substance in our world. The density of water is another example of a "natural" unit of measure like the atmosphere (atm) and the acceleration of gravity (g). As a matter of fact, the original definition of the unit of mass in the metric system, the gram, was based on the density of water (see the Physics Potpourri, "The Metric System," on p. 11). The gram was defined to be the mass of 1 cubic centimeter of water at 0°C. This is why the density of water is exactly 1,000 kg/m^3.

One final note on the concept of density in general: We have considered only *volume* density—the mass or weight per unit volume. For matter that is primarily two-dimensional, such as carpeting or a drumhead, it is often convenient to use *surface* density—the mass or weight per unit area. Similarly, for strings, ropes, and cables, one can use a *linear* density— the mass or weight per unit length. The basic concept of density is that it is a measure of the concentration of matter. It relates mass or weight to physical size.

LEARNING CHECK

1. The mass density of an object equals its _____ divided by its _____.
2. (True or False.) An alcoholic beverage is a mixture of water and ethyl alcohol. A "stronger" beverage (one with more alcohol in it) is likely to have a higher mass density than a "weaker" one.
3. Solid object A weighs more than solid object B. Consequently, we can conclude that
   a) A's weight density must be greater than B's.
   b) A's volume must be greater than B's.
   c) A's weight density and volume must both be greater than B's.
   d) measurements would be required to determine the cause of the weight difference.
4. The specify gravity of a substance is computed by dividing its density by the density of _____.

ANSWERS: 1. mass, volume   2. False   3. (d)   4. water

4.4 Fluid Pressure and Gravity

A fluid is any substance that flows readily. All gases and liquids are fluids, as are plasmas. Granulated solids such as salt or grain can be considered fluids in some situations because they can be poured and made to flow.

Fluids are very important to us: Without air and water, life as we know it would be impossible. In the remainder of this chapter, we will discuss some of the properties of fluids in general. Usually, the statements will apply to both liquids and gases, but the fact that gases are compressible and liquids are not is sometimes important.

We live in a sea of air—the atmosphere. Though we are usually not aware of it, the air exerts pressure on everything in it. This pressure varies with altitude and can cause one's ears to "pop" when riding in a fast elevator. When swimming underwater, the same phenomenon can occur if you go deeper under the surface. In both the atmosphere and underwater, the pressure is caused by the force of gravity. Without gravity, air and water would not be pulled to the Earth. They would simply float off into space, just as we would. The pressure in any fluid increases as you go deeper in it and decreases when you rise through it.

Before considering the exact relationship between depth and pressure, we state two general properties of pressures in fluids. First, fluid pressures act in all directions. When you put a hand under water, the pressure acts not only on the top of your hand, but also on the sides and bottom. Second, the force of gravity causes the pressure in a fluid to vary with depth only, not with horizontal position. In liquids, the pressure depends on the vertical distance from the surface and is independent of the shape of the container.

Explore It Yourself 4.4

Take an empty water or soft drink container (plastic or aluminum) and punch three round holes in the side, one near the bottom, another halfway up the side, and the third near the top. Hold it under a faucet over a sink and run water into it just fast enough to keep it full as water runs out through the holes. What is different about the streams of water coming from the three holes? What does that tell you about the pressure inside the container at the levels of these holes?

One can illustrate both of these principles by filling a rubber boot with water. When holes are punched in the boot, the pressure causes water to run out (● Figure 4.24). Water will run out of holes in the top, sides, and bottom of the toe because the pressure acts in all directions on the inner surface. Water comes out faster from holes that are farther below the surface because the pressure is greater. The speed is the same for all holes that are at the same level, regardless of their orientation (up, down, or sideways) and their location (toe, heel, etc.). This is because the pressure in a particular fluid depends only on the depth, not on the lateral position.

The following law explains how the pressure in a fluid is related to gravity.

● Figure 4.24 A rubber boot is filled with water. The pressure of the water acts on all parts of the inner surface of the boot and forces water out of any hole punched in the boot. The speed of the water coming out of a hole depends on how far the hole is below the surface of the water.

© Vern Ostdiek

LAWS

Law of Fluid Pressure The (gauge) pressure at any depth in a fluid at rest equals the weight of the fluid in a column extending from that depth to the "top" of the fluid divided by the cross-sectional area of the column.

This law is as much a prescription for determining the pressure in a fluid as it is a description of what causes pressure in a fluid. For liquids, we can use it to derive the simple relationship between pressure and depth. Let's say a tank is filled with a liquid so that the bottom is some distance h below the liquid's surface. On the bottom, we look at a rectangular area that has length l and width w (● Figure 4.25). All of the liquid directly above the rectangle is in a column with dimensions l by w by h. The weight of this liquid pushes down on the rectangle. This causes a pressure on the bottom that is equal to the weight of the liquid in the column divided by the area of the rectangle.

$$p = \frac{F}{A} = \frac{W}{A} = \frac{\text{weight of liquid}}{\text{area of rectangle}}$$

● Figure 4.25 A tank is filled with a liquid to a depth h. Any rectangular area on the bottom supports the weight of all of the fluid directly above it. So the pressure on the bottom equals the weight of the liquid in the column divided by the area of that rectangle.

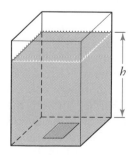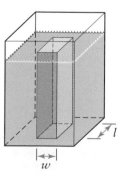

It does not matter what the actual area is: A larger rectangle will have a proportionally larger amount of liquid in the column above. The actual height of the column of liquid is what determines the pressure. We compute the pressure using the fact that the weight of the liquid equals the weight density D_W of the liquid times the volume V of the column.

$$F = W = D_W \times V = D_W \times l \times w \times h$$

$$A = \text{area of rectangle} = l \times w$$

So:

$$p = \frac{W}{A} = \frac{D_W \times l \times w \times h}{l \times w}$$

$$p = D_W h \qquad \text{(gauge pressure in a liquid)}$$

Or, because $D_W = Dg$:

$$p = Dgh \qquad \text{(gauge pressure in a liquid)}$$

This gives the pressure due to the liquid above. If there is also pressure on the liquid's surface (like atmospheric pressure), this will be passed on to the bottom as well. Also, there is nothing special about the bottom: At any level above the bottom, the liquid above exerts a force and, therefore, pressure on the liquid below that level. We can summarize our result as follows.

PRINCIPLES

In a liquid, the absolute pressure at a depth h is greater than the pressure at the surface by an amount equal to the weight density of the liquid times the depth.

$$p = D_w h = Dgh \qquad \text{(gauge pressure in a liquid)}$$

Example 4.6

Let's calculate the gauge pressure at the bottom of a typical swimming pool—one that is 10 feet (3.05 meters) deep. The gauge pressure at that depth, using Table 4.4, is

$$p = D_W h = 62.4 \text{ lb/ft}^3 \times 10 \text{ ft}$$

$$= 624 \text{ lb/ft}^2$$

To convert this to psi, we use the fact that 1 square foot equals 144 square inches.

$$p = 624 \frac{\text{lb}}{\text{ft}^2} = 624 \frac{\text{lb}}{144 \text{ in.}^2}$$

$$= 4.33 \text{ lb/in.}^2 = 4.33 \text{ psi (gauge pressure)}$$

The absolute pressure is this pressure plus the atmospheric pressure, 14.7 psi at sea level. (Absolute pressure = 4.33 psi + 14.7 psi = 19.03 psi.)

At a depth of 20 feet, the gauge pressure would be twice as large, 8.66 psi. If we do this calculation in SI units:

$$p = Dgh = 1{,}000 \text{ kg/m}^3 \times 9.8 \text{ m/s}^2 \times 3.05 \text{ m}$$

$$= 29{,}900 \text{ Pa}$$

The general result for the increase in pressure with depth in water is

$$p = 0.433 \text{ psi/ft} \times h \qquad \text{(for water, } h \text{ in ft, } p \text{ in psi)}$$

For every 10 feet of depth in water, the pressure increases 4.33 psi. ● Figure 4.26 shows a graph of the pressure underwater versus the depth. It is a straight line, since the pressure is proportional to the depth. In seawater, the density is slightly higher, and the pressure increases 4.47 psi for every 10 feet. Submarines and other devices that operate underwater

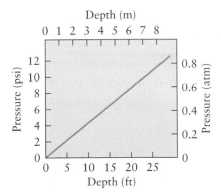

● **Figure 4.26** Graph of the (gauge) pressure underwater versus the depth. The pressure increases 0.433 psi with every foot of depth. All liquids have graphs with the same shape, but the slope is greater if the density is higher.

must be designed to withstand high pressures. For example, at a depth of 300 feet in sea-water, the pressure is 134 psi ($p = 0.447 \times 300$). The force on each square foot of a surface is over 9 tons.

Example 4.7   At what depth in pure water is the gauge pressure 1 atmosphere?

$$p = 0.433 \text{ psi/ft} \times h$$

$$14.7 \text{ psi} = 0.433 \text{ psi/ft} \times h$$

$$\frac{14.7 \text{ psi}}{0.433 \text{ psi/ft}} = h$$

$$h = 33.9 \text{ ft} = 10.3 \text{ m}$$

Since mercury is 13.6 times as dense as water, the gauge pressure is 1 atmosphere at a depth of:

$$h = \frac{33.9 \text{ ft}}{13.6} = 2.49 \text{ ft} = 29.9 \text{ in.} = 0.76 \text{ m} \qquad (\bullet \text{ Figure 4.27})$$

Fluid Pressure in the Atmosphere

The law of fluid pressure has a simple form in liquids. For gases, things are a bit more complicated. We know that the density of a gas depends on the pressure. At greater depths in a gas, the increased pressure causes increased density. The total weight of a vertical column of gas can be computed only with the aid of calculus.

The Earth's atmosphere is a relatively thin layer of a mixture of gases. The decrease in air pressure with altitude is further complicated by variations in the temperature and composition of the gas. The heating of the air by the Sun, the rotation of the Earth, and other factors cause the air pressure at a given place to vary slightly from hour to hour. Also, the pressure is not always the same at points with the same altitude.

In spite of the complexity of the situation, we can make some statements about the general variation of air pressure with altitude. Because the atmosphere is a gas, there is no upper surface or boundary. The air keeps getting progressively thinner—there are fewer

● **Figure 4.27** The pressure at the bottom of each column is 1 atmosphere.

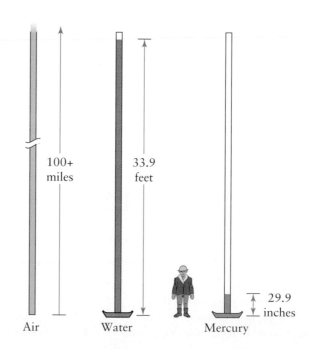

100+ miles — Air

33.9 feet — Water

29.9 inches — Mercury

molecules per unit volume—as you go higher in the atmosphere. At 9,000 meters (30,000 feet) above sea level, the density of the air is only about 35% of that at sea level. At this altitude, the average person cannot remain conscious because there is not enough oxygen in each breath. At about 160 kilometers (100 miles) above sea level, the density is down to one-billionth of the sea-level density. This is often regarded as the effective upper limit of the atmosphere. Spacecraft can remain in orbit at this altitude because the thin air causes very little air resistance.

● Figure 4.28 is a graph of the air pressure versus altitude for the lower atmosphere. (When comparing this graph with Figure 4.26, remember that the depth in a liquid is measured downward from the surface, but the height in the atmosphere is measured upward from sea level. The pressure rapidly decreases with increasing height at low altitudes where the density is still fairly high. Remember that the pressure at any elevation depends on the weight of all of the air above. At 9,000 meters, the pressure is about 0.3 atmospheres. This means that only 30% of the air is above 9,000 meters. Even though the atmosphere extends over 100 miles up, most of the air (70%) is less than 6 miles up.

Air pressure is measured with a *barometer.* The simplest type is the mercury barometer. It consists of a vertical glass tube with its lower end immersed in a bowl of mercury. All of the air is removed from the tube, so there is no air pressure acting on the surface of the mercury in the tube (see ● Figure 4.29). The air pressure on the mercury in the bowl is transmitted to the mercury in the tube. This forces the mercury to rise up in the tube. (When drinking through a straw, you reduce the pressure in your mouth, and the atmospheric pressure forces the drink up the straw.) The mercury will rise in the tube until the pressure at the base of the tube equals the air pressure. Hence, the air pressure is determined by measuring the height of the column of mercury.

When the air pressure is 1 atmosphere (14.7 psi), the column of mercury is 760 millimeters (29.9 inches) long. At lower pressures, the column is shorter, so we can use the length of the column of mercury as a measure of the pressure (760 millimeters of mercury equals 1 atmosphere). Other liquids will work, but the tube has to be much longer if the liquid's density is small. For example, it would take a column of water 10.3 meters (33.9 feet) long to produce, and therefore measure, a pressure of 1 atmosphere (Figure 4.27).

This unit of measure is not limited to measuring air pressure. Blood pressure is also given in millimeters of mercury. If your blood pressure is 100 over 60, it means that the pressure drops from 100 millimeters of mercury during each heartbeat to 60 millimeters of mercury between beats.

A more portable type of barometer consists, more or less, of a very short metal can with the air removed from the inside (see ● Figure 4.30). The air pressure causes the ends to be squeezed inward. The higher the pressure, the greater the distortion of the ends. A pointer is attached to one end and moves along a scale as the distortion of the end changes. This is called an *aneroid barometer.*

The variation of air pressure with height is used to measure the altitude of aircraft. An *altimeter* is an aneroid barometer with a scale that registers altitude instead of pressure (top

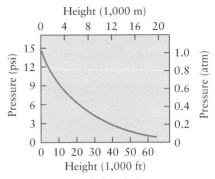

● **Figure 4.28** Graph of the absolute air pressure versus height above sea level. The graph is not a straight line because the density decreases with altitude. The pressure never quite reaches zero.

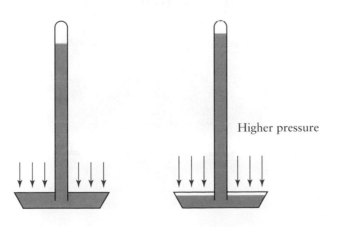

Higher pressure

● **Figure 4.29** A mercury barometer. If there is no air in the tube, the air pressure on the mercury in the bowl forces the mercury to rise up the tube. The pressure at the bottom of the tube equals the air pressure. The higher the air pressure, the higher the column of mercury.

Figure 4.30 Schematic of an aneroid barometer. With higher atmospheric pressure, the top of the can is pushed in more, causing the pointer to indicate higher pressure.

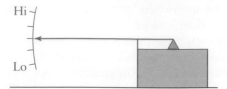

Higher pressure

of Figure 4.31). For example, if an aneroid barometer in an airplane indicated the pressure was 0.67 atmospheres, one could infer from Figure 4.28 that the altitude was 10,000 feet. Each pressure reading on the barometer would correspond to a different altitude.

A more sophisticated instrument is used to measure the vertical speed of an aircraft—how fast it is going up or down. This device, called a *vertical airspeed indicator* in airplanes (bottom of Figure 4.31) and a *variometer* in gliders, senses changes in the air pressure. When the aircraft is going up, the measured air pressure decreases. The instrument converts the rate of change of the air pressure into a vertical speed.

The law of fluid pressure applies to granulated solids in much the same way that it does to liquids. The walls of storage bins and grain silos are reinforced near the bottom because the pressure is higher there. Again, it is the force of gravity pulling on the material above that causes the pressure.

Explore It Yourself 4.5

Fill a glass with water, hold a piece of cardboard or stiff paper on the top with your hand, and invert it. You might want to do this over a sink. Make sure there is a good seal all around the top of the glass before you turn it over. When you remove your hand, why doesn't the water fall out?

Figure 4.31 An altimeter (top) measures altitude by measuring the air pressure. A vertical airspeed indicator (bottom) measures vertical speed by measuring how rapidly the air pressure is increasing or decreasing.

Courtesy of Dakota Ridge Aviation, Inc., Boulder, CO

1. (True or False.) Gravity causes the pressure in the ocean to vary with depth.
2. Two aquariums are completely full of water, but the pressure at the bottom of aquarium A is greater than at the bottom of aquarium B. The cause of this could be
   a) the water in A is deeper than it is in B.
   b) A contains freshwater and B contains seawater.
   c) A is shaped like a box and B is shaped like a cylinder.
   d) any of the above.

3. (True or False.) The atmospheric pressure at 5,000 m above the ground is exactly twice as large as it is at 10,000 m above the ground.
4. A barometer is a device that measures

   _____ .

4.5 Archimedes' Principle

The force of gravity causes the pressure in a fluid to increase with depth. This in turn causes an interesting effect on substances partly or totally immersed in a fluid. In some cases, the substance floats—wood or oil on water, blimps or hot-air balloons in air. In other cases, the substance seems lighter—a 100-pound rock can be lifted with a force of about 60 pounds when it is underwater. Obviously, some force acts on the foreign substance to oppose the downward force of gravity (weight). This force is called the **buoyant force.**

> **DEFINITION**
>
> **Buoyant Force** The upward force exerted by a fluid on a substance partly or completely immersed in it.

This force acts on anything immersed in any gas or liquid. As long as there are no other forces acting on the substance, there are three possibilities (● Figure 4.32). If the buoyant force is *less than* the weight of the substance, it will *sink*. If the buoyant force is *equal* to the weight, the substance will *float*. If the buoyant force is *greater than* the weight of the substance, it will *rise* upward. Examples of each case (in the same order) are a rock in water, a piece of wood floating on water, and a helium-filled balloon rising in air. Both the rock and the balloon experience a net force because the weight and the buoyant force do not cancel each other. This net force causes each to accelerate momentarily until the force of friction offsets the buoyant force and a terminal speed is reached.

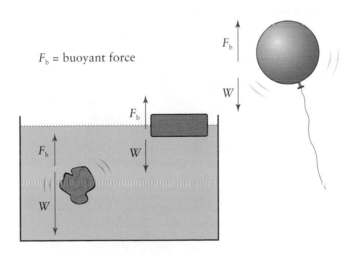

F_b = buoyant force

● **Figure 4.32** The rock sinks because the buoyant force on it is smaller than its weight. The wood floats because the buoyant force on it is equal to its weight. The helium balloon rises because the buoyant force on it is greater than its weight.

● **Figure 4.33** Setup for Explore It Yourself 4.6.

Explore It Yourself 4.6

For this you need a glass or ceramic coffee mug, a rubber band, and a sink with several inches of water standing in it. Loop the rubber band around the handle so that the cup hangs sideways (Figure 4.33). The rubber band should be strong enough to support the mug while also stretching noticeably.

1. Measure how long the stretched rubber band is.
2. Lower the mug into water, making sure no air is trapped in the cup and it doesn't touch anything but water. Measure how long the rubber band is now. What has happened?

At this point, we might pose two questions. First, what causes the buoyant force? Second, what determines the magnitude of the buoyant force? The answer to the first question can be arrived at rather simply in light of the previous section. Consider an object that is completely immersed (● Figure 4.34). Since its bottom surface is deeper in the fluid than its top surface, the pressure on the bottom surface will be greater. Hence the upward force on its bottom surface caused by this pressure is greater than the downward force on its top surface. In short, the difference in fluid pressure acting on the surfaces of the object causes a net upward force. (The forces on the sides of the object are equal and opposite, so they cancel out.)

When something floats on the surface of a liquid, only its lower surface experiences the fluid pressure of the liquid. This causes an upward force.

As for the second question, the size of the buoyant force is determined by a law formulated in the third century B.C. by the Greek scientist Archimedes (see Physics Family Album, Chapters 3 and 4).

PRINCIPLES

> **Archimedes' Principle** The buoyant force acting on a substance in a fluid at rest is equal to the weight of the fluid displaced by the substance.
>
> $$F_b = \text{weight of displaced fluid*}$$

When a piece of wood is placed in water, it displaces some of the water: Part of the wood occupies the same space or volume formerly occupied by water. The weight of this displaced water equals the buoyant force acting on the wood. Obviously, any object that is completely submerged in a fluid will displace a volume of fluid equal to its own volume.

---

* For the mathematically inclined—in the simple case of a box-shaped object immersed (but level) in a liquid, we can prove Archimedes' principle. Assume the box's dimensions are l, w, and h. The downward force on the top of the box is

$$F_{\text{top}} = p_{\text{top}} \times A_{\text{top}} = p_{\text{top}} \times l \times w$$

Here p_{top} is the pressure at the top of the box. At the bottom of the box, the pressure is higher because it is deeper in the fluid. Specifically, the pressure there, p_{bottom}, is

$$p_{\text{bottom}} = p_{\text{top}} + (D_W \times h)$$

where D_W is the weight density of the liquid. Therefore, the upward force on the bottom is

$$F_{\text{bottom}} = p_{\text{bottom}} \times l \times w = [p_{\text{top}} + (D_W \times h)] \times l \times w$$
$$= (p_{\text{top}} \times l \times w) + (D_W \times h \times l \times w)$$

The buoyant force F_b is the upward force minus the downward force

$$F_b = F_{\text{bottom}} - F_{\text{top}} = (p_{\text{top}} \times l \times w) + (D_W \times h \times l \times w) - (p_{\text{top}} \times l \times w)$$
$$= D_W \times h \times l \times w = D_W \times V = W$$
$$= \text{weight of fluid displaced}$$

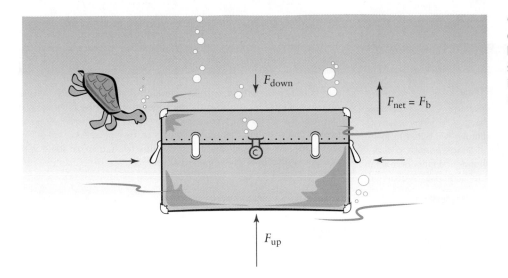

● **Figure 4.34** The fluid pressure causes a force on each surface of an immersed object. The pressure on the lower surface is greater than the pressure on the upper surface. Consequently, the upward force on the bottom is greater than the downward force on the top. The net upward force is the buoyant force.

If the buoyant force on an object is less than its weight, it sinks, but the *net* downward force is reduced. ● Figure 4.35 shows an object hanging from a scale. Its weight is 10 newtons. As the object is lowered into a beaker of water, it displaces some of the water. The scale reading is reduced by an amount equal to the buoyant force. When the scale shows 6 newtons, the buoyant force is 4 newtons.

$$\text{scale reading} = \text{weight} - \text{buoyant force}$$

$$6\,\text{N} = 10\,\text{N} - 4\,\text{N}$$

The weight of the water that spills over the side of the beaker is also 4 newtons. If the object is lowered farther into the water, the buoyant force will increase, and the scale reading will decrease.

Notice that the buoyant force acting on a substance doesn't depend on what the substance is, only on how much fluid it displaces when immersed. Identical balloons filled to the same size with helium, air, and water all have the same buoyant force acting on them. Only the helium-filled balloon floats in air because its weight is less than the buoyant force. Also, the weight of a substance alone does not determine whether or not it will float. A tiny pebble will sink in water, but a 2-ton log will float.

● **Figure 4.35** The weight of an object is 10 newtons. When the object is immersed in water, the scale reads a smaller force because of the upward buoyant force. The buoyant force is equal to the weight of the water that the object displaces.

Superfluids—Friction-free Flow

Supermarkets. Superbowls. Superglue. Superhighways. Superstars. Superfluids? The prefix *super-* has been attached to many nouns to indicate qualities that go beyond those ordinarily attributed to them. Its use draws attention to the fact that these things possess a *super-*abundance of the usual kinds of properties that define them or that the defining characteristics are somehow *super*ior to those of common, garden-variety examples of these entities. Indeed, the word super comes from the Latin, meaning "over" or "above."

Taken at face value, then, we might expect that a *super*fluid would be one whose "fluidity" is more pronounced than in ordinary cases, whose ability to flow, for example, is heightened over that of common fluids. And this is basically true. In technical terms, we speak of a superfluid as having no friction or resistance to flow—that is, no *viscosity.* Unlike ordi-nary fluids, a superfluid can flow through narrow channels with no pressure difference. It also apparently defies gravity by flowing as a film over the edges of containers and by spouting dramatically when heated (the so-called *thermomechanical* or *fountain effect;* see ● Figure 4.36). As we shall discuss later, superfluids display some very interesting rotational phenomena as well.

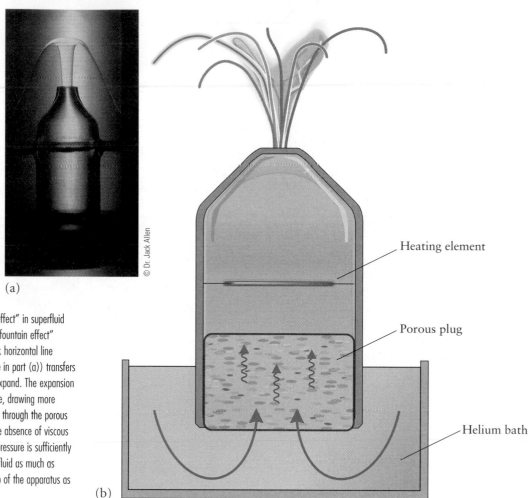

(a)

© Dr. Jack Allen

● **Figure 4.36** (a) The "fountain effect" in superfluid helium. (b) Simplified diagram of the "fountain effect" apparatus. A thin heating element (dark horizontal line crossing the middle of the capillary tube in part (a)) transfers energy to the superfluid, causing it to expand. The expansion pushes the liquid out the top of the tube, drawing more liquid in from the surrounding fluid bath through the porous plug near the bottom of the tube. In the absence of viscous friction with the walls of the tube, the pressure is sufficiently large that a continuous stream of superfluid as much as 15 cm high can be ejected from the top of the apparatus as long as sufficient heat is supplied.

(b)

Heating element

Porous plug

Helium bath

The key is density. Let's take the simple case of a solid object submerged in a liquid. The volume of the liquid displaced by the object is, of course, equal to the object's volume. So:

weight of object = weight density (of object) × volume

$$W = D_W(\text{object}) \times V$$

The buoyant force is

buoyant force = weight of displaced fluid   (Archimedes' principle)

$$F_b = \text{weight density (of fluid)} \times \text{volume}$$

$$= D_W (\text{fluid}) \times V$$

Discovered in 1938 by P. Kapitza (who received a 1978 Nobel Prize for his work) and J. Allen, superfluids are a subclass of what are called *quantum fluids*. (Superconductors, discussed in Chapter 7, are also members of this broad class.) They exhibit striking macroscopic effects due to special microscopic ordering. To see these effects requires that the substance under investigation remain in a liquid state at temperatures that are low enough to permit the ordering to occur. For most elements and compounds, the needed temperatures are *so* low that the substances condense into the solid phase of matter before reaching these critical temperatures. An exception to this behavior is the element helium, which condenses into a normal liquid at around −269°C (and 1 atmosphere pressure) and then undergoes another "phase" transition to the superfluid state below about −271°C.

The approach of a system of particles to superfluidity is associated with the onset of what is termed *Bose–Einstein condensation* in the liquid, in which a macroscopic quantity of liquid particles all come to occupy the same microscopic (quantum) energy state. (For more on quantum physics, see Chapter 10.) An analogy given by University of Illinois physicist Anthony Leggett, who shared the 2003 Nobel Prize in physics, may be helpful here in understanding what is going on.

Imagine you are atop a high mountain overlooking a distant city square on market day. As you examine the shopping crowd, you see people milling about in all directions seemingly at random: Each individual is doing his or her "own thing." Suppose you return to the mountain and inspect the town square on the day of a parade when the milling crowd is replaced by a marching band or a precision drill team. Now every member of the group is doing the same thing at the same time, and it's much easier to see and understand what is going on. The comparable situation in fluid physics is one in which a normal liquid plays the role of the market-day crowd—every atom doing something different—and where a superfluid is like the drill team—all the atoms forced into the same state, every atom doing exactly the same thing at the same time. A superfluid, then, acts like one large collective atom instead of billions of independent individual atoms. This collective action makes the effects of this microscopic phase change noticeable on macroscopic levels.

How does this collective condensation account for flow without friction? Basically, viscosity is absent in a superfluid because it costs the fluid too much energy to have all of its atoms respond together to the frictional forces. Imagine the following experiment: Put some liquid helium at a temperature above the superfluid transition temperature in a cup, and then place the container on the axis of a phonograph turntable. Start the turntable spinning. The normal helium liquid will,

after a few minutes, begin to rotate with the container. Next, cool the liquid below the transition temperature to a superfluid state. The fluid will continue to rotate, with all the helium atoms occupying the same energy state, equal to the rotational energy of the cup. Now stop the rotation of the container. The superfluid helium will continue to spin, unlike a normal fluid for which friction with the walls of the cup would gradually bring it to a halt. This is a graphic illustration of the ineffectiveness of friction to act on a superfluid. The Bose–Einstein condensed helium atoms remain rotating because to collectively pass to a nonrotating state would require an enormous investment of additional energy, which the system doesn't have. The key to this behavior is the cooperative interaction of all the helium atoms. If only a few individual atoms were involved, the energy required to respond to the frictional forces could be found, but for 10^{23} atoms it cannot!

Superfluidity is an important topic of pure research, and increasingly for practical application. For example, in the former category, the study of superfluids in the laboratory will have significance for our understanding of the interior structure of neutron stars, which are thought to be in superfluid states (see p. 120). In addition, because superfluids represent systems where gravitational and electromagnetic effects are minimized, investigating them may elucidate the characteristics of the weak nuclear force (see Chapter 12) in macroscopic domains. The discovery and characterization of superfluidity, in a form of helium called *helium-3* at the incredibly low temperature of only 0.002 degrees above the lowest possible temperature by David Lee, Robert Richardson, and Douglas Osherhoff won them the Nobel Prize in physics in 1996.

On the practical side, the excellent heat transfer properties of superfluids have already been exploited in cooling other systems to temperatures below −270°C. Moreover, the friction-free flow characteristics of superfluid helium have been used to measure the rotation of the Earth and may soon be used in gyroscopes as part of spacecraft navigation systems. Recent studies at the University of California at Berkeley have also opened up the possibility of using oscillations in superfluid helium to define an international standard of pressure based on precisely known quantum-mechanical constants.

Check out the World Wide Web to see what other potential applications exist to justify the use of the prefix *super-* to describe these fluids using some of the key words provided in this Physics Potpourri. For an interesting place to start, see the on-line article in *Physics Today* (Vol. 48, No. 7, July 1995, p. 30) by R. J. Donnelly, who gives a highly readable account of the discovery and early history of superfluids. Do it now! Don't let the opportunity slip away!

From this, we conclude that if the weight density of the object is greater than the weight density of the fluid, the object's weight is greater than the buoyant force. The object sinks. If its density is less than the fluid's density, it floats. By simply comparing the densities (or specific gravities) of the substance and the fluid, one can determine whether or not the substance will float. Any substance with a smaller density than that of a given fluid will float in (or on) that fluid.

In Table 4.4 we see that juniper, ice, gasoline, ethyl alcohol, and all of the gases will float on water because their densities are less than that of water. (Actually, the alcohol would simply mix in with the water. In the case of a fluid being immersed in another fluid, it is best to imagine the first being enclosed in a balloon or some other container that has negligible weight but keeps the fluids from mixing.)

● **Figure 4.37** A steel ball floats in mercury because the density of mercury is greater than that of steel.

The density of mercury is so high that everything in the table except gold and uranium will float on it (● Figure 4.37). Hydrogen and helium both float in air, while radon and carbon dioxide sink.

Ships and blimps float, even though they are constructed out of materials with higher densities than the fluids in which they float. They are composites of substances: Most of the volume of a ship is occupied by air, and most of the volume of a blimp is occupied by helium. In both cases, the *average* density of the object is less than the density of the fluid.

As a ship is loaded with cargo, it sinks lower into the water. This makes it displace more water, thereby increasing the buoyant force and countering the added weight (see ● Figure 4.38).

Archimedes' principle is routinely used to measure the densities or specific gravities of solids that sink. The object is hung from a scale, and the reading is recorded. (It does not matter whether the scale gives mass or weight. Here we assume it reads weight.) Then the object is completely immersed in water, and the new scale reading is recorded (see Figure 4.35). The *difference* in these two readings equals the weight of the displaced water, by Archimedes' principle.

W of displaced water = scale reading out of water − scale reading in water

W of displaced water = scale (out) − scale (in)

So the weight of the object is known, and the weight of an *equal* volume of water is also known. Density is weight divided by the volume. Since the volumes are the same, the density of the object divided by the density of the water *equals* the weight of the object divided by the weight of the water.

$$\frac{\text{density of object}}{\text{density of water}} = \frac{\text{scale reading out of water}}{\text{scale (out)} - \text{scale (in)}}$$

● **Figure 4.38** A loaded ship rides lower in the water because it must displace more water to have a larger buoyant force.

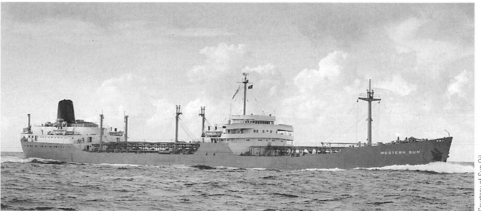

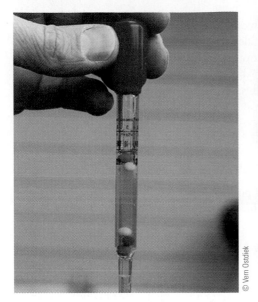
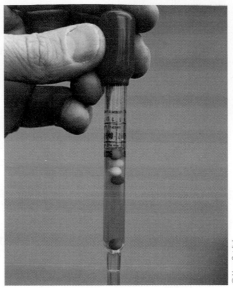

• **Figure 4.39** The number of balls that float in this antifreeze tester depends on the density of the coolant. Higher density coolant (right) has a lower freezing point.

This ratio is just the specific gravity of the object.

$$\text{specific gravity} = \frac{\text{scale (out)}}{\text{scale (out)} - \text{scale (in)}}$$

In Section 4.3, we described how the density of the coolant in an automobile engine indicates the antifreeze content. The density is measured with a simple device that employs Archimedes' principle. • Figure 4.39 shows an antifreeze tester that consists of a narrow glass tube containing five balls. The balls are specially made so that each one has a slightly higher density than the one above it. The coolant is drawn up into the glass tube with a suction bulb. Each ball will float if its density is less than the density of the coolant. The number of balls that float depends on the density of the coolant. If the antifreeze content is high, the coolant's density is high, and all of the balls float.

To check the hemoglobin content of blood, a drop of it is placed in a liquid that has the correct minimum density. If the blood sinks, its density is greater than that of the liquid, and the hemoglobin content is high enough.

The following examples use Archimedes' principle.

Example 4.8

A contemporary Huckleberry Finn wants to construct a raft by attaching empty, plastic, 1-gallon milk jugs to the bottom of a sheet of plywood. The raft and passengers will have a total weight of 300 pounds. How many jugs are required to keep the raft afloat on water?

The buoyant force on the raft must be at least 300 pounds. Consequently, the raft must displace a volume of water that weighs 300 pounds.

$$F_b = \text{weight of water displaced}$$

$$= D_W \,(\text{water}) \times \text{volume of water displaced}$$

$$300\text{ lb} = 62.4\text{ lb/ft}^3 \times V$$

$$V = \frac{300\text{ lb}}{62.4\text{ lb/ft}^3}$$

$$1.0\text{ ft}^3$$

To keep 300 pounds afloat, 4.8 cubic feet of water must be displaced. One cubic foot equals 7.48 gallons. So $4.8 \times 7.48 =$ about 36 one-gallon jugs.

Example 4.9

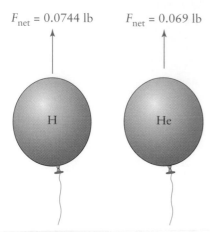

$F_{net} = 0.0744$ lb $F_{net} = 0.069$ lb

H   He

● **Figure 4.40** One cubic foot of hydrogen can lift 0.0744 pounds. One cubic foot of helium can lift 0.069 pounds.

Before the spectacular and tragic destruction of the German airship *Hindenburg* on 6 May 1937, blimps, zeppelins, and balloons were filled with hydrogen. Now helium is used. Let's compare the two gases in terms of their buoyancy in air. Each cubic foot of hydrogen gas weighs 0.0056 pounds at 0°C and 1 atm (Table 4.4). When in an airship, this hydrogen will displace a cubic foot of air. So each cubic foot of hydrogen sustains a buoyant force of 0.08 pounds, the weight of 1 cubic foot of air. Therefore, the net force on each cubic foot of hydrogen gas is

$$\text{net force} = 0.08 \text{ lb} - 0.0056 \text{ lb}$$

$$F = 0.0744 \text{ lb} \qquad \text{(hydrogen)}$$

Each cubic foot of hydrogen can lift 0.0744 pounds (see ● Figure 4.40). If helium is used instead, the buoyant force is still the same, but the weight of each cubic foot of helium is 0.011 pounds. So:

$$\text{net force} = 0.08 \text{ lb} - 0.011 \text{ lb}$$

$$F = 0.069 \text{ lb} \qquad \text{(helium)}$$

Each cubic foot of hydrogen can lift 8% more than a cubic foot of helium. A balloon filled with hydrogen can lift 8% more than an identical balloon filled with helium. However, the big factor that tilts the scale in favor of helium is that it does not burn, as hydrogen does.

The air exerts a buoyant force on everything in it, not just balloons. This force, 0.08 pounds for each cubic foot, is so small that it can be ignored except when dealing with other gases. For example, the volume of a person's body is typically around 2 or 3 cubic feet. (The density of the human body is about the same as that of water. That is why we just barely float in water. So the approximate volume of your body is your weight divided by the weight density of water.) This means that the buoyant force on a person due to the air is between 0.16 and 0.24 pounds.

Explore It Yourself 4.7

(You have to do this one outside—and no smoking!) Fill a glass about two-thirds full with equal amounts of water and gasoline. These two liquids do not mix, and you can clearly see that the gasoline does float on the water. Now put an ice cube into the glass. What does the ice cube do? Does this agree with what you might have predicted using information in Table 4.4?

LEARNING CHECK

1. The _____ acting on a boat in water at rest is equal to the weight of the water displaced by the boat.
2. One balloon is filled with pure water and a second is filled with antifreeze. What does each balloon do after being released by a scuba diver beneath the surface of the ocean?
3. As a person climbs into a rowboat, it must sink lower into the water so that
   a) the net force on the boat remains zero.
   b) it displaces more water, consequently increasing the buoyant force on it.

   c) its bottom is deeper in the water where the higher pressure causes a larger upward force.
   d) All of the above.
4. (True or False.) The buoyant force on a hydrogen-filled balloon is larger than that on a helium-filled balloon with the same size.

ANSWERS: 1. buoyant force 2. pure water balloon rises, the other sinks 3. (d) 4. False

4.6 Pascal's Principle

When a force acts on a surface of a solid, the resulting pressure is "transmitted" through the solid only in the original direction of the force. When you sit on a stool, your weight causes pressure in the legs that is transmitted to the floor. This pressure doesn't act to the side of the legs, only downward.

In a fluid, any pressure caused by a force is transmitted everywhere throughout the fluid and acts in all directions. When you squeeze a tube of toothpaste, the pressure is passed on to all points in the toothpaste. An inward force on the sides of the tube can cause the toothpaste to come out of the end. This property of fluids should be familiar to you and might even be used to distinguish fluids from solids. Pascal's principle is a formal statement of this phenomenon.

Pascal's Principle Pressure applied to an enclosed fluid is transmitted undiminished to all parts of the fluid and to the walls of the container.

This property of fluids is exploited in widely used hydraulic systems. Hydraulic jacks and the brake systems in automobiles are common examples. The basic components of such systems are piston and cylinder combinations. ● Figure 4.41 shows a small piston and cylinder on the left connected via a tube to a larger piston and cylinder on the right. (The effect of gravity can be ignored.) A liquid is in both cylinders and the connecting tube. When a force acts on the left piston, the resulting pressure is passed on throughout the liquid. This pressure causes a force on the right piston. Since this piston is larger than the one on the left, the force ($F = pA$) will also be larger. For example, if the area of the right piston is five times that of the left piston, the force will be five times as large. This system behaves like a lever (Figure 3.10). A small force acting at one place causes a larger force at another place. As with the lever, the smaller piston will move a correspondingly greater distance than the larger piston. The work done is the same for both pistons.

In automotive brake systems, the brake pedal is connected to a piston that slides in the "master" cylinder. This cylinder is connected via tubing to "wheel" cylinders on the brakes on the wheels (● Figure 4.42). (Actually, there are two cylinders in tandem in the master cylinder. Each is connected to two wheels so that if one subsystem fails, the other two wheels will still have braking power.)

Brake fluid fills the cylinders and tubing. When the brake pedal is pushed, the piston produces pressure in the master cylinder that is transmitted to the wheel cylinders. The piston in each wheel cylinder is attached to a mechanism that applies the brakes. In disc brakes, used on the front wheels of most cars, the piston squeezes disc pads against the sides of a rotating disc attached to the wheel. This action is very similar to that of rim brakes on bicycles. The mechanism in drum brakes, used on the rear wheels of most cars, is a bit more complicated.

Using a hydraulic brake system serves two purposes. Using wheel cylinders with larger diameters than those of the master cylinder gives a mechanical advantage: The forces on the wheel-cylinder pistons are larger than the force applied to the master-cylinder piston. Also, this is an efficient way to transmit a force from one location in the car to four other locations.

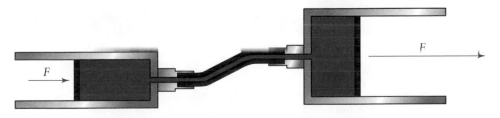

● **Figure 4.41** The pressure caused by the force on the small piston is transmitted throughout the fluid and acts on the larger piston. The resulting force on this piston is larger than the force on the smaller piston.

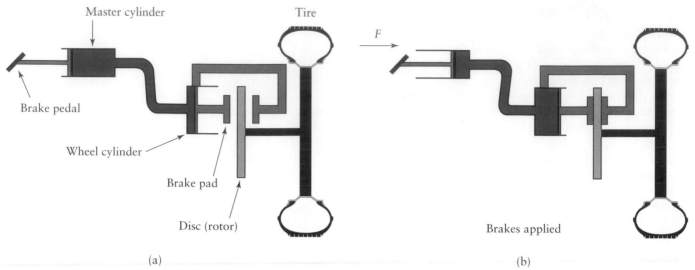

Master cylinder   Tire

Brake pedal

Wheel cylinder

Brake pad

Disc (rotor)

(a)

F

Brakes applied

(b)

● **Figure 4.42** (a) Simplified diagram of an automotive hydraulic brake system. Only one wheel, equipped with disc brakes, is shown. (b) When the brake pedal is pushed, the pressure increase in the master cylinder is passed on through the fluid to the wheel cylinder. The piston is pushed outward, and the brake pads squeeze the disc.

LEARNING CHECK

1. Squeezing the sides of a flexible water bottle to make water come out of the top is an example of exploiting _____ principle.

2. (True or False.) When the pressure in the fluid in two cylinders in a hydraulic system is the same, the forces on the two pistons are not necessarily equal.

ANSWERS: 1. Pascal's   2. True

4.7 Bernoulli's Principle

In this section, we present a simple principle that applies to moving fluids. Water flowing in a stream or through pipes to a water faucet, air moving through heating ducts or as a cool breeze in the summer, blood flowing through your arteries and veins—these are all examples of fluids flowing. It is common for each of these fluids to speed up and slow down as it flows. Accompanying any change in the speed of the fluid is a change in pressure of the fluid. This is stated in the following principle, named after the Swiss physicist and mathematician Daniel Bernoulli.

PRINCIPLES

Bernoulli's Principle For a fluid undergoing steady flow,* the pressure is lower where the fluid is flowing faster.

---

* Steady flow means that there is no random swirling of the fluid and that no outside forces increase or decrease the rate of flow.

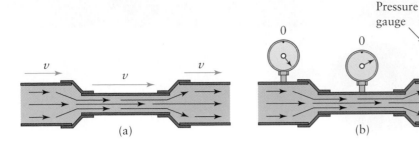

Pressure gauge

(a)   (b)

This principle is based on the conservation of energy. A fluid under pressure has what can be called *pressure potential energy*. The higher the pressure, the greater the potential energy of any given volume of fluid. (When you open a faucet, water rushes out as the potential energy due to pressure is converted into kinetic energy of running water. Low water pressure makes the water come out slowly, with low kinetic energy, because it starts with low potential energy.) Moving fluids have both kinetic and potential energy. When a fluid speeds up, its kinetic energy increases. Its total energy remains constant, so its potential energy and, therefore, its pressure decrease.

One of the best examples of Bernoulli's principle is that depicted in ● Figure 4.43. Water flowing through a pipe passes through a smaller section spliced into the pipe. The water speeds up when it enters the narrow region, and then slows down when it reenters the wide region. This occurs because the volume of fluid passing through each part of the pipe each second is the same. Where the cross-sectional area is smaller, the fluid must flow faster if the same number of cubic inches of fluid is to get through each second. You've seen this already if you've ever used your thumb to partially block water coming out of the end of a garden hose.

Bernoulli's principle tells us that the pressure in the moving water is *smaller in the narrow section* than in the wide sections upstream and downstream. Pressure gauges placed in the pipe show this to be so. This is one of those rare situations in which physical fact runs counter to one's intuition. On first thought, most people would predict that the pressure should be higher in the narrow part of the pipe because the fluid is "squeezed into a smaller stream." But that is not the case. The pressure is actually lower in the narrow part.

Atomizers on perfume bottles utilize Bernoulli's principle. When a bulb is squeezed, air moves through a horizontal tube (● Figure 4.44). The air moves fast, so the air pressure is low. A small tube runs from the horizontal tube down into the perfume. Since the air pressure is reduced, the normal air pressure acting on the surface of the perfume forces the liquid to rise upward in the tube and to enter the moving air. Carburetors used on lawn mowers and the automatic shutoff mechanism on gas pump nozzles also make use of Bernoulli's principle.

Concept Map 4.2 summarizes this chapter's concepts related to pressure.

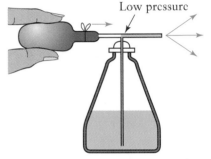

Low pressure

● **Figure 4.44** Perfume atomizers use Bernoulli's principle. Air is forced to move through the horizontal tube when the bulb is squeezed. In this tube, the air has to move fast, so the pressure is low. Normal air pressure on the perfume in the bottle forces it to rise upward into the stream.

Explore It Yourself 4.8

Hold the top of a piece of paper horizontally just below your lips, so that the paper hangs limp (see ○ Figure 4.45). Blow hard over the top of the paper. What does the paper do? What causes this?

© Vern Ostdiek

● **Figure 4.45** Angie is set to illustrate Bernoulli's principle.

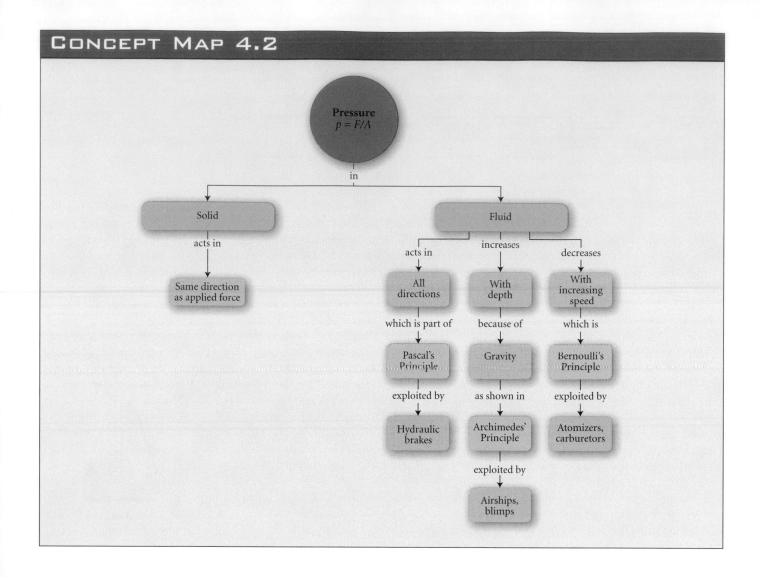

Pressure
$p = F/A$

in

Solid — acts in → Same direction as applied force

Fluid

acts in → All directions → which is part of → Pascal's Principle → exploited by → Hydraulic brakes

increases → With depth → because of → Gravity → as shown in → Archimedes' Principle → exploited by → Airships, blimps

decreases → With increasing speed → which is → Bernoulli's Principle → exploited by → Atomizers, carburetors

LEARNING CHECK

1. Bernoulli's principle is exploited by devices such as _____.

2. (True or False.) At a place in a particular ventilation duct where the air pressure is lowest, the air is likely to be moving fastest.

ANSWERS: 1. perfume atomizer or carburetor 2. True

The Greek philosopher Democritus, who lived about a century before Aristotle, developed the theory that all matter is composed of particles too small to be seen. He called these particles *atoms*, from the Greek word for *indivisible*, because he thought them to be the final result of any repeated subdivision of matter. The atomic theory as expounded by Democritus and other Greek philosophers was not accepted by Aristotle. For this and other reasons, the atomic theory was abandoned for more than 2,000 years.

The successes of Galileo and Newton opened a new era of experimentation and advancement in all of the sciences. The development of chemistry in the seventeenth and eighteenth centuries revived the atomic theory because it could explain many of the new discoveries. Robert Boyle (1626–1691), an Irish-born physical scientist, identified "elements" as those substances that could not be further decomposed by chemical analysis. A century later, the French chemist Antoine Lavoisier (1743–1794; see ● Figure 4.46)

● **Figure 4.46** Lavoisier with his wife Marie, who was also his assistant.

refined the concept of the element and showed that mass is conserved during chemical reactions. When paper burns, the total mass of the combustion products, ashes, and smoke is the same as the total mass of the paper and oxygen that were combined during the burning. (Lavoisier was guillotined during the French Revolution because he worked for King Louis XVI.)

During the eighteenth century, several of the elements were correctly classified, iron, gold, mercury, carbon, and oxygen to name a few. Salt and several other compounds were mistakenly identified as elements because they could not be decomposed into their constituent elements with the techniques then available.

In 1808, John Dalton (1766–1844), an English schoolteacher (● Figure 4.47), presented what was essentially the modern atomic theory. He correctly stated that the properties of each element are determined by the properties of its atoms, the atoms of different

elements have different masses, and two or more different atoms in combination form the basic units of chemical compounds. The actual structure of atoms and the fact that they aren't indestructible were only determined during the last 100 years.

Fluids

The scientific study of fluids probably began with Archimedes' discoveries about buoyancy. In his book *On Floating Bodies*, he proves the famous principle that has immortalized his name. Apparently, Archimedes' interest in the topic began when King Hiero of Syracuse presented him with a problem. The king had given a certain weight of gold to a craftsman to be made into a crown. The finished product was presented to the king, but he suspected that some of the gold had been replaced by an equal weight of silver. King Hiero asked Archimedes to find a way to determine the composition of the crown without damaging it. The story goes that Archimedes thought of the solution when he noticed how water was displaced as he sank into a bathtub. He supposedly rushed home naked, shouting, "Eureka!" ("I have found it!").

Historians do not agree on the exact method used by Archimedes to prove that there was indeed some silver in the crown. He could have compared the volumes of the crown, an equal weight of gold, and an equal weight of silver by measuring the water each displaced when submerged. Or he could have compared the weights of the three objects when immersed in water (as in Figure 4.35). Either method could have been used to determine the amount of silver in the crown.

The concept of specific gravity was developed in Arabia approximately 1,000 years ago. Arabian scientists also described how to use Archimedes' principle to measure the specific gravities of solids and estimated how high the atmosphere extends.

The French mathematician Blaise Pascal (1623–1662) made several discoveries about pressure in liquids, the most famous being

● **Figure 4.47** English scientist John Dalton developed the concepts of atoms and molecules and their relationship to elements and compounds.

● Figure 4.48 A large barometer built by Pascal in Rouen, France, to measure air pressure.

at the bottom of a narrow tube filled with water to the same level.

For 2,000 years after Aristotle's time, the philosophical principle *horror vacui*—that "nature abhors a vacuum"—obscured the understanding of air pressure. It was used to explain why a liquid could be drawn up into a tube by removing some of the air. When a straw is being used, the drink is supposedly pulled into the straw because nature does not allow the creation of a vacuum. Galileo was puzzled by reports that a suction pump could not raise water in a pipe higher than about 33 feet. This indicated to him that there seemed to be a limit to *horror vacui*.

The invention of the mercury barometer by the Italian mathematician Evangelista Torricelli (1608–1647), a student of Galileo's, proved to be the key to discovering the nature of atmospheric pressure. Torricelli correctly suggested that the pressure of the "sea of air" (the atmosphere) forces liquids to rise in a tube when air is removed from it. Pascal learned of Torricelli's work and performed many experiments with his own barometers (see ● Figure 4.48). In one of them, he used red wine in a glass tube 46 feet long. Pascal reasoned that the column of mercury in a barometer should be shorter at higher elevations. Yet he

the principle named after him (Section 4.6). He described how this principle implies that a force can be "amplified" using combinations of different-sized pistons and cylinders (Figure 4.41). Pascal also showed experimentally that the pressure in a liquid depends only on the depth and not on the shape of the container. The pressure at the bottom of an inverted cone filled with water is the same as the pressure

could not detect a difference in the length when he moved his barometer to the top of a church tower. Pascal's brother-in-law took a barometer to the Puy de Dôme, a mountain in south-central France, and did observe the effect.

The German physicist Otto von Guericke (1602–1686) performed several experiments that dramatically illustrated that atmospheric pressure can cause large forces. In one such experiment, he filled a copper sphere with water, sealed it, and then pumped out the water. The unbalanced air pressure on the outside caused the sphere to implode. Von Guericke invented air pumps and used them to produce near-vacuums. In 1654, he demonstrated his *Magdeburg hemispheres* to Emperor Ferdinand III. He fitted two hemispheres together and pumped the air out of the spherical cavity (see ● Figure 4.49). The force of the air pressure acting on the two hemispheres was so great that eight horses pulling on each one were unable to separate them.

Robert Boyle, a mentor and colleague of Isaac Newton, used a modification of the barometer to discover the relationship between pressure and volume in a gas (see the end of Section 4.2). Boyle's apparatus was a U-shaped glass tube partially filled with mercury. One side was sealed off so that some air was trapped above the mercury. By pouring more mercury into the other side or by removing some, Boyle was able to vary the pressure on the trapped air. The difference in height between the mercury in the two sides indicated the pressure on the air. Boyle showed that the volume occupied by the trapped air was inversely proportional to the pressure on it. This law, pV = a constant, is known as Boyle's law.

● Figure 4.49 Otto von Guericke demonstrating that air is stronger than horses.

SUMMARY

- The matter that makes up the material universe can be classified in different ways. In terms of the external, observable properties of matter, we can identify four **phases: solid, liquid, gas,** and **plasma.**

- In terms of the submicroscopic composition of ordinary matter, there are **elements** (consisting of **atoms**), **compounds** (consisting of **molecules**), and combinations of these.

- The nature of the constituent particles—atoms and molecules—and the forces that act between them determine the physical properties of a given element or compound.

- **Pressure, mass density,** and **weight density** are extensions of the concepts of force, mass, and weight.

- Pressure is a measure of the "concentration" of any force spread over an area.

- Mass density is the mass of a substance per unit volume. Weight density is the weight per unit volume.

- Liquids and gases are **fluids:** They flow readily and take the shapes of their containers.

- The force of gravity causes the pressure in a fluid to increase with depth. The **law of fluid pressure** states how the pressure at any point in a fluid is determined by the weight of the fluid above.

- In liquids, the pressure is proportional to the distance below the liquid's surface. Gases are compressible, and the pressure increases with depth in a different way because the density is not constant.

- Anything partly or completely immersed in a fluid experiences an upward **buoyant force.** By **Archimedes' principle,** this force is equal to the weight of the fluid that is displaced. Ships, boats, submarines, blimps, and hot-air balloons all rely on this principle. Archimedes' principle is also routinely used to measure the densities of solids and liquids.

- **Pascal's principle** states that any additional pressure in a fluid is passed on uniformly throughout the fluid.

- **Bernoulli's principle** is exploited by carburetors, atomizers, and other devices. It states that the pressure in a moving fluid decreases when the speed increases, and vice versa.

IMPORTANT EQUATIONS

Equation	Comments	Equation	Comments
Fundamental Equations		*Special-Case Equations*	
$p = \dfrac{F}{A}$	Definition of pressure	$p = D_w h = Dgh$	Pressure (gauge) at a depth h in a liquid
$D = \dfrac{m}{V}$	Definition of mass density	$p = 0.433 \text{ psi/ft} \times h$	Pressure (gauge) in psi underwater at a depth h in feet
$D_W = \dfrac{W}{V}$	Definition of weight density		
$F_b = W_{\text{fluid displaced}}$	Archimedes' principle		

MAPPING IT OUT!

1. In Section 4.1 in the description of matter, the following terms were introduced: *elements, atoms, electrons, protons, neutrons, nucleus, molecules, compounds, mixtures,* and *solutions*. Create a concept map explaining the composition of matter by appropriately organizing and linking these concepts to form meaningful propositions. After completing your concept map, compare your map with that of a classmate or the instructor. Are they the same? Should they be? Discuss the similarities and differences that you find between the maps.

2. Review Section 4.3 carefully. Based on your understanding of this material, develop a concept map to distinguish and relate the concepts of *mass density, weight density,* and *specific gravity*. Your map should address at least such issues as the basic definitions of these concepts and their applications in physics and in everyday life.

(► Indicates a review question, which means it requires only a basic under-standing of the material to answer. The others involve integrating or extending the concepts presented thus far.)

1. ► Describe the four phases of matter. Compare their external, observable properties. Compare the nature of the forces between atoms or molecules (or both) in the solid, liquid, and gas phases.

2. Identify some of the elements that exist in pure form (not in compounds) around you.

3. ► What is the difference between a mixture of two elements and a compound formed from the two elements?

4. If you classify everything around you as an element, a compound, or a mixture, which category would have the largest number of entries?

5. ► Why can gases be compressed much more readily than solids or liquids?

6. You find a sealed, rigid box. How might you go about determining what phase or phases of matter are inside the box without opening it? Can you make the determination in all cases?

7. ► Use the concept of pressure to explain why snowshoes are better than regular shoes for walking in deep snow.

8. The same bicycle tire pump is used to inflate a mountain bike tire to 40 psi and then a road bike tire to 100 psi. What difference would the user notice when using the pump on the two tires?

9. ► Explain the difference between gauge pressure and absolute pressure.

10. ► How can you use the volume of some quantity of a pure substance to calculate its mass?

11. The mass density of a mixture of ethyl alcohol and water is 950 kg/m³. Is the mixture mostly water, mostly alcohol, or about half and half? What is your reasoning?

12. Believe it or not, canoes have been made out of concrete (and they actually float). But even though concrete has a lower density than aluminum, a concrete canoe weighs a lot more that an aluminum one of the same size. Why is that?

13. Would the weight density of water be different on the Moon than it is on Earth? What about the mass density?

14. The way pressure increases with depth in a gas is different from the way it does in a liquid. Why?

15. Workers are to install a hatch (door) near the bottom of an empty storage tank. In choosing how strong to make the hatch, does it matter how tall the tank is? How wide it is? Whether it is going to hold water or mercury? Explain.

16. If the acceleration due to gravity on the Earth suddenly increased, would this affect the atmospheric pressure? Would it affect the pressure at the bottom of a swimming pool? Explain.

17. If the Earth's atmosphere warmed up and expanded to a larger total volume but its total mass did not change, would this affect the atmospheric pressure at sea level? Would this affect the pressure at the top of Mt. Everest? Explain.

18. Is there a pressure variation (increase with depth) in a fuel tank on a spacecraft in orbit? Why or why not?

19. ► Explain how a barometer can be used to measure altitude.

20. Why does the buoyant force always act upward?

21. ► What substances would sink in gasoline but float in water?

22. It is easier for a person to float in the ocean than in an ordinary swimming pool. Why?

23. ► A ship on a large river approaches a bridge and the captain notices that the ship is about a foot too tall to fit under the bridge. A crew member suggests pumping water from the river into an empty tank on the ship. Would this help?

24. In "The Unparalleled Adventure of One Hans Pfaall" by Edgar Allen Poe, the hero discovers a gas whose density is "37.4 times" less than that of hydrogen. How much better at lifting would a balloon filled with the new gas be compared to one filled with hydrogen?

25. A brick is tied to a balloon filled with air and is then tossed into the ocean. As the balloon is pulled downward by the brick, the buoyant force on it decreases. Why?

26. Venus's atmosphere is much more dense than the Earth's while that of Mars is much less dense. Suppose it is decided to send a probe to each planet that, once it arrived, would be carried around in the planet's atmosphere by a helium-filled balloon. How would the size of each balloon compare to the size that would be needed on the Earth?

27. ► What is Pascal's principle?

28. ► How does a car's brake system make use of Pascal's principle?

29. ► What important thing happens when the speed of a moving fluid increases?

30. The pressure in the air along the upper surface of an aircraft's wing (in flight) is lower than the pressure along the lower surface. Compare the speed of the air flowing over the wing to that of the air flowing under the wing.

31. ► How does a perfume atomizer make use of Bernoulli's principle?

PROBLEMS

1. A grain silo is filled with 2 million pounds of wheat. The area of the silo's floor is 400 ft². Find the pressure on the floor in pounds per square foot and in psi.
2. A bicycle tire pump has a piston with area 0.44 in.². If a person exerts a force of 30 lb on the piston while inflating a tire, what pressure does this produce on the air in the pump?
3. A large truck tire is inflated to a gauge pressure of 80 psi. The total area of one sidewall of the tire is 1,200 in.². What is the outward force on the sidewall due to the air pressure?
4. The water in the plumbing in a house is at a gauge pressure of 300,000 Pa. What force does this cause on the top of the tank inside a water heater if the area of the top is 0.2 m²?
5. A box-shaped metal can has dimensions 8 in. by 4 in. by 10 in. high. All of the air inside the can is removed with a vacuum pump. Assuming normal atmospheric pressure outside the can, find the total force on one of the 8-by-10-in. sides.
6. A viewing window on the side of a large tank at a public aquarium measures 50 in. by 60 in. (● Figure 4.50). The average gauge pressure due to the water is 8 psi. What is the total outward force on the window?

● **Figure 4.50**

7. A large chunk of metal has a mass of 393 kg, and its volume is measured to be 0.05 m³.
   a) Find the metal's mass density and weight density in SI units.
   b) What kind of metal is it?
8. A small statue is recovered in an archaeological dig. Its weight is measured to be 96 lb and its volume 0.08 ft³.
   a) What is the statue's weight density?
   b) What substance is it?
9. A large tanker truck can carry 20 tons (40,000 lb) of liquid.
   a) What volume of water can it carry?
   b) What volume of gasoline can it carry?
10. A manufacturer orders 10,000 kg of copper ingots. What volume does the copper occupy?
11. A large balloon used to sample the upper atmosphere is filled with 900 m³ of helium. What is the mass of the helium?

12. A certain part of an aircraft engine has a volume of 1.3 ft³.
    a) Find the weight of the piece when it is made of iron.
    b) If the same piece is made of aluminum, find the weight and determine how much weight is saved by using aluminum instead of iron.
13. The volume of the Drop Tower "Bremen" (a 100-meter-tall tube used to study processes during free fall) is 1,700 m³.
    a) What is the mass of the air that must be removed from it to reduce the pressure inside to nearly zero (1 Pa compared to 100,000 Pa)?
    b) What is the weight of the air in pounds?
14. It is determined by immersing a crown in water that its volume is 26 in.³ = 0.015 ft³.
    a) What would its weight be if it were made of pure gold?
    b) What would its weight be if half of its volume were gold and half lead?
15. Find the gauge pressure at the bottom of a swimming pool that is 12 ft deep.
16. The depth of the Pacific Ocean in the Mariana Trench is 36,198 ft. What is the gauge pressure at this depth?
17. Calculate the gauge pressure at a depth of 300 m in seawater.
18. A storage tank 30 m high is filled with gasoline.
    a) Find the gauge pressure at the bottom of the tank.
    b) Calculate the force that acts on a square access hatch at the bottom of the tank that measures 0.5 m by 0.5 m.
19. The highest point in North America is the top of Mt. McKinley in Alaska, 20,320 ft above sea level. Using the graph in Section 4.4, find the approximate air pressure there.
20. The Concorde, a supersonic passenger jet, flew at an altitude of 15,000 m. What is the approximate air pressure at that height?
21. An ebony log with volume 12 ft³ is submerged in water. What is the buoyant force on it?
22. An empty storage tank has a volume of 1,500 ft³. What is the buoyant force exerted on it by the air?
23. A blimp used for aerial camera views of sporting events holds 200,000 ft³ of helium.
    a) How much does the helium weigh?
    b) What is the buoyant force on the blimp at sea level?
    c) How much can the blimp lift (in addition to the helium)?
24. A modern-day zeppelin holds 8,000 m³ of helium. Compute its maximum payload at sea level.
25. A piece of concrete measures 3 ft by 2 ft by 0.5 ft.
    a) What is its weight?
    b) Find the buoyant force that acts on it when it is submerged in water.
    c) What is the net force on the concrete piece when it is under water?
26. A juniper-wood plank measuring 0.25 ft by 1 ft by 16 ft is totally submerged.
    a) What is its weight?
    b) What is the buoyant force acting on it?
    c) What is the size and the direction of the net force on it?

27. The volume of an iceberg is 100,000 ft³ (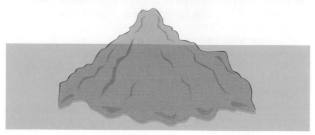Figure 4.51).
    a) What is its weight, assuming it is pure ice?
    b) What is the volume of seawater it displaces when floating? (*Hint:* You know what the weight of the seawater is.)
    c) What is the volume of the part of the iceberg out of the water?

● **Figure 4.51**

28. A boat (with a flat bottom) and its cargo weigh 5,000 N. The area of the boat's bottom is 4 m². How far below the surface of the water is the boat's bottom when it is floating in water?
29. A scale reads 100 N when a piece of aluminum is hanging from it. What does it read when it is lowered so that the aluminum is submerged in water?
30. A rectangular block of ice with dimensions 2 m by 3 m by 0.2 m floats on water. A person weighing 600 N wants to stand on the ice. Would the ice sink below the surface of the water?

CHALLENGES

1. When exactly 1 cup of sugar is dissolved in exactly 1 cup of water, *less* than 2 cups of solution result. Why?
2. Near the end of Section 4.3, we stated that, because the gram was defined to be the mass of 1 cm³ of water, the density of water is exactly 1,000 kg/m³. Verify this.
3. What would be the mass density, weight density, and specific gravity of aluminum on the Moon? The acceleration of gravity there is 1.6 m/s².
4. The mass of the Earth is 6×10^{24} kg, and its radius is 6.4×10^6 m. What is the average mass density of the Earth? The density of the rocks comprising the Earth's outermost layer (its "crust") ranges from 2,000 to 3,500 kg/m³. Based on your answer, what can you conclude about the material deep inside the Earth's interior?
5. In 1993, a football coach in the United States accused an opposing team's punter of using a football inflated with helium instead of air. Estimate how much lighter a football would be if inflated to a gauge pressure of 1 atm with helium instead of air. The volume of a football is approximately 0.1 ft³.
6. Two swimming pools are 8 ft deep, but one measures 20 ft by 30 ft, and the other measures 40 ft by 60 ft. Identical drain valves at the bottom of each pool are 10 in.² in area. Compare the force on each valve.
7. Scuba divers take their own supply of air with them when they go underwater. Why couldn't they just take a long hose with them from the surface and breathe through it (● Figure 4.52)?

● **Figure 4.52**

8. A motorist driving through Colorado checks the tire pressure in Denver (elevation 5,000 ft), then again at the Eisenhower Tunnel (elevation 11,000 ft). Would the pressures be the same? What two main factors affect the tire pressure as the car climbs?
9. A glass contains pure water with ice floating in it. After the ice melts, will the water level be higher, lower, or the same? (Ignore evaporation.)
10. At one point in the novel *Slapstick* by Kurt Vonnegut, the force of gravity on Earth suddenly "increased tremendously." The result:

    . . . elevator cables were snapping, airplanes were crashing, ships were sinking, motor vehicles were breaking their axles, bridges were collapsing, and on and on.

    Would a ship really sink if the force of gravity were increased?
11. A brick rests on a large piece of wood floating in a bucket of water. The brick slides off and sinks. Does the water level in the bucket go up, go down, or stay the same?
12. As a helium balloon rises up in the air, work is done on it against the force of gravity. What is doing the work? What energy transfer or transformation is taking place?

SUGGESTED READINGS

Armbruster, Peter, and Fritz Peter Hessberger. "Making New Elements." *Scientific American* 279, no. 3 (September 1998): 72–77. Description of the techniques used to produce and detect superheavy elements (110 and beyond) by leading scientists in the field working at the Institute for Heavy Ion Research in Darmstadt, Germany.

Cornell, Eric A., and Carl E. Wieman, "The Bose-Einstein Condensate." *Scientific American* 278, no. 3 (March 1998): 40–45. Co-leaders of the team that isolated the first Bose-Einstein condensate in a gas discuss the nature of this material, as well as the technology required to produce it.

Curl, Robert F., and Richard E. Smalley. "Fullerenes." *Scientific American* 265, no. 4 (October 1991): 54–63. Describes the structure, method of production, and possible uses of buckministerfullerene (C_{60}) and other large carbon molecules.

Donnelly, Russell J. "The Discovery of Superfluidity." *Physics Today* 48, no. 7 (July 1995): 30–36. Describes events during the 27-year period between the discoveries of superconductivity and superfluidity.

Ebbesen, Thomas W. "Carbon Nanotubes." *Physics Today* 49, no. 6 (June 1996): 26–32. An account of the history, properties, and possible uses of carbon nanotubes.

Hemley, Russell J., and Neil W. Ashcroft. "The Revealing Role of Pressure in the Condensed Matter Sciences." *Physics Today* 51, no. 8 (August 1998): 26–32. Describes some of the tools, techniques, and discoveries of modern high-pressure physics.

Huffman, Donald R. "Solid C_{60}." *Physics Today* 44, no. 11 (November 1991): 22–29. Describes how crystals of C_{60} are produced.

Moon, Richard E., Richard D. Vann, and Peter B. Bennett. "The Physiology of Decompression Illness." *Scientific American* 273, no. 2 (August 1995): 70–77. A look at the dangers of deep diving, along with the techniques and equipment used to reduce them.

Scerri, Eric R. "The Evolution of the Periodic System." *Scientific American* 279, no. 3 (September 1998): 78–83. Describes the history of the periodic table and includes alternate ways of displaying the elements.

Scientific American 281, no. 3 (November 1999): 98–113. This issue has three articles about lighter-than-air craft: "Floating in Space," "A Zeppelin for the 21st Century," and "The Balloon That Flew Round the World."

Sorbjan, Zbigniew. *Hands-On Meteorology.* American Meteorology Society, 1996. Includes many do-it-yourself experiments on such topics as gases, pressure, and buoyancy.

Weast, Robert C., ed. *Handbook of Chemistry and Physics,* Boca Raton, Florida: CRC Press. Gives a short history of each element.

For additional readings, explore InfoTrac® College Edition, your online library. Go to http://www.infotrac-college.com/wadsworth and use the passcode that came on the card with your book. Try these search terms: blimp, fullerene, carbon nanotubes, aneroid barometer, superfluid, Bernoulli's principle.

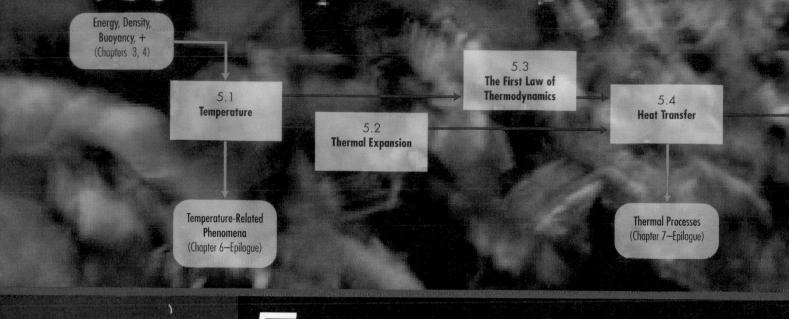

Energy, Density, Buoyancy, +
(Chapters 3, 4)

5.1 Temperature

Temperature-Related Phenomena
(Chapter 6—Epilogue)

5.2 Thermal Expansion

5.3 The First Law of Thermodynamics

5.4 Heat Transfer

Thermal Processes
(Chapter 7—Epilogue)

5 TEMPERATURE AND HEAT

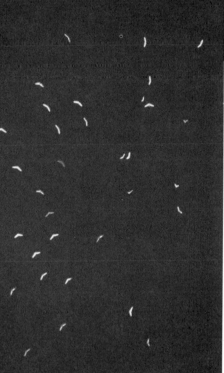

Soaring hanggliders viewed from below. (For more information call 1-800-HANGGLIDE or go to www.ushga.org.)

Photo by Gerry Charlebois

Soaring

Solar-powered flying. It's not usually called that, but that's what it is: Sustained flight using energy from the Sun instead of buoyancy or onboard engines. Sailplanes, hang gliders, and paragliders do it. Many bird species—including some eagles, hawks, vultures, pelicans, and cranes—do it too, although they can flap their wings for locomotion when they need to.

Soaring flights are often local—the pilot or bird doesn't venture far from the starting point and only flies for fun, or for the view, or to look for something to eat. But truly impressive "cross-country" flights are possible, some hundreds of miles long. It is common in some places for hang-glider pilots to fly 100 miles or more without an engine on aircraft that they can carry on their shoulders to a takeoff site. Some migratory birds save valuable energy by soaring for part of their journey, gaining altitude without having to flap their wings, and then gliding downwind.

The key to soaring is finding air that is moving upward faster than the aircraft is moving downward. Gliders flying in still air move like paper airplanes gliding to the ground. Different models lose anywhere from 2 to 50 feet of altitude for every 100 feet they go forward.

If the downward component of a glider's speed is 2 mph, it will gain altitude in air that is rising 3 mph. Why would air be moving upward this fast? Solar generated *thermals*—rising bubbles or columns of air—are the most commonly exploited means of soaring. How does the Sun cause these invisible elevators of rising air? The basics of heat and temperature are all that's needed to understand how, and that's what this chapter is about.

We take a formal look at the concepts of temperature and heat. First, we consider temperature—what it is, how it is measured, and how matter is affected when it changes. Then we refine the concepts of internal energy and heat by showing how they are involved in changing the temperature or phase of matter. The chapter concludes with a discussion of heat engines and heat movers.

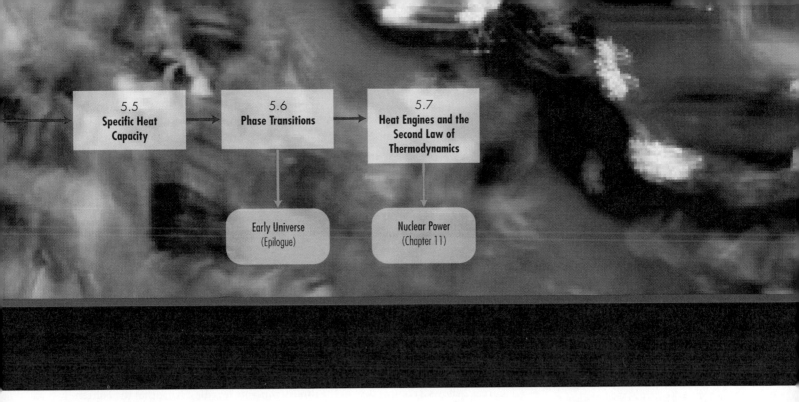

5.5 Specific Heat Capacity	5.6 Phase Transitions	5.7 Heat Engines and the Second Law of Thermodynamics

Early Universe (Epilogue)

Nuclear Power (Chapter 11)

5.1 Temperature

Temperature may be the most commonly used physical quantity in our daily lives, next to time. We might loosely define temperature as a measure of hotness or coldness. But the concepts of hot and cold are themselves rather vague, subjective, and relative. In the summer, an air temperature of 70°F feels cool, while the same temperature in the winter feels warm.

The idea of temperature is distinct from that of heat or internal energy. For example, when one drop is removed from a cup of hot water, both the drop and the cup have the *same* temperature, but the amount of internal energy associated with each is very different (● Figure 5.1). Placing the drop in the palm of your hand would not have nearly the same effect (pain) as pouring the cup of water into it.

All thermometers depend on some physical property that changes with temperature. A common type exploits the fact that a liquid, usually mercury or red-colored alcohol, will expand and contract when heated or cooled, rising or falling in a glass tube as the temperature varies. Some thermometers are based on other temperature-dependent physical properties, including the volume of solids, the pressure or volume of a gas, the electrical properties of metals, the amount and frequency of radiated energy, and the speed of sound in a gas.

There are three different temperature scales in common use: Fahrenheit, Celsius, and Kelvin. The normal freezing and boiling temperatures of water—called the *phase transition* temperatures—may be used to compare the three scales. In the Fahrenheit scale, the

● **Figure 5.1** The cup of water and the drop have the same temperature, but the cup of water can transfer much more internal energy to its surroundings.

© Vern Ostdiek

Explore It Yourself 5.1

Go to the thermometer display in a department store, hardware store, or other retail outlet. Look carefully at the temperatures displayed by the different thermometers. Do they all show exactly the same temperature? Should they be the same? What does this tell you about the accuracy of these instruments?

boiling point of water under a pressure of 1 atmosphere is 212°—designated 212°F. The freezing point of water under this pressure is 32°F. So there are 180 units, called *degrees*, that separate the two temperatures. The Celsius scale, formerly called the centigrade scale, is metric based; it uses 100 degrees between the freezing and boiling points of water. Zero degrees Celsius—designated 0°C—is the freezing temperature, and 100°C is the boiling temperature.

Most of us spend our lives subjected to temperatures within the range of −60 to 120°F (−51 to 49°C). Much higher temperatures exist in common places: the interiors of stoves and automobile engines, the filaments of light bulbs, the flame of a candle, and so on. The Sun's surface temperature is about 10,000°F (5,700°C), and its interior is at about 27,000,000°F (15,000,000°C). Temperatures this high have been produced on the Earth in experiments with plasmas and in nuclear explosions. There is no upper limit on temperature.

At the other extreme, there *is* a limit on cold temperatures. The coldest temperature, called **absolute zero,** is −459.67°F (−273.15°C). Since it is impossible to go below this temperature, the Kelvin scale is a convenient one because it uses absolute zero as its starting point (zero). (This scale is also referred to as the "absolute temperature scale.") The size of the unit in the Kelvin scale is the same as that in the Celsius scale, except it is called a kelvin (K) instead of a degree. Any temperature in the Kelvin scale equals the corresponding Celsius value plus 273.15. The normal boiling and freezing temperatures of water are 373.15 K and 273.15 K, respectively.

● Figure 5.2 shows a comparison of the three temperature scales. Note that the Fahrenheit and Celsius scales agree at −40°. ● Table 5.1 lists some representative temperatures. As a physical quantity, temperature is represented by *T*.

Unfortunately, *T* is used to represent both temperature and period (Section 1.1).

● **Figure 5.2** Comparison of the three temperature scales. The Celsius and Kelvin scales have the same sized unit, or degree. The Fahrenheit degree is five-ninths the Celsius degree.

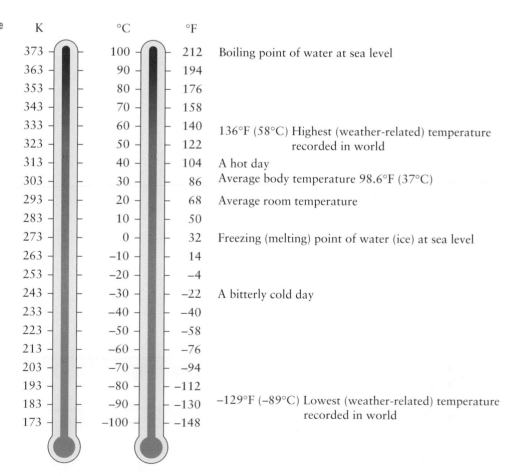

K	°C	°F	
373	100	212	Boiling point of water at sea level
363	90	194	
353	80	176	
343	70	158	
333	60	140	136°F (58°C) Highest (weather-related) temperature
323	50	122	recorded in world
313	40	104	A hot day
303	30	86	Average body temperature 98.6°F (37°C)
293	20	68	Average room temperature
283	10	50	
273	0	32	Freezing (melting) point of water (ice) at sea level
263	−10	14	
253	−20	−4	
243	−30	−22	A bitterly cold day
233	−40	−40	
223	−50	−58	
213	−60	−76	
203	−70	−94	
193	−80	−112	
183	−90	−130	−129°F (−89°C) Lowest (weather-related) temperature
173	−100	−148	recorded in world

• Table 5.1 Representative Temperatures in the Three Temperature Scales

Description	°F	°C	K
Absolute zero	−459.67	−273.15	0
Helium boiling point	−452	−268.9	4.25
Nitrogen boiling point	−320.4	−195.8	77.35
Oxygen boiling point	−297.35	−182.97	90.18
Alcohol freezing point	−175	−115	158
Mercury freezing point	−37.1	−38.4	234.75
Water freezing point	32	0	273.15
Normal body temperature	98.6	37	310.15
Water boiling point	212	100	373.15
"Red hot" (approx.)	800	430	700
Aluminum melting point	1,220	660	933
Iron melting point	2,797	1,536	1,809
Sun's surface (approx.)	10,000	5,700	6,000
Sun's interior (approx.)	27×10^6	15×10^6	15×10^6
Highest laboratory temperature	900×10^6	500×10^6	500×10^6

What determines the temperature of matter? In other words, what is the difference between a cup of coffee when it is hot (200°F) and the same cup of coffee when it is cold (70°F)? The atoms and molecules that compose matter have kinetic energy. In gases, they move about randomly with high speed. In liquids and solids, they vibrate much like a mass on a spring or an object oscillating in a hole (Section 3.5). At higher temperatures, the atoms and molecules in matter move faster and have higher kinetic energies.

PRINCIPLES

The Kelvin temperature of matter is proportional to the average kinetic energy of the constituent particles.

Kelvin scale temperature $\propto$ average *KE* of atoms and molecules

Because of collisions between the particles, during which energy is exchanged, the particles do not all have exactly the same kinetic energy at each instant, nor does the energy of a given particle stay exactly the same from one moment to the next. But the *average* kinetic energy of all of the particles is constant as long as the temperature stays constant.

So when a cup of coffee is hot, the molecules in it have higher average kinetic energy than when it is cold. If you put your finger into hot coffee, the atoms and molecules in the coffee pass on their higher kinetic energy to the atoms and molecules in your finger by way of collisions: Your finger is warmed.

This fact—that temperature depends on the average kinetic energy of atoms and molecules—is very important. It should help you understand many of the phenomena we will discuss in this chapter. In gases, higher kinetic energy means that the atoms and molecules move about with higher speeds. For liquids and solids, the molecules vibrate through a greater distance like a pendulum swinging through a larger arc. You may recall that when particles oscillate like this, they also have potential energy. This is the case here, too, but the potential energy of "bound" atoms and molecules in liquids and solids is not directly related to the temperature. This potential energy is important when a substance undergoes a change of phase—like freezing or boiling. More on this later.

This principle also accounts for the existence of an absolute zero. At colder temperatures, the average kinetic energy of the particles is smaller. If they stopped moving altogether, the average kinetic energy would be zero. This would be the lowest possible

To Breathe or Not to Breathe, That Is the Question

As mentioned in Chapter 4, we live at the bottom of a "sea" of air that provides us with one of the basic ingredients of sustained life: oxygen. Indeed, had the Earth been unable to retain an atmosphere for more than several billion years, the evolution of life as we know it could not have taken place at all. But what physical conditions or quantities determine whether or not a planet is capable of holding an appreciable atmosphere? More specifically, why does the Earth possess an atmosphere, while Mercury does not (see ● Figure 5.3)? Is the composition of such an atmosphere related to these physical conditions? That is, is it possible to understand, in terms of our physics, why the atmosphere of Jupiter is primarily hydrogen, while that of the Earth is made up principally of heavier gases like nitrogen and oxygen (see Table 4.2)?

Every planetary body in our solar system receives some radiation from the Sun. Those closer to the Sun receive more radiation and hence have higher surface temperatures than those farther away. Given the definition of temperature, it is clear that atoms and molecules in the atmosphere of a planet near the Sun will generally have higher average kinetic energies (and higher average speeds) than those in the atmosphere of a planet far from the Sun.

The ability of a planet to retain these atmospheric atoms and molecules depends primarily on the strength of its gravitational field. If the average speeds of the atmospheric particles are high and the planet's gravity is low, the atmosphere will gradually escape. Conversely, if the average speeds of the atmospheric particles are low and the gravity is high, the atmosphere will be retained. This is why Mercury and the Moon do not possess primordial atmospheres. Mercury is a small planet very close to the Sun: It is hot (having a surface temperature over 400 K) and has a low surface gravity (1 *g* on Mercury equals 3.7 m/s² as opposed to 9.8 m/s² on the Earth). Any atmosphere it may once have had escaped long ago because the particles were far too energetic to have been held in by the planet's weak gravity. Similarly, any original atmosphere the Moon may have had must have dissipated billions of years ago despite its cooler temperature (about 300 K, like that of the Earth) because its gravitational field is only about one-sixth that of the Earth.

For planets with moderate temperatures and gravities, some gases can be retained, while others cannot. It is possible to predict which gases will be retained in a given circumstance by comparing the average molecular speed of a gas with the escape velocity of the planet. The *escape velocity* (as discussed in Section 3.5) is the minimum speed needed to overcome the gravitational pull of the planet. The higher the gravity of a planet, the greater its escape velocity. In general, a planet will retain a particular gas in its atmosphere over millions of years only if its escape velocity is four to six times larger than the average molecular speed of that gas. Only under these circumstances will the planet's gravity be high enough to prevent the gas from eventually leaking off into space.

● Figure 5.3 The planet Earth (top) has an atmosphere, but the planet Mercury (bottom) does not.

Figure 5.4 is a plot of characteristic velocity versus temperature and displays this rule. The labeled points show the average surface temperature and escape velocity (divided by six) for the major planets and three satellites: Earth's moon; Titan, a moon of Saturn; and Triton, a moon of Neptune. Superimposed on this plot are lines for different gases tracing the average particle speeds of these gases as a function of temperature.

Several things can be noted: (a) For a given gas, as the temperature increases (far left), the average molecular speed increases. This is to be expected, of course, on the basis of what is meant by the Kelvin temperature of a gas. (b) At a given temperature, the speeds of the lighter species, like hydrogen and helium, are greater than those of the heavier ones, like nitrogen and carbon dioxide. This, too, is easily understood in terms of what we know about the physics of gases. For a fixed temperature, the average kinetic energy of, say, a helium atom will be equal to that of a carbon dioxide molecule. But the kinetic energy of any particle is proportional to the product of its mass and its velocity squared. For a particular value of kinetic energy, higher masses correspond to smaller velocities. Hence, at any temperature, the more massive carbon dioxide molecule will be moving more slowly than the lighter He atom. (c) For a planet to retain a particular gas, its position on the graph must lie above the line for that gas. Notice that the points corresponding to the planets Jupiter and Saturn lie well above all the lines. These planets are so massive and so cold (being far from the Sun) that they can retain all the gases considered here in their atmospheres—even hydrogen, which has the lightest atoms. Earth, on the other hand, can hold the heavier gases like nitrogen and oxygen but not hydrogen. And, as discussed above, the Moon and the planet Mercury are incapable of holding onto any of the common atmospheric constituents, while Titan and Triton possess atmospheres dominated by nitrogen.

The analysis presented here has ignored many factors that complicate the question of planetary atmosphere retention (such as interactions between planetary atmospheres and the "solar wind," a continuous stream of electrons and other charged particles that leave the Sun and propagate outward into the interplanetary medium, which are particularly important for a nearby planet like Mercury), but it is correct in its broad outline. The important point to keep in mind is that, armed only with the physics we have developed so far, we are in a position to make some important predictions about the likely existence of atmospheres on other planetary bodies within the solar system. Such predictions figure prominently in attempts to identify locations beyond the Earth where life may have evolved.

The search for planets orbiting Sun-like stars that may harbor conditions appropriate for the development of carbon-based life forms is one of the most active areas in astronomy today. During the past decade, more than a hundred *extrasolar planets* have been discovered using a variety of astronomical techniques. Most of these systems have orbital characteristics quite different from those of the planets revolving around the Sun, but a few exhibit properties similar to our own solar system. These are clearly the most exciting because they offer the best prospects for locales where other forms of life may exist. For those interested in learning more about the search for extrasolar planets, two good, authoritative places to start are "The Extrasolar Planets Encyclopedia," maintained by the Observatoire Bio du Midi in France (http://www.obspm.fr/encycl/encycl.html), and the California and Carnegie Planet Search site at http://exoplanets.org. Good hunting!

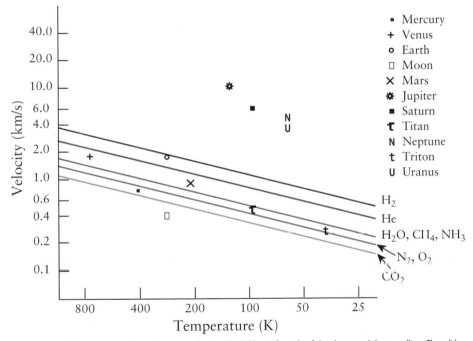

- **Figure 5.4** Surface temperatures and escape velocities (divided by 6) for eight of the planets and three satellites. The solid lines show the variation of molecular speed as a function of temperature for several common atmospheric gases.

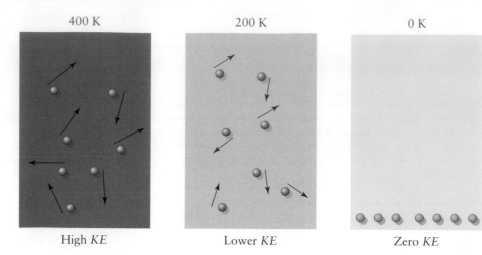

Figure 5.5 At lower temperatures, atoms and molecules have lower average kinetic energy (*KE*). At absolute zero, their kinetic energies would be zero: They would be stationary.

400 K   200 K   0 K

High *KE*   Lower *KE*   Zero *KE*

temperature—absolute zero (see ● Figure 5.5). The word "would" is used because, as it turns out, absolute zero can never be reached. The atoms and molecules in a substance cannot be completely stopped. Researchers do get very close to absolute zero within a billionth of a degree—but they cannot reach it exactly.

At very low temperatures, many substances acquire unusual properties. Plastic and rubber become as brittle as glass. Below about 2 K, helium is a "superfluid" liquid; it flows without friction. (Recall the Physics Potpourri in Section 4.5.) Some materials become "superconductors." They conduct electricity without resistance (more on this in Chapter 7).

The nature of matter is also different at very high temperatures. Above about 6,000 K, the kinetic energies of atoms are so high that they cannot bind together. Hence there can be no solids or liquids—or even molecules. Above about 20,000 K, the electrons break free from atoms, and only plasmas can exist.

LEARNING CHECK

1. The average kinetic energy of the atoms in a gold ring determines its _____.
2. (Choose the *incorrect* statement.) Absolute zero is
   a) the lowest possible temperature.
   b) 0° on the Celsius scale.
   c) the temperature at which atoms and molecules would not be moving.
   d) more than 400 degrees below zero on the Fahrenheit scale.
3. (True or False.) If the temperature of something was raised 50°C, that means its temperature was raised 90°F.

ANSWERS: 1. temperature 2. (b) 3. True

5.2 Thermal Expansion

Thermal expansion is an important phenomenon that is exploited by the common types of thermometers and by a variety of other useful devices. In almost all cases, substances that are not constrained expand when their temperatures increase. (Exceptions include water below 4°C and some compounds of tungsten.) The air in a balloon expands when heated, the mercury in a thermometer expands upward in the glass tube when placed in a hot liquid, and sections of bridges become longer in the summer. If the substance is constrained sufficiently, it will not expand, but forces and pressures will be created in response to the constraint. For example, if an empty pressure cooker is sealed and then heated, the air inside is prevented from expanding. But the pressure inside increases and causes larger and larger outward forces on the inner surfaces of the pressure cooker.

We can see qualitatively why this expansion occurs. At higher temperatures, the atoms and molecules in a solid or a liquid vibrate through a larger distance and so push each other apart slightly. In gases, they move with higher and higher speeds as the temperature rises. A balloon will expand when heated because the higher-speed air molecules will cause higher pressure and push the balloon's surface outward as they collide with it. In these cases, the expansion occurs in all three dimensions. For example, as a brick is heated, its length, width, and thickness all increase proportionally.

We can use logic and basic mathematics to predict the amount of expansion that occurs. Let us first consider the simplest case—the thermal expansion of a long, thin solid such as a metal rod. The main expansion will be an increase in its length l (see ● Figure 5.6). This increase, designated Δl, depends on three factors:

1. The *original length l*. The longer the rod is to begin with, the greater the change in length will be.
2. The *change in temperature*, designated ΔT. The larger the increase in temperature, the greater the increase in length.
3. The *substance*. For example, the increase in length of an aluminum rod will be more than twice that of an identical iron rod under the same conditions.

Point 3 can be tested through experimentation. The expansions of different solids are measured under similar conditions. The results are used to assign a **coefficient of linear expansion** to each material. The value of this coefficient is a fixed parameter of each substance, much like mass density or weight density. It is represented by the Greek letter alpha, α. Since aluminum expands more than iron under the same circumstances, the coefficient of linear expansion of aluminum is larger than that of iron (● Figure 5.7).

The equation that gives the change in length in terms of the change in temperature and the coefficient of linear expansion is

$$\Delta l = \alpha \, l \, \Delta T$$

The change in length is proportional to the change in temperature and to the original length. The coefficient of linear expansion, α, is the constant of proportionality. ● Table 5.2 lists α for several different solids. In the equation, the units of l and Δl must be the same. The units of temperature, usually °C, are canceled by those of α, which are their reciprocal, that is to say, the units of α may be expressed as 1/°C.

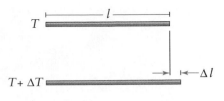

● **Figure 5.6** A metal rod has length l when the temperature is T. When the temperature is increased by an amount ΔT, the length of the rod increases by a proportional amount Δl.

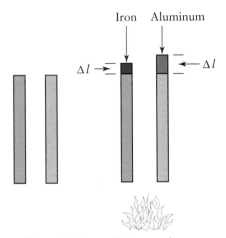

● **Figure 5.7** Because aluminum expands more than iron, given the same increase in temperature, its coefficient of linear expansion is larger.

The center span of a steel bridge is 1,200 meters long on a winter day when the temperature is −5°C. How much longer is the span on a summer day when the temperature is 35°C?

First, the change in temperature is the final temperature minus the initial temperature:

$$\Delta T = 35 - (-5) = 35 + 5 = 40°C$$

From Table 5.2, the coefficient of linear expansion for steel is

$$\alpha = 12 \times 10^{-6}/°C$$

So:

$$\Delta l = \alpha \, l \, \Delta T$$
$$= (12 \times 10^{-6}/°C) \times 1{,}200 \text{ m} \times 40°C$$
$$= (12 \times 10^{-6}/°C) \times 48{,}000 \text{ m-°C}$$
$$= 576{,}000 \times 10^{-6} \text{ m}$$
$$= 0.576 \text{ m}$$

Example 5.1

● **Table 5.2** Some Coefficients of Linear Expansion

Solid	$\alpha \ (\times 10^{-6}/°C)$
Aluminum	25
Brass or bronze	19
Brick	9
Copper	17
Glass (plate)	9
Glass (Pyrex)	3
Ice	51
Iron or steel	12
Lead	29
Quartz (fused)	0.4
Silver	19

The change in length in Example 5.1 is considerable and must be allowed for in the bridge design. Expansion joints, which act somewhat like loosely interlocking fingers, are

● **Figure 5.8** Expansion joints allow for the thermal expansion of bridges and other elevated roadways. Each end of a section of the roadway is connected to a metal "comb." The teeth of the two combs fit together and can move back and forth as the lengths of the sections change. For comparison, the thermometer is 30 centimeters (1 foot) wide.

placed in bridges, elevated roadways, and other such structures to allow thermal expansion to occur (● Figure 5.8).

The equation also works when the temperature decreases. When this occurs, the change in temperature is negative, so the change in length is also negative: The solid becomes shorter. Another way of saying this is that thermal expansion is a reversible process. If something becomes longer when heated, it will get shorter when cooled. Most of the phenomena discussed in this chapter are reversible. If something happens when the temperature increases, the reverse will happen when the temperature decreases.

Explore It Yourself 5.2

If you live near a bridge or elevated roadway on which you can walk safely, find an expansion joint. (Figure 5.8 shows one type.) Photograph it or sketch what it looks like on a hot afternoon and then sometime when it is much cooler. What is different?

The *bimetallic strip* is an ingenious and widely used application of thermal expansion. As the name implies, a bimetallic strip consists of two strips of different metals bonded to one another (● Figure 5.9). The two metals have different coefficients of linear expansion, so they expand by different amounts when heated. The result is that the bimetallic strip bends—one way when heated and the other way when cooled. The greater the change in temperature, the greater the bending. For example, if brass and iron are used, the brass will expand and contract more than the iron will. The brass will be on the outside of the curve when the strip is hot and on the inside of the curve when it is cold.

Thermostats, thermometers, and choke-control mechanisms on automobiles often contain a bimetallic strip that is curled into a spiral. The coil will either partly unwind or wind up more tightly when the temperature changes. To make a thermometer, a pointer is attached to one end of the spiral, and the other end is held fixed. As the temperature varies, the pointer moves over a scale that indicates the temperature (● Figure 5.10). In thermostats, the movement of the end of the spiral is used to turn a switch on or off. The switch might turn on a heater, an air conditioner, a fire alarm, or the cooling unit in a refrigerator.

As mentioned, thermal expansion occurs in all three dimensions. The bridge becomes longer, and also wider and thicker. The area of any surface increases, as does the volume of the solid. If a solid has a hole in it, thermal expansion will make the hole bigger, contrary to most people's intuition (● Figure 5.11). This is because thermal expansion causes every point in a solid to move *away* from every other point. (It is similar to what happens

Normal

Hot

Cold

● **Figure 5.9** In this bimetallic strip, the metal composing the upper layer has a larger coefficient of thermal expansion than the metal composing the lower layer. When heated, the strip curves downward because the upper layer undergoes a greater change in length. When cooled, the strip curves upward.

Figure 5.10 This thermometer uses a bimetallic strip in the shape of a spiral. Even small temperature changes cause the pointer attached to the outer end of the spiral to rotate noticeably.

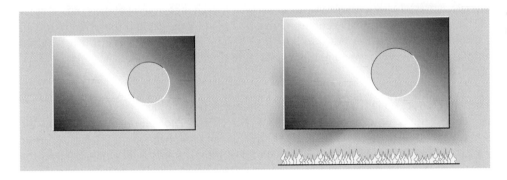

Figure 5.11 The hole becomes larger when the object is heated.

when a photograph is enlarged.) A point on one side of a hole moves away from any point on the other side of the hole.

Liquids

The behavior of liquids is quite similar to that of solids. Since liquids do not hold a certain shape, it is best to consider the change in volume caused by thermal expansion. In general, liquids expand considerably more than solids. This means that when a container holding a liquid is heated, the level of the liquid will usually rise because the increase in volume of the liquid exceeds the increase in volume of the container. When a mercury thermometer is heated, the mercury in the tube and in the glass bulb at the bottom expands more than does the glass, so the level of mercury rises. If the glass expanded more than the mercury, the column would go down at higher temperatures instead of up.

At the beginning of this section, we implied that there are exceptions to the general rule that matter expands when heated. The most important example of this is water that is near its freezing temperature. Above 4°C (39°F), water expands when heated like ordinary liquids. But between 0°C and 4°C, water actually contracts when heated and expands when cooled. The volume of a given amount of water at 3°C is *less* than the volume of the same water at 1°C. This anomaly accounts for the fact that lakes, ponds, and other bodies of water freeze on top first. As long as the average water temperature is above 4°C, the warmer (less dense) water is buoyed to the surface, and the cooler water is at the bottom (see ● Figure 5.12). As the air gets cooler in autumn, the water at the surface is cooled. When the average temperature of the water is below 4°, the cooler water (closer to freezing) is now less dense and rises to the surface. Consequently, the surface water freezes first because it is cooler than the water below, and it is in contact with the cold air.

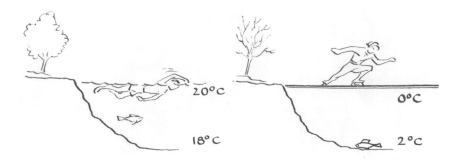

● **Figure 5.12** Above 4°C, the warmer water rises to the surface. Below 4°C, the cooler water rises to the surface.

$T = 273$ K

V

$T = 303$ K

$1.1\ V$

● **Figure 5.13** If the pressure in a gas is kept constant, the volume that the gas occupies is proportional to the Kelvin temperature. A 10% increase in temperature will cause the volume of a balloon to increase by 10%.

© Vern Ostdiek

● **Figure 5.14** Illustrating thermal expansion. (See Explore It Yourself 5.3).

Water is also unusual in that its density when in the solid phase (ice) is less than its density when in the liquid phase. Because of this, ice floats in water, whereas most solids (for example, candle wax) sink in their own liquid.

Gases

The volume expansion of gases is larger than that of solids and liquids. Also, the amount of expansion does not vary with different gases (except at very low temperatures or very high pressures). Instead of relating the expansion to a change in temperature, it is simpler to state the relationship between the volume occupied by the gas and the temperature. In particular, as long as the pressure remains constant, the volume occupied by a given amount of gas is *proportional* to the temperature (in kelvins).

$$V \propto T \quad \text{(gas at fixed pressure)}$$

The temperature must be in kelvins. When the temperature of a gas is increased by some percentage, the volume increases by the same percentage. The volume of a balloon at 303 K (86°F) is about 10% larger than the volume of the same balloon at 273 K (32°F) (see ● Figure 5.13). By comparison, volume change of typical solids is less than 1% over the same temperature range. In liquids, it is a maximum of 5%.

Explore It Yourself 5.3

For this you need a balloon and a pitcher or some other container. It's easiest if the container is transparent enough to see the level of water in it from the outside. See ● Figure 5.14.

1. Inflate the balloon to a point where it will fit completely inside the pitcher but with some space around it and below it, then tie it so it won't deflate. The balloon should not be fully inflated: The air inside must be able to expand.
2. Slowly fill the pitcher with cold tap water as you hold the balloon down, its top level with the pitcher's top, until the pitcher overflows. Hold the balloon completely submerged for a minute or so to allow the air inside the balloon to reach the temperature of the water, adding more water if necessary to keep the pitcher full.
3. Remove the balloon and mark the level of the water.
4. Repeat #2 and #3 using warm or tolerably hot water.
5. Is the water level the same both times? Why? What happened to the air in the balloon when it was in warm water? (See Question 11 at the end of this chapter.)

The general equation that relates the pressure, temperature, and volume of a gas is

$$pV = (\text{a constant})\ T$$

The constant depends on the quantity of gas. A given amount of gas can have any combination of pressure, volume, and temperature as long as the three values satisfy the equation. This equation, known as the *ideal gas law*, expresses the interdependence of p, V, and

© Carlyn Galati/Visuals Unlimited

T. For example, if the volume of the gas is fixed, the pressure will increase whenever the temperature increases. In other words, the pressure is proportional to the temperature (in kelvins) as long as the volume stays constant. In the earlier example of heating air in a pressure cooker, the volume of the air inside would be nearly constant. (This is because the volume expansion of the metal is quite small compared to that of a gas.) So the pressure inside would increase proportionally with the temperature.

Regardless of which phase of matter is involved, changing the temperature of a substance will not change its mass or its weight. Since thermal expansion causes the volume to increase while the mass and weight stay the same, the mass density and weight density decrease. The mass density of a hot piece of iron is slightly less than the mass density of the same piece when it is cold. The mass density of the balloon referred to earlier is about 10% less when its temperature is 303 K than when its temperature is 273 K.

The reduction in density of a gas at constant pressure because of heating is used in hot-air balloons. The air in the balloon, which is basically a large bag with an opening at the bottom, is heated with a burner (see ● Figure 5.15). The pressure inside remains equal to the atmospheric pressure because of the opening. Consequently, the air inside the balloon expands and its density decreases. The balloon can float in the air because it is filled with a gas (hot air) that has a smaller density than the surrounding fluid (cooler air). As the air inside cools, the balloon will sink toward the Earth until the burner heats the air again.

Refer to Archimedes' principle, Section 4.5

LEARNING CHECK

1. A key is placed in a hot oven and warmed to 100°C. As a result
   a) the hole in the key becomes smaller.
   b) the key is bent into an arc.
   c) the key becomes longer.
   d) All of the above.
2. (True or False.) A bimetallic strip is made out of two metals that have different coefficients of linear expansion.

3. What odd thing happens to water when its temperature is increased from 1°C to 2°C?
4. Heating the air in a hot-air balloon reduces the air's _____ .

ANSWERS: 1. (c) **2.** True **3.** It contracts (volume decreases) **4.** density

5.3 The First Law of Thermodynamics

We have discussed what temperature is and how changes in temperature can affect the physical properties of matter—density in particular. The next item to consider is how the temperature of matter is changed. This will lead us back to the concept of energy.

There are two general ways to increase the temperature of a substance:

1. By exposing it to something that has a higher temperature
2. By doing work on it in certain ways

The first way is very familiar to you. When you heat something on a stove, warm your hands over a heater, or feel the Sun warm your face, the temperature increase is caused by exposure to something that has a higher temperature: the stove, the heater, or the Sun. (More on this in Section 5.4.) As mentioned before, the atoms and molecules in the substance being warmed gain kinetic energy from those in the hotter substance. The temperature of a substance will rise only if its atoms and molecules gain kinetic energy.

Friction is an effective means of raising the temperature of a substance by the second way (doing work on it). When something is heated by kinetic friction, work is done on it. This was discussed in more detail at the end of Section 3.4.

Explore It Yourself 5.4

For this, you need a bicycle tire pump and a flat bicycle tire that you can inflate. First touch the lower end of the pump (or the metal fittings on the hose if it has one). Pump up the tire to the recommended pressure, and then touch the lower end (or metal fittings) again. What is different? What has happened?

Another example of raising the temperature of a substance by doing work on it is the compression of a gas. When a gas is squeezed (quickly) into a smaller volume, its temperature increases (see ● Figure 5.16). In diesel engines, air is compressed so much that its temperature is raised above the combustion temperature of diesel fuel. When the fuel is injected into the compressed hot air, it ignites.

Computation of the temperature rise of a gas when it is compressed a certain amount is beyond the scope of this text. But we can illustrate the type of heating that occurs by stating the results of one example: If air at 27°C is compressed to one-twentieth of its initial volume in a diesel engine, its temperature will increase to over 700°C (● Figure 5.17).

Both processes can be reversed to cause the temperature of a substance to decrease. The cooler air in a refrigerator lowers the temperature of a pitcher of tea. Air escaping from a tire is cooled as it expands.

Temperature depends on the average kinetic energy of atoms and molecules. For matter to undergo a change in temperature, its atoms and molecules must gain energy (increase temperature) or lose energy (decrease temperature). In gases, the constituent

● **Figure 5.16** Work is done on a gas when it is compressed in a cylinder. This work causes the temperature of the gas to increase.

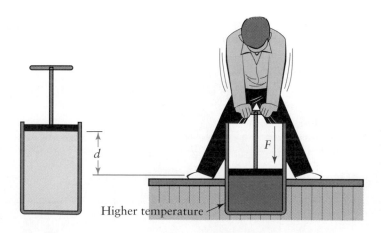

Higher temperature

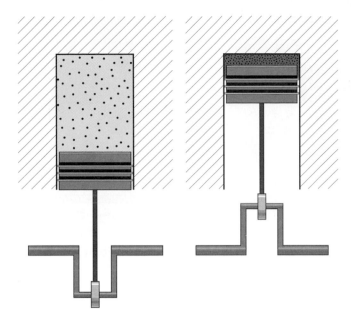

● **Figure 5.17** The air in a diesel engine has an initial temperature of 27°C. It is compressed by the piston until it occupies one-twentieth of its original volume. This raises the temperature to 721°C.

particles have kinetic energy only. All of the energy given to the atoms and molecules acts to increase the temperature of the gas. Things are different in solids and liquids. The atoms and molecules have kinetic energy *and* potential energy because they are bound to each other and oscillate. Energy given to the particles goes to increase both their *PE* and their *KE*. The concept of **internal energy**, which we introduced in Section 3.4, incorporates both forms of energy.

> **DEFINITION**
>
> **Internal Energy** The sum of the kinetic energies and potential energies of all the atoms and molecules in a substance.*

Internal energy is represented by *U*. Its units of measure are the same as those of energy and work.

In gases, the internal energy is the total of the kinetic energies only: The atoms and molecules do not have potential energy. (Gravitational potential energy is not included in internal energy.) In solids and liquids, both the kinetic energy and the potential energy of the particles contribute to the internal energy.

As the temperature of a substance rises, its internal energy increases. If this is accomplished by exposure to a hotter substance, we say that **heat** has flowed from the hotter substance into the cooler substance.

> **DEFINITION**
>
> **Heat** The form of energy that is transferred between two substances because they have different temperatures.

Heat is represented by *Q*. Its units are the same as those of work and energy. Traditionally, the calorie, kilocalorie (also written Calorie), and British thermal unit (Btu) were used exclusively as units of heat. The joule and the foot-pound were used for work and the other forms of energy. Now the joule is becoming the standard unit for heat as well.

---

* Sometimes the concept of internal energy is expanded to include other forms of energy possessed by atoms and molecules. For example, batteries and gasoline have chemical energy that could be included in internal energy.

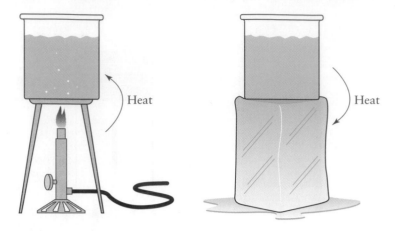

The internal energy of something can increase when heat flows into it and decrease when heat flows out of it (● Figure 5.18). Heat flow is the transfer of energy from a hotter substance to a cooler substance. In this respect, heat is much like work in mechanics. As work is being done, energy is transferred from one thing to another or is transformed from one form to another. A hot pizza does not contain heat, just as a battery and a wound-up spring do not contain work. The battery and the spring contain potential energy, which means they can be used to do work (given the appropriate motor or other mechanism). In the same manner, the hot pizza contains internal energy, and it can therefore transfer heat to something that is cooler. Heat and work are energy in transition while internal energy and potential energy are stored energy.

Two substances with the same temperature are said to be in *thermal equilibrium.* No heat transfers between them. An abandoned cup of hot coffee cools off as heat is transferred from it to the air. Once the coffee and cup reach the same temperature as the air, the transfer stops and they are in thermal equilibrium.

The first law of thermodynamics is a formal summary of the preceding statements.

LAWS

First Law of Thermodynamics The change in internal energy of a substance equals the work done on it plus the heat transferred to it.

$$\Delta U = \text{work} + Q$$

The work referred to in this law must be the type that transfers energy directly to atoms and molecules, such as compressing a gas. Work is done on an object when it is lifted, but this does not affect its internal energy—just its gravitational potential energy.

The work is positive if it is done *on* the substance and negative if the substance does work on something else. When gas in a cylinder is compressed by a piston, the work that is done is positive. If the piston is then released, the gas will expand and push the piston out. In this case, the gas does work on the piston, and the work is negative (● Figure 5.19). Similarly, Q is positive when heat flows into the substance and negative when heat flows out of it. If you place a brick in a hot oven, heat flows into it, and Q is positive. Placing the brick in a freezer would result in a flow of heat out of it and a negative Q.

The first law of thermodynamics is nothing more than a restatement of the law of conservation of energy as it applies to thermodynamic systems. Work done on, or heat transferred to, a substance is "stored" in it as internal energy. In addition to its theoretical significance, the first law of thermodynamics is an important tool used in the analysis of things like internal combustion engines and air conditioners.

This section began with a consideration of how the temperature of matter is changed. How does internal energy fit in with this? Temperature depends on the kinetic energies of the atoms and molecules in a substance. Since these kinetic energies are part of the internal energy of the substance, raising the temperature of something increases its internal energy because it raises the kinetic energies of the atoms and molecules.

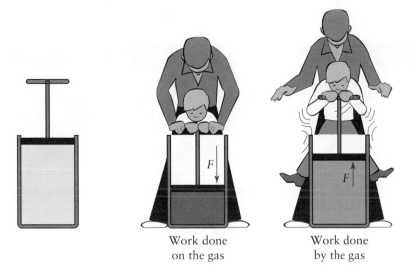

Work done
on the gas

Work done
by the gas

Perhaps you are wondering why the concept of internal energy is used at all, since temperature is determined only by the kinetic energies of particles. Internal energy is useful when considering phase transitions. When water boils on a stove, for example, heat is transferred to it, but the temperature stays the same. This means that the average kinetic energy of the water molecules also stays the same. The heat transferred during a change in phase increases the internal energy by increasing only the potential energies of the molecules. The energy is used to break the bonds between the water molecules and to free them. We will take a closer look at this in Section 5.6.

Concept Map 5.1 summarizes the concepts presented in this section.

CONCEPT MAP 5.1

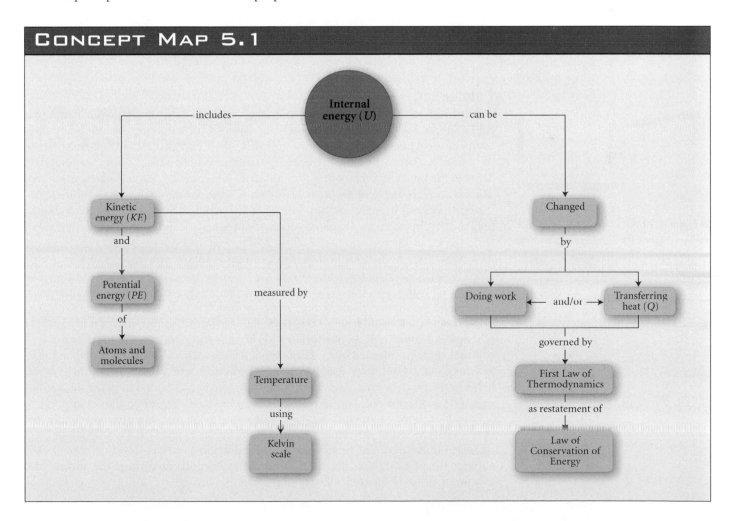

1. (True or False.) The internal energy of a solid body equals the sum of the kinetic energies of all of its constituent atoms and molecules.
2. _____ is transferred from a substance if it is exposed to something that has a lower temperature.
3. The internal energy of air can be increased by
   a) compressing it.
   b) allowing heat to flow out of it.
   c) lifting it.
   d) allowing it to expand.
4. The first law of thermodynamics is a special version of the law of conservation of _____ .

ANSWERS: 1. False 2. Heat 3. (a) 4. energy

5.4 Heat Transfer

Transferring heat is the more common of the two ways to change the temperature of something. Heat transfer occurs whenever there is a temperature difference between two substances or between parts of the same substance. In this section we discuss the three different mechanisms for heat transfer: **conduction, convection,** and **radiation.**

Conduction The transfer of heat between atoms and molecules in direct contact.

Convection The transfer of heat by buoyant mixing in a fluid.

Radiation The transfer of heat by way of electromagnetic waves.

Conduction

Conduction occurs when a pan is placed on a hot stove, when you put your hands into cold water, and when an ice cube comes into contact with warm air. The atoms and molecules in the warmer substance transfer some of their energy directly to the particles in the cooler substance. In these examples, the conduction takes place across the boundary between the two substances, where the atoms and molecules collide with each other. Conduction also is responsible for the transfer of heat from one part of a solid to another part. (Conduction also happens in fluids but is not as important as convection.) Even though only the bottom of a pan is in contact with a hot burner, the heat flows through the metal and soon raises the temperature of all parts of the pan (● Figure 5.20). As the atoms and molecules on the bottom are heated, their constant jostling passes some of their increased speed to neighboring atoms and molecules.

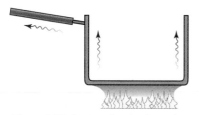

● **Figure 5.20** Heat is conducted to the pan from the flame in contact with it. Conduction also takes place within the pan: Heat flows from the hot bottom up the sides and into the handle.

The ease with which heat flows within matter varies greatly. Materials through which heat moves slowly are called *thermal insulators.* Wool, Styrofoam, and bundles of fiberglass strands are all good insulators because they contain large amounts of trapped air or other gases. Conduction is poor within a gas because the atoms and molecules are not in constant contact with each other. Diamond and metals such as iron and copper are *thermal conductors:* Heat flows readily through them. Concrete, stone, wood, and glass are between the two extremes. A vacuum completely prevents conduction because no atoms or molecules are present.

Metals are good conductors of heat for the same reason that they are good conductors of electricity. Some of the electrons in the atoms in metals are free to move about from one atom to the next. The motion of these "conduction electrons" constitutes an electric current (Chapter 7). These electrons can also carry internal energy from the warmer part of a metal object to the cooler parts.

The conduction of heat within matter is similar to the flow of a fluid, except that nothing material moves from one place to another. The rate at which heat flows from a hot part of an object to a cold part depends on several things—the difference in temperature between the two places, the distance between them, the cross-sectional area through which the heat flows, and how good a thermal conductor the substance is.

Heat conduction is important in our daily lives. In cold weather, we wear clothes that slow the conduction of heat from our warm bodies to the cold air. Handles on metal pans are often made of wood or plastic to reduce the conduction of heat from the burner to your hand. One reason carpets and rugs are used is that they feel warm when you step on them with bare feet. A rug is no warmer than the bare floor next to it, but it is a poorer conductor. When you step on the rug, very little heat is conducted away from your feet, so they stay warm. When you step on bare wood or tile, materials that are better thermal conductors, your feet are cooled more because heat is conducted from them more rapidly (● Figure 5.21).

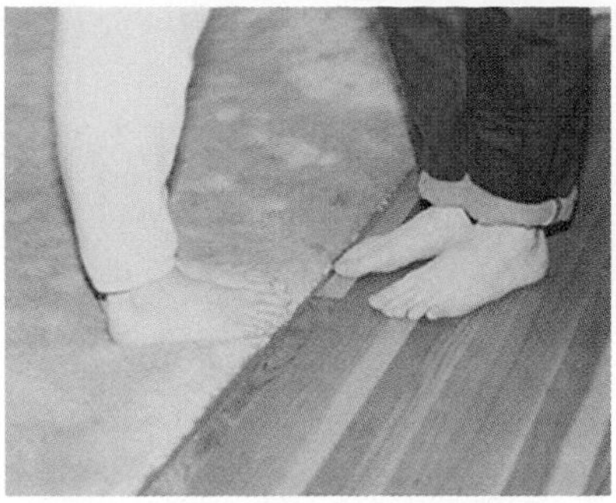

● **Figure 5.21** A rug feels warmer than bare floor because it cuts down on the conduction of heat from your feet.

Convection

Convection is the dominant mode of heat transfer within fluids. Whenever part of a fluid is heated, its thermal expansion causes its density to decrease, so it rises. (Water below 4°C is an exception.) The result is a natural mixing of the fluid. Conduction can then occur between the warmer fluid and the cooler fluid around it.

A room with a wood stove or heater is warmed by convection. The air that is heated is less dense and rises to the ceiling, and cooler, denser air near the floor moves toward the heat source to replace the rising air. The result is a natural circulation of air along the ceiling, floor, and walls of the room (● Figure 5.22). The warmed air near the ceiling cools when it contacts the walls and then sinks to the floor. The same type of circulation can occur in heated aquariums and swimming pools.

You may have heard the statement "heat rises." This is *not* physically correct. Heat is not something material that can rise or fall. "Heated air rises" or "heated fluids rise" are better statements for conveying the idea. Remember also that it is the denser surrounding fluid that pushes the warmer, less dense fluid upwards.

Mechanical mixing of a fluid that causes heat transfer is an example of what is called *forced* convection. Stirring cool cream into hot coffee with a spoon is forced convection— the mixing is not caused by thermal buoyancy. Another example is hot or cold air being blown around the interior of a building or vehicle, causing heat transfer by the mixing of the warmer air with the cooler air.

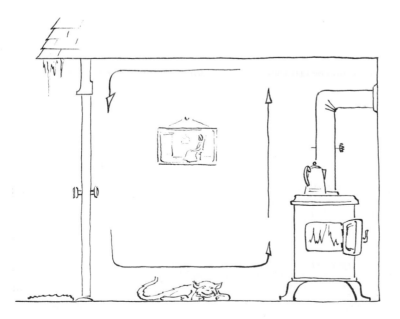

● **Figure 5.22** Convection causes a natural circulation of air in a room heated with a wood stove. Air heated by contact with the stove rises to the ceiling. Air cooled by contact with an outside wall sinks to the floor. This results in lateral movement of air along the ceiling and the floor.

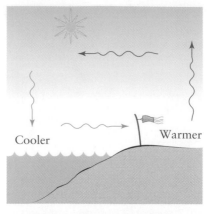

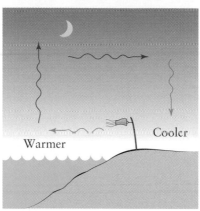

● **Figure 5.23** (top) Sea breeze. Warmed air rises from the heated land, causing cooler air to be drawn in from the sea. (bottom) The land cools at night, so the flow is reversed.

● **Figure 5.24** A glowing light bulb transfers heat to your hand by both radiation and convection.

Convection in the Earth's atmosphere is a major cause of clouds, wind, thunderstorms, and other meteorological phenomena. White, puffy, cumulus clouds are formed when warm air rises into cooler air above.

Sea breezes—steady winds blowing into shore along coasts—are caused by convection. Sunshine warms the land more than the sea, so the air over the ground is heated and rises upward.* This reduces the air pressure over the land, so the higher pressure over the sea forces air to move inland (● Figure 5.23, top). At night, the land cools off more than the sea, so the process is reversed, and a *land breeze* is produced. Air over the cooler ground sinks and forces air to move out to sea (Figure 5.23, bottom).

Large-scale convection takes place within the ocean itself. Water near the equator is heated by the Sun, rises to near the ocean's surface, and flows towards the poles. There, the water cools and sinks deeper, and then flows back towards the equator.

Radiation

Radiation is the transfer of heat via electromagnetic waves. We've all felt the warmth of the Sun on our faces and the heat radiating from a hot fire. This is heat radiation—a type of wave related to radio waves and x rays. (We will discuss electromagnetic waves in more detail in Chapters 8 and 10.) This is the only one of the three types of heat transfer that can operate through a vacuum. The Sun's radiation passes through 150 million kilometers (93 million miles) of empty space and heats the Earth and everything on it. Without the Sun's radiation, the Earth would be a cold, lifeless rock. (More on this in Section 8.7.)

Explore It Yourself 5.5

1. Hold your hand a few inches from the side of a bare light bulb.
2. Remove your hand, let it cool off a bit, then place it the same distance directly above the bulb (● Figure 5.24). What is different? Why?

Infrared heat lamps warm things by emitting radiation. From a distance, you can feel the heat from a camp fire or other heat source because the radiation from it warms your hands and face.

You might think of the radiation as a "vehicle" or "carrier" of internal energy. Internal energy of atoms and molecules is converted into electromagnetic energy—radiation. The radiation then carries the energy through space until it is absorbed by something. When absorbed, the energy in the radiation is converted into internal energy of the atoms and molecules of the absorbing substance.

Everything emits electromagnetic radiation: the Sun, the Earth, your body, this book. The amount and type of radiation depend on the temperature of the emitter. The hotter something is, the more electromagnetic radiation it emits. Things below about 800°F (430°C) emit mostly infrared light, which we cannot see. Hotter substances emit *more* infrared light and also visible light. That is why we can see things in the dark that are "red hot" or "white hot." Things that are even hotter, like the Sun at 10,000°F, emit more infrared and visible light but also emit ultraviolet light. We will take a closer look at heat radiation in Chapter 8.

Everything around you is emitting radiation and also absorbing the radiation emitted by other things. How does this lead to a net transfer of heat? Emission of radiation cools an object, while absorption of radiation warms it. If something absorbs radiation faster

---

* Only the surface of the ground is heated by the Sun, because the soil is a poor conductor. Heat transferred to the sea is quickly spread deeper below the surface by currents and convective mixing. We will also see in Section 5.5 that water requires a great deal of heat to raise its temperature.

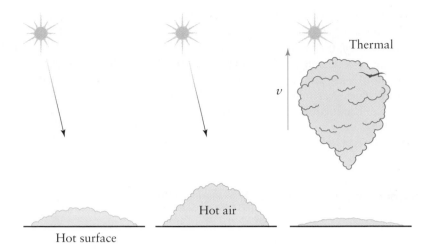

● **Figure 5.25** Thermals are rising "bubbles" of air formed on surfaces heated by the Sun. They give a free upward ride to soaring birds and aircraft. Finding a thermal and then staying in it are a bit difficult because thermals are invisible.

than it emits radiation, it is heated. Your face is warmed by the Sun because it absorbs more radiation than it emits.

Combinations

All three mechanisms of heat transfer are involved in soaring (see p. 172). The Sun warms the Earth via radiation. Air in contact with hot ground is heated by conduction and expands. If conditions are favorable, the hot air will form an invisible "bubble" that breaks free from the surface and rises upward into the air, causing convection. (This is similar to the formation of steam bubbles on the bottom of a pan of boiling water.) These bubbles of rising heated air are called *thermals* (● Figure 5.25). Hang-glider and sailplane pilots and soaring birds such as eagles, hawks, and vultures seek out thermals and circle around in them. The upward speed of a typical thermal is around 5 m/s (11 mph), so they provide an easy, free ride upward.

The different mechanisms of heat transfer become important in reducing heating and cooling costs in buildings. The more heat that flows out of a building in cold weather or into a building in warm weather, the more it costs to heat or cool the building. Reducing this flow of heat saves money (● Figure 5.26). Putting insulation in the walls and ceilings reduces

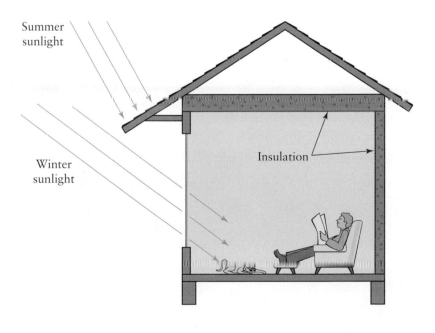

● **Figure 5.26** Heating and cooling costs for a home can be lowered by keeping in mind the mechanisms of heat transfer.

conduction. Using thicker insulation in the ceiling counteracts one effect of convection: The warmest air in the room is at the ceiling, so that is where heat would be transferred most rapidly out of the room during the winter. Using window shades, blinds, and canopies to keep direct sunlight out during the summer reduces heat transfer by radiation.

Thermos bottles are also designed to limit the flow of heat into or out of their contents. They use a near-vacuum between their inner and outer walls that almost eliminates heat flow due to conduction. The inner glass chamber is coated with silver or aluminum to reflect radiation.

The basic concepts of heat transfer are summarized in Concept Map 5.2.

CONCEPT MAP 5.2

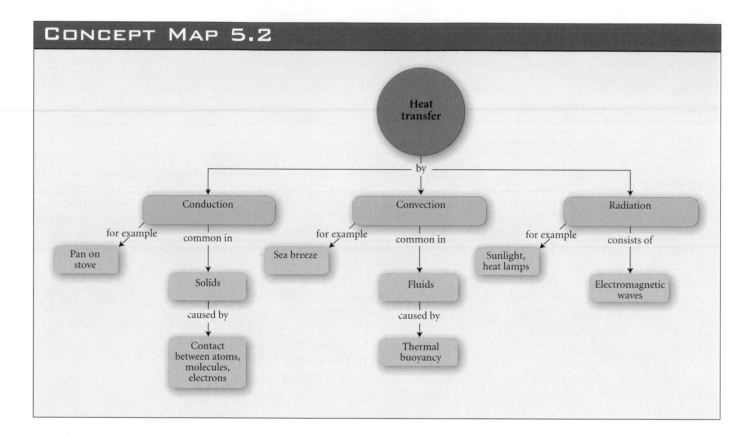

LEARNING CHECK

1. A material through which heat conducts slowly is called a _____.
2. (True or False.) Convection is the only mechanism of heat transfer than can take place in a vacuum.
3. For each of the following situations, indicate which mechanism of heat transfer is the dominant one.
   a) Clothes being dried by hanging them well above a campfire.
   b) A piece of chocolate melting in your mouth.

   c) The surface of the Moon cooling off after the Sun has set.
   d) After a fire alarm sounds, touching a door knob to see if there is fire on the other side of the door.
4. (True or False.) A liquid cannot be heated simultaneously by conduction, convection, and radiation.

ANSWERS: 1. thermal insulator 2. False 3. a) convection b) conduction c) radiation d) conduction 4. False

5.5 Specific Heat Capacity

Transferring heat to a substance or doing work on it increases its internal energy. In this section, we describe how the temperature of the substance is changed as a result. To simplify matters, we assume that no phase transitions take place.

We might state the topic now under consideration in the form of a question: To increase the temperature of a substance by some amount ΔT, what quantity of heat Q must be transferred to it? (We could just as well ask how much work must be done on it.)

The amount of heat needed is proportional to the *temperature increase*. It takes twice as much heat to raise the temperature 20°C as it does to raise it 10°C. So:

$$Q \propto \Delta T$$

The amount of heat needed also depends on the *quantity* (mass) of the substance to which the heat is transferred. For a given increase in temperature, 2 kilograms of water will require twice as much heat as 1 kilogram. So:

$$Q \propto m$$

The quantity of heat required also depends on the *substance*. It takes more heat to raise the temperature of water 1°C than it does to raise the temperature of an equal mass of iron 1°C (● Figure 5.27). As with thermal expansion (Section 5.2), a number can be assigned to each substance indicating the relative amount of heat needed to raise its temperature. This number, called the *specific heat capacity C*, is determined experimentally for each substance. The larger the specific heat capacity of a substance, the greater the amount of heat needed to raise its temperature by a given amount. Thus:

$$Q \propto C$$

We can combine the three proportionalities into the following equation:

$$Q = C m \Delta T$$

The amount of heat required equals the specific heat capacity of the substance times the mass of the substance times the temperature increase. The SI unit of specific heat capacity is the joule per kilogram-degree Celsius (J/kg-°C). If the specific heat capacity of a substance is 1,000 J/kg-°C, it takes 1,000 joules of energy to raise the temperature of 1 kilogram of that substance 1°C. (The kelvin is the same size unit as the °C and can be used, too.) ● Table 5.3 lists the specific heat capacities of some common substances.

● **Table 5.3** Some Specific Heat Capacities

Substance	C (J/kg-°C)
Solids	
Aluminum	890
Concrete	670
Copper	390
Ice	2,000
Iron and steel	460
Lead	130
Silver	230
Liquids	
Gasoline	2,100
Mercury	140
Seawater	3,900
Water (pure)	4,180

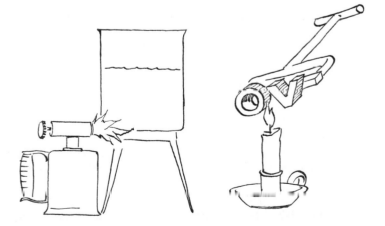

● **Figure 5.27** It takes about nine times as much energy to warm water one degree as it takes to warm the same mass of iron one degree.

Energy Flow in Stars

As discussed in the text, energy can be transported from one location to another by three mechanisms, conduction, convection, and radiation. Usually, the bulk of the energy in any system is transferred by the mechanism that is the most efficient. In a star like our own Sun, energy is moved primarily by radiation throughout most of its interior but by convection during the last 30% or so of its journey from the Sun's core. The specific mechanism at work depends on the local physical conditions in the star.

The Sun is an average star, one of over 10 billion in the Milky Way Galaxy. It is in what is called the *main sequence* phase of its lifetime. We might call this the Sun's "adulthood." In this period, energy is produced at the center, or core, of the star by nuclear fusion reactions involving hydrogen nuclei (protons) (see Chapter 11 for more on fusion). This energy is transported from the core out to regions just beneath the solar surface—a distance of some 70% of the Sun's radius—by radiation, that is, in the form of electromagnetic waves, working its way outward until it reaches what is known as the convection zone. This process is exceedingly slow. The electromagnetic waves travel less than 1 centimeter before they undergo collisions with particles in the solar interior that randomly scatter them, altering their directions of motion. It requires about a million years for energy produced in the Sun's core to reach its outer parts.

When the energy arrives at the outer 30% of the Sun's interior, radiation becomes less efficient as a transport mechanism than convection. The reason for this has to do with what is called the *opacity* of the gas. The opacity is a measure of how effectively the gas absorbs electromagnetic waves. Although the gas in the solar interior is fairly opaque, radiation is still more efficient than convection at moving the energy outward. However, in the very outermost parts of the Sun, the gas temperatures are low enough that hydrogen atoms can exist. The formation of these atoms suddenly makes the gas extremely opaque.

The energy is effectively "dammed up," creating a locally overheated region beneath the Sun's surface. This condition starts the upward movement of the heated, low-density gas and its subsequent mixing with the cooler, overlying fluid. A convection zone develops. From here, the energy finally bursts forth into space at the top of the solar atmosphere.

Photographs of the solar surface taken from high-altitude balloons, rockets, and satellites permit us to see the top of this convection zone (• Figure 5.28). The rolling, bubbling gases can be clearly seen in the form of light and dark convection cells. Such mottling has been variously described as "granulation" or "salt-and-pepper" patterns or as "oatmeal-like" in texture. The brighter areas are warmer, updrafting material. The entire pattern continually changes as the cells rise and fall on time scales of about 10 minutes. Each bubble has an irregular shape and varies in size from 800 to 1200 kilometers across—nearly one-quarter the size of the United States! These are no small cauldrons of boiling brew!

Conduction does not play a significant role in the transport of energy in stars like the Sun because the Sun's material is gaseous and not extremely dense. Conduction does become an important mechanism in the terminal stages of most stars' lives. In this final phase of their evolution, all but the most massive stars end up as small (Earth-sized or less), dense (more than a billion times as dense as water) objects called *white dwarfs*. Because of the compactness of these common end points of stellar evolution, their properties are in some ways similar to those of ordinary solid matter. It may not be too surprising then, given our discussion of heat flow in solids, to find that conduction is an important process controlling their structure and the time it takes for these stars to cool off and finally go dark. In fact, conduction in these terminal stars is so efficient at transporting energy that their interiors have nearly uniform temperatures throughout. Were it not for the thin insulating blanket of normal atmospheric material covering

---

Example 5.2 Let's compute how much energy it takes to make a cup of coffee or tea. Eight ounces of water has a mass of about 0.22 kilograms. How much heat must be transferred to the water to raise its temperature from 20°C to the boiling point, 100°C? The change in temperature is

$$\Delta T = 100 - 20 = 80°C$$

The specific heat capacity C of water is 4,180 J/kg-°C (from Table 5.3). Therefore,

$$Q = C\,m\,\Delta T$$
$$= 4{,}180 \text{ J/kg-°C} \times 0.22 \text{ kg} \times 80°C$$
$$= 73{,}600 \text{ J}$$

It takes an enormous amount of energy to heat water. You may recall that in Example 3.6 we computed the kinetic energy of a small car traveling at highway speed. The answer was 364,500 joules. This much energy is only enough to bring about 5 cups of water to the boiling point from 20°C (• Figure 5.29). Usually, a relatively large amount of mechanical energy does not produce a large temperature change when it is converted into heat. Example 5.3 shows this another way.

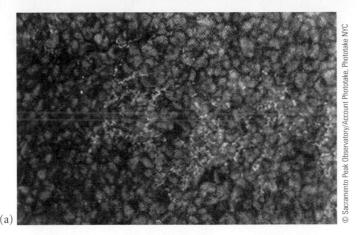

(a)

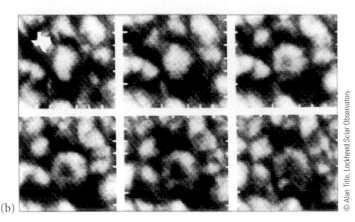

(b)

© Sacramento Peak Observatory/Account Phototake, Phototake NYC

© Alan Title, Lockheed Solar Observatory

● **Figure 5.28** (a) High resolution image of the surface of the Sun showing a pattern of convection cells called *granulation*. (b) Time variations in the pattern of solar granulation. These images were taken at two-minute intervals and show how the granulation pattern on the Sun changes with time. Note particularly the explosive behavior of the central granule. For scale, the upper left panel shows the size of the state of Texas.

them, they would quickly lose all their residual heat and become blackened stellar cinders.

At this stage, you might ask how we know that stars like the Sun transport energy in the manner we've described. Up until recently, *direct* evidence to support the accuracy of this picture was hard to come by. However, with the *Global Oscillations Network Group* (GONG) project and the *Solar and Heliospheric Observatory* (SOHO) spacecraft in the late 1990s, the science of *helioseismology,* the study of surface vibrations on the Sun to understand the nature of the solar interior, reached maturity and provided just the data needed to check our models. It turns out that the roiling motion of the granulation cells near the Sun's surface generates low-amplitude sound waves that propagate through our star. These vibrations can be detected as tiny Doppler shifts in the light emitted by the Sun. (For more on sound waves and the Doppler effect, see Chapter 6, and for a great movie showing the bubbling granulation pattern, check out http:// homepages.wmich.edu/.korista/startstruct.html.) Comparisons between the observed oscillations and the predictions of our best models show extraordinarily good agreement (within ~0.2%!) and place the boundary between the radiative and convection zones at 71.3% of the Sun's radius. For more about helioseismology, the latest data and images from the GONG and SOHO programs may be found by visiting their homepages at www.gong.noao.edu and sohowww .nascom.nasa.gov.

60 mph

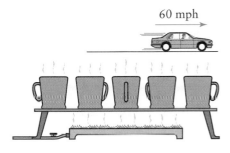

● **Figure 5.29** The energy needed to heat 5 cups of water to boiling is about the same as the energy needed to accelerate a small car to 60 mph.

Example 5.3

A 5-kilogram concrete block falls to the ground from a height of 10 meters. If all of its original potential energy goes to heat the block when it hits the ground, what is its change in temperature?

There are two energy conversions. The block's gravitational potential energy, $PE = mgd$, is converted into kinetic energy as it falls. When it hits the ground, the kinetic energy is converted into internal energy in the inelastic collision (● Figure 5.30). Actually, this internal energy would be shared between the block and the ground, but we assume that

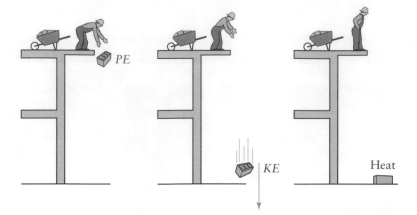

it all goes to the block. So the equivalent amount of heat transferred to the block equals the original potential energy.

$$Q = PE = mgd = 5 \text{ kg} \times 9.8 \text{ m/s}^2 \times 10 \text{ m}$$

$$= 490 \text{ J}$$

The increase in temperature, ΔT, of the block is

$$Q = C m \, \Delta T$$

$$490 \text{ J} = 670 \text{ J/kg-°C} \times 5 \text{ kg} \times \Delta T$$

$$490 \text{ J} = 3,350 \text{ J/°C} \times \Delta T$$

$$\frac{490 \text{ J}}{3,350 \text{ J/°C}} = \Delta T$$

$$\Delta T = 0.15°C$$

At the end of Section 3.4, we discussed how mechanical energy is often converted into internal energy. At that time you may have wondered why you usually don't notice a temperature increase when you slide something across a floor or drop a book on a table. The quantities of energy or work that we typically deal with do not go far in changing the temperatures of the objects involved.

Let's consider one last example in which the amount of mechanical energy is so large that considerable heating does take place.

Example 5.4   A satellite in low Earth orbit experiences slight air resistance and eventually reenters the Earth's atmosphere. As it moves downward through the increasingly dense air, the frictional force of air resistance converts the satellite's kinetic energy into internal energy. If the satellite is mostly aluminum and all of its kinetic energy is converted into internal energy, what would be its temperature increase?

Again, some of the heat is transferred to the air and some to the satellite. Just to get some idea of the potential heating, we assume that all of the heat from the friction flows into the satellite. We do not need to know the mass of the satellite, because m divides out, as we will see.

The heat transferred to the satellite equals its original kinetic energy. In Section 2.8 we computed the speed of a satellite in low Earth orbit—7,900 m/s. So:

$$Q = KE = \frac{1}{2}mv^2 = \frac{1}{2} \times m \times (7{,}900 \text{ m/s})^2$$

$$= (31{,}200{,}000 \text{ J/kg}) \times m$$

The increase in temperature caused by this much heat is

$$Q = (31{,}200{,}000 \text{ J/kg}) \times m = C\,m\,\Delta T$$

$$31{,}200{,}000 \text{ J/kg} = 890 \text{ J/kg-°C} \times \Delta T$$

$$\frac{31{,}200{,}000 \text{ J/kg}}{890 \text{ J/kg-°C}} = \Delta T$$

$$\Delta T = 35{,}000°C$$

Of course, its temperature would not actually increase this much: The satellite would start to melt. Even if 90% of the heat went to the air, the remaining 10% would still be enough to melt the satellite. (This would make $\Delta T = 3{,}500°C$.)

The purpose of Example 5.4 was to show you why satellites and meteoroids usually disintegrate when they enter the Earth's atmosphere. The friction from the air transforms their kinetic energy into enough internal energy to raise their surface temperature by several thousand degrees. The objects simply start melting on the outside. Meteors can be seen at night because of their extreme temperatures: They leave behind a trail of hot, glowing air and meteoroid fragment (see Figure 3.24).

Before the middle of the nineteenth century, the concepts of heat and mechanical energy were not connected to each other. It was thought that heat dealt with changes in temperature only and had nothing directly to do with mechanical energy. (The explanations of heating caused by friction were a bit strange as a result.) Water once again was used as a basis for defining a unit of measure: The calorie was defined to be the amount of heat needed to raise the temperature of 1 gram of water 1°C. Similarly, the British thermal unit (Btu) was defined to be the amount of heat needed to raise the temperature of 1 pound of water 1°F. This made the specific heat capacity of water equal to 1 cal/g-°C = 1 Btu/lb-°F.

In 1843, James Joule announced the results of his experiments in which a measurable amount of mechanical energy was used to raise the temperature of water by stirring it. The amount of internal energy given to the water equaled the mechanical energy expended. The result allowed Joule to calculate what was called the "mechanical equivalent of heat"—the relationship between the unit of energy and the unit of heat:

$$1 \text{ cal} = 4{,}184 \text{ J}$$

It was only later that the metric unit of energy was renamed in honor of Joule.

You may have noticed that the specific heat capacity of water is quite high, nearly twice as high as that of anything else in Table 5.3. This ability to absorb (or release) large amounts of internal energy is another property of water that adds to its uniqueness—and its usefulness. Water is used as a coolant in automobile engines, power plants, and countless industrial processes partly because it is plentiful and partly because its specific heat capacity is so high. Engine parts near where the fuel burns are exposed to very high temperatures. The metal would be damaged, or even melt, if there were no way of cooling the parts. Water is circulated to these areas of the engine, where it absorbs heat from the hotter metal, thereby cooling the parts. The heated water flows to the radiator, where it transfers the heat to the air.

The "food calorie" that dieters count is actually the kilocalorie. So:

$$1 \text{ food calorie} = 1{,}000 \text{ calories}$$

It is used as a measure of the energy content of foods.

5.6 Phase Transitions

A phase transition or "change of state" occurs when a substance changes from one phase of matter to another. ● Table 5.4 lists the common phase transitions, the phases involved, and the effect of the transition on the internal energy of the substance.*

Let's say that a pan of water is placed on a stove and that heat is transferred to the water at some rate. This causes the temperature to rise steadily until the water starts to boil. Then the temperature stays the same (100°C = 212°F at sea level) even though heat is being transferred to the water. The heat is no longer increasing the kinetic energy of the water molecules: It is breaking the "bonds" that hold the molecules in the liquid state. The molecules are given enough energy to break free from the water's surface and become free molecules of steam (● Figure 5.31).

Below the boiling temperature, the water molecules are bound to each other and so have negative potential energies. During boiling, each molecule in turn is given enough energy to break free of the bonds. The average kinetic energy of the molecules stays the same during boiling, and the temperature of the water therefore stays constant.

A similar process occurs when a solid melts. The atoms or molecules go from being rigidly bound to each other in the solid state to being rather loosely bound in the liquid state. As in boiling, the increase in internal energy that occurs during melting goes to increase *only* the potential energies of the atoms or molecules. So the temperature of ice remains at 0°C (32°F) while it is melting.

Condensation and freezing are simply the reverse processes of boiling and melting, respectively. Here the potential energies of the atoms and molecules decrease, but the kinetic energies and the temperature stay the same.

---

* Sublimation is a fairly rare phase transition in which a solid goes directly to a gas, bypassing the liquid phase. The atoms or molecules break free from the rigid forces binding them in a crystal, and move off in the gas phase. Dry ice (solid carbon dioxide) undergoes sublimation at temperatures above −78.5°C under 1 atmosphere. This means that carbon dioxide cannot exist in the liquid phase under normal pressure. Mothballs also undergo sublimation at room temperature.

● Table 5.4 Common Phase Transitions

Name	Phases Involved	Effect
Boiling	Liquid to gas	Increases U
Melting	Solid to liquid	Increases U
Condensation	Gas to liquid	Decreases U
Freezing	Liquid to solid	Decreases U

● **Figure 5.31** The temperature of boiling water stays constant even though heat is flowing into it.

● **Figure 5.32** Higher pressure inside the pressure cooker raises the boiling point of water to about 120°C.

The temperature at which a particular phase transition occurs depends on the properties of the atoms or molecules in the substance—particularly the masses of the particles and the forces acting between them. When these forces are very strong, such as in table salt, the melting and boiling temperatures are quite high. When the forces are very weak, such as in helium, the phase-transition temperatures are very low—near absolute zero.

The boiling temperature of each liquid varies with the pressure of the air (or other gas) that acts on its surface. When the pressure is 1 atmosphere, water boils at 100°C. At an elevation of 3,000 meters (10,000 feet) above sea level, where the pressure is about 0.67 atmospheres, the boiling point of water is reduced to 90°C. That is why it takes longer to cook food by boiling at higher elevations. The temperature of the boiling water is lower, so conduction of heat into the food is slower. If the pressure is *increased* to 2 atmospheres such as in a pressure cooker, the boiling point of water is 120°C and food cooks faster (● Figure 5.32). One type of nuclear power plant, called a *pressurized water reactor*, uses high pressure to keep water from boiling even at several hundred degrees Celsius.

This dependence of the boiling temperature on the pressure makes it possible to induce boiling or condensation simply by changing the pressure. For example, water exists as a liquid at 110°C when under 2 atmospheres of pressure. If the pressure is reduced to 1 atmosphere, the boiling temperature is lowered to 100°C, and the water begins to boil. In the same manner, steam at 110°C and 1 atmosphere pressure will start to condense if the pressure is increased to 2 atmospheres. As we will see in Section 5.7, such pressure-induced phase transitions are the basis for the operation of refrigerators and other "heat movers."

A specific amount of internal energy must be added or removed from a particular substance to complete a phase transition. For example, 334,000 joules of heat must be transferred to each kilogram of ice at 0°C to melt it. This quantity is called the **latent heat of fusion** of water. A much larger amount, 2,260,000 joules, must be transferred to each kilogram of water at 100°C to convert it completely into steam. This is the **latent heat of vaporization** of water. During the reverse processes, freezing and condensation, the same amounts of internal energy must be extracted from the water and steam, respectively.

Example 5.5 Ice at 0°C is used to cool water from room temperature (20°C) to 0°C. How much water can be cooled by 1 kilogram of ice?

Heat flows into the ice as it melts, cooling the water in the process. The maximum amount of water it can cool to 0°C corresponds to all of the ice melting. So the question is, How much water will be cooled by 20°C when 334,000 joules of heat are transferred from it?

$$Q = C m \, \Delta T$$

$$-334,000 \text{ J} = 4,180 \text{ J/kg-°C} \times m \times -20°C$$

$$-334,000 \text{ J} = -83,600 \text{ J/kg} \times m$$

$$\frac{-334,000 \text{ J}}{-83,600 \text{ J/kg}} = m$$

$$m = 4.0 \text{ kg}$$

Ice at 0°C can cool about four times its own mass of water from 20°C to 0°C—provided this takes place in a well-insulated container to minimize heat conduction from the outside.

These large latent heats of ice and water have common applications. The reason that ice is so good at keeping drinks cold is that it absorbs a large amount of internal energy while melting. As long as there is ice in the drink, the temperature remains near 0°C. One reason water is ideal for extinguishing fires is that it absorbs a huge amount of internal energy when it vaporizes. When poured on a hot, burning substance, the water absorbs heat as it boils, thereby cooling the substance (● Figure 5.33).

Explore It Yourself 5.6

You need an electric kitchen range, an empty aluminum soft drink can, an oven mitt or glove, and some kind of clock for timing.

1. Put a tablespoon or so of water into the can.
2. Turn one of the surface heating units to high and wait a minute for it to heat up.
3. Place the can on the unit and time how long it takes before the water starts boiling (you should be able to hear it and see vapor coming out). This might be about 1 minute.
4. "Restart" the clock and time how long it takes for all of the water to be boiled away (when you can no longer hear water boiling). Turn the unit off and **immediately** remove the can using the mitt or glove. If you don't, the can will quickly get hot enough for the paint to start smoking.
5. How do the times compare? What does this tell you about the amount of internal energy it takes to heat up water compared to the amount it takes to convert water at the boiling point to steam?

Here's an example that summarizes the relationship between the temperature of water, its phase, and its internal energy. A block of ice at a temperature of −25°C is placed in a special chamber. The pressure in the chamber is kept at 1 atmosphere while heat is transferred to the ice at a fixed rate. ● Figure 5.34 shows a graph of the temperature of the water versus the amount of heat transferred to it.

Between points *a* and *b* on the graph, the heat transferred to the ice simply raises its temperature. At point *b*, the ice starts to melt, and the temperature stays fixed at 0°C until all of the ice is melted, point *c*. From *c* to *d*, the heat that is transferred to the water goes to increase its temperature. Between *d* and *e*, the water boils while the temperature remains at 100°C. As soon as all of the water is converted into steam, at *e*, the temperature of the steam starts to rise.

We could reverse the process by placing steam in the chamber and transferring heat from it. The result would be like moving along the graph from right to left.

● **Figure 5.33** Exploiting water's high latent heat of vaporization.

Humidity

At temperatures below their boiling points, liquids can gradually go into the gas phase through a process known as **evaporation.** Water left standing will eventually "disappear" because of this. How can this phase transition occur at temperatures below the boiling point? Individual atoms or molecules in a liquid can go into the gas phase if they have enough energy. At temperatures below the boiling point, some of the atoms or molecules do have enough energy to do this. Even though the *average* energy of the particles is too low for boiling to occur, some have more energy than the average, and some have less. Atoms or molecules with higher-than-average energy can break free from the liquid if they are near the surface. Once in the air, the atoms or molecules can remain in the gas phase even though the temperature is below the boiling point.

Because of evaporation, water vapor is always present in the air. The amount varies with geographic location (proximity to large bodies of water), climate, and weather. **Humidity** is a measure of the amount of water vapor in the air.

> **DEFINITION**
>
> **Humidity** The mass of water vapor in the air per unit volume. The density of water vapor in the air.

The unit of humidity is the same as that of mass density. The humidity generally ranges from about 0.001 kg/m³ (cold day in a dry climate) to about 0.03 kg/m³ (hot,

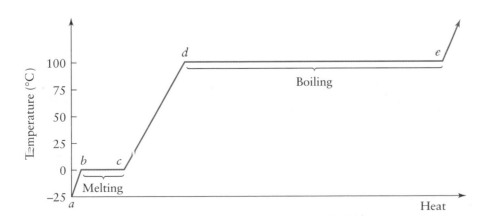

● **Figure 5.34** Graph of the temperature of water versus the heat transferred to it. The phase transitions correspond to the two places where the graph is flat: The internal energy increases, but the temperature stays constant.

● **Figure 5.35** Fog in the cooler air in a valley. Water droplets form when the humidity exceeds the saturation density—when there is too much water vapor in the air.

NOAA Photo Library, NOAA Central Library; OAR/ERL/National Severe Storms Laboratory (NSSL)

humid day). Note that these densities are much less than the normal density of the air, 1.29 kg/m³. Even in humid conditions, water vapor is only a small component of the air—less than 5%.

At any given temperature, there is a maximum possible humidity, called the *saturation density*. This upper limit exists because the water molecules in the air "want" to be in the liquid phase. If there are too many molecules in the air, the probability is high that several will get close enough to each other for the attractive force between them to take over. They begin to form droplets and condense onto surfaces. This is what happens during the formation of fog and dew, or the mist in a shower (● Figure 5.35). The saturation density is *higher* at higher temperatures because the water molecules are moving faster and are less likely to "stick" together when they collide. Hence, more molecules can be present in any volume of air without droplet formation. ● Table 5.5 lists the saturation density of water vapor in the air at several different temperatures.

The reverse of evaporation also takes place: Water molecules in the air near the surface of water can be deflected into the water and "captured." When the humidity is well below the saturation density at that temperature, evaporation occurs faster than reabsorption of water molecules. If the humidity increases to the saturation density, water molecules are reabsorbed into the water at the same rate that they evaporate. The water does not disappear. This is why damp towels dry slowly in humid environments.

The upshot of this is that the humidity alone doesn't determine how rapidly water evaporates. What matters is how close the humidity is to the saturation density. The **relative humidity** is a good indicator of this relation.

DEFINITION	**Relative Humidity** The humidity expressed as a percentage of the saturation density. $$\text{relative humidity} = \frac{\text{humidity}}{\text{saturation density}} \times 100\%$$

When the relative humidity is 40%, this means that 40% of the maximum amount of water vapor is present.

Example 5.6

What is the relative humidity when the humidity is 0.009 kg/m³ and the temperature is 20°C?

From Table 5.5, the saturation density at 20°C is 0.0173 kg/m³. Therefore,

$$\text{relative humidity} = \frac{0.009 \text{ kg/m}^3}{0.0173 \text{ kg/m}^3} \times 100\%$$

$$= 52\%$$

The same humidity in air at 15°C would make the relative humidity 70%. If the air were cooled to 9°C, the relative humidity would be 100%.

● **Table 5.5** Saturation Density of Water Vapor in the Air

Temperature		Saturation Density	Temperature		Saturation Density
(°C)	(°F)	(kg/m³)	(°C)	(°F)	(kg/m³)
−15	5	0.0016	15	59	0.0128
−10	14	0.0022	20	68	0.0173
−5	23	0.0034	25	77	0.0228
0	32	0.0049	30	86	0.0304
5	41	0.0068	35	95	0.0396
10	50	0.0094	40	104	0.0511

When air is cooled and the water vapor content stays constant, the relative humidity *increases*. When air is heated and the humidity stays constant, the relative humidity *decreases*. This is why heated buildings often feel dry in the winter. Cold air from the outside enters the building and is heated. Unless water vapor is artificially added to this air with a humidifier, the relative humidity will be very low.

If air is cooled while the humidity stays constant, eventually condensation begins to occur. The temperature when this happens is called the *dew point* temperature. Often on clear nights, the air temperature drops until the dew point is reached and condensation begins. The result is dew on plants and other surfaces—hence the name "dew point." The dew point is easily predicted if the humidity is known: It is the temperature at which that humidity is the saturation density. Table 5.5 or ● Figure 5.36 can be used for this purpose. For example, when the humidity is 0.0128 kg/m³, the dew point is 15°C (59°F).

Figure 5.36b shows how the dew point can be estimated graphically. The point ○ represents the air described in Example 5.6. After the air is cooled from 20°C to 15°C, its state is marked by ●. The point moved horizontally to the lower temperature. Continued cooling would cause the air to reach the saturation density curve at a temperature of about 9°C, so that is the dew point.

Water droplets often form on the sides of cans and other containers holding cold drinks. This happens when the temperature of the container's sides is below the dew point temperature. Air in contact with the surface is cooled until the dew point is reached, and condensation begins. (This process also occurs when car windows "fog over" in cold weather.) If the air is very dry (low relative humidity), condensation doesn't occur because the dew point is below the temperature of the cold drink.

The process of evaporation cools a liquid. Atoms or molecules in the gas phase have more energy than when in the liquid phase. When water molecules evaporate, they take away some internal energy from the water, thereby cooling it. Here is another way of looking at it: Since the molecules with the higher kinetic energies are the ones that evaporate, the average kinetic energy of the water molecules that remain is lowered.

Our bodies are cooled by the evaporation of perspiration. On hot days, we feel hotter if the humidity is high because evaporation is inhibited. A breeze feels cool because it removes air near our skin that has a higher humidity due to perspiration. The drier air that replaces it allows for more rapid evaporation—and cooling.

Concept Map 5.3 summarizes the basic phase transitions and their effect on the potential energy of atoms and molecules.

● **Figure 5.36** (a) Graph of the saturation density versus temperature (data in Table 5.5). (b) Same graph, showing the situation described in Example 5.6.

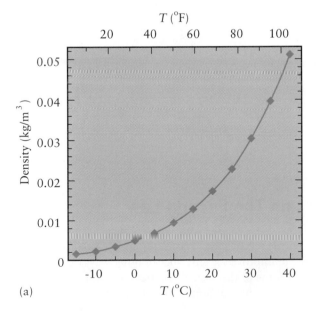

(a)

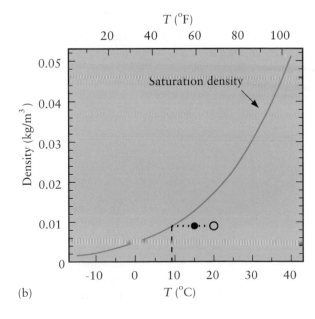

(b)

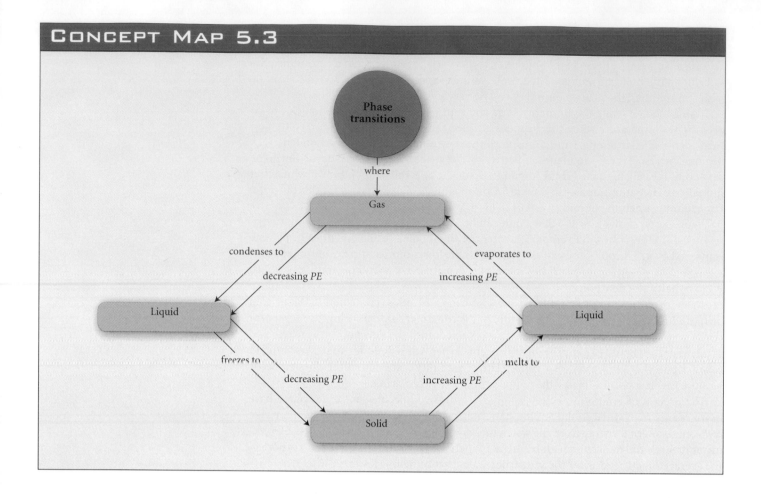

Phase transitions

where

Gas

condenses to

decreasing *PE*

evaporates to

increasing *PE*

Liquid

Liquid

freezes to

decreasing *PE*

melts to

increasing *PE*

Solid

LEARNING CHECK

1. If heat is flowing into water but its temperature is not changing, what can you conclude is happening?
2. (True or False.) Water's very high latent heat of vaporization is a major reason it is very effective at extinguishing fires.
3. The relative humidity of the outside air usually increases during the night. Why?
4. When droplets form on the outside of a glass containing ice water, this means that in the air right next to the glass

a) the humidity is approximately equal to the saturation density.
b) the relative humidity is about 100%.
c) the temperature is approximately equal to the dew point.
d) All of the above.

ANSWERS: 1. It is boiling. **2.** True **3.** Air temperature decreases. **4.** (d)

5.7 Heat Engines and The Second Law of Thermodynamics

Along with the efforts made in the recent past to conserve energy used to heat and cool buildings, much has been done to improve the efficiency of other ways that we use energy. In this section, we will consider some of the basic theoretical principles that are involved in energy-conversion devices that use heat and mechanical energy.

Most of the energy used in our society comes from fossil fuels—coal, oil, and natural gas. Part of these fuels is burned directly for heating, such as in gas stoves and oil furnaces. But most of these fuels are used as the energy input for devices that are classified together as **heat engines.**

Heat Engine A device that transforms heat into mechanical energy or work. It absorbs heat from a hot source such as burning fuel, converts some of this energy into usable mechanical energy or work, and outputs the remaining energy as heat to some lower-temperature reservoir.

Gasoline engines, diesel engines, jet engines, and steam-electric power plants are all heat engines. In gasoline and diesel engines, some of the heat from burning fuel is converted into mechanical energy. The remainder of the heat is ejected to the air from the exhaust pipe, the radiator, and the hot surfaces of the engine. Coal, nuclear, and some types of solar power plants produce electricity by using steam to turn a generator (see ● Figure 5.37). Heat from burning coal, fissioning nuclear fuel, or the Sun is used to boil water. The steam is piped to a turbine (basically a propeller) that is given rotational energy by the steam. A generator connected to the turbine converts rotational energy into electrical energy. After the steam leaves the turbine, it is condensed back into the liquid phase via cooling with water or the air. Cooling towers, which function somewhat like automobile radiators, are used in the latter case. Most of the heat transferred to the water to boil it is ejected from the plant as waste heat when the steam is condensed.

Even though the actual inner workings of heat engines are quite complicated, from an energy point of view we can represent them with a simple schematic diagram (see ● Figure 5.38). Energy in the form of heat from some source is input into a mechanism. The mechanism converts some of the heat into mechanical energy and rejects the remainder. In this simplified representation of a heat engine, the heat source has some fixed (high) temperature T_h and the reservoir that absorbs the rejected heat has some fixed (lower) temperature T_l.

During a given period while the engine is operating, some quantity of heat Q_h is absorbed from the heat source. Some of this energy is converted into usable work, and the

● **Figure 5.37** Simplified sketch of an electric power plant showing the energy (heat) inputs and outputs.

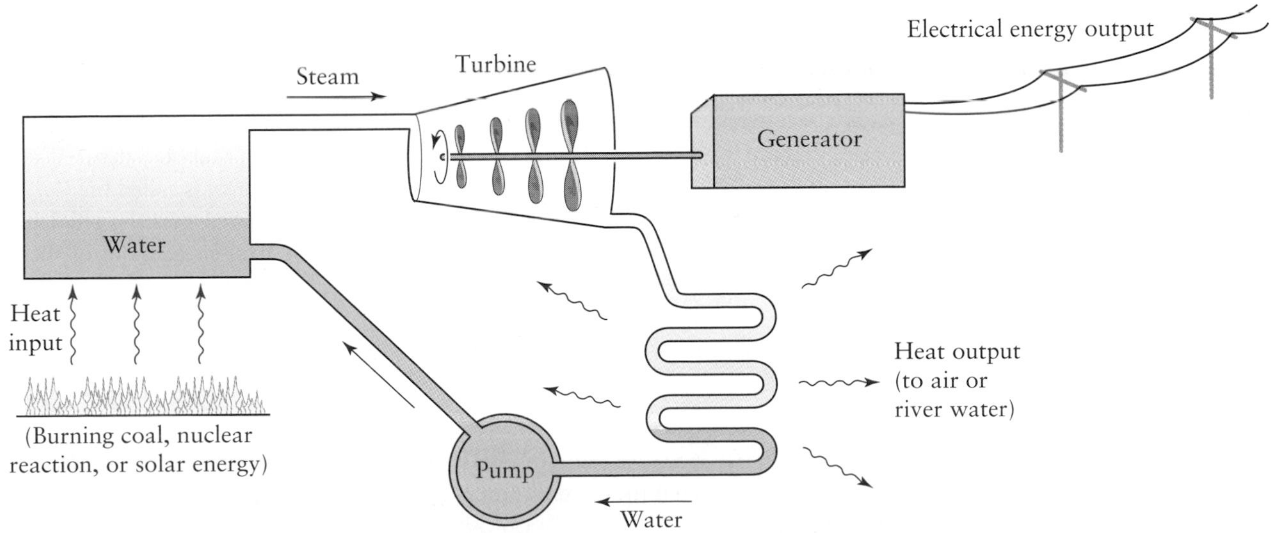

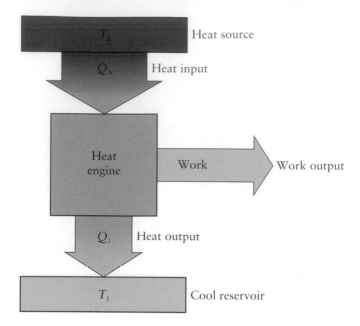

● **Figure 5.38** Diagram of a heat engine. The mechanism absorbs heat from the heat source, uses some of it to do work, and releases heat to some lower-temperature reservoir.

remainder of the original energy input is ejected as waste heat Q_l. This wasted heat is unavoidable. The following law is a formal statement of this.

LAWS

Second Law of Thermodynamics No device can be built that will repeatedly extract heat from a source and deliver mechanical work or energy without ejecting some heat to a lower-temperature reservoir.

The energy efficiency of any device or process is the usable output divided by the total input, times 100%.

$$\text{efficiency} = \frac{\text{energy or work output}}{\text{energy or work input}} \times 100\%$$

If the efficiency of a device is 25%, then one-fourth of the input energy is converted into usable form. The remaining three-fourths are "lost"—released as waste heat.

For heat engines, the input energy is Q_h, and the output is the work. Therefore,

$$\text{efficiency} = \frac{\text{work}}{Q_h} \times 100\% \qquad \text{(heat engine)}$$

As we have seen, there are many different types of heat engines. Some of them use processes that are inherently more efficient than others. However, there is a theoretical upper limit on the efficiency of a heat engine. This maximum efficiency is called the *Carnot efficiency* after the French engineer Sadi Carnot. Carnot discovered that the efficiency of a perfect heat engine is limited by the temperatures of the heat source and of the lower-temperature reservoir. In particular:

$$\text{Carnot efficiency} = \frac{T_h - T_l}{T_h} \times 100\%$$

(T_h and T_l must be in kelvins.)

Real heat engines have friction, imperfect insulation, and other factors that reduce their efficiencies. But even if these were completely eliminated, a heat engine could not have an efficiency of 100%.

Example 5.7

A typical coal-fired power plant uses steam at a temperature of 1,000°F (810 K). The steam leaves the turbine at a temperature of about 212°F (373 K). What is the theoretical maximum efficiency of the power plant? Here, $T_h = 810$ K and $T_1 = 373$ K. So:

$$\text{Carnot efficiency} = \frac{810 - 373}{810} \times 100\%$$

$$= 54\%$$

This is the ideal efficiency. The actual efficiency of a coal power plant is usually around 35 to 40%.

Nearly two-thirds of the energy input to a power plant is lost as waste heat (● Figure 5.39). The energy not used causes thermal pollution by artificially heating the environment. The actual top efficiencies of the other heat engines in common use are similar: diesel engine, 35%; jet engine, 23%; gasoline (piston) engine, 25%.

There are two general ways to improve the efficiency of a heat engine. One is to improve the process and reduce energy losses so that the efficiency gets closer to the Carnot efficiency. The other way is to increase the Carnot efficiency by raising T_h or lowering T_1. The efficiencies of steam-based heat engines have been improved a great deal through the use of higher-temperature steam.

Heat Movers

Refrigerators, air conditioners, and heat pumps are devices that act much like heat engines in reverse. They use an input of energy to cause heat to flow from a cooler substance to a warmer substance—opposite the natural flow from hot to cold. The purpose of refrigerators and air conditioners is to extract heat from an area and thereby cool it. Heat pumps are designed to heat an interior space in the winter as well as cool it in the summer. In these devices, as heat is transferred from a substance being cooled, a larger amount of heat flows into a different substance that is then warmed. A refrigerator removes heat from its interior and ejects heat into the room. We will call these devices **heat movers,** since the term describes what they do.

The mechanisms in the three devices are much the same. A gas, called the *refrigerant,* is forced to change phases in a cyclic process. The gas must have a fairly low boiling point and be condensed easily when the pressure on it is increased. Freon (CCl_2F_2) was the most commonly used refrigerant, although it is being replaced by other compounds that do not destroy ozone after being released into the atmosphere. The gas is compressed into the liquid phase by a pump and forced to flow through a small opening called an *expansion valve* (● Figure 5.40). The pressure on the other side of the valve is kept low so that the refrigerant quickly goes into the gas phase. This phase transition cools the refrigerant, and in the process, it absorbs heat from the surroundings. The gas flows back to the pump, where it is recompressed into the liquid phase. This raises the temperature of the refrigerant, causing heat to flow from it to the surroundings. After the refrigerant is cooled and liquefied, it flows through the expansion valve, and the cycle repeats.

Although the net result is a flow of heat from something cooler to something warmer, the actual heat flow into, and then out of, the refrigerant is always a flow from something warmer to something cooler. Basically, the refrigerant is first cooled to a temperature below that of the substance being cooled and then heated to a temperature above that of

● **Figure 5.39** For every three railroad cars of coal burned at an electric power plant, the energy contained in about two of them goes to waste heat. This is a consequence of the second law of thermodynamics, as well as an imperfect heat engine.

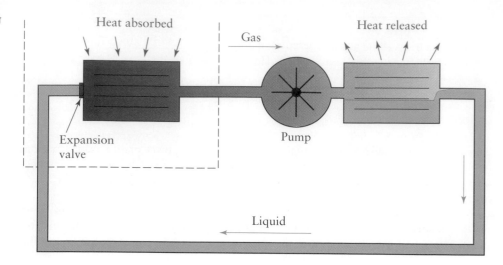

Figure 5.40 Refrigerators remove heat from their interiors by exploiting the phase transition of the refrigerant. As the refrigerant vaporizes, it absorbs heat from the refrigerator's interior. As it condenses, it releases heat to the air in the room.

Heat absorbed

Gas

Heat released

Expansion valve

Pump

Liquid

the substance being warmed. Also note that energy must be supplied to the pump to make the process work.

Heat movers can be represented with a diagram similar to that of heat engines (● Figure 5.41). Unlike heat engines, the mechanism in heat movers has an *input* of energy or work. During a given period of time, a quantity of heat Q_l is absorbed from a cool reservoir with temperature T_l. An amount of work is done on the refrigerant by the pump, and an amount of heat Q_h is released to a warm reservoir with temperature T_h. In refrigerators, the cool reservoir is the air in their interiors. The warm reservoir is the air in the room. Air conditioners absorb heat from the air that they circulate inside a building, a vehicle, and so on, and transfer heat to the warmer outside air. When in the "heating mode," heat pumps remove heat from the outside air, ground, or groundwater and release heat inside a building. In the "cooling mode," heat pumps reverse the heat flow: Heat is transferred from the inside of the building and ejected to the outside.

The law of conservation of energy tells us that the amount of heat released, Q_h, is equal to the amount of heat absorbed, Q_l, plus the energy input to the mechanism. The relative values of Q_h, Q_l, and the energy input depend on the efficiency of the pumping mechanism, the difference between the two temperatures, T_h and T_l, and other factors. In the average refrigerator, the amount of heat transferred from the interior is about *three times* the amount of energy input. In the heating mode, the amount of heat delivered by a good heat pump is about *three times* the amount of energy consumed by the mechanism.

Figure 5.41 Diagram of a heat mover. A mechanism uses an energy input to extract heat from a cool source and to release heat to a higher-temperature reservoir.

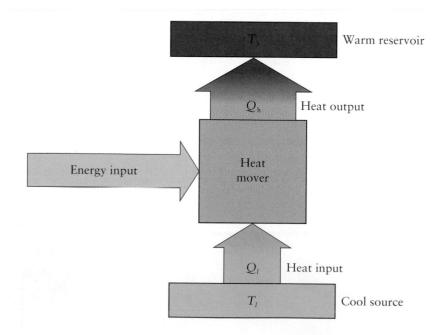

T_h Warm reservoir

Q_h Heat output

Energy input

Heat mover

Q_l Heat input

T_l Cool source

Energy is not being "created": The heat pump is simply extracting heat from a cooler substance and then transferring a larger amount of heat to a warmer substance.

Usable Energy

The outputs of a heat engine are heat flowing into a low-temperature reservoir, thereby increasing its internal energy, and useful work or energy. The latter eventually becomes internal energy as well because of friction and other processes. So the overall effect of a heat engine is to convert internal energy in a higher-temperature source into internal energy in a lower-temperature reservoir. This internal energy, in turn, is no longer available to do as much useful work: Carnot tells us that lower temperatures would give us lower efficiencies. Even though energy is not "lost" or "destroyed" by a heat engine, it is made unavailable for similar use again.

This is a general result of energy transformations: The energy that remains is less usable or less organized than the original energy. Electricity is "high-quality" energy: It is very organized and can be used in many different ways. Light bulbs, heaters, and blow dryers convert electrical energy into lower-quality energy—heat. (Even the light from a light bulb eventually becomes internal energy.) There are many other examples of this. Even the high-quality energy in fine Swiss chocolate becomes low-quality internal energy after we eat it and exercise.

Observations like these lead us to the following conclusion: Natural processes tend to increase the disorder of systems. This tendency is not limited to energy. When you let the air out of a tire, you reduce the order of the system: low-pressure air is less useful and more disordered than high-pressure air. Refined and manufactured metal products, like cars, become less ordered as they rust and fall apart. A fallen tree in a forest decays into a more disordered state.

The physical quantity **entropy** is a measure of the amount of disorder in a system. Natural processes tend to *increase* the entropy of a system because they increase the disorder (● Figure 5.42). This is another way of stating the second law of thermodynamics.

The entropy of part of a system can be decreased temporarily at the expense of the rest of the system. As living things grow, they become more highly ordered, and their entropy decreases. But in so doing, they increase the entropy of the food they consume. After they die, the entropy of their remains once again increases. As a car battery is charged, its entropy decreases, but the charging system is consuming energy and increasing entropy elsewhere.

The term "energy shortage" is somewhat misleading. There is as much energy around now as ever before. But our society uses up high-quality, conveniently stored energy—fossil fuels, uranium, forests—and leaves low-quality internal energy. We are increasing the entropy of the world, leaving less *usable* energy available for future generations. However, by tapping renewable energy sources—solar, wind, geothermal, and tidal—we can use continuous supplies of energy that would otherwise go to waste. Solar energy eventually becomes high-entropy internal energy, whether it is absorbed by the ground or by solar cells. If absorbed by solar cells, we get low entropy electricity as an intermediate step

© John Kaprielian/Photo Researchers

● **Figure 5.42** Devices like this stove transform energy from one form to another, leaving the total amount of energy in the universe unchanged. The output energy (low-temperature heat), however, is in a less usable form than the input energy (high-quality chemical energy in the fuel). The stove increases the entropy of the universe.

LEARNING CHECK

1. (True or False.) The efficiency of a heat engine could be increased by lowering the temperature of its cool reservoir.
2. Which of the following devices is *not* a heat mover?
   a) air conditioner   b) heat engine   c) heat pump
   d) refrigerator
3. (True or False.) For each joule of electrical energy that a refrigerator uses, more than 1 joule of heat can be transferred from the stuff inside it.
4. As a heat engine operates, it does not change the amount of energy there is in the universe, but it does increase the _____ of the universe.

ANSWERS: 1. True   2. (b)   3. True   4. entropy

The first primitive thermometer was invented by none other than Galileo in 1592. It made use of the expansion and contraction of air with temperature. A glass bulb equipped with a long, thin neck was partially filled with water and then inverted into a container of water (● Figure 5.43). The water level would move up and down in the neck as changes in temperature caused the air in the bulb to expand and contract. The device was also affected by changes in atmospheric pressure, so it was a combination thermometer and barometer. Galileo did not equip his device with a temperature scale.

In 1657, a number of disciples of Galileo formed an "academy of experiment" in Florence and developed functional thermometers. These were based on the thermal expansion of alcohol in sealed tubes and were not affected by changes in atmospheric pressure. The academicians established temperature scales for their thermometers by choosing two fixed points and dividing the interval into a number of degrees.

The Florentine thermometers became quite popular in Europe and stimulated many scientists to attempt improvements. An astronomer in Paris named Ismaël Boulliau was the first to use mercury in a thermometer. Perhaps the most famous name in the history of the development of the modern thermometer is Gabriel Daniel Fahrenheit (1686–1736), a scientist and manufacturer of meteorological instruments. Fahrenheit experimented extensively with both alcohol and mercury thermometers, as well as with fixed points and temperature scales. Fahrenheit made the important discovery that the boiling temperature of liquids varies with atmospheric pressure. This accounted for the confusion that other experimenters encountered when trying to use the boiling point of a liquid as a fixed point on a temperature scale. The scale that he developed used as its fixed points the temperature of an ice-salt-water mixture and the temperature of the human body. Fahren-

● **Figure 5.43** Galileo's "thermometer."

heit divided the interval into 96 steps and made the first fixed point 0 degrees. The scale was later adjusted after measurement errors were discovered. The result is our modern Fahrenheit scale with the temperature of melting ice 32° and that of boiling water (at normal atmospheric pressure) 212°.

Several experimenters sought to establish a scale based on 100° between fixed points—a "centigrade" scale. The Swedish astronomer Andreas Celsius (1701–1744) introduced a scale that used 100 degrees between the freezing and boiling points of water, but it read backward: 100° was the freezing temperature and 0° was the boiling temperature. This scale was later inverted and is our modern Celsius scale. But no one temperature scale has ever quite gained universal acceptance. During the eighteenth century, there were more than a dozen different scales in use.

The existence of an "absolute zero" was first indicated in experiments with air thermometers like Galileo's. The volume of a fixed amount of a gas was known to decrease with temperature. One could simply graph the volume of the gas versus the temperature and extrapolate the straight line to the point where the volume would be zero (see ● Figure 5.44). The corresponding temperature would be the coldest possible temperature—absolute zero. It was known to exist even before it was possible to get within 200 degrees of it. This temperature is the zero point of the Kelvin scale, named after the British physicist Lord Kelvin, who first established an absolute temperature scale.

Heat

The development of the understanding of the basic nature of heat presents an interesting counterexample to the general progress of science. Originally, Democritus and other Greeks thought that heat was some kind of material substance that flowed into objects when

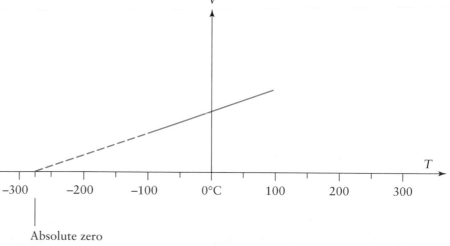

● **Figure 5.44** Lowering the temperature of a gas under constant pressure lowers the volume it occupies. The dotted line indicates that the volume would go to zero at some very low temperature—absolute zero.

they were heated. This view was challenged by many leading scientists of the seventeenth century, including such familiar ones as Isaac Newton and Robert Boyle, who considered heat to involve some kind of internal motion. Although this was the correct approach, it was replaced in the eighteenth century by a reversion to the material model of heat. Joseph Black (1728–1799), a Scottish scientist, and Antoine Lavoisier were two of the strongest proponents of the *caloric theory* of heat. Heat was supposedly an invisible, massless fluid called "caloric." This model, though incorrect, was adequate for explaining many thermodynamic phenomena. Lavoisier measured the specific heat capacities of many substances, and Black successfully measured the quantity of heat needed to melt a given amount of ice. The science of thermodynamics progressed quite nicely during the next century, even though the basic concept of heat was wrong.

The Achilles' heel of the caloric theory was friction. How was heat, and therefore caloric, produced by friction? Among the first to dispute the caloric theory was Benjamin Thompson, later titled Count Rumford (● Figure 5.45). Rumford (1753–1814), one of

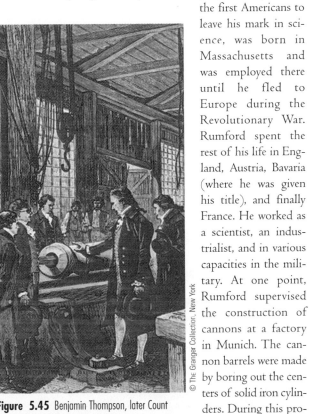

● **Figure 5.45** Benjamin Thompson, later Count Rumford, at a cannon factory.

the first Americans to leave his mark in science, was born in Massachusetts and was employed there until he fled to Europe during the Revolutionary War. Rumford spent the rest of his life in England, Austria, Bavaria (where he was given his title), and finally France. He worked as a scientist, an industrialist, and in various capacities in the military. At one point, Rumford supervised the construction of cannons at a factory in Munich. The cannon barrels were made by boring out the centers of solid iron cylinders. During this process, friction between the drill bit and the

cannon barrel produced heat. The caloric theory implied that the friction released caloric from the metal. Rumford performed many experiments and found that the amount of heat that could be produced when a dull bit was used was unlimited. How could the iron

● **Figure 5.46** James Prescott Joule.

contain an infinite quantity of a material substance (caloric)? Rumford reasoned that the heat was produced by the action of the drill bit and that caloric did not exist.

Several experiments by the English brewer and amateur scientist James Prescott Joule (1818–1889; ● Figure 5.46) indicated that heat is a form of energy. His most famous experiment consisted of a paddle wheel that was made to churn water, thereby heating it, by the action of a falling weight. Joule measured the temperature change of the water and used its specific heat capacity to relate the amount of work done by the falling weight to the heat transferred to the water. The result is the famous "mechanical equivalent of heat" (Section 5.5). In other experiments, Joule measured the heat produced by the compression of air and by electrical currents. This, along with the work of many other scientists, led to the first law of thermodynamics.

A great impetus to the science of thermodynamics was provided by attempts to improve the performance of steam engines. The SI unit of power is named after James Watt, a Scottish engineer who was responsible for several design changes. Nicolas Léonard Sadi Carnot (1796–1832; ● Figure 5.47) was a French engineer who introduced the theoretical analysis of heat engines. Initially a follower of the caloric theory, Carnot was nonetheless able to determine what factors affect the efficiency of a heat engine. His analysis was later modified to include the fact that heat is associated with internal energy, not caloric.

Research in thermodynamics has received new impetus in recent decades as concerns about energy supplies and costs have made energy efficiency ever more important. Success in this area is manifest by the fact that most of today's cars, buildings, and devices use less energy to perform the same function than those of 30 years ago.

● **Figure 5.47** Sadi Carnot.

- **Temperature** is the basis of our sense of hot and cold. Physically, the temperature of a substance is proportional to the average kinetic energy of its atoms and molecules.

- Three different temperature scales are in common use today—the **Fahrenheit, Celsius,** and **Kelvin** scales. The Kelvin temperature scale uses the lowest possible temperature, **absolute zero,** as its zero point.

- In most cases, matter expands when its temperature is raised. The amount of expansion depends on the substance and the temperature change. This property is exploited in mercury and alcohol thermometers and in bimetallic strips.

- The **internal energy** of a substance is the total potential and kinetic energies of its atoms and molecules. It can be increased by doing work on the substance and by transferring **heat** to it.

- The **first law of thermodynamics** states that the change in internal energy of a substance equals the work done on it plus the heat transferred to it.

- Heat can be transferred from one place to another three ways: **Conduction** is the transfer of heat via contact between atoms or molecules. **Convection** is the transfer of heat via buoyant mixing in a fluid. **Radiation** is the transfer of heat via electromagnetic radiation. In some situations, all three processes occur at the same time.

- Except during phase transitions, the temperature of matter increases whenever its internal energy increases. The specific heat capacity is a characteristic of each substance that relates the mass and temperature increase to the heat transferred.

- During phase transitions, the potential energies of the atoms and molecules change while the average kinetic energy remains constant. So the internal energy increases while the temperature stays constant.

- **Evaporation** is a liquid-to-gas transition that occurs below the boiling temperature. It is responsible for the water vapor present in the air.

- **Humidity** and **relative humidity** are two measures of the water-vapor content.

- **Heat engines** are devices that use heat from a hot source to do work. They release heat at a cooler temperature in the process. The maximum efficiency of a theoretically "perfect" heat engine is determined by the two temperatures.

- **Heat movers** use an energy input to remove heat from a cool substance and to transfer it to a warmer substance.

- **Entropy** is a measure of the disorder in a system. Heat engines do not "consume" energy, but they do increase entropy, thus reducing the energy's availability. Increasing the entropy (disorder) of the universe is a result of all natural processes.

IMPORTANT EQUATIONS

Equation	Comments
$\Delta l = \alpha \, l \, \Delta T$	Thermal expansion of a rod
$pV = (\text{a constant}) \, T$	Pressure, volume, and temperature of a fixed amount of a gas (ideal gas law)
$\Delta U = \text{work} + Q$	First law of thermodynamics
$Q = C \, m \, \Delta T$	Heat needed to raise the temperature by ΔT
$\text{relative humidity} = \dfrac{\text{humidity}}{\text{saturation density}} \times 100\%$	Definition of relative humidity
$\text{efficiency} = \dfrac{\text{energy or work output}}{\text{energy or work input}} \times 100\%$	Definition of efficiency of a heat engine
$\text{efficiency} = \dfrac{T_h - T_l}{T_h} \times 100\%$	Carnot efficiency

MAPPING IT OUT!

1. In Section 5.2, we discussed the phenomenon of thermal expansion. The concept map on page 213 was to have been used to help you organize and understand this material better. When the map was printed, however, we realized that several key concepts and linking words/phrases had been left out. Because it was too late to make any changes in the book, it now becomes your task to complete the map, entering the missing information in the appropriate places. If you have any trouble doing so, review the reading on thermal expansion and make a rank-ordered list of the important concepts you find there. Use the list in addressing the deficiencies in the map provided.

2. Figures 5.38 and 5.41 schematically show the operation/function of *heat engines* and *heat movers*, respectively. An alternative way to capture the information and meaning contained in these "flow charts" is through the use of concept maps. Pick one of these figures, and, based on your understanding of it, produce a proper concept map that represents the concepts involved in either heat engines or heat movers.

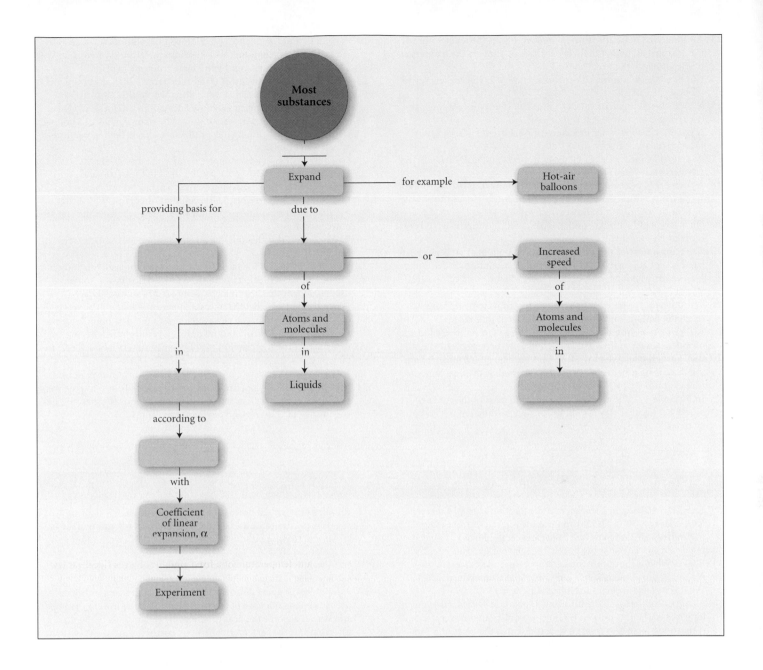

(▶ *Indicates a review question, which means it requires only a basic understanding of the material to answer. The others involve integrating or extending the concepts presented thus far.*)

1. ▶ What are the three common temperature scales? What are the normal boiling and freezing points of water in each scale?

2. The Fahrenheit and Celsius temperature scales agree at −40° (−40°C = −40°F). Do the Fahrenheit and Kelvin temperature scales ever agree? How about the Celsius and Kelvin scales?

3. ▶ What is the significance of absolute zero?

4. ▶ What happens to the atoms and molecules in a substance as its temperature increases?

5. ▶ In the most general terms, why do the Moon and the planet Mercury not have atmospheres?

6. A special glass thermometer is manufactured using a liquid that expands less than glass when the temperature increases. Assuming the thermometer does indicate the correct temperature, what is different about the scale on it?

7. ▶ Explain what a bimetallic strip is and how it functions.

8. Refer to the thermometer in Figure 5.10. Which metal in the bimetallic strip has the larger coefficient of linear expansion, the metal on the outer side of the spiral or the metal on the inner side? How can you determine this?

9. ▶ A certain engine part made of iron expands 1 mm in length as the engine warms up. What would be the approximate change in length if the part were made of aluminum instead of iron?

10. ▶ What is unusual about water below the temperature of 4°C?

11. In Explore It Yourself 5.3, the change in the volume of the air in the balloon is also affected by the fact that an enlarged balloon extends deeper under the water. What effect does this have on the pressure inside the balloon? Does this increase or decrease the observed change in the balloon's volume (compared to the volume change that would occur if the balloon were not submerged)?

12. ▶ What are the two general ways to increase the internal energy of a substance? Describe an example of each.

13. Air is allowed to escape from an inflated tire. Is the temperature of the escaping air higher than, lower than, or equal to the temperature of the air inside the tire? Why?

14. Is it possible to compress air without causing its internal energy to increase? If so, how?

15. ▶ Describe the three methods of heat transfer. Which of these are occurring around you at this moment?

16. A potato will cook faster in a conventional oven if a large nail is inserted into it. Why?

17. A coin and a piece of glass are both heated to 60°C. Which will feel warmer when you touch it?

18. A submerged heater is used in an aquarium to keep the water above room temperature. Should it be placed near the surface of the water or near the bottom to be most effective?

19. On a cool night with no wind, people facing a campfire feel a breeze on their backs. Why?

20. The temperature of the air in a one-room building that is not insulated is kept at exactly 22°C year round. A person inside the building feels cooler when it is cold outside than when it is warm outside. Why?

21. When heating water on a stove, a full pan of water takes longer to reach the boiling point than a pan that is half full. Why?

22. A 1-kg piece of iron is heated to 100°C, and then submerged in 1 kg of water initially at 0°C. The iron cools and the water warms until they are at the same temperature (in thermal equilibrium). Assuming there is no other transfer of heat involved, is the final temperature closer to 0°C, 50°C, or 100°C? Why?

23. In Example 5.3, is it necessary to know the mass of the concrete block? Put another way, would the answer be different if it was a 10-kg block?

24. A piece of aluminum and a piece of iron fall without air resistance from the top of a building and stick into the ground on impact. Will their temperatures change by the same amount? Explain.

25. The specific heat capacity of water is extremely high. If it were much lower, say, one-fifth as large, what effect would this have on processes like fire fighting and cooling automobile engines?

26. ▶ Why does the temperature of water not change while it is boiling?

27. ▶ Describe how changing the air pressure affects the temperature at which water boils.

28. One way to desalinate seawater—remove the dissolved salts so that the water is drinkable—is to distill it: Boil the seawater and condense the steam. The salts stay behind. This technique has one major disadvantage. It consumes a large amount of energy. Why is that?

29. ▶ What is saturation density? How does it change when the temperature increases?

30. ▶ What effect does heating the air in a room have on the relative humidity?

31. *Wood's metal* is an alloy of the elements bismuth, lead, tin, and cadmium that has a melting point of 70°C. Describe how it might be used in an automatic sprinkler system for fire suppression.

32. When trying to predict the lowest temperature that will be reached overnight, forecasters pay close attention to the dew point temperature. Why is the air temperature unlikely to drop much below the dew point? (The high latent heat of vaporization of water is important.)

33. ▶ Explain what a heat engine does and what a heat mover does.

34. ▶ In the winter, the amount of internal energy a heat pump delivers to a house is greater than the electrical energy it uses. Does this violate the law of conservation of energy? Explain.

PROBLEMS

1. Your jet is arriving in London, and the pilot informs you that the temperature is 30°C. Should you put on your jacket? Use Figure 5.2 to determine the temperature in degrees Fahrenheit.

2. On a nice winter day at the South Pole, the temperature rises to −60°F. What is the approximate temperature in degrees Celsius?

3. An iron railroad rail is 700 ft long when the temperature is 30°C. What is its length when the temperature is −10°C?

4. A copper vat is 10 m long at room temperature (20°C). How much longer is it when it contains boiling water at 1 atm pressure?

5. A machinist wishes to insert a steel rod with a diameter of 5 mm into a hole with a diameter of 4.997 mm. By how much would the machinist have to lower the temperature of the rod to make it fit the hole?

6. A brick oven 2 m long (at 20°C) is designed to allow for an expansion of 5 mm without damage. To what temperature would the oven have to be heated to expand that much?

7. A gas is compressed inside a cylinder (see Figure 5.16). An average force of 50 N acts to move the piston 0.1 m. During the compression, 2 J of heat are conducted away from the gas. What is the change in internal energy of the gas?

8. Air in a balloon does 50 J of work while absorbing 70 J of heat. What is its change in internal energy?

9. How much heat is needed to raise the temperature of 5 kg of silver from 20°C to 960°C?

10. A bottle containing 3 kg of water at a temperature of 20°C is placed in a refrigerator where the temperature is kept at 3°C. How much heat is transferred from the water to cool it to 3°C?

11. Figure 5.34 shows a graph of the temperature of ice as heat is transferred to it.
    a) How much heat does each kilogram of ice absorb from point *a* to point *b*?
    b) How much heat does each kilogram of water absorb from point *c* to point *d*?

12. Aluminum is melted during the recycling process.
    a) How much heat must be transferred to each kilogram of aluminum to bring it to its melting point, 660°C, from room temperature, 20°C?
    b) About how many cups of coffee could you make with this much heat (see Example 5.2)?

13. A 1,200-kg car going 25 m/s is brought to a stop using its brakes. Let's assume that a total of approximately 20 kg of iron in the brakes and wheels absorbs the heat produced by the friction.
    a) What was the car's original kinetic energy?
    b) After the car has stopped, what is the change in temperature of the brakes and wheels?

14. A 0.02-kg lead bullet traveling 200 m/s strikes an armor plate and comes to a stop. If all of its energy is converted to heat that it absorbs, what is its temperature change?

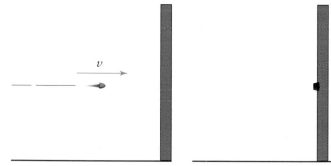

● **Figure 5.48**

15. A 10-kg lead brick is dropped from the top of a 629-m-tall television transmitting tower in North Dakota and falls to the ground.

Assuming all of its energy goes to heat it, what is its temperature increase?

16. Water flowing over the Lower Falls in Yellowstone National Park drops 94 m. If all of the water's energy goes to heat it, what is its temperature increase?

17. On a winter day, the air temperature is −15°C, and the humidity is 0.001 kg/m³.
    a) What is the relative humidity?
    b) When this air is brought inside a building, it is heated to 20°C. If the humidity isn't changed, what is the relative humidity inside the building?

18. On a summer day in Houston, the temperature is 35°C and the relative humidity is 77%.
    a) What is the humidity?
    b) To what temperature could the air be cooled before condensation would start to take place? (i.e., What is the dew point?)

19. Inside a building, the temperature is 20°C, and the relative humidity is 40%. How much water vapor is in each cubic meter of air?

20. On a hot summer day in Washington, D.C., the temperature is 86°F, and the relative humidity is 70%. How much water vapor does each cubic meter of air contain?

21. An apartment has the dimensions 10 m by 5 m by 3 m. The temperature is 25°C, and the relative humidity is 60%. What is the total mass of water vapor in the air in the apartment?

22. The total volume of a new house is 800 m³. Before the heat is turned on, the air temperature inside is 10°C, and the relative humidity is 50%. After the air is warmed to 20°C, how much water vapor must be added to the air to make the relative humidity 50%?

23. The temperature of the air in thermals decreases about 10°C for each 1,000 m they rise (see Challenge 5). If a thermal leaves the ground with a temperature of 30°C and a relative humidity of 31%, at what altitude will the air become saturated and a cloud form? (In other words, at what altitude does the temperature equal the dew point?)

24. In cold weather, you can sometimes "see" your breath. What you are seeing is a mist of small water droplets, the same as in clouds and fog. Suppose air leaves your mouth with temperature 35°C and humidity 0.035 kg/m³ and mixes with an equal amount of air at 5°C and humidity 0.005 kg/m³.
    a) What is the relative humidity of the mixed air if its temperature and humidity equal the averages of those of the two original air masses?
    b) Represent what happens by plotting three points in a graph like Figure 5.36.

25. What is the Carnot efficiency of a heat engine operating between the temperatures of 300°C (573 K) and 100°C (373 K)?

26. What is the maximum efficiency that a heat engine could have when operating between the normal boiling and freezing temperatures of water?

27. As a gasoline engine is running, an amount of gasoline containing 15,000 J of chemical potential energy is burned in 1 s. During that second, the engine does 3,000 J of work.
    a) What is the engine's efficiency?
    b) The burning gasoline has a temperature of about 4,000°F (2,500 K). The waste heat from the engine flows into air at about 80°F (300 K). What is the Carnot efficiency of a heat engine operating between these two temperatures?

28. A proposed ocean thermal-energy conversion (OTEC) system is a heat engine that would operate between warm water (25°C) at the ocean's surface and cooler water (5°C) 1,000 m below the surface. What is the maximum possible efficiency of the system?

CHALLENGES

1. A solid cube is completely submerged in a particular liquid and floats at a constant level. When the temperature of the liquid and the solid is raised, the liquid expands more than the solid. Will this make the solid rise upward, remain floating, or sink? Explain why.

2. Pyrex glassware is noted for its ability to withstand sudden temperature changes without breaking. Explain how its coefficient of linear expansion contributes to this ability.

3. Is it possible to transfer heat to air without changing its temperature? Explain.

4. A heater is placed in water at 1°C.
   a) Sketch the convection circulation that would be produced in the water.
   b) Where in the water should the heater be placed? Should it be moved at some later point in time? If so, where to?

5. As air rises in the atmosphere, its temperature drops, even if no heat flows out of it.
   a) Based on what you learned in Sections 4.4 and 5.3, explain why this is so.
   b) Cumulus clouds form when rising air is cooled to the point where water droplets form because of condensation. Why are these clouds usually much higher above the ground in dry climates than in wet ones?

6. If air at 35°C and 77% relative humidity is cooled to 25°C, what mass of water would condense out of the air in a room that measures 5 m by 4 m by 3 m? (See Problem 18.)

7. The door of a refrigerator is left open. Assuming the refrigerator is in a closed room, will the air in the room eventually be cooled?

SUGGESTED READINGS

Baldwin, Samuel F. "Renewable Energy: Progress and Prospects." *Physics Today* 55, no. 4 (April 2002): 62–67. An update on the current state of renewable energy technologies.

Crane, H. Richard, ed. "How Things Work." *Physics Teacher* 36, no. 5 (May 1998): 302–303. A closer look at bimetallic switches.

Dannen, Gene. "The Einstein-Szilard Refrigerators." *Scientific American* 276, no. 1 (January 1997): 90–95. Describes refrigerators designed by famous physicists Albert Einstein and Leo Szilard in the 1920s.

DeCicco, John, and Marc Ross. "Improving Automobile Efficiency." *Scientific American* 271, no. 6 (December 1994): 52–57. Describes various aspects of automobile efficiency and ways to improve it.

Dostrovsky, Israel. "Chemical Fuels from the Sun." *Scientific American* 265, no. 6 (December 1991): 102–107. The production of hydrogen and other fuels using solar energy is described.

Hunter, Lloyd. "The Art and Physics of Soaring." *Physics Today* 37, no. 4 (April 1984): 34–41. Includes a detailed look at soaring in thermals.

Ouboter, Rudolf de Bruyn. "Heike Kamerlingh Onnes's Discovery of Superconductivity." *Scientific American* 276, no. 3 (March 1997): 98–103. Describes the techniques devised by Kamerlingh Onnes to reach temperatures near absolute zero.

Sorbjan, Zbigniew. *Hands-On Meteorology.* American Meteorology Society, 1996. Includes many do-it-yourself experiments on such topics as temperature, heat, and moisture.

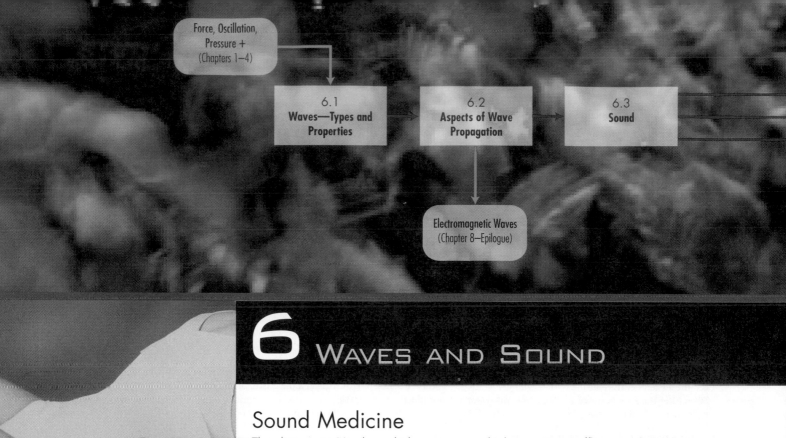

ESWL® being used on a kidney stone. The wave pulse is focused onto the stone because the stone and the spark producing the pulse are located at the foci of the ellipsoidal reflector.

6 WAVES AND SOUND

Sound Medicine

The diagnosis: You have kidney stones, which sometimes afflict people in their twenties and thirties. The condition is very painful, and it can kill you. Your options? One is surgery, but the operation itself is dangerous, and then there is a long and uncomfortable recuperation. But since the early 1980s, there has been an alternative that works in most cases: You can have the stones pulverized with sound. No scalpels involved.

The process is called *extracorporeal shock-wave lithotripsy* (ESWL)®. A device called a *lithotripter* focuses intense sound waves on the stones, which are broken into tiny fragments that can pass out of the patient's system. The sound is produced and focused outside of the body—hence the term *extracorporeal*. Some ESWL® systems make use of a reflector based on the shape of an ellipse (Figure 2.41). An intense sound pulse (shock wave) produced at one focus bounces off the reflector and converges on the other focus. The reflector is positioned so that the stone is at that second focus. Other lithotripters focus the sound with an "acoustic lens," much the way a magnifying glass can be used to focus sunlight to start a fire. In both systems, the sound is produced and focused inside a water-filled "cushion," similar to a balloon, that is pressed against the patient's body. The sound never travels in air.

Why does the ESWL® sound wave continue on target after it passes from water into a patient? What is it about sound waves that makes them useful for this and many other medical applications? Answers to such questions are found in the study of waves.

Waves in general and sound in particular are the main topics of this chapter. Waves are an integral part of our everyday lives. Whether playing a guitar, listening to a radio, clocking the speed of a thrown baseball, or having a kidney stone shattered, we are using a wave of some kind. The two most often used senses—sight and hearing—are highly developed wave-detection mechanisms. In the first part of this chapter, we look at simple waves and examine some of their general properties. The remainder of the chapter is about sound—how it is produced, how it travels in matter, and how it is perceived by humans.

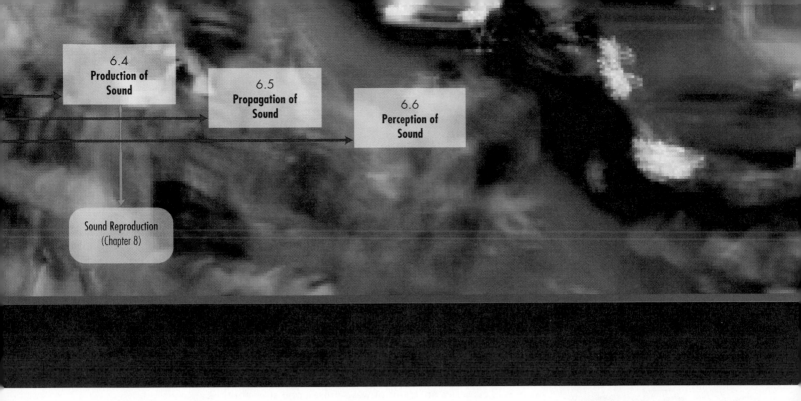

6.4 Production of Sound	**6.5** Propagation of Sound	**6.6** Perception of Sound

Sound Reproduction (Chapter 8)

6.1 Waves—Types and Properties

Ripples moving over the surface of a still pond, sound traveling through the air, a pulse "bouncing" back and forth on a piano string, light from the Sun illuminating and warming the Earth—these are all waves (see ● Figure 6.1). We can see or feel the effects of some waves, such as water ripples and earthquake tremors (called seismic waves), as they pass. Others, such as sound and light, we sense directly. Technology has given us numerous devices that produce or detect waves that we cannot sense (radio, ultrasound, x rays).

What are waves? Though many and diverse, they share some basic features. They all *involve vibration or oscillation* of some kind. Floating petals show the vibration of the water's surface as ripples move by. Our ears respond to the oscillation of air molecules and give us the perception of sound. Also, waves move and *carry energy* yet do not have mass. The sound from a loudspeaker can break a wineglass even though no matter moves from the speaker to the glass. We can define a wave as follows.

● **Figure 6.1** Like all waves, these water ripples involve oscillation.

DEFINITION

> **Wave** A traveling disturbance consisting of coordinated vibrations that transmit energy with no net movement of matter.

Sound, water ripples, and similar waves consist of vibrations of matter—air molecules or the water's surface, for example. The substance through which such waves travel is called the *medium* of the wave. Particles of the medium vibrate in a coordinated fashion to form the wave.

A rope stretched between two people is a handy medium for demonstrating a simple wave (see ● Figure 6.2). A flick of the wrist sends a wave down the rope. Each short segment of the rope is pulled upward in turn by its neighboring segment. The forces between the parts of the medium are responsible for "passing along" the wave. This kind of wave is not unlike a row of dominoes knocking each other over, except that the medium of a wave does not have to be "reset" after a wave goes by.

Many waves—sound, water ripples, waves on a rope—require a material medium. They cannot exist in a vacuum. On the other hand, light, radio waves, microwaves, and x rays can travel through a vacuum because they do not require a medium for their propagation. We will take a close look at these special waves—called *electromagnetic waves*—in Chapter 8.

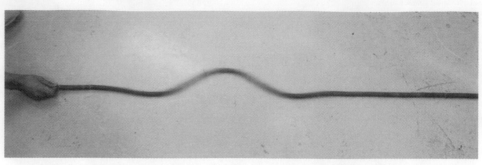

© Vern Ostdiek

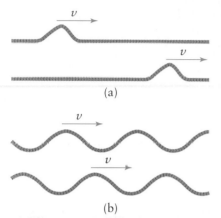

(a)

(b)

● **Figure 6.3** (a) Successive views of a wave pulse as it travels on a rope. (b) A continuous wave traveling on a rope.

Waves occur in a great variety of substances: in gases (sound), in liquids (water ripples), and in solids (seismic waves through rock). Some travel along a line (a wave on a rope), some across a surface (water ripples), and some throughout space in three dimensions (sound). Many more examples could be listed. Clearly, waves are everywhere, and they are diverse.

A wave can be short and fleeting, called a *wave pulse*, or steady and repeating, called a *continuous wave*. The sound of a bursting balloon, a tsunami (large ocean wave generated by an earthquake), and the light from a camera flash are examples of wave pulses. The sound from a tuning fork and the light from the Sun are continuous waves.* ● Figure 6.3 shows a wave pulse and a continuous wave on a long rope. You can see that a continuous wave is like a series or "train" of wave pulses, one after another.

If we take a close look at many different types of waves, we find that they can be classified according to the orientation of the wave oscillations. There are two main wave types, **transverse** and **longitudinal.**

Transverse Wave A wave in which the oscillations are perpendicular (transverse) to the direction the wave travels. Examples: waves on a rope, electromagnetic waves, some seismic waves.

Longitudinal Wave A wave in which the oscillations are along the direction the wave travels. Examples: sound in the air, some seismic waves.

Both types of waves can be produced on a Slinky—a short, fat spring that you may have seen "walk" down steps. If a Slinky is stretched out on a flat, smooth tabletop, a transverse wave is produced by moving one end from side to side, perpendicular to the Slinky's length (● Figure 6.4a). A longitudinal wave is produced by pushing and pulling one end back and forth, first toward the other end, then back (see Figure 6.4b). For each type of wave, one can produce either a wave pulse or a continuous wave.

---

* For now we can treat light as a simple wave. In Chapter 10, we will present the modern view of light as developed by Albert Einstein and others.

Explore It Yourself 6.1

For this, you need a Slinky, preferably the metal kind. (If you don't already have one, get one. You won't regret it.) On a smooth table, large desk, or bare floor, stretch the Slinky out about 5 feet, with a partner holding the other end—or attach it to something heavy so it stays put.

1. Send a transverse pulse down the Slinky by quickly moving your hand to the side and back. Send a longitudinal pulse by quickly moving your hand toward the other end and back. Do the two pulses seem to travel at the same speed? (Do they take about the same amount of time to get to the other end?)
2. Again send a transverse pulse down the Slinky. Watch what happens when it reaches the other end. Does it reflect? If so, is the reflected pulse identical to the original pulse?

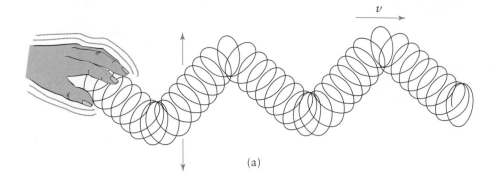

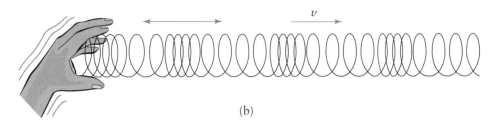

● **Figure 6.4** (a) A continuous transverse wave on a Slinky. Each coil oscillates up and down as the wave travels to the right. (b) A continuous longitudinal wave on a Slinky. Each coil oscillates left and right as the wave travels to the right.

(a)

(b)

A Slinky is not the only medium that can carry both transverse and longitudinal waves. Both kinds of waves can travel in any solid. Earthquakes and underground explosions produce both longitudinal and transverse seismic waves that travel through the Earth. Simple waves that involve oscillation of atoms and molecules must be longitudinal to travel in liquids and gases because of the absence of rigid bonds between the particles.

Many waves are neither purely longitudinal nor purely transverse. Although a water ripple appears to be a simple transverse wave, individual parcels of water actually move in circles or ellipses—they oscillate forward and backward as well as up and down. Waves in plasmas and in the atmosphere are even more complicated. But the two simple types of waves described here are common and well suited for illustrating wave phenomena.

The speed of a wave is the rate of movement of the disturbance. (Do not confuse this with the speed of individual particles as they oscillate.) For a given type of wave, the speed is determined by the properties of the medium. In the waves that we have been discussing, the masses of the particles that oscillate and the forces that act between them affect the wave speed. As a wave travels on a Slinky, for example, each coil is accelerated back and forth by its neighbors. Basic mechanics tells us that the mass of each coil and the size of the force acting on it will determine how quickly it—and therefore the wave—moves. In general, weak forces or massive particles in a medium cause the wave speed to be low.

Often, the speed of waves in a medium can be predicted by measuring some other properties of the medium. After all, the factors that affect wave speed—masses of particles and the forces between them—also affect other properties of a substance. For example, the speed of waves on a stretched rope or a Slinky, or on a taut wire can be computed by using the force F, which must be exerted to keep it stretched, and its **linear mass density** ρ, which equals its mass m divided by its length l. (ρ is the Greek letter rho, pronounced like row.) In particular:

$$v = \sqrt{\frac{F}{P}} \quad \left(\text{wave on a rope or spring; } \rho = \frac{m}{l} \right)$$

Increasing this force, also called the *tension,* will cause the waves to move faster. This is how stringed instruments like guitars and pianos are tuned. (More on this in Section 6.4.)

A student stretches a Slinky out on the floor to a length of 2 meters. The force needed to keep the Slinky stretched is measured and is found to be 1.2 newtons. The Slinky's mass is 0.3 kilograms. What is the speed of any wave sent down the Slinky by the student?

Example 6.1

First, we compute the Slinky's linear mass density.

$$\rho = \frac{m}{l} = \frac{0.3 \text{ kg}}{2 \text{ m}}$$

$$= 0.15 \text{ kg/m}$$

The speed of waves on the Slinky is

$$v = \sqrt{\frac{F}{\rho}} = \sqrt{\frac{1.2 \text{ N}}{0.15 \text{ kg/m}}}$$

$$= \sqrt{8 \text{ m}^2/\text{s}^2} = 2.8 \text{ m/s}$$

The speed of sound in air or any other gas depends on the ratio of the pressure of the gas to the density of the gas. But for each gas, this ratio depends only on the temperature. In particular, the speed of sound in a gas is proportional to the square root of the Kelvin temperature. For air:

$$v = 2.01 \times \sqrt{T} \qquad \text{(speed of sound in air; } v \text{ in m/s, } T \text{ in kelvins)}$$

The speed of sound is lower at high altitudes, not because the air is thinner but because it is colder.

Example 6.2

What is the speed of sound in air at room temperature (20°C = 68°F)? The temperature in kelvins is

$$T = 273 + 20 = 293 \text{ K}$$

Therefore:

$$v = 20.1 \times \sqrt{T}$$

$$= 20.1 \times \sqrt{293} = 20.1 \times 17.1$$

$$= 344 \text{ m/s (770 mph)}$$

The numerical factor (20.1) in the equation in Example 6.2 is determined by the properties of the molecules that comprise air and therefore applies to air only. The speed of sound in any other gas will be different, and the corresponding equation for v will have a different numerical factor. Two examples:

Speed of sound in helium: $v = 58.8 \times \sqrt{T}$ (SI units)

Speed of sound in carbon dioxide (CO_2): $v = 15.7 \times \sqrt{T}$ (SI units)

Explore It Yourself 6.2

(Note: The idea for this one came from the Constructing Physics Understanding Web site at San Diego State University.)

For this, you need two identical narrow-necked glass bottles, the ability to produce a tone in a bottle by blowing across its opening, and a seltzer tablet.

1. Put water into both bottles until they are about half full. "Tune" them so they produce the same tone or note when you blow across their openings by adding or removing water in one of them.
2. Add a seltzer tablet to one of them and wait a minute for the fizzing to die down. Carbon dioxide (CO_2) gas that is produced during the process replaces most of the air in the bottle.
3. See if they still produce the same tone. What is different about the bottle with the carbon dioxide in it? What causes this?

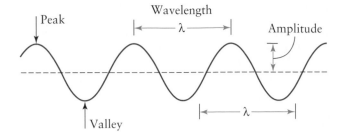

Peak

Wavelength

λ

Amplitude

Valley

λ

For the remainder of this section, we will take a look at some of the properties of a continuous wave. A convenient example is a transverse wave on a Slinky produced by moving one end smoothly side to side. ● Figure 6.5 shows a "snapshot" of such a wave. It shows the shape of the Slinky at some instant in time. Note that the wave has the same sinusoidal shape you've seen before (Figure 2.25).

The high points of the wave are called *peaks* or *crests,* and the low points are called *valleys* or *troughs.* The straight line through the middle represents the equilibrium configuration of the medium—its shape when there is no wave.

In addition to wave speed, there are three other important parameters of a continuous wave that can be measured. They are **amplitude, wavelength,** and **frequency.** At any moment, the different particles of the medium are displaced from their equilibrium positions by different amounts. The maximum displacement is called the *amplitude* of the wave.

DEFINITION

Amplitude The maximum displacement of points on a wave, measured from the equilibrium position.

The amplitude is just a distance equal to the height of a peak or the depth of a valley, which are the same. The amplitude of a particular type of wave can vary greatly. For water waves, it can be a few millimeters for ripples to tens of meters for ocean waves (● Figure 6.6). When we hear a sound, its loudness depends on the amplitude of the sound wave: louder sounds have larger amplitudes.

DEFINITION

Wavelength The distance between two successive "like" points on a wave. For example, the distance between two adjacent peaks or two adjacent valleys. Wavelength is represented by the Greek letter lambda (λ).

There is also a large variation in the wavelengths of particular types of waves. The wavelengths of sound (in air) that can be heard by humans range from about 2 centimeters (very high pitch) to about 17 meters (very low pitch). Typical wavelengths for radio waves are 3 meters for FM stations and 300 meters for AM stations.

Any segment of a wave that is one wavelength long is called one *cycle* of the wave. As each cycle of a wave passes by a given point in the medium, that point makes one complete oscillation—up, down, and back to the starting position. Figure 6.5 shows three complete cycles of a wave.

Amplitude and wavelength are independent features of a wave: A short-wavelength wave can have a small or a large amplitude (see ● Figure 6.7).

To understand what the frequency of a wave is, we must "unfreeze" the wave and imagine it as it moves along. The rate at which the wave cycles pass a point is the frequency of the wave. Recall from Section 1.1 that the unit of measure of frequency is the hertz (Hz).

● **Figure 6.6** Waves with large amplitudes.

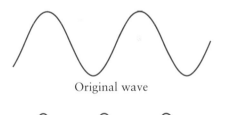

Original wave

Shorter wavelength, same amplitude

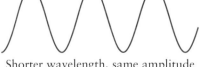

Same wavelength, smaller amplitude

Shorter wavelength, smaller amplitude

● **Figure 6.7** Transverse waves with different combinations of wavelength and amplitude.

Frequency The number of cycles of a wave passing a point per unit time. The number of oscillations per second in the wave.

If you move the end of a Slinky back and forth three times each second, you will produce a wave with a frequency of 3 hertz. The note A above middle C on the piano has a frequency of 440 hertz. This means that 440 cycles of the sound wave reach your ear each second. The piano wires producing the sound and the air molecules in the room all vibrate with the same frequency, 440 hertz.

Under ideal conditions, a person with good hearing can hear sounds with frequencies as low as 20 hertz or as high as 20,000 hertz. Frequency is important in other kinds of waves as well. Each radio station broadcasts a radio wave with a specific frequency, for example, 1,100 kilohertz = 1,100,000 hertz, or 92.5 megahertz = 92,500,000 hertz.

Amplitude, wavelength, and frequency can be identified for both transverse waves and longitudinal waves, although the amplitude of a longitudinal wave is a bit difficult to visualize. It is still the maximum displacement from the equilibrium position, but in this case, the displacement is along the direction the wave is traveling. ● Figure 6.8 shows a close-up of a Slinky with no wave and then with a longitudinal wave traveling on it. The amplitude is the farthest distance that any coil is displaced to the right or left of its equilibrium position. The regions where the coils are squeezed together are called **compressions,** and the regions where they are spread apart are called **expansions** or rarefactions. The wavelength is the distance between two adjacent compressions or two adjacent expansions.

The speed of a wave, its wavelength, and its frequency are related to each other in a simple way. Imagine a continuous wave passing by a point, perhaps ripples moving by a plant stem. The speed of the wave equals the number of cycles that pass by each second, multiplied by the length of each cycle. For example, if five cycles pass the stem each second and the peaks of the ripples are 0.03 meters apart, the wave speed is 0.15 m/s (see ● Figure 6.9). In general:

$$\text{wave speed} = \text{number of cycles per second} \times \text{length of each cycle}$$

The two quantities on the right of the equal sign are the frequency of the wave and the wavelength, respectively. Therefore:

$$v = f\lambda$$

The velocity of a continuous wave is equal to the frequency of the wave times the wavelength.

In many cases, all waves that travel in a particular medium have the same speed. Wave pulses, low-frequency continuous waves, and high-frequency continuous waves all travel with the same speed. Sound is an important example of this; sound pulses, low-frequency sounds, and high-frequency sounds travel through the air with the same speed, 344 m/s

● **Figure 6.8** The amplitude of a longitudinal wave on a Slinky equals the greatest lateral displacement of the coils.

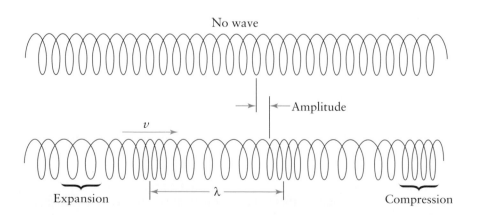

No wave

Amplitude

Expansion

λ

Compression

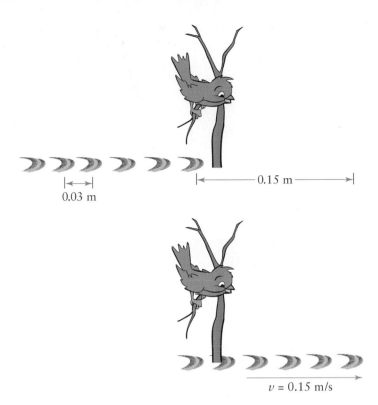

0.03 m

0.15 m

$v = 0.15$ m/s

at room temperature. Similarly, light, radio waves, and microwaves travel with the same speed—3×10^8 m/s—in a vacuum. According to the equation $v = f\lambda$, when the wave speed is the same for all waves, *higher* frequency waves must have proportionally *shorter* wavelengths. A 20-hertz sound wave has a wavelength of about 17 meters, while a 20,000-hertz sound wave has a wavelength of about 1.7 centimeters.

Example 6.3

Before a concert, musicians in an orchestra tune their instruments to the note A, which has a frequency of 440 hertz. What is the wavelength of this sound in air at room temperature? The speed of sound at this temperature is 344 m/s. So:

$$v = f\lambda$$

$$344 \text{ m/s} = 440 \text{ Hz} \times \lambda$$

$$\frac{344 \text{ m/s}}{440 \text{ Hz}} = \lambda$$

$$\lambda = 0.78 \text{ m} = 2.6 \text{ ft}$$

The wavelength of sound with a frequency of 220 hertz is twice as large—1.56 meters.

Not all continuous waves have the simple sinusoidal shape shown in Figure 6.5. In fact, waves with precisely that shape are relatively rare. Any continuous wave that does not have a sinusoidal shape is called a *complex wave*. ● Figure 6.10 shows two examples. Note that there are three different-sized peaks in each cycle of the upper wave. The shape of a wave is called its *waveform*. The two complex waves in the figure have about the same wavelength and amplitude, but they have very different waveforms. The waveform is another feature that is needed when comparing complex waves. We will take a closer look at this in Section 6.6.

Concept Map 6.1 summarizes the characteristics of waves.

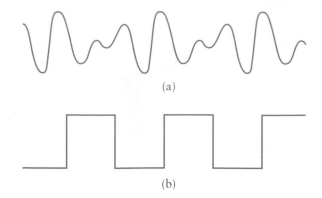

Figure 6.10 Two examples of complex waves. The lower wave, a "square wave," is used in many electronic devices.

(a)

(b)

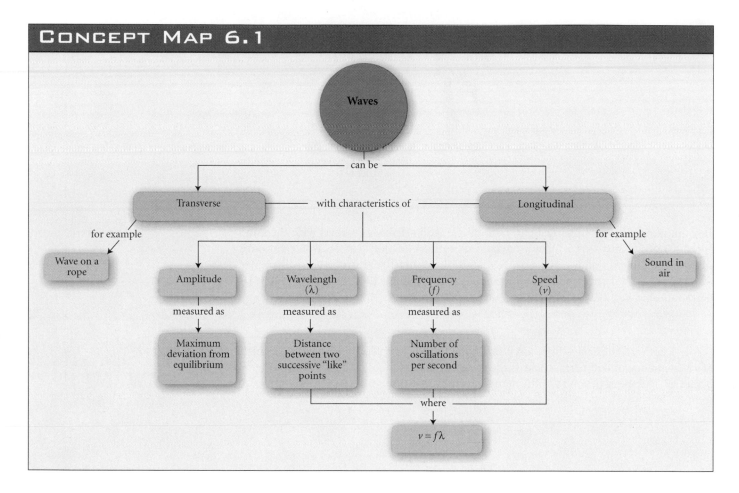

CONCEPT MAP 6.1

Waves

can be

Transverse ——— with characteristics of ——— Longitudinal

for example

Wave on a rope

for example

Sound in air

Amplitude

Wavelength (λ)

Frequency (f)

Speed (v)

measured as

Maximum deviation from equilibrium

measured as

Distance between two successive "like" points

measured as

Number of oscillations per second

where

$v = f\lambda$

LEARNING CHECK

1. (True or False.) The flash from a camera is an example of a continuous wave.
2. A _____ wave is one with the oscillations perpendicular to the direction of propagation.
3. The only way to make sound travel faster in air is to change the air's _____.
4. (True or False.) Regions in a longitudinal wave where the particles of the medium are squeezed together are called compressions.

5. Two sound waves traveling through the air have different frequencies. Consequently, they must also have different
   a) amplitudes.
   b) speeds.
   c) wavelengths.
   d) All of the above.

ANSWERS: 1. False   2. transverse   3. temperature   4. True   5. (c)

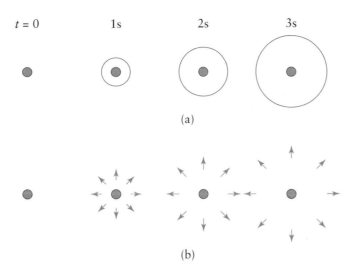

(a)

(b)

● **Figure 6.11** (a) A wavefront is used to show how a pulse spreads over water. (b) The same wave pulse, at the same times, represented with wave rays. The rays point in the direction the wave travels and are perpendicular to the wavefront.

6.2 Aspects of Wave Propagation

In this section, we consider what waves do as they travel. For waves traveling along a surface or throughout space in three dimensions, it is convenient to use two different ways to represent the wave. We will call these the **wavefront model** and the **ray model.** ● Figure 6.11 shows how each is used to illustrate a wave pulse on water as it travels from the point where it was produced. The wavefront is a circle that shows the location of the peak of the wave pulse. A ray is a straight arrow that shows the direction a given segment of the wave is traveling. A laser beam and sunlight passing through a hole in a window shade are like individual rays of light that we can see if there is dust in the air. (See the lower part of Figure 2.48.) On the other hand, the rays of water ripples are not visible, but we do see the wavefronts (Figure 6.1).

For a continuous water wave, the wavefronts are concentric circles about the point of origin that represent individual peaks of the wave (● see Figure 6.12). The largest circle shows the position of the first peak that was produced. Each successive wavefront is smaller because it came later and has not traveled as far. The distance between adjacent wavefronts is equal to the wavelength of the wave. Again, a continuous wave is like a series of wave pulses produced one after another. The rays used to represent a continuous wave are lines radiating from the source of the wave (the blue arrows in Figure 6.12). The wavefronts arriving at a point far from the source are nearly straight lines (far right in Figure 6.12). The corresponding rays are nearly parallel.

For a wave moving in three-dimensional space, like the sound traveling outward from you in all directions as you whistle, the wavefronts are spherical shells around the source of the wave. The wavefront of a wave pulse, such as the sound from a hand clap, expands like a balloon that is being inflated very fast. For continuous three-dimensional waves, like

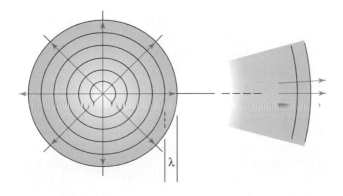

● **Figure 6.12** Wavefronts and rays for a continuous wave on a surface. Far away from the wave source, the wavefronts are nearly straight, and the rays are nearly parallel.

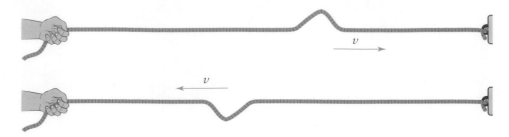

Figure 6.13 A wave pulse traveling on a rope is reflected at a fixed end. In this case, the pulse is inverted.

a steady whistle, the wavefronts form a series of concentric spherical shells that expand like the circular wavefronts of a wave on a surface. A 440-hertz tuning fork produces 440 of these wavefronts each second. The surface of each wavefront expands outward with a speed of 344 m/s (at room temperature). As with waves on a surface, the rays used to represent a continuous wave in three dimensions are lines radiating outward from the wave source.

One inherent aspect of the propagation of waves on a surface or in three dimensions is that the amplitude of the wave necessarily decreases as the wave gets farther from the source. A certain amount of energy is expended to create a wave pulse or each cycle of a continuous wave. This energy is distributed over the wavefront and determines the amplitude of the wave: The greater the amount of energy given to a wavefront, the larger the amplitude. As the wavefront moves out, it gets larger, so this energy is spread out more and becomes less concentrated. This attenuation accounts for the decrease in loudness of sound as a noisy car moves away from you and for the decrease in brightness of a lightbulb as you move away from it.

One can infer when the amplitude of a wave is changing by noting changes in the wavefronts or the rays. If the wavefronts are growing larger, the amplitude is getting smaller. The same thing is indicated when the rays are diverging (slanting away from each other).

At great distances from the source of a three-dimensional wave, the wavefronts become nearly flat and are called *plane waves*. The corresponding rays are parallel, and the wave's amplitude stays constant. The light and other radiation we receive from the Sun come as plane waves because of the great distance between the Earth and the Sun.

With this background, we will look at several phenomena associated with wave propagation.

Reflection

Think about how many times you looked in a mirror today. That's a very common use we make of the reflection of waves, but it's not the only one. As we will see, the sound we hear inside rooms is affected by reflection, musical instruments like guitars make use of it when producing sound, radar and sonar systems use it for a variety of purposes (like checking how fast we are driving), and so on.

A wave is reflected whenever it reaches a boundary of its medium or encounters an abrupt change in the properties (density, temperature, and so on) of its medium. A wave pulse traveling on a rope is reflected when it reaches a fixed end (see ● Figure 6.13). It "bounces" off the end and travels back along the rope. Notice that the reflected pulse is inverted. When the end of the rope is attached to a very light (but strong) string instead, the reflected pulse is not inverted. (The incoming pulse causes two pulses to leave the junction, a reflected pulse and a pulse that continues into the light string.) This reflection occurs because of an abrupt change in the density of the medium from high density (for the heavy rope) to low density (for the light string).

Similarly, a wave on a surface or a wave in three dimensions is reflected when it encounters a boundary. The wave that "bounces back" is called the *reflected wave*. Rays are more commonly used to illustrate reflection because they show nicely how the direction of each part of the wave is changed. When a wave is reflected from a straight boundary (for surface waves) or a flat boundary (in three dimensions), the reflected wave appears to be expanding out from a point behind the boundary (● Figure 6.14). This point is

Image (apparent source of reflected wave)

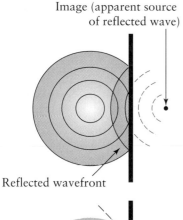

Reflected wavefront

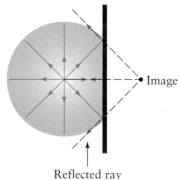

● Image

Reflected ray

● **Figure 6.14** The reflection of water ripples off the side of a pool. Both models of the wave show that the reflected wave appears to diverge from a point behind the wall, called the *image*.

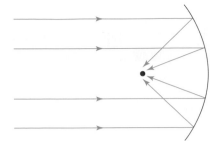

called the *image* of the original wave source. An echo is a good example: Sound that encounters a large flat surface, such as the face of a cliff, is reflected and sounds like it is coming from a point behind the cliff.

Our most common experience with reflection is that of light from a mirror. The image that you see in a mirror is a collection of reflected light rays originating from the different points on the object you see. (More on this in Section 9.2.)

Reflection from surfaces that are not flat (or straight) can cause interesting things to happen to waves. ● Figure 6.15 shows a wave being reflected by a curved surface. Note that the rays representing the reflected part of the wave are converging toward each other. This means that the amplitude of the wave is increasing—the wave is being "focused." Parabolic microphones seen on the sidelines of televised football games use this principle to reinforce the sounds made on the playing field. Satellite receiving dishes do the same with radio waves (● Figure 6.16).

A reflector in the shape of an ellipse has a useful property. (We saw in Section 2.8 that the orbits of satellites, comets, and planets can be ellipses.) An ellipse has two points in its interior called *foci* (plural of focus). If a wave is produced at one focus, it will converge on the other focus after reflecting off the ellipse. All rays originating from one focus reflect off the ellipse and pass through the other focus (● Figure 6.17). A room shaped like an ellipse is called a *whispering chamber* because a person standing at one focus can hear faint sounds—even whispering—produced at the other focus. This property of the ellipse is also used in the medical treatment of kidney stones (see "Sound Medicine," p. 216).

© Vern Ostdiek

● **Figure 6.16** Parabolic "dish" antennas focus radio waves by reflection.

Explore It Yourself 6.3

Locate an isolated building with a large, flat side. Stand more than 15 meters (50 feet) away, and clap your hands once: You should hear an echo shortly after the clap. Slowly walk toward the building, clapping your hands occasionally. What is different about the time between a clap and when you hear its echo? At some point you should notice that you no longer hear the echo because it comes too soon after the direct sound.

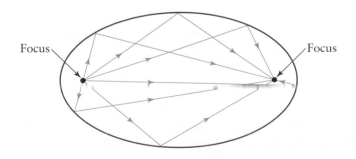

Focus   Focus

● **Figure 6.17** The focusing property of an ellipse. Each ray of a wave produced at one focus reflects off the ellipse and passes through the other focus.

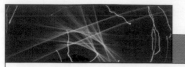

The Hubble Relation— Expanding Our Horizons

We mention in this section that the Doppler effect can be used to determine the line-of-sight speeds for individual stars bright enough to allow accurate measurement of the shifts in frequency of the light emitted by them. The question of whether or not the speeds of approach or recession of large aggregates of stars (galaxies) could be measured from similar shifts in the combined light emitted by all the members of the galaxy was first investigated by Vesto Slipher beginning in about 1912. By 1925, Slipher had carefully studied the radiation emitted by 40 galaxies and found that he could indeed determine the speeds with which they were moving. What he discovered was quite surprising: A number of the galaxies had very high speeds, up to 5,700 km/s (13 million mph), and 38 out of the 40 showed frequency shifts that indicated they were moving away from us.

During the last 75 years, much has been accomplished in extragalactic astronomy, but the essentials of Slipher's results remain true.

The vast majority of galaxies—and certainly all of those farther from us than about 3 million light-years (MLY)—are moving away from us and are doing so with velocities that are, in many cases, significant fractions of the speed of light. One of the more important additional facts to be discovered about galaxies during this period concerns the relationship between the recessional speed of a galaxy and its distance from us. Shortly after Slipher reported his results, Edwin Hubble noticed that galaxies with low recessional speeds were all relatively close to us, whereas those with high recessional speeds were more remote. By 1929, Hubble, in whose honor the Hubble Space Telescope was named, had amassed enough data to show a direct proportionality between the speeds with which galaxies are receding from us and their distances. This connection between speed and distance for all but the nearest galaxies has been verified by modern observations and is shown in ● Figure 6.18. It is now referred to as the *Hubble relation*.

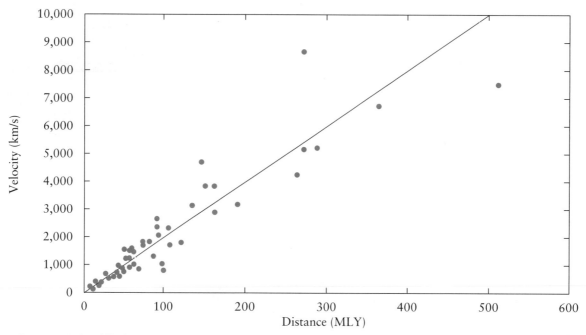

● **Figure 6.18** The Hubble relation. Plotted here are the recessional velocities of galaxies versus their distances. (1 light-year = 9.46×10^{12} km.) The (red) line represents a best fit to the data and yields a slope of about 20 km/s/MLY. More recent measurements including many more galaxies out to distances of more than 2000 MLY give a linear relation with a slightly larger slope of ~22 km/s/MLY.

Doppler Effect

Can you recall the last time a loud, fast-moving car or train passed near you? The pitch or tone of its sound dropped suddenly as it went by—although you may be so used to this you didn't notice. This is the Doppler effect—an apparent change in the frequency of a wave due to motion of the source of the wave or the receiver. ● Figure 6.20 shows the wavefronts emitted by a moving source—perhaps a bug moving over the surface of a pond or a train blowing its whistle. (Compare this to Figure 6.12 where the source is not moving.) Each wavefront expands outward from the point where the source was when it emitted that wavefront. (The speed of a wave in a medium is constant and is not affected

What is the significance of the Hubble relation? Only that the universe is expanding! To see this, we must recognize that galaxies (or more properly, clusters of galaxies) are the "atoms" of the universe, and just as one speaks of an "expanding gas" when the atoms of the gas are moving away from one another, so one can talk of an "expanding universe" where the atoms of the universe, the galaxies, are observed to be systematically separating from each other. And this is precisely what the Hubble relation is telling us. No matter in what direction we look, we find galaxies rushing away from us with speeds proportional to their distances: The farther away they are, the faster they are receding from us. The "raisin bread" analogy depicted in ● Figure 6.19 may be helpful to you in visualizing what is happening here.

The Hubble relation (and its interpretation) has had a profound influence on *cosmology,* the study of the structure and evolution of the universe. Indeed, by studying the motion of the most remote galaxies observable and checking for departures from a strictly linear trend in the Hubble relation, it is just now becoming possible to determine whether the universe will continue to expand forever or gradually slow to a stop and then begin to collapse in on itself. For example, recent data (c. 2003) seem to show that at large distances the Hubble relation curves in such a way as to indicate that the universe will expand forever. For the latest on cosmology and the expansion of the universe, see the Epilogue, beginning on p. 508, and the references cited there. The important point, however, to keep firmly in mind now is that were it not for the Doppler effect, the existence of the Hubble relation might never have become known.

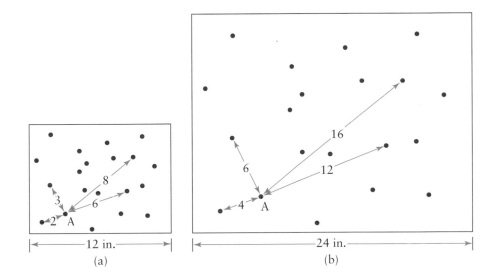

● **Figure 6.19** The "raisin bread" analogy for the expansion of the universe. As the bread rises (through the action of the yeast in the dough), it doubles in size in a time equal to, say, 1 hour. Consequently, all the raisins move farther apart. Some characteristic distances from one of the raisins, say, A, to several others are shown. Since each distance doubles during an hour, each raisin must move away from the one selected at a speed proportional to its distance. This remains true no matter which raisin is chosen as the origin. The point is that a uniform expansion leads naturally to a Hubble-like relationship between velocity and distance.

by any motion of the wave source.) In front of the moving source, the wavefronts are bunched together. This means that the wavelength is *shorter* than when the source is at rest and therefore that the frequency of the wave is *higher.* Behind the moving source, the wavefronts are spread apart: The wavelength is *longer,* and the frequency is *lower* than when the source is at rest. In both places, the higher the speed of the wave source, the greater the frequency shift.

The frequency of sound that reaches a person in front of a moving train is higher than that perceived when the train is not moving. A person behind the moving train hears a lower frequency. As a train or a fast car moves by, you hear the sound shift from a higher

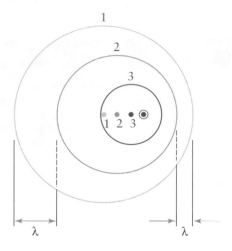

Figure 6.20 The Doppler effect with a moving source. A wave source moves to the right with constant speed. Each dot indicates the source's location when a wavefront was emitted. Wavefront 1 was emitted when the source was at position 1, and similarly for 2 and 3. Ahead of the source (to the right), the wavelength of the wave is decreased. Behind the source (to the left), the wavelength is increased.

frequency (pitch) to a lower frequency. The change in the loudness of the sound, which you also hear, is *not* part of the Doppler effect: It involves a separate process.

A similar shift in frequency of sound occurs if you are moving toward a stationary sound source (see ● Figure 6.21). This Doppler shift happens because the speed of the wave relative to you is higher than that when you are not moving. The wavefronts approach you with a speed equal to the wave speed plus your speed. Since the wavelength is not affected, the equation $v = f\lambda$ tells us that the frequency of the wave is increased in proportion to the speed of the wave relative to you. By the same reasoning, when one is moving away from the sound source, the frequency is reduced.*

The Doppler effect is routinely taken into account by astronomers. The frequencies of light emitted by stars that are moving toward or away from the Earth are shifted. If the speed of the star is known, the original frequencies of the light can be computed. If the frequencies are known instead, the speed of the star can be computed from the amount of the Doppler shift. Such information is essential for determining the motions of stars in our galaxy or of entire galaxies throughout the universe (see Physics Potpourri on p. 228).

Echolocation is the process of using the reflection of waves off an object to determine its location. *Radar* and *sonar* are two examples. Basic echolocation uses reflection only: A wave is emitted from a point, reflected by an object of some kind, and detected on its return to the original point. The time between the emission of the wave and the detection of the reflected wave (the round-trip time) depends on the speed of the wave and the distance to the reflecting object. For example, if you shout at a cliff and hear the echo 1 second later, you know that the cliff is approximately 172 meters away. This is because the sound travels a total of 344 meters (172 meters each way) in 1 second (at room temperature). If it takes 2 seconds, the cliff is approximately 344 meters away, and so on (see ● Figure 6.22).

With sonar, a sound pulse is emitted from an underwater speaker, and any reflected sound is detected by an underwater microphone. The time between the transmission of

---

* For the musically inclined: To shift the frequency of sound by a musical half-step, the speed of the source of the sound (or the listener, if the source is at rest) must be about 20 m/s (45 mph). For example, if the sound from a car's horn is the note A when the car and the listener are at rest, it will be heard as an A sharp when the car is approaching a listener at 20 m/s and as an A flat when it is moving away at 20 m/s. (The occupants of the car always hear an A.) This value applies when the temperature is around 20°C. When it is colder, the required speed would be smaller. (Why?)

Figure 6.21 The Doppler effect with moving observers. The person in the car on the left hears a higher frequency than the pedestrian. The person in the car on the right hears a lower frequency.

the pulse and the reception of the reflected pulse is used to determine the distance to the reflecting object. Basic radar uses a similar process with microwaves that reflect off aircraft, raindrops, and other things.

Incorporating the Doppler effect in echolocation makes it possible to immediately determine the speed of an approaching or departing object. A moving object causes the reflected wave to be Doppler shifted. If the frequency of the reflected wave is *higher* than that of the original wave, the object is moving *toward* the source. If the frequency is *lower*, the object is moving *away*.

Doppler radar uses this combination of echolocation and the Doppler effect. The time between transmission and reception gives the distance to the object, while the amount of frequency shift is used to determine the speed. Law-enforcement officers use Doppler radar to check the speeds of vehicles (see ● Figure 6.23), and Doppler radar is also used in baseball, tennis, and other sports to clock the speed of a ball. Dust, raindrops, and other particles in air reflect microwaves, making it possible to detect the rapidly swirling air in a tornado with Doppler radar. Another potentially life-saving application is the detection of wind shear—drastic changes in wind speed near storms that have caused low-flying aircraft to crash. (There are a few more examples at the end of Section 6.3.)

● **Figure 6.23** Radar unit in use.

Bow Waves and Shock Waves

If you've ever heard a sonic boom, or been jostled by the wake of a passing watercraft while floating in the water, you've had experience with the topic of this section. ● Figure 6.24a shows another series of wavefronts produced by a moving wave source. This time the speed of the wave source is *greater than* the wave speed. The wavefronts "pile up" in the forward direction and form a large-amplitude wave pulse called a **shock wave.** This is what causes the V-shaped bow waves produced by boats.

Aircraft flying faster than the speed of sound produce a similar shock wave. In this case, the three-dimensional wavefronts form a conical shock wave, with the aircraft at the cone's apex. This conical wavefront moves with the aircraft and is heard as a sonic boom (a sound pulse) by persons on the ground.

Diffraction

Think about walking down a street and passing by an open door or window with sound coming from inside. You can hear the sound even before you get to the opening, as well as after you've passed it. The sound doesn't just go straight out of the opening like a beam, it spreads out to the sides. This is **diffraction.** ● Figure 6.25 shows wavefronts as they

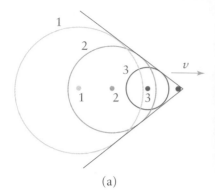

(a)

(b)

© Marilyn Newton/Reno Gazette-Journal

● Figure 6.24 (a) A shock wave is produced when the speed of a wave source exceeds the wave speed. Parts of the wavefronts combine along the two black lines to form a V-shaped wavefront. (b) The jet-powered car *Thrust SSC* traveling faster than sound on 25 September 1997. Dust can be seen being kicked up by the shock wave extending away from the car on both sides.

reach a gap in a barrier. These might be sound waves passing through a door or ocean waves encountering a breakwater. The part of the wave that passes through the gap actually sends out wavefronts to the sides as well as ahead. The rays that represent this process show that the wave "bends" around the edges of the opening.

The extent to which the diffracted wave spreads out depends on the ratio of the size of the opening to the wavelength of the wave. When the opening is much larger than the wavelength, there is little diffraction: The wavefronts remain straight and do not spread out to the sides appreciably. This is what happens when light comes in through a window. The wavelength of light is less than a millionth of a meter, and consequently, there is little diffraction. When the wavelength is roughly the same size as the opening, the diffracted wave spreads out much more (see ● Figure 6.26). The sizes of windows and doors

● Figure 6.25 The diffraction of a wave as it passes through an opening in a barrier. The wavefronts spread out to the sides after passing through.

● Figure 6.26 Diffraction of water waves passing through a narrow gap in a barrier. Short-wavelength waves (left) are not diffracted as much as long-wavelength waves (right) passing through an opening of the same size.

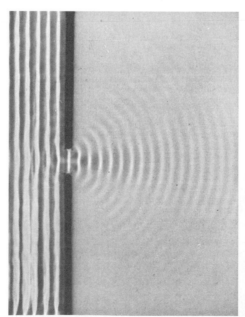

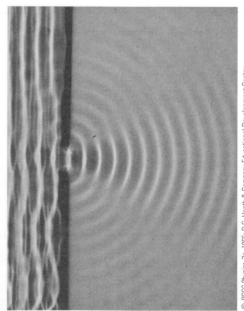

© PSSC *Physics*, 2e, 1965, D.C. Heath & Company Educational Development Center

are well within the range of the wavelengths of sound waves, so sound diffracts a great deal after passing through them. Higher frequencies (shorter wavelengths) are not diffracted as much as the lower frequencies.

Explore It Yourself 6.4

Stand near an open window or door of a building in which a continuous sound, such as recorded music, is being produced (see ○ Figure 6.27).

1. Move back and forth past the opening, and notice that you can hear sound when you are well off to one side. Where is the sound the loudest?
2. Move back and forth past the opening again, and this time pay close attention to how much treble and bass there are in the sound. Is there a difference depending on where you are?

● **Figure 6.27** Experiencing diffraction of sound.

Interference

Interference arises when two continuous waves, usually with the same amplitude and frequency, arrive at the same place. The sound from a stereo with the same steady tone coming from each speaker is an example of this situation. Another way to cause interference is to direct a continuous wave at a barrier with two openings in it. The two waves that emerge from the two openings will diffract (spread out), overlap each other, and undergo interference.

Consider the case of identical, continuous water waves produced by two small objects made to oscillate up and down in unison on the surface of the water. As these two waves travel outward, each point in the surrounding water moves up and down under the influence of both waves. If we move around in an arc about the wave sources, we find that at some places the water is moving up and down with a large amplitude. At other places, the water is actually still—it is not oscillating at all (● Figure 6.28).

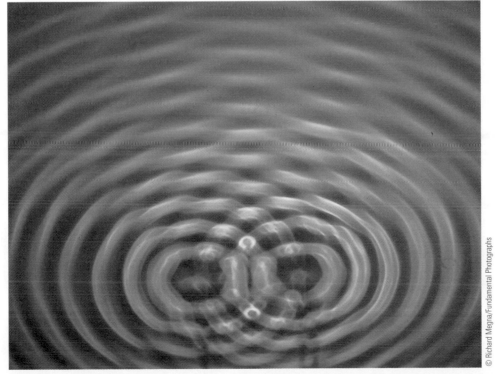

● **Figure 6.28** Interference pattern of water waves from two nearby sources. The thin lines of calm water indicate destructive interference. Between these lines are regions of large-amplitude waves, caused by constructive interference.

© Richard Megna/Fundamental Photographs

To see why this characteristic pattern of large-amplitude and zero-amplitude motion arises, consider ● Figure 6.29a—a sketch showing two waves at one moment in time. The thicker lines represent peaks of the waves, and the thinner lines represent the valleys. In Figure 6.29b, the straight lines labeled C indicate the places where the two waves are "in phase"—the peak of one wave matches the peak of the other and valley matches valley. The two waves reinforce each other and the amplitude is large. This is called **constructive interference.** On the straight lines labeled D, the waves are "out of phase"—the peak of one wave matches the valley of the other. The two waves cancel each other. (Whenever one wave has upward displacement, the other has downward displacement, and vice versa. Therefore, the net displacement is always zero.) This is called **destructive interference.** Figure 6.29c shows the same waves a short time later, after the waves have traveled one-half of a wavelength. The pattern of constructive and destructive interference is not altered as the waves travel outward. If the photograph shown in Figure 6.28 had been taken earlier or later, it would look the same.

Whether the two waves are in phase or out of phase depends on the relative distances they travel. To reach any point on line C_1 in Figure 6.29b, the two waves travel the same distance and consequently arrive with peak matching peak and valley matching valley. Along the line C_2, the wave from the source on the left must travel a distance equal to one wavelength farther than the wave from the source on the right. The reverse is true along the line on the left labeled C. In general, there is *constructive interference* at all points where one wave travels one, or two, or three . . . wavelengths farther than the other wave.

On the other hand, along the line of destructive interference labeled D_1, the wave from the source on the left has to travel one-half wavelength farther than the wave from the source on the right. They arrive with peak matching valley and cancel each other. Along

● **Figure 6.29** Interference of two waves.

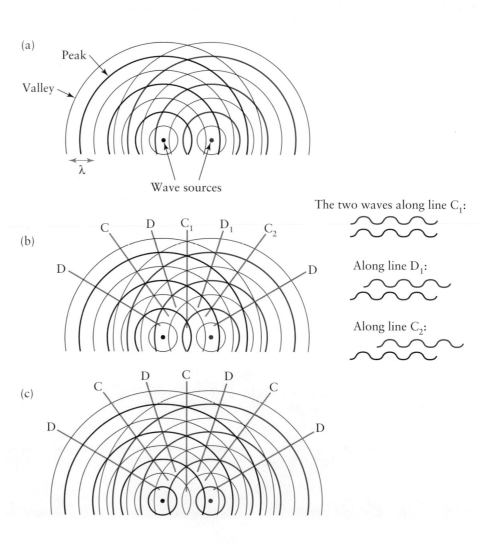

(a)

Peak

Valley

λ

Wave sources

(b)

C   D C_1 D_1 C_2

D   D

(c)

D   C   D

C   C

D   D

The two waves along line C_1:

Along line D_1:

Along line C_2:

the far right line labeled "D," the wave from the source on the left has to travel $1\frac{1}{2}$ wavelengths farther, so the two waves again arrive out of phase. The reverse is true for the lines showing destructive interference on the left. In general, there is *destructive interference* at all points where one wave travels $\frac{1}{2}$, or $1\frac{1}{2}$, or $2\frac{1}{2}$, . . . wavelengths farther than the other wave. At places in between constructive and destructive interference, the waves are not completely in phase or out of phase so they partially reinforce or cancel each other.

Sound and other longitudinal waves can undergo interference in the same way. We can imagine Figure 6.29 representing sound waves with the peaks corresponding to compressions and the valleys corresponding to expansions. Along the lines of constructive interference, one would hear a loud, steady sound. Along the lines of destructive interference, one would hear no sound at all. In Chapter 9, we will look at the interference of light waves.

These are some of the more important phenomena associated with waves as they propagate. Later in this chapter and in Chapter 9, we will take a closer look at some of these and introduce others that are particularly important for light.

LEARNING CHECK

1. As a sound wave or water ripple travels out from its source, its _____ decreases.
2. Which of the following can change the frequency of a wave?
   a) interference
   b) Doppler effect
   c) diffraction
   d) All of the above
3. (True or False.) The amplitude of a sound wave can be increased by making it reflect off a curved surface.

4. A sophisticated echolocation system can determine the following about an object by reflecting a wave off of it:
   a) in what direction it is located
   b) how far away it is
   c) how fast it is approaching or moving away
   d) All of the above
5. When two identical waves undergo _____ interference, the net amplitude is zero.

ANSWERS: 1. amplitude 2. (b) 3. True 4. (d) 5. destructive

6.3 Sound

Our most common experience with sound is in air, but it can travel in any solid, liquid, or gas. For example, when you speak, much of what you hear is sound that travels to your ears through the bones and other tissues in your head. That is why a recording of your voice does not sound the same to you as what you hear when you are talking. ● Table 6.1 lists the speed of sound in some common substances.

Explore It Yourself 6.5

You might want to do this when no one is around to see you or hear you.

1. Speak a sentence in a normal way, and pay attention to how your voice sounds.
2. Now block the sound into both ears. (One way is to cover both of them with the palms of your hands.) Speak the sentence again. What is different?
3. Record your voice (with a tape recorder, computer, or video camera), and then listen to the recording. What is different?

● Table 6.1	Speed of Sound in Some Common Substances	
Substance	**Speed***	
	(m/s)	(mph)
Air		
At −20°C	320	715
At 20°C	344	770
At 40°C	356	795
Carbon dioxide	269	600
Helium	1,006	2,250
Water	1,440	3,220
Human tissue	1,540	3,450
Aluminum	5,100	11,400
Granite	4,000	9,000
Iron and steel	5,200	11,600
Lead	1,200	2,700

*At room temperature (20°C) except as indicated.

The speed of sound in any substance depends on the masses of its constituent atoms or molecules and on the forces between them. The speed of sound is generally higher in solids than in liquids and gases because the forces between the atoms and molecules in solids are very strong. Sound in gases and liquids is a longitudinal wave, whereas in solids it can be either longitudinal or transverse. In the rest of this chapter, we will concentrate mainly on sound in air.

Sound is produced by anything that is vibrating and causing the air molecules next to it to vibrate. ● Figure 6.30 shows a representation of a sound wave that was emitted by a vibrating tuning fork. The shading represents the air molecules that we, of course, cannot see. The wave looks very much like a longitudinal wave on a Slinky (Figure 6.8). These compressions and expansions travel at 344 m/s (at room temperature).

The air pressure in each compression is higher than normal atmospheric pressure because the air molecules are squeezed closer together. Similarly, the pressure in each expansion is below atmospheric pressure. Beneath the sketch is a graph of the air pressure along the direction the wave is traveling. Note that it has the characteristic sinusoidal shape. So a sound wave can be represented by a series of pressure peaks and valleys—a pressure wave. It is more convenient to think of sound as regular fluctuations of air pres sure than as vibrations of molecules, although it is both. The amplitude of a sound wave is the maximum pressure change. For a very loud sound, this is only about 0.00002 atmospheres.

It is these pressure variations that our ears detect and convert into the sensation of sound. The eardrum is a flexible membrane that responds to pressure changes. That is why your ears "pop" when you ride in a fast elevator. The oscillating pressure of a sound wave forces the eardrum to vibrate in and out. A remarkable series of organs converts this oscillation of the eardrum into an electrical signal to the brain that is perceived as sound.

● **Figure 6.30** A representation of part of the sound wave emitted by a tuning fork. The air pressure is increased in each compression and reduced in each expansion, as shown in the graph. (From *Physics in Everyday Life* by Richard Dittman and Glenn Schmeig. Used by permission.)

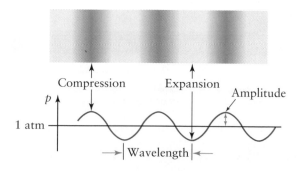

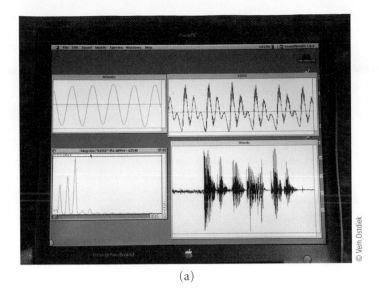

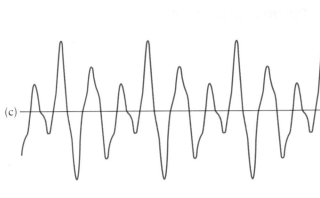

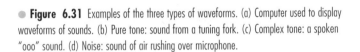

(a)

(b)

(c)

(d)

● **Figure 6.31** Examples of the three types of waveforms. (a) Computer used to display waveforms of sounds. (b) Pure tone: sound from a tuning fork. (c) Complex tone: a spoken "ooo" sound. (d) Noise: sound of air rushing over microphone.

The **waveform** of a sound wave is the graph of the air-pressure fluctuations caused by the sound wave. The easiest way to display the waveform of sound is to connect a microphone to an oscilloscope, an electronic device often seen displaying heartbeats in television hospital shows. (Most personal computers have this capability as well ● Figure 6.31a.) The oscilloscope shows a graph of the pressure variations detected by the microphone.

We can classify sounds by their waveforms (see Figure 6.31b, c, d). A **pure tone** is a sound with a sinusoidal waveform. A tuning fork produces a pure tone, as does a person carefully whistling a steady note. A **complex tone** is a complex sound wave. The waveform of a complex tone repeats itself but is not sinusoidal. Therefore, any complex tone also has a definite wavelength and a definite frequency. Most steady musical notes are complex tones.

The third type of sound is called **noise.** Noise has a random waveform that does not repeat over and over. For this reason, noise does not have a definite wavelength or frequency. The sound of rushing air is a good example of noise. (In everyday speech, "noise" is often used to describe any unwanted sound, even if it is a pure tone or a complex tone.)

Sound with frequencies outside the range of 20 to 20,000 hertz cannot be heard by people. Inaudible sound with frequency less than 20 hertz is called **infrasound.** High-amplitude infrasound can be felt, rather than heard, as periodic pressure pulses. The hearing ranges of elephants and whales extend into the infrasound region. Sound with frequencies higher than 20,000 hertz (20 kilohertz) is called **ultrasound.** The audible ranges of dogs, cats, moths, mice, and bats extend into ultrasound frequencies; they can hear very high frequency sounds that humans cannot.

Sound Applications

Before going into greater detail about sounds mainly meant for human hearing, we consider some of the many other uses of sound.

A variety of animals use sound for echolocation—to "see" their surroundings and to find prey. Dolphins and some other marine animals emit clicking sounds that reflect off

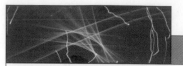
Putting Sound to Work

Perhaps you would like to know if it is windy enough to go sailing or too windy to launch a hot-air balloon. Or maybe you want to see if the snow is deep enough for skiing or if the water is high enough in a river to go kayaking. So you go online or make a phone call to get the information, but who or what actually takes the measurement of the wind speed or snow depth? It may be an automatic device that uses sound.

The desire for such information is not limited to those with recreation in mind. An electric utility may want to monitor the wind a hundred meters above the ground at some location, day and night for months, to see if it would be a good place to install a wind-energy "farm." The stakes are much higher for a pilot coming in for a landing, wanting to know if there will be a drastic change in wind speed or direction, called **wind shear,** that could bring a jumbo jet crashing to the ground. Meteorologists whose job might be to forecast what the weather will be like tomorrow or to predict how the Earth's climate might change over the next century need all of the information they can get about what the atmosphere is doing, both near the Earth's surface and extending upward hundreds or thousands of meters. Increasingly, measurements of wind velocity, air temperature, and other quantities are being made by instruments that utilize sound.

Probably the simplest such device is a **sonic depth gauge,** basically a sonar unit pointed downward. It sends out a pulse of ultrasound that reflects off snow or water and uses the time it takes for the reflected wave to return to compute the distance to the reflecting surface—similar to Figure 6.22—and from that, the depth. (The air temperature must be measured and used in the calculation because the speed of sound changes as the air temperature changes.)

Measuring wind speed and direction with sound is a bit more complicated. Traditional **anemometers**—wind meters—use propellers or "cups" on spokes that are spun by the wind. But these devices wear out and they can be frozen in ice in cold weather. Hundreds of weather stations spread across the United States are replacing these with **sonic anemometers,** which have no moving parts. They are also critical components of wind-shear alert systems at some major airports. Sound is carried along as the air moves, so it takes sound less time to travel a given distance downwind than it takes it to travel the same distance upwind. (Just like a boat motoring down a river versus up a river.) A sonic anemometer sends a pulse of ultrasound in one direction and records the time it takes to travel a short distance (roughly 1 foot). It repeats this with a pulse sent in the opposite direction (● Figure 6.32). The component of the wind's velocity in the given direction is computed using these two times. One benefit of using the times for the sound going both ways is that the actual speed of sound cancels out: The anemometer gives the correct wind speed regardless of the air temperature.

Direct measurement of the air temperature and wind velocity above the ground is done twice a day, at about 1,000 sites worldwide. A lightweight instrument pack called a **radiosonde** is carried upward tens of kilometers by a helium-filled balloon. It radios back to a ground station the measurements it takes as it is rising. This system has been in place for decades and is still one of the most important sources of vital information about the atmosphere. But the instruments are expensive and they are usually not recovered for reuse, so only two "profiles" of the atmosphere are measured every 24 hours at each site.

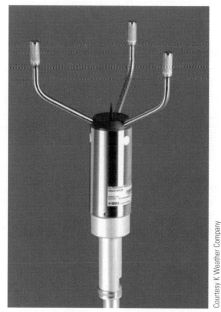

● **Figure 6.32** Sonic anemometer, consisting of three ultrasonic transducers at the corners of a triangle. Each transducer is used alternately as a transmitter and as a detector. Sound pulses are sent between each pair of transducers, in both directions, and the times for each of the 6 one-way trips are measured. These measurements are used to compute the component of the wind's velocity along the direction of each of the three sides of the triangle. These in turn are used to find the wind's speed and direction.

● **Figure 6.34** Bats use ultrasonic echolocation to navigate in the dark.

fish and other objects. By paying attention to how long it takes for reflected sound to return, to the direction from which it comes, and to how strong the reflected sound is, a dolphin can get a very good idea of the sizes and locations of nearby objects. Bats use very high frequency sound, usually ultrasound, in a highly sophisticated echolocation system that employs the Doppler effect (● Figure 6.34). Each species uses a characteristic range of frequencies, anywhere from 16 to 150 kilohertz. The bat emits a short burst of sound that reflects off surrounding objects like the ground, trees, and flying insects. The bat detects these echoes and uses the time it takes the sound to make the roundtrip to determine the distances to the objects. The shift in the frequency of sound reflected off moving objects is used by the bat to track down its dinner—flying insects. The bat even compensates for the Doppler shift in the frequency of the emitted sound caused by its own motion.

There are two types of sound-based, "remote-sensing" instruments that measure wind velocity or air temperature at many levels far above the ground, day and night: **sodar** (**so**nic **d**etection **a**nd **r**anging) and **RASS** (**r**adio **a**coustic **s**ounding **s**ystem). Both are essential tools in atmospheric research programs, and a few units are in regular use to supplement data from radiosondes.

Sodar functions somewhat like the sonic depth meter, but it sends sound upward. The returning sound that it detects is that scattered by the air itself. The atmosphere is generally filled with turbulent eddies, like the small vortices seen in smoke rising from burning incense—except the eddies are usually invisible. The twinkling of stars and the apparent shimmering of things seen far in the distance on a hot day are caused by these temperature-induced eddies affecting light traveling through them. When sound encounters these eddies, some of it is scattered in all directions, including back toward the sound source. Sodar uses a very sensitive receiver to detect this returning sound. Any motion of the air toward or away from the sodar will cause a Doppler shift in the sound frequency, which is used to measure the wind speed. A typical sodar instrument will send sound pulse "beams" in three directions (vertical, and two near-vertical) and use the Doppler shifts of the return sound to compute the wind speed and direction. By waiting a specific period of time before accepting the return sound—half a second, for example—the unit can select which elevation above the ground to measure the wind velocity. A common hourly wind report from a sodar unit consists of wind speeds and directions at 10, 20, and 30, and so on, meters above the ground, up to perhaps several hundred meters, depending on the system and the conditions.

RASS measures air temperatures aloft using the Doppler effect and an interaction between sound waves and high-frequency radio waves (● Figure 6.33). A sound wave and a radio wave are emitted upward, with the radio wave's wavelength twice that of the sound wave's. (This makes the frequency of the radio wave over 400,000 times that of the sound wave. Why?) The radio wave is scattered by the higher-density air in the sound-wave compressions, and some of this scattered energy is detected back on the ground. Since the radio-wave signal that is received is reflecting off sound-wave compressions moving upward, the decrease in the radio wave's frequency because of the Doppler effect can be used to determine the speed of the sound wave. Then the equation for the speed of sound given in Section 6.1 is used to calculate the air temperature. Incidentally, the two-to-one ratio of the wavelengths of the two waves is chosen so that interfer-

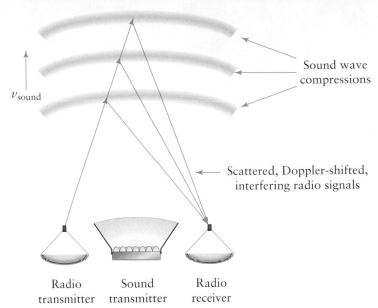

● **Figure 6.33** Simplified diagram of an RASS system. Radio waves (blue) are scattered downward by the compressions of the upward-moving sound wave (pink), and undergo constructive interference. The Doppler shift of the radio signal is used to compute the speed of sound, and from that, the air temperature.

ence enhances the return signal: Radio waves reflected downward from two successive sound-wave compressions are in phase and undergo constructive interference.

Research on new ways to use sound to probe the atmosphere continues. For example, studies have shown that severe tornado-producing thunderstorms generate infrasound that might be detected and used to warn of approaching tornadoes. The devices described here and others like them indicate that, just as from one person to another, sound turns out to be a useful way for the atmosphere to communicate information to us about what it is doing.

For more information about sodar and RASS, and to look at actual wind and temperature measurements, go to the ABLE (Atmospheric Boundary Layer Experiment) Web-site, sponsored by Argonne National Laboratory. Other good sources can be found by doing a web search for terms such as sodar, RASS, and sonic anemometer.

Although most applications of sound in science and technology use ultrasound, a few interesting devices have been developed that use lower-frequency sound. Prototype refrigerators have been built that use sound to pump the refrigerant (refer to Figure 5.40). A very large amplitude sound wave inside a specially designed chamber produces huge pressure oscillations. The chamber is fitted with valves that, during the low-pressure part of the sound-wave cycle, allow gas that has just absorbed heat to enter it and then, during the high-pressure part of the sound-wave cycle, allow the compressed gas to escape to be recondensed. Such a sonic pump or compressor has far fewer moving parts than a conventional motor-driven pump, and there is no need for a lubricant. This makes it easier to use ozone-safe refrigerants that can be incompatible with lubricants now in use.

During the 20th century, many useful applications of ultrasound were developed. Ultrasound is used in motion detectors that turn on the lights when a person enters a

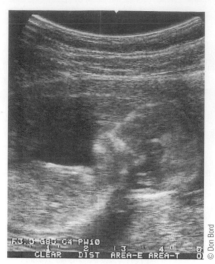

● **Figure 6.35** Ultrasonic image of author Don Bord's son Jeffrey at 13 weeks' gestation.

room or that set off an alarm when an intruder enters an area. It is also used to control rodents and insects; clean jewelry and intricate mechanical and electronic components; weld plastics; sterilize medical instruments; enhance certain chemical reactions; and measure the speed of the wind (see the Physics Potpourri on p. 238.)

Ultrasound can also be used to produce light. First discovered in the 1930s, **sonoluminescence** (from the Latin words for sound and light) has been the subject of intense research in recent years. A bubble inside water emits flashes of light as pressure oscillations caused by low-frequency ultrasound make the bubble expand and collapse. The temperature inside the bubble rises to over 10,000 K during collapse—hotter than the surface of the sun—and the light pulse lasts less than a billionth of a second. Recent research suggests that the light-producing process is similar to that occurring inside x-ray tubes (discussed in Chapter 8).

Ultrasound has several uses in medicine. It is routinely used to form images of internal organs and fetuses (see ● Figure 6.35). High-frequency ultrasound, typically 3.5 million hertz, is sent into the body and is partially reflected as it encounters different types of tissue. These reflections are analyzed and used to form an image on a television monitor. Some sophisticated ultrasonic scanning devices also use the Doppler effect. The beating heart of a fetus and the flow of blood in arteries can be monitored by detecting the frequency shift of the reflected ultrasound.

Recently developed *acoustic surgery* uses ultrasound for tasks such as destroying tumors. Focused, high-intensity sound causes heating that destroys tissue. The precision of such an "acoustic scalpel" can exceed that of a conventional knife.

Another use of ultrasound in medicine is *ultrasonic lithotripsy,* a procedure that breaks up kidney stones that have migrated to the bladder. (This is not the same as the ESWL® described at the beginning of this chapter.) A large-amplitude 27,000-hertz sound wave travels through a steel tube inserted into the body and placed in contact with the stone. The ultrasound breaks the stone into small pieces, somewhat like a singer breaking a wineglass. A procedure similar to ultrasonic lithotripsy has recently been developed to break up blood clots.

LEARNING CHECK

1. (True or False.) Sound cannot travel through solid steel.
2. We are able to hear because our ears respond to the changes in _____ associated with a sound wave.

3. Which of the following is *not* one of the ways sound can be used?
   a) to make a bubble in water produce light
   b) to relay information to orbiting satellites
   c) to form images of internal organs
   d) to measure the air temperature aloft

ANSWERS: 1. False 2. pressure 3. (b)

6.4 Production of Sound

In the remaining sections of this chapter, we will take a brief look at the three P's of acoustics: the production, propagation, and perception of sound.

The sounds that we hear range from simple pure tones like a steady whistle to complicated and random waveforms like those found on a noisy street corner. Most of the sound we hear is a combination of many sounds from different sources. The loudness usually fluctuates, as do the frequencies of the component sounds.

Sound is produced when vibration causes pressure variations in the air. Any flat plate, bar, or membrane that vibrates produces sound. The tuning fork shown in Figure 6.30 is a nice example. A dropped garbage-can lid, a vibrating speaker cone, and a struck drumhead produce sound the same way. The tuning fork executes simple harmonic motion and produces a pure tone. The garbage-can lid and the drumhead have more complicated motions and so produce complex tones or noise.

The various musical instruments represent some of the basic types of sound producers. Drums, triangles, xylophones, and other percussion instruments produce sound by direct vibration. Each is made to vibrate by a blow from a mallet or drumstick.

Guitars, violins, and pianos use vibrating strings to produce sound (● Figure 6.36). By itself, a vibrating string produces only faint sound because it is too thin to compress and expand the air around it effectively. These instruments employ "soundboards" to increase the sound production. (The electric versions of guitars and violins pick up and amplify the string vibrations electronically.) One end of the string is attached to a wooden soundboard, which is made to vibrate by the string. The vibrating soundboard, in turn, produces the sound. ● Figure 6.37 shows a simplified diagram of the process used in pianos. When a note is played, a hammer strikes the piano wire and produces a wave pulse. The pulse travels back and forth on the wire, being reflected each time at the ends. The soundboard receives a "kick" each time the pulse is reflected at that end. This makes the soundboard vibrate at a frequency equal to the frequency of the pulse's back-and-forth motion. The sound dies out because the pulse loses energy to the soundboard during each cycle.

The strings on guitars are plucked instead of struck, giving the pulses a different shape. This is part of the reason why the sound of a guitar is different from that of a piano. Violin strings are bowed, resulting in even more complicated wave pulses. In all three instruments, the frequency of the pulse's motion depends on the speed of waves on the string and on the length of the string. When a string is tuned by being tightened, the wave speed is increased. The pulse moves faster on the string and makes more "round-trips" each second—the frequency of the sound is raised. Different notes are played on the same guitar or violin string by using a finger to hold down the string some distance from its fixed end. The pulse travels a shorter distance between reflections, makes more round trips per second, and produces a higher-frequency sound.

A similar process is used in flutes, trumpets, and other wind instruments. Here it is a pressure pulse in the air inside a tube that moves back and forth (● Figure 6.38). Initially, a sound pulse is produced at one end of the tube by the musician. This pulse travels down the tube and is partially reflected and partially transmitted at the other end. The transmitted part spreads out into the air, becoming the sound that we hear. The reflected part returns to the mouthpiece end, where it is reflected again and reinforced by the musician. (The sound pulse is also inverted at each end: It goes from a compression to an expansion, and vice versa.) The musician must supply pressure pulses at the same frequency that the pulse oscillates back and forth in the tube. This, of course, is the frequency of the sound that is produced.

Different notes are played by changing the length of the tube—by opening side holes in woodwinds and by using valves or slides in brasses. The speed of the pulses is determined by the temperature of the air. This is one reason why musicians "warm up" before a performance. The air inside the instrument is warmed by the musicians' breath and hands. Hence the frequencies of the notes are higher than when the air inside is cool.

● **Figure 6.36** Guitarist Bonnie Raitt makes the strings vibrate by plucking them. Most of the sound that we hear comes from the front plate of the guitar, which is made to vibrate by the strings.

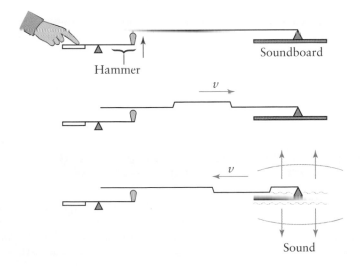

● **Figure 6.37** Sound production in a piano. The hammer creates a wave pulse that oscillates back and forth on the piano wire. (The amplitude is not to scale.) The pulse causes the soundboard to vibrate at the frequency of the pulse's oscillation.

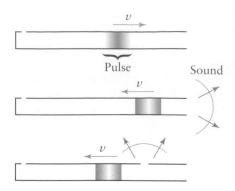

Pulse   Sound

Figure 6.38 Sound production in a flute. The musician causes a sound (pressure) pulse to oscillate back and forth in the tube. Each time the pulse reaches the right end, it sends out a compression—part of the sound that we hear. Opening a side hole shortens the path of the pulse, thereby increasing the frequency.

Figure 6.39 All singers produce sound by making their vocal cords vibrate. The position of the jaw, the size of the mouth opening, and other factors also influence the sound.

The human voice uses several types of sound production and modification mechanisms. Some consonant sounds like "*sss*" and "*fff*" are technically noise: They are hissing sounds produced by air rushing over the teeth and lips. The randomly swirling air produces sounds with random, changing frequencies. The vocal cords, located inside the Adam's apple in the throat, are the primary sound producers for singing and for spoken vowel sounds (see • Figure 6.39). When air is blown through the vocal cords, they vibrate and produce pressure pulses (sound) much like the reed of a saxophone. This sound is modified by the shapes of the air cavities in the throat, mouth, and nasal region. Muscles in the throat are used to tighten and to loosen the vocal cords, thereby changing the pitch of the sound. Moving the tongue or jaw changes the shape of the mouth's air cavity and allows for different sounds to be produced. A sinus cold can change the sound of one's voice because swelling changes the configuration of the nasal cavity.

Explore It Yourself 6.6

Two simple exercises illustrate how complicated human speech is.

1. Using a mirror, look carefully at your jaw, tongue, and lips as you recite the alphabet slowly. Note just how much their positions change as you speak different sounds.
2. Speak the first four vowel sounds (a, e, i, o). Now speak one of these sounds, "lock" your jaw, tongue, and lips in place (don't let them move), and then try to make the other three sounds. What happens?

Perhaps you've heard someone speak who had inhaled helium. (This is *not* a recommended exercise. It is possible to suffocate because of lack of oxygen in the lungs.) The speed of sound in helium is nearly three times that in air (refer to Table 6.1). This raises the frequencies of the sounds and gives the speaker a falsetto voice.

Sound waves carry energy, as do all waves. This means that the source of the sound must supply energy. Speaking loudly or playing an instrument for extended periods of time can tire you out for this reason. For continuous sounds, it is more relevant to consider the power of the source, since the energy must be supplied continuously. Most instruments, including the human voice, are very inefficient; typically, only a small percentage of the energy output of the performer is converted into sound energy.

LEARNING CHECK

1. Increasing the speed of waves on a guitar string increases the _____ of the sound that it produces.

2. (True or False.) The vocal cords are used to make all sounds that are produced in human speech.

ANSWERS: 1. frequency 2. False

6.5 Propagation of Sound

Once a sound has been produced, what factors affect the sound as it travels to our ears? The general aspects of wave propagation discussed in Section 6.2 of course apply to sound waves. Of these, reflection, diffraction, and the reduction of amplitude with distance from the sound source are most important in influencing the sound that actually reaches us.

The simplest situation is a single source of sound in an open space—such as a person talking in an empty field. The sound travels in three dimensions, and its amplitude decreases as the wavefronts expand. In particular, the amplitude is inversely proportional to the distance from the sound source.

$$\text{amplitude} \propto \frac{1}{d}$$

When you move to twice as far away from a steady sound source, the amplitude of the sound is decreased by one-half. The sound becomes quieter as you move away from the sound source.

Sound propagation is more complicated inside rooms and other enclosures. First, diffraction and reflection of sound allow you to hear sound from sources that you can't see because they are around a corner. We are so accustomed to this phenomenon that it doesn't seem mysterious. Second, even when the source is inside the room with you, most of the sound that you hear has been reflected one or more times off the walls, ceiling, floor, and any objects in the room. This has a large effect on the sound that you hear.

• Figure 6.40 shows that sound emitted by a source in a room can reach your ears in countless ways. Consider a single sound pulse like a hand clap. In an open field, you would hear only a single, momentary sound as the pulse moves by you. A similar pulse produced in a room is heard repeatedly: You hear the sound that travels directly to your ears; then sound that reflects off the ceiling, floor, or a wall before reaching your ears; then sound that is reflected twice, three times, and so on. These reflected sound waves travel greater and greater distances before reaching your ears and are heard successively later than the direct sound. This process of repeated reflections of sound in an enclosure is called **reverberation.** The single hand clap is heard as a continuous sound that fades quickly.

• Figure 6.41 compares the sound from a hand clap as it is heard in an open field and in a room. Each graph shows the amplitude of the sound that is heard versus the time after the sound pulse is produced. The reverberation causes the sound to "linger" in the room. The indirect sound that one hears after the initial direct pulse is called the *reverberant sound.* The amount of time it takes for the reverberant sound to fade out depends on the size of the room and the materials that cover the walls, ceiling, and floor. Sound is

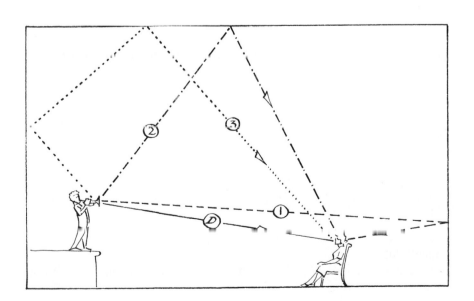

● **Figure 6.40** Some possible pathways for the sound in a room to travel from the source to your ears; *D* represents the sound that reaches your ears directly. Rays 1 and 2 are reflected once. Ray 3 is reflected twice. In most rooms, the sound can be reflected more than a dozen times and still be heard.

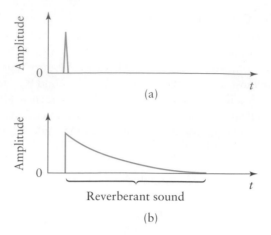

● **Figure 6.41** Comparison of the sound heard from a hand clap in (a) an open field and in (b) a room. A single pulse is heard in the field, whereas a continuous, fading sound is heard in the room.

never completely reflected by a surface: Some percentage of the energy in an incoming wave is absorbed by the surface, leaving the reflected wave with a reduced amplitude. (Concrete absorbs only about 2% of the incident sound's energy, whereas carpeting and acoustical ceiling tile can absorb around 90%.) A room with a large amount of sound-absorbing materials in it will have little reverberation. After a few reflections, the sound loses most of its energy and cannot be heard.

The *reverberation time* is used to compare the amount of reverberation in different rooms. It is the time it takes for the amplitude of the reverberant sound to decrease by a factor of 1,000. It varies from a small fraction of a second for small rooms with high sound absorption to several seconds for large, brick-walled gymnasiums and similar enclosures. The Taj Mahal is made of solid marble, which absorbs very little sound. The reverberation time of its central dome is over 10 seconds.

Explore It Yourself 6.7

1. You can make a rough measurement of the reverberation time of a large room or a racquetball court with a stopwatch. Have a friend produce a single loud handclap while you start the stopwatch at the same instant. Stop it the moment that the reverberant sound can no longer be heard. You may want to do the exercise several times and compute the average of the values.
2. If the room has a great deal of reverberation, you can easily illustrate the impact it has on speech communication. Stand next to a friend, near the middle of the room, and begin a normal conversation (or simply repeat a sentence to each other). Move a few steps apart, and exchange sentences again. What is different? Continue moving farther apart and conversing until you can no longer understand each other.
3. Repeat the process while speaking more slowly. What is different about your experiences this time? Explain.
4. Find a room that is about the same size but with carpeting, drapes, or other sound absorbing material. Repeat steps 1 to 3. What is different?

When a steady sound is produced in a room, such as a trumpet playing a long note in an auditorium, the sound that one hears is affected by reverberation in a number of ways:

1. The sound is louder than it would be if you had heard it at the same distance in an open field.
2. The sound "surrounds" you; it comes from all directions, not just straight from the source (Figure 6.40).

3. Beyond a short distance from the sound source, the loudness does not decrease as rapidly with distance as it would in an open field. That is why one can often hear as well near the back of an auditorium as in the middle.

4. Not only does the sound fade gradually when the source stops, the sound also "builds" when the source starts.

Moderate reverberation has an overall positive effect on the sound that we hear, particularly music (see ● Figure 6.42). However, excessive reverberation adversely affects the clarity of both speech and music. Speech and music are a series of short, steady sounds interspersed with short moments of silence. Each note, word, or syllable is followed by a brief pause. If we again graph the amplitude of sound versus time, we can see the effect of reverberation (see ● Figure 6.43). In an open field, one hears each syllable or note as a distinct, separate sound. In a room, the individual sounds begin to merge. As a new note is played or a new word is spoken, the reverberant sound from the preceding one can still be heard. The longer the reverberation time, the more the sounds overlap each other and the harder it is to understand speech. Racquetball courts have hard, smooth walls and very high reverberation times; that is why it is very difficult for players to converse unless they are close to each other. It is recommended that the reverberation time of rooms used for oral presentations and lectures should be around 0.5–1 second. For concert halls, it should be from 1 to 3 seconds, depending on the type of music being performed.

Many other factors besides reverberation time must be taken into account by architects and building designers. For example, a balcony must be high above the main floor and not extend out too far, or little reverberant sound will reach the seats below it. Also, sound is focused by concave walls and ceilings. Building an elliptical auditorium could result in the sound being concentrated in a small area of the room (Figure 6.17).

● **Figure 6.42** Symphony Hall in Boston is very successful at using reverberation to enhance sound.

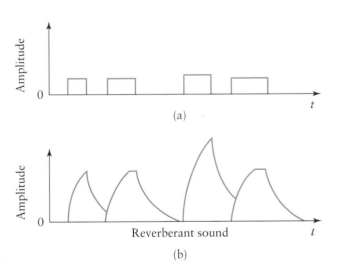

(a)

(b)

Reverberant sound

● **Figure 6.43** A series of steady sounds, such as spoken syllables, is heard (a) in an open field and (b) in a room. Reverberation in the room causes the sounds to merge. This tends to blend musical notes and makes speech more difficult to understand.

LEARNING CHECK

1. A stationary lawnmower produces a steady sound in an open field. If a listener moves to a distance one-half as far away from the mower, the amplitude of the sound would be
   a) one-half as large
   b) one-fourth as large
   c) twice as large
   d) four times as large

2. (True or False.) Reverberation can both inhibit conversation and enhance music.

3. The _____ of a room with walls that are good reflectors of sound is longer than that of a similar room with walls that are good absorbers of sound.

ANSWERS: 1. (c) 2. True 3. reverberation time

6.6 Perception of Sound

In this section, we consider some aspects of sound perception—how the physical properties of sound waves are related to the mental impressions we have when we hear sound. We will be comparing psychological sensations, which can be quite subjective, to measurable physical quantities. (A similar situation: "hot" and "cold" are subjective perceptions that are related to temperature, which is a measurable physical quantity.) To make things simple, we will limit ourselves to steady, continuous sounds. This frees us from having to include such effects as reverberation in a room—that is, how the sound builds up, how it decays, and so on.

The main categories that we use to describe sounds subjectively are **pitch, loudness,** and **tone quality.**

> The *pitch* of a sound is the perception of highness or lowness. The sound of a soprano voice has a high pitch and that of a bass voice has a low pitch. The pitch of a sound depends primarily on the frequency of the sound wave.

> The *loudness* of a sound is self-descriptive. We can distinguish among very quiet sounds (difficult to hear), very loud sounds (painful to the ears), and sounds with loudness somewhere in between. The loudness of a sound depends primarily on the amplitude of the sound wave.

> The *tone quality* of a sound is used to distinguish two different sounds even though they have the same pitch and loudness. A note played on a violin does not sound quite like the same note played on a flute. The tone quality of a sound (also referred to as the *timbre* or *tone color*) depends primarily on the waveform of the sound wave.

Pitch

Pitch is perhaps the most accurately discriminated of the three categories, particularly by trained musicians. It depends almost completely on the frequency of the sound wave: The higher the frequency, the higher the pitch. Noise does not have a definite pitch, because it does not have a definite frequency.

Pitch is essential to nearly all music. There is a great deal of arithmetic in the musical scale; each note has a particular numerical frequency. ● Figure 6.44 shows the frequencies of the notes on the piano keyboard. There are seven octaves on the piano, each consisting of 12 different notes. Within each octave, the notes are designated A through G, plus five sharps and flats. Each note in a given octave has exactly twice the frequency of the corresponding note in the octave below. For example, the frequency of the lowest note on the piano, an A, is 27.5 hertz; for the A in the next octave, it is 55 hertz; it is 110 hertz for the third A, and so on. The frequency of middle C is 261.6 hertz.*

Certain combinations of notes are pleasing to the ear, whereas others are not. Nearly 2,500 years ago, Pythagoras indirectly discovered that two different notes are in harmony when their frequencies have a simple whole-number ratio. For example, a musical fifth is any pair of notes whose frequencies are in the ratio 3 to 2. Any E and the first A below it have this ratio of frequencies, as do any G and the first C below it.

Figure 6.44 also shows the approximate ranges of singing voices and some instruments. For normal speech, the ranges are approximately 70 to 200 hertz for men and 140 to 400 hertz for women. When whispering, you do not use your vocal cords, and you produce much higher-frequency "hissing" sounds.

Loudness

The loudness of a sound is determined mainly by the amplitude of the sound wave. The greater the amplitude of the sound wave that reaches your eardrums, the greater the per-

---

* For the mathematically inclined: The frequencies of the notes in the most commonly used tuning scheme are based on the twelfth root of 2 ($\sqrt[12]{2}$ = 1.05946). The frequency of each note equals the frequency of the note a half-step lower multiplied by $\sqrt[12]{2}$. (Put another way, the frequency of each note is about 5.9% higher than that of the next lower one.) Going up the scale, 12 half-steps yield a note whose frequency is $(\sqrt[12]{2})^{12}$ = 2 times that of the initial note.

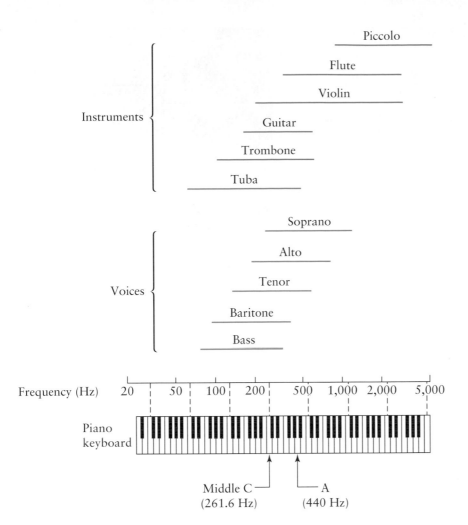

● **Figure 6.44** The frequencies of the notes on a piano are indicated and compared with the approximate frequency ranges of different singing voices and musical instruments. This frequency scale, and, therefore, the piano keyboard, is a logarithmic scale: The interval between 50 and 100 hertz is the same as the interval between 500 and 1,000 hertz. (From *Physics in Everyday Life* by Richard Dittman and Glenn Schmieg. Used by permission.)

ceived loudness of the sound. The actual pressure amplitudes of normal sounds are extremely small, typically around one-millionth of 1 atmosphere. This causes the eardrum to vibrate through a distance of around 100 times the diameter of a single atom. An extremely faint sound has an amplitude of less than one-billionth of 1 atmosphere, and it makes the eardrum move *less* than the diameter of an atom. The ear is an amazingly sensitive device.

There is a specially defined physical quantity that depends on the amplitude of sound but is more convenient for relating amplitude to perceived loudness. This is the sound pressure level or simply the **sound level.**

The standard unit of sound level is the decibel (dB). The range of sounds that we are normally exposed to has sound levels from 0 decibels to about 120 decibels. ● Figure 6.45 shows some representative sound levels along with their relative perceived loudness. Sound level does not take into account the irritation of the sound. Your favorite music played at 100 decibels may not sound as loud as an annoying screech of fingernails on a blackboard with a sound level of 80 decibels.

The relationship between the amplitude of a steady sound and its sound level is based on factors of 10. A sound with 10 times the amplitude of another sound has a sound level that is 20 decibels higher. A 90-decibel sound has 10 times the amplitude of a 70-decibel sound.

The following five statements describe how the perceived loudness of a sound is related to the measured sound level. These are general trends that have been identified by researchers after testing large numbers of people. For a particular person, the actual numerical values of the sound levels can vary somewhat from those listed. Also some of

Figure 6.45 The decibel scale with representative sounds that produce approximately each sound level. Sound levels above about 85 decibels pose a danger to hearing. Levels above 120 decibels cause ear pain and the potential for permanent hearing loss.

Jet takeoff (60 m)	120 dB	
Construction site	110 dB	*Intolerable*
Shout (1.5 m)	100 dB	
Heavy truck (15 m)	90 dB	*Very loud*
Urban street	80 dB	
Automobile interior	70 dB	*Noisy*
Normal conversation (1 m)	60 dB	
Office, classroom	50 dB	*Moderate*
Living room	40 dB	
Bedroom at night	30 dB	*Quiet*
Broadcast studio	20 dB	
Rustling leaves	10 dB	*Barely audible*
	0 dB	

the values given, particularly in numbers 1 and 3 below, depend on the frequencies of the sounds.

1. The sound level of the quietest sound that can be heard under ideal conditions is 0 decibels. This is called the *threshold of hearing.*

2. A sound level of 120 decibels is called the *threshold of pain.* Sound levels this high cause pain in the ears and can result in immediate damage to them.

3. The minimum increase in sound level that makes a sound noticeably louder is approximately 1 decibel. For example, if a 67-decibel sound is heard and after that a 68-decibel sound, we can just perceive that the second sound is louder. If the second sound had a sound level of 67.4 decibels, we could not notice a difference in loudness.

4. A sound is judged to be twice as loud as another if its sound level is about 10 decibels higher. A 44-decibel sound is about twice as loud as a 34-decibel sound. A 110-decibel sound is about twice as loud as a 100-decibel sound. This is a cumulative factor: A 110-decibel sound is about four times as loud as a 90-decibel sound, and so on.

5. If two sounds with equal sound levels are combined, the resulting sound level is about 3 decibels higher. If one lawn mower causes an 80-decibel sound level at a certain point nearby, starting up a second identical lawn mower next to the first will raise the sound level to about 83 decibels. It turns out that 10 similar sound sources are perceived to be twice as loud as a single source (see • Figure 6.46).

Figure 6.46 Ten equivalent sources sound about twice as loud as one.

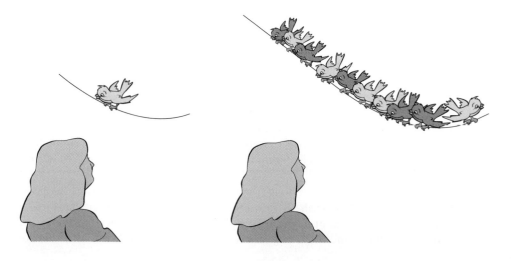

The loudness of pure tones and, to a lesser degree, of complex tones, also depends on the frequency. This is because the ear is inherently less sensitive to low- and high-frequency sounds. The ear is most sensitive to sounds in the frequency range of 1,000 to 5,000 hertz. For example, a 50-hertz pure tone at 78 decibels, a 1,000-hertz pure tone at 60 decibels, and a 10,000-hertz tone at 72 decibels all sound equally loud. (At very high sound levels, 80 decibels and above, the ear's sensitivity does not vary as much with frequency as it does at lower sound levels.) One reason for this variation in sensitivity is that a considerable amount of low-frequency sound is produced inside our bodies by flowing blood and flexing muscles. The ear is less sensitive to low-frequency sounds so these internal sounds do not "drown out" the external sounds that we need to hear.

Sound levels are measured with sound-level meters (● Figure 6.47). Most sound-level meters are equipped with a special weighting circuit (called the A scale) that allows them to respond to sound much as the ear does. The response to low and high frequencies is diminished. The readings on the A scale are designated dBA. When operating in the normal mode (called the C scale) a sound-level meter measures the sound level in decibels, treating all frequencies equally. When in the A scale mode, a sound-level meter responds like the human ear and therefore indicates the relative loudness of the sound.

Loud sounds can not only damage your hearing; they can also affect the physiological and psychological balance of your body. Since the beginning of humankind, the sense of hearing has been used as a warning device: Loud sounds often indicate the possibility of danger, and the body automatically reacts by becoming tense and apprehensive. Constant exposure to loud or annoying sounds puts the body under stress for long periods of time and consequently jeopardizes the physical and mental well-being of the individual.

The Occupational Safety and Health Administration (OSHA) has established standards designed to protect workers from excessive sound levels. Workers must be supplied with sound-protection devices such as earplugs if they are exposed to sound levels of 90 dBA or higher (● Table 6.2). Some communities also have noise ordinances designed to reduce the sound levels of traffic and other activities.

Tone Quality

The tone quality of a sound is not as easily described as loudness or pitch. Comparisons such as full versus empty, harsh versus soft, or rich versus dry are sometimes used. The tone quality of a sound is very important to our ability to identify what produced the sound. The sound of a flute is different from the sound of a clarinet, and we notice this even if they produce the same note at the same sound level. The tone quality of a person's voice helps us to identify the speaker.

The tone quality of a sound depends primarily on the waveform of the sound wave. If two sounds have different waveforms, we usually perceive different tone qualities. The simplest waveform is that of a pure tone: sinusoidal. Pure tones have a soft, pleasant tone quality (unless they are very loud or high pitched). Complex tones with waveforms that are nearly sinusoidal share the same characteristics. Unlike frequency and sound level, a waveform cannot be expressed as a single numerical factor.

What determines the waveform of a sound wave? The following mathematical principle gives us a way to comparatively analyze waveforms.

● **Figure 6.47** A sound-level meter.

© Vern Ostdiek.

● Table 6.2 OSHA Noise Limits	
Sound Level (dBA)	**Daily Exposure (hours)**
90	8
92	6
95	4
97	3
100	2
102	1.5
105	1
110	0.5
115	0.25

Any complex waveform is equivalent to a combination of two or more sinusoidal waveforms with definite amplitudes. These component waveforms are called *harmonics*. The frequencies of the harmonics are whole-number multiples of the frequency of the complex waveform.

For our purposes, this means that any complex tone is equivalent to a combination of pure tones. These pure tones, called the **harmonics,** have frequencies that are equal to 1, 2, 3, . . . times the frequency of the complex tone. For example, ● Figure 6.48 shows the

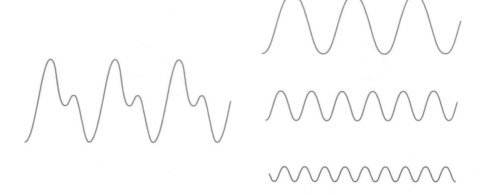

Figure 6.48 The complex-tone waveform on the left is a combination of the three pure-tone waveforms on the right. The pure tones (harmonics) have frequencies that are 1, 2, and 3 times the frequency of the complex tone.

waveform of a complex tone whose frequency is, let's say, 100 hertz. It is equivalent to a combination of the three pure tones shown whose frequencies are 100, 200, and 300 hertz. This means that we could artificially produce this complex tone by carefully playing the individual pure tones simultaneously. This is one method that electronic music synthesizers can use to create sounds.

The tone quality of a complex tone depends on the number of harmonics that are present and on their relative amplitudes. These two factors give us a quantitative way of comparing waveforms. A spectrum analyzer is a sophisticated electronic instrument that indicates which harmonics are present in a complex tone and what their amplitudes are. In general, complex tones with a large number of harmonics have a rich tone quality. Notes played on a recorder or a flute contain only a couple of harmonics, so they sound similar to pure tones. Violins and clarinets, on the other hand, have over a dozen harmonics in their notes and consequently have richer tone qualities.

The waveform of noise is a random "scribble" that doesn't repeat. This is because noises are composed of large numbers of frequencies that are not related to each other: They are not harmonics. "White noise" contains equal amounts of all frequencies of sound. (It is so named because one way to produce "white light" is to combine equal amounts of all frequencies of light.) The sound of rushing air approximates white noise.

Specifications

CD player section

Frequency Response
20 – 20,000 Hz
.
.
.

Cassette section

Frequency Response
80 – 12,000 Hz
.
.
.

Figure 6.49 Excerpts from the instruction booklet for a portable stereo. Even though the frequency of the highest note played by musical instruments is usually less than 4,000 Hz, high-fidelity sound equipment must be able to reproduce the high-frequency harmonics present in the sound.

Explore It Yourself 6.8

For this one, you have to be able to whistle.

1. Do as in Explore It Yourself 6.5 Step 1 and Step 2, but with a vowel sound or a sung note. Is the tone quality different when your ears are blocked?
2. Repeat while whistling a steady tone. Is the tone quality different when your ears are blocked? Is the effect as pronounced as in Step 1?

The existence of higher-frequency harmonics in musical notes explains why high-fidelity sound-reproduction equipment must respond to frequencies up to about 20,000 hertz. Even though the frequencies of musical notes are generally less than 4,000 hertz, the frequencies of the harmonics *do* go up to 20,000 hertz and higher. (Since our ears can't hear any harmonic with a frequency above 20,000 hertz, it isn't necessary to reproduce them.) To accurately reproduce a complex tone, each higher-frequency harmonic must be reproduced (see ● Figure 6.49).

The study of sound perception brings together the fields of physics, biology, and psychology. The mechanical properties of sound waves interact with the physiological mechanisms in the ear to produce a psychological perception. One of the challenges is to relate subjective descriptions of different sound perceptions to measurable aspects of the sound waves. The hearing apparatus itself is one of the most remarkable in all of nature. It responds to an amazing range of frequencies and amplitudes and still captures the beauty and subtlety of music. At the same time, it is very fragile. We need to learn more about how our sense of hearing works and how it can be protected.

Concept Map 6.2 summarizes the general aspects of sound perception.

CONCEPT MAP 6.2

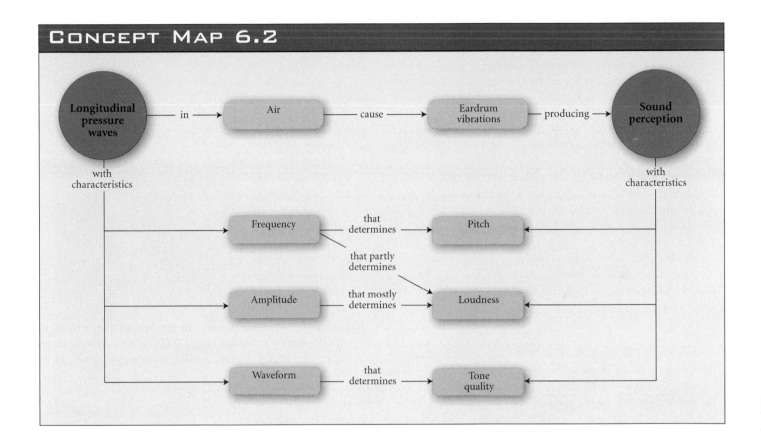

LEARNING CHECK

1. Choose the *correct* statement.
   a) The perceived pitch of a sound usually doesn't depend on the sound's frequency.
   b) The frequency of each musical note is 10 hertz greater than that of the next lower note.
   c) The decibel is the standard unit of measure of tone quality.
   d) The perceived loudness of a sound depends somewhat on its frequency.

2. (True or False.) The sound from two identical sound producers will sound twice as loud as that from just one.

3. A complex tone can be produced by combining two or more _____.

4. Telephones do not transmit sounds with frequencies above 3,400 Hz. But all musical notes with frequencies below 3,400 Hz should sound normal when they are heard over a telephone.

ANSWERS: 1. (d) 2. False 3. pure tones 4. False

The struggle to understand the nature of waves and wave propagation was historically waged on two fronts, optics and acoustics. Optics is the formal study of light, and acoustics is the study of sound and vibration. Progress in both areas over the last four centuries or so has produced a nearly complete picture of wave phenomena. The two areas of investigation complemented each other well because some wave properties are more easily exhibited with light, whereas others are more easily illustrated with sound. For example, although diffraction of sound is a very common occurrence, the diffraction of light was "discovered" first by the Italian mathematician Francesco Maria Grimaldi (1618–1663). A beam of light passing through a tiny hole in a plate produces a projected circle of light on a screen that is larger than the size of the hole. This is because the light spreads out as it passes through the hole. Isaac Newton also experimented with light diffraction.

Similarly, the Doppler effect was first investigated with sound. Predicted for both light and sound by the Austrian physicist Christian Johann Doppler (1803–1853) in 1842, the Doppler effect is not easily observed with light because of its extreme speed. In 1845, a Dutch meteorologist named Christoph Heinrich Dietrich Buys-Ballot (1817–1890) illustrated the Doppler effect using trains to provide the motion for a sound source, trumpets. He also used the trumpets as a stationary source and put listeners on the moving train.

Sound

Acoustics is one of the oldest branches of physics and yet one of the newest. It is one of the oldest because the first recorded discovery in acoustics occurred 200 years before Aristotle. It is one of the newest because most of our present knowledge of sound has been developed only in the last 150 years. Why did acoustics progress more slowly than, say, mechanics and thermodynamics? One main reason is that fairly sophisticated equipment is needed for experimentation because of the relatively high speed and rapid fluctuations (high frequencies) of sound. The development of calculus in the eighteenth century allowed for greater theoretical understanding of general wave phenomena, but experimental verification and investigation were limited by the crude technology.

Another reason for the slow development of acoustics is that certain areas, most notably the design of musical instruments, defied the complete mathematical analysis made possible in mechanics by Newton's work. To this day there is no "magic equation" that can be used to design a flute, for example, based solely on mathematical principles. The exact length, location of tone holes, and sizes of the tone holes have been determined as much by trial and error on the part of instrument makers as by mathematical analysis.

The earliest discoveries in acoustics were apparently made around 550 B.C. by Pythagoras, a Greek philosopher and mathematician whose name is immortalized by a theorem in geometry.

Although Pythagoras left no record of his work, the accounts of his successors indicate that he used a stretched string to discover that two pitches sound good together (are in harmony) if the ratio of their frequencies is simple: 2 to 1, 3 to 2, and so on. A treatise written by Aristotle indicates that he had a good understanding of the nature of sound and sound production.

A rare contribution to science by the Romans was made by the architect Vitruvius around 50 B.C. In a treatise on the design of theaters, he discussed the effects of reverberation and echoes on sound clarity.

The next notable discoveries in acoustics were made by the father of experimental science, Galileo. He completed the analysis of the vibrations of strings and correctly determined the relationship between the frequency of vibrations and the string's length, mass density, and tension. Galileo also showed that the pitch of a sound is determined by its frequency. At about this same time, a French Franciscan friar named Marin Mersenne (1588–1648; ● Figure 6.50) made fairly good measurements of the speed of sound in air and of the actual frequencies of musical notes. Other experimenters measured the speed of sound quite accurately by measuring the time between the flash and the arrival of the sound from a distant gun.

● **Figure 6.50** Marin Mersenne.

In his *Principia*, Isaac Newton showed how the properties of air can be used to theoretically predict the speed of sound. However, his predicted value was markedly lower than the measured value because he did not take into account the fact that the air is rapidly heated and cooled by the compressions and expansions associated with a sound wave. Newton then "fudged" his analysis by introducing erroneous compensating factors that raised his theoretical value so as to agree with the measured one. Even the great master was not correct on all matters.

A major step toward understanding the nature of complex tones was made by the French mathematician Joseph Fourier (1768–1830; ● Figure 6.51). He discovered that any complex waveform is actually a composite of simple, sinusoidal waveforms (Section 6.6). Fourier did not apply this principle to sound. The German physicist Georg Ohm (1787–1854), famous for his work on the conduction of electricity, used Fourier's discovery to distin-

● **Figure 6.51** Joseph Fourier.

guish pure tones and complex tones and to show that complex musical tones are combinations of pure tones.

Perhaps the greatest contributions to the science of acoustics, particularly sound perception, were made by Hermann von Helmholtz (1821–1894; ● Figure 6.52). Helmholtz was a man of prodigious analytical talent, which he applied as a surgeon, physiologist, physicist, and mathematician. In addition to his work in acoustics, Helmholtz made important contributions to electricity and magnetism and to the principle of the conservation of energy. It has been said that as a doctor he began to study the eye and the ear but found that he needed to learn more physics. Then he found that he needed to study mathematics to learn physics. In this way, he became a noted physiologist, a noted physicist, and a noted mathematician.

Helmholtz used his medical understanding of the ear to make several discoveries about sound perception. He distinguished the sensations of loudness, pitch, and tone quality in a sound. He showed that the tone quality of a complex tone depends on the harmonics that it contains. To detect these harmonics, he designed a set of "Helmholtz resonators"—hollow spheres tuned to particular frequencies that would amplify an individual pure

tone in a complex tone. Most of his discoveries are beyond the level of this text. Helmholtz's book, *Sensations of Tone*, was perhaps *the* musical acoustics textbook for nearly a century. Only with the advent of modern electronic instrumentation in the 1900s has it been possible to improve upon Helmholtz's findings about sound perception.

In recent decades, there has been a resurgence of interest in the field of musical acoustics. One example of this is the establishment of the Institut de Recherche et Coordination Acoustique/Musique in Paris in the late 1970s. Here musicians and scientists work together, using computers and other sophisticated electronic instruments, to study and create musical sounds. Many physicists find musical acoustics to be a particularly gratifying field of research because it combines the art and beauty of music with the elegance and depth of physical science.

● **Figure 6.52** Hermann von Helmholtz.

SUMMARY

- Waves are everywhere. They can be classified as **transverse** or **longitudinal** according to the orientation of the wave oscillations.

- The speed of a wave depends on the properties of the medium through which it travels. For continuous periodic waves, the product of the **frequency** and the **wavelength** equals the wave speed.

- Once waves are produced, they are often modified as they propagate. **Reflection** and **diffraction** cause waves to change their direction of motion when they encounter boundaries.

- **Interference** of two waves produces alternating regions of larger amplitude and zero-amplitude waves.

- The **Doppler effect** and the formation of **shock waves** are phenomena associated with moving wave sources. The former also occurs when the receiver of the waves is moving.

- Although sound can refer to a broad range of mechanical waves in all types of matter, we often restrict the term to the longitudinal waves in air that we can hear. Sound waves in air are generally represented by

the air-pressure fluctuations associated with the **compressions** and **expansions** in the sound wave.

- Sound with frequency too high to be heard by humans, above about 20,000 hertz, is called **ultrasound.** Ultrasound is used for echolocation by bats and for a variety of procedures in medicine.

- The sounds that we can hear can be divided into **pure tones, complex tones,** and **noise.**

- Sound is produced in many different ways, all resulting in rapid pressure fluctuations that travel as a wave. Different musical instruments employ diverse and sometimes multiple sound-production mechanisms, including vibrating plates or membranes, vibrating strings, or vibrating columns of air in tubes.

- Sound propagation inside rooms and other enclosures is dominated by repeated reflection, called **reverberation.** This causes individual sounds to linger after they are produced. Moderate reverberation causes a positive blending of successive sounds, such as musical notes, but can adversely affect the clarity of speech.

- We use three main characteristics to classify a steady sound we perceive: **pitch, loudness,** and **tone quality.** The pitch of a sound depends mainly on the frequency of the sound wave.

- The loudness of a sound depends mainly on the amplitude (or **sound level**) of the sound, but it is also affected by the frequency.

- The tone quality of a sound depends on the **waveform** of the sound wave. The waveform of a complex tone depends on the number and amplitudes of the separate pure tones, called **harmonics,** that comprise it.

IMPORTANT EQUATIONS

Equation	Comments
$\rho = \dfrac{m}{l}$	Linear mass density of rope, wire, string, Slinky, and so on.
$v = \sqrt{\dfrac{F}{\rho}}$	Speed of waves on a rope, wire, string, Slinky, and so forth

Equation	Comments
$v = 20.1 \times \sqrt{T}$	Speed of sound waves in air (SI units)
$v = f\lambda$	Relates frequency, wavelength, and speed for continuous waves

MAPPING IT OUT!

1. Review Section 6.2 on wave propagation. Identify at least five new concepts introduced in this section and devise a way to integrate these concepts in a meaningful way into Concept Map 6.1 on p. 224.

QUESTIONS

(▶ *Indicates a review question, which means it requires only a basic understanding of the material to answer. The others involve integrating or extending the concepts presented thus far.*)

1. Take a close look at the pulse traveling on the rope in Figure 6.2. It was produced by moving the hand quickly to the left (up in the photo), then right. Notice that the rope is blurred ahead of the peak of the pulse and behind it but that the peak itself is not. Explain why.
2. ▶ Give an example of a wave that does not need a medium in which to travel and a wave that does need a medium.
3. ▶ What is the difference between a longitudinal wave and a transverse wave? Give an example of each.
4. A popular distraction in large crowds at sporting events during the 1980s and 1990s was the "wave." The people in one section would quickly stand up, and then sit down; the people in the neighboring sections would follow suit in succession, resulting in a visible pattern in the crowd that would travel around the stadium. Which of the two types of waves is this? Compared to the waves described in this chapter, how is the stadium wave different?
5. A long row of people are lined up behind one another at a service window. Joe E. Clumsy stumbles into the back of the person at the end and pushes hard enough to generate a wave in the people waiting. What type of wave is produced?
6. A person attaches a paper clip to each coil of a Slinky—about 90 in all—in such a way that waves will still travel on it. What effect, if any, will the paper clips have on the speed of the waves on the Slinky?
7. The speed of an aircraft is sometimes expressed as a Mach number: Mach 1 means that the speed is equal to the speed of sound. If you wish to determine the speed of an aircraft going Mach 2.2, for

example, in meters per second or miles per hour, what additional information do you need?
8. Based on information in Section 4.1 and in the periodic table of the elements (back inside cover), why is it not surprising that sound travels faster in helium than in air?
9. If you were actually in a battle fought in space like the ones shown in science fiction movies, would you hear the explosions that occur? Why or why not?
10. ▶ Explain what the amplitude, frequency, and wavelength of a wave are.
11. ▶ A low-frequency sound is heard, and then a high-frequency sound is heard. Which sound has a longer wavelength?
12. The rays that could be used to represent the steady sound emitted by a warning siren on the top of a pole would look similar to the gravitational field lines (Section 2.8) representing the field around an object. Explain why they are alike, and point out the one important difference.
13. When trying to hear a faint sound from something far away, we sometimes cup a hand behind an ear. Explain why this can help.
14. ▶ What useful thing happens to a wave when it encounters a concave reflecting surface?
15. ▶ A person riding on a bus hears the sound from a horn on a car that is stopped. What is different about the sound when the bus is approaching the car compared to when the bus is moving away from the car?
16. ▶ Explain the process of echolocation. How is the Doppler effect sometimes incorporated?
17. In the past, ships often carried small cannons that were used when approaching shore in dense fog to estimate the distance to the hidden land. Explain how this might have been done.

18. You can buy a measuring device billed as a "sonic tape measure." Describe how a device equipped with an (ultrasonic) speaker, a microphone, and a precision timer could be used to measure the distance from the device to a wall (for example).

19. While a shock wave is being generated by a moving wave source, is the Doppler effect also occurring? Explain.

20. Describe some of the things that would happen if the speed of sound in air suddenly decreased to, say, 20 m/s. What would it be like living next to a freeway?

21. ▶ If a boat is producing a bow wave as it moves over the water, what must be true about its speed?

22. When a wave passes through two nearby gaps in a barrier, interference will occur, provided that there is also diffraction. Why must there be diffraction?

23. A recording of a high-frequency pure tone is played through both speakers of a portable stereo placed in an open field. A person a few meters in front of the stereo walks slowly along an arc around it. How does the sound that is heard change as the person moves?

24. As a loud, low-frequency sound wave travels past a small balloon, the balloon's size is affected. Explain what happens. (The effect is too small to be observed under ordinary circumstances.)

25. ▶ Describe the waveforms of pure tones, complex tones, and noise.

26. ▶ What is ultrasound, and what is it used for?

27. ▶ Describe how sound is produced in string instruments. Why does tightening a string change the frequency of the sound it makes?

28. A conditioning drill consists of repeatedly running from one end of a basketball court to the other, turning around and running back.

Sometimes the drill is changed and the runner turns around at half court, or perhaps at three-fourths of the length of the court. Describe how the number of round trips a runner can do each minute changes when the distance is changed and how this is related to a guitarist changing the note generated by a string by pressing a finger on it at some point.

29. A special room contains a mixture of oxygen and helium that is breathable. Two musicians play a guitar and a flute in the room. Does each instrument sound different from when it is played in normal air? Why or why not?

30. ▶ What is reverberation? How does reverberation affect how we hear sounds?

31. ▶ What are the three categories used to describe our mental perception of a sound? Upon what physical properties of sound waves does each depend?

32. A 100-Hz pure tone at a 70-dB sound level and a 1,000-Hz pure tone at the same sound level are heard separately. Do they sound equally loud? If not, which is louder, and why?

33. A sound is produced by combining three pure tones with frequencies of 200 Hz, 400 Hz, and 600 Hz. A second sound is produced using 200 Hz, 413 Hz, and 600 Hz pure tones. What important difference is there between the two sounds?

34. ▶ The highest musical note on the piano has a frequency of 4,186 Hz. Why would a tape of piano music sound terrible if played on a tape player that reproduces frequencies only up to 5,000 Hz?

PROBLEMS

1. Two children stretch a jump rope between them and send wave pulses back and forth on it. The rope is 3 m long, its mass is 0.5 kg, and the force exerted on it by the children is 40 N.
   a) What is the linear mass density of the rope?
   b) What is the speed of the waves on the rope?

2. The force stretching the D string on a certain guitar is 150 N. The string's linear mass density is 0.005 kg/m. What is the speed of waves on the string?

3. What is the speed of sound in air at the normal boiling temperature of water?

4. The coldest and hottest temperatures ever recorded in the United States are −80°F (211 K) and 134°F (330 K), respectively. What is the speed of sound in air at each temperature?

5. A 4-Hz continuous wave travels on a Slinky. If the wavelength is 0.5 m, what is the speed of waves on the Slinky?

6. A 500-Hz sound travels through pure oxygen. The wavelength of the sound is measured to be 0.65 m. What is the speed of sound in oxygen?

7. A wave travelling 80 m/s has a wavelength of 3.2 m. What is the frequency of the wave?

8. What frequency of sound traveling in air at 20°C has a wavelength equal to 1.7 m, the average height of a person?

9. Verify the figures given in Section 6.1 for the wavelengths (in air) of the lowest and highest frequencies of sound that people can hear (20 and 20,000 Hz, respectively).

10. What is the wavelength of 3.5 million Hz ultrasound as it travels through human tissue?

11. The frequency of middle C on the piano is 261.6 Hz.
    a) What is the wavelength of sound with this frequency as it travels in air at room temperature?
    b) What is the wavelength of sound with this frequency in water?

12. A steel cable with total length 30 m and mass 100 kg is connected to two poles. The tension in the cable is 3,000 N, and the wind makes

the cable vibrate with a frequency of 2 Hz. Calculate the wavelength of the resulting wave on the cable.

13. In a student laboratory exercise, the wavelength of a 40,000-Hz ultrasound wave is measured to be 0.868 cm. Find the air temperature.

14. A 1,720-Hz pure tone is played on a stereo in an open field. A person stands at a point that is 4 m from one of the speakers and 4.4 m from the other. Does the person hear the tone? Explain.

15. A person stands directly in front of two speakers that are emitting the same pure tone. The person then moves to one side until no sound is heard. At that point, the person is 7 m from one of the speakers and 7.2 m from the other. What is the frequency of the tone being emitted?

16. A radio acoustic sounding system (RASS) instrument uses 915-MHz radio waves (wavelength = 0.328 m).
    a) What is the wavelength of the sound waves used to measure air temperature?
    b) If the Doppler shift of the radio wave reflected from a region several hundred meters above the ground indicates a sound speed of 330 m/s, what is the air temperature?

17. A sonic depth gauge is placed 5 m above the ground. An ultrasound pulse sent downward reflects off snow and reaches the device 0.03 seconds after it was emitted. The air temperature is −20°C.
    a) How far is the surface of the snow from the device?
    b) How deep is the snow?

18. Redo Problem 17, using 0°C for the air temperature. Does it appear that the accuracy of the measured snow depth depends on the device using the correct air temperature?

19. The huge volcanic eruption on the island of Krakatoa, Indonesia, in 1883 was heard on Rodrigues Island, 4,782 km (2,970 miles) away. How long did it take the sound to travel to Rodrigues?

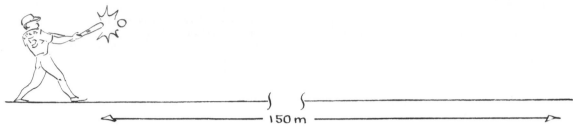

● **Figure 6.53** Problem 20.

20. A baseball fan sitting in the "cheap seats" is 150 m from home plate. How much time elapses between the instant the fan sees a batter hit the ball and the moment the fan hears the sound?

21. A geologist is camped 8,000 m (5 miles) from a volcano as it erupts.
    a) How much time elapses before the geologist hears the sound from the eruption?
    b) How much time does it take the seismic waves produced by the eruption to reach the geologist's camp, assuming the waves travel through granite as sound waves do?

22. A person stands at a point 300 m in front of the face of a sheer cliff. If the person shouts, how much time will elapse before an echo is heard?

23. A sound pulse emitted underwater reflects off a school of fish and is detected at the same place 0.01 s later. How far away are the fish?

● **Figure 6.54** Problem 23.

24. The sound level in a quiet room is 45 dB. Approximately how many times louder is it when a TV is turned on and the sound level is 75 dB?

25. Approximately how many times louder is a 100-dB sound than a 60-dB sound?

26. What are the frequencies of the first four harmonics of middle C (261.6 Hz)?

27. The frequency of the highest note on the piano is 4,186 Hz.
    a) How many harmonics of that note can we hear?
    b) How many harmonics of the note one octave below it can we hear?

CHALLENGES

1. Redraw Figure 6.15 for a convex boundary (one with the curve in the opposite direction). What happens to the amplitude of the reflected wave?

2. Perform the Explore It Yourself exercise on p. 227, and measure the distance to the wall when you could no longer distinguish the echo. Use this distance to determine the amount of time between the direct sound that you heard and the echo.

3. Jack and Jill go for a walk along an abandoned railroad track. Jack puts one ear next to a rail, while Jill, 200 m away, taps on the rail with a stone. How much sooner does Jack hear the sound through the steel rail than through the air?

4. A stationary siren produces a sound with a frequency of 500 Hz. A person hears the sound while riding in a car traveling 30 m/s toward the siren. Compute the frequency of the sound that is heard by using the relative speed of the listener and the sound wave. Assume the air temperature is 20°C.

5. An entrepreneur decides to invent and market a device that will fool the Doppler radar units used to detect cars that are speeding. The device would be placed at the very front of the car and would detect the radar signal, determine its frequency, and transmit back its own radar signal that would make the radar unit register a legal speed. What would the frequency of the "fake" signal have to be in comparison to the original? If a second unit were designed to be placed at the back end of the car, what would be different about the frequency it would have to use compared to that used by the unit at the front?

6. The guitar string described in Problem 2 is approximately 0.6 m long. What is the frequency of oscillation of a pulse on the string? (*Hint:* The frequency equals 1 divided by the period, the amount of time it takes the pulse to go back and forth once.)

7. Use the relationship between amplitude and sound level described in Section 6.6 to determine how many times larger the amplitude of a 120-dB sound is compared to the amplitude of a 0-dB sound.

8. The frequency of the lowest note played on a flute at room temperature (20°C) is 262 Hz. What would be the frequency of the same note when the flute is played outside in the winter and the air temperature in the flute is 0°C?

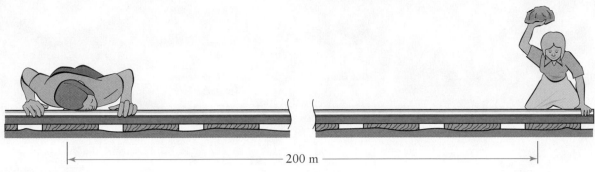

── 200 m ──

● **Figure 6.55** Challenge 3.

SUGGESTED READINGS

Bedard, Alfred J., Jr. and Thomas M. Georges. "Atmospheric Infrasound." *Physics Today* 53, no. 3 (March 2000): 32–37. An overview of natural and human-made sounds present in the atmosphere with frequencies below 20 Hz, plus a note on "animal infrasound."

El-Baz, Farouk. "Space Age Archaeology." *Scientific American* 277, no. 2 (August 1997): 60–65. Describes some of the modern tools of archaeologists, including ground-penetrating radar and underground sonar.

Freedman, Wendy L. "The Expansion Rate and Size of the Universe." *Scientific American* 267, no. 5 (November 1992): 54–60. Describes observational studies related to the Hubble relation.

Ladbury, Ray. "Ultrahigh-Energy Sound Waves Promise New Technologies." *Physics Today* 51, no. 2 (February 1998): 23–24. A report about a breakthrough in the production of very large amplitude sound waves.

Morrison, Phillip. "Wonders: Double Bass Redoubled." *Scientific American* 278, no. 5 (May 1998): 109–111. An anecdotal description of some of the uses of infrasound.

Neuweiler, Gerhard. "How Bats Detect Flying Insects." *Physics Today* 33, no. 8 (August 1980): 34–40.

Osterbrock, Donald E., Joel A. Gwinn, and Ronald S. Brashear. "Edwin Hubble and the Expanding Universe." *Scientific American* 269, no. 1 (July 1993): 84–89. A look at the life and work of astronomer Edwin Hubble.

Pohlmann, Ken C. "How Things Work." *Scientific American* 279, no. 3 (September 1998): 109. A brief look at the technology behind CDs and DVDs.

Rossing, Thomas D. *The Science of Sound*, 3rd ed. Reading, Mass: Addison–Wesley, 2001. A standard introductory acoustics textbook.

Stix, Gary. "Technology and Business: Road Warriors." *Scientific American* 277, no. 6 (December 1997): 38–40. A report on the first car to go faster than the speed of sound.

Suslick, Kenneth S. "The Chemical Effects of Ultrasound." *Scientific American* 260, no. 2 (February 1989): 80–86. Describes the use of ultrasound to produce cavities in liquids, resulting in sonoluminescence and other effects.

Swift, Gregory W. "Thermoacoustic Engines and Refrigerators." *Physics Today* 48, no. 7 (July 1995): 22–28. Describes heat engines and refrigerators based on sound.

ter Haar, Gail. "Acoustic Surgery." *Physics Today* 54, no. 12 (December 2001): 29–34. Describes the use of ultrasound in cancer treatment and other medical procedures.

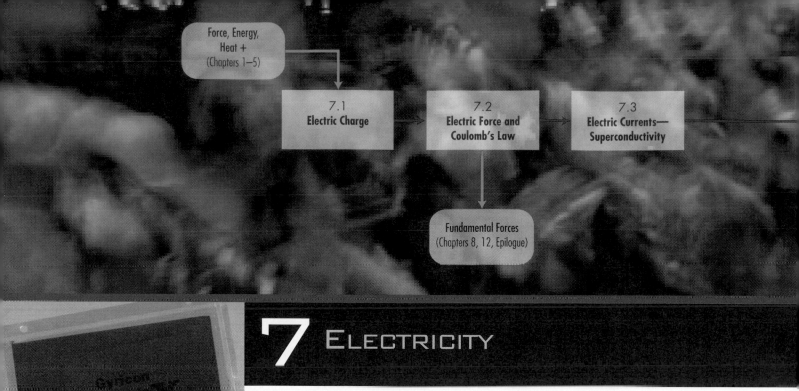

An earlier prototype of electronic paper.

7 ELECTRICITY

Electronic Paper

They first showed up in a few stores around the turn of the century: display signs that look something like a cross between paper and a flat-panel computer monitor. They represent the first use of electronic paper, a new technology that may finally have combined many of the key advantages of paper and thin video displays. Like regular paper, e-paper is flexible and relatively thin. It has high contrast and uses reflected light. It can easily be read from wide viewing angles. But like video screens, it is erasable and writeable, so it is not discarded or recycled when the information on it is obsolete. It is simply updated electronically.

Sometime in the near future, you might roll up your e-newspaper after reading it, and then use it again when the newspaper's new edition is released and downloaded to it. If only a fraction of the newspapers, magazines, and disposable documents currently produced could be replaced by reusable electronic paper, there would be environmental benefits like reduced consumption of landfill space and trees.

One of the frontrunners in this developing technology consists of millions of tiny beads sandwiched between two thin plastic sheets. Half of the surface (a hemisphere) of each bead is one color and the other a contrasting color (for example, black and white). Each bead can be rotated using a small force, but otherwise they stay in place. Letters are formed on the electronic paper by rotating some of the beads so we see their dark sides. How are the beads rotated in a controlled fashion? The answer lies in electrostatics, the topic of the following two sections.

In this chapter, we consider some of the basic aspects of electricity, starting with *electrostatics*—phenomena involving electric charges that are not moving. In the remainder of the chapter, we discuss the physics of moving charges—*electric current*. We introduce the important quantities *voltage, resistance,* and *current* and show how they are involved in many devices around us and in living things. The last two sections deal with power and energy in electric circuits and the two types of electric current—*AC* and *DC*.

7.1 Electric Charge

How many electrical devices have you used so far today? How many are operating around you right now? Do you realize that as you read these words, the information is sent to

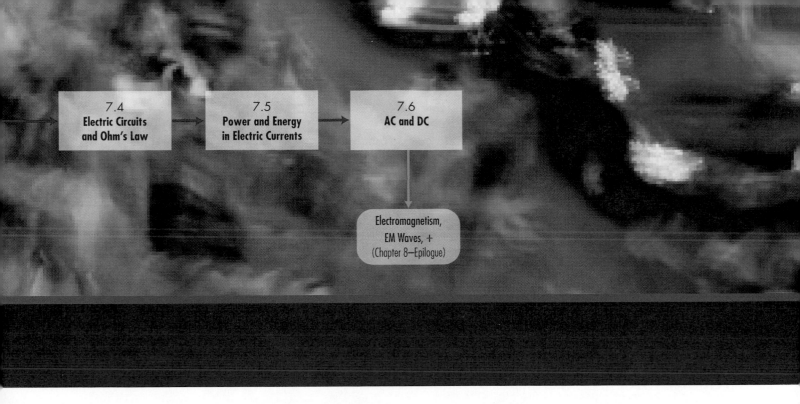

| 7.4 Electric Circuits and Ohm's Law | → | 7.5 Power and Energy in Electric Currents | → | 7.6 AC and DC |

Electromagnetism, EM Waves, + (Chapter 8—Epilogue)

your brain and processed there in the form of electrical signals? When you turn this page, your brain will communicate to the muscles in your hand in the same way. Electricity governs your life in ways you probably rarely think about. Even the properties of all of the matter you rely on—the air you breathe, the water you drink, the chair you sit in—are largely determined by electrical forces acting in and between atoms. From the latest electronic gizmo you bought to the "glue" that holds matter together, electricity is inextricably woven into your life. Most of the material in the remainder of this text is connected to electricity to one degree or another. So let's take a closer look at what electricity is.

Explore It Yourself 7.1

Note: Exercises like this sometimes do not work well when the relative humidity is high.

You need a Styrofoam cup (or inflated balloon), short pieces of thread, and fingernail-sized pieces of paper. "Charge" the cup by rubbing its sides back and forth in your hair or on fur-like material. Lower the cup to a few centimeters above the pieces of paper and thread. What happens? Does electricity appear to be stronger than gravity? (The cup can be "recharged" as needed. To "discharge" a cup, you can rub its outside on a metal water faucet.)

The word *electricity* comes from *electron*, which is based on the Greek word for amber. Amber is a fossil resin that attracts bits of thread, paper, hair, and other things after it has been rubbed with fur. You may have noticed that a comb can do the same after you use it (Figure 7.1). This phenomenon, known as the "amber effect," was documented by the ancient Greeks, but its cause remained a mystery for more than two millennia. The results of numerous experiments, some conducted by the American statesman Benjamin Franklin, indicated that matter possessed a "new" property not connected to mass or gravity. This property was eventually traced to the atom and is called **electric charge.**

 Figure 7.1 The amber effect. The comb attracts bits of paper because it was charged by rubbing.

© Fundamental Photographs

DEFINITION

Electric Charge An inherent physical property of certain subatomic particles that is responsible for electrical and magnetic phenomena. Charge is represented by q, and the SI unit of measure is the *coulomb* (C).

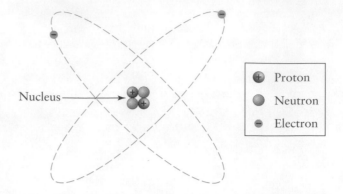

Nucleus

⊕ Proton

● Neutron

⊖ Electron

Early on, we stated that there are three fundamental things that physicists can quantify or measure: space, time, and properties of matter (Section 1.1). Until now, mass has been the only fundamental property of matter that we have used. Electric charge is another basic property of matter, but it is intrinsically possessed only by electrons, protons, and certain other subatomic or "elementary" particles (more on this in Chapter 12). Unlike mass, there are two different (and opposite) types of electric charge, appropriately named *positive* and *negative* charge. One coulomb of positive charge will "cancel" 1 coulomb of negative charge. In other words, the *net* electric charge would be zero ($q = -1 \text{ C} + 1 \text{ C} = 0 \text{ C}$).

Recall that every atom is composed of a nucleus surrounded by one or more electrons (Section 4.1). The nucleus itself is composed of two types of particles, protons and neutrons (● Figure 7.2). Every electron has a charge of -1.6×10^{-19} C, and every proton has a charge of $+1.6 \times 10^{-19}$ C. (To have a total charge of -1 C, over 6 billion billion electrons are needed.) Neutrons are so named because they are neutral; they have no net electric charge. Normally, an atom will have the same number of electrons as it has protons, which means the atom as a whole is neutral. This number is the *atomic number*, which determines the atom's identity. For example, an atom of the element helium has two protons in the nucleus and two electrons in orbit around the nucleus. The positive charge possessed by the two protons is exactly balanced by the negative charge possessed by the two electrons, so the net charge is zero. Most of the substances that we normally encounter are electrically neutral simply because the total number of electrons in all of the atoms is equal to the total number of protons.

A variety of physical and chemical interactions can cause an atom to gain one or more electrons or to lose one or more of its electrons. In these cases, the atom is said to be *ionized*. For example, if a helium atom gains one electron, it has three negative particles (electrons) and two positive particles (protons). This atom is said to be a **negative ion,** since it has a net negative charge (● Figure 7.3a). The value of its net charge is just the charge on the "extra" electron, $q = -1.6 \times 10^{-19}$ C. Similarly, if a neutral helium atom

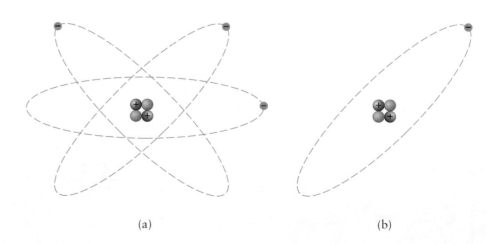

(a)

(b)

loses one electron, it becomes a **positive ion,** since it has two positive particles and only one negative particle. Its net charge is $+1.6 \times 10^{-19}$ C (Figure 7.3b).

In many situations, ions are formed on the surface of a substance by the action of friction. When a piece of amber, plastic, or hard rubber is rubbed with fur, negative ions are formed on its surface. The contact between the fur and the material causes some of the electrons in the atoms of the fur to be transferred to some of the atoms on the surface of the solid. The fur acquires a net positive charge, because it has fewer electrons than protons. Similarly, the amber, plastic, or hard rubber acquires a net negative charge since it has an excess of electrons (● Figure 7.4). Combing your hair can charge the comb in the same way. Rubbing glass with silk causes the glass to acquire a net positive charge. Some of the electrons in the surface atoms of the glass are transferred to the silk, which becomes negatively charged.

Explore It Yourself 7.2

The "popcorn" exercise. Note: Exercises like this sometimes do not work when the relative humidity is high.

For this you need a Styrofoam cup, several fingernail sized pieces of aluminum foil wadded up into little balls, and a horizontal metal surface (a baking pan or the space between heating elements on a kitchen range work). Place the foil wads on the metal surface and charge the cup (as in Explore It Yourself 7.1). Slowly lower the cup to a few centimeters above the foil wads. What happens? (If all goes well, things happen fast—so you have to watch carefully.) What causes this?

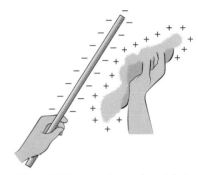

● **Figure 7.4** The contact between fur and plastic when they are rubbed together causes some of the electrons in the fur to be transferred to the plastic. This leaves the plastic with a net negative charge and the fur with a net positive charge.

Ion formation by friction is a complicated phenomenon that is not completely understood. It is affected by many factors such as which materials are used and the relative humidity.

LEARNING CHECK

1. The physical property possessed by certain particles that is responsible for electrical and magnetic phenomena is _____.

2. (True or False.) All three types of particles found in atoms are electrically charged.

3. A positive ion is formed when
   a) a neutral atom gains an electron.
   b) a neutral atom loses an electron.
   c) a negative ion loses an electron.
   d) All of the above.

ANSWERS: 1. electric charge 2. False 3. (b)

7.2 Electric Force and Coulomb's Law

The original amber effect illustrates that electric charges can exert forces. You may have noticed hair being pulled toward a charged comb or "static cling" between items of clothes removed from a dryer. These are the most common situations—two objects with opposite charges attracting each other (● Figure 7.5). The negatively charged comb exerts an attractive force on the positively charged hair. In addition, two objects with the same kind of charge (both positive or both negative) repel each other. When two similarly charged combs are suspended from threads, they push each other apart. Just remember this simple rule: *Like charges repel, unlike charges attract.*

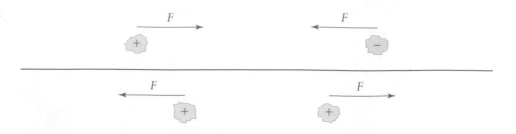

Explore It Yourself 7.3

Note: Exercises like this sometimes do not work when the relative humidity is high.

Attach a thread (roughly an arm's length) to a Styrofoam cup. Charge it (as in Explore It Yourself 7.1), and then let it dangle from the thread. Charge a second Styrofoam cup, and then bring it close to the hanging one. What happens? The size of the force on the cup is related to how far the thread is angled from vertical. Does the force on the cup seem to be affected by how far apart the two cups are?

Sodium ion

Chlorine ion

● **Figure 7.6** Ordinary table salt consists of positive sodium ions and negative chlorine ions. The strong attractive forces between the oppositely charged ions hold them in a rigid crystalline array.

This force between charged objects is extremely important in the physical world, particularly at the atomic level. It holds atoms together and makes it possible for them to exist. In each atom, the positively charged protons in the nucleus exert attractive forces on the negatively charged electrons. The electric force on each electron keeps it in its orbit, much as the gravitational force exerted by the Sun keeps the Earth in its orbit. The forces between the atoms in many compounds arise because opposite charges attract. For example, when salt is formed from the elements sodium and chlorine, each sodium atom gives up an electron to a chlorine atom. The resulting ions exert attractive forces on one another because they are oppositely charged (● Figure 7.6). Matter as we know and experience it would not exist without the electrical force.

Recall Newton's law of universal gravitation in Section 2.8, which gives the size of the force acting between two masses. The corresponding law for the electrical force, **Coulomb's law,** is very similar.

LAWS

Coulomb's Law The force acting on each of two charged objects is directly proportional to the net charges on the objects and inversely proportional to the square of the distance between them.

$$F \propto \frac{q_1 q_2}{d^2}$$

The constant of proportionality in SI units is 9×10^9 N-m^2/C^2. Therefore:

$$F = \frac{(9 \times 10^9) q_1 q_2}{d^2}$$

SI units—F in newtons, q_1 and q_2 in coulombs, and d in meters

The force on q_1 is equal and opposite to the force on q_2, by Newton's third law of motion. Note that if both objects have the same kind of charge (both positive or both negative), the force F is positive. This indicates a repulsive force. If one charge is negative and the other positive, the force F is negative, indicating an attractive force. If the distance between two charged objects is doubled, the forces are reduced to one-fourth their original values (● Figure 7.7).

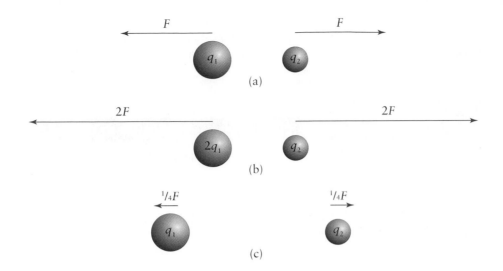

(a)

(b)

(c)

● **Figure 7.7** Coulomb's law. Doubling the charge on one object, (b), doubles the force on both. Doubling the distance between the charges, (c), reduces the forces to one-fourth their original values.

Perhaps it is not a surprise that Coulomb's law has the same form as Newton's law of universal gravitation. After all, mass and charge are both fundamental properties of the particles that comprise matter. We must remember, however, that the (gravitational) force between two bodies due to their masses is *always* an attractive force, while the (electrostatic) force between two bodies due to their electric charges can be attractive or repulsive, depending on whether they have opposite charges or not. Also, all matter has mass and so experiences and exerts gravitational forces, whereas the electrostatic force normally acts between objects only when there is a net charge on one or both of them. Generally, when two objects have electric charges, the electrostatic force between them is much stronger than the gravitational force. For example, the electrostatic force between an electron and a proton is about 10^{39} times as large as the gravitational force between them.

For more information on the properties and relative strengths of the forces in nature, see Section 12.2.

It is possible for a charged object to exert a force of attraction on a second object that has no net charge. This is what happens when a charged comb is used to pick up bits of paper or thread. Here, the negatively charged comb attracts the nuclei of the atoms and repels the electrons. The orbits of the electrons are distorted so that the electrons are, on the average, farther away from the charged comb than the nuclei (● Figure 7.8). This results in a net attractive force because the repulsive force on the slightly more distant, negatively charged electrons is smaller than the attractive force on the closer, positively charged protons.

Some molecules naturally have a net negative charge on one side and a net positive charge on the other. They are called *polar molecules*. Think of the similarity to the two poles of a magnet. Water molecules have this property. (As we will see in Chapter 8, that's why microwave ovens work.) If a polar molecule is free to rotate—as in a liquid—it will

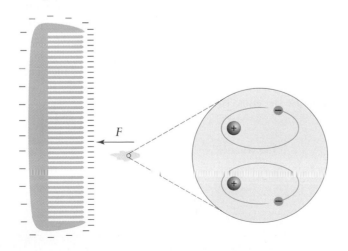

● **Figure 7.8** A charged comb exerts a net attractive force on a neutral piece of paper because the electrons are displaced slightly away from the negatively charged comb. (The distorted electron orbits are exaggerated in this drawing.) The attractive force on the closer nuclei is stronger than the repulsive force on the electrons.

be attracted to a charged object. Its side with the charge opposite that on the object will turn toward the object, and the attractive force on that side will be stronger than the repulsive force on the other side, as with the atoms above.

Explore It Yourself 7.4

Note: Exercises like this sometimes do not work when the relative humidity is high.
    Turn on a water faucet just enough to have a very small, but steady, stream trickling out. Bring a charged Styrofoam cup near the stream. What happens?

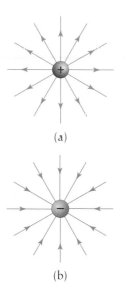

(a)

(b)

● **Figure 7.9** Electric field lines in the space around a positive charge (a) and around a negative charge (b). The arrows show the direction of the force that would act on a positive charge placed in either field.

The electrostatic force is another example of "action at a distance." As with gravitation, the concept of a field is useful. In the space around any charged object, there is an *electric field*. This field is the "agent" of the electrostatic force: It will cause any charged object to experience a force. The electric field around a charged particle is represented by lines that indicate the direction of the force that the field would exert on a positive charge. Thus the *electric field lines* around a positively charged particle point radially outward, and the field lines around a negatively charged particle point radially inward (● Figure 7.9).

The strength of an electric field at a point in space is equal to the size of the force that it would cause on a given charged object placed at that point, divided by the size of the charge on the object.

$$\text{electric field strength} = \frac{\text{force on a charged object}}{\text{charge on the object}}$$

Where the field is strong, a charged object will experience a large force. The strength of the electric field is indicated by the spacing of the field lines: Where the lines are close together, the field is strong. The electric field around a charged particle is clearly weaker at greater distances from it.

Any time a positive charge is in an electric field, it experiences a force in the *same* direction as the field lines. A negative charge in an electric field feels a force in the *opposite* direction of the field lines (● Figure 7.10). Remember that the electric field exerts the force on a charged object; in Chapter 8, we will show that electric fields can be produced without directly using electric charges.

Perhaps you have had the experience of walking across a carpeted floor and receiving a shock when you touched a metal doorknob. This is more likely to happen in winter than in summer because the relative humidity is usually lower, and electrostatic charging takes place more readily. The shock is due to charges flowing between you and the doorknob, and it may be accompanied by a visible spark. Air normally does not allow charges to flow through it. A spark occurs when there is an electric field strong enough to ionize atoms in the air. Freed electrons accelerate in a direction opposite to the direction of the electric field, and positive ions accelerate in the same direction as the field. The electrons and ions pick up speed and collide with other atoms and molecules, ionizing them or causing them to emit light (see ● Figure 7.11). Lightning is produced in this same way on a much larger scale as Benjamin Franklin demonstrated using kites, keys, and metal rods in the middle of the 18th century (see Physics Potpourri on p. 266). In Chapter 10, we will discuss how atomic collisions cause the emission of light.

● **Figure 7.10** In an electric field, the force on a positively charged body is parallel to the field, while the force on a negatively charged body is opposite to the direction of the field.

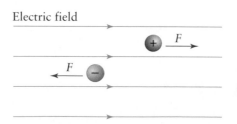

Electric field

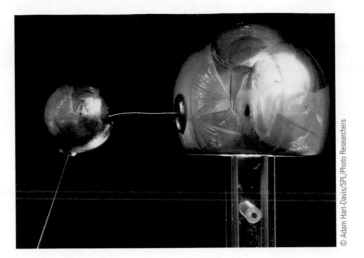

© Adam Hart-Davis/SPL/Photo Researchers

● **Figure 7.11** A large spark created in a lab. The spark is produced because there is a very strong electric field in the space between the two objects.

Explore It Yourself 7.5

Note: Exercises like this sometimes do not work when the relative humidity is high.
For this you need burning incense in a room where the air is still. Bring a charged Styrofoam cup near the rising stream of smoke. What happens? Can you think of a practical use for this phenomenon?

While most of the electrical devices we rely on make use of electric currents, some depend primarily on electrostatics. One important example of the latter is the *electrostatic precipitator,* an air-pollution control device (● Figure 7.12). Tiny particles of soot, ash, and dust are major components of the airborne emissions from fossil-fuel-burning power plants and from many industrial processing plants. Electrostatic precipitators can remove nearly all of these particles from the emissions. The flue gas containing the particles is passed between a series of positively charged metal plates and negatively charged wires (● Figure 7.13). The strong electric field around the wires creates negative ions in the particles. These negatively charged particles are attracted by the positively charged plates and collect on them. Periodically, the plates are shaken so the collected soot, ash, and dust slide down into a collection hopper. This "fly ash" must then be disposed of, but in some cases, it has uses—for example, as a filler in concrete.

Courtesy of Iatan Generating Station

● **Figure 7.12** Electrostatic precipitator at a large coal-fired power plant near Iatan, Missouri.

● **Figure 7.13** Schematic of an electrostatic precipitator. (a) The emissions (flue gas) flow around negatively charged wires hanging between positively charged plates. (b) Top view of one channel. The particles are charged negatively by the strong electric field around the wires. Consequently, they are attracted to, and collect on, the plates.

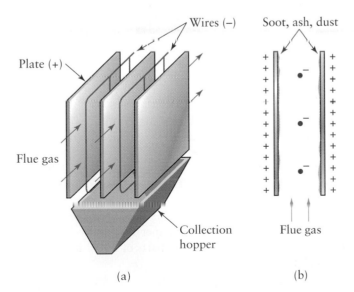

(a)   (b)

Electrifying Sights and Sounds: A Thunderstorm Primer

Few of us have not witnessed the fearsome power and awesome beauty of a thunderstorm, complete with torrential rains and hail, ear-splitting claps of thunder, and brilliant, sometimes deadly, strokes of lightning (● Figure 7.14). At any given time, there are roughly 2000 thunderstorms in progress around the world, producing some 30 to 100 cloud-to-ground lightning flashes each second, totaling about 5 million lightning strikes a day. But what are these lightning strikes? How are they produced and what makes them so dangerous? The answers, as Benjamin Franklin first discovered in the early 1750s (cf. Physics Family Album, p. 283), are rooted in the physics of electricity.

Lightning is akin to the short sparks we experience when we touch a metal doorknob after walking across a wool rug in a dry environment in the winter—only on a much grander scale! Cumulonimbus clouds, the most common thunderstorm cloud, and the Earth effectively acquire opposite electrical charges (like your finger and the doorknob), the air between serving as an insulating material. When the separated charge grows sufficiently large, the strong electric field between the cloud and the Earth, typically some 300,000–400,000 volts/meter, causes an electrical breakdown of the insulating air creating an ionized path between the cloud and the ground along which charge can flow, that is, a lightning discharge. Usually during this discharge, negative charge is transferred from the lower part of the cloud to neutralize the induced positive charge on the Earth below. The average maximum current in such lightning flashes is about 30,000 amperes, lasting for only about 30 microseconds and delivering about a coulomb of charge. It is the large currents carried by lightning discharges that make them so dangerous to human beings (see Physics Potpourri on p. 276). Every year an average of 200 persons in the United States die from injuries sustained after being struck by lightning. Lightning is the leading cause of weather-related personal injuries. Thus, while the overall functional explanation for lightning strikes is well-understood, many of the details have only been elucidated relatively recently.

For example, one major puzzle surrounding cloud-to-ground lightning, the most important type of lightning in terms of its impact on human life, has been how the cloud acquires charge—or more properly, a charge separation—in the first place. One popular theory takes advantage of the mixed phases of water typically found in large (3–4 kilometers horizontally by 5–8 kilometers vertically) thunderstorm clouds: liquid water droplets, ice crystals, and *graupel* or soft hail. It is believed that, as the faster-falling graupel collides with the smaller ice crystals, positive charge is transferred from the porous hail to the ice crystals, so that the former acquire a net negative charge and the latter a net positive charge. (This is a bit like rubbing a plastic rod with a piece of fur; see Figure 7.4.) As the graupel continues to move downward relative to the ice crystals, a charge separation develops within the cloud with the negative charges concentrated in the lower portions of the cloud and the positive charges residing higher up (as much as a kilometer or more) in the cloud (● Figure 7.15). Although this scenario for producing a

● **Figure 7.14** Lightning and sparks are caused by very strong electric fields.

© Tony Stone Worldwide

The electronic paper described at the beginning of this chapter uses electric fields to form letters and other images. One type, called SmartPaper™, consists of millions of tiny beads between two thin plastic sheets (see ● Figure 7.16).

One side of each bead is some color and negatively charged, and the other side is a contrasting color and positively charged. An electric field exerts opposite forces on the two sides (recall Figure 7.10), causing the beads to rotate until they are aligned with the field. (As we will see in Chapter 8, this is just like what a compass needle does in a magnetic field.) Letters are formed on the electronic paper by selectively applying upward and downward electric fields at different places, so parts of the display are one color and the rest are the other color.

charge separation shows promise in explaining lightning in common terrestrial thunderstorms, it is not the only theory of thundercloud electrification, and it may not be able to explain other types of discharges called *warm cloud* lightning.

The average lightning stroke is between 6 and 8 miles long and consists of a large surge of current usually moving upward from the ground into the cloud along a path of ionized air. (*Intracloud lightning* is also quite common wherein the discharge occurs along an ionized track connecting a positive portion of a cloud with a negatively charged region of the same cloud.) The characteristic lightning flash is produced by atmospheric ions recombining with electrons and relaxing to their normal energy states by emitting light (Section 10.3). What appears as a single lightning stroke may actually be composed of several individual strokes, the average for most displays being about 3 or 4, each one separated by 40 to 80 milliseconds. The average peak power delivered by such lightning flashes is about 10^{12} watts, equivalent to 10 billion 100-watt lightbulbs!

Like lightning, thunder, which accompanies most displays, is another natural phenomenon about which humans have speculated for centuries. The consensus view today, however, remains that described by M. Hirn in 1888, namely that air along the lightning train is heated to a high temperature ($\sim$20,000°C) and rapidly expands to form a shock wave, which soon decays to a sound wave we hear. However, even under the best of conditions, thunder generally cannot be heard at all at distances of more than about 25 kilometers from the lightning channel that produced it. This is due to the upward refraction of the sound waves resulting from vertical temperature variations and wind shear. But that's another story. (As noted in the *Explore It Yourself* on page 21, you can often estimate the distance to a lightning strike from the time difference between the visible flash and the audible thunder associated with it.)

So, next time you find yourself in the middle of one of nature's fantastic fireworks displays, as you contemplate its majesty and beauty, think too about the awesome physics that makes it all possible!

For those interested in learning more about thunderstorms and lightning and how these phenomena are currently being studied using rockets, high-altitude airplanes, and spacecraft, check out NASA's Global Hydrology and Climate Center website on lightning and atmospheric electricity research at http://thunder.msfc.nasa.gov.

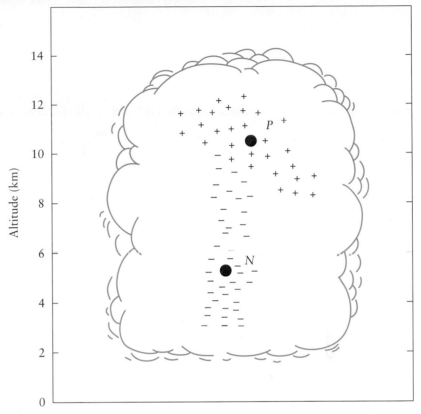

● **Figure 7.15** Schematic drawing showing the distribution of the main concentrations of positive and negative charge in a large thundercloud. The solid black dots indicate the effective centers of charge for each distribution. If each cloud of charge were replaced by an equivalent point charge located at these centers of charge, then the size of the charge at *P* would be about $+40$ C, while that at *N* would be -40 C. (After M. A. Uman, *Lightning*, p. 3 (New York, Dover, 1984).)

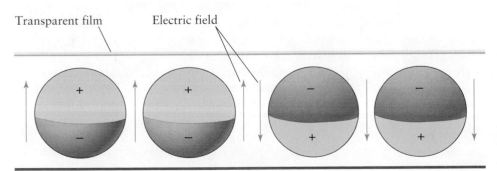

● **Figure 7.16** Simplified edge view of a segment of SmartPaper™ electronic paper. One half of each tiny bead has a negative charge and is some color (red in this example), and the other half has a positive charge and is a contrasting color. An electric field exerts opposite forces on the two sides of each bead, so the bead rotates and aligns with the field. Letters are formed by using electric fields to turn the red sides up in parts of the display and the blue sides up elsewhere.

The type of transistor used most widely in computers and similar devices is the field-effect transistor (FET). In FETs, an electric field controls the flow of electricity through the transistor. Electric fields play crucial roles in the operation of LCDs (liquid crystal displays) and touchpads on laptop computers as well.

Concept Map 7.1 summarizes the ideas presented in these first two sections of Chapter 7.

CONCEPT MAP 7.1

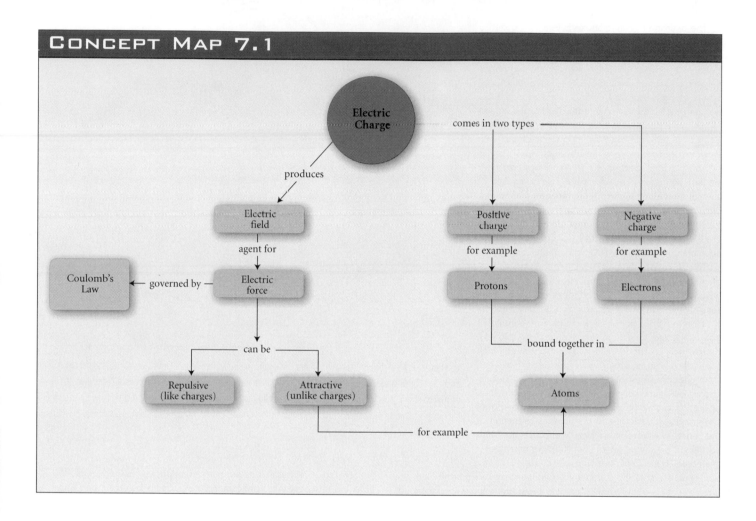

LEARNING CHECK

1. Like charges _____ ; unlike charges _____ .

2. Two charged objects exert forces on each other. The sizes of both forces will increase if
   a) the objects are brought closer together.
   b) the charge on one of them is increased.
   c) the charges on both of them are increased.
   d) All of the above.

3. A charged particle creates an electric _____ in the space around it.

4. (True or False.) An electrostatic precipitator uses an electric field to make it rain.

ANSWERS: 1. repel, attract   2. (d)   3. field   4. False

7.3 Electric Currents—Superconductivity

An **electric current** is a flow of charged particles. The cord on an electrical appliance encloses two separate metal wires covered with insulation. When the appliance is plugged in and operating, electrons inside each wire move back and forth. Inside a television picture tube, free electrons are accelerated from the back of the tube to the screen at the front. There is a near vacuum inside the picture tube, so the electrons can travel without colliding with gas molecules (● Figure 7.17). When salt is dissolved in water, the sodium and chlorine ions separate and can move about just like the water molecules. If an electric field is applied to the water, the positive sodium ions will flow one way (in the direction of the field), and the negative chlorine ions will flow the other way.

Regardless of the nature of the moving charges, the quantitative definition of electric current is as follows.

<table>
<tr><td rowspan="6" style="writing-mode: vertical-lr">DEFINITION</td><td>

Current The rate of flow of electric charge. The amount of charge that flows by per second.

$$\text{current} = \frac{\text{charge}}{\text{time}} \qquad I = \frac{q}{t}$$

The SI unit of current is the *ampere* (A or amp), which equals 1 coulomb per second. Current is measured with a device called an *ammeter*.

</td></tr>
</table>

A current of 5 amperes in a wire means that 5 coulombs of charge flow through the wire each second. (● Table 7.1 lists some representative currents.) Either positive charges

● **Table 7.1** Typical Currents in Common Devices

Device	Current (A)
Calculator	0.0001
Spark plug	0.001
Clock radio	0.03
60-watt light bulb	0.5
Color television	2
1,200-watt hair dryer	10

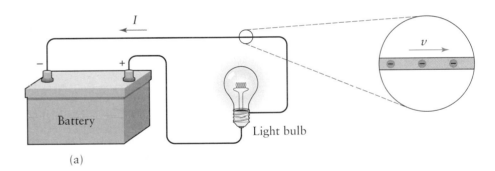

(a)

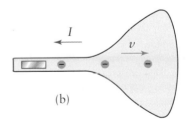

(b)

● **Figure 7.17** Examples of electric current. (a) Electrons flowing inside a metal wire. (b) Electrons moving through the near-vacuum inside a television picture tube. (c) Positive ions and negative ions, from dissolved salt, flowing through water.

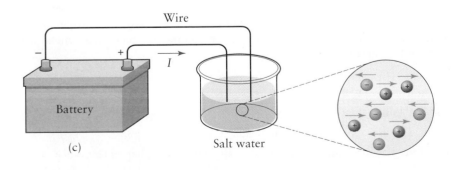

(c)

or negative charges can comprise a current. The effect of a positive charge moving in one direction is the same as that of an equal negative charge moving in the opposite direction. Originally, it was believed that positive charges flow through metals. Even though it was determined later that negative electrons flow, this convention was retained: A current is represented as a flow of positive charges, even if it is actually due to a flow of negative charges in the opposite direction. If positive ions are flowing to the right in a liquid, then the current is to the right. If negative charges (like electrons) are flowing to the right, the direction of the current is to the left. In Figure 7.17, the current is to the left in (a) and (b), and to the right in (c).

The ease with which charges move through different substances varies greatly. Any material that does not readily allow the flow of charges through it is called an electrical *insulator*. Substances like plastic, wood, rubber, air, and pure water are insulators because the electrons are tightly bound in the atoms, and electric fields are usually not strong enough to rip them free so they can move. Our lives depend on insulators: The electricity powering the devices in our homes could kill us if insulators, like the covering on power cords, didn't keep it from entering our bodies.

An electrical *conductor* is any substance that readily allows charges to flow through it. Metals are very good conductors because some of the electrons are only loosely bound to atoms and so are free to "skip along" from one atom to the next when an electric field is present. In general, solids that are good conductors of heat are also good conductors of electricity. As mentioned before, liquids such as water are conductors when they contain dissolved ions. Most drinking water has some natural minerals and salts dissolved in it and so conducts electricity. Solid insulators can become conductors when wet because of ions in the moisture. The danger of being electrocuted by electrical devices increases dramatically when they are wet.

Semiconductors are substances that fall in between the two extremes. The elements silicon and germanium, both semiconductors, are poor conductors of electricity in their pure states, but they can be modified chemically ("doped") to have very useful electrical properties. Transistors, solar cells, and numerous other electronic components are made out of such semiconductors. The electronic revolution in the second half of the twentieth century, including the development of inexpensive calculators, computers, sound-reproduction systems, and other devices, came about because of semiconductor technology (● Figure 7.18).

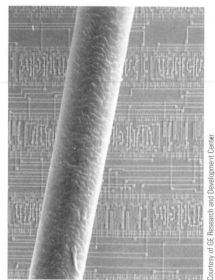

● **Figure 7.18** Greatly enlarged image of electronic circuits etched on a semiconductor chip. The human hair lying on the chip gives you an idea of the chip's small size.

Courtesy of GE Research and Development Center

Resistance

What makes a 100-watt light bulb brighter than a 60-watt bulb? The size of the current flowing through the filament determines the brightness. That, in turn, depends on the filament's **resistance.**

In general, a conductor will have low resistance and an insulator will have high resistance. The actual resistance of a particular piece of conducting material—a metal wire, for example—depends on four factors:

Composition. The particular metal making up the wire affects the resistance. For example, an iron wire will have a higher resistance than an identical copper wire.

Length. The longer the wire is, the higher its resistance.

Diameter. The thinner the wire is, the higher its resistance.

Temperature. The higher the temperature of the wire, the higher its resistance.

The filament of a 100-watt bulb is thicker than that of a 60-watt bulb, so its resistance is lower. As we will see in later sections, this means a larger current normally flows through the 100-watt bulb, and consequently, it is brighter.

Resistance can be compared to friction. Resistance inhibits the flow of electric charge, and friction inhibits relative motion between two substances. In metals, electrons in a current move among the atoms and in the process collide with them and give them energy. This impedes the movement of the electrons and causes the metal to gain internal energy. The consequence of resistance is the same as that of kinetic friction—heating. The larger the current through a particular device, the greater the heating.

Superconductivity

In 1911, the Dutch physicist Heike Kamerlingh Onnes made an important discovery while measuring the resistance of mercury at extremely low temperatures. He found that the resistance decreased steadily as the temperature was lowered, until at 4.2 K ($-452.1°F$) it suddenly dropped to zero (● Figure 7.19). Electric current flowed through the mercury with *no* resistance. Onnes named this phenomenon *superconductivity* for good reason: Mercury is a perfect conductor of electric current below what is called its *critical temperature* (referred to as T_c) of 4.2 K. Subsequent research showed that hundreds of elements, compounds, and metal alloys become superconductors, but only at very low temperatures. Until 1985, the highest known T_c was 23 K for a mixture of the elements niobium and germanium.

Superconductivity seems too good to be true: electricity flowing through wires with *no* loss of energy to heating. Once a current is made to flow in a loop of superconducting wire, it can flow *for years* with no battery or other source of energy because there is no energy loss due to resistance. A great deal of the electrical energy that is wasted as heat in wires could be saved if conventional conductors could be replaced with superconductors. But the superconducting state for a given material has limitations. Resistance returns if the temperature is raised above the superconductor's T_c, if the current in it becomes too large, or if it is placed in a magnetic field that is too strong.

Practical superconductors were developed in the 1960s and are now widely used in science and medicine. Most of them are compounds of the element niobium. *Superconducting electromagnets,* the strongest magnets known, are used to study the effects of magnetic fields on matter and to direct high-speed charged particles. The huge particle accelerator at the Fermi National Accelerator Laboratory near Chicago, shown in Figure 8.16, uses superconducting electromagnets to guide protons as they are accelerated to nearly the speed of light. An entire experimental passenger train was built that levitated by superconducting electromagnets. *Magnetic resonance imaging* (MRI) uses superconducting electromagnets to form incredibly detailed images of the body's interior. (Magnetism and electromagnets are discussed in Chapter 8.)

Widespread practical use of these superconductors is severely limited because they must be kept cold using liquefied helium. Helium is very expensive and requires sophisticated refrigeration equipment to cool and to liquefy. Once a superconducting device is cooled to the temperature of liquid helium, bulky insulation equipment is needed to limit the flow of heat into the helium and the superconductor. These factors combine to make the so-called low-T_c superconductors unwieldy or uneconomical except in certain special applications when there are no alternatives.

But hope for wider use of superconductivity blossomed in 1987 when a new family of "high-T_c" superconductors was developed with critical temperatures as high as 90 K, since then pushed upward to about 140 K. This was an astounding breakthrough because these materials can be made superconducting through the use of liquid nitrogen (boiling point 77 K). This liquid is widely available, it is cheap compared to liquid helium, and it can be used with much less sophisticated insulation. However, the new high-T_c superconductors are handicapped by a couple of unfortunate properties: They are brittle and consequently are not easily formed into wires, and they aren't very tolerant of strong magnetic fields or large electric currents. If these problems can be overcome, a new revolution in superconducting technology will occur.

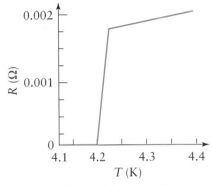

● **Figure 7.19** Graph of the resistance of a sample of mercury versus temperature, showing the transition to superconductivity.

7.4 Electric Circuits and Ohm's Law

An electric current will flow in a light bulb, a radio, or other such device only if an electric field is present to exert a force on the charges. A flashlight works because the batteries produce an electric field that forces electrons to flow through the light bulb. An *electric circuit* is any such system consisting of a battery or other electrical *power supply*, some electrical device like a light bulb, and wires or other conductors to carry the current to and from the device (see ● Figure 7.20). The power supply acts like a "charge pump": It forces charges to flow out of one terminal, go through the rest of the circuit, and flow into the other terminal. Electrons typically move through a circuit quite slowly, about 1 millimeter per second. In this respect, an electric circuit is much like the cooling system in a car in which the water pump forces coolant to flow through the engine, radiator, and the hoses connecting them.

The concepts of energy and work are used to quantify the effect of a power supply in a circuit. In a flashlight, for instance, the batteries cause electrons to flow through the bulb's filament. Since a force acts on the electrons and causes them to move through a distance, *work is done* on the electrons *by the batteries.* In other words, the batteries give the electrons energy. This energy is converted into internal energy and light as the electrons go through the light bulb. This leads to the concept of electric **voltage.**

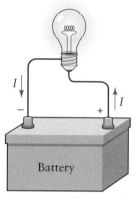

● **Figure 7.20** A simple electric circuit. The battery supplies the energy or "pressure" needed to move the charges through the circuit. Charge does not build up anywhere; for each coulomb of charge that leaves the positive terminal, 1 coulomb enters the negative terminal.

DEFINITION

Voltage The work that a charged particle can do divided by the size of the charge. The energy per unit charge given to charged particles by a power supply.

$$V = \frac{work}{q} \qquad V = \frac{E}{q}$$

The SI unit of voltage is the *volt* (V), which is equal to *1 joule per coulomb.* Voltage is measured with a device called a *voltmeter.*

● **Table 7.2** Examples of Common Voltages	
Description	**Voltage (V)**
Nerve impulse	0.1
D-cell battery	1.5
Car battery	12
Wall socket (varies)	120
TV or computer monitor picture tube (typical)	20,000
Power plant generator (typical)	24,000
High-voltage transmission line (typical)	345,000

A 9-volt battery gives 9 joules of energy to each coulomb of electric charge that it moves through a circuit. Each coulomb does 9 joules of work as it flows through the circuit. (● Table 7.2 lists some typical voltages.)

If we return to the analogy of a battery as a charge pump, the voltage plays the role of pressure. A high voltage causing charges to flow in a circuit is similar to a high pressure causing a fluid to flow (● Figure 7.21). Even when the circuit is disconnected from the power supply and there is no charge flow, the power supply still has a voltage. In this case, the electric charges have potential energy. (Voltage is also referred to as electric *potential.*)

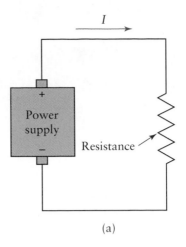
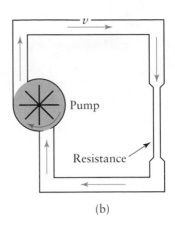

(a) (b)

● **Figure 7.21** (a) The flow of charge in an electric circuit is much like (b) the flow of water through a closed path. The power supply corresponds to the water pump, and the resistance corresponds to the narrow segment of pipe. The pressure on the output side of the pump is much like the voltage on the "+" terminal of the power supply. The electric current corresponds to the rate of flow of the water.

The size of the current that flows through a conductor depends on its resistance and on the voltage causing the current. **Ohm's law,** named after its discoverer, Georg Simon Ohm, expresses the exact relationship.

Ohm's Law The current in a conductor is equal to the voltage applied to it divided by its resistance.

$$I = \frac{V}{R} \quad \text{or} \quad V = IR$$

The units of measure are consistent in the two equations: If I is in amperes and R is in ohms, V will be in volts.

By Ohm's law, the higher the voltage for a given resistance, the larger the current. The larger the resistance for a given voltage, the smaller the current. By applying different-sized voltages to a given conductor, one can produce different-sized currents. A graph of the voltage versus the current will be a straight line whose slope is equal to the conductor's resistance (● Figure 7.22). Reversing the polarity of the voltage (switching the "+" and "−" terminals) will cause the current to flow in the opposite direction.

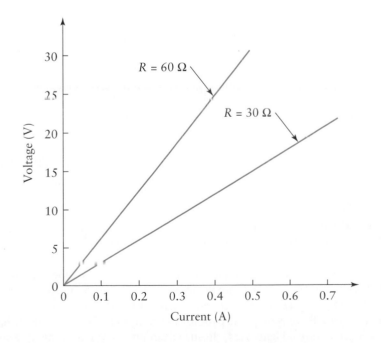

● **Figure 7.22** Graph of voltage versus current for two conductors with different resistances (R). For each resistance, the voltage needed to produce a given current is proportional to the current.

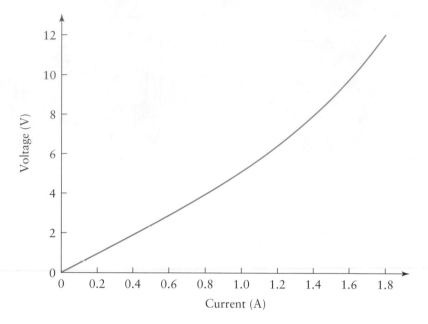

● **Figure 7.23** Graph of voltage versus current for a brake light bulb on a car. The graph curves upward because the resistance is higher with larger currents. This is because the filament is hotter.

Example 7.1 A light bulb used in a 3-volt flashlight has a resistance equal to 6 ohms. What is the current in the bulb when it is switched on? By Ohm's law:

$$I = \frac{V}{R} = \frac{3\,\text{V}}{6\,\Omega}$$

$$= 0.5\,\text{A}$$

Example 7.2 A small electric heater has a resistance of 15 ohms when the current in it is 2 amperes. What voltage is required to produce this current?

$$V = IR = 2\,\text{A} \times 15\,\Omega$$

$$= 30\,\text{V}$$

Often, the resistance of a conductor changes when the voltage changes. At higher voltages, a larger current flows through the filament of a light bulb, so its temperature is also higher. The resistance of the hotter filament is consequently greater (● Figure 7.23). Some semiconductor devices, called *diodes,* are designed to have very low resistance when current flows through them in one direction but very high resistance when a voltage tries to produce a current in the other direction. Water with salt dissolved in it generally has lower resistance when higher voltages are applied to it: Doubling the voltage will more than double the current. A graph of *V* versus *I* for ordinary tap water is less steep at higher voltages.

Many electrical devices are controlled by changing a resistance. The volume control on a radio or a television simply varies the resistance in a circuit. Turning up the volume reduces the resistance, so more current flows in the circuit, resulting in louder sound. A dimmer control used to change the brightness of the lights in a room works the same way.

Series and Parallel Circuits

In many situations, several electrical devices are connected to the same electrical power supply. A house may have a hundred different lights and appliances all connected to one cable entering the house. An automobile has dozens of devices connected to its battery. There are two basic ways in which more than one device can be connected to a single electrical power supply—by a series circuit and by a parallel circuit.

In a *series circuit,* there is only one path for the charges to follow, so the same current flows in each device (see ● Figure 7.24). In such a circuit, the voltage is divided among the

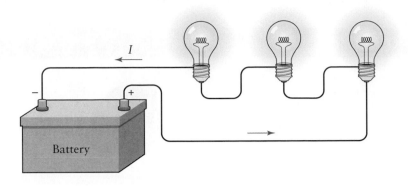

devices: The voltage on the first device plus the voltage on the second device, and so on, equals the voltage of the power supply. For example, if three light bulbs with the same resistance are connected in series to a 12-volt battery, the voltage on each bulb is 4 volts. If the bulbs had different resistances, each one's "share" of the voltage would be proportional to its resistance.

Notice that the current in a series circuit is stopped if any of the devices breaks the circuit (● Figure 7.25). A series circuit is not normally used with, say, a number of light bulbs because if one of them burns out, the current stops and all of the bulbs go out. A string of Christmas lights that flash at the same time uses a series circuit so that all the bulbs go on and off together.

In a *parallel circuit,* the current through the power supply is "shared" among the devices while each has the same voltage (● Figure 7.26). The current flowing in the first device plus the current in the second device, and so on, equals the current put out by the power supply. There is more than one path for the charges to follow—in this case, three. If one of the devices burns out or is removed, the others still function. The light bulbs in multiple-bulb light fixtures are in parallel so that if one bulb burns out, the others remain lit. Often, the two types of circuits are combined: One switch may be in series with several light bulbs that are in parallel.

Three light bulbs are connected in a parallel circuit with a 12-volt battery. The resistance of each bulb is 24 ohms. What is the current produced by the battery?

Example 7.3

The voltage on each bulb is 12 volts. Therefore, the current in each bulb is

$$I = \frac{V}{R} = \frac{12\ V}{24\ \Omega}$$

$$= 0.5\ A$$

The total current supplied by the battery equals the sum of the currents in the three bulbs.

$$I = 0.5\ A + 0.5\ A + 0.5\ A$$

$$= 1.5\ A$$

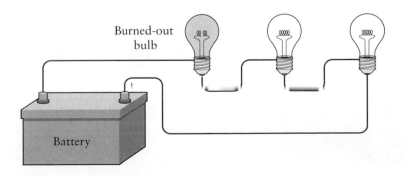

Burned-out bulb

Battery

● **Figure 7.25** If one device in a series circuit fails, such as a bulb burning out, the current stops and all of the devices go off.

Electricity and the Human Body

As in all living organisms, from tiny single-celled protozoans to huge whales and sequoia trees, electricity plays vital roles in the normal functioning of the human body. Our nervous system, sketched in ● Figure 7.27, is a highly sophisticated internal information network composed of different types of *neurons* (nerve cells). Sensory neurons provide input to the brain, associative neurons process information in the brain, and motor neurons carry signals from the brain to muscles and some glands. There are approximately 12 billion nerve cells in the human body, some up to a meter long.

A signal is transmitted along an individual neuron in the form of a small voltage change between the inside and the outside of the cell. A neuron at rest has an excess of negative ions at its inner surface, resulting in a voltage of about -0.07 volts. A nerve signal involves a momentary flow of positive sodium ions into the neuron until the polarity is reversed and the voltage is about $+0.03$ volts. That part of the neuron quickly returns to the resting state, but the small voltage "pulse" travels along the neuron at up to 100 m/s.

The specialized *sensory cells* associated with touch, taste, sight, and the other senses produce electrical signals when they are stimulated by the outside environment. These signals are passed on from neuron to neuron, through the spinal cord, and into various regions of the brain. Command signals from the brain are transmitted by motor neurons in the opposite direction to muscles, which contract after the electrical signal is received.

Recently, researchers have developed ways to control computers using electrical signals associated with muscle contraction. One goal is to provide physically disabled individuals full use of computers without the need of keyboards and pointing devices.

The brain is constantly swarming with neural signals as it processes information and formulates response signals. (As you read these words, think of the miracles taking place in your brain as the signals from the nerves in your eyes are transformed into a mental image and from that, information is extracted.) An *electroencephalogram* (EEG) is a record of the periodic voltage changes associated with brain activity. EEGs are made by monitoring the minute voltages that appear on the scalp near different parts of the brain. EEGs show characteristic patterns for different levels of mental activity (fully alert wakefulness, asleep, at rest with the eyes closed, and so on) and can indicate certain disorders such as epilepsy.

The heart pumps blood by coordinated muscle contraction followed by relaxation. Unlike other muscles, such as those in your arms, the heart does not rely on a signal from the brain to contract. A small

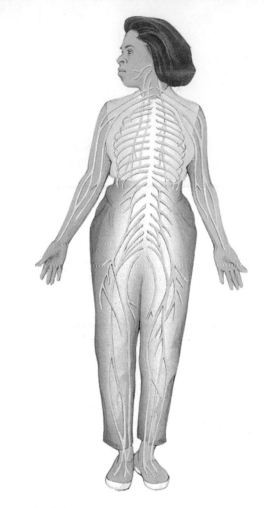

● **Figure 7.27** The human nervous system.

group of cells in the heart produces electrical impulses spontaneously at a rate of around 70 per minute. (The actual frequency of the signal, and therefore the heart rate, is regulated by the brain and autonomic nervous system. For example, the heart rate is increased during exer-

● **Figure 7.26** A simple parallel circuit. Each bulb has the same voltage (the voltage of the power supply). If one bulb burns out, current can still flow through the other bulbs.

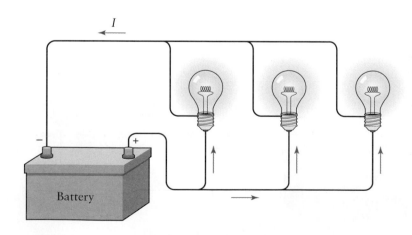

cise.) This impulse initiates a wave of contraction of heart muscle in the required coordinated fashion. An *electrocardiogram* (ECG, more commonly called EKG) is similar to an EEG in that it is a record of the voltages produced by various parts of the heart as it beats (● Figure 7.28). EKGs are routinely used to detect irregularities in the heart's operation, such as a heart murmur.

Often, with age, the intrinsic ability of the heart to control its beating diminishes. A common problem is that different parts of the heart start to beat at different rates because the controlling signal is too weak. A pacemaker can often be implanted near the heart to remedy this problem. This device supplies a small electrical shock to the heart at regular intervals to reestablish coordinated heart contraction.

Because the body relies so heavily on small voltages for its normal function, it is extremely vulnerable to electricity from the outside. Interestingly, it is the *current* that actually flows through the body, not the voltage, that is most important in determining the physiological effect. A single shock of static electricity received from a doorknob may be caused by several thousand volts, but the effect is localized (a sharp pain in your finger) because only a small amount of charge flows through the interior of your body.

The effect of a steady current on your body depends on the size of the current and on the part of the body through which it flows. Currents less than about 1 milliampere usually cannot be felt. A current of around 7 milliamperes or above through your hand (for example) would block signals to your muscles, and you would not be able to move your fingers. A current of 100 milliamperes or more through the heart can induce uncoordinated contractions, called *fibrillation,* leading to death. Steady currents above 300 milliamperes through the heart are usually fatal.

You should realize that a current cannot flow through your body unless it is part of a complete circuit. A person hanging by the arms from a 240-volt power line will have no current flowing through his or her body. But touching the ground or a different wire would provide a complete circuit for the charge to flow through.

In some cases, sending a current through part of the body is beneficial—even life-saving. The pacemaker is one example. Another is the defibrillator: For over 50 years, they have been used to save lives by sending a momentary current through fibrillating hearts, thereby restoring normal heartbeat. Some individuals with high risk of fibrillation can have a miniature, automatic defibrillator surgically implanted. Over the past decade, another implanted device has been developed that controls tremors by sending electrical pulses into part of the brain.

From the very "glue" that holds together atoms and molecules to the vehicle for thought and communication in the human body, electricity is essential to the existence of matter and life itself.

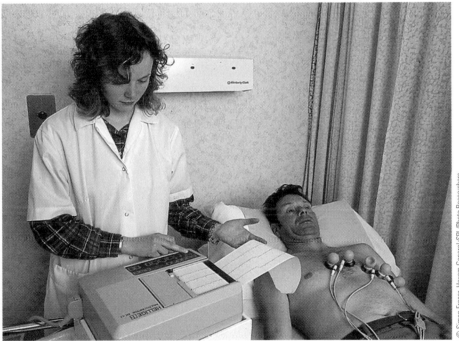

● **Figure 7.28** An electrocardiogram being recorded.

The concept of voltage is quite general and is not restricted to electrical power supplies and electric circuits. Whenever there is an electric field in a region of space, a voltage exists because the field has the potential to do work on electric charges. The strength of an electric field can be expressed in terms of the voltage change per unit distance along the electric field lines. For example, air conducts electricity when the electric field is strong enough to ionize atoms in the air. The minimum electric field strength required for this to happen is between 10,000 and 30,000 volts per centimeter, depending on the conditions. This means that if there is a spark one-fourth of an inch long between your finger and a doorknob, the voltage that causes the spark is at least 7,500 volts.

As transistors and other components on integrated circuit chips (ICs) are made smaller, even the low voltages that are used to make them operate (typically around 1 volt) produce very strong electric fields. Inside modern ICs, electric field strengths can reach 400,000 V/cm. Designers of ICs must keep this in mind because electric fields only about 25% stronger than this can disrupt circuit processes.

LEARNING CHECK

1. (True or False.) A battery stores electric charge in the same way that a bottle stores water.
2. Which of the following is *not* true about voltage?
   a) It is analogous to fluid pressure.
   b) It equals the energy per unit charge.
   c) It equals the work that can be done per unit charge.
   d) Its unit of measure is equivalent to a joule per ampere.

3. The size of the current flowing through a conductor equals the _____ divided by the _____ .
4. (True or False.) In a parallel circuit, the size of the current flowing through each device may be different.

ANSWERS: 1. False   2. (d)   3. voltage, resistance   4. True

7.5 Power and Energy in Electric Currents

Review power and energy in Chapter 3

Since a battery or other electrical power supply must continually put out energy to cause a current to flow, it is important to consider the **power output**—the rate at which energy is delivered to the circuit. The power is determined by the voltage of the power supply and the current that is flowing. Think of it this way: The power output is the amount of energy expended per unit amount of time. The power supply gives a certain amount of energy to each coulomb of charge that flows through the circuit. Consequently, the energy output per unit time equals the *energy given to each coulomb of charge* multiplied by the *number of coulombs that flow through the circuit per unit time:*

energy per unit time = energy per coulomb × number of coulombs per unit time

These three quantities are just the power, voltage, and current, respectively. Consequently, the power output of an electrical power supply is

$$\text{power} = \text{voltage} \times \text{current}$$

$$P = VI$$

The units work out correctly in this equation also: joules per coulomb (volts) multiplied by coulombs per second (amperes) equals joules per second (watts).

The power output of a battery is proportional to the current that it is supplying: the larger the current, the higher the power output.

Example 7.4

In Example 7.1, we computed the current that flows in a flashlight bulb. What is the power output of the batteries?

Recall that the batteries produce 3 volts and that the current in the light bulb is 0.5 amperes. The power output is

$$P = VI = 3\,\text{V} \times 0.5\,\text{A}$$

$$= 1.5\,\text{W}$$

The batteries supply 1.5 joules of energy each second.

What happens to the energy delivered by an electrical power supply? In a light bulb, less than 5% is converted into visible light, and the rest becomes internal energy. Even the visible light emitted by a light bulb is absorbed eventually by the surrounding matter and transformed into internal energy. (Interior lighting is actually used to heat some buildings.) Electric motors in hair dryers, vacuum cleaners, and the like convert about 60% of their energy input into mechanical work or energy while the remainder goes to internal energy. The mechanical energy is generally dissipated as internal energy through friction. In a similar way, we can trace the energy conversions in other electrical devices and the outcome is the same: Most electrical energy eventually becomes internal energy.

Ordinary metal wire converts electrical energy into internal energy whenever there is a current flowing. You may have noticed when using a hair dryer that its cord becomes warm. This heating, called **ohmic heating,** occurs in any conductor that has resistance, even when the resistance is quite small. The huge cables used to conduct electricity from power plants to cities are heated by this effect. This heating represents a loss of usable energy.

The temperature that a current-carrying wire reaches due to ohmic heating depends on the size of the current and on the wire's resistance. Increasing the current in a given wire will raise its temperature. Many devices utilize this effect. The resistances of heating elements in toasters and electric heaters are chosen so that the normal operating current is large enough to heat them until they glow red hot and can toast bread or heat a room. The filament in an incandescent light bulb is made so thin that ohmic heating causes it to glow white hot and emit enough light to illuminate a room (see ● Figure 7.29).

Ohmic heating is a major consideration in the design of sophisticated integrated circuit chips. Even though the currents flowing through the tiny transistors are extremely small, there are so many circuits in such a small space that special steps must be taken to make sure the heat that is produced is conducted away. Since a superconductor has zero resistance, there is no ohmic heating. The overall efficiencies of most electrical devices could be improved if regular wires could be replaced by superconductors. Superconducting transmission lines would allow electricity to be carried from a power plant to a city with no loss of energy. The limitations of currently known superconductors make such uses impracticable.

A sufficiently large current in any wire can cause it to become very hot—hot enough to melt any insulation around it or to ignite combustible materials nearby. Fuses and circuit breakers are put into electric circuits as safety devices to prevent dangerous overheating of wires. If something goes wrong or if too many devices are plugged into the circuit and the current exceeds the recommended safe limit for the size of wire used, the fuse or circuit breaker will automatically "break" the circuit and the current will stop. (A fuse is a fine wire or piece of metal inside a glass or plastic case. When the current exceeds the fuse's design limit, the metal melts away, and the circuit is broken; see ● Figure 7.30.) Designers of electric circuits in cars, houses, and other buildings must choose wiring that is large enough to carry the currents needed without overheating. They must also include fuses or circuit breakers that will disconnect a circuit if it is overloaded.

Most electrical devices are rated by the power that they consume in watts. The equation $P = VI$ can be used to determine how much current flows through the device when it is operating.

● **Figure 7.29** The current causes the thin filament wire to become white hot, even as the larger wires connected to it stay much cooler.

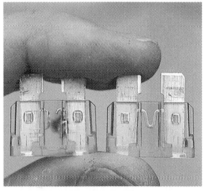

● **Figure 7.30** Two 20-ampere fuses used in automobiles. At some point, the current in the fuse on the left exceeded 20 amperes so it burned out.

An electric hair dryer is rated at 1,875 watts when operating on 120 volts. What is the current flowing through it?

Example 7.5

$$P = VI$$

$$1{,}875\ \text{W} = 120\ \text{V} \times I$$

$$\frac{1{,}875\ \text{W}}{120\ \text{V}} = I$$

$$I = 15.6\ \text{A}$$

The wires in the electric cord must be large enough to allow 15.6 amperes to flow through them without becoming dangerously hot.

Figure 7.31 An electric meter registers the amount of energy consumed in kilowatt-hours.

The highest current that can flow in a particular wire without causing excessive heating depends on the size of the wire. This is one reason why electric utilities use high voltages in their electrical power supply systems. The electricity delivered to a city, subdivision, or individual house must be transmitted with wires. Since $P = VI$, using a large voltage makes it possible to transmit the same power with a smaller current. If low voltages were used, say, 100 volts instead of the more typical 345,000 volts, much larger cables would have to be used to handle the larger currents.

Customers pay for the electricity supplied to them by electric companies based on the amount of energy they use. An electric meter keeps track of the total energy used by monitoring the power (rate of energy use) and the amount of time each power level is maintained (see ● Figure 7.31). Recall the equation used to define power:

$$P = \frac{E}{t}$$

Therefore:

$$E = Pt$$

The amount of energy used is equal to the power times the time elapsed. If P is in watts and t is in seconds, E will be in joules.

Example 7.6   If the hair dryer discussed in Example 7.5 is used for 3 minutes, how much energy does it use?

The power is 1,875 watts. To get E in joules, we must convert the 3 minutes into seconds.

$$t = 3 \text{ min} = 3 \times 60 \text{ s} = 180 \text{ s}$$

So:

$$E = Pt = 1,875 \text{ W} \times 180 \text{ s}$$

$$= 340,000 \text{ J}$$

This is a large quantity of energy—about the same as the kinetic energy of a small car going 60 mph (Example 3.6). Another comparison: A 150-pound person would have to climb 1,700 feet (about 170 floors) to gain 340,000 joules of potential energy.

A typical household can consume more than a billion joules of electrical energy each month. For this reason, a more appropriately sized unit of measure is used for electrical energy—the kilowatt-hour (kWh). Energy in kilowatt-hours is computed by expressing power in kilowatts (kW) and time in hours. The conversion factor between joules and kilowatt-hours is

$$1 \text{ kWh} = 1 \text{ kW} \times 1 \text{ h}$$

$$= 1,000 \text{ W} \times 3,600 \text{ s}$$

$$= 3,600,000 \text{ J}$$

In Example 7.6, the energy used by the hair dryer is about 0.1 kilowatt-hours. The cost of electricity varies from region to region, but it is typically around 10 cents per kilowatt-hour. This means that it costs about 1 cent to run the hair dryer 3 minutes. (Would you climb 1,700 feet for 1 cent?)

Perhaps you have wondered why a common 1.5-volt D-cell battery is larger than a 9-volt battery. The voltage of a battery really has nothing to do with its physical size. Different 1.5-volt batteries range from the size of a button (for wristwatches) to larger than a beer can. The size is more of an indication of the amount of electrical energy stored in the battery. A large battery can supply the same current (and the same power) for a longer time than a small battery with the same voltage.

1. An automobile headlight and a dashboard light use the same voltage, but the power input to the headlight is much larger because
   a) its resistance is lower.
   b) the current in it is larger.
   c) it uses energy faster.
   d) All of the above.
2. Name one or more devices in which ohmic heating is desirable.

3. (True or False.) A fuse is designed to stop the flow of charge in a circuit if the current is too large.
4. The standard unit of energy used in electric utility bills is the _____.

ANSWERS: 1. (d)   2. toaster, electric heater   3. True   4. Kilowatt-hour

7.6 AC and DC

The electric current supplied by a battery is different from the current supplied by a normal household wall socket. Batteries supply **direct current (DC),** and household outlets supply **alternating current (AC).** A DC power supply, such as a battery, causes a current to flow in a fixed direction in a circuit (● Figure 7.32). The current flows out of the positive (+) terminal of the power supply, moves through the circuit, and flows into the negative (−) terminal of the power supply. If the total resistance in the circuit doesn't change, the size of the current remains constant (as long as the battery doesn't run down). A graph of the current I versus time t is simply a horizontal line.

In an AC power supply, the polarity of the two output terminals switches back and forth—the voltage alternates. This causes the current in any circuit connected to the power supply to alternate as well. It flows counterclockwise, then clockwise, then back to counterclockwise, and so on. All the while the size of the current is increasing, then decreasing, and so forth. A graph of the current in an AC circuit shows this variation in the size and direction of the current. (When I goes below zero, it means that the direction has reversed; see ● Figure 7.33).

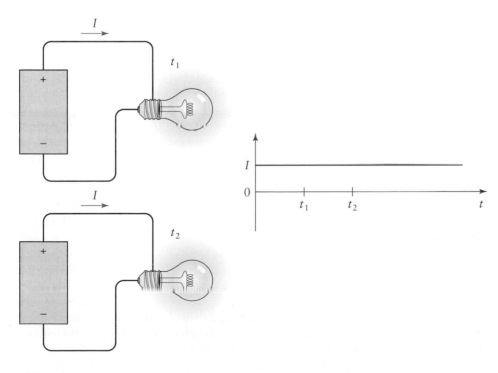

● **Figure 7.32** Direct current. The current flows in one direction and doesn't increase or decrease, as shown in the graph of current versus time.

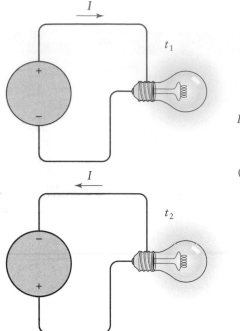

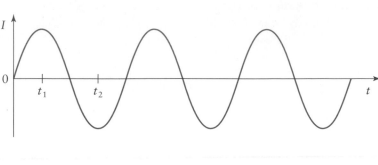

● **Figure 7.33** Alternating current. The direction of the current switches back and forth, and the size of the current varies continuously.

Explore It Yourself 7.6

You will have to be in a room lit only by tube-type fluorescent lights (compact fluorescent lights don't work as well), and have a pen, pencil, or other thin object that is white or light colored. Position yourself so that the fluorescent light is behind or above you. Hold the pen in such a way that the light shines on it and there is a dark background behind it. Move the pen rapidly back and forth (sideways) with your hand. You should see faint lines parallel to the pen. What causes this? (Hint: AC is involved.) Why doesn't this work nearly as well with incandescent light bulbs?

We have seen this kind of oscillation before in Section 2.6 and throughout Chapter 6. One can even think of AC as a kind of "wave" causing the charges in a conductor to oscillate back and forth. Almost all public electric utilities in the United States supply 60-hertz AC. The voltage between the two slots in a wall outlet oscillates back and forth 60 times per second. (In Europe, the standard frequency of AC is 50 hertz.)

Some electronic devices (such as light bulbs) can operate on AC or DC, while others require one or the other. Electric motors and generators must be designed to operate on or to produce either AC or DC. There are devices that can convert an AC voltage to a DC voltage, and vice versa. Batteries can produce direct current only. For this reason, automobiles have DC electrical systems. (The alternator in an automobile generates AC, which is then converted into DC to be compatible with the battery.)

Alternating current has one distinct advantage over DC: Simple, highly efficient devices called *transformers* can "step up" or "step down" AC voltages. This makes it possible to generate AC at a power plant at some intermediate voltage, step it up to a very high voltage (typically over 300,000 volts) for economical transmission, and then step it down again to lower voltages for use in homes and industries. There is no counterpart of the transformer for DC. Another important use of AC is in electronic sound equipment. One example: If a 440-hertz tone is recorded on tape and then played back, the "signal" going to the speaker will be an alternating current with a frequency of 440 hertz. (We will discuss transformers and sound reproduction in Chapter 8.)

LEARNING CHECK

1. Direct current (DC)
   a) is produced by batteries.
   b) is used by most electric utilities.
   c) carries the audio signal to speakers.
   d) oscillates.

2. (True or False.) An incandescent light bulb normally used with AC would not work if connected to DC (with the proper voltage).

3. AC voltages can be easily increased or decreased by using a _____ .

ANSWERS: 1. (a) 2. False 3. transformer

PHYSICS FAMILY ALBUM

The first person to undertake a successful systematic analysis of electric as well as magnetic effects was William Gilbert (1544–1603), a contemporary of Galileo. Born into an English family of comfortable means, Gilbert studied medicine at Cambridge and was later appointed physician to Queen Elizabeth I. This position left him with sufficient time and financial resources to pursue his studies of electricity and magnetism (see ● Figure 7.34). Gilbert showed that the two types of effects are distinct, and he dispensed with a number of misconceptions about them. He carefully tested various substances to see which exhibited the amber effect. Some of the terminology that we use today, such as "pole," originated with Gilbert.

After Gilbert published his findings in the book *De Magnete* in 1600, nothing new was discovered for some 60 years. Otto von Guericke, famous for his experiments with atmospheric pressure (see page 166), constructed a huge electrostatic machine to illustrate the amber effect on a large scale. It consisted of a large, rotating sulfur ball that was charged by friction. It could attract feathers, bits of paper, and other things from considerable distances. He was also the first to record electrostatic repulsion. During the next 200 years, demonstrations of electrostatic effects with such machines became very popular as parlor amusements and as topics of public lectures. Benjamin Franklin's interest in electricity began after he viewed such a demonstration in Boston in 1746.

The findings of two other experimenters, the Englishman Stephen Gray (1666-1736) and the Frenchman Charles Dufay (1698–1739), are also noteworthy. Gray discovered, somewhat by accident, that the electrostatic charge could flow through some substances but not through others. He thus discovered electrical conductors and insulators. Materials that are good at showing the amber effect turn out to be insulators. Conductors like iron cannot be charged in the usual way because the charge will simply flow into the person holding it. Conductors can be charged by keeping them out of contact with other conductors. Gray showed that the human body is itself a conductor by suspending a person by nonconducting strings from a framework (● Figure 7.35).

Dufay studied electrostatic attraction and repulsion and concluded that there must be two different kinds of electric charge to account for them. The two types, which he called *vitreous* and *resinous*, obeyed the rule "like charges repel and unlike charges attract."

Our current names for the two types of charges, positive and negative, come from Benjamin Franklin (● Figure 7.36). Franklin performed his electrical experiments during the late 1740s and early 1750s. In describing his results, he imagined that electricity (or "electrical fire") is a single fluid found in all

● **Figure 7.34** A 1903 painting of William Gilbert demonstrating electrostatics to Queen Elizabeth I.

Painting by A. A. Hunt, 1903, Townhall of Colchester, England. From Bundy Library, Norwalk, CT

© Johann Gabriel Doppelmayr, Neu-endeckte Phaenoma von dewundernswurdigen Wurckungen der Natur, Nuremberg

● **Figure 7.35** Stephen Gray's demonstration that the human body conducts electricity.

objects and that it is capable of being circulated among objects. The flow of electrical fluid is initiated by rubbing (friction), a process in which one object, say B, acquires an excess of electrical fire while another object, say A, receives a deficit. Franklin described this circumstance as follows: "Hence have arisen some new terms among us: we say, B . . . is electrised positively; A negatively. Or rather, B is electrised plus; A, minus." Thus Franklin introduced the terms *positive* and *negative* to describe electricity—not to distinguish different types of charges but to describe the states of objects having a greater or lesser amount of electrical fire in proportion to their normal share.

Coulomb's law, the force law for static electric charges, seems to have been discovered independently by three different men: Charles Coulomb (1736-1806), John Robison (1739–1805), and Henry Cavendish (1731–1810). (Cavendish, whom we encountered in Section 2.8, was a noted recluse who amassed a wealth of experimental findings that were not published until long after his death.) But the Frenchman Coulomb is credited with discovering the law. His experience as an engineer, including a 9-year stint in Martinique, provided him with the skills he needed to construct precision force balances.

The investigation of electric currents was triggered by a chance observation of an Italian biologist, Luigi Galvani (1737–1798). Frog legs that he was preparing twitched when touched by a charged scalpel. A careful experimenter, Galvani undertook an extensive study of the phenomenon and thus made the first discov-

© CORBIS

● **Figure 7.36** Benjamin Franklin (1706–1790).

© CORBIS

● **Figure 7.37** Alessandro Volta, inventor of the battery.

eries of the role of electricity in living systems. However, Galvani clung to the mistaken notion that the electricity in living organisms was somehow different from ordinary electricity.

Galvani's work came to the attention of the physicist Alessandro Volta (1745–1827) (see ● Figure 7.37) in northern Italy. Volta was already an accomplished experimenter in electricity: He was the first to devise a way to measure what was later named voltage. In the process of verifying and extending the discoveries of Galvani, Volta invented a type of battery. This was a pivotal discovery: Batteries were able to supply much larger currents than were possible with electrostatic generators. The invention brought immediate acclaim to Volta. In 1801, Napoleon viewed a demonstration of the device and was very impressed.

Experimenters throughout Europe quickly began building their own "Voltaic cells," and the understanding of moving electricity advanced rapidly. The foremost researcher in this area was a German schoolteacher named Georg Simon Ohm (1787–1854). Working in isolation from the more famous physicists of the time, Ohm introduced the important quantities of voltage, current, and resistance. He carefully studied the factors that affect the resistance of conductors and discovered the relationship that bears his name. Ohm's discoveries, unlike Volta's, were first met with skepticism and

even contempt, perhaps because he was not a member of the traditional scientific community. It was only a few years before his death that Ohm was finally given the respect and recognition that he deserved.

The large currents that could be produced by batteries were a boon to science. Some of the larger batteries built had power outputs of several thousand watts. It was soon discovered that sending a current through water decomposed it into hydrogen and oxygen. This process, known as *electrolysis*, was applied to many other substances and led to the discovery of several new elements. It also showed that electricity is involved in the very structure of matter.

The explosion of electrical devices that started appearing in the late nineteenth century, and continues to this day, revolutionized nearly every aspect of the lives of billions of people worldwide. Our dependence on electricity is perhaps best illustrated by what happens when it is unexpectedly cut off. The great blackouts that occurred in the northeastern United States in 2003, 1977, and 1965 brought commerce, transportation, science—nearly every aspect of modern society—to almost a complete halt. One could argue that the discoveries of researchers into the basics of electricity affect how we live more than those in any other field of physics.

SUMMARY

- Electrons, protons, and certain other subatomic particles possess a physical property, called **electric charge,** that is the basic source of electrical and magnetic phenomena.

- Forces act between any objects that possess net electric charge, positive or negative. Like charges repel, unlike charges attract.

- The electrostatic force, expressed by **Coulomb's law,** is responsible for binding electrons to the nucleus in atoms, for the amber effect (such as static cling), and for a number of other phenomena.

- Electric charges produce electric fields in the space around them. This field is the agent for the electrostatic force, just as the gravitational field is the agent of the gravitational attraction between objects.

- Most useful applications of electricity involve electric currents. They most often consist of electrons being made to flow through metal wires by an electrical power supply, such as a battery.

- The flow of charge is analyzed using **voltage, current,** and **resistance.**

- **Ohm's law** states that the current in a circuit equals the voltage divided by the resistance.

- The power consumption in a circuit depends on the voltage and the current.

- The electrical energy needed to cause a current to flow through a resistance is converted into internal energy. This **ohmic heating** is exploited by incandescent light bulbs to produce light.

- Fuses and circuit breakers are used to automatically disconnect a circuit if the current is large enough to cause excessive ohmic heating.

- At extremely low temperatures, many materials become superconductors—they have zero resistance. Consequently, no energy is lost to heating when electric currents flow through them. Superconductors now in use are limited to special-purpose scientific and medical instruments.

- There are two types of electric current: **alternating (AC)** and **direct (DC).** Since batteries produce DC, battery-powered devices generally employ DC.

- Transformers can be built to "step up" or to "step down" an AC voltage from one value to another. This makes AC particularly convenient for electrical supply networks such as electric utilities.

Equation	Comments	Equation	Comments
$F = \dfrac{(9 \times 10^9)q_1 q_2}{d^2}$	Coulomb's law (in SI units)	$I = \dfrac{V}{R}$	Ohm's law
$I = \dfrac{q}{t}$	Definition of current	$V = IR$	Ohm's law
		$P = VI$	Electrical power consumption
$V = \dfrac{E}{q} = \dfrac{work}{q}$	Definition of voltage	$E = Pt$	Energy used during time t

MAPPING IT OUT!

1. Sections 7.3 and 7.4 introduce several important concepts for understanding electric circuits. These include, but are not limited to, *current, resistance,* and *voltage.* Using these fundamental concepts as a starting point, develop a concept map that captures your understanding of them and that properly establishes their relationships to one another and to electrical circuits (another concept). Is there one main concept that unites your map? If so, which one is it? If not, how did/does that fact influence the structure of your concept map?

2. Concept maps can be used as substitutes for traditional note-taking when you read articles in magazines and newspapers. This question illustrates the point.

   Chapter 7 describes a number of interesting devices based on the concepts of electric charge and force, electric currents, and electric circuits. Some of these applications of the principles of electricity include semiconductor devices, electrostatic precipitators, superconducting devices (like MRI machines), transformers, and so on. Using the bibliography provided at the end of this chapter as a starting point, locate and read an article on one of these devices. After reading the article, look back through it and circle/identify the key concepts in it. Then, using only the concepts you have identified, try to construct a concept map that accurately relates the major ideas/issues raised in the article. After completing the map, reflect for a moment on the degree of difficulty of this exercise. Are some major "concepts" missing from the article that you, the reader, had to "add," based on your own knowledge, to make sense of the article and, hence, your map? What does this suggest to you about the quality and/or level of sophistication of the piece you read?

QUESTIONS

(▶ Indicates a review question, which means it requires only a basic understanding of the material to answer. The others involve integrating or extending the concepts presented thus far.)

1. ▶ All matter contains both positively and negatively charged particles. Why do most things have no net charge?
2. A particular solid is electrically charged after it is rubbed, but it is not known whether its charge is positive or negative. How could you determine which charge it has by using a piece of plastic and fur?
3. ▶ What is a positive ion? A negative ion?
4. What remains after a hydrogen atom is positively ionized?
5. ▶ Describe the similarities and the differences between the gravitational force between two objects and the electrostatic force between two charged objects.
6. If the electrostatic force suddenly became much weaker (the constant of proportionality in Coulomb's law became much smaller), what sort of things would happen?
7. a) A negatively charged iron ball (on the end of a plastic rod) exerts a strong attractive force on a penny even though the penny is neutral. How is that possible?
   b) The penny accelerates toward the ball, hits it, and then is immediately repelled. What causes this sudden change to a repulsive force?

8. ▶ What is an electric field? Sketch the shape of the electric field around a single proton.
9. ▶ At one moment during a storm, the electric field between two clouds is directed toward the east. What is the direction of the force on any electron in this region? What is the direction of the force on a positive ion in this region?
10. ▶ Explain what an electrostatic precipitator is and how it works.
11. Salt water contains an equal number of positive and negative ions. When salt water is flowing through a pipe, does it constitute an electric current?
12. ▶ Materials can be classified into four categories based on the ease with which charges can flow through them. Give the names of these categories and describe each one.
13. A solid metal cylinder has a certain resistance. It is then heated and carefully stretched to form a longer, thinner cylinder. After it cools, will its resistance be the same as, greater than, or less than what it was before?
14. A student using a sensitive meter that measures resistance finds that the resistance of a thin wire is changed slightly when it is picked up with a bare hand. What causes the change in the resistance, and does it increase or decrease?

15. If a new material is found that is a superconductor at all temperatures, what parts of some common electric devices would definitely *not* be made out of it?

16. ▶ Explain what current, resistance, and voltage are.

17. What causes the "popcorn" effect in Explore It Yourself 7.2 to stop?

18. ▶ Describe Ohm's law.

19. A power supply is connected to two bare wires that are inserted into a glass of salt water. The resistance of the water decreases as the voltage is increased. Sketch a graph of the voltage versus the current in the water showing this type of behavior.

20. ▶ There are two basic schemes for connecting more than one electrical device in a circuit. Name, describe, and give the advantages of each.

21. Make a sketch of an electric circuit that contains a switch and two light bulbs connected in such a way that if either bulb burns out the other still functions, but if the switch is turned off, both bulbs go out.

22. Two 9-volt batteries are connected in series in an electric circuit. Use the concept of energy to explain why this combination is equivalent to a single 18-volt battery. When connected in parallel, what are two 9-volt batteries equivalent to?

23. An electrical supply company sells two models of 100-watt power supplies (the maximum power output is 100 W), one with an output of 12 V and the other 6 V. What can you conclude about the maximum current that the two power supplies can produce?

24. A simple electric circuit consists of a constant-voltage power supply and a variable resistor. What effect does reducing the resistance have on the current in the circuit and on the power output of the power supply?

25. ▶ What is the purpose of having fuses or circuit breakers in electric circuits? How should they be connected in circuits so they will be effective?

26. A 20-A fuse in a household electric circuit burns out. What catastrophe could occur if it is replaced by a 30-A fuse?

27. ▶ Why is it economical to use extremely high voltages for the transmission of electrical power?

28. ▶ Explain what AC and DC are. Why is AC used by electric utilities? Why is DC used in flashlights?

29. If AC in a circuit can be thought of as a wave, which kind is it, longitudinal or transverse?

30. If the electric utility company where you live suddenly changed the frequency of the AC to 20 Hz, what problems might this cause?

PROBLEMS

1. Five coulombs of charge flow through a car stereo every 20 s. What is the electric current?

2. A lightning stroke lasts 0.05 s and involves a flow of 100 C. What is the current?

3. A current of 0.7 A goes through an electric motor for 1 min. How many coulombs of charge flow through it during that time?

4. A calculator draws a current of 0.0001 A for 5 min. How much charge flows through it?

5. A current of 12 A flows through an electric heater operating on 120 V. What is the heater's resistance?

6. A 120-V circuit in a house is equipped with a 20-A fuse that will "blow" if the current exceeds 20 A. What is the smallest resistance that can be plugged into the circuit without causing the fuse to blow?

7. The resistance of each brake light bulb on an automobile is 6.6 Ω. Use the fact that cars have 12-V electrical systems to compute the current that flows in each bulb.

8. The light bulb used in an overhead projector has a resistance of 24 Ω. What is the current through the bulb when it is operating on 120 V?

9. The resistance of the skin on a person's finger is typically about 20,000 Ω. How much voltage would be needed to cause a current of 0.001 A to flow into the finger?

10. A 150-Ω resistor is connected to a variable-voltage power supply.
    a) What voltage is necessary to cause a current of 0.3 A in the resistor?
    b) What current flows in the resistor when the voltage is 18 V?

11. Compute the power consumption of the electric heater in Problem 5.

12. An electric eel can generate a 400-V, 0.5-A shock for stunning its prey. What is the eel's power output?

13. An electric train operates on 750 V. What is its power consumption when the current flowing through the train's motor is 2,000 A?

14. All of the electrical outlets in a room are connected in a single parallel circuit (● Figure 7.38). The circuit is equipped with a 20-A fuse, and the voltage is 120 V.

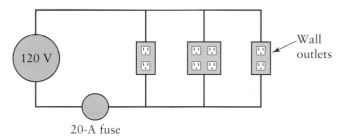

● **Figure 7.38**

    a) What is the maximum power that can be supplied by the outlets without blowing the fuse?
    b) How many 1,200-W appliances can be plugged into the sockets without blowing the fuse?

15. A car's headlight consumes 40 W when on low beam and 50 W when on high beam.
    a) Find the current that flows in each case ($V = 12$ V).
    b) Find the resistance in each case.

16. Find the current that flows in a 40-W bulb used in a house ($V = 120$ V). Compare this with the answer to the first part of Problem 15.

17. An electric clothes dryer is rated at 4,000 W. How much energy does it use in 40 min?

18. A clock consumes 2 W of electrical power. How much energy does it use each day?

19. Which costs more, running a 1,200-W hair dryer for 5 min or leaving a 60-W lamp on overnight (10 h)?

20. A representative lightning strike is caused by a voltage of 200,000,000 V and consists of a current of 1,000 A that flows for a fraction of a second. Calculate the power.

21. A toaster operating on 120 V uses a current of 9 A.
    a) What is the toaster's power consumption?
    b) How much energy does it use in 1 min?

22. A certain electric motor draws a current of 10 A when connected to 120 V.
    a) What is the motor's power consumption?
    b) How much energy does it use during 4 h of operation? Express the answer in joules and in kilowatt-hours.

23. The generator at a large power plant has an output of 1,000,000,000 W at 24,000 V.
    a) If it were a DC generator, what would be the current in it?
    b) What is its energy output each day—in joules and in kilowatt-hours?
    c) If this energy is sold at a price of 10 cents per kilowatt-hour, how much revenue does the power plant generate each day?

24. A light bulb is rated at 60 W when connected to 120 V.
    a) What current flows through the bulb in this case?
    b) What is the bulb's resistance?
    c) What would be the current in the bulb if it were connected to 60 V, assuming the resistance stays the same?
    d) What would be its power consumption in this case?

25. An electric car is being designed to have an average power output of 4,000 W for 2 h before needing to be recharged. (Assume there is no wasted energy.)
    a) How much energy would be stored in the charged-up batteries?
    b) The batteries operate on 30 V. What would the current be when they are operating at 4,000 W?
    c) To be able to recharge the batteries in 1 h, how much power would have to be supplied to them?

26. The resistance of an electric heater is 10 Ω when connected to 120 V. How much energy does it use during 30 min of operation?

CHALLENGES

1. Compute the electric force acting between the electron and the proton in a hydrogen atom. The radius of the electron's orbit around the proton is about 5.3×10^{-11} m.

2. Use the result from Challenge 1 and the equation for centripetal force from Chapter 2 to compute the speed of the electron as it moves around the proton. The electron's mass is 9.1×10^{-31} kg.

3. Compute the number of electrons that flow through a wire each second when the current in the wire is 0.2 A.

4. Using your understanding of the nature of internal energy and temperature, explain why one might expect the resistance of a solid to increase if its temperature increases.

5. The current that flows through an incandescent light bulb immediately after it is turned on is higher than the current that flows moments later. Why?

6. An electrical device called a *diode* is designed to have very low resistance to current flowing through it in one direction but very large resistance to current flow in the other direction. Sketch a graph of the voltage versus current for such a device.

7. Imagine a company offering a line of hair dryers that operate on different voltages, say, 12, 30, 60, and 120 V. If all are rated at 1,200 W, find the current that would flow in each as it operates. What would be different about the heating filament wires and the motors in the various hair dryers?

8. Perform the calculation referred to in the last sentence of Example 7.6.

9. Combine Ohm's law and the equation for power consumption to derive the equation that gives the power in terms of current and resistance. Use the result to answer the following: A cable carrying electrical energy wastes 10 kWh of energy each day because of ohmic heating. If the current in the cable is doubled but the cable's resistance remains the same, how much energy will it waste each day?

10. A defibrillator sends approximately 0.1 C of charge through a patient's chest in about 2 ms. The average voltage during the discharge is approximately 3,000 V. Compute the average current that flows, the average power output, and the total energy consumed.

Bering, Edgar A., III, Arthur A. Few, and James R. Benbrook. "The Global Electric Circuit." *Physics Today* 51, no. 10 (October 1998): 24–30. Describes the main features of the 1,000-A current that flows from the Earth to the upper atmosphere and back.

Ditlea, Steve. "The Electronic Paper Chase." *Scientific American* 285, no. 5 (November 2001): 50–55. Describes competing technologies in the race to create digital paper.

Eisenberg, Mickey S. "Defibrillation: The Spark of Life." *Scientific American* 278, no. 6 (June 1998): 86–90. A look at the 50-year evolution of this life-saving technology.

Geballe, Theodore H. "Superconductivity: From Physics to Technology." *Physics Today* 46, no. 10 (October 1993): 52–56. A brief history of superconductivity.

MacIsaac, Dan, Gary Kanner, and Graydon Anderson. "Basic Physics of the Incandescent Lamp (Lightbulb)." *The Physics Teacher* 37, no. 9 (December 1999): 520–525.

Mende, Stephen B., Davis D. Sentman, and Eugene M. Wescott. "Lightning between Earth and Space." *Scientific American* 277, no. 2 (August 1997): 56–59. A report on types of recently discovered electrical discharges that go upward from thunderstorms.

Ouboter, Rudolf de Bruyn. "Heike Kamerlingh Onnes's Discovery of Superconductivity." *Scientific American* 276, no. 3 (March 1997): 98–103. Describes the techniques devised by Kamerlingh Onnes to reach temperatures near absolute zero and discover superconductivity.

Schecter, Bruce. *The Path of No Resistance: The Story of the Revolution in Superconductivity.* New York: Simon and Schuster, 1989. A captivating account written for nonscientists by a physicist.

Segrè, Emilio. *From X-Rays to Quarks.* New York: W. H. Freeman and Co., 1980. A history of modern physics, by a Nobel laureate.

"Special Issue: The Ubiquitous Electron." *Physics Today* 50, no. 10 (October 1997): 25–61. Five articles commemorating the hundredth anniversary of the discovery of the electron.

"Special Report: Fuel Cells." *Scientific American* 281, no. 1 (July 1999): 72–93. Three articles on the electrochemical alternative to the battery.

Williams, Earle R. "The Electrification of Thunderstorms." *Scientific American* 259, no. 5 (November 1988): 88–99. Describes models that have been developed to explain the cause of lightning.

For additional readings, explore InfoTrac® College Edition, your online library. Go to http://www.infotrac-college.com/wadsworth and use the passcode that came on the card with your book. Try these search terms: electronic paper, Benjamin Franklin, electrostatic precipitator, lightning, superconductivity, electroencephalogram, electrocardiogram, defibrillator.

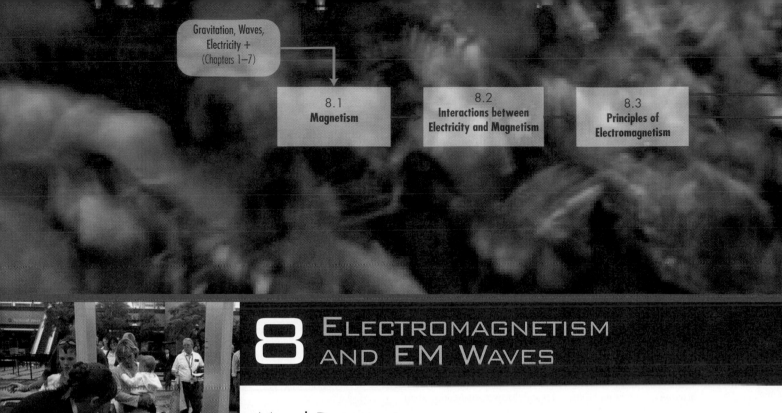

Gravitation, Waves,
Electricity +
(Chapters 1–7)

8.1	8.2	8.3
Magnetism	**Interactions between Electricity and Magnetism**	**Principles of Electromagnetism**

8 ELECTROMAGNETISM AND EM WAVES

Airport metal detector.

© Andre Lambertson/CORBIS

Metal Detectors

They are the first line of defense against people trying to smuggle weapons onto passenger planes or into schools, government buildings, and many other places. Metal detectors probe your clothing and body without physically touching you, looking for metal that could be part of a gun, a knife, or other dangerous object. A device that can find hidden items on a person walking through an arch seems like something from science fiction. But in today's world, it is routine.

How do these devices work their magic? Although metal detectors operate on electricity, it is magnetism that probes you. Brief magnetic pulses are sent through you and around you, typically about 100 times a second. (These would make a compass needle twitch back and forth slightly at the same rate.) The device carefully monitors how swiftly each magnetic pulse dies out. Any metal object encountered by a pulse is induced to produce its own magnetic pulse, which affects how rapidly the total pulse dies out. Sophisticated electronics in the metal detectors sense this change and signal that metal is present. They detect iron and other metals that ordinary magnets attract as well as metals like aluminum and gold that do not respond to magnets.

This explanation might raise some questions in your mind. How are the magnetic pulses produced? How does the metal detector monitor how the pulses die out? Why do they cause "nonmagnetic" metals like aluminum to produce magnetic pulses? The answers lie in the key concepts presented in this chapter, the fundamental ways in which electricity and magnetism interact with each other.

Magnetism and its useful interrelationship with electricity are the subjects of this chapter. First, the properties of permanent magnets and the Earth's magnetic field are described. Next, we demonstrate how electric fields and magnetic fields intertwine whenever motion or change is involved. These concepts are used to explain how many common electrical devices operate. They also suggest the existence of electromagnetic (EM) waves. The properties and uses of the different types of EM waves are the main topics of the latter half of this chapter.

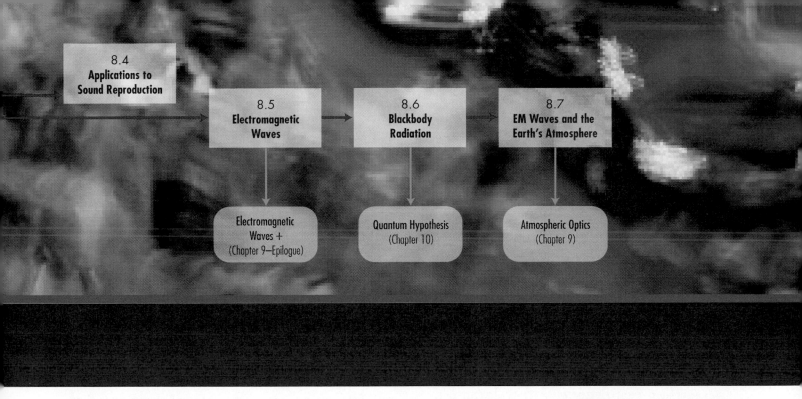

8.4
Applications to Sound Reproduction

8.5
Electromagnetic Waves

8.6
Blackbody Radiation

8.7
EM Waves and the Earth's Atmosphere

Electromagnetic Waves +
(Chapter 9–Epilogue)

Quantum Hypothesis
(Chapter 10)

Atmospheric Optics
(Chapter 9)

8.1 Magnetism

Magnetism was first observed in a naturally occurring ore called *lodestone*. Lodestones were fairly common around Magnesia, an ancient city in Asia Minor. Small pieces of iron, nickel, and certain other metals are attracted by lodestones, much as pieces of paper are attracted by charged plastic (● Figure 8.1). The Chinese were probably the first to discover that a piece of lodestone will orient itself north and south if suspended by a thread or floated in water on a piece of wood. The *compass* revolutionized navigation because it allowed mariners to determine the direction of north even at sea in cloudy weather. It was also one of the few useful applications of magnetism up to the nineteenth century.

Now magnets are made into a variety of sizes and shapes out of special alloys that exhibit much stronger magnetism than lodestone. All simple magnets exhibit the same compass effect—one end or part of it is attracted to the north, and the opposite end or part is attracted to the south. The north-seeking part of a magnet is called its **north pole,** and the south-seeking part is its **south pole.** All magnets have both poles. If a magnet is broken into pieces, each part will have its own north and south poles. The south pole of one magnet exerts a mutually attractive force on the north pole of a second magnet. The south poles of two magnets repel each other, as do the north poles (● Figure 8.2). Simply: *Like poles repel, unlike poles attract* (just as with electric charges).

Metals that are strongly attracted by magnets are said to be *ferromagnetic.* Such materials have magnetism induced in them when they are near a magnet. If a piece of iron is brought near the south pole of a magnet, the part of the iron nearest the magnet has a north pole induced in it, and the part farthest away has a south pole induced in it (● Figure 8.3). Once the iron is removed from the vicinity of the magnet, it loses most of the induced magnetism. Some ferromagnetic metals actually retain the magnetism induced in them—they become *permanent magnets.* Regular magnets and compass needles are made of such metals. Ferromagnetism is also the basis of magnetic data recording. More on this later.

As with gravitation and electrostatics, it is useful to employ the concept of a field to represent the effect of a magnet on the space around it. A *magnetic field* is produced by a magnet and acts as the agent of the magnetic force. The poles of a second magnet experience forces when in the magnetic field: Its north pole has a force in the same direction as the magnetic field, but its south pole has a force in the opposite direction. A compass

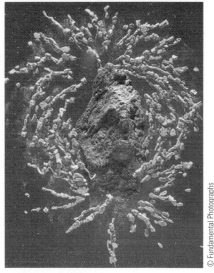

● **Figure 8.1** Lodestone, a natural magnet, attracting pieces of metallic ore.

© Fundamental Photographs

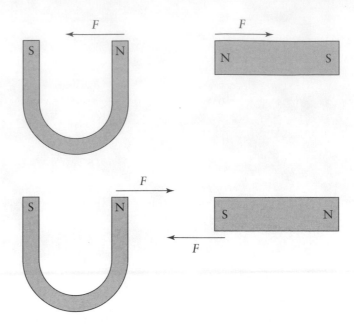

● **Figure 8.2** The poles of two magnets exert forces on each other. Like poles repel each other, and unlike poles attract each other.

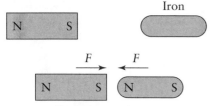

● **Figure 8.3** When a piece of ferromagnetic material (like iron) is brought near a magnet, it has magnetism induced in it. That is why it is attracted by the magnet.

can be thought of as a "magnetic field detector" because its needle will always try to align itself with a magnetic field (● Figure 8.4). The shape of the magnetic field produced by a magnet can be "mapped" by noting the orientation of a compass at various places nearby. Magnetic *field lines* can be drawn to show the shape of the field, just as electric field lines are used to show the shape of an electric field (● Figure 8.5). The direction of a field line at a particular place is the direction that the north pole of a compass needle will point.

Because magnets respond to magnetic fields, the fact that compass needles point north indicates that the Earth itself has a magnetic field. The shape of the Earth's field has been mapped carefully over the course of many centuries because of the importance of compasses in navigation. The Earth's magnetic field has the same general shape as the field around a bar magnet with its poles tilted about 12° with respect to the axis of rotation (● Figure 8.6). The direction of "true north" shown on maps is determined by the orientation of the Earth's axis of rotation. (The axis is aligned closely with Polaris, the North Star.)

Because of the tilt of the Earth's "magnetic axis," at most places on Earth, compasses do not point to true north. For example, in the western two-thirds of the United States, compasses point to the right (east) of true north, while in New England compasses point to the left (west) of true north. The difference, in degrees, between the direction of a compass and the direction of true north varies from place to place and is referred to as the *magnetic declination.* (In parts of Alaska, the magnetic declination is as high as 25° east.) This must be taken into account when navigating with a compass.*

---

* The Earth's magnetic poles actually move around: The south magnetic pole is not exactly where it was 20 years ago. At most places on Earth, a compass does not point exactly in the same direction it did 20 years ago. Detailed maps used for navigation are corrected periodically to show any change in magnetic declination.

● **Figure 8.4** The forces on the poles of a compass placed in a magnetic field are in opposite directions. This causes the needle to turn until it is aligned with the field.

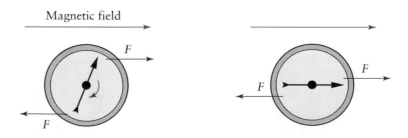

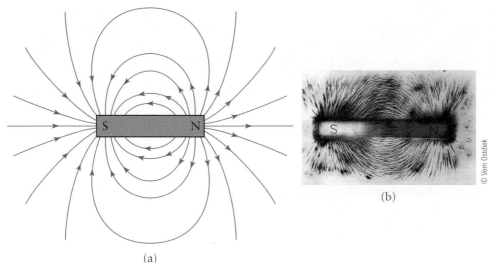

(a)

(b)

© Vern Ostdiek

● **Figure 8.5** (a) Sketch of the magnetic field in the space around a bar magnet. Note that the field lines point toward the south pole and away from the north pole. (b) Photograph of iron filings around a magnet. Each tiny piece of iron becomes magnetized and aligns itself with the magnetic field.

Explore It Yourself 8.1

Determine what the magnetic declination is where you live. You can do this by finding it on a topographic map or other high-precision navigation map, or by searching for a Web site that finds it for you based on your ZIP code or latitude and longitude. If its magnitude is a large number, say, 10° or so, get a compass and find a place where you can see Polaris (the North Star) at night. Compare the direction the compass points to the direction of Polaris. Does the difference agree with the reported magnetic declination?

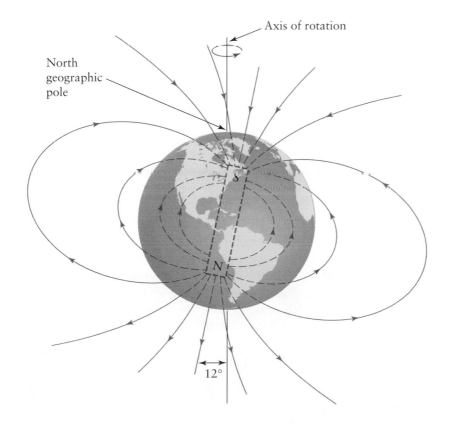

● **Figure 8.6** The Earth's magnetic field. It is shaped as if there were a huge bar magnet deep inside the Earth, tilted 12° relative to the Earth's axis of rotation.

Incidentally, the Earth's field is responsible for the magnetism in lodestone. This naturally occurring ferromagnetic ore is weakly magnetized by the Earth's magnetic field. Another thing to note about the Earth's magnetic field: The Earth's *north* magnetic pole is at (near) its *south* geographic pole, and vice versa. Why?

1. The north pole of a magnet is attracted to the south pole of a second magnet.
2. The north pole of a compass needle points to the north.

Therefore, a compass's north pole points at the Earth's south magnetic pole. This is not a physical contradiction: It is a result of naming the poles of a magnet after directions instead of, say, + and −, or A and B.

Some organisms use the Earth's magnetic field to aid navigation. Although the biological mechanisms that they employ have not yet been identified, certain species of fish, frogs, turtles, birds, newts, and whales are able to sense the strength of the Earth's field or its direction (or both). The former allows the animal to determine its approximate latitude (how far north or south it is) because the Earth's magnetic field is stronger near the magnetic poles (Figure 8.6). Some migratory species travel thousands of miles before returning home, guided—at least in part—by sensing the Earth's magnetic field.

Superconductors, so named because of their ability to carry electric current with zero resistance, react to magnetic fields in a rather startling fashion. When in the superconducting state, the material will expel any magnetic field from its interior. This phenomenon, known as the *Meissner effect,* is why strong magnets are levitated when placed over a superconductor (see ● Figure 8.7). When trying to determine whether a material is in the superconducting state, it is easier to test for the presence of the Meissner effect than it is to see if the resistance is exactly zero.

You have probably noticed that magnetism and electrostatics are very similar: There are two kinds of poles and two kinds of charges. Like poles repel as do like charges. There are magnetic fields and electric fields. However, there are some important differences. Each kind of charge can exist separately, while magnetic poles always come in pairs. (Modern theory indicates the possible existence of a particular type of subatomic "elementary particle" that does have a single magnetic pole. As of this writing, such a "magnetic monopole" has not been found.) Furthermore, all conventional matter contains positive and negative charges (protons and electrons) and can exhibit electrostatic effects by being "charged." But, with the exception of ferromagnetic materials, most matter shows very little response to magnetic fields.

We should also point out that the electrostatic and magnetic effects described so far are completely independent. Magnets have no effect on pieces of charged plastic, for instance, and vice versa. This is the case as long as there is no motion of the objects or changes in the strengths of the electric and magnetic fields. As we shall see in the following sections, a number of fascinating and useful interactions between electricity and magnetism take place when motion or change in field strength occurs.

Concept Map 8.1 summarizes the similarities and differences between electrostatics and magnetism.

● **Figure 8.7** The Meissner effect. A magnet levitating above a high-T_c superconductor cooled with liquid nitrogen.

© David Parker/Photo Researchers

LEARNING CHECK

1. An iron nail is brought near a magnet. Which of the following is *not* true?
   a) The nail has both a north pole and a south pole.
   b) The nail becomes either a north pole or a south pole.
   c) The magnet exerts an attractive force on the nail.
   d) The nail exerts an attractive force on the magnet.

2. (True or False.) The shape of the magnetic field around a bar magnet is almost exactly the same as the shape of the electric field around a positive charge.

3. Magnetic _____ indicates how far away from true north a compass needle points.

ANSWERS: 1. (b) **2.** False **3.** declination

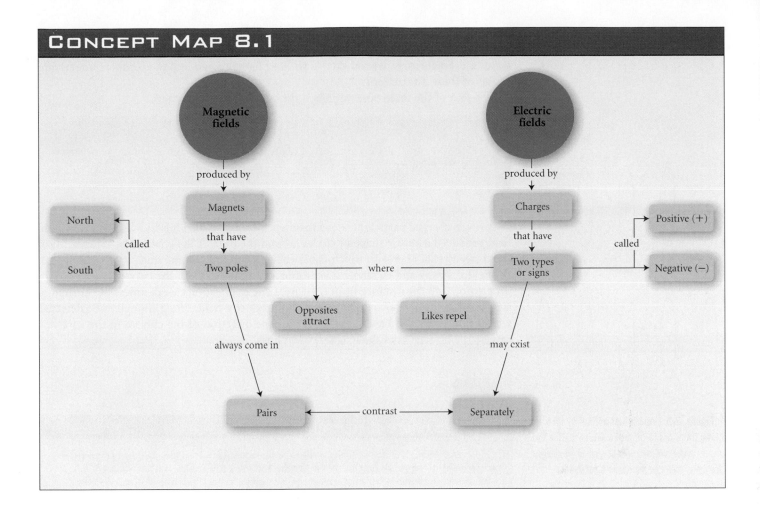

8.2 Interactions Between Electricity and Magnetism

Consider the following items that we usually take for granted: electric motors in hair dryers, vacuum cleaners, computer disk drives, elevators, and countless other devices; generators that produce most of the electricity we use; speakers, audio, and videotape recorders, and high-fidelity microphones; and the waves that make radios, wireless telephones, radar, microwave ovens, medical x rays, and our eyes work. What do all of these have in common? They all are possible because electricity and magnetism interact with each other in basic—and very useful—ways. The word *electromagnetic*, which appears dozens of times in this chapter, is perhaps the best indication of just how intertwined these two phenomena are.

Before we delve into these interactions, let's summarize and review the key aspects of electrostatics and magnetism presented in Sections 7.1, 7.2, and 8.1:

- Electric charges produce electric fields in the space around them (see Figure 7.9).
- An electric field, regardless of its origin, causes a force on any charged object placed in it (see Figure 7.10).
- Magnets produce magnetic fields in the space around them (see Figure 8.5).
- A magnetic field, regardless of its origin, causes forces on the poles of any magnet placed in it (see Figure 8.4).

These statements have been worded in a particular way because, as we shall see, it is the electric and magnetic *fields* that are involved in the interplay between electricity and magnetism. In this section, we describe three basic observations of these interactions and discuss some useful applications of them. In Section 8.3, we summarize the underlying concepts in the form of two principles like the four statements above. We emphasize the ways in which electricity and magnetism interact and how these help us understand such things as how many electrical devices work and what light and other electromagnetic waves are. Fortunately, we can do this without having to go into the complex underlying causes of these interactions.

The first of the three observations is the basis of electromagnets.

Observation 1: A moving electric charge produces a magnetic field in the space around it. An electric current produces a magnetic field around it.

A single charged particle creates a magnetic field only when it is moving. The magnetic field produced is in the shape of circles around the path of the charge (see ● Figure 8.8). For a steady (DC) current, which is basically a succession of moving charges, in a wire, the field is steady and its strength is proportional to the size of the current and inversely proportional to the distance from the wire. (The field is quite weak unless the current is large. A current of 10 amperes or more will produce a field strong enough to be detected with a compass; see ● Figure 8.9.) Reversing the direction of the current in the wire will reverse the directions of the magnetic field lines.

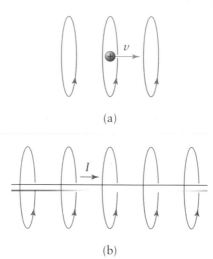

(a)

(b)

● **Figure 8.8** Magnetic field produced by (a) a moving charge and by (b) a wire carrying DC. The field lines are circles concentric with the path of the charges. (The power source for the current is not shown.)

Explore It Yourself 8.2

For this, you will need a car or similar vehicle and a compass. Open the hood of the car (when the engine is not running), and locate the battery and the large cables that carry current from it to the starter. Close the hood, and hold a compass over the hood just above where a cable is located. Have a friend start the car, and watch the compass while the starter is engaged. What happens? What causes this?

Most applications of this phenomenon use coils—long wires wrapped in the shape of a cylinder, often around an iron core. The magnetism induced in the iron greatly enhances the magnetic field of the coil. The magnetic field of such a coil (when carrying a direct current) has the same shape as the field around a bar magnet. (See ● Figure 8.10 and compare it to Figure 8.5.) This device is an *electromagnet*. It behaves just like a permanent magnet as long as there is a current flowing. One end of the coil is a north pole, and the other is a south pole. Electromagnets have an advantage over permanent magnets in that the magnetism can be "turned off" simply by switching off the current. (Large electromagnets are routinely used to pick up scrap iron.)

● **Figure 8.9** (a) The compass needles align with the Earth's magnetic field when no current is in the wire. (b) The compass needles show the circular shape of the magnetic field produced by a large current (15 amperes) flowing in the wire.

(a)

(b)

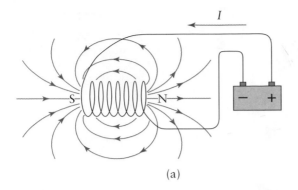

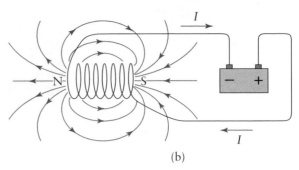

(a)

(b)

● **Figure 8.10** (a) The magnetic field produced by a current in a coil of wire. The field has the same shape as that produced by a bar magnet. When the direction of the current is reversed (b), the polarity of the magnetic field is also reversed.

A coil whose length is much greater than its diameter is called a *solenoid*. If an iron rod is partially inserted into a solenoid with a hollow core, the rod will be pulled in when the current is switched on (see ● Figure 8.11). Solenoids are used in common devices for striking doorbell chimes, opening valves to allow water to enter and to leave washing machines, withdrawing deadbolts in electric door locks, and engaging starter motors on car and truck engines.

Electromagnets are used to produce the strongest magnetic fields on Earth. Two factors contribute to stronger fields, wrapping more coils around the cylinder and using a larger electric current. The former suggests the use of thinner wire so that more coils can fit into the same amount of space. But smaller wire requires smaller electric current so the wire does not overheat and melt. This limitation is overcome in *superconducting electromagnets* (● Figure 8.12). When the wire used in an electromagnet is a superconductor, it can carry huge electric currents with no ohmic heating because there is no resistance. Very small superconducting electromagnets can generate very strong magnetic fields while using much less electrical energy than a conventional electromagnet. (We describe some uses of superconducting electromagnets in Section 7.3 and later in this section.)

Refer to the discussion on superconductivity in Section 7.3.

Superconducting electromagnets do have limitations, though. The superconducting state is lost if the temperature, electric current, or magnetic field strength exceeds certain values. Most superconducting electromagnets now found in laboratories throughout the world use a compound of niobium and tin that must be kept cold with liquid helium ($T = 4$ K). The added cost of the liquid helium system is offset by the high magnetic fields achieved and the great reduction in use of electric energy compared to conventional electromagnets.

The polarity of an electromagnet is reversed if the direction of the current is reversed (see Figure 8.10b). An alternating current in a coil will produce a magnetic field that oscillates: It increases, decreases, and switches polarity with the same frequency as the current. Such an oscillating magnetic field will cause a nearby piece of iron to vibrate. The oscillating magnetic field of a coil with AC in it is used in many common devices, as we shall see in the following sections.

● **Figure 8.11** (a) The iron rod is pulled into the coil (solenoid) when the current flows. (b) The solenoid (red) in this doorbell chime pulls in the iron rod when a current flows in it. The rod strikes the black bar, producing sound.

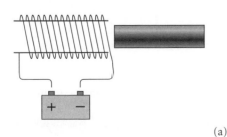

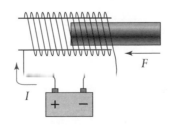

(a)

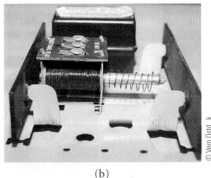

(b)

Courtesy of GE Research Laboratory

● **Figure 8.12** The pattern of magnetic field lines surrounding a pair of powerful superconducting electromagnets is made visible by scattering iron nails on a piece of white plywood.

Aluminum

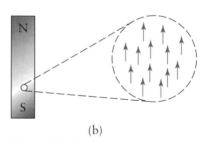

(a)

(b)

● **Figure 8.13** The arrows represent the magnetic fields of electrons in individual atoms. (a) The fields remain randomly oriented in nonferromagnetic materials. (b) Inside ferromagnetic material that is magnetized, the individual magnetic fields are aligned.

Not only does this first interaction explain how electromagnets work, it gives us new insight into permanent magnets as well. Since electrons in atoms are charged particles in motion about the nucleus, they produce magnetic fields. Also, the electrons have their own magnetic fields associated with their spin (more on this in Chapter 12). In any unmagnetized material, the individual magnetic fields of the electrons are randomly oriented and cancel each other out (● Figure 8.13). In ferromagnetic materials, these fields can be aligned with one another by an external magnetic field; the material then produces a net magnetic field. So we can conclude that moving electric charges are the causes of magnetic fields even in ordinary bar and horseshoe magnets.

This brings us back to a statement made at the beginning of Chapter 7: Electric charges are the cause of both electrical and magnetic effects. We might regard electricity and magnetism as two different manifestations of the same thing—charge.

The second observation helps us understand how things like electric motors and speakers work.

> **Observation 2:** A magnetic field exerts a force on a moving electric charge. Therefore, a magnetic field exerts a force on a current-carrying wire.*

A stationary electric charge is not affected by a magnetic field, but a moving charge usually is. Note that this second observation is a logical consequence of the first: Anything that produces a magnetic field will itself be affected by other magnetic fields.

A curious characteristic of electromagnetic phenomena is that the effects are often perpendicular to the causes. The direction of the magnetic field from a current-carrying

---

* If a charge's velocity or the direction of a current is parallel to the magnetic field or in the opposite direction, the magnetic field does not exert this force.

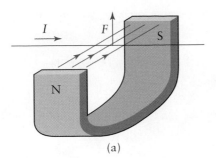

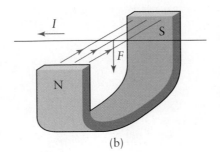

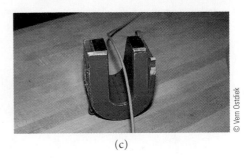

(a)   (b)   (c)

● **Figure 8.14** (a) The force on a current-carrying wire in a magnetic field. (b) When the direction of the current is reversed, the direction of the force is also reversed. (c) A current-carrying wire levitating in a magnetic field.

wire is perpendicular to the direction the current is flowing (Figure 8.8). Similarly, the force that a magnetic field exerts on a moving charge or on a current-carrying wire is perpendicular to both the direction of the magnetic field and the direction the charge is flowing. For example, if a horizontal magnetic field is directed away from you and a wire is carrying a current to your right, the force on the wire is *upward* (● Figure 8.14). If the direction of the current is reversed, the direction of the force is reversed (downward). An alternating current would cause the wire to experience a force that alternates up and down.

Electric motors—like those in hair dryers and elevators—exploit this electromagnetic interaction. The simplest type of electric motor consists of a coil of wire mounted so that it can rotate in the magnetic field of a horseshoe-shaped magnet (● Figure 8.15). A direct current flows through the coil, the magnetic field causes forces on the sides of the coil, and it rotates. Once the coil has completed half of a rotation, a simple mechanism reverses the direction of the current. This reverses the force on the coil, causing it to rotate another half-turn. This process is repeated, and the coil spins continuously. Motors designed to run on AC can exploit the fact that the direction of the current is automatically reversed 120 times each second (sixty-cycles-per-second AC, with two reversals each cycle).

Liquid metals, such as the molten sodium used in certain nuclear reactors, can be moved through pipes using an electromagnetic pump that has no moving parts. If the metal has to be moved in a pipe that is oriented north-south, for example, a large electric current can be sent across the pipe—east to west perhaps. Then, if a strong magnetic field is directed downward through the same section of pipe, the current-carrying metal will be forced to move southward.

Several large-scale devices used in experimental physics make use of the effect of magnetic fields on moving charged particles. High-temperature plasmas cannot be kept in any conventional metal or glass container because the container would melt. Since plasmas are composed of charged particles, magnetic fields can be used to contain them in what is known as a "magnetic bottle." This is one approach being employed in the attempt to harness nuclear fusion as an energy source (Section 11.7).

In the absence of other forces, a charged particle moving perpendicularly to a magnetic field will travel in a circle: The force on the particle is always perpendicular to its

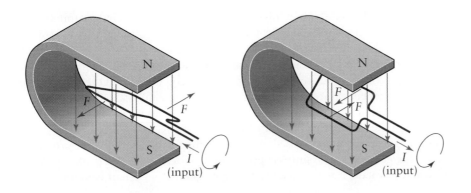

● **Figure 8.15** Simplified sketch of an electric motor. The loop of wire rotates because of the forces on its sides. Each time the loop becomes horizontal, the direction of the current is reversed, and the rotation continues.

● **Figure 8.16** Fermi National Accelerator Laboratory, near Batavia, Illinois. Inside the huge ring, charged particles are accelerated to nearly the speed of light. Magnetic fields are used to keep the charges moving in a circle.

velocity and is therefore a centripetal force. An electron, proton, or other charged particle can be forced to move in a circle by a magnetic field and then gradually accelerated during each revolution. Particle accelerators used for experiments in atomic, nuclear, and elementary particle physics, as well as for producing radiation for cancer treatments at some large hospitals, operate on this principle. The world's highest-energy particle accelerator currently in operation, the Tevatron at the Fermi National Accelerator Laboratory (Fermilab) near Batavia, Illinois, is 2 kilometers (1.2 miles) in diameter (● Figure 8.16). The Large Hadron Collider (LHC) near Geneva, Switzerland, scheduled to become operational in 2007, is even larger than the Tevatron. With a diameter of 8.6 kilometers (5.3 miles), LHC is the largest scientific instrument built so far (see page 86). Both accelerators are designed to use superconducting electromagnets to keep positively charged particles circulating in one direction and negatively charged particles circulating in the opposite direction. The head-on collisions of these oppositely charged particles yield information about the fundamental forces and particles in nature (more on this in Chapter 12).

The third observed interaction between electricity and magnetism is used by electric generators. Recall that the first observation tells us that moving charges create magnetic fields. The third one is a similar statement about moving magnets.

Observation 3: A moving magnet produces an electric field in the space around it. A coil of wire in motion relative to a magnet has a current induced in it.

The electric field around a moving magnet is in the shape of circles around the path of the magnet. This circular electric field will force charges in a coil of wire to move in the same direction—as a current (● Figure 8.17). The process of inducing an electric current with a magnetic field is known as **electromagnetic induction.** All that is required is that the magnet and coil move relative to each other. If the coil moves and the magnet remains stationary, a current is induced. If the motion is steady in either case, the induced current is in one direction. If either the coil or the magnet oscillates back and forth, the current alternates with the same frequency—it is AC.

Electromagnetic induction is used in the most important device for the production of electricity, the generator. The simplest generator is basically an electric motor. When the coil is forced to rotate, it moves relative to the magnet, so a current is induced in it (see

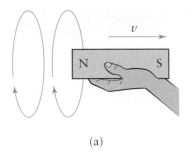

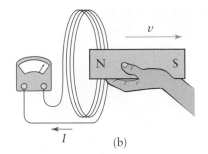

(a)

(b)

(c)

(d)

© Vern Ostdiek

© Vern Ostdiek

● Figure 8.17 (a) The electric field produced by a moving magnet is circular. (The magnetic field of the magnet is not shown.) (b) A magnet will induce a current in a coil as it passes through. (c) The ammeter shows there is no current in the coil of wire (yellow) when the horseshoe magnet (blue) is not moving. (d) A current flows when the magnet is moving.

● Figure 8.18). We might call this device a "two-way energy converter." When electrical energy is supplied to it, it is a motor. It converts this electrical energy into mechanical energy of rotation. When it is mechanically turned (by hand cranking, by a fan belt on a car engine, or by a turbine in a power plant), it is a generator. It converts mechanical energy into electrical energy.

This motor-generator duality is used in dozens of *pumped-storage* hydroelectric power stations. During the night when there is a surplus of electrical energy available from other power stations, the motor mode is used to pump water from one reservoir to another that is at a higher elevation. Most of the electrical energy is converted into "stored" gravitational potential energy. During the peak time of electric use the next day, water flows in the opposite direction, and the generator mode is used as the moving water turns the pumps (now acting as turbines) that turn the motors (now acting as generators), thereby producing electricity. You might say that the system functions like a rechargeable gravitational battery.

Another application of this technology is *regenerative braking,* used in electric and hybrid vehicles. While accelerating and cruising, electric motors turn the wheels using electricity from batteries. During braking, the motors function as generators: The wheels turn them, and the electricity that is generated can partially recharge the batteries. Instead of all of the vehicle's kinetic energy being converted into wasted heat—the case with conventional friction brakes—some of it is saved for reuse.

In summary, when electric charges or magnets are in motion, electricity and magnetism are no longer independent phenomena. The three observations given here are statements of experimental facts that illustrate this interdependence. They can be demonstrated easily using a battery, wires, a compass, a large magnet, and a sensitive ammeter. The fact that electricity and magnetism interact only when there is motion (and then the effects are perpendicular to the causes) is somewhat startling when compared to, say, gravitation and electrostatics. As we saw in Chapters 2 and 7, gravitational and electrostatic forces are always toward or away from the objects causing them, and they act whether or not anything is moving or changing. These basic yet surprising interactions between electricity and magnetism are crucial to our modern electrified society.

Concept Map 8.2 summarizes the interactions between electricity and magnetism.

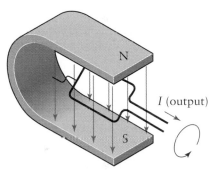

● Figure 8.18 Simplified sketch of a generator. As the loop of wire rotates relative to the magnetic field, a current is induced in it.

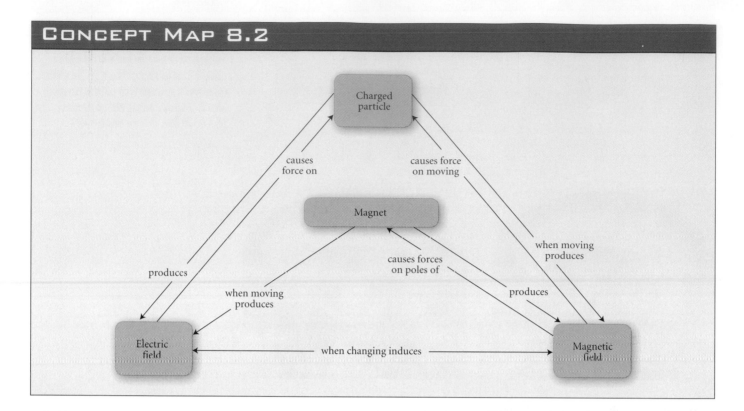

LEARNING CHECK

1. (True or False.) A stationary electric charge produces a magnetic field around it.
2. How does one reverse the polarity (switch the north and south poles) of an electromagnet?
3. A magnetic field exerts a force on an electric charge if the charge is _____.
4. A current can be made to flow in a coil of wire if
   a) it is connected to a battery.
   b) it is put into motion near a magnet.

   c) a magnet is put into motion near it.
   d) All of the above.
5. When the coil in a simple electric motor is turned—by a crank, for example—the motor functions as a _____.

ANSWERS: 1. False   2. reverse the current   3. moving   4. (d)   5. generator

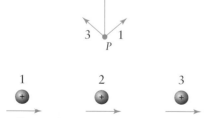

8.3 Principles of Electromagnetism

The interactions between electricity and magnetism described in the previous section, along with other similar observations, suggest the following two general statements. We might call these the **principles of electromagnetism:**

1. An electric current or a changing electric field induces a magnetic field.
2. A changing magnetic field induces an electric field.

These two statements summarize the previous observations and also emphasize the symmetry that exists. In both cases, a "changing" field means that the strength or the direction of the field is changing. The first principle can be used to explain the first observation: As a charge moves by a point in space, the strength of the electric field increases and then decreases. All the time the direction of the field is changing as well (● Figure 8.19). The effect of this is to cause a magnetic field to be produced. Similarly, the second principle explains electromagnetic induction.

● **Figure 8.19** The electric field at the point *P* changes as the charged particle moves by. The upper blue arrows indicate the magnitude and direction of the electric field for three different locations of the particle. The magnetic field that is induced at point *P* is directed straight out of the paper.

As mentioned in Section 7.6, a transformer is a device used to step up or step down AC voltages. It represents one of the most elegant applications of electromagnetism. In essence, a transformer consists of two separate coils of wire in close proximity. An AC voltage is applied to one of the coils, called the "input" or "primary" coil, and an AC voltage appears at the other coil, called the "output" or "secondary" coil (● Figure 8.20). The AC in the primary coil produces an oscillating magnetic field through both coils. Most transformers have both coils wrapped around a single ferromagnetic core to intensify the magnetic field and guide it from one coil to the other. This oscillating (and therefore changing) magnetic field induces an AC current in the output coil. Note that a DC input would produce a steady magnetic field that would not induce a current in the output coil. Transformers do not work with DC.

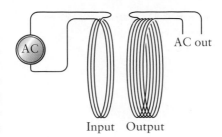

● **Figure 8.20** Simplified diagram of a transformer. The alternating magnetic field produced by the AC in the input coil induces an alternating current in the output coil. In this case, the output voltage would be higher than the input voltage.

Now, how can the voltage of the output be different from the voltage of the input? Each "loop" or "turn" of the output coil has the same voltage induced in it. The voltages in all of the turns add together so that the more turns there are in the output coil, the higher the total voltage. The ratio of the number of turns in the two coils determines the ratio of the input and output voltages. In particular:

$$\frac{\text{voltage of output}}{\text{voltage of input}} = \frac{\text{number of turns in output coil}}{\text{number of turns in input coil}}$$

$$\frac{V_o}{V_i} = \frac{N_o}{N_i}$$

If there are twice as many turns in the output coil as in the input coil, the output voltage will be twice the input voltage. If there are one-third as many turns in the output coil, the output voltage will be one-third the input voltage. Thus, the AC voltage can be stepped up or stepped down by any desired amount by adjusting the ratio of the number of turns in the two coils.

Example 8.1

A transformer is being designed to have a 600-volt output with a 120-volt input. If there are to be 800 turns of wire in the input coil, how many turns must there be in the output coil?

$$\frac{V_o}{V_i} = \frac{N_o}{N_i}$$

$$\frac{600 \text{ V}}{120 \text{ V}} = \frac{N_o}{800}$$

$$800 \times 5 = N_o$$

$$N_o = 4{,}000 \text{ turns}$$

In addition to being used to change voltages in electrical distribution systems, transformers are used in a wide variety of electrical appliances. Most electrical components used in radios, calculators, and the like require voltages much smaller than 120 volts. Appliances designed to operate on household AC must include transformers to reduce the voltage accordingly. High-intensity desk lamps also use transformers; that is what makes their bases so heavy (● Figure 8.21). The spark used to ignite gasoline in automobile engines is generated using a type of transformer called a "coil." The number of turns in the output coil is many times the number of turns in the input coil. A spark is produced by first sending a brief current into the input. A magnetic field is produced that quickly disappears. This induces a very high voltage (around 25,000 volts) in the output, which is conducted to the spark plugs to ignite the fuel.

Understanding electromagnetism allows us to address some of the questions raised at the beginning of the chapter about how metal detectors work. The magnetic pulses are produced by sending an electric current through a coil of wire for a short period of time. When the current stops, the magnetic field that was created dies out quickly, and this

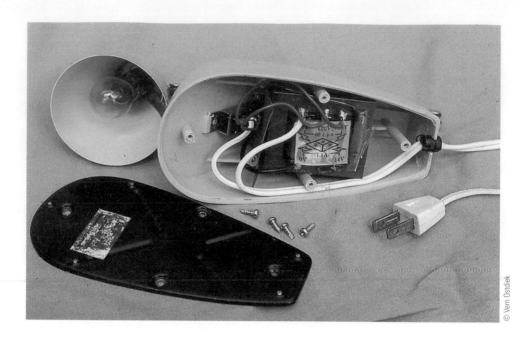

© Vern Ostdiek

● **Figure 8.21** The transformer in this lamp converts 120 volts AC into 14 volts AC ("Hi") or 12 volts AC ("Lo").

decreasing field induces an electric current in the coil. This current is used to monitor how swiftly the magnetic pulse dies out.

Metals are detected because the rapidly changing magnetic field of each pulse induces electrons in the metal to move—as in the secondary coil in a transformer—and this current produces an opposite magnetic pulse. This change in the total magnetic field affects the current induced in the coil. The electronics is designed to detect any such change and signal an alarm.

LEARNING CHECK

1. A magnetic field will be produced at some point in space if the electric field at that point
   a) gets stronger.
   b) gets weaker.
   c) changes direction.
   d) All of the above.
2. (True or False.) A transformer works with AC but not DC.

ANSWERS: 1. (d) 2. True

8.4 Applications to Sound Reproduction

A hundred years or so ago, the only people who listened to music performed by world-class musicians were those few who could attend live performances. Today, people in the most remote corners of the world can hear concert-quality sound from large home entertainment systems, pocket-sized MP3 players, and many devices in between. The first Edison phonographs were strictly mechanical and did a fair job of reproducing sound. It was the invention of electronic recording and playback machines that brought true high fidelity to sound reproduction. The sequence that begins with sound in a recording studio and ends with the reproduced sound coming from a speaker in your home or car includes components that use electromagnetism.

The key to electronic sound recording and playback is first to translate the sound into an alternating current and then later translate the AC back into sound. The first step requires a microphone, and the second step requires a speaker. Although there are several different types of microphones, we will take a look at what is called a dynamic micro-

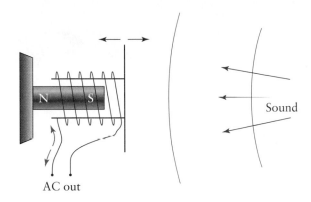

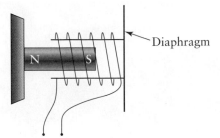

Diaphragm

Sound

AC out

● **Figure** **8.22** Simplified diagram of one type of microphone. Sound waves cause the diaphragm and coil to oscillate relative to the magnet. This induces AC in the coil.

phone. It consists of a magnet surrounded by a coil of wire attached to a diaphragm (see ● Figure 8.22). The coil and diaphragm are free to oscillate relative to the stationary magnet. When sound waves reach the microphone, the pressure variations in the wave push the diaphragm back and forth, making it and the coil oscillate. Since the coil is moving relative to the magnet, an oscillating current is induced in it. The frequency of the AC in the coil is the same as the frequency of the diaphragm's oscillation, which is the same as the frequency of the original sound. That is all it takes. This type of dynamic microphone is referred to as a "moving coil" microphone. The alternative is to attach a small magnet to the diaphragm and keep the coil stationary—a "moving magnet" microphone.

Let's skip ahead now to when the sound is played back. The output of the CD player, radio, or other audio component is an alternating current that has to be converted back into sound by a speaker. The basic speaker is quite similar to a dynamic microphone. In this case, the coil (called the "voice coil") is connected to a stiff paper cone instead of to a diaphragm (● Figure 8.23). Recall from Section 8.2 that an alternating current in the voice coil in the presence of the magnet will cause the coil to experience an alternating force. The voice coil and the speaker cone oscillate with the same frequency as the AC input. The oscillating paper cone produces a longitudinal wave in the air—sound.

Microphones and speakers are classified as *transducers*. They convert mechanical oscillation due to sound into AC (microphone), or they convert AC into mechanical oscillation and sound (speaker). They are almost identical. In fact, a microphone can be used as a speaker, and a speaker can be used as a microphone. But, as with motors and generators, each is best at doing what it is designed to do.

Explore It Yourself 8.3

For this, you will need a pair of headphones and a stereo or tape recorder with a microphone input jack that matches the headphone plug. Plug the headphones into the microphone jack, and talk into them as if they were microphones. Does it work?

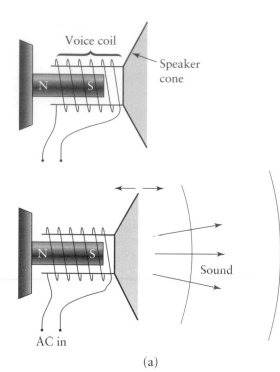

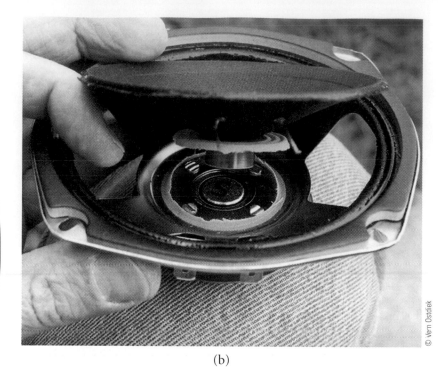

(a)

(b)

● **Figure 8.23** (a) Simplified sketch of a speaker. AC in the voice coil causes the cone to be forced in and out, thereby producing sound. (b) Photograph of a speaker with the speaker cone (between fingers) and voice coil (red band) assembly removed. The voice coil fits around the magnet (labeled "S").

© Vern Ostdiek

Most sound recording, from simple cassette recorders to sophisticated studio tape machines, is done on magnetic tape. The tape is a plastic film coated with a thin layer of fine ferromagnetic particles that retain magnetism. Sound is recorded on the tape using a recording head, a ring-shaped electromagnet with a very narrow gap (● Figure 8.24). During recording, an AC signal (from a microphone, for example) produces an alternating magnetic field in the gap of the recording head. As the tape is pulled past the gap, the particles in each part of the tape are magnetized according to the polarity of the head's magnetic field at the instant they are in the gap. The polarity of the particles changes from north-south to south-north, and so on, along the length of the tape.

To play back the recording, the tape is pulled past a playback head, often the same head used for recording. The magnetic field of the particles in the tape oscillates back and forth and induces an oscillating magnetic field in the tape head (● Figure 8.25). This oscillating magnetic field induces an oscillating current (AC) in the coil—electromagnetic induction again.

Magnetic recording is not limited to sound reproduction. Television video recorders (VCRs) record both sound and visual images on magnetic tape. Computers store information magnetically on tapes, floppy disks, and hard disks (● Figure 8.26).

● **Figure 8.24** (a) Simplified sketch of a tape head. During recording, (b) and (c), AC in the coil induces alternating magnetism in the tape.

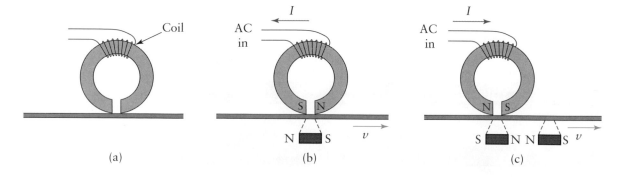

(a)

(b)

(c)

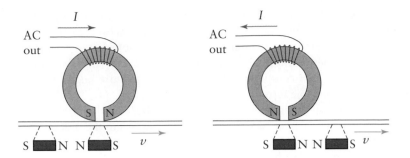

● **Figure 8.25** During playback, the alternating magnetism in the tape induces AC in the coil on the tape head.

The AC signals produced by microphones, CD players, and tape playback heads are quite weak. Amplifiers are used to increase the power of these signals before they are sent to speakers. Amplifiers also allow the listener to modify the sound by adjusting its loudness with the volume control and its tone quality with the bass and treble controls.

Digital Sound

A revolution in sound reproduction occurred in the 1980s with the advent of digital sound reproduction, the method used in *compact discs* (CDs) and various computer sound file formats, including *MP3*. In a process known as analog-to-digital conversion, the sound wave to be recorded is measured and stored as numbers. For CDs, the actual voltage of the AC signal from a microphone is measured 44,100 times *each second* (● Figure 8.27). Note that this frequency is more than twice the highest frequency that people can hear. The waveform of the sound is "chopped up" into tiny segments and then recorded as numerical values. These numbers are stored as binary numbers using 0s and 1s, just as information is stored in computers. To play back the sound, a digital-to-analog conversion process reconstructs the sound wave by generating an AC signal whose voltage at each instant in time equals the numerical value originally recorded. After being "smoothed" with an electronic filter, the waveform is an almost perfect copy of the original.

A huge amount of data is associated with digital sound reproduction—millions of numbers for each minute of music. CDs (and DVDs) store these data in the form of

● **Figure 8.26** Examples of magnetic storage media used by computers. The data tape at lower right can store over 1,000 times as much information as the floppy disk at upper left.

Refer to the discussion on waveforms in Section 6.3.

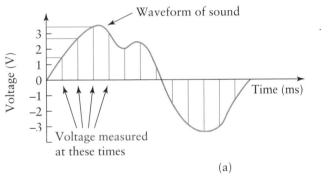

Time (ms)	Voltage (V)
0.0227	1.5
0.0454	2.8
0.0680	3.5
.	.
.	.
.	.

(a)

(b)

● **Figure 8.27** Digital sound reproduction in CDs. (a) To record the sound, the voltage of the waveform is measured 44,100 times each second. The resulting numbers are stored on magnetic tape for later use. (b) During playback, the voltage of the output at each time is set equal to the numerical value that was stored originally. The reconstructed waveform is then "smoothed" using an electronic filter. The resulting waveform is an almost perfect reproduction of the original.

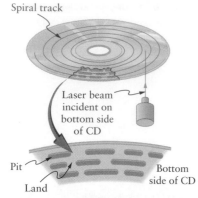

Spiral track

Laser beam
incident on
bottom side
of CD

Pit

Land

Bottom
side of CD

● **Figure 8.28** Information is stored on CDs and DVDs in digital form, as microscopic pits used to represent 0s and 1s. It is read from the spinning disk by the beam from a tiny laser. (From *College Physics,* 2e, by Urone. Used by permission.)

microscopic pits in a spiral line several miles long (see ● Figure 8.28). A tiny laser focused on the pits reads them as 0s and 1s. The amount of information stored on a 70-minute CD is equivalent to more than a dozen full-length encyclopedias. A standard DVD can store about seven times as much data. Little wonder that CDs and DVDs have also been embraced by the personal-computer industry as a way to store huge amounts of information in durable, portable form.

The superior quality of digital sound comes about because the playback device looks only for numbers. It can ignore such things as imperfections in the disk or tape, the weak random magnetization in a tape that becomes tape hiss on cassettes, and the mechanical vibration of motors that we hear as a rumble on phonographs. A sophisticated error-correction system can even compensate for missing or garbled numbers. Because the pickup device in a CD player does not touch the disc, each CD can be played over and over without the slow deterioration that a needle in a phonograph groove causes. This combination of high fidelity and disk durability made the CD system an immediate hit with consumers.

This is just a glimpse of some of the factors in state-of-the-art high-fidelity sound reproduction. Perhaps we are all so accustomed to it that we cannot appreciate how much of a technological miracle it really is. The next time you listen to high-quality recorded music, remember that it is all possible because of the basic interactions between electricity and magnetism described in Section 8.2.

LEARNING CHECK

1. Which of the following does *not* use electromagnetic induction?
   a) speaker
   b) transformer
   c) dynamic microphone
   d) cassette tape player
2. (True or False.) A speaker can be used as a crude microphone.

3. The process of making a digital recording of a sound wave involves
   a) digital-to-analog conversion.
   b) analog-to-digital conversion.
   c) electromagnetic induction.
   d) All of the above.

ANSWERS: 1. (a)   2. True   3. (b)

8.5 Electromagnetic Waves

Eyes, radios, televisions, radar, x-ray machines, microwave ovens, heat lamps, tanning lamps. . . . What do all of these things have in common? They all use **electromagnetic waves (EM waves).** EM waves occupy prominent places both in our daily lives and in our technology. These waves are also involved in many natural processes and are essential to life itself. In the rest of this chapter, we will discuss the nature and properties of electromagnetic waves and will look at some of their important roles in today's world.

As the name implies, EM waves involve both electricity and magnetism. The existence of these waves was first suggested by the nineteenth-century physicist James Clerk Maxwell while he was analyzing the interactions between electricity and magnetism. Consider the two principles of electromagnetism stated in Section 8.3. Let's say that an oscillating electric field is produced at some place. The electric field switches back and forth in direction while its strength varies accordingly. This oscillating electric field will induce an oscillating magnetic field in the space around it. But the oscillating magnetic field will then induce an oscillating electric field. This will then induce an oscillating magnetic field and so on in an endless "loop": The principles of electromagnetism tell us that a continuous succession of oscillating magnetic and electric fields will be produced. These fields travel as a wave, an EM wave.

DEFINITION

Electromagnetic Wave A transverse wave consisting of a combination of oscillating electric and magnetic fields.

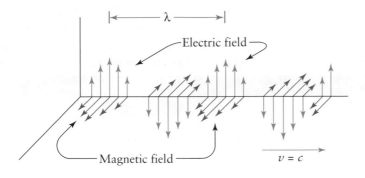

Electromagnetic waves are transverse waves because the oscillation of both of the fields is perpendicular to the direction the wave travels.* ● Figure 8.29 shows a "snapshot" of an EM wave traveling to the right. (The three axes are perpendicular to each other.) In this particular case, the electric field is vertical. As the wave travels by a given point in space, the electric field oscillates up and down, the way a floating petal oscillates on a water wave. The magnetic field at the point oscillates horizontally.

Figure 8.29 should remind you of the transverse waves we described in Chapter 6 (Figure 6.5, for example). Electromagnetic waves do differ from mechanical waves in two important ways. First, they are a combination of two waves in one: an electric field wave and a magnetic field wave. These cannot exist separately. Second, EM waves do not require a medium in which to travel. They can travel through a vacuum: The light from the Sun does this. They can also travel through matter. Light through air and glass, and x rays through your body, are common examples.

Electromagnetic waves travel at an extremely high speed. Their speed in a vacuum, called the "speed of light" because it was first measured using light, is represented by the letter c. Its value is

$$c = 299{,}792{,}458 \text{ m/s} \qquad \text{(speed of light)}$$

or

$$c = 3 \times 10^8 \text{ m/s} \qquad \text{(approximately)}$$
$$= 300{,}000{,}000 \text{ m/s}$$
$$= 186{,}000 \text{ miles/s} \qquad \text{(approximately)}$$

All of the parameters introduced for waves in Chapter 6 apply to EM waves. The wavelength can be readily identified in Figure 8.29. The amplitude is the maximum value of the electric field strength. The equation $v = f\lambda$ holds with v replaced by c. There is an extremely wide range of wavelengths of EM waves, from the size of a single proton, about 10^{-15} meters, to almost 4,000 kilometers for one type of radio wave. The corresponding frequencies of these extremes are about 10^{23} hertz and 76 hertz, respectively. Most EM waves used in practical applications have extremely high frequencies compared to sound.

Example 8.2

An FM radio station broadcasts an EM wave with a frequency of 100 megahertz. What is the wavelength of the wave?

The prefix "mega" stands for 1 million. Therefore, the frequency is 100 million hertz.

$$c = f\lambda$$

$$300{,}000{,}000 \text{ m/s} = 100{,}000{,}000 \text{ Hz} \times \lambda$$

$$\frac{300{,}000{,}000 \text{ m/s}}{100{,}000{,}000 \text{ Hz}} = \lambda$$

$$\lambda = 3 \text{ m}$$

---

* Some EM waves in plasmas can be longitudinal.

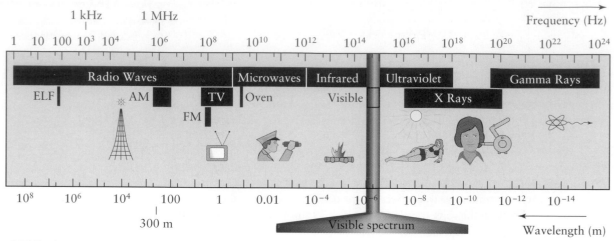

● **Figure 8.30** The electromagnetic (EM) spectrum.

Electromagnetic waves are named and classified according to frequency. In order of increasing frequency, the groups, or "bands," are **radio waves, microwaves, infrared radiation, visible light, ultraviolet radiation, x rays**, and **gamma rays (γ rays).** (Use of the word "radiation" instead of "waves" is not significant here.) ● Figure 8.30 shows these groups along with frequency and wavelength scales. This is called the **electromagnetic spectrum.** Notice that the groups overlap. For example, a 10^{17}-hertz EM wave could be ultraviolet radiation or an x ray. In cases of overlap, the name applied to an EM wave depends on how it is produced.

We will briefly discuss the properties of each group of waves in the electromagnetic spectrum—how they are produced, what their uses are, and how they can affect us. The great diversity of uses of EM waves arises from the variety of ways in which they can interact with different kinds of matter. All matter around us contains charged particles (electrons and protons), so it seems logical that EM waves can affect and be affected by matter. The oscillating electric field can cause AC currents in conductors; it can stimulate vibration of molecules, atoms, or individual electrons; or it can interact with the nuclei of atoms. Which sort of interaction occurs, if any, depends on the frequency (and wavelength) of the EM wave and on the properties of the matter through which it is traveling—its density, molecular and atomic structure, and so on.

In principle, an electromagnetic wave of any frequency could be produced by forcing one or more charged particles to oscillate at that frequency. The oscillating field of the charges would initiate the EM wave. The "lower-frequency" EM waves (radio waves and microwaves) are produced this way: A transmitter generates an AC signal and sends it to an antenna. At higher frequencies, this process becomes increasingly difficult. Electromagnetic waves above the microwave band are produced by a variety of processes involving molecules, atoms, and nuclei. Note that charged particles are present in all of these processes.

There is one other factor to keep in mind: Electromagnetic waves are a form of energy. Energy is needed to produce EM waves, and energy is gained by anything that absorbs EM waves. The transfer of heat by way of heat radiation is one example.

Radio Waves

Radio waves, the lowest frequency EM waves, extend from less than 100 hertz to about 10^9 Hz (1 billion hertz or 1,000 megahertz; ● Figure 8.31). Within this range are a number of frequency bands that have been given separate names—for example, ELF (extremely low frequency), VHF (very high frequency), and UHF (ultra high frequency). Most frequencies are given in kilohertz (kHz) or megahertz (MHz). Sometimes radio waves are classified by wavelength: long wave, medium wave, or short wave.

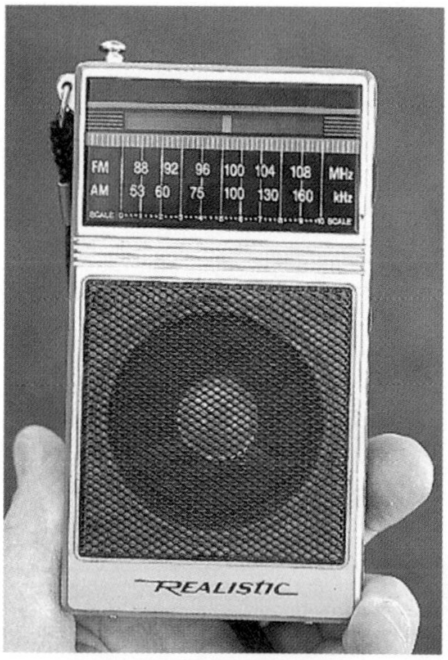

As mentioned earlier, radio waves are produced using AC with the appropriate frequency. Radio waves propagate well through the atmosphere, which makes them practi-

● **Figure 8.31** A radio is an EM wave detector that can select a single-frequency radio wave.

© Vern Ostdiek

cal for communication. Lower-frequency radio waves cannot penetrate the upper atmosphere, so higher frequencies are used for space and satellite communication. Only the very lowest frequencies can penetrate ocean water.

By far the main application of radio waves is in communication. The process involves broadcasting a certain frequency of radio wave with sound, video, or other information "encoded" in the wave. The radio wave is then picked up by a receiver, which recovers the information. Sometimes, this is a one-way process (commercial AM and FM radio and television), but in most other applications, it is two-way: Each party can broadcast as well as receive. Narrow frequency bands are assigned for specific purposes. For example, frequencies from 88 to 108 megahertz (88 million hertz to 108 million hertz) are reserved for commercial FM radio. There are dozens of bands assigned to government and private communication.

Microwaves

The next band of EM waves, with frequencies higher than those of radio waves, is the microwave band. The frequencies extend from the upper limit of radio waves to the lower end of the infrared band, about 10^9–10^{12} hertz. The wavelengths range from about 0.3 m to 0.3 mm.

One use of microwaves is in communication. For example, most satellite-based communication systems use frequencies in the microwave band. Early experiments with microwave communication led to the most important use of microwaves, *radar* (*ra*dio *d*etection *a*nd *r*anging), after the discovery that microwaves are reflected by the metal in ships and aircraft. As we discussed in Section 6.2, radar is echolocation using microwaves. The time it takes microwaves to make a round-trip from the transmitter to the reflecting object and back is used to determine the distance to the object. Radar systems are quite sophisticated: Doppler radar can determine the speed of an object moving toward or away from the transmitter by measuring the frequency shift of the reflected wave. Such radars are essential tools for air traffic control and monitoring severe weather. In the early 1990s, the *Magellan* spacecraft used imaging radar to penetrate Venus's permanent cloud cover and map the planet's surface (● Figure 8.32). Similar radar equipment placed in orbit around the Earth is used to form images of its surface, for such purposes as monitoring changes in the global environment and searching for archaeological sites.

● **Figure 8.32** Three-dimensional view of Venus's surface, generated using radar data from the *Magellan* spacecraft.

Microwaves have gained wide acceptance as a way to cook food. The goal of cooking is to heat the food, in other words, increase the energies of the molecules in the food. Conventional ovens heat the air around the food and rely on conduction (in solids) or convection (in liquids) to transfer the heat throughout the food. Microwave ovens send microwaves (typically with f = 2,450 megahertz and λ = 0.122 meters) into the food. The microwaves penetrate the food and raise the energies of the molecules directly. Recall from Section 7.2 that water consists of polar molecules—they have a net positive charge on one side and a net negative charge on the other side (● Figure 8.33). The electric field of a microwave exerts forces on the two sides of the water molecules in food. These forces are in opposite directions and twist the molecule. Since the electric field is oscillating, the molecules are alternately twisted one way and then the other. This process increases the

● **Figure 8.33** (a) Simplified sketch of a water molecule showing the net charges on its sides. (b, c) The oscillating electric field of a microwave twists the molecule back and forth, giving it energy.

Electric field

(a)

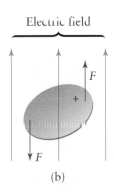

F

F

(b)

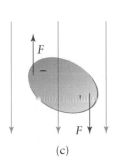

F

F

(c)

kinetic energy of the molecules and thereby raises the temperature of the food. Cooking with microwaves is fast because energy is given directly to all of the molecules. It does not rely completely on the conduction of heat from the outside to the inside of the food—a much slower process.

Infrared

Infrared radiation (IR; also called *infrared light*) occupies the region between microwaves and visible light in the electromagnetic spectrum. The frequencies are from about 10^{12} hertz to about 4×10^{14} hertz (400,000,000 megahertz). The wavelengths of IR range from approximately 0.3 to 0.00075 millimeters.

Infrared radiation is ordinarily the main component of heat radiation (introduced in Section 5.4). Everything around you is both absorbing and emitting infrared radiation, just as you are. The warmth you feel from a fire or heat lamp is due to your skin absorbing the IR. Infrared radiation is constantly emitted by atoms and molecules because of their thermal vibration. Absorption of IR by a cooler substance increases the vibration of the atoms and molecules, thus raising the temperature. We will take a closer look at heat radiation and its uses in Section 8.6.

Infrared radiation is commonly used in wireless remote-control units for televisions, and for short-distance wireless data transfer between such devices as personal digital assistants (PDAs) and laptop computers (see ● Figure 8.34). These units emit coded IR that is detected by other devices. In this capacity, IR is used much like radio waves. Another use of IR is in lasers; some of the most powerful ones in use emit infrared light (see Section 10.8).

Visible Light

Visible light is a very narrow band of frequencies of EM waves that happens to be detectable by human beings. Certain specialized cells in the eye, called *rods* and *cones,* are sensitive to EM waves in this band. They respond to visible light by transmitting electrical signals to the brain, where a mental image is formed. (The visible ranges of some animals such as hummingbirds and bees extend into the ultraviolet band. Some flowers that seem plain to humans are quite attractive to these nectar eaters.)

Visible light is a component of the heat radiation emitted by very hot objects. About 44% of the Sun's radiation is visible light: It glows white hot. Incandescent light bulbs produce visible light in the same way. Fluorescent and neon lights use excited atoms that emit visible light. In Chapter 10, we will discuss this process and describe how infrared and ultraviolet light and even x rays are emitted by excited atoms.

Within the narrow band of visible light, the different frequencies are perceived by people as different colors. The lowest frequencies of visible light, next to the infrared band, are perceived as the color red. The highest frequencies are perceived as violet. ● Table 8.1 shows the approximate frequencies and wavelengths of the six main colors in the rainbow.

Note how narrow the frequency band is: The highest frequency of light we can see is less than twice the lowest. By comparison, the range of frequencies of sound that can be heard is huge: The highest is 1,000 times the lowest.

● **Figure 8.34** This remote-control unit and the data window on this computer (tip of pencil) use infrared radiation to transmit information.

© Vern Ostdiek

● **Table 8.1** Approximate Frequencies and Wavelengths of Different Colors		
Color	Frequency Range ($\times 10^{14}$ Hz)	Wavelength Range ($\times 10^{-7}$ m)
Red	4.0–4.8	7.5–6.3
Orange	4.8–5.1	6.3–5.9
Yellow	5.1–5.4	5.9–5.6
Green	5.4–6.1	5.6–4.9
Blue	6.1–6.7	4.9–4.5
Violet	6.7–7.5	4.5–4.0

Most colors that you see are combinations of many different frequencies. White represents the extreme: One way to produce white light is to combine equal amounts of all frequencies (colors) of light. Rainbow formation involves reversing the process: White light is separated into its component colors. When no visible light reaches the eye, we perceive black.

In our daily lives, visible light is the most important of all electromagnetic waves. The entire next chapter is dedicated to optics, the study of visible light.

Ultraviolet Radiation

Ultraviolet radiation (UV), also called *ultraviolet light,* is a band of EM waves that begins just above the frequency of violet light and extends to the x-ray band. The frequency range is from about 7.5×10^{14} hertz to 10^{18} hertz.

Ultraviolet light is also part of the heat radiation emitted by very hot objects. About 7% of the radiation from the Sun is UV. This part of sunlight is responsible for suntans and sunburns. Ultraviolet radiation does not warm the skin as much as IR, but it does trigger a chemical process in the skin that results in tanning (● Figure 8.35). Overexposure leads to sunburn as a short-term effect, and repeated overexposure during a person's lifetime increases the chances of developing skin cancer. In Section 8.7, we describe how the ozone layer protects us from excessive UV in sunlight.

Some substances undergo "fluorescence" when irradiated with UV: They emit visible light. The inner surfaces of fluorescent lights are coated with such a substance. The UV emitted by excited atoms in the tube strikes the fluorescent coating, and visible light is produced. The same process is used in plasma TVs. Some fluorescent materials appear to be colorless under normal light and can be used as a kind of invisible ink. They can be seen under a UV lamp but are invisible otherwise.

X Rays

The next higher frequency electromagnetic waves are x rays. They extend from about 10^{16} to 10^{20} hertz. An important feature of x rays is that their range of wavelengths (about 10^{-8} to 10^{-11} meters) includes the size of the spacing between atoms in solids. X rays are partially reflected by the regular array of atoms in a crystal and so can be used to determine the arrangement of the atoms. X rays also travel much greater distances through most types of matter compared to UV, visible light, and other lower-frequency EM waves.

● **Figure 8.35** Tanning requires a source of ultraviolet light.

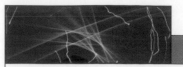
N Rays: "C'est une erreur."

The years before the beginning of the twentieth century were truly revolutionary in the history of modern physical science. Starting in 1893 with Sir William Crookes's experiments with cathode-ray tubes (the forerunners of television picture tubes) and continuing through 1895 with Roentgen's discovery of x rays, the world of physics was turned topsy-turvy. Virtually every physicist in Europe was conducting experiments on these new phenomena. The period from 1895 to 1905 was one of intense activity, intense excitement, and intense rivalry—among individuals *and* nations—in physics.

One of the early experimenters with x rays was René Blondlot (● Figure 8.36a), a professor of physics at the University of Nancy in France. Blondlot had already established a solid reputation as a physicist when, in 1903, in connection with his research on x rays, he reported the discovery of a new type of radiation, which he dubbed "N rays" after the place of their discovery, Nancy. Using first a small spark and later a low-intensity gas flame, whose increase in brightness gave witness to the presence of the N rays, Blondlot found that these rays were emitted from a variety of sources: x-ray tubes, ring-shaped gas burners (but not ordinary Bunsen burners), sheets of iron and silver heated to glowing, and the Sun.

Originally thought to be similar to infrared radiation, N rays possessed a number of unique qualities: They passed through plates made of platinum but not chunks of rock salt; they passed through dry but not wet paper; they could be produced by objects on which stresses were applied (like a stick bent by the hands) or by hardened metal (as in a steel file). The marvels, mysteries, and means of manu-

● **Figure 8.36** (a) René Prosper Blondlot (1849–1930). (b) Title page from Blondlot's book on N rays.

© Jean Loup Charmet Photographe, Paris, courtesy AIP Emilio Segrè Visual Archives, Physics Today Collection

"N" RAYS

A COLLECTION OF PAPERS COMMUNICATED
TO THE ACADEMY OF SCIENCES

*WITH ADDITIONAL NOTES AND INSTRUCTIONS FOR
THE CONSTRUCTION OF PHOSPHORESCENT
SCREENS*

BY
R. BLONDLOT

CORRESPONDENT OF THE INSTITUTE OF FRANCE
PROFESSOR IN THE UNIVERSITY OF NANCY

TRANSLATED BY
J. GARCIN

INGÉNIEUR E.S.E., LICENCIÉ-ÈS-SCIENCES

*WITH PHOSPHORESCENT SCREEN AND OTHER
ILLUSTRATIONS*

LONGMANS, GREEN, AND CO.
39 PATERNOSTER ROW, LONDON
NEW YORK AND BOMBAY
1905

© Photo Researchers

(b)

X rays are produced by smashing high-speed electrons into a "target" made of tungsten or some other metal (● Figure 8.37). The electrons spontaneously emit x rays as they are rapidly decelerated on entering the metal. X rays are also emitted by some of the atoms excited by the high-speed electrons.

Medical and dental "x-ray" photographs are made by sending x rays through the body. Typically, x rays with frequencies between 3.6×10^{18} hertz and 12×10^{18} hertz are used. As x rays pass through the body, the degree to which they are absorbed depends on the material through which they pass. Tissue containing elements with relatively large atomic

● **Figure 8.37** Simplified sketch of an x-ray tube. Electrons are accelerated to a very high speed by the high voltage. X rays are emitted as the electrons enter the metal target.

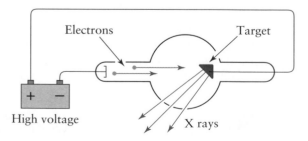

Electrons   Target

High voltage   X rays

facturing and manifesting these rays were described by Blondlot in great detail in 26 articles and a book (see ● Figure 8.36b).

Once the discovery of N rays was announced, experimenters around the world rushed to reproduce and to extend Blondlot's work. The results were mixed. Some workers reported success, but most reported failure, including such renowned physicists as Rayleigh, Langevin, and Rubens, a German physicist who pioneered studies in infrared radiation. Suspicion about the reality of N rays began to grow, reaching a high point in the summer of 1904 when a group of concerned scientists decided to send an envoy to Blondlot's laboratory in Nancy to investigate the activities there firsthand. The spy was Professor Robert L. Wood of Johns Hopkins University, a well-known expert in optical phenomena and debunker of numerous spiritualist scams.

Upon arriving in Nancy, Wood was treated to the complete gamut of N-ray phenomena by none other than Blondlot himself. And having witnessed the demonstrations, Wood parted on good terms with his host and published his report. In it Wood recounted how, when asked to hold a steel file (a well-known emitter of N rays) near Blondlot's forehead so as to enhance the latter's ability to see a dimly lit clock face, he instead substituted a piece of wood (one of the few objects known *not* to emit N rays), with no adverse effects on the results of the experiment as reported by Blondlot. Wood himself reported no improvement in the clock's visibility when the file was placed near his line of vision. If this were not enough, Wood noted that during a critical experiment in which the spectrum of N rays was to be produced by diffracting them through an aluminum prism, Wood pocketed the prism in the dark with no apparent alteration of the successful outcome of the demonstration as described by Blondlot. After Wood's exposé, the issue of N rays was dead.

To this day, explanations of the N-ray affair are unsatisfying. Probably the best that can be said is that problems associated with the subjective observation of low-intensity sources with varying energy output, coupled with the failure to perform a well-controlled experiment and the desire to achieve personal prestige and to foster national pride,

led to spurious results. What does seem clear is that the story of N rays is not a tale of deliberate fraud on the part of Blondlot or his associates. Moreover, it was not a hoax. It was, quite simply, a mistake. In the words of Professor Josef Bolfa, when interviewed on the subject, *"C'est une erreur."*

Perhaps one point to take from all of this is that physics is a human endeavor carried out by human beings, and, as in all areas of human activity, mistakes are made. Physicists, like all scientists, are not infallible. Nor are they always as honest, unbiased, or objective as we might wish. In the first three years of this century, scandals involving two physicists, one at Lucent Technologies' Bell Laboratories and another at the Lawrence Berkeley National Laboratory, showed all too clearly that physicists are subject to the same temptations and have the same character flaws as any other mortal. In each of these instances, an expert panel of investigators at each of the two affected institutions concluded that the physicists had fabricated and/or falsified data relating to the alleged discovery of superconductivity in buckyballs (page 135) on the one hand and the existence of element 118 in the periodic table (page 136) on the other. Reports issued by these review committees led to the dismissal of the two individuals and to the retraction by co-authors or publishers of many of the scientific articles based on the fraudulent data.

A second lesson to emerge from sagas like these is that physics, as a scientific discipline, is self-correcting. The inability of other experimenters at other, independent laboratories around the world to duplicate or corroborate the results reported by these miscreants ultimately led to their discovery and downfall. Thus, although a few physicists may be guilty of cheating—either subconsciously as in the case of N rays or deliberately as in the more recent cases—to obtain the answers they desire, the collective action of many physicists over a period of years generally produces results that are reliable, unbiased, and reflective of how the natural world truly operates. As Pier Oddone, deputy director of the Berkeley Lab, has said, "In the end, nature is the checker. Experiments have to be reproducible."

numbers (Z), like calcium ($Z = 20$), tend to absorb x rays more effectively than those that contain predominantly light elements like carbon ($Z = 6$), oxygen ($Z = 8$), or hydrogen ($Z = 1$). Lead, with atomic number 82, is a particularly good shield for blocking x radiation. Bones, which are rich in calcium, absorb x rays better than soft tissue such as muscle or fat, and hence show up more clearly on x rays (● Figure 8.38).

An x ray, or, more properly, a *roentgenogram,* named after Wilhelm C. Roentgen (pronounced *rent' gen*), who discovered x rays in 1895, is really an image of the x-ray shadows cast on film by various structures of the body. The greater the absorption of x rays by the body tissue or structure, the darker the shadow and the darker the image on the developed x ray. (Originally, x-ray images were called *skiagraphs* from the Greek meaning "shadow graphs.") Today, most x-ray images are made using a special film sandwiched between two intensifying screens. The latter are pieces of cardboard covered with crystals that absorb x rays well and give off visible or UV light in response. The film is coated on both sides with a light-sensitive emulsion, and each side of the film produces a picture of the light emitted by the intensifying screen in contact with it. Because the intensifying screens are more efficient at producing x-ray images than a single piece of x-ray–sensitive film alone, the x-ray dosages required to give well-exposed images using these screens are about one-tenth those needed without them. This is an important consideration, because medical x rays are the largest source of artificially produced radiation in the United States, comprising about 10% of the total radiation received per year by the average resident in this country. Protecting the public against unnecessary exposure to damaging radiation in diagnostic radiology is one of the greatest challenges to health and radiological physicists. Little wonder that such specialists recommend, when possible, the use of x-ray images produced with intensifying screens, even though they are somewhat blurrier (that is, less well defined) than those made directly on film.

X rays and gamma rays can be harmful because they are **ionizing radiation**—radiation that produces ions as it passes through matter. Such radiation can "kick" electrons out of atoms, leaving a trail of freed electrons and positive ions. This process can break chemical bonds between atoms in molecules, thereby altering or breaking up the molecule. Living cells rely on very large, sophisticated molecules for their normal functioning and reproduction. Disruption of such molecules by ionizing radiation can kill the cell or cause it to mutate, perhaps into a cancer cell. The human body can (and does) routinely replace dead cells, but massive doses of x rays or other ionizing radiation can overwhelm this process and cause illness, cancer, or death. We will discuss this more in Section 11.3.

Gamma Rays

The highest-frequency EM waves are gamma rays (γ rays). The frequency range is from about 3×10^{19} hertz to beyond 10^{23} hertz. The wavelength of higher-frequency gamma rays is about the same distance as the diameter of individual nuclei. Gamma rays are emitted in a number of nuclear processes: radioactive decay, nuclear fission, and nuclear fusion, to name a few. We will study these processes in detail in Chapter 11.

This concludes our brief look at the electromagnetic spectrum. Even though the various types of waves are produced in different ways and have diverse uses, the only real difference in the waves themselves is their frequency and, therefore, their wavelength.

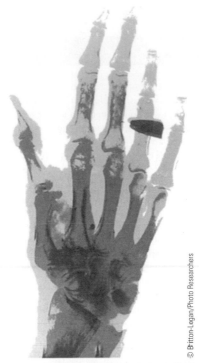

● **Figure 8.38** X-ray photograph of a human hand. In this negative image, areas that appear dark are those that strongly absorbed the incident x radiation. Bones are much more efficient at absorbing x rays because of their calcium content. Elements like gold and silver found in most jewelry are even better absorbers of x rays because of their higher atomic numbers.

© Britton-Logan/Photo Researchers

LEARNING CHECK

1. What simple thing can you do with a charged object to make it generate an electromagnetic wave?
2. Which of the following is *not* a common use of microwaves?
   a) cooking
   b) radar
   c) medical imaging
   d) communication
3. If we see two objects that have different colors, the light waves coming from them have different _____.
4. Tanning and sunburn are caused by the _____ component of sunlight.
5. (True or False.) Bones show up in x-ray images because they don't absorb x rays as effectively as muscle and other tissue does.

ANSWERS: 1. make it oscillate 2. (c) 3. frequencies 4. ultraviolet 5. False

8.6 Blackbody Radiation

Every object emits electromagnetic radiation because of the thermal motion of its atoms and molecules. We have already seen how this *heat radiation* is one method of transferring heat. Without radiation from the Sun, the Earth would be a frozen rock. In this section, we take a closer look at heat radiation and consider some of its uses.

The nature of the heat radiation emitted by a given object—the range of frequencies or wavelengths of EM waves present and their intensities—depends on the temperature of the object and on the characteristics of its surface (for example, its color). A hypothetical object that is perfectly black—one that absorbs all EM waves that strike it—would actually be the best at emitting heat radiation. Referred to as a *blackbody*, it would emit radiant energy at a higher rate than any other object at the same temperature, and the intensities of all of the wavelengths of EM waves emitted could be predicted quite accurately (Figure 8.39). The heat radiation emitted by such an object is referred to as **blackbody radiation (BBR).**

Blackbody radiation, an idealized representation of heat radiation, has been analyzed thoroughly and is well understood. The actual heat radiation emitted by real objects usually is not much different from BBR, so we can use it as a model of heat radiation.

The heat radiation emitted by any object (such as your own body, the Sun, a blackbody) is a broad band of electromagnetic waves. Within this band, some wavelengths are emitted more strongly than others: The intensity of the different wavelengths, the amount of energy released per square meter per second, varies with wavelength. For example, the heat radiation from the Sun contains more energy in each wavelength of visible light than in each wavelength of IR radiation. The intensity of the visible wavelengths is higher than that of the IR wavelengths.

A graph showing the intensity of each wavelength of radiation emitted by a blackbody is called a *blackbody radiation curve* (Figure 8.40). The size and shape of the graph change with the object's temperature. The graph for a real object (not a blackbody) would be similar.

The total amount of radiation emitted per second by an object obviously depends on how large it is. A 100-watt light bulb is brighter (it emits more light) than a 10-watt light bulb because its filament is larger. Aside from this factor, it is the object's *temperature* that has the greatest influence on the amount and types of radiation emitted. Three aspects of heat radiation are affected by the object's temperature.

1. The amount of each type of radiation (such as microwave and IR) emitted increases with temperature.

An incandescent light bulb fitted with a dimmer control illustrates this well. Dimming the light causes the filament temperature to decrease. This reduces the amount of visible light emitted: The bulb's brightness is decreased. The amount of infrared is also reduced.

2. The total amount of radiant energy emitted per unit area per unit time increases rapidly with any increase in temperature. For a particular blackbody, the total radiant energy emitted per second (power) is proportional to the temperature (in kelvins) raised to the fourth power.

$$P \propto T^4 \qquad (T \text{ in kelvins})$$

Doubling the Kelvin temperature of an object will cause it to emit 16 times as much radiant energy each second. The human body at 310 K (98.6°F) radiates about 25% more

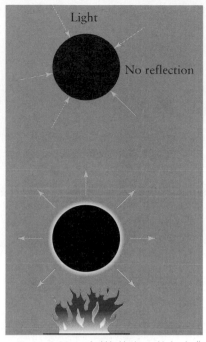

 Figure 8.39 An ideal blackbody would absorb all EM radiation that strikes it. It would also radiate more heat radiation than any other object with the same size and the same temperature.

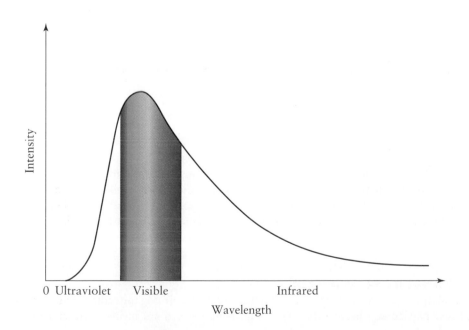

 Figure 8.40 Typical blackbody radiation curve. It indicates the amount of energy emitted at each wavelength of the EM spectrum.

power than it would at room temperature, 293 K (68°F). If you could see infrared, humans would appear to glow more brightly than their cooler surroundings.

3. At higher temperatures, more of the power is emitted at successively shorter wavelengths (higher frequencies) of electromagnetic radiation. For a blackbody, the wavelength that is given the maximum power (the peak of the blackbody radiation curve) is inversely proportional to its temperature.

$$\lambda_{max} = \frac{0.0029}{T} \ (\lambda_{max} \text{ in meters, } T \text{ in kelvins})$$

Objects cooler than about 700 K (about 800°F) emit mostly IR with smaller amounts of microwaves, radio waves, and visible light. There is not enough visible light to be detected by the human eye. Above this temperature, objects emit enough visible light to glow. They appear red hot because more red (longer wavelength) radiation is emitted than any other visible wavelength. The peak wavelength is still in the infrared region. (Several factors influence the minimum temperature needed to cause an object to glow. These include the size and color of the object, the brightness of the background light, and the acuity of the observer's eyesight.)

Explore It Yourself 8.4

For this, you need a room lit by incandescent lights equipped with a dimmer. It must also be dark when the lights are turned off. With the lights at their brightest, note the appearance of a white piece of paper. Turn the lights down, and again note the appearance of the paper. Does it appear to be exactly the same color? What is different?

The filament of an incandescent light bulb can be as hot as 3,000 K. The peak wavelength of its heat radiation curve is in the infrared region, not far from the visible band. The visible light emitted is a bit stronger in the longer wavelengths, so it has a slightly reddish tint. At 6,000 K, the Sun's surface emits heat radiation that peaks in the visible band (the wavelength of the peak is one-half that of the light bulb's). It appears to be white hot (● Figure 8.41). When comparing the curves, keep in mind that the Sun's is higher *not* because the Sun is bigger than a light bulb but because it is hotter.

Example 8.3 Assuming that the Sun is a blackbody with a temperature of 6,000 K, at what wavelength does it radiate the most energy?

$$\lambda_{max} = \frac{0.0029}{T}$$

$$= \frac{0.0029}{6,000 \text{ K}}$$

$$= 4.8 \times 10^{-7} \text{ m}$$

Table 8.1 shows this to be in the blue part of the visible band.

Some stars are hot enough to appear bluish. Sirius and Vega are two examples. The peaks of their radiation curves are in the UV, so they emit more of the shorter wavelength visible light (blue) than the longer wavelengths (see Explore It Yourself 8.5 on p. 322).

The temperature dependence of blackbody radiation is responsible for a number of interesting phenomena, and it is used in some ingenious ways. The following are examples.

Temperature Measurement

The temperature of an object can be determined by examining the radiation that it emits. This is particularly useful when very high temperatures are involved, as in a furnace,

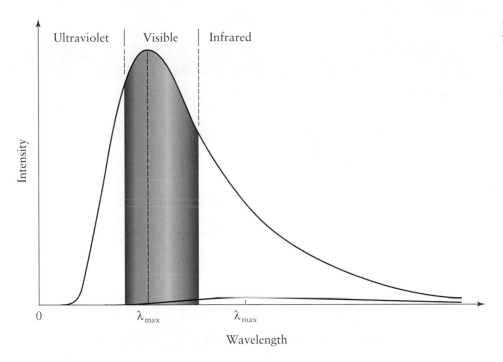

● **Figure 8.41** Blackbody radiation curves for the Sun (upper) and a light bulb (lower).

because nothing has to come into contact with hot matter. Special devices called *pyrometers* measure the amount and types of radiation emitted and use the rules mentioned above to determine the temperature. This process is used to measure the temperature of the Sun and other stars.

The electronic ear thermometer works in a similar way: It determines the patient's body temperature by measuring the intensity of infrared radiation emitted by an eardrum.

Detection of Warm Objects

Most things on Earth have temperatures that cause them mainly to emit infrared light. Anything that can detect IR can use this fact to locate warmer-than-average objects, since they will emit more IR. Rattlesnakes and certain other snakes use IR to hunt mice and other warm-blooded animals at night. These snakes have sensitive organs that detect the higher-intensity infrared emitted by objects warmer than their surroundings.

Infrared-sensitive photographic film, video cameras, and other detection devices have many practical uses. For instance, IR photographs, called *thermograms,* can show where heat is escaping from a poorly insulated house and can detect ohmic heating caused by a short circuit in an electrical substation (● Figure 8.42). They can locate warmer areas on the human body that might be caused by tumors. The military uses IR detectors to locate soldiers at night. Heat-seeking missiles automatically steer themselves toward the hot exhaust of aircraft.

● **Figure 8.42** Thermograms—infrared photographs—show regions that are warmer or cooler than the surroundings.

Cosmic Background Radiation—A Relic of the Big Bang

The next time you turn on your TV to watch the news or your favorite "reality" program, tune the set to a channel that is not used by your local cable company or network affiliates and observe the "snow" on the screen. A part of the "signal" you are receiving is from radiation that was produced during the formation of the universe in a catastrophic event popularly referred to as the *Big Bang*.* Indeed, our belief that the universe was in fact born in a fiery explosion of space and time nearly 14 billion years ago is intimately connected to the existence and character of the weak microwave radiation picked up by your TV as "noise."

The description of what is now called the *cosmic background radiation* as "noise" is an apt one, for it was the search for the persistent source of radio interference in their 20-foot horn antenna (● Figure 8.43) in 1965 that eventually led Bell Laboratory researchers Arno Penzias and Robert Wilson to the startling conclusion that the sky is filled with microwaves. Regardless of the direction in which a suitably tuned receiver is pointed, it will detect such radiation and with nearly the same intensity. The equivalent temperature of the interfering radiation discovered by Penzias and Wilson was found to be about 3.5 K. Because of the limitations of their original equipment, they were not able to fully describe the wavelength dependence of this "noise." In the years following their report, measurements were made by other groups at other wavelengths, and all yielded temperatures of between about 2.7 and 3.0 K. The spectrum was thus shown to be that of a blackbody, characteristic of matter in thermodynamic equilibrium with radiation at a temperature of a bit less than 3 K. The best and most recent demonstration of the blackbody nature of this cosmic radiation comes from data obtained by the *Cosmic Background Explorer* (COBE) satellite; ● Figure 8.44a shows the results of measurements made by this instrument, which yield an excellent fit to a blackbody radiation curve for a temperature of 2.726 K. Using the method of Example 8.3, this temperature gives a wavelength of 1.06×10^{-3} meters (1.06 millimeters) for the peak in the cosmic blackbody radiation curve; this is squarely in the microwave region of the EM spectrum (see Figure 8.30). For their careful work in tracking down and drawing attention to the significance of this 3 degree radiation, Penzias and Wilson were awarded the Nobel Prize in 1978.

Just exactly *how* are the existence and character of this microwave radiation connected to the creation of the universe, and *what* can it tell us about the conditions present during this birthing event? One of the implications of the discovery that the galaxies are rushing away from one another with speeds that are proportional to their separations is that at earlier times in the history of the universe, the galaxies were all closer together than they are now (see the Physics Potpourri on the Hubble relation in Chapter 6). The earlier the epoch, the more densely packed were the galaxies, until at some point all the matter and energy in the universe were concentrated in an infinitely small volume. Insofar as the universe comprises everything, nothing can exist outside it; thus this period of infinite density marks not only the temporal, but the spatial beginning of the universe. Space and time were created simultaneously in an explosion of mass and energy from this singular condition. This *is* the Big Bang.[†]

● **Figure 8.43** Arno Penzias (left) and Robert Wilson. In the background is the Bell Labs microwave antenna that was used to detect the cosmic background radiation.

If we treat the early universe like a highly compressed gas, then we expect it to have a high temperature according to the laws of thermodynamics discussed in Chapter 5. As this hot, dense "gas" expanded, it thinned out and cooled off. In the process, matter, initially in the form of elementary particles (see Chapter 12), began to condense out of the sea of pure radiant energy that was the universe at that time. Once this condensation took place, matter began to interact strongly with the radiation, absorbing, scattering, and emitting it profusely. The distance traveled by any given beam of radiation was very small before it encountered particles that delayed and deflected it. In this way, the matter acted as a very effective dam to the free propagation of radi-

---

* In 1993, *Sky and Telescope* magazine held a competition among its readership to provide a better name for this event, one more in keeping with its seriousness and significance. After reviewing more than 13,000 submissions, a panel of judges determined that *none* of the entries served to capture the essence of this creation event any better than the Big Bang, a term coined by astronomer Fred Hoyle in 1950.

† The Big Bang model of the universe is not the only one that has been proposed by scientists. There have been several variants on this theme that have appeared in the literature, as well as one fundamentally different model called the **steady-state theory,** which alleges that the gross properties of the universe remain fixed throughout all space and for all time. This model has fallen from favor precisely because it cannot easily account for the microwave background radiation observed by Penzias and Wilson.

ation at this phase in the development of the universe.

There came a time, however, when the continued expansion of the universe rendered the density and temperature of matter low enough that the interactions between the radiation and the matter became far less pronounced. At this stage, believed to have occurred some 400,000 to a million years after the initial explosion when electrons combined with protons to form hydrogen atoms, the radiation "decoupled" from the matter in the universe and was free to move through it virtually unimpeded. The universe became transparent to radiation. This radiation then filled the universe nearly uniformly and isotropically. Of course, just after its separation from matter, the temperature of the radiation was still quite high. Current theories place it at about 3,000 K. As the universe inexorably expanded and cooled, though, this temperature dropped until at the present time we expect it to have reached about 3 K. Since the radiation was originally in equilibrium with the matter in the universe, it remained so after decoupling and exhibits today the spectrum of a blackbody. Thus, one prediction of the Big Bang model arising from the expansion of the universe from an infinitely hot, dense state is the existence of a low-temperature cosmic background radiation with a blackbody distribution. And this is exactly what Penzias and Wilson observed.

The discovery of the microwave background radiation was one of the most important pieces of physical evidence leading to the widespread acceptance by scientists of the Big Bang model as the correct description of the formation of the universe. Continuing studies of the cosmic microwave background radiation by COBE and more recently by the Wilkinson Microwave Anisotropy Probe (WMAP) launched by NASA in June 2001 have now produced highly detailed maps of the sky revealing microkelvin fluctuations in temperature (Figure 8.44b). Such displays are the "baby pictures" of the universe when it was only 379,000 years old. The size and scale of these minute temperature variations are helping scientists to understand how galaxy formation, driven by such fluctuations, began in the early universe. But that's another story.

For more information on the latest developments from the WMAP, go to http://map.gsfc.nasa.gov.

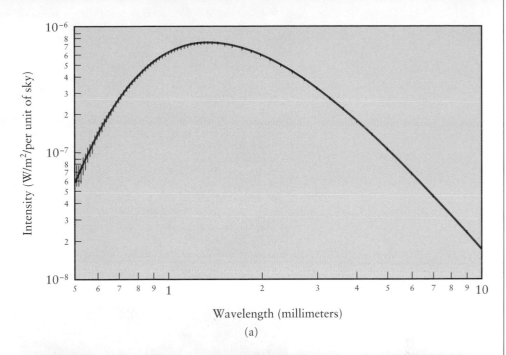

(a)

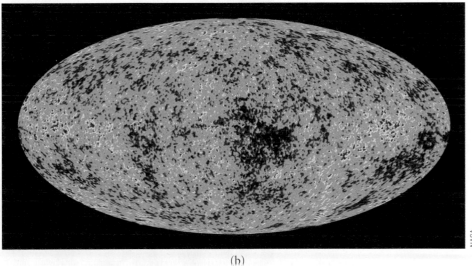

NASA

(b)

● **Figure 8.44** (a) COBE data showing the blackbody spectrum of cosmic radiation. (b) WMAP all-sky image showing microkelvin temperature variations in the cosmic background radiation. Regions shown in red are warmer, and those in blue-violet are cooler.

Explore It Yourself 8.5

Next time you're outside on a clear night in a place well removed from city lights and free of obstructions, take a careful look at the sky. Do the stars you see all appear equally bright? Do they all appear the same color? The color (and to some extent the brightness) of a star is directly related to its surface temperature, as described earlier in this section. You can establish the relative temperatures for many of the brightest stars by assessing their colors and applying the physics of blackbodies that you have learned thus far.

To get started, use the star chart in ● Figure 8.45 to locate the prominent wintertime constellation of Orion. Identify the bright stars Rigel and Betelgeuse. Notice the differences in their colors. Betelgeuse has a surface temperature of only about 3,200 K (quite low by stellar standards but hot enough to melt iron). It should appear distinctly red to you. By contrast, Rigel is a much hotter star ($T = 10,000$ K) and should look bluish-white to your eyes.

Scan the sky for other bright stars in the vicinity of Orion, like Sirius in Canis Major or Aldebaran in Taurus. Based on their colors, how do their surface temperatures compare to those of Rigel and Betelgeuse? How would you describe the colors of the majority of the brightest-appearing stars in the sky? Suppose you were told that most of the stars in our Galaxy near the Sun have rather low surface temperatures and should appear red-orange in color. Is this statement at odds with your answer to the previous question? Can you offer an explanation for any discrepancy between your observations and the true state of stars in the solar neighborhood?

● **Figure 8.45** Winter constellations, in the evening at mid-northern latitudes.

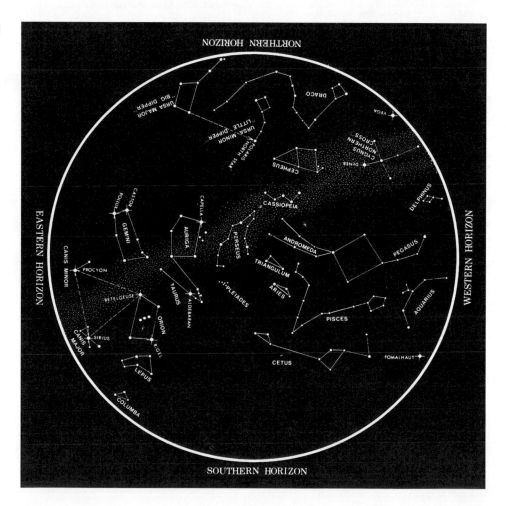

8.7 EM Waves and the Earth's Atmosphere

Many substances in the atmosphere surrounding the Earth interact with EM waves in important ways. Some of these interactions are crucial to the existence of life on this planet, another helps us communicate, and some add to the beauty that characterizes life on Earth. The visible phenomena, rainbows, for example, are described in the chapter dedicated to visible light, Chapter 9. Some of the others are discussed here.

Ozone Layer

The sunlight that keeps the Earth warm and provides the energy for plants to grow also contains ultraviolet radiation that is harmful to living things. But life has evolved on this planet because the atmosphere has protected it from this UV. In the region between about 20 and 40 kilometers above the Earth, known as the *ozone layer,* there is a comparatively high concentration of *ozone* (O_3), the form of oxygen with three atoms in each molecule. This ozone absorbs most of the harmful UV in sunlight. The ozone layer has been a shield protecting living things on Earth.

In 1974, it was reported that *chlorofluorocarbons* (CFCs), chemical compounds such as Freon used in refrigerators, air conditioners, and as an aerosol propellant in spray cans, could be depleting the ozone layer. The CFCs that are released into the air drift upward to the ozone layer and chemically break up the ozone molecules with alarming efficiency. The 1995 Nobel Prize in chemistry was awarded to the discoverers of this effect. Because of this, CFCs were banned for use as aerosol propellants in the United States. Then in 1985 it was discovered that a "hole" developed in the ozone layer over Antarctica during the later part of each year. The concentration of ozone in a huge section of the atmosphere was reduced by about one-half (● Figure 8.46). A review of old satellite measurements revealed that this hole had developed in previous years as well and that the one in 1982 was twice the area of the United States. Scientists now believe that the ozone hole is produced by a complex set of processes involving chlorine that originates in CFCs. During the southern winter, chlorine molecules (Cl_2) are released by chemical reactions that take place in extremely high clouds, called *polar stratospheric clouds* (PSCs), over the sunless South Pole. The return of sunlight in spring then triggers the reactions that cause chlorine to break up ozone molecules.

Until recently it was thought that a similar ozone hole would not form over the North Pole. But in March 1997 the ozone level over the Arctic dropped to a record low for that time of year. The depletion was not as severe as in the southern ozone hole, but it suggests that the damage to the ozone layer is increasing.

Global monitoring of the ozone layer revealed an overall decline during the latter part of the twentieth century, not just over the poles. Continued reduction in ozone levels could have tragic consequences. Rates of occurrence of skin cancer could rise. Crop yields could decrease, since increased UV adversely affects many plant species. But unprecedented international cooperation led to the banning of CFCs in developed countries. Levels of

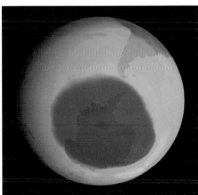

● **Figure 8.46** False-color image of a recent "ozone hole" over the southern hemisphere, based on satellite measurements taken on 11 September 2003. The region of greatly reduced ozone, shown in blue, covers most of Antarctica and reaches the southern tip of South America.

CFCs in the atmosphere seem to be declining, but the extremely high chemical stability of these compounds means they will continue to do damage for many years to come.

Greenhouse Effect

The **greenhouse effect** is so named because it is partly responsible for keeping greenhouses warm in cold weather. Glass and certain other materials allow visible light to pass through them while they absorb or reflect the longer-wavelength infrared radiation. A glass wall or roof on a building allows visible light to enter and to warm the interior. As the temperature of things inside increases, they emit more IR. Without the glass, this IR would escape from the enclosure and carry away the added energy. But the glass blocks the IR, so the added internal energy is trapped and the enclosure is warmed. (The glass not only reduces heat loss by radiation, it also eliminates convection: The heated air that rises cannot leave and take internal energy with it.) A car parked in the sun with the windows up is much warmer than the outside air because of this heating. Windows on the sunny side of a building help to keep it warm in the same way.

The greenhouse effect occurs naturally in the Earth's atmosphere. Water vapor, carbon dioxide (CO_2), and other gases in the air act somewhat like glass in that they allow visible light from the Sun to pass through to the Earth's surface while they absorb part of the infrared radiation emitted by the warmed surface (● Figure 8.47a). The atmosphere is heated by the IR that is absorbed; it is about 35°C (65°F) warmer than it would be without this effect.

From 1959 to 2002, the carbon-dioxide content of the atmosphere rose 18% (Figure 8.47b). Evidence points to human activity as the likely cause. During the last century, huge quantities of fossil fuels—coal, oil, and natural gas—have been burned. This has released vast amounts of carbon dioxide into the air. At the same time, forests have been

● **Figure 8.47** (a) The Earth's atmosphere produces a greenhouse effect. Increased carbon dioxide content in the air could cause too much heating. (b) Graph of CO_2 content in the air versus time, measured at the observatory atop Mauna Loa in Hawaii. The concentration varies throughout each year because plants in the northern hemisphere take in more CO_2 in the summer when they are most active.

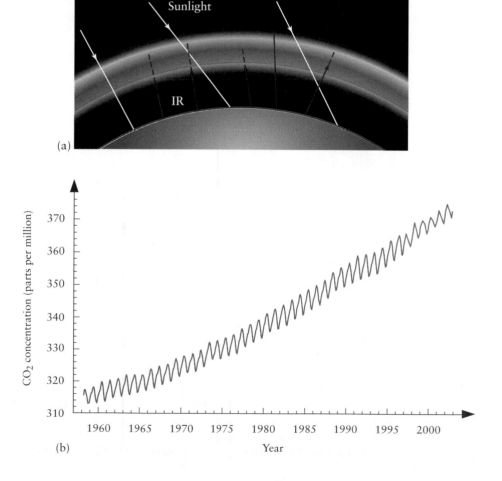

cut down for building materials and cleared for farming and human occupation. This contributes to the problem because trees and other plants take carbon dioxide out of the air. The concentration of methane is also increasing. Methane gas, released into the air by some animals and by rice as it grows, and the ozone-threatening chlorofluorocarbons are also "greenhouse gases" like water vapor and CO_2. The result of this is the possibility of global warming—the entire atmosphere heating up.

Anyone who has lived through a harsh winter might think that the warming of the atmosphere is not such a bad idea. The main concern is that we might be starting a runaway greenhouse effect that could lead to global disaster. The temperature of the Earth's atmosphere could rise enough to trigger changes in the weather patterns. Once-verdant areas could become deserts, and vice versa. The polar ice caps could start melting, thereby raising the level of the oceans and flooding heavily populated coastal areas around the world. A runaway greenhouse effect did occur naturally on the planet Venus, where the surface temperature is now 460°C. Conditions on Venus prevented excessive atmospheric carbon dioxide from being trapped in carbonate rocks or dissolved in oceans, as is the case on Earth.

The atmospheric greenhouse effect is such a complex phenomenon that it is nearly impossible to predict exactly what will happen. Different scenarios have been proposed: As the Earth heats up, huge amounts of CO_2 dissolved in the oceans could be released and could exacerbate the global warming. Increased evaporation of water would raise the water-vapor content of the atmosphere and possibly cause more heating. Or the higher humidity could lead to more cloud cover, which might cool the Earth as less sunlight reaches its surface. About all we can be sure of is that we are altering the life-sustaining blanket in which we live even though we cannot predict what the effects will be.

The Ionosphere

Did you ever wonder why you can pick up AM radio stations that are several hundred kilometers away but usually cannot receive FM radio or television more than about 80 kilometers away? About 50–90 kilometers above the Earth's surface, in a region of the atmosphere known as the *ionosphere*, there is a relatively high density of ions and free electrons. Radio waves from transmitters travel upward into the ionosphere. Higher-frequency radio waves, such as those used in FM radio and television, pass through the ionosphere and out into space (● Figure 8.48b). Lower-frequency radio waves, such as the 500–1,500 kilohertz waves used in AM radio, reflect off the ionosphere and return to Earth (Figure 8.48a). The range for high-frequency radio waves is limited to "line-of-sight" reception. The curvature of the Earth eventually blocks the signal from the transmitting tower. Low-frequency radio waves can skip off the ionosphere and travel farther around the planet. The radio waves used to communicate with spacecraft must have high frequency to pass through the ionosphere.

The ionosphere is also the home of the *auroras*—the northern lights and the southern lights. Charged particles from the Sun excite atoms and molecules in the ionosphere, causing them to emit light. (More on this in Chapter 10.)

Astronomy

Stars, galaxies, and other objects in space emit all types of electromagnetic radiation. Astronomers were originally limited to studying only visible light through optical telescopes. Now, they examine the entire spectrum of EM radiation to gain a more complete understanding about the universe. The atmosphere is a hindrance to some of these investigations. Ultraviolet light from stars and galaxies is absorbed by ozone and other gases. Infrared light is also absorbed, mainly by water vapor. Lower-frequency radio waves are absorbed in the ionosphere. Even visible light is affected by the random swirling of the air which causes stars to twinkle and degrades the images formed in large telescopes. Microwaves and higher-frequency radio waves are about the only EM waves from space that are not affected by the atmosphere.

Since the early 1960s, dozens of telescopes and other astronomical instruments have been placed in orbit to overcome the deleterious effects of Earth's atmosphere. Among the

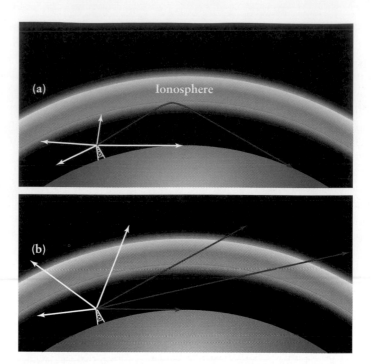

● Figure 8.48 (a) Low-frequency radio waves are bent back to the Earth's surface by the ionosphere. This allows commercial AM radio to be transmitted hundreds of miles. (b) High-frequency radio waves pass through the ionosphere. Because of this, transmission of commercial FM and television is limited to about 100 km.

most important and sophisticated of recent space missions are the four that comprise NASA's Great Observatories Program. Each was designed to examine different parts of the EM spectrum, from infrared light for studying cool stars and interstellar dust, to gamma radiation emitted in high-energy processes associated with supernova explosions and neutron-star collisions. The most famous of these, the *Hubble Space Telescope* (HST), was launched in 1990 and is equipped with instruments that analyze visible, ultraviolet, and shorter-wavelength infrared light (see ● Figure 8.49). After a shaky start, HST has achieved enormous success: Research astronomers, as well as millions of people worldwide, have been captivated by the stunning images it has returned. (There is more on HST in Section 9.2.) The other spacecraft in the program, along with the year in which each was launched, are the *Compton Gamma-Ray Observatory* (1991), the *Chandra X-Ray Observatory* (1999), and the *Spitzer Space Telescope* (infrared, 2003). Astronomy textbooks for years to come are likely to contain illustrations and findings supplied by these observatories. Space exploration continues to give astronomers access to the entire electromagnetic spectrum.

● Figure 8.49 (Left) The *Hubble Space Telescope* during deployment from the space shuttle *Discovery*. (Right) *Hubble Space Telescope* photograph of the Eagle Nebula.

Space Telescope Science Institute/NASA/SPL/Photo Researchers

NASA

1. Why would a decrease in the amount of ozone in the atmosphere increase the likelihood of people getting sunburn?
2. Of the following gases, which is the most important contributor to the Earth's greenhouse effect?
   a) water vapor
   b) ozone
   c) CFCs
   d) methane
3. (True or False.) The *Hubble Space Telescope* and devices that detect other segments of the EM spectrum are sometimes placed in orbit because the Earth's atmosphere affects EM waves passing through it.

ANSWERS: 1. Increases solar UV   2. (a)   3. True

PHYSICS FAMILY ALBUM

The first recorded investigation of magnetism was carried out by William Gilbert (see Chapter 7 Physics Family Album) at the same time he was studying electricity. In the course of his work, Gilbert built a sphere out of lodestone to model the effects of the Earth on a compass needle. With this *terrella* (little Earth), he accounted for a number of phenomena, such as magnetic declination, that had been observed by navigators as they traveled around the globe. One of these was an effect observed by Columbus on his famous voyage: The declination angle changes with longitude.

As recently as 200 years ago, electricity and magnetism seemed to be similar but independent phenomena. Reports that compasses were affected by lightning storms suggested that some kind of interaction between the two existed, but nothing concrete was observed until 1820. In that year, Hans Christian Oersted (1777–1851; ● Figure 8.50), a physics professor at the University of Copenhagen, announced that he had observed a compass needle being deflected by an electrical current in a nearby wire. Oersted had begun experimenting with electricity after he heard of Volta's invention of the battery. His first attempts to observe the effect failed because, understandably, he was not expecting the deflection to be perpendicular to the current-carrying wire.

Within months, news of the discovery had spread throughout the European scientific community. The exact relationship between an electric current and the strength and configuration of the magnetic field that it produced were quickly deduced. Most of this work was done by the French physicist André Marie Ampère (1775–1836). The son of a prosperous merchant in a village near Lyons, Ampère showed signs of genius at an early age. His path to greatness was sidetracked when his father was executed during the French Revolution. It took Ampère a year to recover from the shock and to continue his education.

Before his studies of electromagnetism, Ampère had gained fame for his work in mathematics and chemistry. His talents and interests spread to many areas: During his career he held professorships in mathematics, philosophy, astronomy, and physics. (It seems he was also a classic absent-minded professor.)

Ampère's greatest work was triggered by Oersted's discovery. Within a week of hearing about it, Ampère had thoroughly investigated it and prepared his own paper on the topic. Ampère showed that the magnetic field of a current could be concentrated by bending the wire into a loop

● **Figure 8.50** Hans Christian Oersted.

or wrapping it up as a solenoid. He correctly suggested that the magnetism in permanent magnets is caused by tiny currents in the molecules. He made the second important observation of electromagnetism: A magnetic field exerts a force on a current-carrying wire. Ampère's most famous finding is that two current-carrying wires exert forces on each other (● Figure 8.51). The force is due to magnetism even though no permanent magnets are involved. This simple setup elegantly illustrates how magnetism is an inextricable part of moving electric charges.

An American physicist, Joseph Henry (1797–1878), is notable for his research on electromagnets. He improved their efficiency to

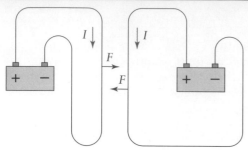

● **Figure 8.51** Two parallel current-carrying wires exert forces on each other. The magnetic field produced by the wire on the left exerts a force on the wire on the right, and vice versa.

the point of making an electromagnet capable of lifting 50 times its own weight when connected to a small battery. Henry also had the idea of using electromagnets for long-distance communication—the telegraph. He freely mentioned this idea to others, who later patented the process and became wealthy because of it.

The final major discoveries in electromagnetism were made by two of the greatest names in physics, Michael Faraday and James Clerk Maxwell. These men present interesting contrasts in their backgrounds and in the way in which they approached physics. Faraday was born into a poor family and was, for the most part, self-educated. He became the greatest experimenter in physics of the time. Maxwell came from a well-to-do family and was educated in Cambridge. His approach was highly theoretical: He incorporated sophisticated mathematics in his explanations of electromagnetism.

Faraday (1791–1867), son of a blacksmith, grew up near London. His apprenticeship as a bookbinder afforded him the opportunity to read extensively about the great discoveries in electricity. His break came when he was able to attend some lectures given by the great chemist Sir Humphry Davy (1778–1829). Davy gained fame for his use of electrolysis to discover sodium, potassium, and several other elements. Faraday took detailed notes of the lectures, carefully bound them, and showed them to Davy. Davy was sufficiently impressed to hire Faraday and to launch him on his historic career. As Davy's assistant, Faraday came into contact with some of the greatest scientists of the time. He soon developed into a first-rate chemist and made some noteworthy discoveries.

Faraday first experimented with electricity in 1821, but his greatest work was done in the 1830s. Like many other experimenters, he assumed that magnetism could be used to produce electricity because the reverse was true. His initial attempts failed because he used steady magnetic fields. But on 29 August 1831, using a device similar to a transformer, Faraday noted that a current

was momentarily induced in one coil as the other was connected to or disconnected from a battery (● Figure 8.52). Electromagnetic induction occurred in the coil only if the magnetic field in it was changing.

In the following months, Faraday developed a keen understanding of electromagnetic induction. He constructed primitive electric motors, generators, and transformers. In addition to these practical inventions, he made important contributions to the theoretical understanding of electricity and magnetism. He introduced the concept of "lines of force" (field lines) to model what were later called electric and magnetic fields. Faraday's successors marveled at his ability to understand sophisticated phenomena without the aid of higher mathematics. Faraday fell short of predicting the existence of electromagnetic waves.

Faraday's scientific career was interrupted by a mental breakdown in 1840—perhaps caused by mercury poisoning. Five years later, he resumed his work. Faraday was renowned as a great lecturer throughout his career. He filled lecture halls with excited patrons who paid to hear him explain the beauty of science (● Figure 8.53). Eventually, his mental problems, particularly loss of memory, forced him to resign his professorship at the Royal Institution, London. He left a monumental diary of several thousand pages detailing his experimental findings.

One of Faraday's greatest admirers was the young James Clerk Maxwell (1831–1879; ● Figure 8.54). Maxwell grew up on his family's estate in rural Scotland. He quickly showed his intellectual ability in school, and he published a mathematics paper by the age of 15. At Cambridge, he did some interesting work on the perception of color and proved mathematically that the rings around the planet Saturn could not be rigid solids. The latter established his prowess as a mathematical physicist.

Maxwell's work in electricity and magnetism began in 1855, about the time he accepted a physics professorship in Aberdeen,

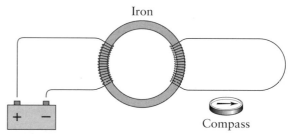

● **Figure 8.52** Apparatus used by Faraday to discover electromagnetic induction. A current is induced in the right coil only as the direct current in the left coil is either switched on or switched off. The compass is used to detect the induced current.

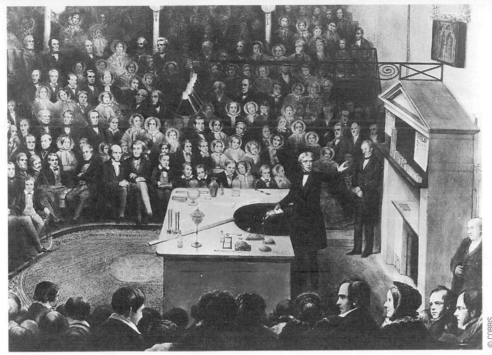

● **Figure 8.53** Michael Faraday delivering a public lecture on electricity and magnetism at the Royal Institution, London, in 1856. Faraday devoted a great deal of effort to his lectures and drew large crowds.

Scotland. He synthesized the findings of Faraday and others and provided a concrete mathematical framework for describing electrical and magnetic phenomena. He solidified Faraday's concept of fields, which was later applied to gravity as well as electricity and magnetism, and introduced an important clarification of the idea of current. He is most famous for a set of equations, called Maxwell's equations, that summarize the basics of electromagnetism. These equations are on the same level of importance in physics as Newton's laws.

Maxwell saw that his mathematical description of electricity and magnetism indicated that traveling waves of electric and magnetic fields could exist. That in itself was not particularly startling, but when he computed the speed of these waves, it turned out to be nearly equal to the measured speed of light. The conclusion was inescapable: Light was one type of "electromagnetic wave." Not only had Maxwell integrated electricity and magnetism, he had also brought light into the same realm.

Maxwell went on to make important discoveries in thermodynamics. He did not gain the level of recognition during his life that he deserved, partly because his highly mathematical writings were rather inaccessible. In addition to his theoretical discoveries, Maxwell left his mark on the way physics is done. Like Newton nearly 200 years earlier, he showed that advanced mathematics plays a vital role in physics.

Some nine years after Maxwell's death, his prediction of electromagnetic waves was proved correct. The German physicist Heinrich Hertz (1857–1894), a former assistant to Hermann von Helmholtz, produced radio waves and showed that they had properties similar to light. In the century that has followed, our technological society has filled the skies with electromagnetic waves.

● **Figure 8.54** James Clerk Maxwell in 1855, while at Cambridge University.

- Every simple magnet has a **north pole** and a **south pole,** so named because the two parts of the magnet are naturally attracted to the north and to the south, respectively.

- When two magnets are near each other, the like poles repel and the unlike poles attract.

- The magnetic field produced by a magnet causes forces on the poles of any other magnet.

- A compass is simply a small magnet that is free to rotate when in the presence of a magnetic field. It can be used to determine the direction of the magnetic field at any point in space.

- The Earth has its own magnetic field that causes compasses to point toward the north. The Earth's magnetic poles do not coincide with its geographic poles, so compasses do not point exactly north at most places on Earth.

- Many phenomena depend on the interactions between electricity and magnetism. These interactions occur only when there is some kind of change taking place, such as motion of charges or a magnet.

- The basic electromagnetic interactions can be stated in the form of three simple observations:
  (1) Moving charges produce magnetic fields.
  (2) Magnetic fields exert forces on moving charges.
  (3) Moving magnets induce currents in coils of wire (**electromagnetic induction**).
  These processes are exploited in a variety of useful devices, from electromagnets and electric motors to microphones and speakers.

- The **principles of electromagnetism** express the fundamental relationship between electricity and magnetism: "An electric current or a changing electric field induces a magnetic field," and "a changing magnetic field induces an electric field."

- These principles of electromagnetism summarize the three observations and predict the existence of **electromagnetic (EM) waves**—traveling combinations of oscillating electric and magnetic fields.

- EM waves can be classified according to frequency. From low to high, the bands are **radio waves, microwaves, infrared radiation, visible light, ultraviolet radiation, x rays,** and **gamma rays.** The different waves are involved in a diverse number of natural processes and technological applications.

- **Blackbody radiation** is a broad band of electromagnetic waves emitted by an object because of the thermal motion of atoms and molecules. The amount of radiation emitted and the intensities of different wavelengths depend on an object's temperature. This makes it possible to locate warmer objects even in the dark.

- The ozone layer in the Earth's atmosphere absorbs most of the harmful UV present in sunlight. Compounds known as CFCs drift upward and reduce the concentration of ozone.

- Carbon dioxide, water vapor, methane gas, and CFCs contribute to a **greenhouse effect** in the Earth's atmosphere. They allow sunlight to pass through the Earth's surface and to warm it while they absorb much of the IR emitted by the heated surface. The result is a warming of the atmosphere that may increase because of heightened concentrations of these gases.

- The ionosphere, a region in the upper atmosphere containing ions and free electrons, reflects lower-frequency radio waves back to the Earth. This greatly increases the range of radio communication for these lower frequencies.

- Astronomers have overcome the absorption of different bands of EM waves by the atmosphere by placing telescopes and other instruments in space.

IMPORTANT EQUATIONS

Equation	Comments
$\dfrac{V_o}{V_i} = \dfrac{N_o}{N_i}$	Input and output voltages of a transformer
$c = f\lambda$	Relates speed, wavelength, and frequency of EM waves
$\lambda_{max} = \dfrac{0.0029}{T}$	Peak wavelength of BBR curve (SI units)

MAPPING IT OUT!

1. Reread Section 8.5 on electromagnetic waves. Make a list of the main concepts introduced in this section, and then, taking *electromagnetic waves* as the organizing concept, develop a concept map that includes the basic properties of EM waves, their relationship to electric and magnetic fields, as well as the principal types of EM waves that scientists have distinguished. How could your map be integrated or connected with Concept Map 8.2 on p. 302? That is, how and where would you link your map to Concept Map 8.2?

2. As mentioned in Exercise 2, in Mapping It Out! in Chapter 7, concept maps can often take the place of more traditional note-taking when reading magazine or newspaper articles. Here is another opportunity for you to practice the advantages of this technique.

    Using the reference list at the end of this chapter as a start, locate and read *two* articles on either global warming *or* the ozone-layer

problem. After reading them, review both again, and circle what you believe to be the key concepts introduced in each. Then, using *only* the concepts you circled, construct a concept map for each that represents the major points or positions discussed in the articles. After finishing the task, consider, as you did in Chapter 7, the difficulty of this exercise. Are some important concepts missing from the articles that you had to supply from your own background or knowledge in order to make sense of the articles? Evaluate how conceptually complete the articles were. Consider how your answer to the previous question might affect your judgment of the quality of the articles. Then use your concept maps to reconsider the degree to which you might be inclined to trust the accuracy or fairness of the articles. Comment on these issues as part of your report.

(▶ Indicates a review question, which means it requires only a basic understanding of the material to answer. The others involve integrating or extending the concepts presented thus far.)

1. Three bar magnets are placed near each other along a line, end to end, on a table. The net magnetic force on the middle one is zero, and its north pole is to the left. Make two sketches showing the possible arrangements of the poles of the other magnets.
2. ▶ Sketch the shape of the magnetic field around a bar magnet.
3. ▶ What happens to a ferromagnetic material when it is placed in a magnetic field?
4. What causes magnetic declination? Is there a place where the magnetic declination is 180° (a compass points south)? If so, approximately where?
5. ▶ Describe the three basic interactions between electricity and magnetism.
6. ▶ Explain what *superconducting electromagnets* are. What advantages do they have over conventional electromagnets? What disadvantages do they have?
7. ▶ Name five different *basic* devices that use at least one of the electromagnetic interactions.
8. ▶ In many cases, the effect of an electromagnetic interaction is perpendicular to its cause. Describe two different examples that illustrate this.
9. To test whether a material is a superconductor, a scientist decides to make a ring out of the material and then to see whether a current will flow around in the ring with no steady energy input.
   a) Explain how a magnet could be used to initiate the current.
   b) At some later time, how could the scientist check to see whether the current is still flowing in the ring but without touching the ring?
10. ▶ A coil of wire has a large alternating current flowing in it. A piece of aluminum or copper placed near the coil becomes warm even if it does not touch the coil. Explain why.
11. In the particular accelerator shown in Figure 8.16, what are the possible directions of the magnetic field that keeps the particles traveling around the circular path?
12. ▶ What is the "motor-generator duality"? Explain how it is used.
13. Explain why two wires, each with a current flowing in it, exert forces on each other even when they are not touching each other.
14. ▶ Explain why a transformer doesn't work with DC.
15. What would a speaker do at the instant a low-voltage battery is connected to it?
16. If a magnetic tape or floppy disk is placed near a strong magnet, the information on it is erased. Why is that? What happens?
17. ▶ What is *analog-to-digital conversion*, and how is it used in sound reproduction?

18. ▶ Explain how electromagnetic waves are a natural outcome of the principles of electromagnetism.
19. An electromagnetic wave travels in a region of space occupied only by a free electron. Describe the resulting motion of the electron.
20. ▶ List the main types of electromagnetic waves in order of increasing frequency. Give at least one useful application for each type of wave.
21. ▶ Alternating current with a frequency of 1 million Hz flows in a wire. What in particular could be detected traveling outward from the wire?
22. ▶ What are the main uses of microwaves? Explain how each process works.
23. A liquid compound is not heated by microwaves the way water is. What can you conclude about the nature of the compound's molecules?
24. Aircraft equipped with powerful radar units are forbidden from turning them on when on the ground near people. Explain why this is so.
25. ▶ Which type of EM wave does your body emit most strongly?
26. ▶ What is different about our perceptions of the different frequencies within the visible light band of the EM spectrum?
27. A heat lamp is designed to keep food and other things warm. Would it also make a good tanning lamp? Why or why not?
28. ▶ How are x rays produced?
29. ▶ Why are x rays more strongly absorbed by bones than by muscles and other tissues?
30. ▶ What is *blackbody radiation*? How does the radiation emitted by a blackbody change as its temperature increases?
31. A light bulb manufacturer makes bulbs with different "color temperatures," meaning that the spectrum of light they emit is similar to a blackbody with that temperature. What would be the difference in appearance of the light from bulbs with color temperatures of 2,700 K and 4,000 K?
32. Explain how infrared light can be used to detect some types of animals in the dark. Can you think of a situation in which this would not work?
33. ▶ What is different about a star that appears reddish compared to one that appears bluish?
34. ▶ What effect does the ozone layer have on the EM waves from the Sun? What is currently threatening the ozone layer?
35. ▶ Describe the greenhouse effect that is occurring in the Earth's atmosphere.
36. ▶ How does the ionosphere affect the range of radio communications?

1. A radio is designed to operate on a 9-V battery or on household current. It contains a transformer that reduces 120 V AC to 9 V AC. What is the ratio of the number of turns in the output coil to the number of turns in the input coil?
2. The generator at a power plant produces AC at 24,000 V. A transformer steps this up to 345,000 V for transmission over power lines. If there are 2,000 turns of wire in the input coil of the transformer, how many turns must there be in the output coil?
3. Compute the wavelength of the carrier wave of your favorite radio station.
4. What is the wavelength of the 76-Hz ELF wave (used to communicate with submarines)?
5. Compute the frequency of an EM wave with a wavelength of 1 in. (0.0254 m).

6. The wavelength of an electromagnetic wave is measured to be 600 m.
   a) What is the frequency of the wave?
   b) What type of EM wave is it?
7. Determine the range of wavelengths in the UV radiation band.
8. A piece of iron is heated with a torch to a temperature of 900 K. How much more energy does it emit as blackbody radiation at 900 K than it does at room temperature, 300 K?
9. The filament of a light bulb goes from a temperature of about 300 K up to about 3,000 K when it is turned on. How many times more radiant energy does it emit when it is on than when it is off?
10. What is the wavelength of the peak of the blackbody radiation curve for the human body ($T = 310$ K)? What type of EM wave is this?

11. What wavelength EM wave would be emitted most strongly by matter at the temperature of the core of a nuclear explosion, about 10,000,000 K? What type of wave is this?

12. What is the frequency of the EM wave emitted most strongly by a glowing element on a stove with temperature 1,500 K?

13. The blackbody radiation emitted from a furnace peaks at a wavelength of 1.2×10^{-6} m (0.0000012 m). What is the temperature inside the furnace?

14. What is the lowest temperature that will cause a blackbody to emit radiation that peaks in the infrared?

CHALLENGES

1. The Earth's magnetic field lines are not parallel to its surface except in certain places: They actually "dip" downward at some angle to the ground. (An ordinary compass does not show this.)
   a) At what places on Earth would the "magnetic dip" be the greatest?
   b) Where on the Earth would the dip be zero?

2. A solenoid connected to a 60-Hz AC source will produce an oscillating magnetic field, as we have seen. If a permanent magnet is inserted into the solenoid, it will oscillate, but *not* with the same frequency that an unmagnetized piece of iron would. Why? (This is why the "hum" or "buzz" from electrical devices is sometimes 60-Hz sound and sometimes 120-Hz sound.)

3. The "right-hand rule" is a way to determine the direction of the magnetic field produced by moving charges. Imagine wrapping your right hand around the path of the charges so that the positive charges (or the current) flow from the "little finger" side of your fist to the thumb side (● Figure 8.55). Then your fingers circle the path in the same direction as the magnetic field lines. Use this rule to verify the directions of the magnetic fields shown in Figures 8.8 and 8.9. How would you use the rule to find the direction of the magnetic field lines around a moving negative charge?

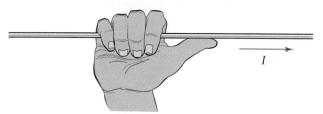

● **Figure 8.55**

4. Sketch a diagram of an atom with one electron orbiting the nucleus. Use the right-hand rule (see Challenge 3) to determine the direction of the magnetic field produced by the electron.

5. If a coil of wire is connected to a very sensitive ammeter and then waved about in the air, a current will be induced in it even if there are no magnets around. Why?

6. Even though the output voltage of a transformer can be much larger than the input voltage, the power output is nearly the same as the power input. (There is some energy loss due to ohmic heating.) Use this to determine the relationship between the input and output currents and the number of turns in the input and output coils.

7. The highest frequency sound that can be recorded by a tape recorder depends on the size of the gap in the recording head. Why? Would a wider or a narrower gap be capable of recording higher frequencies?

8. Describe what happens to an electric charge as an electromagnetic wave passes through the region around it. Explain why the charge will produce another electromagnetic wave. (This process occurs in the ionosphere.)

9. Would an electromagnetic pump be able to move a liquid superconductor? Why or why not?

10. The nuclei of carbon atoms that are found in nature come in two main "varieties" called *isotopes* (more on this in Chapter 11). The carbon-14 nuclei have a greater mass than the carbon-12 nuclei. The two types of atoms can be separated from each other by ionizing them, accelerating them all to some uniform velocity, and then passing them between the poles of a magnet. Why does this separate the two isotopes? Where does each type go? What would happen if all the atoms did not have the same speed?

SUGGESTED READINGS

"Cover Suite: X-Ray Astronomy Takes Off." *Sky and Telescope* 98, no. 2 (August 1999): 44–57. Three articles describing the dawn of a new era in high-resolution x-ray astronomy following the launch of the Chandra X-Ray Observatory.

George, Mark S. "Stimulating the Brain." *Scientific American* 289, no. 3 (September 2003): 66–69. "Activating the brain's circuitry with pulsed magnetic fields may help ease depression, enhance cognition, even fight fatigue."

Lederman, Leon M. "The Tevatron." *Scientific American* 264, no. 3 (March 1991): 48–55. A close look at the huge proton-antiproton collider near Chicago, written by a 1988 Nobel laureate.

Levi, Barbara Goss. "Nobel Chemistry Prize Gives a Stratospheric Boost to Atmospheric Scientists." *Physics Today* 48, no. 12 (December 1995): 21–22. An account of the discovery of the threat CFCs pose to atmospheric ozone.

Livingston, James D. "100 Years of Magnetic Memories." *Scientific American* 279, no. 5 (November 1998): 106–111. A history of magnetic recording, including its role in some major events of the twentieth century.

Post, Richard F. "Maglev: A New Approach." *Scientific American* 282, no. 1 (January 2000): 83–87. Describes the promise of "a safer, cheaper system for magnetically levitating trains."

Schneider, David. "Profile: Raymond V. Damadian." *Scientific American* 276, no. 6 (June 1997): 32–34. A brief account of the invention of MRI and its inventor.

Schwarzchild, Bertram. "WMAP Maps the Entire Cosmic Microwave Sky with Unprecedented Precision." *Physics Today* 56, no. 4 (April 2003): 21–24. A report on the findings of the Wilkinson Microwave Anisotropy Probe during its first year in orbit.

Smith, Chris Llewellyn. "The Large Hadron Collider." *Scientific American* 283, no. 1 (July 2000): 32–34. Describes the LHC near Geneva, Switzerland, which "will be a particle accelerator of unprecedented energy and complexity, a global collaboration to uncover an exotic new layer of reality."

Toigo, Jon William. "Avoiding a Data Crunch." *Scientific American* 282, no. 5 (May 2000): 58–74. Describes the amazing evolution of magnetic computer hard drives, future limitations to continued improvements, and possible alternative technologies.

Weart, Spencer R. "The Discovery of the Risk of Global Warming." *Physics Today* 50, no. 1 (January 1997): 34–40. A look at the 100-year history of scientific concerns about the possibility of global warming.

For additional readings, explore InfoTrac® College Edition, your online library. Go to http://www.infotrac-college.com/wadsworth and use the passcode that came on the card with your book. Try these search terms: metal detector, geomagnetism, superconducting electromagnets, Tevatron, digital sound, ozone hole, global warming, cosmic background radiation, WMAP, Hubble Space Telescope, Great Observatories.

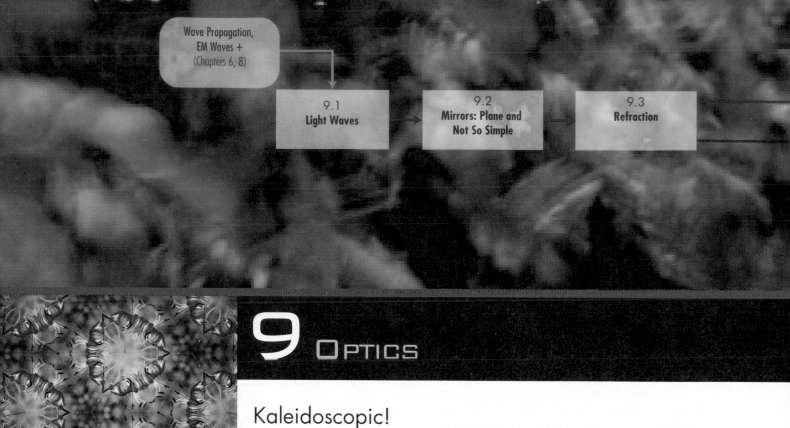

Wave Propagation, EM Waves + (Chapters 6, 8)

| 9.1 Light Waves | → | 9.2 Mirrors: Plane and Not So Simple | → | 9.3 Refraction |

9 OPTICS

Kaleidoscopic image

Royalty Free/CORBIS

The kaleidoscope was invented by Sir David Brewster in 1816. The name derives from the Greek words *kalos* meaning *beautiful*, *eîdos* meaning "form," and *skopeîn* meaning "to view" or "to see."

Kaleidoscopic!

We've all heard the expression, "It's all done with mirrors!" Often, the phrase is used pejoratively to describe an event or occurrence that seems magical if not unbelievable, a conjurer's trick, defying logic and common sense. Yet applying this descriptor to the breathtaking array of colors and shapes produced by the *kaleidoscope* would be both literally and figuratively correct.

In this instrument, the arrangement of plane mirrors produces highly symmetric displays of great complexity and apparent delicacy. If the kaleidoscope includes a chamber containing bits of colored glass or stones that may be rotated, the patterns that it is capable of creating can be a nearly endless source of amusement to children and adults. Take a close look at the chapter-opening photograph.

Can you tell how many mirrors were used to create the pattern shown? Can you say how those mirrors were oriented with respect to one another?

The wonder and fascination afforded by what has today become as much an art form as a toy can be enhanced by understanding the physical principle—the law of reflection—that underlies how it works. And by applying this principle in the study of plane mirrors, you should be able to frame answers to the questions posed above. The law of reflection and similar relationships that we investigate in this chapter provide the basis for many practical devices, like cameras, telescopes, and liquid-crystal displays, and for many of the most colorful and striking natural phenomena, like rainbows, blue skies, and soap-bubble iridescence. The subject of this chapter is *optics*, the study of light and its interaction with matter.

9.1 Light Waves

Light generally refers to the narrow band of electromagnetic (EM) waves that can be seen by human beings. These are transverse waves with frequencies from about 4×10^{14} hertz to 7.5×10^{14} hertz (see Table 8.1 on p. 312). The corresponding wavelengths are so small that we will find it useful to express them in *nanometers* (nm). One nanometer is one-billionth of a meter.

$$1 \text{ nanometer} = 10^{-9} \text{ meter} = 0.000000001 \text{ meter}$$

$$1 \text{ nm} = 10^{-9} \text{ m}$$

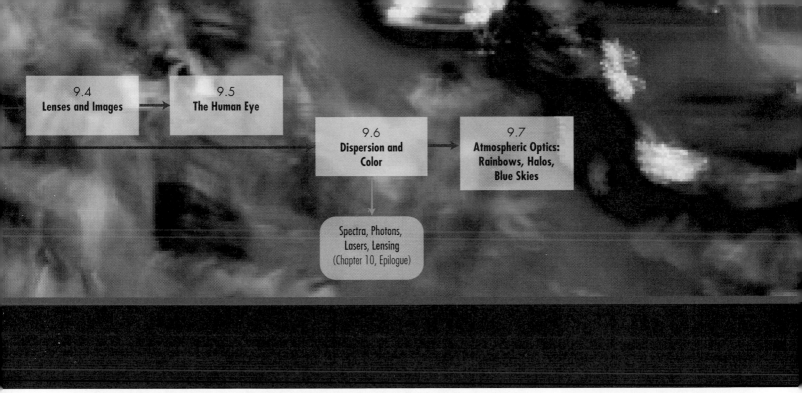

The wavelengths of visible light (in a vacuum or in air) range from about 750 nanometers for low-frequency red light to about 400 nanometers for high-frequency violet light. Two important points from Section 8.5 that you should keep in mind are (1) different frequencies of light are perceived as different colors, and (2) white light is typically a combination of all frequencies in the visible spectrum.

The various properties of waves described in the first part of Chapter 6 also apply to light. (You may find it useful to review Sections 6.1 and 6.2 at this time.) As with sound and water ripples, we will use both wavefronts and rays to represent light waves. Recall that a wavefront shows the location in space of one particular phase (peak or valley, for example) of the wave. For a light bulb, the wavefronts are spherical shells (not unlike balloons) expanding outward at the speed of light (see ● Figure 9.1). A *light ray* is a line drawn in space representing a "pencil" of light that is part of a larger beam. Rays are represented as arrows and indicate the direction the light is traveling. A laser beam can often be thought of as a single light ray. The light from a light bulb can be represented by light rays radiating outward in all directions. (Be careful not to confuse these light rays with the electric and magnetic field lines discussed in the previous chapters.)

Some of the general characteristics of wave propagation, like reflection, are readily observed with light waves. But other phenomena are more rare in everyday experience because of two factors:

1. The speed of light is extremely high (3×10^8 m/s in a vacuum).
2. The wavelengths of light are extremely short.

For example, we must turn to distant galaxies moving away from us at high speeds to easily observe the Doppler effect with light (see the Physics Potpourri, "The Hubble Relation—Expanding Our Horizons," on p. 228). As we saw in Section 6.2, the Doppler effect with sound is, on the other hand, quite common because the speed of sound is only about 350 m/s and the wavelengths of sound (in air) are in the centimeter to meter range.

In these first two sections, we will describe some of the phenomena that can occur when light encounters matter. The remaining sections of the chapter deal with important things that occur after light has traveled inside transparent material.

(a)

(b)

● **Figure 9.1** The light from a light bulb represented (a) by wavefronts and (b) by light rays.

Reflection

Reflection of light waves is extremely common: Except for light sources like the Sun and light bulbs, everything we see is reflecting light to our eyes. There are two types of reflection, specular and diffuse.

Refer to the discussion of reflection in Section 6.2.

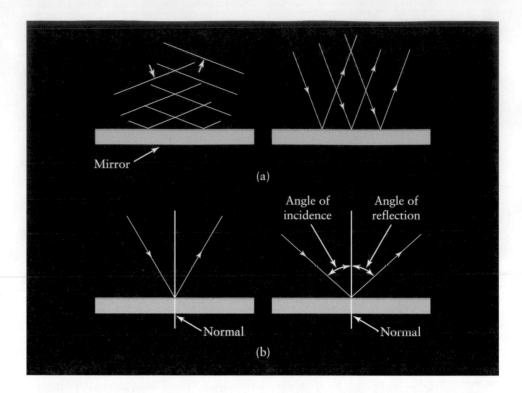

● **Figure 9.2** (a) Specular reflection using wavefronts and light rays. (b) Specular reflection of a single light ray with different angles of incidence.

Specular reflection is the familiar type that we see in a mirror or in the surface of a calm pool of water. A mirror is a very smooth, shiny surface, usually made by coating glass with a thin layer of aluminum or silver. Specular reflection occurs when the direction the light wave is traveling changes (● Figure 9.2). By changing the angle of the incident (incoming) light ray and observing the reflected ray, we see that the light behaves somewhat like a billiard ball bouncing off a cushion on a pool table.

Figure 9.2b shows an imaginary line drawn perpendicular to the mirror and touching it at the point where the incident ray strikes it. This line is called the **normal.** The angle between the incident ray and the normal is called the **angle of incidence,** and the angle between the reflected ray and the normal is called the **angle of reflection.** Our observations indicate that these angles are always equal. The following law, first described in a book entitled *Catoptrics,* thought to have been written by Euclid in the third century B.C., states this formally.

LAWS

Law of Reflection The angle of incidence equals the angle of reflection.

So specular reflection of light is much like sound echoing off a cliff, as described in Chapter 6.

● **Figure 9.3** By placing a pair of pins in front of a plane mirror and then aligning their images with another pair of pins, you can easily convince yourself of the validity of the law of reflection.

© Don Bord

Explore It Yourself 9.1

Stand one edge of a small plane mirror on a flat, horizontal surface (a piece of cardboard or foam board will do nicely) so that its reflecting side is upright (vertical). Position four straight pins so that they form an upside-down "V" with its "corner" approximately in the middle of the mirror (● Figure 9.3). Look into the mirror and move your head until you can align the pair of pins on one side with the images of the other pair in the mirror. Devise a way to use this arrangement to verify the law of reflection. That is, how might you show that the angle of incidence equals the angle of reflection by this demonstration?

The other type of reflection, **diffuse reflection,** occurs when light strikes a surface that is not smooth and polished but rough like the bottom of an aluminum pan or the surface of this paper. The light rays reflect off the random bumps and nicks in the surface and scatter in all directions (see ● Figure 9.4). The law of reflection still applies, but the rays encounter segments of the irregular surface oriented at different angles and therefore leave the surface with different directions. That is why you can shine a flashlight on the aluminum and see the reflected light from different angles around the pan. With specular reflection off a mirror, you could see the reflected light from only one direction.

Except for light sources and smooth, shiny surfaces like mirrors, every object we see is reflecting light diffusely. This diffuse reflection of ambient light results in light radiating outward from each point on a surface. You can see every point on your hand as you turn it in front of your face because each point on your skin is reflecting light in all directions.

Things can have color because light actually penetrates into the material and is partially reflected and partially absorbed along its way into and out of the material. The reflected light that leaves the surface will have color if pigments in the material absorb some frequencies (colors) more efficiently than others. A white surface, like this paper, reflects all frequencies of light. If you shine just red light on it, it will appear red. With just blue light, it will appear blue. A colored surface, like that of a red fire extinguisher, "removes" some frequencies of the light. A red surface reflects the lower-frequency light (red) most effectively and absorbs much of the rest (see ● Figure 9.5). If you shine red light on it, it will appear red. With blue or any other single color, it will appear black: Very little of the light will be reflected.

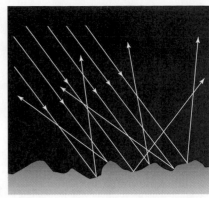

● **Figure 9.4** Diffuse reflection of light from a rough surface.

Explore It Yourself 9.2

Check out your supply of holiday wrapping materials. You're likely to find some transparent, colored-plastic gift wrap in the collection. Common hues include red, green, yellow, and blue. (If you don't have a ready supply of gift wrap, plastic food wrap now comes in several colors and may be purchased at your local grocery store.) Cover a flashlight with one color of plastic, say, red, and a second flashlight with another, say, green. Shine the two beams at the same spot on a white piece of paper. What color do you see? Try the experiment with two filters of other colors. Again, what color results from adding the two beams together? This type of color production is called *additive mixing.* Now look at a brightly lit window through one of the colored sheets, say, the red one. What do you see? Without changing your perspective, interpose a second filter, say, the green one, along your sight line. What do you see now? Is the effect different from what you saw when you combined the two flashlight beams with the same two filters? If so, how has it changed? Can you account for the differences based on your understanding of how color is produced in common objects? Try sandwiching together two of the other colored sheets. Do you find a similar result? This type of color production is called *subtractive mixing* because, typically, all but one color is subtracted (absorbed) from an incident beam of white light, leaving the residual color to be passed to the eye.

Diffraction

As with all waves, diffraction of light as it passes through a hole or slit is observable only when the width of the opening is not too much larger than the wavelength of the light.

● **Figure 9.5** A red surface reflects the red contained in white light much more effectively than it does blue, green, and other colors.

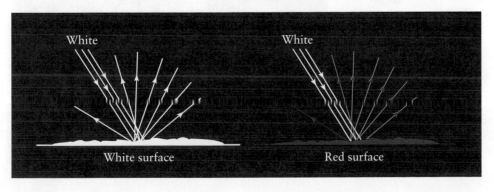

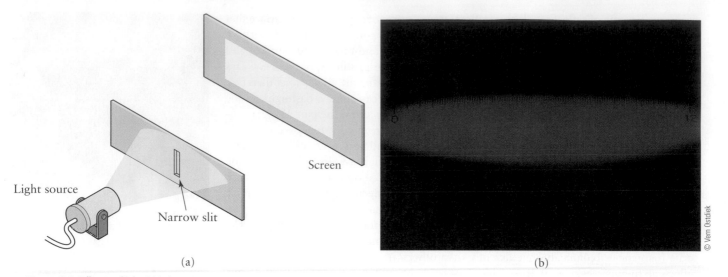

(a)

(b)

© Vern Ostdiek

● **Figure 9.6** Diffraction of light. (a) Light passing through a narrow slit spreads out, as shown on the screen. (b) Photograph of laser light projected onto a screen after passing through a slit 0.008 centimeter wide. The screen was 10 meters from the slit.

This means that light doesn't spread out after passing through a window nearly as much as sound does, but diffraction is observed when a very narrow slit (about the width of a human hair) is used (● Figure 9.6). The narrower the slit, the more the light spreads out.

Interference

In Section 6.2, we described how waves can undergo interference. Recall that when two identical waves arrive at the same place, they add together. If the two waves are "in phase"—peak matches peak—the resulting amplitude is doubled. This is called **constructive interference.** At any point where the two waves are "out of phase"—peak matches valley—they cancel each other. This is **destructive interference.**

Interference of light waves is an important phenomenon for two reasons. First, in experiments conducted around 1800, the British physician Thomas Young used interference to prove that light is indeed a wave. Second, interference is routinely used to measure the wavelength of light. We will consider two types of interference, *two-slit* interference and *thin-film* interference.

When a light wave passes through two narrow slits that are close together, the two waves emerging from the slits diffract outward and overlap. If the light consists of a single frequency (color), a screen placed behind the slits where the two light waves overlap will show a pattern of bright areas alternating with dark areas (● Figure 9.7). At each

● **Figure 9.7** Two-slit interference. (a) Light passing through two narrow slits forms an interference pattern on the screen. The bright areas occur at places where the light waves from the two slits arrive in phase. (b) Photograph of an interference pattern formed by laser light passing through two narrow slits 0.025 centimeter apart. The screen was 10 meters from the slits.

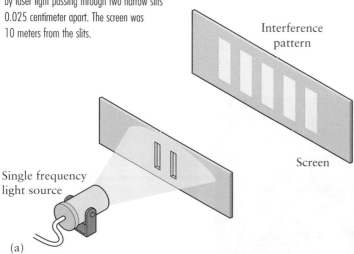

(a)

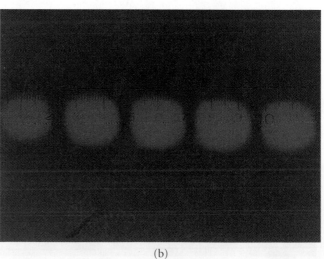

(b)

© Vern Ostdiek

bright area, the two waves from the slits are in phase and undergo constructive interference. Conversely, at each dark area the two waves are out of phase and undergo destructive interference—they cancel each other. There is a bright area at the center of this *interference pattern* because the two waves travel exactly the same distance in getting there, so they are in phase. At the first bright area to the left of center, the wave from the slit on the right has to travel a distance exactly equal to one wavelength farther than the wave from the slit on the left. This puts them in phase also. Similarly, at each successive bright area on the left side, the wave from the right slit has to travel 2, 3, 4, . . . wavelengths farther than the wave from the left slit. At each bright area on the right side of the pattern, it is the wave from the left slit that has to travel a whole number of wavelengths farther.

At the first dark area to the left of the center of the pattern, the wave from the right slit travels one-half wavelength farther than the wave from the left slit. The two waves are out of phase and interfere destructively. At the next dark area on the left, the additional distance is $1\frac{1}{2}$ wavelengths, then $2\frac{1}{2}$ wavelengths at the next, and so on. True constructive and destructive interference actually occurs only at the centers of the bright and dark areas. At points in between, the waves are neither exactly in phase nor exactly out of phase, so they partially reinforce or partially cancel each other.

The distance between two adjacent bright or dark areas is determined by the distance between the two slits, the distance between the screen and the slits, and the wavelength of the light. Since the first two can be measured easily, their values can be used to compute the wavelength of the light. (See Challenge exercise 2 on p. 386 for more details.)

The swirling colors you see in oil or gasoline on wet pavement are caused by *thin-film interference.* Part of the light striking a thin film of oil reflects off it, and part of it passes through to be reflected off the water (● Figure 9.8). The light wave that passes through the film before being reflected travels a greater distance than the wave that reflects off the upper surface of oil. If the two waves emerge in step, there is constructive interference. If they emerge out of step, there is destructive interference.

The wavelength of the light, the thickness of the film, and the angle at which the light strikes the film combine to determine whether the interference is constructive, destructive, or in between. With single-color (one wavelength) light, one would see bright areas and dark areas at various places on the film. With white light, one sees different colors at different places on the film. At some places, the film thickness and angle of incidence will cause constructive interference for the wavelength of red light, at other places for the wavelength of green light, and so on.

Interference in thin films in hummingbird and peacock feathers is the cause of their iridescent colors. Soap bubbles are also colored by interference of light reflecting off the two surfaces of the soap film (● Figure 9.9).

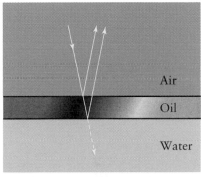

● **Figure 9.8** Interference of light striking a thin film of oil. The light reflecting from the upper surface of the film undergoes interference with the light that passes through to, and is reflected by, the lower surface. Part of the light continues on through the lower surface.

● **Figure 9.9** Examples of colors generated by thin-film interference.

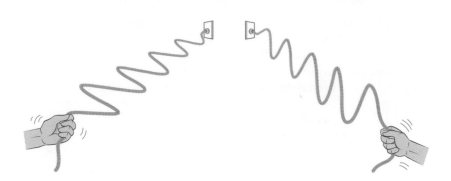

Polarization

The fact that light could undergo diffraction and interference convinced Young and other scientists of his time that light can behave like a wave. The other model of light elaborated by Newton (see Physics Family Album, p. 380), which suggested that light is a stream of tiny particles, could not account for these distinctively wavelike phenomena. **Polarization** reveals that light is a *transverse* wave rather than a longitudinal wave.

A rope secured at one end can be used to demonstrate polarization. If you pull the free end tight and move it up and down, a wave travels on the rope that is *vertically polarized*. Each part of the rope oscillates in a vertical plane (● Figure 9.10). In a similar manner, moving the free end horizontally produces a *horizontally polarized* wave on the rope. Moving the free end at any other angle with the vertical will also produce a polarized wave. Polarization is possible only with transverse waves.

Explore It Yourself 9.3

Go to the sunglasses display in a department or drug store and choose a polarized pair. Examine an overhead light fixture through one lens of the glasses. Rotate the lens as you look at the light. Does the intensity (brightness) of the source change? Now find a place where bright light is reflected off the top of a display case or a highly polished floor. Again view the light through one lens of the glasses, and rotate the lens through at least 90°. Is there any change in the image brightness now? Can you relate the image intensity to the angle of rotation of the lens? Pick up another pair of polarized sunglasses. Arrange them so that light passes through the right lens of one and the left lens of the other before reaching your eye. Rotate one pair with respect to the other. What happens? Can you relate what you see now to what you saw with the reflected light?

Light, being a transverse wave, can be polarized. A *Polaroid filter*, like the lenses of Polaroid sunglasses, absorbs light passing through it unless the light is polarized in a particular direction. This direction is coincident with the *transmission axis* of the filter. Light polarized in this direction passes through the Polaroid largely unaffected, light polarized perpendicular to this direction is blocked (absorbed), and light polarized in some direction in between is partially absorbed (● Figure 9.11).

The light that we get directly from the Sun and from light fixtures is a mixture of light waves polarized in all different directions. The light is said to be "natural" or "unpolarized" because it has no preferred plane of vibration. When natural light encounters a Polaroid filter, it emerges polarized along the transmission axis (see ● Figure 9.12). The filter allows only those parts of the light wave that are polarized in this direction to pass through; the rest are absorbed.

Now, if this light encounters a second Polaroid filter, the amount of light that emerges will depend on the orientation of the transmission axis of the second filter. If the axis of the second filter is aligned with that of the first, the light will pass through. If the axis of the second is perpendicular to that of the first, all of the light will be blocked by the sec-

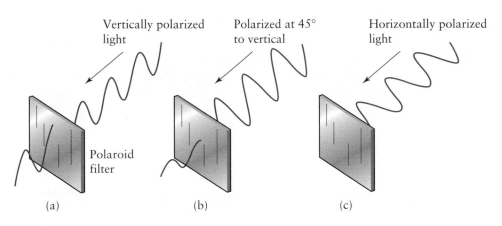

Vertically polarized light

Polarized at 45° to vertical

Horizontally polarized light

Polaroid filter

(a)   (b)   (c)

ond filter. This is referred to as "crossed Polaroids" (● Figure 9.13). When the angle between the transmission axes of the two filters is other than zero or 90°, some of the light will pass through, with the intensity becoming progressively less for angles closer to 90°.

Polaroid sunglasses are very useful because light is polarized to some extent when it reflects off a surface like water, asphalt, or the paint on the hood of a car. In particular, the reflected sunlight is partially polarized horizontally (● Figure 9.14). This reflected light, called *glare*, is usually bright and annoying. Sunglasses using Polaroid lenses with their transmission axes vertical will block most of this reflected light, which makes it easier to see the surface itself.

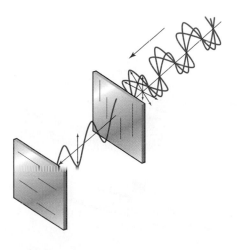

Explore It Yourself 9.4

Grab those polarizing sunglasses again, and this time examine a liquid-crystal display on a watch or a calculator or the gas pump the next time you fill 'er up. Rotate the polarizing lens as you did in the previous Explore It Yourself exercise. Describe what happens. What does this tell you about the light emanating from the LCD? Try the same experiment with light from the sky on a clear sunny day. When the Sun is near the horizon, look through the polarizing sunglasses at the blue sky in directions north or south (approximately 90° away from the Sun). Rotate the glasses as before. What do you see? Can you draw any conclusions about the nature of the scattered sunlight in these directions? (See Challenge exercise 12 on p. 387 for more.)

Liquid-crystal displays (LCDs) used in calculators, digital watches, laptop computers, and video games also use polarization. The liquid crystal is sandwiched between crossed Polaroids, and this assembly is placed in front of a mirror. Without the liquid crystal present, the display would be dark: Light passing through the first Polaroid is polarized

Figure 9.14 Light reflected off water is partially polarized horizontally. Polaroid sunglasses with their transmission axes vertical block this glare.

vertically and would be totally absorbed when it reached the second Polaroid. No light would reach the mirror, so none would be reflected back from the display. The specially chosen liquid crystal material between the Polaroids is arranged so that in its normal state its molecules change the polarization of the light passing through it. The liquid crystal twists the polarized light 90°. The polarization is changed from vertical to horizontal for light passing through from the front and from horizontal to vertical for light passing through from the rear. The light that was vertically polarized by the first Polaroid is made horizontally polarized by the liquid crystal, passes through the second Polaroid, is reflected back through the display, and emerges polarized vertically (see ● Figure 9.15a).

To make images on the display, segments of it are darkened. Here is where the liquid crystal property is used. An electric field is switched on in the parts of the display to be darkened. This causes the molecules of the liquid crystal to rotate and to become aligned in one direction. As a result, they no longer change the polarization of light passing through. Light going through this segment is absorbed by the second Polaroid, and that

Figure 9.15 Expanded view of part of a liquid-crystal display. (a) The liquid crystal changes the polarization of the light so that it can make a round trip through the crossed Polaroids. (b) The voltage neutralizes this part of the liquid crystal so the light is blocked by the second Polaroid. This part of the display is darkened.

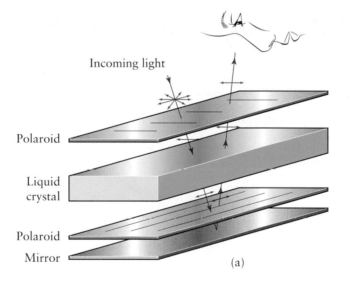

Incoming light

Polaroid

Liquid crystal

Polaroid

Mirror

(a)

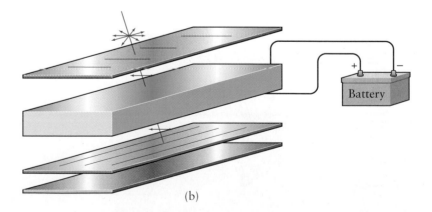

Battery

(b)

part of the display is dark (see Figure 9.15b). By the way, the transmission axis of the first Polaroid is oriented vertically so that you can see the display while wearing Polaroid sunglasses. If you rotate an LCD and look at it while wearing Polaroid sunglasses, at one point all of it will be dark.

It has been discovered that some color-blind animals such as squid can sense the polarization of light. This helps squid see plankton, which are transparent.

LEARNING CHECK

1. (True or False.) According to the law of reflection, if the angle of incidence of a ray striking a mirrored surface is 35°, then the angle of reflection of the ray leaving the surface is 55°.
2. The two types of reflection are _____ and _____.
3. Two otherwise identical light waves arriving at the same point with peak matching valley undergo
   a) constructive interference.
   b) destructive interference.
   c) polarization.
   d) reflection.
   e) None of the above.
4. Light is totally absorbed when sent through two Polaroid filters if their respective transmission axes are _____.

ANSWERS: 1. False **2.** specular, diffuse **3.** (b) **4.** perpendicular

9.2 Mirrors: Plane and Not So Simple

Most mirrors that we use are *plane* mirrors: They are flat and almost perfect reflectors of light. When we use a mirror to "see ourselves," light that is diffusely reflected off our clothes and face strikes the mirror and undergoes specular reflection. Some of the rays leaving the mirror are going in the proper direction to enter our eyes and to give us an image of ourselves. The image appears to be on the other side of the mirror. (We are so accustomed to this that we don't think about it. But imagine your reaction if you had never seen a mirror image before.) ● Figure 9.16 shows a person viewing his image in a plane mirror. Instead of showing every light ray traveling outward from every point on the person, we show selected rays that happen to enter the person's eyes. The dashed lines from the image show the apparent paths taken by the rays when traced back to the image.

A number of important, practical devices employ plane mirrors. They are used as "optical levers" to amplify small rotations in specialized laboratory instruments. For example, as the mirror rotates through an angle θ, the reflected beam will be turned through an angle 2θ. A plane mirror in a single-lens reflex camera (● Figure 9.17) is used to redirect light from the lens to the viewfinder. When the shutter is pressed, the mirror tilts up allowing the light to reach the film. More recently, *micromirrors*, small enough to fit through the eye of needle (0.5 mm or less), have become indispensable parts of modern telecommunications networks, where they are used in high-speed optical switches to control the flow of information through optical fibers (see Section 9.3).

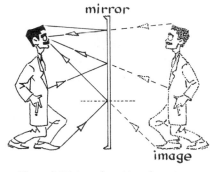

● **Figure 9.16** Image formed by a plane mirror. The rays reaching the viewer's eyes form an image that appears to be on the other side of the mirror. (The normal for the lower ray is shown as a dashed line to illustrate that the law of reflection applies.)

Explore It Yourself 9.5

Hold up your right hand in front of a plane mirror. What do you see? (Be cautious now! Describe the image carefully.) If you're having trouble, place a tube of toothpaste or other object with writing on it in front of the mirror. Now what can you say? The process you're witnessing is called *inversion* and is one of the common characteristics of plane mirrors. Now move your hand slowly toward the mirror with your index finger extended in the direction of the mirror's surface. What happens to the image of your hand in the mirror? Continue advancing your hand. When your finger is just about to touch the mirror, where is the image of your finger in the mirror? What conclusions can you draw about the distances of objects placed in front of a plane mirror relative to those of their images?

Figure

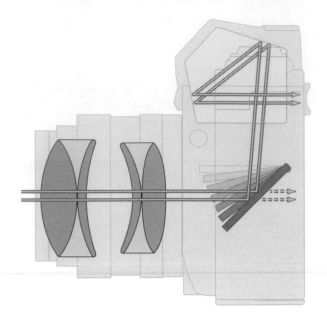

Figure 9.17 A single-lens reflex (SLR) camera. Light enters through the lens, strikes the mirror, travels to the prism and then to the photographer's eye. When the shutter is released, the mirror pops up, allowing light to pass directly to the film. The mirror then drops down.

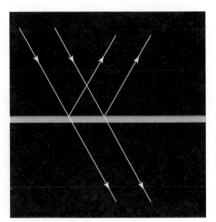

Figure 9.18 Light striking a half-silvered mirror. Part of the light is reflected, and part passes through.

"One-Way Mirror"

A "one-way mirror" is made by partially coating glass so that it reflects some of the light and allows the rest to pass through. This is called a *half-silvered mirror* (● Figure 9.18). When used as a window or wall between two rooms, it will function as a one-way mirror if one of the rooms is brightly lit and the other is dim. It will appear to be an ordinary mirror to anyone in the bright room, but it will appear to be a window to anyone in the dim room. This is because, in the bright room, the light reflected off the half-silvered mirror is much more intense than the light that passes through from the other room. In the dim room, the transmitted light from the bright room dominates (see ● Figure 9.19).

A person in the dim room can see what is happening in the bright room without being seen by anyone in the bright room (see ● Figure 9.20). This device is often used in interview and interrogation rooms and as a means of observing customers in stores and gambling casinos. Note that if a bright light is turned on in the dimmer room, the one-way effect is destroyed. Ordinary window glass is a crude one-way mirror because it does reflect some of the light that strikes it. At night, one can see into a brightly lit room through a window, but anyone in the room has difficulty seeing out.

Figure 9.19 A half-silvered mirror used as a "one-way mirror" between two rooms. In the room on the left, a person sees mostly reflected light (a mirror). In the room on the right, a person sees mostly transmitted light (a window).

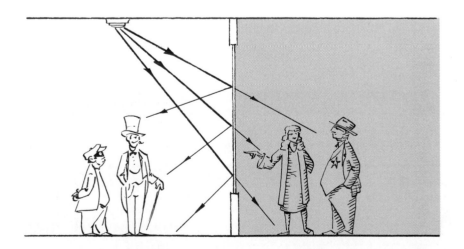

Courtesy of Libby-Owens-Ford

Courtesy of Libby-Owens-Ford

● **Figure 9.20** (Left) In a well-lit room, the half-silvered mirror appears to be an ordinary mirror. (Right) In a dark room on the other side, the half-silvered mirror appears to be a window.

Explore It Yourself 9.6

Stand a small, rectangular plane mirror vertically on a table. Now place another similar plane mirror adjacent to the first with one of its vertical edges running alongside that of the other. Adjust the angle between the mirrors to be about 45°. Place a small object (a coin or a die will do nicely) midway between the mirrors. How many images of the object do you see reflected in the mirrors? Make the angle between the mirrors roughly 60°. How many reflected images do you see now? Set the mirror angle to 30°. Count the number of images in this case. Do you see a pattern developing between the total number of objects (actual plus images) arrayed around the cluster and the angle separating the mirrors? How do you think your observations might be applied to the construction of a kaleidoscope?

Curved Mirrors

As we saw in Section 6.2, reflectors—mirrors in this case—that are curved have useful properties. Parallel light rays that reflect off a properly shaped *concave mirror*—a mirror that is curved inward—are focused at a point called the **focal point** (● Figure 9.21a). The energy in the light is concentrated at that point. Sunlight focused by a concave mirror can heat things to very high temperatures (Figure 9.21b). Even when a mirror's surface is curved, the law of reflection still holds at each point that a ray strikes the mirror. If a normal line is drawn at each point (as was done in Figure 9.2b), the angle of incidence equals the angle of reflection. One such normal line is shown in Figure 9.21a.

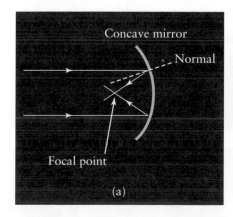

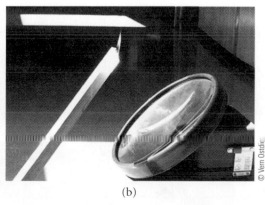

© Vern Ostdiek

● **Figure 9.21** (a) Parallel rays reflecting off a concave mirror converge on a point called the *focal point*. (The normal for the upper ray shows that the law of reflection applies.) (b) Sunlight focused by a concave mirror can generate high temperatures. This piece of wood was in flames a few seconds after being placed at the mirror's focal point.

(a)

(b)

(c)

● **Figure 9.22** (a) Plane mirror image.
(b) Enlarged image in a concave mirror. (c) Reduced image in a convex mirror.

A concave mirror can be used to form images that are enlarged—magnified. Magnifying makeup mirrors are concave mirrors, as are the large mirrors used in astronomical telescopes. ● Figure 9.22b shows a magnified image seen in a concave mirror.

A *convex mirror* is one that is curved outward. The image formed by a convex mirror is *reduced*—it is smaller than the image formed by a plane mirror (Figure 9.22c). The advantage of a convex mirror is that it has a wide field of view—images of things spread over a wide area can be viewed in it. ● Figure 9.23 shows the fields of view for a convex mirror and a plane mirror of the same size. One glance at a well-placed convex mirror on a bike path allows quick surveillance of a large area (● Figure 9.24). Passenger-side rearview mirrors on cars and auxiliary "wide-angle" rearview mirrors on trucks and other vehicles are convex so the driver can view a large region to the rear. Care must be taken when using such a mirror because the reduced image makes any object appear to be farther away than it actually is.

Explore It Yourself 9.7

A shiny metal spoon can serve as either a concave mirror or a convex mirror, although its shape is usually far from ideal. The front side of the spoon is a concave mirror. Place your finger close to the spoon, and look at its image in the spoon. Describe what you see. Now hold the same side of the spoon a comfortable distance in front of your face. What is different about the image of your face that you see? The back side of a spoon is a convex mirror. Describe the image of your face and surrounding environment seen in this side of the spoon.

● **Figure 9.23** The field of view of a convex mirror is much larger than that of a plane mirror.

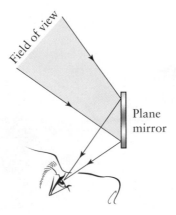

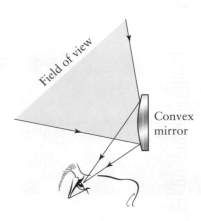

Astronomical Telescope Mirrors

The largest telescopes used by astronomers to examine stars, galaxies, and other celestial objects make use of curved mirrors. ● Figure 9.25 shows a common design for such telescopes. Light from the distant source enters the telescope and reflects off a large concave mirror called the *primary mirror*. The reflected rays converge onto a much smaller convex mirror called the *secondary mirror*. The rays are reflected back toward the primary mirror, pass through a hole in its center, and converge to form an image at the focal point *F*. The primary mirror is the key component of the telescope.

Telescope mirrors have as their basic functions the gathering of light and the concentration of that light to a point. The ability of a mirror to collect light increases with its surface area. To acquire enough radiation to study faint objects adequately, astronomers have sought to build instruments with larger and larger apertures (openings) and, hence, larger light-collecting areas.

The quality of the images produced by telescopes is greatly affected by the shapes of the mirrors. The easiest curved mirror to make is one whose surface has the shape of part of a sphere. But such a *spherical mirror* is not perfect for the task of focusing light rays. ● Figure 9.26a shows that parallel light rays reflecting off a spherical mirror are not all focused at the same point. An image formed using such a mirror will be somewhat blurred. This phenomenon is called **spherical aberration.** We will see in Section 9.4 that the same thing happens with lenses.

As the name implies, spherical aberration is a defect associated with spherical surfaces. A concave mirror in the shape of a parabola does not have this aberration. (You may recall that we saw the parabola in Section 2.6.) A *parabolic mirror* will concentrate all the rays coming from a distant source at the same point (Figure 9.26b). Thus, the ideal surface for a telescope mirror (or for that matter, reflectors in auto headlamps and household flashlights) is one shaped like a parabola. Fabricating very large mirrors, some as big as 8 meters (26 feet) in diameter, with the precise parabolic shape is an enormous technical challenge. One technique called *spin casting* exploits the fact that the surface of a liquid rotating at a steady rate has the required parabolic shape.

● **Figure 9.24** This convex mirror on a bike path at the University of Minnesota allows quick surveillance of a large area that would normally be hidden from view.

Explore It Yourself 9.8

If you have a phonograph turntable, you can make your own spinning parabolic mirror. Place a lightweight soup bowl—dark in color if possible—on the turntable so that *its center matches that of the platter.* (You may have to remove the short spindle from the center or place a couple of useless records on first so that the spindle doesn't hold up the center of the bowl.) Carefully pour cooking oil into the bowl until it is about half full (water works, but it sloshes a lot). Turn on the turntable, and use the $33\frac{1}{3}$ revolutions per minute speed. If there is enough light in the room, place your hand about a foot above the bowl and to one side. Place your head above the bowl and off to the other side, and look for the image of your hand in the oil's surface. How does it appear? If the room is not well lit, replace your hand with a dim lamp or a low-wattage frosted light bulb. (Be *very careful* when using a lamp, because its magnified image can be very bright.) Move your hand or the lamp up and down. How does the image change? Turn the speed up to 45 revolutions per minute. What is different? Watch the image as you turn off the turntable and it slows to a stop. Describe what you see.

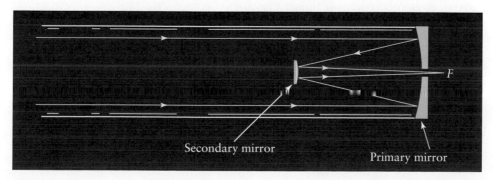

● **Figure 9.25** This is a basic design of a large astronomical telescope. The large, concave primary mirror and the small, convex secondary mirror combine to focus incoming light at the focal point *F*.

Secondary mirror

Primary mirror

● **Figure 9.26** (a) A spherical mirror does not reflect all incoming rays from a distant source to the same point. Rays located well off the symmetry axis of the mirror are brought to a focus closer to the mirror than those rays nearer the axis. This effect is called *spherical aberration.* (b) A parabolic mirror focuses all the light from a distant source to a single point. It is the ideal shape for a telescope mirror.

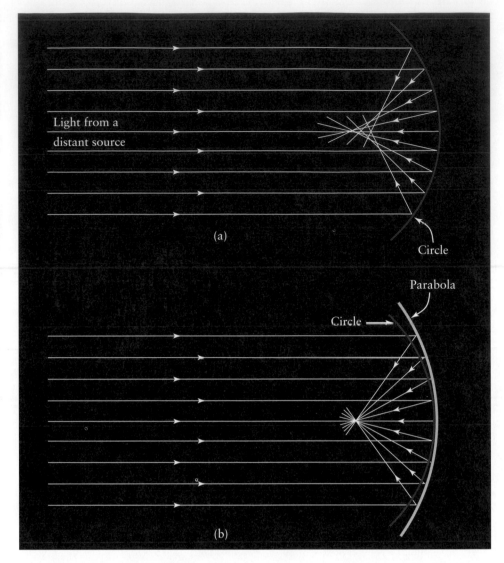

Light from a distant source

(a)

Circle

Parabola

Circle →

(b)

Beginning in the late 1980s, several telescopes were built in Canada that employ rotating liquid mirrors. A bowl of mercury typically more than 1 meter in diameter is spun at the proper rate to give its surface the desired parabolic shape. Although such mirrors can't be tilted and can therefore be used only to look at points very near the zenith, they are useful for certain types of astronomical and atmospheric research.

Shortly after the Hubble Space Telescope (HST; see Figure 8.49) was placed in orbit on 25 April 1990, scientists discovered that its primary mirror was afflicted with a type of spherical aberration. At the edge of the 2.4-meter-diameter mirror, its surface is 0.002 millimeters off of what it is supposed to be. This seemingly minuscule error in the mirror's shape drastically reduced the telescope's ability to form sharp images. In December 1993, space shuttle astronauts installed corrective optics on the instrument platform of the Space Telescope to correct this problem and allow the observatory to perform as designed. ● Figure 9.27 shows the dramatic improvement in the ability of the Space Telescope to resolve fine detail on celestial objects as a result of the repairs. Further improvements in the performance of HST have occurred with the installation of the Advanced Camera for Surveys (ACS) by Space Shuttle astronauts in the spring of 2002. The ACS replaces the nearly 12-year-old Faint Object Camera and, with its higher resolution, larger field of view, and greater light collection efficiency, should provide about a factor-of-ten improvement in the capability of the Space Telescope.

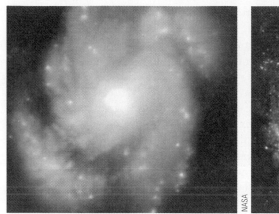

● **Figure 9.27** Hubble Space Telescope images of the core of the galaxy M100 taken before (left) and after (right) the installation of corrective optics in December 1993.

From the ground, recent efforts to combat the blurring effects of the Earth's atmosphere (see the discussion on p. 325) and thereby to increase the resolution of optical instruments, have involved the use of "adaptive optics" (AO). Here, elaborate wavefront sensors, fast computers, and deformable mirrors are used to produce images so sharp that it's as though the atmosphere had disappeared entirely. ● Figure 9.28a shows a typical design of an AO system, similar to those in use with the two 10-m Keck Telescopes in Hawaii, the 8.2-m Gemini North (Hawaii) and South (Chile) instruments, and the Very Large Telescope array (Chile). The key optical element is the "rubber mirror"—a thin glass mirror whose surface can be slightly deformed by up to a hundred tiny actuators attached

● **Figure 9.28** (a) An adaptive optics (AO) system. The distorted wavefronts Σ_1 are analyzed and reshaped by the use of a deformable "rubber" mirror. The corrected wavefronts Σ_2 are transmitted to the scientific instruments and to a wavefront sensor that provides feedback for controlling the deformable mirror. (b) Laser used in conjunction with a telescope equipped with adaptive optics to compensate for atmospheric turbulence.

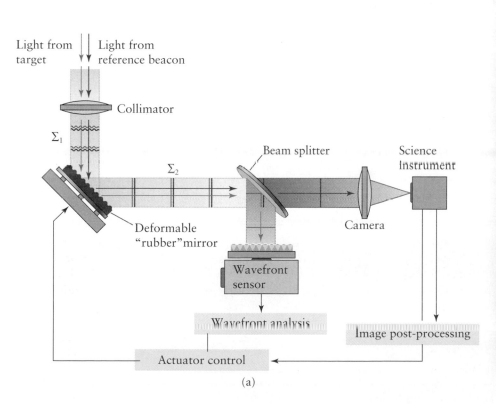

Light from target | Light from reference beacon

Collimator

Σ_1

Σ_2

Beam splitter

Science Instrument

Deformable "rubber" mirror

Camera

Wavefront sensor

Wavefront analysis

Image post-processing

Actuator control

(a)

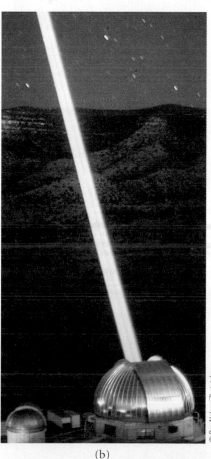

(b)

to the back. If the tiny deformations in the rubber mirror can be made to counteract the deformations in the wavefronts from distant sources caused by turbulence in the Earth's atmosphere, then the original wavefronts can be restored and the blurring in the image removed. Fast computers are required to monitor continuously the rapidly changing conditions in the atmosphere, compute the necessary corrections, and signal the accuators to move.

This technique normally uses a bright star in the field of view of the telescope as the reference beacon for the wavefront assessment. In the absence of such a star, an artificial one can be created using a high-power, highly focused laser beam projected up through the telescope (Figure 9.28b). Called "laser guidestars," the light from the beam that is scattered downward either from air molecules at altitudes of 10 to 40 km or from sodium atoms residing as high as 90 km produces a faint but measurable target for the wavefront sensors. As indispensable as AO systems are for optimizing the light-gathering and resolving power of large, 10-m class telescopes, still greater advances will be needed in this field to harness the full capabilities of the *really* large telescopes with 20-m apertures planned for the future. To meet the challenges, new initiatives described as "atmospheric tomography"—similar to "medical tomography" where a 3–D view of the system is produced—will be required.

Concept Map 9.1 summarizes the types and uses of mirrors.

CONCEPT MAP 9.1

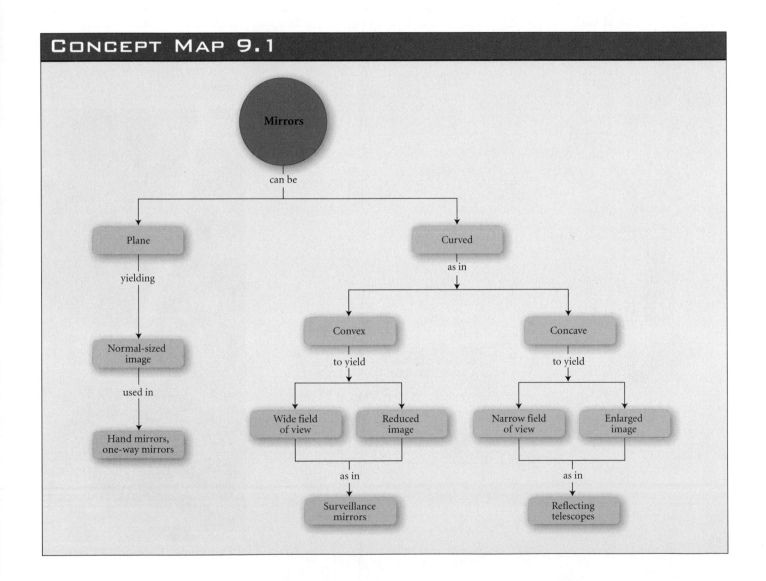

1. An object is placed 20 cm in front of a plane mirror. The image of the object formed by the mirror is located
   a) 10 cm in front of the mirror.
   b) 10 cm behind the mirror's surface.
   c) at the mirror's surface.
   d) 20 cm in front of the mirror.
   e) 20 cm behind the mirror's surface.
2. Parallel light rays from a distant object strike the surface of a concave mirror. After leaving the mirror's surface, the rays will converge toward the _____ of the mirror.

3. (True or False.) A curved mirror that produces an enlarged image will also have a greater field of view than a plane mirror.
4. The primary mirror in an astronomical telescope must be a
   a) half-silvered mirror.
   b) convex mirror.
   c) concave mirror.
   d) plane mirror.
   e) hexagonal mirror.

ANSWERS: 1. (e)   2. focal point   3. False   4. (c)

9.3 Refraction

In this section, we describe what happens when light interacts with glass and other transparent material. Imagine a light ray from some source in air arriving at the surface of a second transparent substance like glass. The boundary between the air and glass is called the *interface*. As mentioned earlier, some of the incident light is reflected back into the air, but the rest of it is transmitted across the interface into the glass (see ● Figure 9.29). The law of reflection gives us the direction of the reflected light ray, but what about the transmitted light ray?

Explore It Yourself 9.9

Fill a large, clear glass beaker or other similar vessel (a clear glass coffeepot works well) about halfway with water. Place a pencil or other long cylindrical object into the container so that it makes an oblique angle with the liquid surface. Look through the vessel from a direction perpendicular to the plane containing the pencil. Keep your eyes about level with the air-water interface. Describe the appearance of the object, paying particular attention to its shape and size in air and in the water.

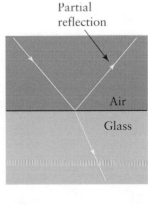

(a)

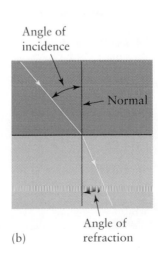

(b)

● **Figure 9.29** (a) Refraction of light as it enters glass, with partial reflection shown. (b) The normal and angles of incidence and refraction. Note that the ray is bent toward the normal.

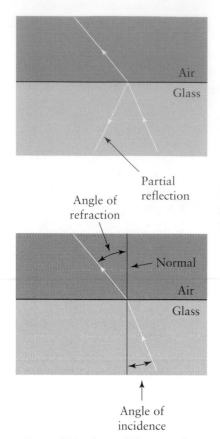

Partial reflection

Angle of refraction

Normal

Air

Glass

Angle of incidence

● **Figure 9.30** Refraction of light as it goes from glass into air. The path of the refracted ray is the reverse of the path shown in Figure 9.29.

The light that passes into the glass is *refracted*—the transmitted ray is bent into a different direction than the incident ray. (If the incident ray is perpendicular to the interface, there is no bending.) This bending is caused by the fact that light travels slower in glass than in air. We again draw in the normal, a line perpendicular to the interface. The angle between the transmitted ray and the normal, called the **angle of refraction,** is smaller than the angle of incidence. We can also have the reverse process: A light ray traveling in glass reaches the interface and is transmitted into air (● Figure 9.30). In this case, the angle of refraction is larger than the angle of incidence. The following law summarizes these observations.

LAWS

Law of Refraction A light ray is bent toward the normal when it enters a transparent medium in which light travels slower. It is bent away from the normal when it enters a medium in which light travels faster.

Notice the symmetry in Figures 9.29 and 9.30. This is one example of the *principle of reversibility:* The path of a light ray through a refracting surface is reversible. The path that a ray takes when going from air into glass is the same path it would take if it turned completely around and went from glass into air. In all of the figures in this and the following sections, the arrows on the light rays could be reversed and the paths would be the same.

Refraction of light affects how you see things that are in a different medium, such as under water. ● Figure 9.31a shows a coin at the bottom of an empty glass; Figure 9.31b shows the same coin at the bottom of a glass full of water. Note the difference in the sizes of the coin in the two photos. The coin in Figure 9.31b appears larger because the image of the coin formed in water is closer to the observer than the image formed in air. ● Figure 9.32 displays a ray diagram for this circumstance revealing how the bending of the rays of light from the edges of the coin as they enter the air from the water causes the coin to appear to be closer to the surface than it is.

The speed of light in any transparent material is less than the speed of light c in a vacuum. For example: $c = 3 \times 10^8$ m/s, while the speed of light in water is 2.25×10^8 m/s, and the speed of light in diamond is only 1.24×10^8 m/s. ● Table 9.1 gives the speed of yellow light in selected transparent media.* (Critical angle and the variation of speed with color will be discussed later.)

---

* For the mathematically inclined: Each transparent material has an *index of refraction, n,* given by the equation

$$n = \frac{c}{v} \quad (v = \text{speed of light in the material})$$

Using the information in Table 9.1, the index of refraction of water is 1.33 and that of diamond is 2.42. The exact mathematical relationship between the angles of incidence and refraction is:

$$n_1 \sin A_1 = n_2 \sin A_2 \quad (\text{Snell's law; see p. 379})$$

where n_1 and A_1 are the index of refraction and the angle between the ray and the normal, respectively, in medium 1, and n_2 and A_2 are the corresponding quantities for medium 2. Looking ahead to Figure 9.36b, for example, $n_1 = 1.00$ (air), $A_1 = 14.7°$ (angle of refraction), $n_2 = 1.46$ (glass), and $A_2 = 10°$ (angle of incidence).

● **Figure 9.31** A coin underwater (b) appears closer and larger to an observer looking down from above than does an identical one in air (a).

© David Rogers

(a)

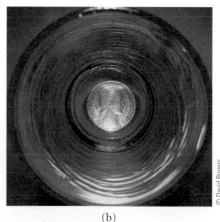

© David Rogers

(b)

● Table 9.1 Speed of (Yellow) Light and Critical Angles for Selected Materials

Phase	Substance	Speed	Critical Angle[a]
Gases at 0°C and 1 atm	Carbon dioxide	2.9966×10^8 m/s	
	Air	2.9970	
	Hydrogen	2.9975	
	Helium	2.9978	
Liquids at 20°C	Benzene	1.997×10^8 m/s	41.8°
	Ethyl alcohol	2.203	47.3°
	Water	2.249	48.6°
Solids at 20°C	Diamond	1.239×10^8 m/s	24.4°
	Glass (dense flint)	1.81	37.0°
	Glass (light flint)	1.90	39.3°
	Glass (crown)	1.97	41.1°
	Table salt	2.00	41.8°
	Fused silica	2.05	43.2°
	Ice	2.29	49.8°

[a] Critical angles are for air as the second medium.

At this point, you may have two questions in mind. Why is the speed of light lower in transparent media such as water and glass, and why does this cause refraction? To answer the first question, we must remember that light is an electromagnetic wave—a traveling combination of oscillating electric and magnetic fields. As light enters glass, the oscillating electric field in the wave causes electrons in the atoms of the glass to oscillate at the same frequency. These oscillating electrons emit EM waves (light) that travel outward to neighboring electrons. However, the wave emitted by each electron is not exactly "in step" with the incident wave that is causing the electron to oscillate: The emitted wave lags the incident wave slightly. As the light wave is relayed from electron to electron, these phase lags accumulate, effectively reducing the speed with which the light propagates through the medium.

As to the second question, we can see why a change in the speed of light causes bending by carefully following wavefronts as they cross an interface. A good analogy to this is a marching band, lined up in rows, crossing (at an angle) the boundary between dry ground and a muddy area (see ● Figure 9.33). Each person in the band is slowed as his or her feet slip and sink in the mud. To remain aligned and properly spaced, each person has to turn a bit, and the whole band ends up marching in a slightly different direction. The slowing of the wavefronts of light entering glass produces the same kind of change in the direction of propagation of the light. Note that the reduction in speed also makes the wavelength *shorter*. This has to be the case because the frequency of the wave remains the same (why?) and we must have $v = f\lambda$.

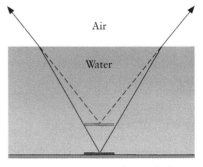

● **Figure 9.32** Ray diagram showing why the object in Figure 9.31b appears closer to the observer when seen through water. The object looks larger because its image lies closer to the observer. (From *Physics: A World View*, 5e, by Kirkpatrick and Francis. Used by permission.)

● **Figure 9.33** (a) Wavefront representation of light refracting as it enters glass. The change in direction occurs because the light travels slower in the glass. (b) For comparison, rows of a marching band alter their direction as they enter muddy ground.

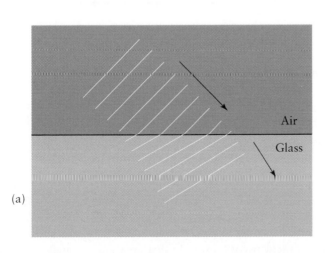

(a)

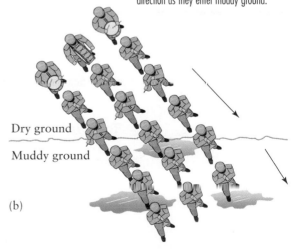

(b)

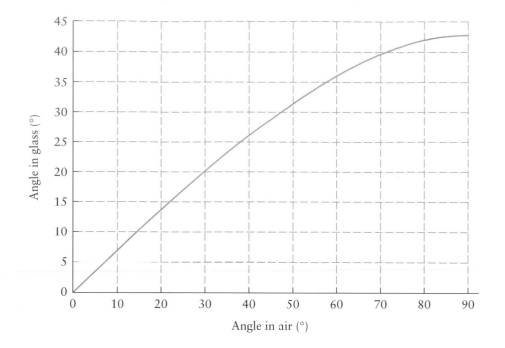

● **Figure 9.34** Graph showing the angles between a light ray and the normal when it passes through an air-glass interface. For a ray passing from air into glass, the angle of incidence is the "angle in air." When the ray passes from glass into air, the angle of incidence is the "angle in glass."

The law of refraction was discovered in 1621 by the Dutch physicist Willebrord Snel and later expressed in its common form by the French mathematician René Descartes (see the footnote on p. 352). It gives the exact value of the angle of refraction when the angle of incidence and the speeds of light in the two media are known.

The graph in ● Figure 9.34 shows the relationship between the two angles for an air-glass boundary. (A similar graph could be constructed for different pairs of media.) The angle between the normal and the light ray in air is plotted on the horizontal axis, and the corresponding angle for the ray in glass is plotted on the vertical axis. Notice that the angle in glass is smaller than the angle in air. For a ray going from air into the glass, this means that the angle of refraction is smaller than the angle of incidence, as shown in Figure 9.29. For the largest possible angle of incidence, 90°, the angle of refraction is about 43°.

Example 9.1

● Figure 9.35 depicts a light ray going from air into glass with an angle of incidence of 60°. Find the angle of refraction.

We use Figure 9.34 with the angle in air equal to 60°. Locate 60° on the horizontal axis, follow the dashed line up to the curve, and then move horizontally to the left. The point on the vertical scale indicates the angle in glass is about 36°. Therefore, the angle of refraction is 36°.

● **Figure 9.35** The light ray is refracted twice as it passes through the block of glass. Its final path is parallel to its initial path if the two surfaces of the glass are parallel. (Note the partial reflection.)

But what happens to light that passes completely through a sheet of glass like a windowpane? Each light ray is bent toward the normal when it enters the glass and then is bent away from the normal when it reenters the air on the other side. No matter what the original angle of incidence is, the two bends exactly offset each other, and the ray emerges from the glass traveling parallel to its original path (see Figure 9.35).

Total Internal Reflection

Consider now the following experiment shown schematically in ● Figure 9.36. Rays of light originating in glass strike the boundary (separating it from the surrounding air) at ever-increasing angles. In this case, since the speed of light in glass is smaller than the speed of light in air, we see that the ray is bent *away* from the normal. In other words, the angle of refraction is *larger* than the angle of incidence, as shown in Figure 9.36b. Moreover, the angle of refraction increases as the angle of incidence does.

Because of reversibility we can use the same graph, Figure 9.34, to find the angle of refraction when the angle of incidence is known. In this situation, the angle of incidence

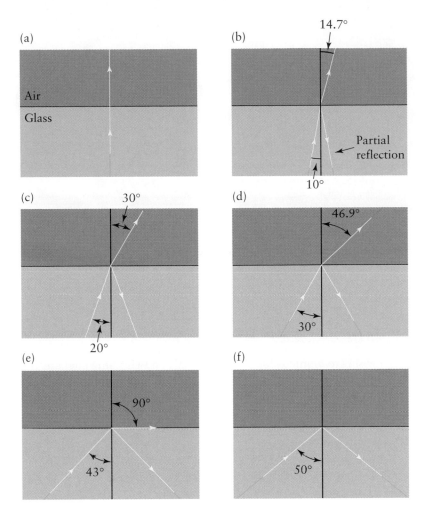

(a)

Air

Glass

(b)

14.7°

Partial
reflection

10°

(c)

30°

20°

(d)

46.9°

30°

(e)

90°

43°

(f)

50°

● **Figure 9.36** A light ray traveling in glass strikes a glass-air interface with different angles of incidence. The angle of incidence in (e) is the *critical angle,* for which the angle of refraction is 90°. Total internal reflection (e and f) occurs when the angle of incidence is equal to, or greater than, the critical angle. No light enters the air.

is the "angle in glass," while the angle of refraction is the "angle in air." For example, when the angle of incidence is 20°, we locate 20° on the vertical scale, move horizontally to the curve, and then drop down to the axis where we see that the angle of refraction would be about 30°. This is shown in Figure 9.36c.

What distinguishes this case from the previous one, in which the incident ray was in air, is the existence of a **critical angle** of incidence for which the angle of refraction reaches 90°. When the angle of incidence equals this critical angle, the transmitted ray travels out *along* the interface between the two media (Figure 9.36e). For angles of incidence greater than the critical angle, the formerly transmitted ray is bent back into the incident medium and does not travel appreciably into the second medium. When this happens, we have a condition called **total internal reflection.** This is shown in Figure 9.36e and f.

Explore It Yourself 9.10

Let's make use of the half-filled beaker again. This time place the container on a sheet of white paper on which you have drawn a circle whose diameter is about two-thirds that of the bottom of the beaker. Make sure the circle and beaker are concentric with one another. Look through the side of the beaker obliquely at the water's surface. Can you see the entire circle? If not, what portion of the circle cannot be observed? Shift your gaze up and down a bit. How does the image of the circle seen through the water's surface change? What do you think is happening to prevent you from seeing the light rays originating from the part of the circle that is hidden from view?

Figure **9.37** Maximizing illumination of the surface of a swimming pool.

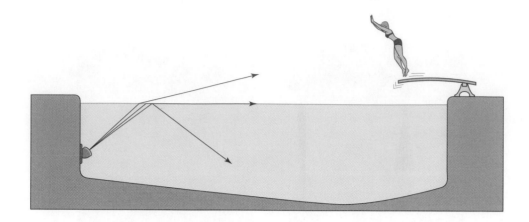

The graph in Figure 9.34 indicates that the critical angle for a glass-air interface, the angle in glass that makes the angle in air equal to 90°, is about 43°. In general, different media have different critical angles. Table 9.1 includes the values of the critical angle for light rays entering air from different transparent media. Notice that the critical angle is smaller for materials in which the speed of light is lower. Diamond, with the lowest speed of light in the list, has a critical angle of less than 25°—little more than half that of glass.

Example 9.2

A homeowner wishes to mount a floodlight on a wall of a swimming pool under water so as to provide the maximum illumination of the surface of the pool for use at night (see ● Figure 9.37). At what angle with respect to the wall should the light be pointed?

To illuminate the surface of the water, the refracted ray at the water-air interface should just skim the water's surface. This means that the angle of refraction must be 90°, so the incident angle must be the critical angle.

Table 9.1 shows that the critical angle for water is about 49°. The homeowner should direct the floodlight upward so that the beam makes an angle of roughly 49° with the vertical side wall of the pool.

Optical fibers are flexible, coated strands of glass that utilize total internal reflection to channel light. In a sense, an optical fiber does to light what a garden hose does to water. ● Figure 9.38 shows the path of a light ray that enters one end of an optical fiber. When the ray strikes the wall of the fiber, it does so with an angle of incidence that is greater than the critical angle, so the ray undergoes total internal reflection. This is repeated each time the ray encounters the fiber's wall, so the light is trapped inside until it emerges from the other end. You may have seen decorative lamps with light bursting from the ends of a "bouquet" of optical fibers. Fiber-optic cables consisting of dozens or even hundreds of individual optical fibers are now in common use to transmit information (● Figure 9.39). Telephone conversations, audio and video signals, and computer information are encoded ("digitized") and then sent through fiber-optic cables as pulses of light produced by tiny lasers. A typical fiber-optic cable can transmit thousands of times more information than a conventional wire cable that is much larger in diameter. For example, the first fiber-optic transatlantic cable, which went into service in 1988, was designed to carry 40,000 simultaneous conversations over just two pairs of glass fibers. By contrast, the last of the large copper bundles installed for overseas communications (1983) could handle only about 8000 calls.

As noted above, one significant advantage of using optical fibers to conduct light from one place to another is their flexibility. If bundles of free fibers are bound together in a

Figure **9.38** Multiple internal reflections within an optical fiber.

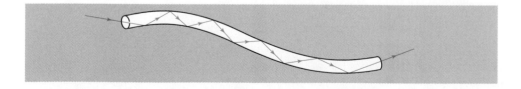

● **Figure 9.39** A fiber-optic light-guide cable (below) can typically carry many more telephone conversations than a much larger wire cable (above).

© Spencer Grant/Photo Researchers

way that preserves the relative positions of adjacent fibers from one end to the other, the bundle may transmit images as well as light. Devices incorporating this technology are routinely used to examine hard-to-reach places like the insides of a nuclear reactor, a jet engine, or the human body. When used in the latter capacity, the instrument is generally called an *endoscope*. Specific examples of endoscopes include bronchoscopes (for examining lung tissue), gastroenteroscopes (for checking the stomach and digestive tract), and colonoscopes (for surveying the bowel).

LEARNING CHECK

1. Upon entering a medium in which the speed of light is 1.5×10^8 m/s from a medium in which the speed of light is 2.0×10^8 m/s, an oblique ray will
   a) be bent away from
   b) be bent toward
   c) travel along
   d) travel perpendicular to
   e) remain undeviated with respect to
   the normal to the interface between the two media.

2. In air, a source of light has a wavelength of 590 nm. The same source under water will have (i) a longer wavelength, (ii) a shorter wavelength, or (iii) the same wavelength.

3. The critical angle is that angle of incidence for which the angle of refraction equals _____.

4. Name an important communication system component that makes use of total internal reflection.

ANSWERS: 1. (b) 2. (ii) 3. 90° 4. optical fiber

9.4 Lenses and Images

In Section 9.2, we described how curved mirrors are used in astronomical telescopes and other devices to redirect light rays in useful ways. Microscopes, binoculars, cameras, and many other optical instruments use specially shaped pieces of glass called *lenses* to alter the paths of light rays. As was the case with mirrors, the key to making glass or other transparent substances redirect light is the use of curved surfaces (interfaces) rather than flat (planar) ones.

Fresnel, Pharos, and Physics

For coastal dwellers, be it along one of the Great Lakes or the eastern or western seaboards, the sight of a lighthouse (or *pharos*) with its sweeping beam warning mariners miles away of the dangers lurking near the shore is not an uncommon one. For inland folk, the beams of the headlamps of oncoming cars offer similar warning to approaching traffic. In both these instances, the measure of safety afforded to sea-farers, drivers, and pedestrians depends on the distance to which these beacons can be seen. This, in turn, depends upon the intensity and degree of focus of the beams. We all owe a debt of gratitude to a nineteenth-century physicist for developing the principles that under-lie the successful function of lighthouses and auto headlights, as well as traffic lights, classroom overhead projectors, stage spotlights, recre-ational vehicle rear-viewing screens, spacecraft solar panels, and many other practical, everyday devices.

Augustin Jean Fresnel (1788–1827; ● Figure 9.40) was a French engineer and physicist who made many important contributions to our understanding of the nature of light. He was a strong proponent of the wave theory and the first to argue persuasively that light was a trans-verse wave based on its polarization. Using the wave theory, he devel-oped mathematical formulas that accurately predicted the diffraction patterns produced by apertures of various shapes and sizes. But Fres-nel was a practical man, and much of his effort throughout his adult life was devoted to public works projects, like road construction, designed to improve the infrastructure in his native France.

In 1824, Fresnel merged his interests in optics with his desire to serve his country by accepting a commission with the Lighthouse Board. In this capacity, Fresnel set about improving the science of pharos by developing compound lenses to capture and focus light instead of using mirrors, as had been done previously. The result was the much-heralded *Fresnel lens*, whose modern usage has spread far beyond the lighthouse for which it was originally intended.

Fresnel realized that the refraction of light by a lens occurs only at the two curved surfaces of the lens. The material between serves no purpose in focusing the light and may, if thick enough, actually reduce the light intensity significantly by absorption. So, to preserve the refractive effects of the surfaces but eliminate the bulk and absorption of the lens, Fresnel subdivided his lens into prismlike segments and then cut out the middle part of each prism (● Figure 9.41). The result-ing wedges have the same refracting properties as the prisms and, hence, the original lens. The main advantages that derive from this design are ones of light weight, low absorption, and, by suitable arrangement of the wedges, high light-collecting efficiency.

When employed in a lighthouse, the various wedge segments are arrayed along an arc. In this configuration, the top and bottom prisms serve to both refract and internally reflect the rays from the source. Moreover, the central wedge is frequently replaced by a very thin, plano-convex lens. Originally, Fresnel lenses for lighthouses were ranked in six main sizes called *orders*; the order is determined by the distance from the light source to the lens. A first-order Fresnel lens might have included up to 1000 prisms, stood nearly 10 feet tall, mea-sured 6 feet in diameter, and weighed up to 3 tons. The first Fresnel lens was installed in the Cardovan Tower Lighthouse on the Gironde River in France. It produced a beam that was five times more power-ful than the reflector system used previously and was visible to the hori-zon more than 20 miles away. Although Fresnel lenses were intro-duced in the United States in 1841 for service in lighthouses in New York harbor, it was not until 1851 that the government authorized the use of this technology throughout its national system of warning bea-cons (● Figure 9.42).

Modern Fresnel lenses are usually made out of plastic with rotary symmetric grooves on one surface only for ease of production. Lenses of this type can be found in many applications where the need to cre-ate bright, focused beams of light exists. Two good examples are the

● **Figure 9.40** Augustin Jean Fresnel. French engineer and physicist whose work in optics included both practical (lighthouse lenses) and theoretical (mathematical description of diffraction) innovations.

Suppose we grind a block of glass so that one end takes the shape of a segment of a sphere as shown in cross section in ● Figure 9.43. Let parallel rays strike the convex spher-ical surface at various points above and below the line of symmetry (called the *optical axis*) of the system. If one applies the law of refraction at each point to determine the angle of refraction of each ray, the results shown in Figure 9.43 are found. In particular, rays trav-eling along the optical axis emerge from the interface undeviated. Rays entering the glass at points successively above or below the optical axis are deviated ever more strongly toward the optical axis. The result is to cause the initially parallel bundle of rays to gradu-ally *converge* together—to become focused—into a small region behind the interface. This point is called the *focal point* and is labeled *F* in the figure. (Compare this to Figure 9.21a in which a concave mirror causes light rays to converge.)

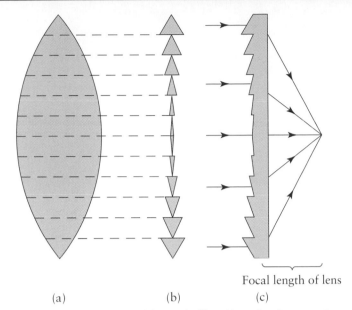

Focal length of lens

(a)　　　　(b)　　　　(c)

● **Figure 9.41** Cut-away view of the principle of Fresnel lenses. An ordinary converging lens may be thought of as a stacked set of prisms (a). If the central portions of each prism are removed and the remaining wedges aligned, the resulting structure (b) has the same refracting characteristics without the added weight or light absorption. Modern Fresnel lenses (c) display a circular symmetric array of grooves on one side of a piece of clear plastic cut to match the shapes of the wedges. This device works to focus light from a distant source just like the original lens.

lenses in many automobile headlamps and in the warning lights on emergency vehicles. Because Fresnel lenses also have a wide field of view, they are ideal for imaging panoramic scenes that often lie hidden behind large vehicles. Perhaps you've seen such lenses in the rear windows of large recreational vehicles. So once again, the theme of safety threads the application of Fresnel's design. One cannot help wondering how many lives have been saved at sea and on land during the past nearly two centuries through the innovation of a young Frenchman with an eye for detail and a commitment to service.

The application of physics principles in service of society, particularly to promote its safety and security, is not a new phenomenon. The early Greeks, notably among them Archimedes (see Physics Family Album in Chapter 3), used then-known laws of mechanics to develop both defensive and offensive weapons to protect their cities from invasion. Among his many accomplishments, Benjamin Franklin's invention of the lightning rod (see Section 7.1) stands as a prime example of the use of physics to safeguard lives and property. Reflect a moment on your own life. Can you think of some other ways in which physical principles have been applied to make you more safe and secure as you work, play, and travel from place to place? Pick a particular area or device that interests you and search the World Wide Web for details on precisely how physics has been used to enhance the quality of life for you and your fellow human beings.

● **Figure 9.42** A second-order Fresnel lighthouse lens from Standard's Rock lighthouse on Lake Superior. This is one of only a few lenses of this size to be used on the Great Lakes; most were much smaller. For example, fourth-order Fresnel lenses were typical in lighthouses along the shores of Lake Michigan. The bull's-eye lenses in the center created a characteristic flash pattern as it rotated.

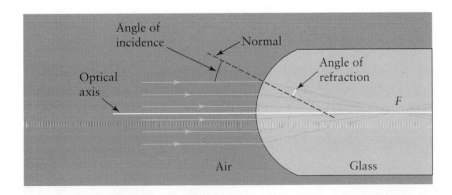

Angle of incidence

Normal

Optical axis

Angle of refraction

F

Air　　　　Glass

● **Figure 9.43** Refraction at a convex spherical surface showing the convergence of light rays. F is the focal point, that is, the point at which the light rays are concentrated after having passed through the surface.

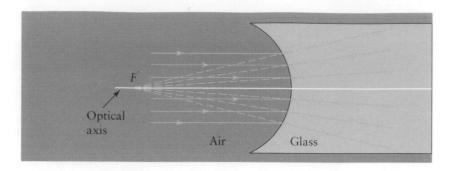

Figure 9.44 Refraction at a concave spherical surface showing the divergence of a beam of parallel light. F' is the focal point, that is, the point from which the rays appear to diverge after having passed through the surface.

● Figure 9.44 shows the behavior of parallel rays refracted across a spherical interface that, instead of bowing outward, curves inward. In this case, the emergent rays *diverge* outward as though they had originated from a point F' to the left of the interface. The ability to either bring together or spread apart light rays is the basic characteristic of lenses, be they camera lenses, telescope lenses, or the lenses in human eyes.

Although the situation shown in Figure 9.43 does occur in the eye, in most devices the light rays must enter and then leave the optical element (lens) that redirects them. Common lenses have two refracting surfaces instead of one, with one surface typically in the shape of a segment of a sphere and the second either spherical as well or flat (planar). The effect on parallel light rays passing through both surfaces is similar to that in the previous examples with one refracting surface. A *converging lens* causes parallel light rays to converge to a point, called the *focal point* of the lens (● Figure 9.45). The distance from the lens to the focal point is called the **focal length** of the lens. A more sharply curved lens has a shorter focal length. Conversely, if a tiny source of light is placed at the focal point, the rays that pass through the converging lens will emerge parallel to each other. This is the principle of reversibility again.

Explore It Yourself 9.11

Look around your home, office, or dorm room for a simple, hand-held magnifying lens. Stand under an overhead light, and holding the lens in one hand, project the image of the source onto your other hand. Move the lens up or down to produce a clearly focused image (not just a blur of light) on your palm. Now estimate the distance from the lens to the image. Congratulations! You've just found the focal length of the lens. This same technique works for any converging lens as long as the source is a good deal farther from the lens than its focal length. Project the image on a white index card or a piece of white paper instead of your hand. Describe the image of the source produced by the lens.

Figure 9.45 Converging lenses focusing parallel light rays at their focal points. A more sharply curved lens has a shorter focal length.

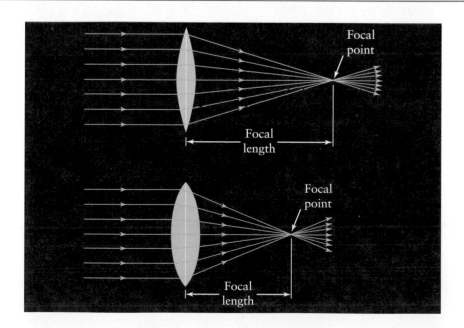

A *diverging lens* causes parallel light rays to diverge after passing through it. These emergent rays appear to be radiating from a point on the other side of the lens. This point is the focal point of the diverging lens (● Figure 9.46). The distance from the lens to the focal point is again called the focal length, but for a diverging lens it is given as a negative number, −15 centimeters, for example. If we reverse the process and send rays converging toward the focal point into the lens, they emerge parallel.

For both types of lenses there are two focal points, one on each side. Clearly, if parallel light rays enter a converging lens from the right side in Figure 9.45, they will converge to the focal point to the left of the lens. Whether a lens is diverging or converging can be determined quite easily: If it is *thicker* at the center than at the edges, it is a converging lens; if it is *thinner* at the center, it is a diverging lens (● Figure 9.47).

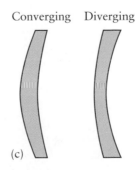

Figure 9.46 Parallel light rays diverging after passing through a diverging lens. The rays appear to radiate from the focal point, to the left of the lens.

Image Formation

The main use of lenses is to form images of things. First, let's consider the basics of image formation when a converging lens is used. Our eyes, most cameras (both still and video), slide projectors, movie projectors, and overhead projectors all form images this way. ● Figure 9.48 illustrates how light radiating from an arrow, called the *object*, forms an *image* on the other side of the lens. One way of setting up this example would be to point a flashlight at the arrow so that light would reflect off the arrow and pass through the lens. The image could be projected onto a piece of white paper placed at the proper location to the right of the lens.

Although each point on the object has countless light rays spreading out from it in all directions, it is easier first to consider only three particular rays from a single point—the arrow's tip. These rays, shown in Figure 9.48, are called the **principal rays.**

1. The ray that is initially parallel to the optical axis passes through the focal point (F) on the other side of the lens.
2. The ray that passes through the focal point (F') on the same side of the lens as the object emerges parallel to the optical axis.
3. The ray that goes exactly through the center of the lens is undeviated because the two interfaces it encounters are parallel.

Note that the image is *not* at the focal point of the lens. Only parallel incident light rays converge to this point.

We could draw principal rays from each point on the object, and they would converge to the corresponding point on the image. This kind of image formation occurs when you take a photograph (● Figure 9.49) or view a slide on a projection screen. In the latter case, light radiating from each point on the slide converges to a point on the image on the screen. Note that the image is inverted (upside down). That's why you load slides in upside down if you want their images to be right side up.

The distance between the object and the lens is called the **object distance,** represented by s, and the distance between the image and the lens is called the **image distance,** p. By convention (with the light traveling from left to right), s is positive when the object is to the left of the lens, and p is positive when the image is to the right of the lens. If we place the object at a different point on the optical axis, the image would also be formed at a different point. In other words, if s is changed, p changes. Using a lens with a different focal

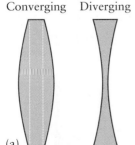
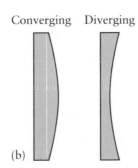

Converging Diverging Converging Diverging Converging Diverging

(a) (b) (c)

Figure 9.47 Examples of different types of lenses: (a) bi-convex (left), bi-concave (right); (b) plano-convex (left), plano-concave (right); (c) meniscus-convex (left), meniscus-concave (right).

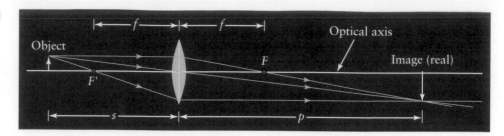

Figure 9.48 Arrangement of a simple converging lens showing the object and image positions as well as the focal points, *F* and *F'* and the focal lengths. Here *s* and *p* are on opposite sides of the lens and are both considered positive. The three principal rays from the arrow's tip are shown.

length would also cause *p* to change. For example, the image would be closer to the lens if the focal length were shorter.

Explore It Yourself 9.12

Let's experiment some more with the magnifying lens. This time use a small electric light as a source; a high-intensity desk lamp works well. Stand as far away from the lamp as possible and again project a focused image of it on a white index card or piece of white paper (the screen; see ● Figure 9.50). Move the lens toward the source, adjusting the position of the screen continuously to produce a clear, focused image. How does the distance of the screen from the lens (the image distance) change as the distance between the source and the lens (the object distance) is reduced? Describe the nature and change in the image as you gradually shorten the object distance. What happens when the lens gets closer to the lamp than about one focal length? At this point, remove the screen and look back through the lens at the source. (Be sure to turn down the lamp intensity or otherwise shield your eyes so that you can view the source comfortably.) How does the lamp appear now? Repeat the experiment, keeping especially careful track of the appearance of the image when the object distance is greater than about two times the focal length of the lens, about equal to two focal lengths, between one and two focal lengths, and, finally, less than one focal length. Record your observations for later study.

The following equation, known as the *lens formula,* relates the image distance *p* to the focal length *f* and the object distance *s.*

$$p = \frac{sf}{s - f} \qquad \text{(lens formula)}$$

The following example shows how the lens formula can be used.

Figure 9.49 Image formation in a camera. The third principal ray from a point at the top and from a point at the bottom of the object are shown.

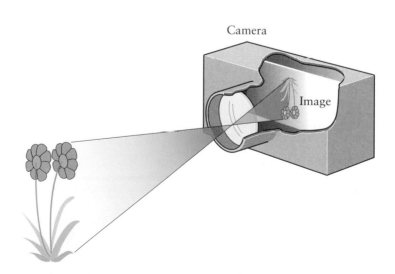

Example 9.3

In a slide projector, a slide is positioned 0.102 meter from a converging lens that has a focal length of 0.1 meter. At what distance from the lens must the screen be placed so that the image of the slide will be in focus?

The screen needs to be placed a distance p from the lens, where p is the image distance for the given focal length and object distance. So:

$$p = \frac{sf}{s - f} = \frac{0.102 \text{ m} \times 0.1 \text{ m}}{0.102 \text{ m} - 0.1 \text{ m}}$$

$$= \frac{0.0102 \text{ m}^2}{0.002 \text{ m}}$$

$$= 5.1 \text{ m}$$

If the slide-to-lens distance is increased to 0.105 meter, the distance to the screen (p) would have to be reduced to 2.1 meters.

If the lens is replaced by one that has a shorter focal length, the distance to the screen would have to be reduced as well.

The images formed in the manner just described are called **real images.** Such images can be projected onto a screen. Our eyes see the image on the screen because the light striking the screen undergoes diffuse reflection. A simple magnifying glass is a converging lens, but the image that it forms under normal use is not a real image—it can't be projected onto a screen. We see the image by looking *into* the lens, just as we see a mirror image by looking *into* the mirror. This type of image is called a **virtual image.** ● Figure 9.51 shows how an image is formed in a magnifying glass. In this case, the object is between the focal point F' and the lens, so the object distance s is *less than* the focal length f of the lens. Note that the image is enlarged and that it is upright. It is also on the same side of the lens as the object, which means that p is negative. This situation is very much like the image formation with a concave mirror (Figure 9.22b).

● **Figure 9.50** Single-lens image formation.

Example 9.4

A converging lens with focal length 10 centimeters is used as a magnifying glass. When the object is a page of fine print 8 centimeters from the lens, where is the image?

$$p = \frac{sf}{s - f} = \frac{8 \text{ cm} \times 10 \text{ cm}}{8 \text{ cm} - 10 \text{ cm}}$$

$$= \frac{80 \text{ cm}^2}{-2 \text{ cm}}$$

$$= -40 \text{ cm}$$

The negative value for p indicates that the image is on the same side of the lens as the object. Therefore, it is a virtual image and must be viewed by looking through the lens.

A virtual image is also formed when you look at an object through a diverging lens (● Figure 9.52). In this case, the image is smaller than the object, as it is with a convex mirror.

● **Figure 9.51** Image formation when the object is between the focal point and the lens. A virtual image of the arrow is formed, seen by looking through the lens toward the object.

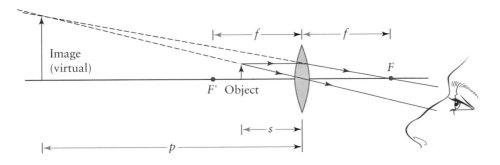

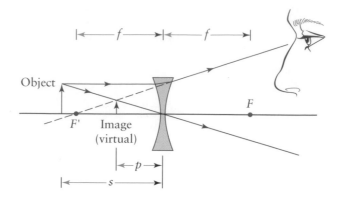

Figure 9.52 Image formation with a diverging lens. The image is virtual, so it must be viewed by looking through the lens.

Magnification

In Figures 9.48 through 9.52, the size of the image is not the same as the size of the object. This is one of the most useful properties of lenses: They can be used to produce images that are enlarged (larger than the original object) or reduced (smaller than the original object). In either case, the **magnification, M,** of a particular configuration is the height of the image divided by the height of the object.

$$M = \frac{\text{image height}}{\text{object height}}$$

If the image is twice the height of the object, the magnification is 2. If the image is upright, the magnification is positive. If the image is inverted, the magnification is negative (because the image height is negative).

The magnification that one gets with a particular lens changes if the object distance is changed. Because of the simple geometry, the magnification also equals *minus* the image distance divided by the object distance.

$$M = \frac{-p}{s}$$

From this we can conclude that:

If p is positive (image is to the right of the lens and real), M is negative; the image is inverted (Figures 9.48 and 9.49).

If p is negative (image is to the left of the lens and virtual), M is positive; the image is upright (see Figures 9.51, 9.52, and ● 9.53).

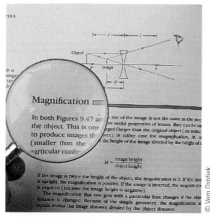

● **Figure 9.53** The image seen through the magnifying glass is magnified, upright, and virtual.

Example 9.5   Compute the magnification for the slide projector in Example 9.3 and for the magnifying glass in Example 9.4.

In the first case in Example 9.3, $s = 0.102$ meters and $p = 5.1$ meters. Therefore:

$$M = \frac{-p}{s} = \frac{-5.1 \text{ m}}{0.102 \text{ m}}$$

$$= -50$$

The image is 50 times as tall as the object, and it is inverted (since M is negative). A slide that is 35 millimeters tall has an image on the screen that is 1,750 millimeters (1.75 meters) tall. When s is 0.105 meters, p is 2.1 meters, and the magnification is -20.

In Example 9.4, $s = 8$ centimeters and $p = -40$ centimeters. Consequently:

$$M = \frac{-p}{s} = \frac{-(-40 \text{ cm})}{8 \text{ cm}}$$

$$= +5$$

The image of the print seen in the magnifying glass is five times as large as the original, and it is upright (since M is positive).

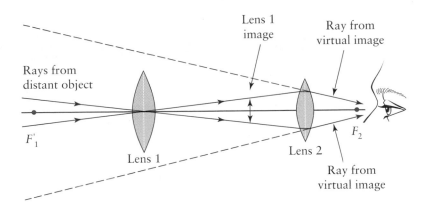

Telescopes and microscopes can be constructed by using two or more lenses together. Figure 9.54 shows a simple telescope consisting of two converging lenses. The real image formed by lens 1 becomes the object for lens 2. The light that could be projected onto a screen to form the image for lens 1 simply passes on into lens 2. In essence, lens 2 acts as a magnifying glass and forms a virtual image of the object. In this telescope, the image is magnified but inverted. Replacing lens 2 in Figure 9.54 with a diverging lens yields a telescope that produces an upright image.

Aberrations

In real life, lenses do not form perfect images. Suppose we carefully apply the law of refraction to a number of light rays, all initially parallel to the optical axis, as they pass through a real lens that has a surface shaped like a segment of a sphere (Figure 9.55). We would find that the lens exhibits the same flaw we saw in Section 9.2 with spherically shaped curved mirrors: *spherical aberration*. Figure 9.55 shows that rays striking the lens at different points do not cross the optical axis at the same place (compare Figure 9.26a). In other words, there is no single focal point. This causes images formed by such lenses to be somewhat blurred. Lens aberrations of this type can be corrected, but this process is complicated and often necessitates the use of several simple lenses in combination.

One type of aberration shared by all simple lenses even when used under ideal conditions is *chromatic aberration*. A lens affected by chromatic aberration, when illuminated with white light, produces a sequence of more or less overlapping images, varying in size and color. If the lens is focused in the yellow-green portion of the EM spectrum where the eye is most sensitive, then all the other colored images are superimposed and out of focus, giving rise to a whitish blur or fuzzy overlay (see Figure 9.56). For a converging lens, the blue images would form closer to the lens than the yellow-green images, while the reddish images would be brought to a focus farther from the lens than the yellow-green ones.

The cause of chromatic aberration has its roots in the phenomenon of *dispersion*, to be discussed in Section 9.6. The remedy for this problem, originally thought to be insoluble by none other than Newton himself, was discovered around 1733 by C. M. Hall and

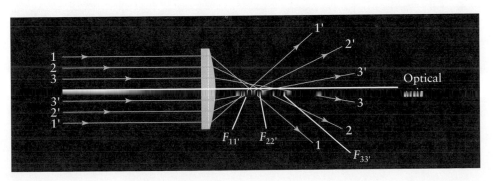

Figure 9.55 Spherical aberration for a convex lens illuminated by a beam of parallel light. Rays 1 and 1′ are brought to a focus at $F_{11'}$, while rays 2 and 2′, and 3 and 3′ are focused at $F_{22'}$, and $F_{33'}$, respectively.

The *Camera Obscura*: A Room with a View

One-way mirrors, discussed in Section 9.2, provide a means of monitoring what goes on in another room without being seen by those present. Peep holes, small apertures in doors, walls, or shades, sometimes fitted with lenses, that give visual access to adjacent spaces, are another common means of surveillance whose origin dates to antiquity and whose use has extended well beyond the realm of "cloak-and-dagger" activities. In their simplest form, these pinholes are the precursors to our modern cameras and have provided the means not only to spy on your neighbors but to render detailed images of natural phenomena like solar eclipses and to create beautiful works of art.

The *camera obscura* (from the Latin, meaning "dark chamber") consists of a small circular hole located in an otherwise opaque screen through which light from a bright area enters a darkened space. The rays from objects outside the room form an inverted image on the wall (or other surface) opposite the aperture (see Figures 9.49 and ● 9.57). The optical principle underlying this effect was known to Aristotle, and the Arab astronomer and scholar Alhazen is said to have used a camera obscura to monitor safely the progress of solar eclipses nearly 1000 years ago. Even Leonardo da Vinci includes some commentary about it in his notebooks. But the first full discussion of the camera obscura and the suggestion that it be used in drawing was

given by Giovanni Battista della Porta in 1558 in his *Magiae Naturalis*. The basic idea in the artistic context was to attach a piece of paper to the screen on which the image was projected and to trace it. Once the outline of the scene had been completed, it could later be "colorized" much as black-and-white films are today. As della Porta writes, "If you cannot paint, you can by this arrangement, draw the outline of the images with a pencil."

Della Porta also promoted the use of a concave mirror in the darkened room to magnify and invert the image formed by the pinhole. He even went so far as to suggest placing a converging lens in the hole to improve the quality of the image. Although popularized by della Porta in the second edition of his book in 1589, both of these innovations may have pre-dated della Porta's book by as much as twenty years. The early mathematical astronomer Johannes Kepler, who discovered that the planets move in elliptical orbits, was reported to have used a tent camera obscura when conducting topographical surveys of the land in and around present-day Austria.

Recently, it has been suggested that some of the great Renaissance painters, notably Jan Vermeer (1632–1675), used the camera obscura to create some of their masterpieces (● Figure 9.58). While this view is hotly contested by some art historians, it is undeniable that the camera obscura was instrumental in permitting artists to produce detailed images of exotic locales and breathtaking natural and man-made structures that eventually were widely published. These early picture postcards began to bring distant places and customs into the homes of people everywhere and thereby initiated the process of globalization that the modern telecommunications revolution has finally completed.

By the middle of the seventeenth century, tiny versions of the camera obscura disguised as drinking goblets and books were in use by the wealthy to surreptitiously monitor events going on elsewhere in a room or banquet hall. In 1669, Robert Boyle (see the Physics Family Album in Chapter 4) described for the Royal Society of London his invention of a portable box camera (see Explore It Yourself 9.13). Within less than twenty years, German inventors had extended Boyle's design to produce the first small, modern reflex camera. It included a simple telephoto lens (a long-focal-length converging lens followed by a short-focal-length diverging one) with a diaphragm to stop down the lens and sharpen the image and a small plane mirror set at 45° to send the image to an oiled paper or opalescent glass screen for viewing or copying (● Figure 9.59). So, by 1685, the camera in the form that we know it today was born, the culmination of several centuries of progress from its humble beginnings in a darkened room with a

● **Figure 9.57** Twelve-foot-wide inverted image of Mount Graham produced by the world's largest known *camera obscura* in the Gov. Aker Observatory at Discovery Park in Arizona. The camera uses a three-lens assembly, 40 inches in diameter, to form the panorama shown.

● **Figure 9.56** Diagram showing the positions of the focal points in various colors for a simple lens. The focal points for blue, green, and red are successively farther from the lens. A screen placed at point *P* will show an image whose outside edges are tinged red-orange, while a screen placed at point *P'* will reveal an image whose outside edges are bluish.

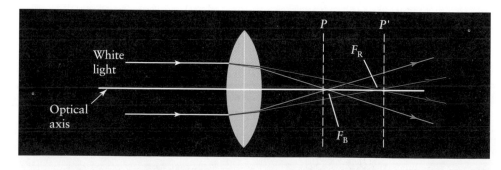

Figure 9.58 On the left is Vermeer's painting *The Music Lesson* (c. 1662–64). On the right is a photographic reconstruction of *The Music Lesson,* based on a full-size scale model of the original room. The precision with which the model could be made owes itself to Vermeer's accuracy in depicting the exact sizes and geometric relations between the various elements of the painting, something likely made possible through the use of the camera obscura.

peephole. All that remained was the development of a convenient, reliable photographic film in the early twentieth century by George Eastman to complete the picture.

As suggested in the foregoing discussion, optical principles have played an important role for centuries in the development of tools for gathering intelligence for both military and civilian use. But they have also contributed to the development of the graphic and visual arts, thereby adding immeasurably to our understanding and enjoyment of our planet. This is no less true today. For example, high-resolution cameras in satellites and aircraft working at optical and infrared wavelengths continuously survey all manner of human activity around the globe. Compact digital video cameras capture for posterity the moments that make life worth living—births, marriages, graduations, sunsets, rainbows. If you're interested, "dissect," figuratively, if not literally, your favorite optical device to see how it works. Search the web for information on the design and function of the instrument and on how it relies on physics principles to produce the images you prize so highly. A good place to start is at www.howthingswork.com. Have fun!

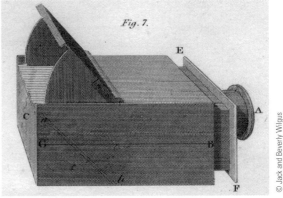

Figure 9.59 Exterior view of a wooden box camera obscura circa 1817. This device has a lens (A) that is moved by means of a drawerlike assembly (EF). The 45° mirror, which reflects the image upward to the screen, is shown in outline (*ab*). A hinged shade with side curtains to aid in viewing the image is shown in its raised position.

Explore It Yourself 9.13

You can easily make a portable pinhole box camera to illustrate some of the features noted above. Take a shoebox or other similar rectangular box, and place a small hole about 0.5 mm in diameter in one end. In one side of the box, cut a rectangular viewing slot to permit you to see the image cast on the end of the box opposite the hole. Now point the end of the device with the pinhole toward a sunlit window or other bright scene and examine the image produced by your camera. (To help improve the contrast of the image you may want to cover the inside screen end of the box with white paper.) How does the image look? Experiment with different-sized pinholes to see which one offers the best compromise between image clarity and image brightness.

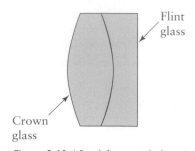

Crown glass
Flint glass

● **Figure 9.60** A Fraunhofer cemented achromatic doublet lens.

later (in 1758) developed and patented by John Dolland, a London optician. It involves using two different types of glass mounted in close proximity. ● Figure 9.60 shows a common configuration called a *Fraunhofer cemented achromat* (meaning "not colored"). The first lens is made of crown glass, the second of dense flint glass (see Table 9.1). These materials are chosen because they have nearly the same dispersion. To the extent to which this is true, the excess convergence exhibited by the first lens at bluish wavelengths is compensated for by the excess divergence produced by the second lens at these same wavelengths. Similar effects occur at the other wavelengths in the visible spectrum, permitting cemented doublets of this type to correct more than 90% of the chromatic aberration found in simple lenses.

Explore It Yourself 9.14

Most households have large, thick magnifying glasses. Using this handy device and a burning candle, you can easily demonstrate some of the basics of chromatic aberration. With the candle as your source of illumination, locate the lens so that it forms a real image of the candle flame on a white card or piece of paper (the *screen*). (If you need help in determining where to place the lens relative to the candle and the screen to accomplish this task, see Explore It Yourself 9.12.) Move the screen closer to the lens. What color is the outside edge of the now-blurred image of the candle flame? Next, move the screen away from the lens past the position of best focus of the candle image. What color does the perimeter of the blurred image appear now? Try examining the source directly by looking back through the lens at the candle. The chromatic effects should be far more apparent.

LEARNING CHECK

1. (True or False.) A light ray can pass through a diverging lens without being deflected.
2. _____ lenses are thicker at their centers than at their edges, while _____ lenses are thinner at their centers than at their edges.
3. An image formed using a converging lens
   a) is always real.
   b) is always virtual.
   c) can always be projected on a screen.
   d) can have a negative magnification.
   e) can never be upright.
4. The magnification of an object by a certain lens is equal to -0.5. The image formed by the lens must be
   a) upright and enlarged.
   b) upright and reduced.
   c) inverted and enlarged.
   d) inverted and reduced.
   e) virtual and unchanged in size.

ANSWERS: 1. True 2. Converging, diverging 3. (d) 4. (d)

9.5 The Human Eye

The eye may be the most sophisticated of our sense organs. But up to the point when the light rays are absorbed and the signal to the brain is formed, the eye is a fairly simple optical instrument.

● Figure 9.61 shows a simplified cross section of the human eye. Light enters from the left and is projected onto the retina at the rear of the eyeball. The iris, the part of the eye that is colored, controls the amount of light that enters the eye. Its circular opening, the pupil, is large in dim light and small in bright light. As light enters the cornea, it converges because of the cornea's convex shape (Figure 9.43 shows this effect). The eye's *auxiliary lens,* simply called the *lens,* makes the light converge even more. This lens is used by the eye to ensure that the image is in focus on the retina. The effect of the cornea and the lens is the same as if the eye were equipped with a single converging lens, so the image formation process shown in Figure 9.48 applies.

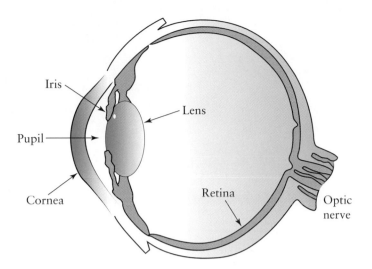

● **Figure 9.61** The human eye. Light passes through the opening in the iris, the pupil, and forms an image on the retina. The cornea and the lens act as a single converging lens.

For an object to be seen clearly, the image must be in focus on the retina. Within the eye, the image distance is always the same—the diameter of the eyeball. We are able to focus on objects near and far (with small and large object distances) because the focal length of the lens can be varied by changing its shape. When the eye is focused on a distant object (farther than, say, 5 meters away) the lens is thin and has a long focal length (● Figure 9.62). For a near object, special muscles make the lens thicker. This shortens the focal length of the lens so that the image of the near object is focused on the retina. Unlike cameras and other optical devices, the eye has a constant value for p, the image distance, but it accommodates different values for s, the object distance, by changing its focal length f.

The two most common types of poor eyesight, nearsightedness and farsightedness, are the result of improper focusing. *Nearsightedness*, or myopia, occurs when close objects are in focus but distant objects are not. The light rays from a distant object are brought into focus before they reach the retina. When the rays do reach the retina, they are out of focus (● Figure 9.63). The problem is remedied by placing a properly chosen *diverging lens* in front of the eye. The corrective lens is either held in place by a frame (glasses) or placed in direct contact with the cornea (contact lenses). This lens makes the light rays diverge slightly before they enter the eye. This moves the point where the rays meet back to the retina.

Farsightedness, or *hyperopia*, is the opposite: Distant objects are in focus, but near objects are not. The cornea and the lens do not make the rays from a near object converge

Humans, like most mammals, accommodate by changing the shape of the eye lens. Fish, however, move the lens itself toward or away from the retina, thus changing the value of p.

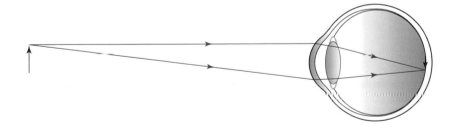

● **Figure 9.62** The eye can form images on the retina of either distant or near objects by changing the thickness (and therefore the focal length) of the lens.

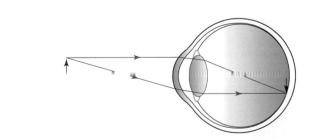

Figure 9.63 (a) A nearsighted eye causes rays from a distant object to converge too quickly. (b) A diverging lens corrects the problem.

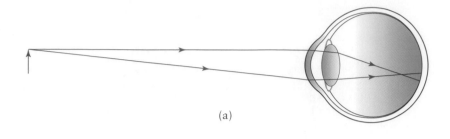

(a)

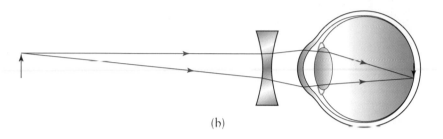

(b)

The first use of specially shaped lenses to correct astigmatism was made by Sir George B. Airy, Astronomer Royal of Great Britain, in 1825 to improve his own poor eyesight.

enough. The rays reach the retina before they meet, and the image is out of focus (● Figure 9.64). The remedy for this condition is a properly chosen *converging lens* placed in front of the eye. This lens makes the light rays converge slightly before they enter the eye, thereby bringing the light rays into focus on the retina.

Another common problem is *astigmatism,* which occurs when the cornea is not symmetric. For example, the cornea's focal length might be shorter for two parallel rays that enter the eye one above the other compared to the focal length for two parallel rays that enter side by side. Objects appear distorted. This condition can often be corrected by using a specially shaped lens that has the opposite asymmetry.

Eye Surgery

During the later part of the twentieth century, several types of corrective eye surgery became commonplace. As described earlier in this section, the convex shape of the cornea

Figure 9.64 (a) A farsighted eye does not cause light rays from a near object to converge enough. (b) A converging lens corrects the problem.

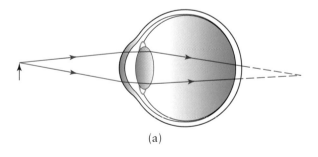

(a)

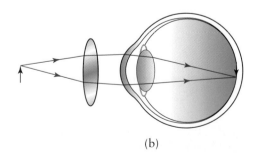

(b)

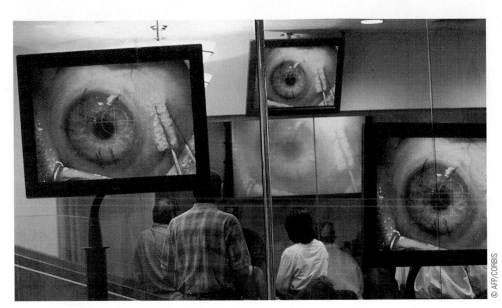

© AFP/CORBIS

causes light rays entering the eye to converge. If the cornea's shape is imperfect (too sharply curved, for example), the person has impaired vision. In the 1970s, ophthalmologists perfected *radial keratotomy* (RK), a method for correcting myopia. The strategy is to precisely flatten the cornea—make it less sharply curved—so that the eye's focal length is increased to the correct value. A surgeon makes several radial (spokelike) incisions in the cornea. As the cornea heals, it becomes flatter, so that the patient no longer needs corrective lenses (in most cases).

In the 1990s, two forms of laser surgery were developed that offer alternatives to RK. *Photorefractive keratectomy* (PRK) is a procedure that reshapes the cornea using an ultraviolet laser controlled by a computer. (For more on lasers, see Section 10.8.) The laser selectively vaporizes (removes) tissue on the cornea's surface until the desired shape is achieved. PRK is used to correct both myopia and astigmatism, and may soon be approved for treating hyperopia. A newer method, called *laser in situ keratomileusis* (LASIK; ● Figure 9.65), is similar to PRK but more complicated. The surgeon cuts a flap of corneal tissue, uses a laser to remove tissue beneath it, and then replaces the flap.

9.6 Dispersion and Color

Most of you at one time or another have seen decorative glass pendulums hanging in windows through which sunlight streamed. If so, you probably noticed patches of bright, rainbow-hued light playing about the room as the pendulums slowly turned in response to air currents. Did you ever wonder how such beauty was produced? Sir Isaac Newton did, and he performed several experiments in an attempt to answer this question. He concluded that sunlight—white light—was a mixture of all the colors of the rainbow and that upon being refracted through transparent substances like glass, it could be *dispersed,* or separated, into its constituent wavelengths (colors). The process of refraction was seen to be color dependent!

This color dependence of refraction comes about because the speed of light in any medium is slightly different for each color. In glass, diamond, ice, and most other common transparent materials, *shorter* wavelengths of light travel slightly *slower* than longer wavelengths. Violet travels a little slower than blue, blue travels slower than green, and so on. In the case of common glass, the speed of violet light is about 1.95×10^8 m/s, while the speed of red light is about 1.97×10^8 m/s. The speeds of the other colors lie between these two values. This slight difference in the speeds of different colors causes dispersion. In diamond, the difference in speeds of violet and red light is comparatively larger—

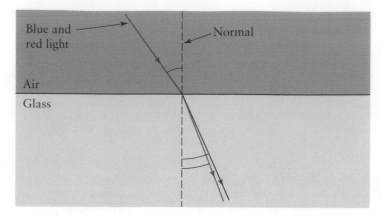

Figure 9.66 Dispersion at an air-glass interface produces a separation of red and blue rays. The angle of refraction of the blue ray is smaller than the angle of refraction of the red ray.

Blue and red light

Normal

Air

Glass

about 2%, compared to about 1% for glass—so the dispersion is greater. That is why one sees such brilliant colors in diamonds. The difference in speeds of violet and red light in water is comparatively smaller than it is in glass.

Up until this time, we have ignored the fact that the speed of light in a medium depends on wavelength. (You can think of our previous study of refraction as having been done using light of only a single color or wavelength, something known as *monochromatic* light.) What effect does this now have on how violet light rays are refracted at an air-glass interface relative to how red rays are refracted?

Consider ● Figure 9.66, showing an incoming ray of light that we will assume is a mixture of only blue and red wavelengths. Because the speeds of both blue light *and* red light are lower in glass than they are in air, we expect that both rays will be bent toward the normal based on our analysis in Section 9.3. But we know that the speed of blue light in glass is lower than the speed of red light in the same medium, so that the blue light will be bent slightly *more* toward the normal than the red light. In Figure 9.66, this results in the angle of refraction for blue light being a bit smaller than the angle of refraction for red light. Thus, although both rays are bent toward the normal upon passing into the glass, the blue ray is refracted more strongly and emerges from the interface along a different path than the red ray. The colors have been dispersed, or separated, as a result of the refraction process because of the wavelength dependence of the speed of light.

If the incoming beam is now allowed to contain the remaining colors between red and blue, the emergent rays for each will fall between the limits set by the red and blue rays. What is produced is a *spectrum*—the different colors spread over a range of angles. It is the process of dispersion, acting to sort out the different colors, that causes the chromatic aberration in simple lenses described at the end of Section 9.4.

The difference between the angles of refraction for the red and blue rays above amounts to less than 0.5°. This may not sound like much, but it is some 30 times the minimum angular separation between rays that the human eye can detect under bright conditions and, therefore, would certainly be noticeable. If additional air-glass surfaces are introduced, more refractions may occur, and the angular spread in the emerging rays may be increased. The light is said to be more highly dispersed in this case.

A *prism* is a common device used to disperse light and form a spectrum (● Figure 9.67). Prisms were well known and highly prized by the Chinese (as indicated in missionary reports dating from the early 1600s) for their ability to generate color. Today, they are highly valued by scientists for much the same reason. For example, one can analyze the radiation emitted by a source of light by dispersing its light into a spectrum and measuring the intensity (amount) of radiation coming off in the various wavelengths (colors). If the source radiates like a blackbody (see Section 8.6), this information might be used to determine the temperature of the source. As we shall see in Chapter 10, it is also possible to determine the chemical composition of a source by examining its spectrum. ● Figure 9.68 shows two common configurations of dispersing prisms and the path a single-color light ray follows through each.

© David Parker/SPL/Photo Researchers

Figure 9.67 Dispersion of white light into a spectrum of colors as it passes through a 60° prism. Each color is bent a different amount because the speed is slightly different for each wavelength.

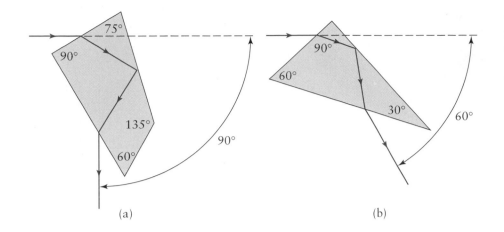

LEARNING CHECK

1. A _____ lens is used to compensate for hyperopia.
2. When myopia is corrected surgically, the shape of the _____ is changed.
3. (True or False.) Blue light travels more slowly through glass than red light does.
4. A ray of violet light and a ray of yellow light enter a block of glass from air with the same angle of incidence. The angle of refraction for the violet ray will be

(a) much larger than that of the yellow ray.
(b) a little larger than that of the yellow ray.
(c) equal to that of the yellow ray.
(d) a little smaller than that of the yellow ray.
(e) much smaller than that of the yellow ray.

ANSWERS: 1. converging 2. cornea 3. True 4. (d)

9.7 Atmospheric Optics: Rainbows, Halos, Blue Skies

Rainbows

"My heart leaps up when I behold a rainbow in the sky." This is how the poet Wordsworth described his reaction to a rainbow, and it is probably not too bad a description of how many of us feel upon seeing a dazzling, colored arc stretching across the sky. Rainbows are both beautiful and puzzling. How do the elements of water and sunlight combine to produce such spectacles? Armed with the information in Sections 9.3 and 9.6, we are in a position to find out.

Explore It Yourself 9.15

Most of us have seen rainbows while washing our cars or watering our lawns or gardens on sunny days, but have you ever considered what goes into creating a rainbow? The next time you're outside using a hose and see a rainbow, ask yourself some of these questions, and experiment to find the answers if they're not obvious to you. Where is the Sun in the sky relative to you and the direction in which the rainbow is seen? How does the production of a rainbow depend upon the size of the water droplets in the mist provided by the hose? How are the colors of the rainbow ordered from inside to outside? How large is the rainbow arc? (To help you with the assessment of the dimensions of the rainbow, you can use the information in ● Figure 9.69, which shows some angle-measuring techniques or tricks long used by astronomy students to establish the angular separations of celestial objects in the sky.) Keep the answers to these questions in mind to reinforce and/or confirm what you read in the remainder of this section.

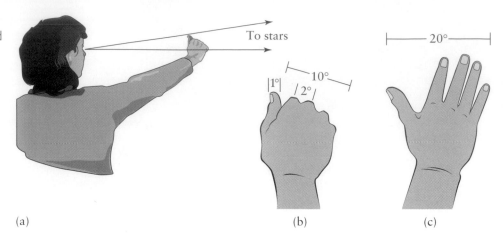

Figure 9.69 Using the human hand to measure angles roughly. (a) Extend the hand at arm's length and sight along it. (b) The thumb subtends about 1°, two knuckles about 2°, and the fist about 10°. (c) The hand subtends about 20°.

(a)

(b)

(c)

Before doing so, however, we need to point out some rainbow basics. First, rainbows consist of arcs of colored light (spectra) stretching across the sky, with the red part of the spectrum lying on the outside of the bow and the blue-violet part lying on the inside. Second, rainbows are always seen against a background of water droplets with the Sun typically at our backs. These two basic characteristics of rainbows are what we seek to understand.

Imagine a beam of light from the Sun striking a raindrop. For simplicity, we will assume that raindrops are spherical, although real falling raindrops are shaped more like the squashed circular pillows that decorate sofas. If we apply the law of refraction at each surface, concentrating only on those rays that make a U-turn in the drop and return in the general direction of the Sun (so as to be consistent with the second rainbow basic above), we find the result shown in ● Figure 9.70. The ray striking the drop at its center (ray 1) returns directly back along its incident direction and defines the *axis* of the drop.

Figure 9.70 Paths of light rays through a water drop. Ray 7 is the Descartes ray.

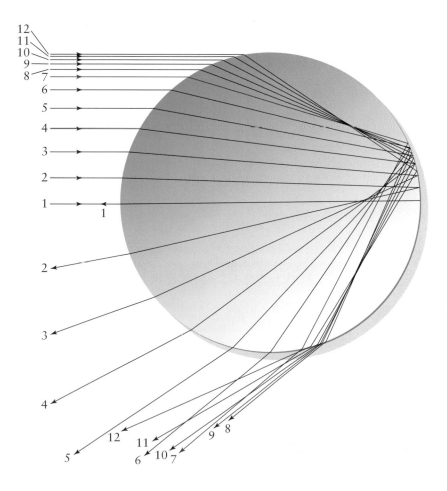

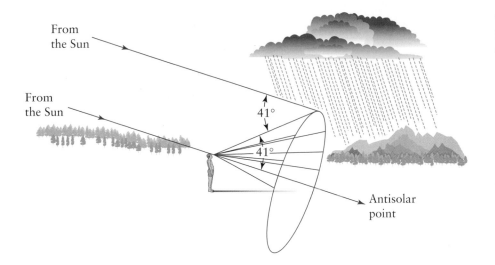

Figure 9.71 From Descartes's construction, the rainbow is predicted to be a circle of angular radius 41°, centered on the antisolar point.

Rays entering above the axis exit below the axis, and vice versa. The farther above the axis the ray enters, the greater its emergent angle up to a point defined by ray 7. This ray is called the *Descartes ray*, after René Descartes, who first suggested this explanation for rainbows in 1637.

For rays entering above the Descartes ray, the exit angles are *less* than that of the Descartes ray. Thus rays entering the drop on either side of the Descartes ray emerge at about the same angle as the Descartes ray itself, leading to a concentration of rays leaving the droplet at a maximum angle corresponding to that of the Descartes ray. This angle is about 41° for rays 6 through 10 in Figure 9.70.

This concentration of reflected and refracted sunlight at exit angles near 41° produces rainbows. The Descartes model predicts that rainbows should consist of circles of light of angular radii equal to 41°, centered on a point opposite the Sun in the sky—the *antisolar point*. Thus, to see a rainbow, we need to look for these concentrated rays in a direction about 41° from the "straight back" direction with the Sun behind us (● Figure 9.71). Notice, if the Sun is above the horizon, the antisolar point will be below the horizon along the direction of your shadow. In this case, the rainbow circle intersects the horizon, and we see only an arc of the circle. For earthbound observers, the best rainbow apparitions occur when the Sun is on the horizon, for then we see half of the rainbow circle. If the Sun is higher in the sky than about 41° above the horizon, then no rainbow can be seen from the ground because the antisolar point lies 41° or more below the horizon, and the rainbow circle never reaches above the horizon. This is why observers throughout most of the continental United States rarely see rainbows at noon. When viewed from an aircraft, a rainbow can form a complete circle.

So far, we have addressed several aspects of the shape and location of rainbows but not their colors. To do so, we must include the phenomenon of dispersion. Recall from Section 9.6 that blue light is more strongly deviated in passing through transparent media than is red light. This means that the maximum emergent angle from the raindrop for blue light will be smaller than the maximum emergent angle for red light (● Figure 9.72). Therefore, the blue light is concentrated at slightly smaller angles than is the red light. Calculations show that blue-violet light is concentrated in a circle of angular radius of about 40°, while red light is concentrated at an angle of about 42°. The other colors of the rainbow fall in between. A more detailed model, including dispersion, then predicts that real rainbows should consist of bands of color in the sky a total of some 2° or so wide, with blue-violet colors on the inside and red-orange colors on the outside. And this is precisely what is seen.

The application of the laws of reflection and refraction (including dispersion) to falling raindrops gives us an explanation for what is called the *primary* rainbow. The primary results from *one* internal reflection of the rays in the drop at the rear surface. Higher-order rainbows may be produced by rays executing two or more internal reflections before leaving the droplet. Some of you no doubt have seen *secondary* rainbows lying outside the primaries along arcs of circles having angular radii of approximately 51°

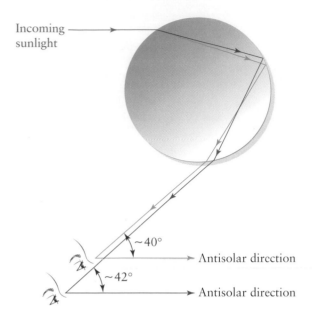

Incoming sunlight

~40°

Antisolar direction

~42°

Antisolar direction

● **Figure 9.73** Primary and secondary rainbows.

(● Figure 9.73). The ordering of the colors of these rainbows is reversed from that of the primaries. All of these properties are explicable in terms of the laws of optics with which we have become familiar (● Figure 9.74).

Halos

Halos, circular arcs of light, often with reddish inner edges, surrounding the Sun or full Moon might be considered winter's answer to rainbows. When the temperature in the upper atmosphere drops below freezing, ice crystals form. At high elevations in the temperate regions of the Earth, one common shape exhibited by such crystals is that of a hexagon, similar to a short, unsharpened pencil (● Figure 9.75a). When seen in cross section, such crystals may be considered as pieces of 60° prisms, and they deviate light like them (see Figure 9.75b).

As in the case of rays entering raindrops, if one traces the paths of rays entering such a crystal at various incident angles, one finds that there is a concentration of exiting rays with deviation angles near 22°. Thus, when light from the Sun or the Moon enters a cloud of such ice crystals having all possible orientations, the emergent rays tend to be clustered into circular arcs having angular radii of 22° centered on the source of illumination.

To see a ray of light forming part of a halo, we should look in a direction 22° away from the Sun or Moon (see ● Figure 9.76). When doing so, you may notice that the inner edge of the halo circle is tinted red. This is again the result of dispersion. At each refraction, the blue component of sunlight (or moonlight, which is merely reflected sunlight) is more

● **Figure 9.74** Schematic diagram showing the production of a secondary rainbow from two internal reflections in a raindrop. The path shown is for a typical ray of yellow light. The ray emerges at an angle of about 51° with respect to the antisolar direction. Red rays emerge with slightly smaller angles, while blue rays emerge with somewhat larger angles.

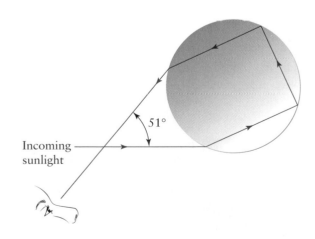

51°

Incoming sunlight

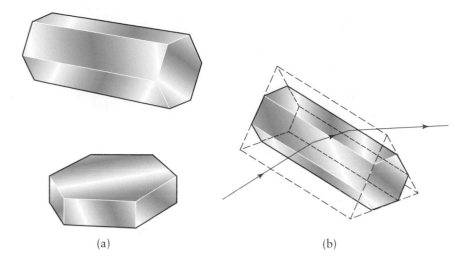

● **Figure 9.75** (a) Two simple ice crystal forms: top, a columnar or pencil crystal; bottom, a plate crystal. (b) A light ray passing through a pencil crystal is refracted as if it were passing through a 60° prism.

(a)   (b)

strongly refracted than is the red component. Consequently, the angle of concentration for the blue light is somewhat greater than it is for red light, and the latter piles up preferentially at the inner edge of the halo, as indicated in ● Figure 9.77.

What we have described is the well-known 22° halo. There are also 46" halos, which result from light entering one face of the pencil crystal and leaving through one end. These halos are much fainter than the 22° halos and are much harder to see—partly because they occupy such a large portion of the sky, having angular diameters of more than 90°! And these are but two of many, many phenomena associated with ice crystal reflection and refraction. Such magnificence surrounds us daily if only we allow our eyes to be open to it. A knowledge of physics can help us to appreciate these natural wonders more deeply.

Another commonly observed atmospheric phenomenon seen mostly surrounding the Moon is a *corona*. This apparition has an angular extent of only a few degrees, is predominantly reddish brown in color, and is caused principally by diffraction, not refraction and dispersion.

Blue Skies

The most common of all atmospheric optical phenomena is the blue sky. It is caused by air molecules *scattering* sunlight in all directions. As a light wave travels through the atmosphere, the wave's oscillating electric field causes the electrons in air molecules to oscillate with the same frequency. From the discussion in Chapter 8, we know that oscillating electric charges (electrons, in this case) emit electromagnetic radiation (light, in this case). This emitted light, which travels outward in all directions (hence the use of the term *scattered*), is what we see.

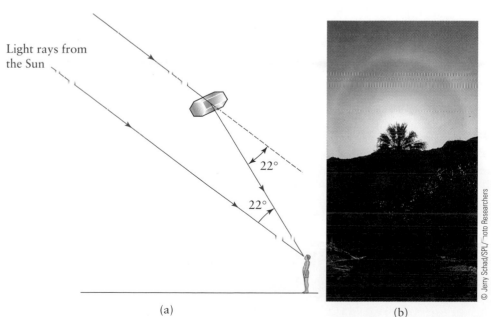

Light rays from the Sun

22°

22°

(a)   (b)

© Jerry Schad/SPL—hoto Researchers

● **Figure 9.76** (a) Schematic representation of how a 22° halo is produced, showing an enlarged image of the ice crystals typically responsible for this phenomenon. To see sunlight that is refracted through an angle of 22° by the ice crystals to form part of the halo, one looks to an angle of 22° away from the sun in the sky. (b) Actual 22° halo around the sun (itself partially obscured by the intervening tree).

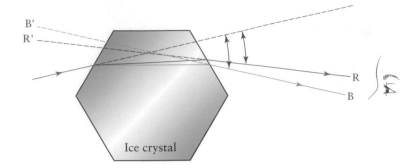

Figure 9.77 Dispersion of sunlight by ice crystals produces halos with reddish inner edges. Because the deviation of blue rays is larger than that of red rays, the blue rays appear to originate along the line $B'B$, farther away from the Sun than the red rays, which appear to come from the direction $R'R$. Thus, the inner edge of the halo—that is, the part nearest the Sun—is tinged reddish.

But why is it blue instead of white like the incident sunlight? It turns out that the electrons in air molecules are much more efficient at absorbing and radiating higher frequencies of light. When blue light makes an electron in an air molecule oscillate, it absorbs and scatters much more of the incident radiant energy than when red light makes it oscillate. In particular, the amount of radiant energy scattered per second by air molecules is inversely proportional to the light wave's wavelength raised to the fourth power.

$$\frac{E_{\text{scattered}}}{t} \propto \frac{1}{\lambda^4}$$

The wavelength of blue light is about 0.7 times that of red light (from Table 8.1). Consequently, there is about 4.2 ($= 1/0.7^4$) times as much blue in scattered sunlight as there is red. The sky is pale blue rather than "pure" blue since all frequencies of light are present.

Not only does molecular scattering give us beautiful blue skies, it is also responsible for the stunning orange and red colors often seen in clouds shortly after sunset and shortly before sunrise (● Figure 9.78a). The sunlight that reaches such clouds must travel hundreds

Figure 9.78 (a) Photo of clouds shortly after sunset. (b) Sunlight on its way to a cloud at sunset has most of the blue removed by scattering.

(a)

Reddish light illuminates cloud

Blue preferentially scattered by air

Sunlight (all frequencies)

Atmosphere

Earth

(Drawing not to scale.)

(b)

of kilometers through the atmosphere. Along the way, more and more of the radiant energy is removed from the sunlight and transferred to scattered light. The higher frequencies (like blue) are much more strongly attenuated by this process than are the lower frequencies (like red). Consequently, the light that reaches the cloud is reddish in color since comparatively less of the red in the original sunlight has been removed (Figure 9.78b).

The color of the sky directly overhead at dusk and at dawn is influenced by the ozone layer. At these times, much of the blue in the sunlight that reaches the air near the zenith has been scattered away. Without the ozone layer, this would cause the sky overhead to be a much more pale blue or even white. Ozone preferentially absorbs lower frequencies (like red) in visible light. Since the amount of ozone in the atmosphere is quite small, this process has little effect on visible sunlight during the day. But at dusk and at dawn, the sunlight that reaches the air molecules overhead travels a great distance almost horizontally through the ozone layer. The absorption of lower frequencies by ozone along the light's path partially balances the loss of higher frequencies due to scattering by air molecules. The light that reaches overhead contains less red than it would without the ozone layer, so the light scattered downward to us also contains less red and therefore appears bluer.

LEARNING CHECK

1. Which of the following is *not* involved in the formation of a typical 41° rainbow? (Indicate all that apply.)
   a) diffraction
   b) dispersion
   c) internal reflection
   d) refraction
   e) interference
2. If the Sun were just rising in the east at the time a rainbow is seen, then in what part of the sky would the rainbow be located?
   a) directly overhead at the zenith point.
   b) toward the northern horizon.
   c) toward the southern horizon.
   d) toward the eastern horizon.
   e) toward the western horizon.
3. Which of the following are responsible for producing halos around the Sun?
   a) water droplets
   b) dust particles
   c) ice crystals
   d) carbon dioxide molecules
   e) ozone molecules
4. (True or False.) In the atmosphere, the intensity of scattered sunlight of wavelength 480 nm is less than that of light of wavelength 590 nm.

ANSWERS: 1. (a),(e)   2. (e)   3. (c)   4. False

PHYSICS FAMILY ALBUM

Although much knowledge about the way light propagates in transparent media had been amassed before 1600, the developments in optics that occurred in the seventeenth century quickly eclipsed all that had been done during the previous 1,500 years. The first decade of that century saw the invention of the microscope and the refracting telescope, the latter being effectively used by Galileo to discover craters on the Moon, the phases of Venus, and four of the moons of Jupiter (● Figure 9.79). By 1611, Johannes Kepler (1571–1630) had discovered total internal reflection. In or about 1621, the law of refraction was found by Willebrord Snel (or in its anglicized form Snell), a professor of mathematics at the University of Leiden, after many years of experimentation. Unfortunately, Snel's results were not publicized until years after his death, and by that time Descartes had already succeeded in getting his version of the law of refraction into print.

René Descartes (1596–1650; ● Figure 9.80), "father of

● **Figure 9.79** Galileo demonstrating his discovery of the satellites of Jupiter to the counselors of Venice in 1610.

© The Granger Collection, New York

● **Figure 9.80** René Descartes at the court of Queen Christina of Sweden.

By permission of the Syndics of the Cambridge University Library · *© The Granger Collection, New York*

modern philosophy" and famous for his statement *Cogito, ergo sum* ("I think, therefore I am"), was also an accomplished mathematician and physicist. He founded the study of analytical geometry and was the first to develop the use of the hypothetical model as a tool of research. For example, his model for the propagation of light may be compared to that of a tennis ball moving uniformly. The law of reflection can be deduced by imagining an elastic collision of the ball with a stationary, impenetrable surface and applying the principle of conservation of linear momentum. Interestingly, Descartes was one of the first to recognize the momentum of a particle as an important physical quantity. These results appeared in 1637 at the beginning of Descartes's *La Dioptrique*.

The laws of reflection and refraction reappear in another of Descartes's works entitled *Météores*. Here he presents a mathematical explanation of primary and secondary rainbows, accounting for their angular dimensions and locations but not for their colors. It remained for Isaac Newton to produce the correct explanation for color in rainbows.

At the end of Chapter 2, several of the accomplishments of Sir Isaac Newton in the field of optics were mentioned, including his invention of the reflecting telescope and his explanation of the phenomenon of dispersion. These two items are very closely connected, the former being a direct consequence of the latter. In about 1666, Newton performed an *experimentum crucis* (a "critical experiment") in which, by a clever arrangement of two prisms, he was able to demonstrate that light of a single color undergoes no further disper-

sion upon being refracted through a prism (● Figure 9.81). From this and similar experiments, Newton concluded, "Light itself is a heterogeneous mixture of differently refrangible rays," asserting that there is an exact correspondence between color and the "degree of refrangibility," the least refrangible rays being "disposed to exhibit a red colour."

Newton's prismatic experiments seem to have convinced him of two things: first, that "light"—white light or sunlight—was not a "body" unto itself but an aggregate of corpuscles (tiny particles). "Each colour is caused by uniformly moving globuli," he wrote. "The uniform motion which gives the sensation of one colour is different from the motion which gives the sensation of any other colour." In particular, Newton believed the particles producing the sensation of blue light moved faster in glass than those associated with red light and that it was for this reason that blue rays were more strongly refracted. Newton's corpuscular theory of light permitted him to account for many of the properties of light, notably its straight-line propagation. But Newton recognized that certain phenomena, for example, diffraction and interference, could best be explained by treating light as a wave. Thus, even before the beginning of the eighteenth century, we see an apparent duality (particle versus wave) in the nature of light emerging. We will have more to say on this matter in Chapter 10.

The second thing of which Newton became convinced as a result of his color experiments was the impossibility of producing refracting telescopes (ones using only combinations of lenses) that were free from chromatic aberrations. It was for this reason that

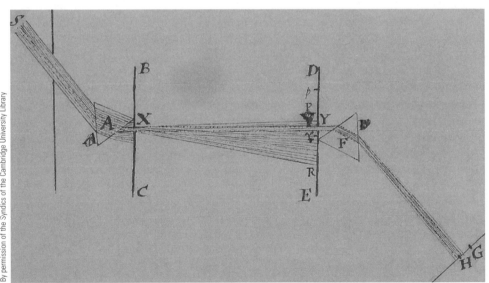

● **Figure 9.81** Newton's original drawing of his experiment with two prisms. The slit Y allowed only one color in the spectrum from prism A to go through prism F. The light leaving the second prism was the single color that entered it, not a complete spectrum. This showed that a prism does not somehow "add" color to light.

Newton turned his attention to the development of reflecting telescopes.

Newton's discoveries about light and color were published in 1704 in his *Opticks*. In one of Newton's investigations, he extended Descartes's model of rainbows, explicitly taking into account dispersion, and calculated the size of the rainbow arcs for each of the different colors in the primary and secondary rainbows. This book ends with a series of "Queries"—questions about the nature of light and related phenomena in need of further explanation and study. It has been argued that these questions formed the most important part of the book insofar as they strongly influenced the course of research on light for the next two centuries.

Newton's particle model of light stood for less than a century before it was dealt a nearly mortal blow by Thomas Young (1773–1829; ● Figure 9.82). Young was born into an English family of Quakers and was quickly seen to be a genius. By the age of 14, he knew several languages and had written an autobiography in Latin. Although Young studied medicine and became a physician, his intellectual pursuits spanned an incredible variety of disciplines, from physics to hieroglyphics (he worked on the Rosetta Stone) to insurance.

● **Figure 9.82** Thomas Young. His work with interference demonstrated that light possesses wavelike properties.

Young championed the wave model of light and showed that it could account for interference and diffraction of light. In his most famous experiment, Young passed light through two pinholes and produced an interference pattern similar to the one in Figure 9.7. From this pattern he was able to measure the wavelengths of different colors of light. Young was also the first to suggest that light could be polarized and was therefore a transverse wave. Young's success, along with that of other physicists, including Fresnel, in the early 1800s, caused the wave model of light to replace Newton's particle model—no small feat considering Newton's great stature. But as we shall see in Section 10.2, the particle model of light was reincarnated a century later.

Optical science in the nineteenth century also profited from the work of another polymath, Karl Friedrich Gauss (1777–1855; ● Figure 9.83). Gauss was a child prodigy whose accomplishments in mathematics, physics, and astronomy mark him as one of the greatest intellects of all time. Among Gauss's many contributions to physics, we mention only his work in magnetism and, of course, optics. In collaboration with another German physicist, Wilhelm Weber, Gauss established an observatory dedicated to the study and mapping of terrestrial magnetism. He also developed a self-consistent set of units (distance, mass, and time) that could be used to measure nonmechanical quantities like those encountered in his electromagnetic investigations.

In the field of optics, Gauss invented the heliotrope, an instrument used in surveying and triangulation work, but only *after* he fully developed the necessary optical theory. In 1841, he published *Dioptrische Untersuchungen*, in which he analyzed the path of light through a system of lenses. In this work, in addition to determining how image distance, object distance, and the focal length of a lens are related to one another (the lens formula), he showed that any system of lenses is equivalent to a properly chosen single lens. This was Gauss's last significant contribution to science, and one of his biographers has called it his greatest work.

Advancements in optics by no means came to an end in the mid-1800s. Beginning shortly after World War II, applied optics began to flourish as the mathematical methods of communication theory and high-speed digital computers were brought into the discipline. Optical technology, driven by the development of the laser in 1960 (see Section 10.8), continues to advance, bringing improve-

● **Figure 9.83** Karl Friedrich Gauss.

ments in our lives and in our understanding of the universe. But if rapid progress has recently occurred, it has happened because modern scientists have, in the words of Newton, "stood on the shoulders of giants," including Descartes, Gauss, and Newton himself.

SUMMARY

- Visible light consists of electromagnetic waves with wavelengths between about 400 and 750 nanometers.

- When two or more light waves overlap in any region of space, **interference** occurs. If the waves are in phase, they reinforce one another, yielding **constructive interference.** If the waves are out of phase, they tend to cancel one another to some degree, producing **destructive interference.**

- Because light is a transverse wave, it can be **polarized,** that is, made to oscillate along some preferred direction. Polaroid sunglasses and LCDs make use of this property of light.

- The **law of reflection** states that the **angle of incidence** equals the **angle of reflection.**

- Mirrors make use of the law of reflection to modify and control the direction of propagation of light rays: Plane mirrors produce undistorted virtual images of the object before them; concave mirrors can concentrate light rays and yield magnified images; convex mirrors disperse light rays and have a wide field of view.

- **Refraction** occurs when light crosses a boundary between two different transparent media. The **law of refraction** connects the size of the **angle of refraction** to that of the angle of incidence: The refraction angle is smaller than the incident angle if the wave speed in the transmitted medium is lower than that in the incident medium. Conversely, it is larger than the incident angle if the wave speed in the transmitted medium is higher than that in the incident medium.

- When the wave speed in the medium in which the transmitted ray travels is higher than that in the medium in which the incident ray travels, there exists a **critical angle** of incidence for which the angle of refraction will be 90°. For angles of incidence larger than the critical angle, incident rays undergo **total internal reflection** and remain trapped in the incident medium. This effect produces the brilliant sparkle in diamonds and underlies the operation of optical fibers.

- Lenses are optical devices that use refraction to control the paths of light rays and to form images. Simple, thin lenses may be either diverging or converging, according to whether their **focal lengths** are negative or positive, respectively. By knowing the focal length of a lens and the position of an object relative to the lens, the location of the image formed by the lens may be determined using the lens formula.

- The **magnification** of the image provides information about the size and orientation of the image. It equals the negative ratio of the image distance to the object distance.

- **Dispersion** describes the process whereby individual wavelengths (colors) comprising a beam of light are separated at the boundary between two transparent media. Dispersion occurs because the speed of light in any medium other than a vacuum is different for different wavelengths. Dispersion is responsible for unwanted chromatic aberrations in simple lenses, but when controlled by the use of prisms, it can be employed to establish the temperature and composition of luminous sources.

- Application of the laws of reflection and refraction, including the effects of dispersion, can lead to an increased understanding and appreciation of many natural phenomena in the area of atmospheric optics, like rainbows and halos.

IMPORTANT EQUATIONS

Equation	Comments
$p = \dfrac{sf}{s - f}$	Lens formula
$M = \dfrac{\text{image height}}{\text{object height}}$	Magnification of a lens system
$M = \dfrac{-p}{s}$	Magnification using image distance and object distance

MAPPING IT OUT!

1. The "shell" of a concept map dealing with lenses and their properties is shown below. Most of the concepts and all of the linking phrases needed to form meaningful propositions have been left out. Complete this map by selecting the appropriate concepts and linkages from the lists that follow. Some connecting words will have to be used more than once to finish the map correctly. If you need to, be sure to review Sections 9.3 through 9.5 for help in properly determining and relating the relevant concepts.

2. Sections 9.6 and 9.7 deal with the phenomena of dispersion and its applications to atmospheric optics. One such application is the explanation of rainbows. Develop a concept map that you could use to teach a friend or family member about the formation and basic characteristics of rainbows. Your map should include links to the fundamental physical processes of reflection, refraction, and dispersion and should distinguish between primary and secondary rainbows. It should also indicate the conditions that must be satisfied to observe rainbows.

Concepts	Linking Words
Altered size	which can be
Real ($p > 0$)	come in
Upright ($M > 0$)	for example
Diverging	form
Ray diagram	using

(continued on next page)

Concepts	Linking Words
Altered orientation	can be combined in
Eye lens	located by
Inverted ($M < 0$)	
Reduced ($|M| < 1$)	
Two principal rays	
Converging	
Telescope	
Enlarged ($|M| > 1$)	
Microscope	
Virtual ($p < 0$)	

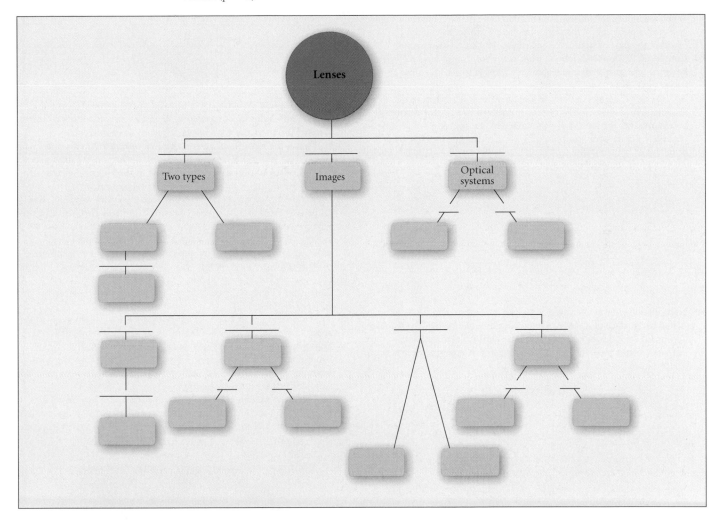

QUESTIONS

(▶ Indicates a review question, which means it requires only a basic understanding of the material to answer. The others involve integrating or extending the concepts presented thus far.)

1. ▶ Why are the Doppler effect and diffraction not as commonly observed with light as they are with sound?
2. ▶ Describe specular reflection and diffuse reflection.
3. ▶ The law of reflection establishes a definite relationship between the angle of incidence of a light ray striking the boundary between two media and its angle of reflection. Describe this relationship.
4. The cover of a book appears blue when illuminated with white light. What color will it appear in blue light? Red light? Explain.

5. A ballet dancer is on stage dressed in a green body suit. How could you light the stage so that the dancer's costume looked black to the audience?
6. ▶ Describe how light passing through two narrow slits can produce an interference pattern.
7. A person looking straight down on a film of oil on water sees the color red. How thick could the film be to appear thin? How thick could it be at another place where violet is seen? (Some useful information is given at the beginning of Section 9.1.)

8. If the wavelength of visible light were around 10 cm instead of 500 nm and we could still see it, what effect would this have on diffraction and interference?

9. An interference pattern is formed by sending red light through a pair of narrow slits. If blue light is then used, the spacing of the bright areas (where constructive interference takes place) won't be the same. How will it be different? Why?

10. ▶ What is polarized light? How do Polaroid sunglasses exploit polarization?

11. Describe how you could use two large, circular Polaroid filters in front of a circular window as a kind of window shade.

12. Just before the Sun sets, a driver encounters sunlight reflecting off the side of a building. Will Polaroid sunglasses stop this glare?

13. You get a new pair of sunglasses as a birthday gift from a friend. When wrapping the present, your friend has removed the price tag and all the other labels from the sunglasses. How can you tell whether the pair of shades has polarizing lenses in them or not?

14. Compared to a person's height, what is the minimum length (top to bottom) of a mirror that will allow the person to see a complete image from head to toe?

15. An observer O stands in front of a plane mirror as shown in the accompanying figure. Which of the numbered locations 1 through 5 best represents the location of the image of the source S seen by the observer? Justify your choice by appealing to the appropriate law(s) of optics.

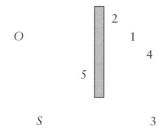

16. Light traveling along direction SO in the accompanying figure strikes the surface of a plane mirror. Which of the paths OP, OQ, OR, or OT best describes the path of the light reflected from the mirror? Defend your choice by appealing to the appropriate law(s) of optics.

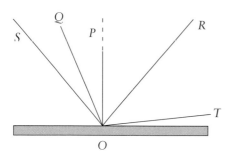

17. Consider the figure below. At which of the lettered positions would an observer be able to see the image of the dot in the mirror?

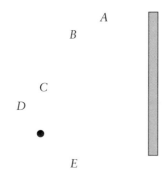

18. Two plane mirrors are hinged along one edge and set at right angles to one another as shown in the figure below. A light ray enters the system, striking mirror 1 with an angle of incidence of 45°. Draw a diagram showing the direction in which the ray will exit the system. This device is called a *corner reflector* and is the basis for the design of bicycle reflectors and some highway signs.

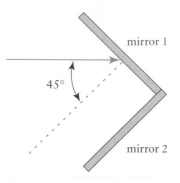

19. ▶ What is different about an image (of a nearby object) formed with a convex mirror compared to an image formed with a concave mirror? What are the advantages of each type of mirror?

20. ▶ What is the ideal shape of concave mirrors used in telescopes?

21. If you were lost in the forest and wanted to start a small fire to keep warm or cook a meal using sunlight and a small lens, what type of lens—converging or diverging—would you use and why?

22. What kind of mirror (concave, convex, or plane) do you think dental care workers use to examine patients' teeth and gums for disease? Explain your choice.

23. ▶ Describe how the path of a ray is deviated as it passes (at an angle) from one medium into a second medium in which the speed of light is lower. Contrast this with the case when the speed of light in the second medium is higher.

24. How would Figure 9.29 be different if the glass were replaced by water or by diamond?

25. The speed of light in a certain kind of glass is exactly the same as the speed of light in benzene—a liquid. Describe what happens when light passes from benzene into this glass, and vice versa.

26. A piece of glass is immersed in water. If a light ray enters the glass from the water with an angle of incidence greater than zero, in which direction is the ray bent?

27. A light ray in air enters a block of clear plastic as shown in the figure. Which of the numbered paths represents the correct one for the ray in the plastic? Justify your choice by appealing to the appropriate law(s) of optics.

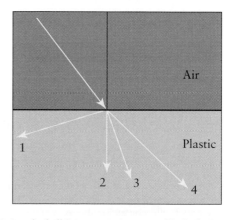

28. After hitting the ball into a water trap, a golfer looks into the pond and spies the ball within apparent easy reach. Reaching in to retrieve the ball, the golfer is surprised to find that it cannot be grasped even with a fully extended arm. Explain why the golfer was deceived into thinking that the location of the ball was close at hand.

29. One sometimes hears the expression, "It was like shooting fish in a barrel!" This usually is taken to mean that the task, whatever it was, was easy to complete. But is it really easy to shoot fish in a barrel? Only if you know some optics! Suppose you're in a boat and spy a large fish a few meters away. If you want to shoot the fish, how should you aim? Above the image of the fish? Below it? Directly at the image? Explain your choice. (You may assume that the path of the bullet you fire will not be deviated from a straight line upon entering the water, unlike light.)

30. Rank (from smallest to largest) the angle of refraction for a light ray in air entering each of the following substances with an angle of incidence equal to 30°: (i) water; (ii) benzene; (iii) dense flint glass; (iv) diamond.

31. ▶ What is total internal reflection, and how is it related to the critical angle?

32. Explain why images seen through flat, smooth, uniform, plate-glass windows are undistorted.

33. ▶ For transparent solids, distinguish between effects of surfaces that curve inward and those that curve outward on the paths of a parallel bundle of light rays incident on each.

34. ▶ Distinguish converging lenses from diverging lenses, and give examples of each type.

35. Of the three converging lenses shown in Figure 9.47, which would you expect to have the shortest focal length?

36. ▶ What are the three principal rays?

37. ▶ Contrast real images with virtual images in as many ways as you can.

38. Indicate whether each of the following is a real image or a virtual image.
    a) Image on the retina in a person's eye.
    b) Image one sees in a rearview mirror.
    c) Image one sees on a movie screen.
    d) Image one sees through eyeglasses or contact lenses.

39. A convex lens forms a clear, focused image of some small, fixed object on a screen. If the screen is moved closer to the lens, will the lens have to be (i) moved closer to the object, (ii) moved farther from the object, or (iii) left at the same location to produce a clear image on the screen at its new location? Explain your answer.

40. Sketch the image of the letter **F** formed on a screen by a pinhole camera or on the film in a simple camera.

41. ▶ How is the magnification of a lens related to the object distance? How is it related to the image distance? What is the significance of the sign of the magnification? What is the significance of its magnitude (size)?

42. Estimate the values of the magnification in Figures 9.48, 9.51, and 9.52.

43. ▶ What is chromatic aberration? How can it be cured?

44. ▶ How is the eye able to form focused images of objects that are different distances away?

45. ▶ When a person is nearsighted, what happens in the eye when the person is looking at something far away? What is used to correct this problem?

46. Make a drawing of the front part of the eye (see Figure 9.61). Indicate where tissue should be removed during laser surgery to correct myopia. Where would it be removed to correct hyperopia?

47. ▶ Describe the phenomenon of dispersion, and explain how it leads to the production of a spectrum.

48. Two light waves that have wavelengths of 700 and 400 nm enter a block of glass (from air) with the same angle of incidence. Which has the larger angle of refraction? Why? Would the answer be different if the light waves were going from glass into air?

49. Suppose a 20-m long tube is filled with benzene and the ends sealed off with thin disks of glass. If pulses of red and blue light are admitted simultaneously at one end of the tube, will they emerge from the opposite end together, that is, at the same time? If not, which pulse will arrive at the far end of the tube first? Why?

50. The difference in speed between red light and violet light in glass is smaller than the difference in speed between the same two colors in a certain type of plastic. For which material, glass or plastic, would the angular spread of the two colored rays after entering the material obliquely from air be the largest? Why?

51. ▶ What is a prism? Why are such devices useful to scientists?

52. Would a prism made of diamond be better than one made of glass? Why or why not?

53. Referring to Figure 9.68, why doesn't the ray inside each of the two prisms get refracted out of the prisms at the second interface instead of turned back into the glass? Explain what must be happening at each of these boundaries to create the ray paths displayed.

54. ▶ Describe the accepted model of rainbows. Specifically, discuss how the model accounts for the size, shape, location, and color ordering of primary rainbows.

55. Suppose an explosion at a glass factory caused it to "rain" tiny spheres made of glass. Would the resulting rainbow be different from the normal one? If so, how might it be different and why?

56. ▶ Compare the primary and secondary rainbows as regards their angular size, color ordering, and number of internal reflections that occur in the rain droplets.

57. Suppose you are told by close friends that they had witnessed a glorious rainbow in the west just as the Sun was setting. Would you believe them? Why or why not?

58. ▶ How is a 22° halo formed? Describe a measurement technique you might use to distinguish a genuine 22° halo from some other "halolike" phenomenon seen surrounding the Sun or the Moon.

59. Water droplets in clouds scatter sunlight similar to the way air molecules do. Since clouds are white (during the day), what can you conclude about the frequency dependence of scattering by cloud particles?

PROBLEMS

1. A light ray traveling in air strikes the surface of a slab of glass at an angle of incidence of 50°. Part of the light is reflected and part is refracted. Find the angles the reflected and refracted rays make with respect to the normal to the air-glass interface.

2. A ray of yellow light crosses the boundary between glass and air, going from the glass into air. If the angle of incidence is 20°, what is the angle of refraction?

3. Using Figure 9.34, find the angles of refraction for a light ray passing from air into glass with the following angles of incidence: 5°, 10°, and 20°. Do you notice a trend in the resulting values? If so, describe it. Based on your observations, what would you predict the angle of refraction to be for an angle of incidence of 40°? How does your value compare with that inferred from Figure 9.34?

4. Using the definition of the index of refraction contained in the footnote on p. 352 and the data in Table 9.1, compute the index of refraction, n, for the following substances: (i) air; (ii) benzene; and (iii) glass (any type).

5. A fish looks up toward the surface of a pond and sees the entire panorama of clouds, sky, birds, and so on, contained in a narrow cone of light, beyond which there is darkness. What is going on here to produce this vision, and how large is the opening angle of the cone of light received by the fish?

6. A camera is equipped with a lens with a focal length of 30 cm. When an object 2 m (200 cm) away is being photographed, how far from the film should the lens be placed?

7. A 2.0-cm-tall object stands in front of a converging lens. It is desired that a virtual image 2.5 times larger than the object be formed by the lens. How far from the lens must the object be placed to accomplish this task, if the final image is located 15 cm from the lens?

8. When viewed through a magnifying glass, a stamp that is 2 cm wide appears upright and 6 cm wide. What is the magnification?
9. A person looks at a statue that is 2 m tall. The image on the person's retina is inverted and 0.005 m high. What is the magnification?
10. What is the magnification in Problem 6?
11. A small object is placed to the left of a convex lens and on its optical axis. The object is 30 cm from the lens, whose focal length is 10 cm. Determine the location of the image formed by the lens. Describe the image.
12. If the object in Problem 11 is moved toward the lens to a position 8 cm away, what will the image position be? Describe the nature of this new image.
13. The focal length of a diverging lens is negative. If $f = -20$ cm for a particular diverging lens, where will the image be formed of an object located 50 cm to the left of the lens on the optical axis? What is the magnification of the image?

14. The equation connecting s, p, and f for a simple lens can be employed for spherical mirrors, too. A concave mirror with a focal length of 8 cm forms an image of a small object placed 10 cm in front of the mirror. Where will this image be located?
15. If the mirror described in the previous problem is used to form an image of the same object now located 16 cm in front of the mirror, what would the new image position be? Assuming that the magnification equations developed for lenses also apply to mirrors, describe the image (size and orientation) thus formed.
16. Compute the approximate ratio of the amount of blue light in blue sky to the amount of orange.
17. If the wavelength of light is doubled, by what fraction does the amount of the light scattered by the Earth's atmosphere change? Is the amount of light scattered at the new wavelength larger or smaller than the amount of light scattered at the old wavelength?

CHALLENGES

1. When white light undergoes interference by passing through narrow slits, dispersion occurs. Why? What is the ordering of the various colors, starting from the middle of the pattern?
2. As the discussion of Young's double slit experiment in Section 9.1 indicates, the separation, Δx, between two adjacent bright areas or *fringes* in the interference pattern is determined by the distance, a, between the slits, the distance, S, between the plane of the slits and the screen on which the pattern is projected, and the wavelength, λ, of light used in the experiment. Specifically, the relationship among these variables is given by:

$$\Delta x = \left(\frac{S}{a}\right)\lambda.$$

Based on this equation, for a given slit spacing and screen distance, for which color of light, red or blue, will the fringe spacing be greater? For a given slit spacing and wavelength, if the screen distance, S, is increased, what happens to the fringe spacing? If the separation between the slits is decreased, for a given value of λ and S, how will Δx change? Suppose that a beam of red light from a He-Ne laser ($\lambda = 633$ nm) strikes a screen containing two narrow slits separated by 0.200 mm. The fringe pattern is projected on a white screen located 1.00 m away. Find the distance in millimeters between adjacent bright areas near the center of the interference pattern that develops. As mentioned in the text, the relationship given above could be used to calculate the wavelength of light in an experiment in which the other three quantities are measured.
3. In Section 9.6, we described how the speed of light varies with wavelength (or frequency) for transparent solids. But the speed of light is also a function of temperature and pressure. This dependence is most marked for gases and is instrumental in producing such things as mirages and atmospheric refraction, the latter phenomenon being the displacement of an astronomical object (like the Sun or another star) from its true position because of the passage of its light through the atmosphere. Because the Earth's atmosphere is a gaseous mixture and easily compressed, its density is highest near the Earth's surface (where the weight of the overlying layers exerts a strong compressional force) and gradually declines with altitude. (Refer to the discussion in Section 4.4 and Figure 4.28.) Thus, the speed of light in the atmosphere is lowest near the surface and gradually gets higher, approaching c as one goes farther and farther into space. Using this fact and the law of refraction, sketch the path a light ray from the Sun would follow upon entering the Earth's atmosphere, and predict the apparent position of the Sun relative to its true position (● Figure 9.89). What does this tell you about the actual location of the Sun's disk relative to your local horizon when you see it apparently setting brilliantly in the west in the evening?

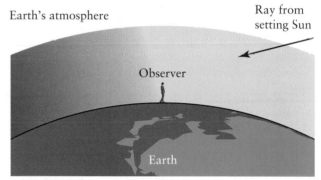

● **Figure 9.89** Challenge 3 (Drawing not to scale.)

4. Would the critical angle for a *glass-water* interface be less than, equal to, or greater than the critical angle for a *glass-air* interface? Why?
5. Light enters diamond from air with angle of incidence 45°. What is the angle of refraction? There is useful information in the footnote on p. 352.
6. While above water, a swimmer's eyes are focused on a nearby boat. When the swimmer submerges, the underwater part of the boat will not be in focus even though it is the same distance away and the swimmer's eyes have not been refocused. Why? Watertight goggles correct this. Why?
7. The form of the lens formula used most commonly in physics is

$$\frac{1}{f} = \frac{1}{s} + \frac{1}{p}$$

   a) Use this to derive the equation that gives s in terms of p.
   b) In a camera equipped with a 50-mm focal-length lens, the maximum distance that the lens can be from the film is 60 mm. What is the closest an object can be to the camera if its photograph is to be in focus? What is the magnification?
   c) An extension tube is added between the lens and the camera body so that the lens can be positioned 100 mm from film. How close can the object be now? What is the magnification?
8. For $f = 10$ cm, use the lens formula (see Challenge 7) to compute the image distance, p, for the values of object distance, s, given in the table below. A few values of p have already been included so you can check your work. Make a graph of p versus s using your results. What does the sign of p signify? For what values of s does p change the most? What happens to p as s gets larger and larger? Can you relate your conclusions to the operation of a simple camera or your

eye? (*Note:* If you are familiar with the use of PC-based spreadsheet programs, the calculations required in this exercise are particularly easy to complete. Even the plot may be done using the graphics capabilities supported by most such programs.)

s (cm)	p (cm)	s (cm)	p (cm)
2.0		25.0	
4.0	−6.67	50.0	12.50
8.0		100	
9.999		200	10.53
12.0		500	
14.0	35.00	1000	
18.0		10,000	
20.0			

9. A 35-mm camera is used to photograph a kitten. The camera is equipped with a standard lens that has a focal length of 50 mm. The camera is focused by moving the lens closer to or farther away from the film at the back of the camera.

   a) First the kitten is photographed when it is 350 mm (13.8 in.) in front of the camera lens. Use the lens formula to compute the distance that the lens must be from the film for the image to be in focus.

   b) Compute the magnification and the height of the image, assuming that the kitten is 100 mm tall.

   c) The kitten is then photographed from a distance of 3,500 mm (11.5 ft). Compute the lens-to-film distance, the magnification, and the image height.

10. A camera is equipped with a telephoto lens that has a focal length of 200 mm. Repeat part (c) of Challenge 9 for this situation.

11. Why can't you go to the end of a rainbow? What happens to the rainbow when you walk toward one of its ends?

12. When the Sun is straight west and low in the sky, the light from blue sky to the north or to the south is polarized. Why?

SUGGESTED READINGS

Greenler, Robert. *Rainbows, Halos, and Glories.* New York: Cambridge University Press, 1990. "This book describes beautiful things that can be seen in the sky, things that can be seen without special equipment or special location, things that can be seen by anyone who sees."

Jayawardhana, Ray. "The Age of Behemoths." *Sky and Telescope* 103, no. 2 (February 2002): 30–37. The author highlights the development of 8-meter-class telescopes and the ways in which these instruments will inaugurate a truly new age of astronomical discovery.

Lynch, David K., and William Livingston. *Color and Light in Nature.* New York: Cambridge University Press, 1995. "An endlessly fascinating exploration of phenomena that are familiar to us all, but are often taken for granted even by trained scientists."

Musa, Samuel. "Active-Matrix Liquid-Crystal Displays." *Scientific American* 277, no. 5 (November 1997): 124. An explanation of how the color LCDs on laptop computers work.

Schilling, Govert. "Adaptive Optics Comes of Age." *Sky and Telescope* 102, no. 4 (October 2001): 30–40. This article discusses the techniques now being employed by astronomers using the world's largest telescopes to cancel out the blurring effects of the Earth's atmosphere and thereby to sharpen their views of the universe.

Sorbjan, Zbigniew. *Hands-On Meteorology.* American Meteorology Society, 1996. Includes many do-it-yourself experiments on such phenomena as rainbows and blue sky.

Stix, Gary. "The Triumph of Light." *Scientific American* 284, no. 1 (January 2001): 80–86. An update on the promise of optical fibers for providing near-infinite carrying capacity for modern communications networks.

Wanjek, Christopher. "Hubble's New Eyes." *Sky and Telescope* 103, no. 3 (March 2002): 34–39. Description of HST's new Advanced Camera for Surveys and the promise it holds for revolutionizing deep-space imaging.

For additional readings, explore InfoTrac® College Edition, your online library. Go to http://www.infotrac-college.com/wadsworth and use the passcode that came on the card with your book. Try these search terms: Fresnel lens, camera obscura, LCD, adaptive optics, optical fibers, LASIK, Thomas Young, K. F. Gauss.

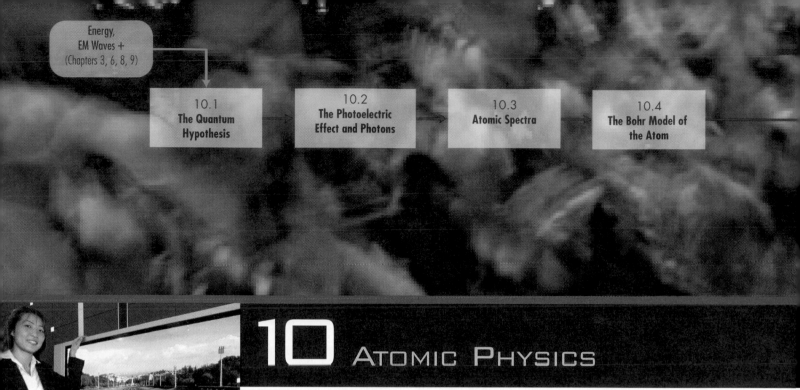

Energy,
EM Waves +
(Chapters 3, 6, 8, 9)

| 10.1 The Quantum Hypothesis | 10.2 The Photoelectric Effect and Photons | 10.3 Atomic Spectra | 10.4 The Bohr Model of the Atom |

10 ATOMIC PHYSICS

Plasma TV

Something Old, Something New

Fluorescent light and plasma TV. One is a familiar, trusted, usually ignored device that's been providing efficient lighting for decades. The other became one of the first status symbols of the twenty-first century. One is typically hidden in fixtures and noticed only if it doesn't work correctly (it flickers or buzzes). The other costs thousands of dollars and is often the center of attention—even becoming part of the studio set for TV news programs and military press briefings. Its inventors shared an Emmy Award in 2002. One is mundane, the other glamorous.

What do these two devices have in common? They produce light using the same process. A plasma display is basically millions of tiny fluorescent lamps emitting light in a coordinated fashion. It is a three-step process: Electrons and ions move through a plasma and give energy to atoms. These atoms in turn emit ultraviolet light, which strikes a special material called a phosphor. The phosphor's atoms absorb the UV, which we could not see, and emit the visible light that we do see.

How is the ultraviolet light produced? How is it absorbed by atoms in the phosphor, and how does that yield visible light? The answers involve the concepts of photons, emission-line spectra, and atomic energy levels—all key ideas presented in this chapter.

Fluorescent lights and plasma TVs are examples of the many devices that utilize the two main topics of this chapter, modern atomic physics and the quantum nature of light. The first part of the chapter describes how the efforts of early twentieth-century physicists to understand three puzzling physical processes led to the concept of photons and the Bohr model of the atom. This is followed by a description of the wave nature of atomic particles and the emergence of the revolutionary new physics known as quantum mechanics. The rest of the chapter shows how the quantum-mechanical model of the atom successfully explains the production of atomic spectra and x rays and the operation of lasers.

10.1 The Quantum Hypothesis

By the end of the nineteenth century, physicists were quite satisfied with the great progress that had been made in the study of physics. Many questions that had been puzzling scientists for thousands of years had been answered. Advances such as Newton's

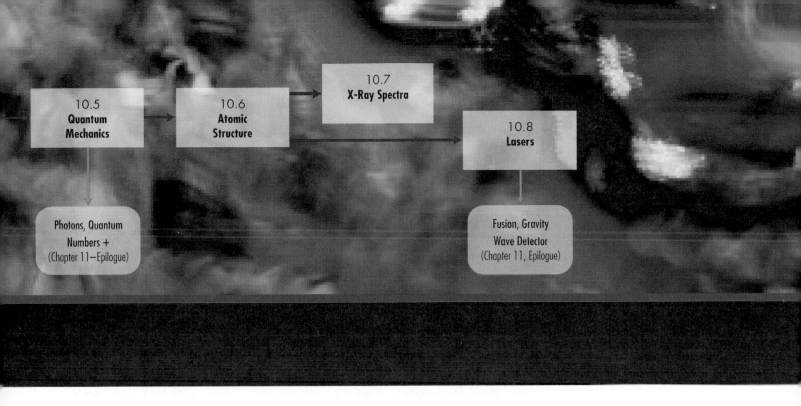

great treatise on mechanics and the clarification of electromagnetism by Maxwell gave physicists cause to celebrate the deep understanding of the physical world that they had acquired. In fact, some physicists worried that the field might be dying; they feared that *all* of their questions might soon be answered.*

But some problems defied solution, and advancements in experimental equipment and techniques led to new discoveries. The implications arising from attempts to explain these phenomena were so revolutionary that by the early 1900s many of these same physicists were bewildered and wondered just how much they really did know.

In the first five sections of this chapter, we take an historical approach to three of these problems. We describe how investigations into them led to a reinterpretation of the very nature of light and the way it interacts with matter. The phenomena are (1) **blackbody radiation,** (2) the **photoelectric effect,** and (3) **atomic spectra.**

Physicists knew a great deal about these phenomena, but the fundamental understanding of their *causes* had eluded them. It was much like the period before Newton: Astronomers knew a lot about the shapes of the orbits of the Moon and the planets, but they didn't know what caused these particular shapes. Newton's mechanics and his law of universal gravitation provided the answer. Similarly, scientists some two centuries later were seeking the theoretical basis for these three phenomena.

Blackbody Radiation

Everything around you, as well as your own body, is constantly emitting electromagnetic (EM) radiation. At normal room temperature, mostly infrared (IR) radiation is emitted. Very hot bodies that are actually glowing, such as the Sun and the heating elements in a toaster, emit visible light and ultraviolet (UV) radiation as well as copious amounts of infrared. Bodies that are very cold, on the other hand, emit very weak, long-wavelength infrared and microwaves. This radiation was studied carefully by scientists in the nineteenth century, and its properties were well-known but not understood. It was determined that a perfectly "black" body, one that would absorb all light and other EM radiation incident upon it, would also be a perfect emitter of EM radiation. Such a body is called a *black-body,* and the EM radiation emitted by it is called *blackbody radiation* (BBR).

This is an overview of the characteristics of blackbody radiation. For more details, including applications, refer to Section 8.6.

---

* With the advent of *grand unified theories* and *theories of everything* (TOEs) (see Section 12.5) within the last three decades or so, similar concerns have been raised again by some physicists. While the views expressed by these scientists may not be universally shared, the issues raised by them concerning our progress toward viable TOEs are important.

Explore It Yourself 10.1

The glowing filament of a regular incandescent lamp acts roughly like a blackbody with a temperature of about 2,500 to 3,000 K. You can view the visible part of the radiation it emits with a compact disc. The spiral line of data on the CD causes light to undergo dispersion—the CD acts somewhat like a prism. (The light undergoes diffraction and interference as it reflects off the closely spaced segments of the line, and different colors (frequencies) of light undergo constructive interference in different directions.)

Use the "shiny" side of the CD (usually the bottom), and position it so you can see the reflection of a glowing light bulb near the hole in the center. (It's OK if the bulb is in a light fixture.) Slowly tilt the CD so the image of the bulb moves to one side of the hole in the center, and note the appearance of colors on the opposite side of the hole. What do the colors look like? Does it appear that all colors of the visible spectrum are present?

The characteristics of the BBR emitted by a given blackbody at a particular temperature can be illustrated with a graph. Imagine examining each wavelength of radiation in turn, from very short wavelength ultraviolet to much longer wavelength microwaves, and measuring the energy that is emitted each second (the power) from the blackbody. The graph of the resulting data, plotted as the intensity of the radiation versus wavelength, is called a *blackbody radiation curve*. ● Figure 10.1 shows two BBR curves for blackbodies at two different temperatures. These graphs illustrate two important ways that BBR changes when the temperature of the body is increased:

1. More *energy* is emitted per second *at each wavelength* of EM radiation.
2. The *wavelength* at which the *most energy* is emitted per second (in other words, the *peak* of the BBR curve) *shifts to smaller values*. That is why toaster elements glow red hot while extremely hot stars glow blue hot.

Why does a blackbody emit different amounts of EM radiation at different wavelengths in precisely this way? Using the principles of electromagnetism, we can easily see why EM waves are emitted. Atoms and molecules are continually oscillating, and they contain charged particles—electrons and protons. We saw in Chapter 8 that this kind of system will produce EM waves. But the exact mechanism involved, one that would account for the two features above, was a mystery to scientists.

● **Figure 10.1** Graph of intensity versus wavelength for the EM waves emitted by blackbodies at two different temperatures.

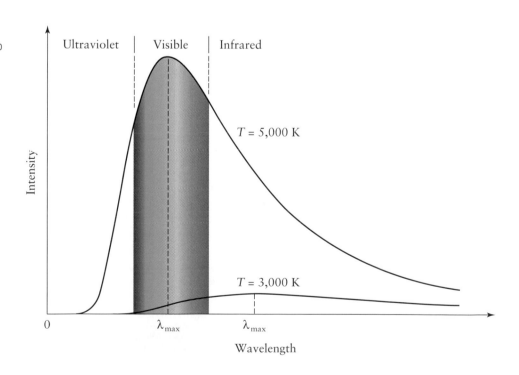

The clue to solving the puzzle was discovered by the German physicist Max Planck (● Figure 10.2) in the year 1900. First, by trial and error, Planck derived a mathematical equation that fit the shape of the BBR curves. This is a common first step in theoretical physics. The mathematical shapes of the planetary orbits—ellipses—were known about a century before Newton explained *why* they were ellipses. The equation did little to increase the understanding of the fundamental process, but Planck also developed a model that would account for his equation.

Planck proposed that an oscillating atom in a blackbody can have only certain fixed values of energy. It can have zero energy or a particular energy E, or 2, 3, 4, 5, and so on, times the energy E. In other words, the energy of each atomic oscillator is **quantized.** The energy E is called the fundamental **quantum** of energy for the oscillator. The allowed values of energy for the atom are integral multiples of this quantum:

allowed energy = 0,   or E,   or 2E,   or 3E,   . . .

This is quite different from an ordinary oscillator, such as a mass hanging from a spring or a child on a swing, which can have a continuous range of energies, not just certain values.

We can illustrate the difference between quantized and continuous energy values by comparing a stairway and a ramp. A cat lying on a stairway has quantized potential energy: It can be on only one of the steps, and each step corresponds to a particular *PE*. A cat lying on a ramp does not have quantized potential energy. It can be anywhere on the ramp. Its height above the ground can have any value within a certain range, and therefore, its *PE* is one of a continuous range of values (● Figure 10.3).

The concept of energy quantization for the oscillating atoms was revolutionary. There seemed to be no logical reason for it. But it worked. Planck's quantized atomic oscillators could emit light only in bursts as they went from a higher energy level to a lower one. He showed that light emitted in this fashion by a blackbody resulted in the correct BBR curves. Moreover, he determined that the basic quantum of energy was proportional to the oscillator's frequency. In particular:

$$E = hf \qquad h = 6.63 \times 10^{-34} \text{ J-s}$$

The constant h is called *Planck's constant.* The allowed energies are:

allowed energy = 0,   or hf,   or 2hf,   or 3hf,   . . .

Planck was unsure of the implication of his model. He regarded it mainly as a helpful gimmick that gave him the correct result. But it turned out to be the first of many scientific discoveries about how things at the atomic level are quantized.

● **Figure 10.2** Max Planck (1858–1947), whose work started what can be called the *quantum revolution.*

LEARNING CHECK

1. The temperature of blackbody A is 2,000 K and that of blackbody B is 1,000 K. Regarding the electromagnetic radiation emitted by the two blackbodies:
   a) A emits more energy per second at any given wavelength than B.
   b) The peak of A's blackbody radiation curve is at a longer wavelength than that of B's.
   c) B emits infrared but A does not.
   d) All of the above.

2. (True or False.) The energy of an oscillating atom in a blackbody can be any value within a certain limited range.

3. When something is restricted to having only certain numerical values, it is said to be _____.

ANSWERS: 1. (a)   **2.** False   **3.** quantized

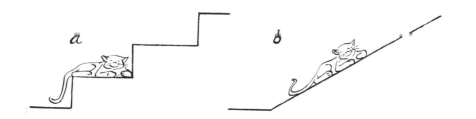

● **Figure 10.3** (a) A "quantized" cat. Its potential energy is restricted to certain values, one for each step. (b) On the ramp the cat can be anywhere, so its potential energy is not quantized.

10.2 The Photoelectric Effect and Photons

The second phenomenon puzzling physicists at the beginning of the twentieth century was the photoelectric effect. It seemed to be unrelated to blackbody radiation, other than that it also involved light, but the concept of quantization turned out to be the key to explaining it, too. The photoelectric effect was accidentally discovered by Heinrich Hertz during his experiments with electromagnetic waves. Hertz was producing EM wave pulses by generating a spark between two conductors. The EM wave would travel out in all directions and induce a spark between two metal knobs used to detect the wave. Hertz noticed that when these knobs were illuminated with ultraviolet light, the sparks were much stronger. The UV was somehow increasing the current in the spark. This is one example of the photoelectric effect.

The photoelectric effect is exhibited by metals when exposed to x rays, UV light, or (for some metals including sodium and potassium) high-frequency visible light. Somehow, the EM waves give energy to electrons in the metal, and the electrons are ejected from the surface (● Figure 10.4). (It was these freed electrons that enhanced the spark in Hertz's apparatus.) Because of the nature of EM waves, we shouldn't be too surprised that this sort of thing can happen. But again, some of the characteristics of the photoelectric effect that scientists saw showed them that the fundamental process was not understood. The biggest puzzle was the relationship between the speed or energy of the electrons and the incident light. It might seem reasonable to suppose that brighter light would cause the electrons to gain more energy and be ejected from the surface with higher speeds. It was found that brighter light caused *more* electrons to be ejected each second, but it did not increase their energies. Even more of a surprise was the finding that only the frequency (color) of the light affected the electron energies. Higher-frequency light ejected electrons with higher energy. Even extremely dim light with high enough frequency would immediately cause electrons to be ejected.

The explanation of the photoelectric effect was supplied in 1905 by Albert Einstein (● Figure 10.5). Planck had suggested that light is emitted in discrete "bursts" or "bundles." Einstein took this idea one step further and proposed that the light itself remains in bundles or "packets" and is absorbed in this form. The electrons in metal can only absorb light energy by absorbing one of these discrete quanta of radiation. The amount of energy in each quantum of light depends on the frequency. In particular:

$$E = hf \qquad \text{(energy of a quantum of EM radiation)}$$

This is the same equation for the energy of Planck's quantized atomic oscillators. Einstein suggested that light and other electromagnetic waves are quantized, just like the energy of oscillating atoms.

This idea, that the energy of an EM wave is quantized, allows us to picture the wave as being made up of individual particles now called **photons.** (The word was coined by G. N. Lewis in 1926.) Each photon carries a quantum of energy, $E = hf$, and propagates at the speed of light. The total energy in an EM wave is just the sum of the energies of all of the photons in the wave. Notice how different this picture is from the wave picture of light developed in Chapter 8. Both pictures are correct and present mutually complementary aspects of EM waves. Under certain circumstances, such as those involving refraction or interference of light, the wave nature of light is manifested. In other circumstances, including those involving the emission or absorption of light, the particle aspects of light are demonstrated. These two different sides of light are like the front and back sides of a person (● Figure 10.6). Sometimes, we see only the front of an individual. We recognize the individual by facial structure, eye color, hair color, and so on. At other times, we see only the back of the person. Here again we may recognize the person, but now by virtue of body structure and gait. In each case, we see different aspects of the same person, but we are still able to recognize the person, although for the most part the clues leading to recognition are not the same in the two instances. The situation with light is analogous. Different experiments reveal different aspects of what we recognize to be the same type of EM radiation.

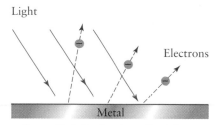

Light

Electrons

Metal

● **Figure 10.4** The photoelectric effect. Light striking a metal surface causes electrons to be ejected.

© AIP Neils Bohr Library

● **Figure 10.5** Albert Einstein (1879–1955) enjoying one of his favorite pastimes—sailing.

● **Figure 10.6** (a) Recognition of an individual as she approaches is often made using such qualities as facial structure, eye color, hair color and style, and so on. (b) Recognition of the same person as she moves away from us can often be made using different qualities, such as body structure, gait, and so on.

(a)   (b)

© Chris Gilbert   © Chris Gilbert

With Einstein's proposal, the observed aspects of the photoelectric effect came together. Higher-frequency light ejects electrons with more energy because each photon has more energy to give. Brighter light simply means that more photons strike the metal each second. This results in more electrons being ejected each second, but it does not increase the energy of each electron.* Einstein's explanation of the photoelectric effect earned him the 1921 Nobel Prize in physics. The new understanding of the nature of light and the way it interacts with matter profoundly altered the course of twentieth-century physics.

Just how much energy does a typical photon have? Not very much. The energy of a photon of visible light is only about 3×10^{-19} joules. On this scale, it is convenient to use a much smaller unit of energy, called the *electronvolt* (eV). One electronvolt is the potential energy of each electron in a 1-volt battery. Since voltage is energy per charge, the charge of an electron multiplied by voltage equals energy:

$$1 \text{ electronvolt} = 1.6 \times 10^{-19} \text{ coulomb} \times 1 \text{ volt}$$

$$= 1.6 \times 10^{-19} \text{ coulomb} \times 1 \text{ joule/coulomb}$$

$$1 \text{ eV} = 1.6 \times 10^{-19} \text{ J}$$

The energy of visible photons is on the order of 2 electronvolts, a much easier number to deal with. In terms of the electronvolt, the value of Planck's constant is

$$h = 6.63 \times 10^{-34} \text{ J-s} = 4.136 \times 10^{-15} \text{ eV/Hz}$$

● Figure 10.7 shows the photon energies corresponding to different types of EM waves.

Compare the energies associated with a quantum of each of the following types of EM radiation.   **Example 10.1**

$$\text{red light: } f = 4.3 \times 10^{14} \text{ Hz}$$

$$\text{blue light: } f = 6.3 \times 10^{14} \text{ Hz}$$

$$\text{x ray: } f = 5 \times 10^{18} \text{ Hz}$$

---

* For each metal, there does exist a threshold (or cutoff) frequency below which no photoelectrons will be emitted, regardless of how intense the light is. Photons with frequencies less than this threshold possess too little energy to eject electrons from the metal.

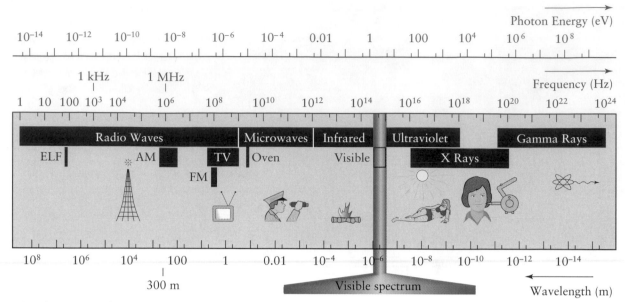

Figure 10.7 Photon energies in the EM spectrum. Visible light photons range from approximately 1.7 to 3.1 electronvolts (red light to blue light).

For each one, we use Planck's original equation for a quantum of energy.

$$E = hf$$
$$= 4.136 \times 10^{-15} \text{ eV/Hz} \times 4.3 \times 10^{14} \text{ Hz}$$
$$= 1.78 \text{ eV} \quad \text{(red light)}$$

Using the same equation for blue light and x rays, we get:

$$E = 2.61 \text{ eV} \quad \text{(blue light)}$$
$$E = 20,700 \text{ eV} \quad \text{(x ray)}$$

Notice how much greater the energy of a quantum of x radiation is compared to that of either red or blue light. Little wonder then that high doses of x rays can be harmful to the human body.

Before we go on to the third puzzle that faced scientists in the early 1900s, let's look at some offshoots of the photoelectric effect. This phenomenon is the key to "interfacing" light with electricity. Just as the principles of electromagnetism make it possible to convert motion into electrical energy and vice versa, the ability of electrons to absorb the energy in photons makes it possible to detect, measure, and extract energy from light. ● Figure 10.8 shows a schematic of a device that can detect light. When no light strikes the metal, no current flows in the circuit because there is nothing to carry the charge through the tube. When light strikes the metal, electrons are ejected and attracted to the positive

Figure 10.8 Schematic of a light detector. (a) No current flows in the circuit because there is nothing to carry charge through the tube. (b) Light releases electrons from the metal, allowing a current to flow. The size of the current indicates the brightness of the light.

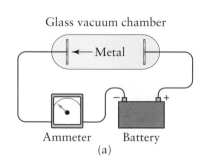

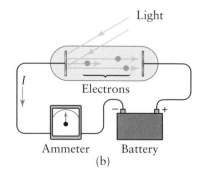

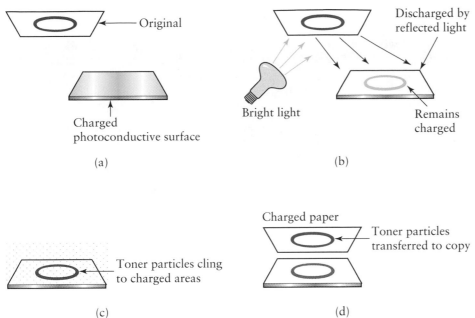

- Original
- Charged photoconductive surface

(a)

- Discharged by reflected light
- Bright light
- Remains charged

(b)

- Toner particles cling to charged areas

(c)

- Charged paper
- Toner particles transferred to copy

(d)

terminal of the tube. The result is a current flowing in the circuit. This sort of detector could be used to automatically count people entering a building. The detector is placed on one side of the door, and a light source is placed on the other side, pointed toward the detector. When something blocks the light going to the detector, the current stops. A counter connected to an ammeter in the circuit then automatically tallies one count. A similar setup could be used to automatically open and close a door.

Photocopying machines and laser printers use the interplay between electricity and light in a process known as *electrophotography*. The key part of the operation is a special *photoconductive* surface. It is normally an insulating material, but it becomes a conductor when exposed to light. Electrons bound to atoms are freed when they absorb photons in the incident light.

The process of forming an image begins when the photoconductive surface is charged electrostatically. The surface retains the charge until light strikes it, and the freed electrons allow the charge to flow off the surface. A mirror image of the material to be printed is formed on the photoconductive surface using light. In the case of a photocopying machine, a bright light shines on the original, and the reflected light strikes the charged surface (● Figure 10.9). White areas on the original reflect most of this light onto the corresponding areas of the photoconductive surface, which are consequently discharged by the large number of incident photons. Dark parts of the original—printed letters, for example—reflect very little light. Consequently, the corresponding regions on the photoconductive surface retain their electrical charge. Then fine particles of toner, somewhat like a solid form of ink, are brought near the surface. The toner particles are attracted to the charged regions and collect on them, while the discharged areas remain clear. A blank piece of paper, also electrically charged, is placed in contact with it. The toner particles are attracted to the paper and collect on it. The final image is "fused" on the paper by melting the toner particles into the paper.

In a laser printer, a laser under computer control illuminates and discharges those parts of the photoconductive surface that will not be dark in the image to be printed. The rest of the process parallels the steps in photocopiers.

Other specialized "photosensitive" materials, many of them semiconductors, have been devised for diverse uses. Light meters in cameras, the light-sensing elements in video and digital cameras that convert optical images into electrical signals, scanning elements in fax machines, sensitive photodetectors used by astronomers to measure the faint light from distant stars and galaxies, and solar cells used to convert energy in sunlight into electricity (● Figure 10.10) are just some of the devices that rely on extracting the energy in photons.

● **Figure 10.10** Author Don Bord (left) and students on top of Emory Peak in Big Bend National Park, Texas. The solar cells absorb photons in sunlight to generate electrical energy that powers a radio transmitter for park personnel.

1. Is the energy associated with a photon of blue light
   a) greater than
   b) less than
   c) equal to
   the energy associated with a photon of red light? Why?
2. (True or False.) In the photoelectric effect, increasing the frequency of the light being used will increase the energy of the electrons that are ejected.

3. (True or False.) In the photoelectric effect, increasing the brightness of the light being used will increase the energy of the electrons that are ejected.
4. Name two common devices that make use of the photoelectric effect or similar process in which electrons absorb photons.

ANSWERS: 1. (a), higher frequency 2. True 3. False 4. photocopier, laser printer, digital camera, solar cell. . . .

10.3 Atomic Spectra

The third problem that defied explanation at the beginning of the twentieth century involved the spectra produced by the various chemical elements. Suppose we use a prism to examine the light produced by the heated filament of an ordinary incandescent bulb. The light from the bulb will be dispersed upon passing through the prism, as described in Chapter 9, and a spectrum will be produced (see Figure 9.67 and Explore It Yourself 10.1 on p. 390). The spectrum will appear as a continuous band of the colors of the rainbow, one color smoothly blending into the next. Such a spectrum is called a **continuous spectrum,** and it is characteristic of the radiation emitted by a hot, luminous solid.

Imagine now a sample of some gas confined in a glass tube and induced to emit light, as by heating (● Figure 10.11). If we examine the light from the luminous gas with a prism, we do not see a continuous spectrum. Instead we see an **emission-line spectrum** consisting of a few, isolated, discrete lines of color. In this case, the source is not emitting radiation at all wavelengths (colors) but only at certain selected wavelengths. If a different type of gas is investigated, one finds that the resultant line spectrum is different. Each type of gas has its own unique set of spectral lines (● Figure 10.12).

Explore It Yourself 10.2

This is just like Explore It Yourself 10.1—only this time you use a compact fluorescent lamp. (The type shown in Figure 4.1 works well.) It's best if the CD is a couple of meters from the lamp and there are no other light sources around. What is different about the spectrum of colors? (The spectrum of light emitted by a compact fluorescent lamp is not a true simple emission-line spectrum, but it is very similar to one.) You might also try this with street lights and neon lights.

● **Figure 10.11** The spectrum of a hot gas consists of several discrete colors (three in this example).

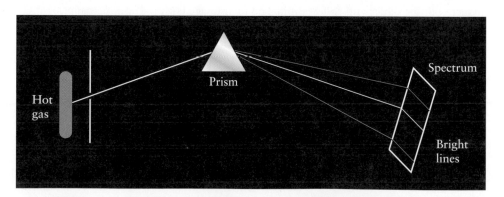

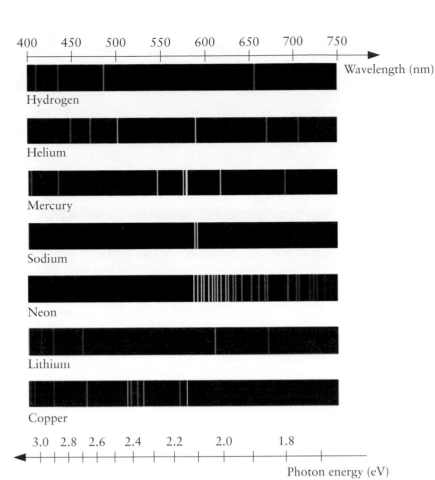

● **Figure 10.12** Emission spectra of selected elements. (Photo of spectral lines Courtesy of Bausch and Lomb.)

The fact that luminous, vaporized samples of material produce line spectra when their light is dispersed (as by a prism) was discovered in the 1850s. Once chemists recognized that each element has its own special spectrum, it became possible for them to determine the compositions of substances in the laboratory by examining their spectra. Robert Bunsen (of Bunsen burner fame) and Gustav Kirchhoff, two German scientists, were pioneers in this new field of *spectroscopy*, the study of spectra, during the last half of the nineteenth century. (See the Physics Potpourri on p. 398.) The problem that remained was to understand why luminous gases produce line spectra and not continuous spectra, and how it is that each element has its own unique spectral "fingerprint" by which it can be identified.

Spectroscopy has grown to be one of the most useful tools for chemical analysis in fields as diverse as law enforcement and astronomy. Suspected poisons or substances found at a crime scene can be identified by comparing their line spectra to those in a catalog of known elements and compounds. Astronomers can determine what chemicals exist in the atmospheres of stars by examining spectra of the starlight with the aid of telescopes.

The explanation of atomic spectra inspired one of the most crucial periods of advancement in physics ever. In the remainder of this chapter, we describe how a picture of the structure of the atom was created and refined to account for atomic spectra. This effort spawned a "new" physics for dealing with matter on the scale of atoms—quantum mechanics.

LEARNING CHECK

1. A hot luminous solid emits a _____ spectrum, while a hot luminous gas produces a(n) _____ spectrum.

2. (True or False.) Two different gases can have the same emission-line spectrum.

ANSWERS: 1. continuous, emission-line 2. False

Cosmic Chemistry, "...To Dream of Such a Thing."

In 1844, the French philosopher Auguste Comte published the following view of the prospects of the science of astronomy:

> The stars are only accessible to us by a distant visual exploration. This inevitable restriction therefore not only prevents us from speculating about life on all these great bodies, but also forbids the superior inorganic speculations relative to their chemical or even their physical natures.

Like many other predictions about the future course of scientific discovery, Comte's was soon proved false. Indeed, within 20 years, two German scientists would pioneer techniques that would allow the composition of substances to be discovered from a long-distance examination of the light they emit.

Beginning in 1859, Robert Bunsen (1811–1899) and Gustav Kirchhoff (1824–1887; see ● Figure 10.13) undertook a series of experiments at the University of Heidelberg that laid the foundations for what is now known as the field of spectroscopy. A key ingredient

to their success was the burner developed by Bunsen a few years earlier. The Bunsen burner, which produces a flame with a very high temperature but low luminosity (brightness), is ideal for examining the distinctive colors imparted by chemical salts to flames containing them (● Figure 10.14). Bunsen had originally used filters to distinguish the various hues emitted by different substances, but Kirchhoff suggested that a much surer distinction between different chemicals might be obtained by examining the spectra of the colored flames. Together they designed and constructed the first laboratory spectroscope using a 60° hollow prism filled with carbon disulfide and then carefully examined the light emitted by various hot gases (Figure 10.11). What they found was that each element produced its own special set of bright colored lines, the line patterns being as unique as a person's fingerprints. The importance of this observation was not lost on Bunsen and Kirchhoff. "We can base a method of qualitative analysis on these lines," Bunsen wrote, "that greatly broadens the field of chemical research and leads to the solution of problems previously beyond our grasp."

One problem thought to be beyond the grasp of scientists in the mid-nineteenth century (as evidenced by Comte's remark) was the determination of the chemical compositions of the stars. Bunsen and Kirchhoff's efforts to do just that are thought to have originated from a fire in the city of Mannheim, some 10 miles west of Heidelberg. Bunsen and Kirchhoff apparently observed the fire from their laboratory window in the evening and turned their spectroscope toward the flames. There they were able to detect barium and strontium. Some time later, while reflecting on this event, Bunsen remarked that if he and Kirchhoff could analyze a fire in Mannheim, might they not do the same for the Sun? "But," he added, "people would think we were mad to dream of such a thing."

But dreams sometimes have a way of coming true, and such was the case here. The realization of the dream began with Kirchhoff's demonstration that a conspicuous dark line (labeled with the letter *D* by Fraunhofer in 1814) in the spectrum of the Sun was unambiguously due to the element sodium. Further experiments convinced Kirchhoff that a substance capable of emitting a certain spectral

● **Figure 10.13(a)** Chemist Robert Bunsen, inventor of the Bunsen burner and codeveloper of the science of spectroscopy.

● **Figure 10.13(b)** Physicist Gustav Kirchhoff, who worked with Bunsen, left his mark in many areas of physics.

10.4 The Bohr Model of the Atom

At the beginning of the twentieth century, little was known about the atom. In fact, many doubted that atoms existed at all. This made the origin of atomic spectra a mystery, in spite of the fact that spectroscopy was a booming field. In 1911, an important experiment performed by Ernest Rutherford in England revealed that the positive charge in an atom is concentrated in a tiny core—the nucleus (more on this in the Physics Family Album in Chapter 11). This result was quickly followed in 1913 by a model of the atom put forth by the great Danish physicist Niels Bohr (● Figure 10.15). **Bohr's model** of the atom, later modified, as we shall see in the next section, successfully explained the nature of atomic line has a strong absorptive power for the same line. He concluded that the dark *D* line in the solar spectrum was produced by absorption by sodium in the atmosphere of the Sun. The Sun, like the Earth, contained sodium! By 1862, Kirchhoff had detected calcium, magnesium, iron, chromium, nickel, barium, copper, and zinc in the Sun, in addition to sodium. The analysis of the composition of celestial bodies had begun.

The excitement felt by Bunsen and Kirchhoff during this period may be best conveyed by quoting from a letter written by Bunsen to an English colleague on 15 November 1859:

> At present, Kirchhoff and I are engaged in an investigation that doesn't let us sleep. Kirchhoff has made a wonderful, entirely unexpected discovery in finding the cause of the dark lines in the solar spectrum, and he can increase them artificially in the Sun's spectrum or produce them in a continuous spectrum and in exactly the same position as the corresponding Fraunhofer lines. Thus, a means has been found to determine the composition of the Sun and the fixed stars with the same accuracy as we determine strontium chloride, etc. with our chemical reagents.

No sooner had Kirchhoff and Bunsen unlocked the door to the chemical secrets of the universe beyond the Earth than they announced (in 1860) a new terrestrial metal, cesium (from the Latin *caesius*, "sky blue"), so named because of its brilliant blue spectral lines. The following year they published the discovery of yet another element, named rubidium (from the Latin *rubidus*, "dark red") for its strong red emission lines. In the years that followed, several other elements were identified spectroscopically: thallium (1861), indium (1863), gallium (1875), scandium (1879), and germanium (1886). In addition, studies of the spectra of the Sun and stars led to the discovery of helium and to the recognition that hydrogen is the most abundant element in the universe.

Although Bunsen and Kirchhoff did not understand *how* the spectra of the different elements were produced, they were highly successful at exploiting the uniqueness of such spectra for the purpose of identifying elements in various sources. The ultimate explanation for atomic spectra did not come until the work of Max Planck and Niels Bohr at the beginning of the twentieth century. Nonetheless, it is hardly an understatement to say that the birth of what might be called "cosmic chemistry" occurred in Heidelberg in 1859 as a result of the blending of the talents of a skilled chemist and a mathematical physicist who dared to dream.

For more on the lives and accomplishments of these two scientists, check out the *Dictionary of Scientific Biography* or explore InfoTrac College Edition, your online library (see page 427).

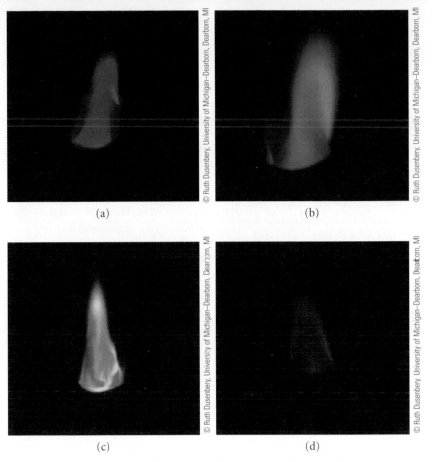

● **Figure 10.14** Characteristic colors for Bunsen burner flame tests for the elements (a) lithium, (b) sodium, (c) copper, and (d) potassium. In each case, the flame color reflects the wavelengths of the dominant emission lines in the spectrum of each element (see Figure 10.12).

spectra. But like Planck's explanation of blackbody radiation, Bohr's model of the atom was based on assumptions that did not seem sensible at the time. The basic features of Bohr's model are the following:

1. The atom forms a miniature "solar system," with the nucleus at the center and the electrons moving about the nucleus in well defined orbits. The nucleus plays the role of the Sun and the electrons are like planets.
2. The electron orbits are quantized, that is, electrons can only be in certain orbits about a given atomic nucleus. Each allowed orbit has a particular energy associated

● Figure 10.15 Niels Bohr (1885–1962), one of the giants in the field of atomic physics.

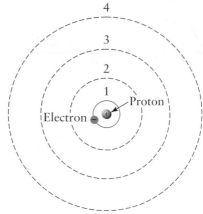

● Figure 10.16 The Bohr model of the hydrogen atom. The electron can have only certain orbits. Four of these are shown. The figure is not drawn to scale: The fourth orbit is actually 16 times the size of the first orbit.

● Figure 10.17 A hydrogen atom with its electron (a) in orbit 2 and (b) in orbit 3. The electron has more energy when it is in orbit 3.

with it, and the larger the orbit, the greater the energy. Electrons do not radiate energy (emit light) while in one of these stable orbits.

3. Electrons may "jump" from one allowed orbit to another. In going from a low-energy orbit to one of higher energy, the electron must *gain* an amount of energy equal to the difference in energy it has in the two orbits. When passing from a high-energy orbit to a lower-energy one, the electron must *lose* the corresponding amount of energy.

● Figure 10.16 shows the Bohr model for the simplest atom, that of the element hydrogen (atomic number = 1). A lone electron orbits the nucleus, in this case a single proton. The electron can be in any one of a large number of orbits (four are shown). In each orbit, the electrical force of attraction between the oppositely charged electron and proton supplies the centripetal force needed to keep the electron in orbit.

When in orbit 1, the electron has the lowest possible energy. The electron has more energy in each successively larger orbit (● Figure 10.17). The electron's energy while in any of the orbits is negative because it is bound to the nucleus (see the end of Section 3.5). To go to a larger orbit, the electron must gain energy. The maximum energy the electron can have and still remain bound to the proton is called the **ionization energy.** If the electron acquires more than this energy, it breaks free from the nucleus, and the atom is ionized. The resulting positive ion in this case is a bare proton.

How does the Bohr model account for the characteristic spectra emitted by luminous hydrogen gas? When an electron in a larger (higher-energy) orbit "jumps" to a smaller orbit, it loses energy. One of the ways it can lose this energy is by emitting light. This process is the origin of atomic spectra.

Imagine that the electron in our hydrogen atom is in the sixth allowed orbit, with its energy represented by E_6. Suppose the electron makes a transition (a jump) to an inner orbit, say, the second one, where its energy is E_2. To do so, the electron must lose an amount of energy, ΔE, equal to:

$$\Delta E = E_6 - E_2$$

In what is called a *radiative transition,* the electron loses this energy through the emission of a photon whose energy is just equal to ΔE (● Figure 10.18). The photon, spontaneously created during the transition of the electron from orbit 6 to orbit 2, carries off the excess energy into space in the form of EM radiation. This is much like Planck's atomic oscillators emitting photons. Because the energy of a photon equals h times its frequency, we have:

$$\text{photon energy} = \Delta E = E_6 - E_2$$

And:

$$\text{photon energy} = hf$$

Therefore:

$$hf = E_6 - E_2$$

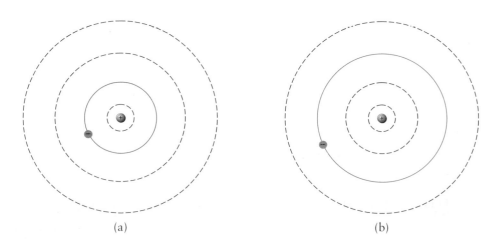

(a)   (b)

The frequency of the emitted light is directly proportional to the difference in the energy of the orbits between which the electron jumped. The larger the energy difference, the higher the frequency of the light given off in the process. A downward transition from orbit 3 to orbit 2 will produce a photon with lower energy and lower frequency, because the energy difference between orbit 3 and orbit 2 is smaller.

$$\Delta E = E_3 - E_2 \quad \text{is smaller than} \quad \Delta E = E_6 - E_2$$

For the 6-to-2 transition in hydrogen, a violet-light photon is emitted. For the 3-to-2 transition, it is a red-light photon.

Each possible downward transition from an outer orbit to an inner orbit results in the emission of a photon with a particular frequency. An appropriately heated sample of hydrogen gas will emit light with these different frequencies but no other radiation. This is the line spectrum of hydrogen. (We will discuss this more in Section 10.6.)

A downward electron transition can also occur without the emission of light through what is called a *collisional transition*. In this case, a collision between a hydrogen atom and another particle (perhaps another hydrogen atom) can induce the electron in the outer orbit to spontaneously jump to an inner orbit. The energy that the electron loses can be transferred to the other particle, or it can be converted into increased kinetic energy of both colliding particles. (This is much like the collision described in Figure 3.34.) Collision-induced transitions are generally important in dense gases where the numbers of atoms or molecules per unit volume of gas are large and the likelihood of two or more gas particles colliding is therefore quite high.

So far, we have focused on how an electron in an outer orbit may lose energy by jumping to an inner orbit. But the reverse of this process also occurs: An electron in an inner orbit can gain just the right amount of energy and jump to an outer orbit. For example, a hydrogen atom in the lowest energy state might gain the amount of energy needed for its electron to jump from orbit 1 to, say, orbit 5. To do this, the electron would have to acquire energy:

$$\Delta E = E_5 - E_1$$

One way to do this would be through a collision: the hydrogen atom could collide with another atom with more energy.

Another way an electron can jump to an outer orbit is by absorbing a photon with the proper energy (● Figure 10.19). For example, an electron in orbit 1 can jump to orbit 5 if it absorbs a photon with energy:

$$\text{photon energy} = \Delta E = E_5 - E_1$$

If a sample of hydrogen gas is irradiated with a broad band of EM waves (like blackbody radiation), many such transitions to outer orbits will occur. Some of the photons in the incident radiation will have just the right energies to induce transitions from inner orbits to outer orbits. In the process, the number of photons with these particular energies will be reduced, and the intensity of the EM radiation at the corresponding frequencies will decrease. This reduction of intensity of light at only certain frequencies after it passes through a gas results in an **absorption spectrum** (● Figure 10.20). For example, if white

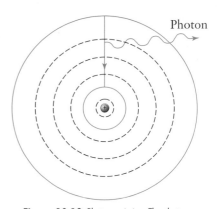

● **Figure 10.18** Photon emission. The electron makes a transition from orbit 6 to orbit 2 and emits a photon. The photon's energy equals the energy lost by the electron.

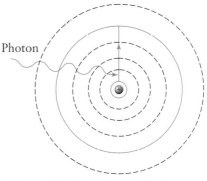

● **Figure 10.19** Photon absorption. The electron makes a transition from orbit 1 to orbit 5 by absorbing a photon.

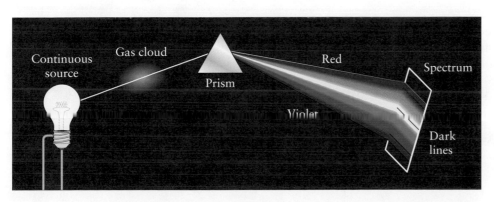

● **Figure 10.20** Absorption spectrum of a gas. The gas absorbs photons in the light passing through it, so at those frequencies (colors) the spectrum is fainter.

light is passed through hydrogen gas and then dispersed with a prism, dark bands appear at certain frequencies. They are exactly the same frequencies that are in the emission spectrum of luminous hydrogen.

This is how the element helium was discovered. In 1868, some of the absorption lines in the spectrum of sunlight were found not to correspond with any elements known at that time. The existence of a new element in the Sun's atmosphere was suggested to account for these lines. This new element was named after the Greek word (*helios*) for Sun (see the Physics Potpourri "What's in a Name" on p. 136).

The Bohr model was successful at explaining the origin of atomic spectra. But Bohr's model of electron orbits rested on two unexplained assumptions. First, there are only certain allowed orbits. While one can place a satellite in orbit about the Earth with any radius, the electron orbits were restricted to specific radii. These radii were determined by a seemingly arbitrary but nonetheless effective condition: that the angular momentum of the electron in its orbit is quantized. Much like the energy of Planck's quantized atomic oscillators, the orbital angular momentum of Bohr's electrons could only have the following values:

Refer to the discussion on angular momentum in Section 3.8.

$$\text{allowed angular momentum} = \frac{h}{2\pi}, \quad \text{or } 2\frac{h}{2\pi}, \quad \text{or } 3\frac{h}{2\pi}, \quad \ldots$$

So quantization was popping up everywhere: in the energy of oscillating atoms, the energy of EM waves, and now in the orbital angular momentum of atomic electrons.

The second assumption that physicists of the time found objectionable was that as long as an electron remained in one of its allowed orbits, it did not emit EM radiation. Maxwell's work had indicated that whenever a charged object undergoes acceleration, including centripetal acceleration, it will radiate. Put another way, an electron in a periodic orbit is much like a charge oscillating back and forth, and the latter results in the production of an EM wave. An orbiting electron should be continually radiating, and in the process it should lose energy and spiral into the nucleus. According to the physical laws known at the time, atoms could not remain stable and should collapse in a fraction of a second!

Even though Bohr had a model that worked, clearly the physics behind it was not understood. What was needed was a revolution in our understanding of how nature works at the atomic level.

Concept Map 10.1 summarizes the Bohr model of the atom.

LEARNING CHECK

1. Which of the following is *not* true in the Bohr model of the atom?
   a) The electrons orbit the nucleus.
   b) The orbits of the electrons are quantized.
   c) While moving in a stable orbit, electrons can emit photons.
   d) Electrons can jump to a higher-energy orbit if they gain the right amount of energy.
2. In the Bohr model of the atom, a photon may be _____ when an electron jumps from a large, high-energy orbit to a smaller, low-energy orbit.

3. An atom is said to have been _____ when it absorbs a photon with sufficient energy to free an electron.
4. (True or False.) There is always a direct correlation between the absorption spectrum of a particular gas and its emission spectrum.

ANSWERS: 1. (c) **2.** emitted **3.** ionized **4.** True

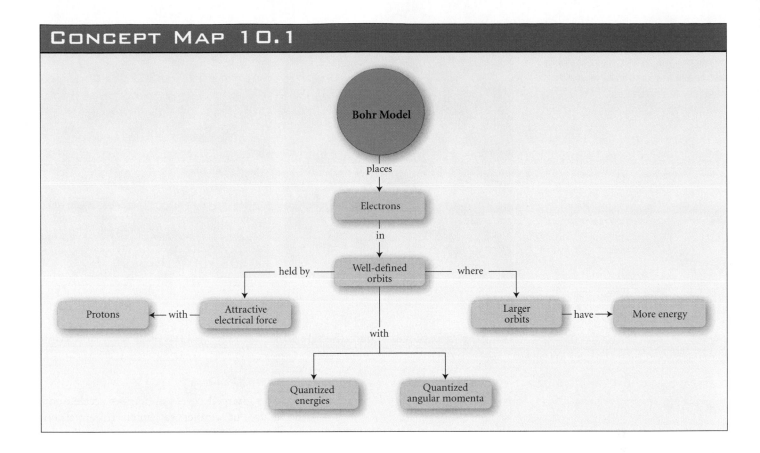

10.5 Quantum Mechanics

The success of Bohr's model of the atom, even though it was at odds with accepted principles of physics at the time, indicated that perhaps a new physics was needed to describe what goes on at the atomic level. The first step in this direction came during the summer of 1923. While working on his doctorate in physics, a French aristocrat named Louis Victor de Broglie (rhymes with "Troy") proposed that electrons and other particles possess wavelike properties (● Figure 10.21). Einstein had shown that light has both wavelike and particlelike properties, so why not electrons too? The wave associated with any moving particle has a specific wavelength (called the **de Broglie wavelength**) that depends on the particle's momentum:

$$\lambda = \frac{h}{mv} \quad \text{(de Broglie wavelength)}$$

The higher the momentum of a particle, the shorter its wavelength. High-speed electrons have shorter wavelengths than low-speed electrons.

Once again, Planck's constant shows up. Since h is such a tiny number, de Broglie wavelengths are extremely small. This means that the wave properties of particles are only manifested at the atomic and subatomic level.

● **Figure 10.21** Louis de Broglie (1892–1987) was the first to suggest that particles can behave like waves.

What is the de Broglie wavelength of an electron with speed 2.19×10^6 m/s? (This is the approximate speed of an electron in the smallest orbit in hydrogen.)

Using the mass given in the back inside cover, the electron's momentum is

$$mv = 9.11 \times 10^{-31} \text{ kg} \times 2.19 \times 10^6 \text{ m/s}$$

$$= 1.995 \times 10^{-24} \text{ kg-m/s}$$

Example 10.2

● **Figure 10.22** These patterns were produced by sending (a) a beam of x rays and (b) a beam of electrons through an aluminum target. The similarity exists because the electrons act like waves as they interact with the aluminum atoms.

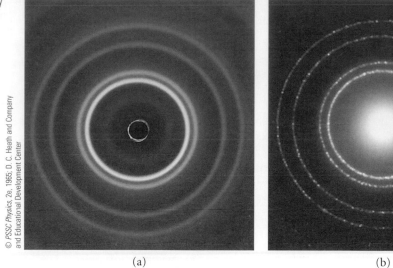

© PSSC Physics, 2e, 1965; D. C. Heath and Company and Educational Development Center

(a)   (b)

© PSSC Physics, 2e, 1965; D. C. Heath and Company and Educational Development Center

Using the value of h in SI units (Section 10.1):

$$\lambda - \frac{h}{mv} = \frac{6.63 \times 10^{-34}\text{J-s}}{1.99 \times 10^{-24}\text{ kg-m/s}}$$

$$= 3.32 \times 10^{-10}\text{ m} = 0.332\text{ nm}$$

This distance is in the same range as the diameters of atoms.

Another radical theory had entered the arena of physics. Although submitted by a newcomer, de Broglie's hypothesis was not completely rejected by the physics community because Einstein himself found it plausible. Then, in 1925, the puzzling results of some experiments were interpreted as proof of the existence of de Broglie's waves. In a series of experiments, the American physicist Clinton Davisson (with various collaborators) showed that a beam of high-speed electrons underwent diffraction when sent into a nickel target. The electrons were behaving just like waves. In fact, one gets the same kind of scattering pattern using x rays or electrons (see ● Figure 10.22).

Other experiments have verified the wave properties of electrons, protons, and other particles. Young's classic two-slit experiment, by which he proved that light is a wave, can be used to make particles undergo interference (● Figure 10.23). The *electron microscope* exploits the wave properties of electrons. Instead of using ordinary light like a conven-

● **Figure 10.23** Particles, such as electrons, undergo interference when passed through narrow slits. They exhibit a property of waves (compare to Figure 9.7).

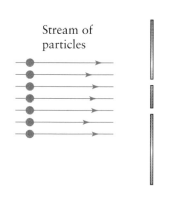

Stream of particles

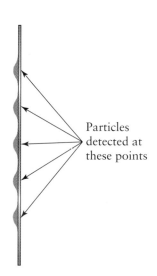

Particles detected at these points

tional microscope, an electron microscope uses a beam of electrons that acts like a beam of electron waves. The magnification can be much higher with an electron microscope because the de Broglie wavelength of the electrons is much shorter than the wavelengths of visible light (● Figure 10.24). Diffraction and interference of waves passing around and between small objects such as cells strongly affect image clarity. Shorter-wavelength waves (electrons) diffract and interfere much less than do longer-wavelength waves (light).

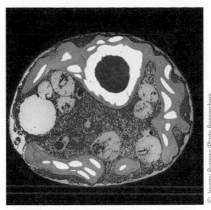

● **Figure 10.24** An image of an alga cell taken with an electron microscope.

Explore It Yourself 10.3

You can investigate in more detail the diffraction and interference of electrons passing through slits using a suite of programs developed at Kansas State University of Professor Dean Zollman and collaborators. Visit the Visual Quantum Mechanics Web site at http://phys.educ.ksu.edu/vqm/ and select one of the available simulations (for example, Double Slit Diffraction). Follow the instructions and explore the dependence of the resulting electron interference patterns on such parameters as slit width, slit spacing, and electron-beam intensity. Compare your findings with the observed characteristics of light-wave diffraction and interference discussed in Chapters 6 and 9.

A more recently developed microscope uses the wavelike nature of electrons to form tantalizing images of individual atoms on the surfaces of solids. The *scanning tunneling microscope* (STM) uses an extremely fine pointed needle that scans back and forth over a surface. A positive voltage maintained on the needle attracts the electrons at the surface of the sample. If electrons were simply particles, they could not traverse the gap—less than 1 nanometer wide—between the surface and the needle. But their wave nature allows the electrons to penetrate or to "tunnel" through the gap and to reach the needle (● Figure 10.25). The tiny current of electrons that flows decreases rapidly if the gap widens. As the needle scans back and forth over the surface (like someone mowing a lawn), a feedback mechanism moves the needle up and down to keep this tunneling current constant. The varying height of the needle is recorded and used to draw a contour map of the surface that clearly shows individual molecules and atoms (● Figure 10.26 and Figure 4.7c). The 1986 Nobel Prize in physics was shared by the inventor of the electron microscope and the codevelopers of the STM.

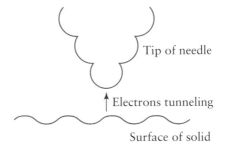

● **Figure 10.25** Simplified sketch of a scanning tunneling microscope (STM). The wavelike nature of the electrons allows them to cross the gap from the surface to the needle.

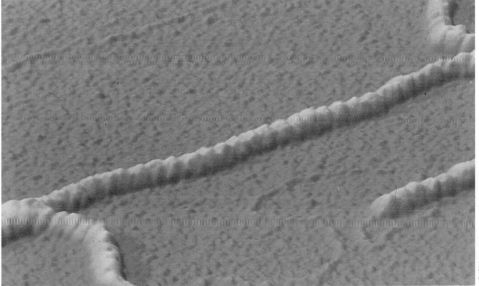

● **Figure 10.26** An STM image (large central object) of DNA-protein complexes (strands of DNA and their protein envelopes). The image was obtained from metal-coated specimens. The structure of the strands is clearly resolved. A DNA strand without its protein envelope, which has a diameter of approximately 1 nanometer, is visible in the lower middle of the image. The images were produced by IBM Zurich Research Center scientists.

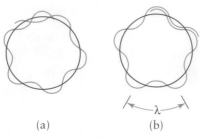

(a)   (b)

● **Figure 10.27** Simplified wave representation of an electron orbiting a nucleus. The orbit in (a) is not possible because the wave interferes destructively with itself. The orbit in (b) is allowed. This corresponds to orbit 5, since it has 5 wavelengths fitted into the orbit.

Another major triumph for de Broglie's wave hypothesis was its ability to explain Bohr's quantized orbits. De Broglie reasoned that since an orbiting electron acts like a wave, its wavelength has to affect the circumference of its orbit. In particular, the electron's wave wraps around on itself, and in doing so it must interfere with itself. Only if the wave interferes constructively (peak matches peak) can the electron's orbit remain stable (● Figure 10.27). This means that the circumference of its orbit must equal exactly one de Broglie wavelength, or two, or three, and so on.

$$\text{circumference of orbit} = \lambda, \quad \text{or } 2\lambda, \quad \text{or } 3\lambda, \quad \dots$$

If r is the radius of a circular orbit, then the circumference is $2\pi r$.

$$2\pi r = \lambda, \quad \text{or } 2\lambda, \quad \text{or } 3\lambda \quad \dots$$

Example 10.3

Using the results of Example 10.2, find the radius of the smallest orbit in the hydrogen atom.

In de Broglie's model, the circumference of the smallest orbit must equal the de Broglie wavelength of the electron. We calculated this wavelength to be 0.332 nanometers for the smallest orbit. So:

$$2\pi r = \lambda = 0.332 \text{ nm}$$

$$r = \frac{0.332 \text{ nm}}{2\pi} = \frac{0.332 \text{ nm}}{2 \times 3.14} = \frac{0.332 \text{ nm}}{6.28}$$

$$= 0.0529 \text{ nm}$$

The allowed values for the circumference of an electron's orbit leads to Bohr's allowed values for the electron's angular momentum. Since the de Broglie wavelength is given by

$$\lambda = \frac{h}{mv}$$

the circumference can have the following values.

$$2\pi r = \frac{h}{mv}, \quad \text{or } 2\frac{h}{mv}, \quad \text{or } 3\frac{h}{mv}, \quad \dots$$

When we divide both sides by 2π and multiply both sides by mv, we get:

$$mrv = \frac{h}{2\pi}, \quad \text{or } 2\frac{h}{2\pi}, \quad \text{or } 3\frac{h}{2\pi}, \quad \dots$$

But mvr is the angular momentum of the electron (see Section 3.8). We saw this same relationship at the end of Section 10.4. De Broglie's condition on the circumference of the electron's orbit turns out to be identical to Bohr's condition on the angular momentum of the electron.

The success of the concepts of quantization and wave-particle duality at the atomic level could not be overlooked. In the later half of the 1920s, a flurry of activity resulted in a formal mathematical model that incorporated these ideas—**quantum mechanics.** The two principal founders were Werner Heisenberg (● Figure 10.28) and Erwin Schrödinger.

One of the main contributions of Heisenberg was the **uncertainty principle.** Because electrons and other particles on the atomic scale have wavelike properties, we can no longer think of them as being like tiny, shrunken marbles. They are a bit spread out in space, more like tiny, fuzzy cotton balls. In the old particle model of the electron, it was possible, at least in theory, to state exactly where an electron is and exactly what its momentum is at any instant in time. Heisenberg stated that the wave nature of particles makes this impossible. One cannot specify both the position and the momentum of an electron to arbitrarily high precision. The more precisely you know the position of the electron, the less precisely you can determine its momentum. If we let Δx represent the

● **Figure 10.28** German physicist Werner Heisenberg (1901–1976), codeveloper of the science of quantum mechanics.

uncertainty in the position of a particle, and Δmv represent the uncertainty in the momentum of the particle, then:

$$\Delta x \, \Delta mv \geq h \qquad \text{(uncertainty principle)}$$

The product of the two uncertainties is greater than or equal to Planck's constant. No matter how good the experimental apparatus, the *best* we can do is limited by this principle. On the atomic scale, particles cannot be localized in the same way that they can be on a large scale.

Schrödinger established a mathematical model for the waves associated with particles. In his model, a simple system like a hydrogen atom can be described by a *wave function*. The wave function gives the wavelength of the particle and carries information about the likelihood of finding the electron at a given location at a specified time. Knowledge of the wave function allows one to calculate the most probable position of the electron with respect to the nucleus, as well as the electron's average energy. Although often difficult to determine in practice, the wave function of a system contains all the information ever needed about the system.

During the early decades of the twentieth century, the view of what matter is like at the submicroscopic level changed dramatically. Electromagnetic waves have particlelike properties, and electrons and other particles have wavelike properties. Clearly, nature is very different at this level.

LEARNING CHECK

1. If the speed of an electron increases, its de Broglie wavelength
   a) increases.
   b) decreases.
   c) stays the same.
   d) may increase or decrease.
2. Name a device that makes use of the wavelike nature of electrons.
3. The wavelike aspect of particles is responsible for which of the following?
   a) the diffraction of electrons as they pass through aluminum
   b) the quantized orbits of the electron in a hydrogen atom
   c) electrons' ability to tunnel through a gap
   d) All of the above.
4. (True or False.) At any instant in time, it is possible to specify simultaneously the position and the momentum of an electron to arbitrarily high precision.

ANSWERS: 1. (b) 2. electron microscope, STM 3. (d) 4. False

10.6 Atomic Structure

The findings of de Broglie, Heisenberg, and Schrödinger force us to revise the simple Bohr model of the atom with its planetary structure. We can no longer represent electrons as little particles moving in nice circular orbits. In reality, each electron is described by a wave that represents the probability of finding it at different locations. The atom is pictured as a tiny nucleus surrounded by an "electron cloud" (● Figure 10.29). The density of the cloud at each point in space indicates the likelihood of finding the electron there. The different allowed orbits of the electrons appear as clouds with different sizes and shapes. So the simple drawings of atoms that we used earlier in this text (such as Figure 7.2) are not strictly correct. They did, however, serve the purpose of presenting the basic structure of the atom without complicating the picture with the wave nature of the electrons.

Because we can't say exactly where the electron is in each of its orbits, it is more useful to concentrate on the electron's energy. After all, when determining the frequency of radiation emitted or absorbed by an atom, the important quantity is the difference between the electron's initial and final energy. In the new model, we describe the electrons as being in certain allowed energy states or **energy levels.** For the hydrogen atom, the lowest energy level (corresponding to the innermost orbit in the Bohr theory) is called the

Figure 10.29 Computer-generated plots of the "electron clouds" for four atomic orbits. (a) $n = 1$ (ground state) probability density. (b)–(d) Three of the four probability distributions for $n = 2$ (first excited state; see the discussion of the Pauli exclusion principle on page 412). The probability of finding the electron at a given location is specified by the color of the display: Red represents the highest probability, followed by yellow, green, light blue, and finally dark blue. The cloud sizes show that the electron has a greater chance of being found farther from the nucleus (the center of the distributions) in the $n = 2$ orbits than in the $n = 1$ orbit. This is consistent with the Bohr model of the atom. (The scale of these images has been magnified about 50 million times.)

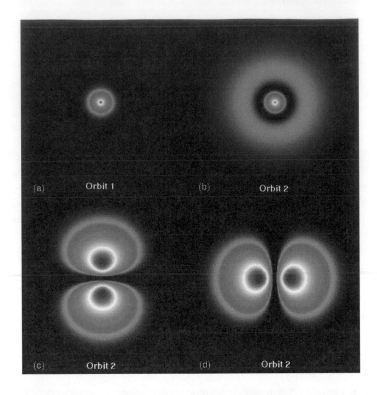

(a) Orbit 1   (b) Orbit 2

(c) Orbit 2   (d) Orbit 2

ground state. The higher energy states are referred to as **excited states.** We now represent the structure of the atom schematically using an **energy-level diagram,** like the one for hydrogen shown in ● Figure 10.30. Each energy level is labeled with a **quantum number,** n, beginning with the ground state having $n = 1$ and continuing on up. As the quantum number increases, so does the energy associated with the state. Moreover, as n gets larger, the difference in energy between the adjacent states becomes smaller. The difference in energy between the $n = 4$ and the $n = 3$ states is smaller than the difference in energy between the $n = 3$ and the $n = 2$ states. Finally, we again note the existence of a maximum allowed energy above which the electron is no longer bound to the nucleus. The state is designated $n = \infty$, and its energy is the ionization energy.

The numbers on the left in the energy-level diagram are the electron energies for each state. (The negative values indicate that the electron is bound to the nucleus, as we mentioned in Section 10.4.) The transition of an electron from one orbit to another corresponds to the atom going from one energy level to another. The change in energy of the electron as a result of the "energy-level transition" is found by comparing the energies of the two states.

Figure 10.30 Energy-level diagram for hydrogen. Each level corresponds to one of the allowed electron orbits.

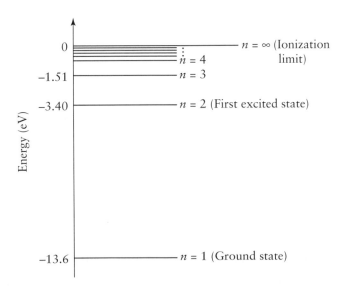

Energy (eV)

0 — $n = \infty$ (Ionization limit)
  $n = 4$
−1.51 — $n = 3$
−3.40 — $n = 2$ (First excited state)

−13.6 — $n = 1$ (Ground state)

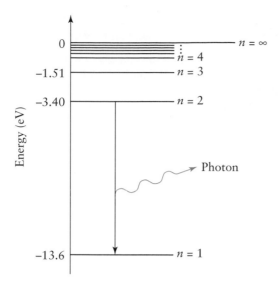

● **Figure 10.31** Downward electronic transition from level $n = 2$ to level $n = 1$ in hydrogen. The transition is accompanied by the emission of a photon with energy 10.2 electronvolts.

Using the energy-level diagram, we can give a more complete picture of the emission spectrum of hydrogen. Suppose an atom in the $n = 2$ state undergoes a transition to the $n = 1$ state by emitting a photon. (Typically, an atom will remain in an excited state for only about a billionth of a second.) The transition is represented by an arrow drawn from the initial level to the final level (● Figure 10.31). The photon that is emitted has an energy equal to the difference in energy between the two levels. Since electron transitions may occur only from one allowed energy state to another, the arrows representing such transitions *must* begin and end on allowed levels within the energy-level diagram. In this case:

$$\text{photon energy} = \Delta E = E_2 - E_1 = -3.4 \text{ eV} - (-13.6 \text{ eV})$$

$$= 10.2 \text{ eV}$$

This is a photon of UV light (refer to Figure 10.7).

Similarly, we can represent all possible downward transitions from higher energy levels to lower energy levels. ● Figure 10.32 is an enlarged energy-level diagram for hydrogen

● **Figure 10.32** Energy-level diagram for hydrogen showing different possible energy-level transitions. The number with each arrow is the wavelength (in nanometers) of the photon that is emitted.

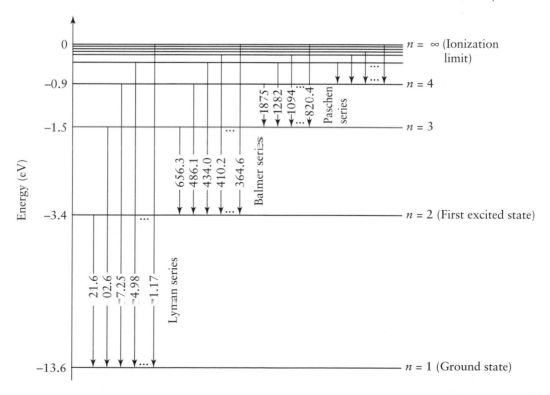

showing various possible energy-level transitions from higher levels to lower ones. The number with each arrow is the wavelength (in nanometers) of the photon that is emitted.

Example 10.4

Find the frequency and wavelength of the photon emitted when a hydrogen atom goes from the $n = 3$ state to the $n = 2$ state.

We first find the energy of the photon, then use that to determine its frequency. The wavelength we get from the equation $c = f\lambda$.

$$\text{photon energy} = hf = E_3 - E_2$$

$$= -1.51 \text{ eV} - (-3.4 \text{ eV})$$

$$= 1.89 \text{ eV}$$

Therefore:

$$f = \frac{1.89 \text{ eV}}{h} = \frac{1.89 \text{ eV}}{4.136 \times 10^{-15} \text{ eV/Hz}}$$

$$= 4.57 \times 10^{14} \text{ Hz}$$

For the wavelength:

$$\lambda = \frac{c}{f} = \frac{3 \times 10^8 \text{ m/s}}{4.57 \times 10^{14} \text{Hz}}$$

$$= 6.56 \times 10^{-7} \text{ m} = 656 \text{ nm}$$

From Figure 10.7, we see that this is a photon of visible light. (To be more precise, Table 8.1 shows that it is red light.)

By doing similar calculations for the other transitions, we can draw the following conclusions about the light that can be emitted by excited hydrogen atoms.

1. Transitions from higher energy levels to the *ground* state ($n = 1$) result in the emission of *ultraviolet* photons. This series of emission lines is referred to as the Lyman series.
2. Transitions from higher energy levels to the $n = 2$ state result in the emission of *visible* photons. This series of emission lines is referred to as the Balmer series (see Figure 10.12 and Example 10.4).
3. Transitions from higher energy levels to the $n = 3$ state result in the emission of infrared photons. This series of emission lines is referred to as the Paschen series.
4. Downward transitions to other states result in the emission of infrared or other lower-energy photons.

An atom in the $n = 3$ state or higher can make transitions to intermediate energy levels instead of jumping directly to the ground state. For example, an atom in, say, the $n = 4$ state may jump to the $n = 1$ state, or it may go from $n = 4$ to $n = 2$ and then from $n = 2$ to $n - 1$. In the latter case, two different photons would be emitted (● Figure 10.33). Atoms in the higher energy levels can undergo different "cascades" in returning to the ground state.

We can envision what happens when hydrogen gas in a tube is heated to a high temperature, as in Figure 10.11, or is excited by passing an electric current through it. The billions and billions of atoms will be excited, some to each of the possible higher energy levels. The excited atoms then undergo transitions to lower levels, with different atoms "stopping" at different levels according to a complex set of quantum-mechanical transition probabilities. The hydrogen gas continuously emits photons corresponding to all of the possible energy-level transitions, although some transitions are more favored than others depending in part upon the gas temperature. This is hydrogen's emission spectrum. (As in the Bohr model, excited atoms can also lose energy via collisions. No photons are emitted in this case.)

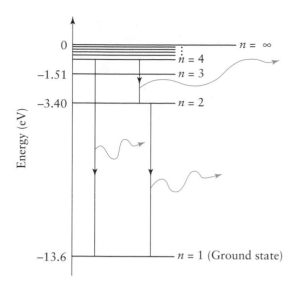

Upward energy-level transitions occur when the atom gains energy, by absorption of a photon with the proper energy, or perhaps by collision. For example, a hydrogen atom in the ground state can jump to the $n = 2$ state by absorbing a photon with energy 10.2 electronvolts. When white light—containing photons with many different energies—passes through hydrogen, those photons with just the right energy can be absorbed, leading to the observed absorption spectrum (see Figure 10.20).

An atom is ionized if it absorbs enough energy to make the electron energy greater than zero. For example, a hydrogen atom in the ground state is ionized if it absorbs a photon with energy greater than 13.6 electronvolts. This process, referred to as "photoionization," is essentially the same as the photoelectric effect. Any excess energy the electron has appears as kinetic energy.

We've presented a fairly complete picture of the hydrogen atom, but what of the other elements? The presence of more than one electron complicates things, but the general principles of atomic structure are much the same. Each element has its own atomic energy-level diagram, with correspondingly different energy values for each level. The various downward transitions between these levels produce the element's characteristic emission spectrum. Atoms with larger atomic numbers have more protons in the nucleus, so the force on the inner electrons is stronger. This means that the electrons are more tightly bound and that their energies have larger magnitudes but are still negative.

One very common device that uses the emission spectra of elements is the neon sign (● Figure 10.34a)—so named because neon is one of the most commonly used elements in them. These signs are made by placing low-pressure gas in a sealed glass tube. A high-voltage alternating current power supply connected across the ends of the tube causes electrons to move back and forth through the tube. The electrons excite the atoms of the gas by collision, and the atoms emit photons as they return to their ground states. Since

(a)

(b)

● **Figure 10.34** Two examples of emission spectra from excited gases. (a) A neon sign. (b) The aurora borealis (northern lights).

different elements have different emission spectra, signs can be made to emit different colors by being filled with different gases. Neon-filled signs are red because there are several bright red lines in neon's emission spectrum (see Figure 10.12).

Fluorescent lights and plasma displays (TVs) make use of ultraviolet emission lines (see p. 388). As in neon lights, the atoms of a gas (mercury in the former and xenon or neon in the latter) are collisionally excited by a current. The photons that are emitted are mostly ultraviolet, so a second substance, called a phosphor, is used to convert these to the visible-light photons that we see. Atoms in the phosphor, which is coated onto the inside surface of the chamber, absorb the UV photons and jump to higher energy levels. They return to the ground state via two steps: First, they jump to intermediate energy levels, which gives energy to neighboring atoms, then they return to the ground state as they emit visible-light photons. Different phosphors produce different frequencies of light. Fluorescent lights use phosphors that emit several different colors, which combine to approximate white light. Each pixel in a plasma display consists of three separate subpixel cells, one with a red phosphor, another with green, and the third with blue. Different colors are produced by varying the brightness of the red, green, and blue light. This is done by varying the current in each cell.

Nature provides us with a spectacular example of emission spectra, the *aurora borealis* (northern lights; Figure 10.34b) and the *aurora australis* (southern lights). These light displays, seen at night mainly in the far north and the far south, arise from a fascinating combination of physical processes. It begins at the Sun where ionized atoms and free electrons are flung out into space as part of the solar wind (see p. 177). After a journey of several days they reach the Earth, where the charged particles are guided by the magnetic field surrounding our planet (recall Figure 8.6). The lightweight electrons spiral around the magnetic field lines and follow them toward the north and south poles. If the conditions are right, these energetic electrons enter the upper atmosphere (down to about 100 kilometers above the Earth) and collide with atoms, molecules, and ions, thereby exciting them to emit spectra. The most common emission is a whitish-green glow from oxygen atoms. Energy-level transitions in excited nitrogen molecules are responsible for a pink that is often seen.* Much of the emitted radiation is in the ultraviolet and infrared bands, which we can't see. In the 1990s, it was discovered that electrical discharges sometimes travel *upward* from thunderstorms, exciting atoms and molecules. These produce faint light displays similar to auroras, between 50 and 100 kilometers above the Earth's surface.

Explore It Yourself 10.4

Go back to the Visual Quantum Mechanics site on the Web at http://phys.educ.ksu.edu/vqm/ and select the simulation called Gas Lamps—Emission. This activity offers you the chance to construct a set of energy levels that can produce (match) the emission spectrum of one of several available elements. Start with a simple spectrum, like that of hydrogen, which you know something about, to get the hang of the "game." Then try some of the less familiar and more complicated examples. After you've explored this simulation to your satisfaction, return to the program menu and select Gas Lamps—Absorption and complete the equivalent exercise. What conclusions can you draw about the relationship between the energy level structure of an atom and the appearance of the absorption and emission spectra it is capable of producing?

The structure of atoms with more than one electron is governed by a principle formulated by Wolfgang Pauli in 1925. Pauli, who was awarded the 1945 Nobel Prize in physics for his work, carefully analyzed the emission spectra of different elements noting that some expected transitions did not occur when the lower energy level was already occupied by a certain number of electrons. From this and other evidence, Pauli concluded

---

* It should be noted that the mechanism that creates the aurorae is much like that used in conventional televisions and computer monitors. In their picture tubes, fast-moving electrons are guided along their paths to the screen by magnetic fields. The image we see on the screen is comprised of light emitted by atoms that are given energy when these electrons collide with them.

Table 10.1 Ground-State Configurations of Several Atoms

Element	Atomic Number	Number of Electrons in Level		
		$n = 1$	$n = 2$	$n = 3$
Hydrogen	1	1	0	0
Helium	2	2	0	0
Lithium	3	2	1	0
Carbon	6	2	4	0
Oxygen	8	2	6	0
Neon	10	2	8	0
Sodium	11	2	8	1

that only a limited, fixed number of electrons could populate each energy level in an atom. (By analogy, we can't sit comfortably on a sofa if it is already full of people.) Pauli stated his conclusion in the **exclusion principle:**

PRINCIPLES

Two electrons cannot occupy precisely the same quantum state at the same time.*
(Pauli exclusion principle)

For each energy level, there exists a set number of quantum states available to the electrons. Once all of the quantum states corresponding to a given energy are filled, any remaining electrons in the atom must occupy other energy levels that have vacancies. The number of quantum states associated with each energy level is related to the electron's orbital angular momentum and spin (see Section 12.2). In the $n = 1$ level, the number of distinct quantum states is 2, so that the maximum number of electrons that can occupy this level is 2. For the $n = 2$ energy level, the maximum number of allowed states and, hence, electrons is 8; for the $n = 3$ level it is 18. The general rule is, for level n, the occupation limit of electrons is $2n^2$. The ground state of a multi-electron atom is one in which all electrons are in the lowest energy levels consistent with the Pauli exclusion principle. If any one electron is in a higher energy level, the atom is in an excited state. • Table 10.1 shows the ground-state energy-level populations for several atoms.

The properties of each element are determined to a great extent by the ground-state configuration of its atoms, particularly the number of electrons in the highest energy level that is occupied. Table 10.1 shows that helium in the ground state has the $n = 1$ energy level filled and that neon has the $n = 2$ energy level filled. Because of this, helium and neon have similar properties: They are both gases at normal room temperature and pressure and are very stable. They don't burn or react chemically in other ways except under special circumstances. Hydrogen, lithium, and sodium have similar properties because they all have one electron in the highest occupied energy level (when in their respective ground states).

The periodic table of the elements (back inside cover) was developed by Dmitri Mendeleev in 1869. He arranged the elements known at the time according to their properties: Elements with similar properties were placed in the same column. After Pauli's discovery, it was determined that the elements in each column have similar ground-state configurations, and that is why they are alike. (For example, hydrogen, lithium, and sodium are in the first column, called Group 1A, because each has one electron in its highest occupied energy level.) This is one indication of how understanding quantum mechanics can play a crucial role in other scientific disciplines like chemistry.

Concept Map 10.2 summarizes the types of transitions atomic electrons can undergo

---

* The exclusion principle applies to an entire class of elementary particles called *fermions*, which includes electrons. More on this in Chapter 12.

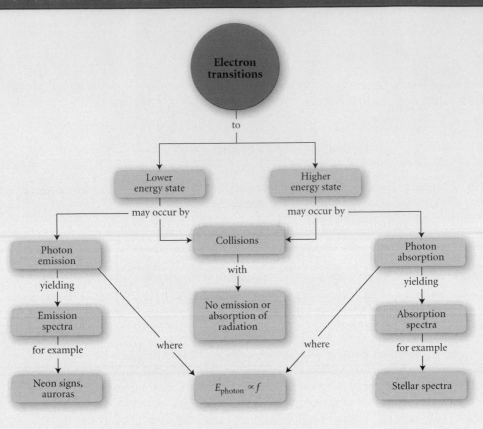

LEARNING CHECK

1. Indicate whether the following transitions within the energy-level diagram of hydrogen are accompanied by the *emission* or *absorption* of photons.
   a) $n = 5$ to $n = 3$
   b) $n = 1$ to $n = 8$
   c) $n = 8$ to $n = 9$
   d) $n = 3$ to $n = 5$
   e) $n = 6$ to $n = 5$

2. Assuming only one photon is emitted or absorbed in the course of each transition given above in Question 1, rank order (from highest to lowest) the frequencies of the photons involved in each transition.

3. The Balmer series is
   a) a set of play-off matches in golf named after the famous professional golfer Arnold Balmer.

   b) a sequence of emission lines in the ultraviolet portion of the electromagnetic spectrum due to hydrogen.
   c) a sequence of emission lines in the visible portion of the electromagnetic spectrum due to hydrogen.
   d) a sequence of emission lines due to hydrogen involving transitions from higher levels down to the $n = 3$ energy level.

4. What exactly is (or are) "excluded" by the Pauli exclusion principle?

5. (True or False.) The chemical properties of the elements in the periodic table are primarily determined by the electronic ground-state configuration of their atoms.

ANSWERS: 1. Emission (a), (e). Absorption (b), (c), (d). 2. (b), (a) & (d), (e), (c) 3. (c) 4. Electrons in the same quantum state. 5. True.

10.7 X-Ray Spectra

We have seen that the Bohr model of the atom (as refined by quantum mechanics) had great success in explaining the spectrum produced by hydrogen. Another early triumph of the Bohr atom came in connection with the study of x rays. Soon after Bohr published his model, a young English physicist named H. G. J. Moseley used it to explain the characteristic spectra of x rays.

In Section 8.5, we described how x rays are produced. Electrons are accelerated to a high speed and then directed into a metal target (Figure 8.35). A band of x rays is emitted, with different wavelengths having different intensities. ● Figure 10.35 shows the x-ray spectra when the elements tungsten (W) and molybdenum (Mo) are used as targets. These graphs are like blackbody radiation curves in that they are plots of intensity versus wavelength. Most of the x rays are produced as the electrons are rapidly decelerated upon entering the target. (This is one example of Maxwell's finding that accelerated charges emit EM waves.) This *bremsstrahlung* (German for "braking radiation") appears as the smooth part of the spectra covering the full range of wavelengths. Bremsstrahlung spectra are much the same for different elements. But the two sharp peaks for the molybdenum target are unique to this element and constitute its characteristic spectrum. Other elements exhibit characteristic spectra but at different wavelengths.

Moseley compared the characteristic x-ray spectra of different elements and found a simple relationship between the frequencies of the peaks and the atomic number of the element. For each peak, a graph of the square root of the frequency versus the atomic number of the element was a straight line (● Figure 10.36). This was a very practical discovery because it allowed him to determine the atomic number, and therefore, the identity, of unknown elements. All he had to do was determine the wavelengths of the characteristic x-ray peaks and use his graph. This method was a key factor in the discovery of several new elements, including promethium (atomic number 61) and hafnium (atomic number 72).

After Bohr developed his model of the atom, Moseley (● Figure 10.37) quickly realized that the characteristic x rays are much like the emission lines of hydrogen. He showed that they are produced when one of the innermost electrons jumps from one orbit (energy level) to another. Different elements have x-ray peaks with different wavelengths because the electron energy levels have different energy values. In particular, atoms with higher atomic numbers have more protons in the nucleus and bind the inner electrons more tightly.

Moseley proposed the following scenario to account for the x-ray spectra of the elements. If a bombarding electron collides with, say, a molybdenum atom and knocks out an electron from the $n = 1$ level, electrons in the upper levels will cascade down to fill the

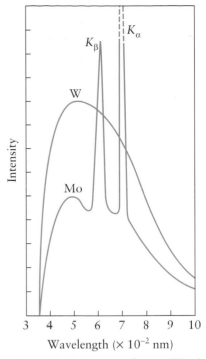

● **Figure 10.35** X-ray spectra of tungsten (W) and molybdenum (Mo) resulting from the bombardment of targets made of each element by electrons having energies of 35,000 electronvolts. Over the wavelength range shown, the spectrum of tungsten is a continuous one produced by bremsstrahlung processes. The spectrum of molybdenum includes two very strong characteristic peaks.

Note how small the wavelengths of these x rays are: ~0.05 nm.

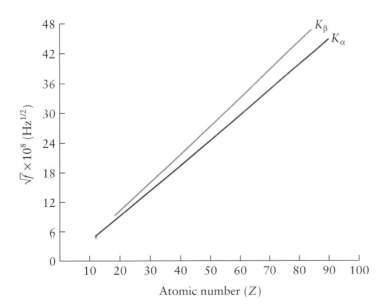

● **Figure 10.36** Moseley's diagram showing the relationship between the square root of the frequency for two characteristic x-ray peaks and the atomic number (Z) of the emitting element.

University of Oxford, Museum of the History of Science. Courtesy of the Niels Bohr Library

● **Figure 10.37** H. G. J. Moseley (1887–1915), whose work with x-ray spectra confirmed Bohr's model of the atom. Moseley's brilliant career was cut short when he enlisted during World War I. He was killed in action.

vacancy left by the ejected electron, and photons will be emitted. The lowest frequency (longest wavelength) x-ray photon corresponds to the transition from $n = 2$ to $n = 1$. Because the lowest energy level was called the "K shell" by x-ray experimenters, this is called the K_α (K-alpha) peak. Electrons jumping from $n = 3$ to $n = 1$ emit photons that form the K_β (K-beta) peak. Notice that the K_α and K_β peaks correspond to the two lowest-frequency lines in hydrogen's Lyman series.

With this knowledge, we can see why high-speed electrons are needed to produce the characteristic x rays of heavy elements. The inner electrons are so tightly bound that it takes very high energy electrons to knock them out.

10.8 Lasers

The word *laser* is an acronym derived from the phrase "<u>l</u>ight <u>a</u>mplification by <u>s</u>timulated <u>e</u>mission of <u>r</u>adiation." By examining this phrase one piece at a time, we will describe how the laser operates. In doing so we will find yet another example of how our theory of atomic structure provides the basis for understanding a process whose practical application has touched nearly every aspect of modern life.

Imagine that an electron in an atom has been excited to some higher energy level by either a collision or the absorption of a photon with the proper energy, E. Generally, such an electron will remain in this excited state for only a short time (around a billionth of a second) before returning to a lower energy level. If this decay to a lower state occurs spontaneously by radiation (which is usually the case in low-density gases where collisions are rare), a photon will be emitted in some random direction.

It turns out, however, that an electron can be *stimulated* to return to its original energy level through the intercession of a second photon with the same energy, E (● Figure 10.38). If a group of atoms all having their electrons in this same excited state is "bathed" in light consisting of photons with energy E, the atoms will be stimulated to decay by emitting additional photons with the same energy E. By this process, the intensity of a beam of light is increased or <u>a</u>mplified as a result of <u>s</u>timulated <u>e</u>mission by atoms in the region through which the <u>r</u>adiation passes. We have a laser.

To achieve this amplification, we must first arrange for the majority of the atoms through which the stimulating radiation travels to have their electrons in the same excited state. Otherwise, the unexcited atoms will just absorb the radiation. This is not an easy situation to arrange because the excited atoms normally don't stay that way for long.

The problem can be overcome because many atoms possess excited states referred to as *metastable*. Once in such a state, the electrons tend to remain there for a relatively long time (perhaps a thousandth of a second instead of a billionth) before spontaneously decaying to a lower state. During the time it takes to excite more than half of the atoms to that state, the ones already excited don't jump back down. This process, called "pumping," can result in the majority of the atoms being in the same metastable, excited state. This condition is called a *population inversion* because there are more atoms with electrons populating an upper energy level than a lower one, the reverse of the usual situation. This condition is necessary for the generation of laser light.

● **Figure 10.38** Stimulated emission. (a) Photon absorption places an atom in an excited state. (b) A photon with the same energy stimulates the excited atom to return to the lower state and emit an identical photon.

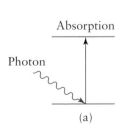

Absorption

Photon

(a)

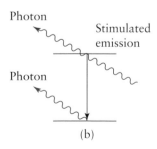

Photon

Stimulated emission

Photon

(b)

If we now irradiate the population-inverted atoms with photons of the correct energy, a chain reaction can be established that greatly amplifies the incident light beam. One photon stimulates the emission of another identical one, these two stimulate the emission of two more, these four stimulate the emission of four more, and so on. In the end, an avalanche of photons is produced, all having the same frequency (color). The beam is said to be *monochromatic*. The light amplification can produce a very high-intensity beam of light.

Besides being intense and monochromatic, laser light has an additional property that makes it extremely useful: *coherence*. This means that the stimulating radiation and the additional emitted laser radiation are in phase: The crests and troughs of the EM waves at a given point all match up (Figure 10.39b). An ordinary light source such as an incandescent bulb emits incoherent light (different parts of the beam are not in phase with one another) because the excited atoms giving off the radiation do so independently of each other. The emitted photons in this case may be considered to be individual, short "wave trains" bearing no constant phase relation to one another (Figure 10.39a). By contrast, stimulated emission of radiation by excited atoms produces photons that are in phase with the stimulating radiation. The coherence of laser light contributes to its high intensity. Coherent sources like lasers are important for producing interference patterns and *holograms* (described at the end of this section).

Although not in widespread use except for some specialized scientific applications, the ruby laser is simple in concept and illustrates all the important design characteristics of most other types of lasers. This device is shown schematically in Figure 10.40. It consists of a ruby rod whose ends are polished and silvered to become mirrors, one of which is partially (1–2%) transparent. The ruby rod is composed of aluminum oxide (Al_2O_3) in which some of the aluminum atoms have been replaced by chromium atoms. The chromium atoms produce the lasing effects. The rod is surrounded by a flash tube capable of producing a rapid sequence of short, intense bursts containing green light with a wavelength of 550 nanometers. The chromium atoms are excited by these flashes from their lower state E_0 to state E in a process referred to as *optical pumping*. (The high-intensity flash tube "pumps" energy into the chromium atoms.) The chromium atoms quickly spontaneously decay back to level E_0, or, in some cases, to the metastable level E_1 (Figure 10.41). With very strong pumping, a majority of the atoms can be forced into state E_1, and a population inversion is produced. Eventually, a few of the chromium atoms in state E_1 decay to the state E_0, thus emitting photons that stimulate other excited chromium atoms to execute the same transition. When these photons strike the end mirrors, most of them are reflected back into the tube. As they move back in the opposite direction, they cause more stimulated emission and increased amplification. A small fraction of the photons oscillating back and forth through the rod is transmitted through the partially silvered end and makes up the narrow, intense, coherent laser beam. The beam produced by a ruby laser has a wavelength of 694.3 nanometers, a deep red color.

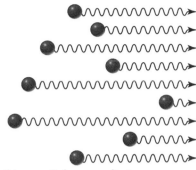

(a) Incoherent radiation from excited atoms

(b) Coherent radiation from excited atoms

 Figure 10.39 (a) Excited atoms in incoherent sources like ordinary light bulbs emit photons independently. The directions of propagation and the relative phases of emitted light are random. (b) Excited atoms in a laser emit coherent light. Photons produced by stimulated emission all travel in the same direction in phase with one another.

 Figure 10.40 Schematic of a ruby laser system. The pumping radiation is provided by a high-intensity flash tube. The stimulating photons are reflected back and forth between the parallel end mirrors to build up a beam of high intensity. The laser beam consists of photons that escape through the partially transmitting end reflector at the right. Lasers of this type must be pulsed to avoid overheating the ruby rod and possibly cracking it.

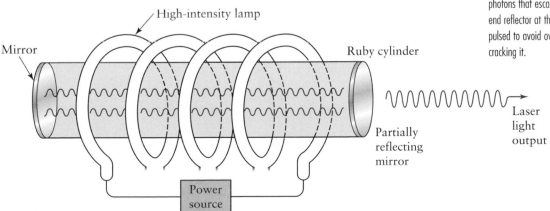

Holograms—3-D Images, No Glasses Required

Increasingly, 3-D imagery is becoming the norm in video gaming and web-based displays for everything from architectural renderings to Martian landforms to zoological nanostructures. For the most sophisticated of these applications, a headset using high-speed shuttered LCD lenses and costing hundreds of dollars is required, but for more common 3-D displays, only an inexpensive ($2–3) pair of glasses with one red lens and one blue one is sufficient to provide a sense of depth—a third dimension—when viewing a properly prepared photograph or cinematic image.* True 3-D objects, however, generally change

---

* We enjoy a sense of depth or three dimensions largely because each eye sees a slightly different view of the same scene. 3-D *anaglyphic* images are created by superimposing two photographs, one taken in red light and the other in blue, which record a particular scene from two slightly different perspectives. When viewing such a composite picture with a pair of red/blue glasses, the red image is seen by one eye while the blue one is seen by the other. The brain combines these two different images together to produce a 3-D image.

appearance as you move around them. You see different parts of an object depending on where you sit or stand. Such is not the case with typical, simple, noninteractive "3-D" images. The same aspects of the scene are observed from the left side of the image as from the right—glasses and all. The images are basically two-dimensional because ordinary photographs capture only the intensity of the light striking the film and nothing about its phase.

The production of true 3-D images that faithfully reproduce all aspects of the original objects was first accomplished in 1947 by Dennis Gabor. Gabor developed a way to preserve information about the relative phases of light beams emitted by or reflected from different points on an object in a complex *interference pattern*. A simple version of the experimental setup is shown in ● Figure 10.42. A beam of

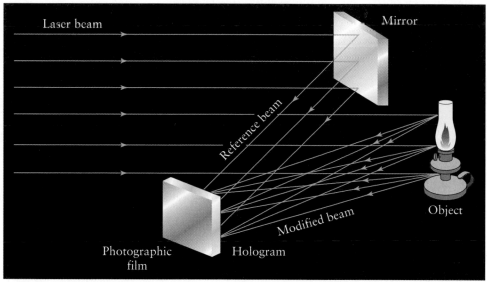

● **Figure 10.42** One method of hologram production. Laser beams reflected from the mirror and from the object combine at the photographic film to produce an interference pattern. The developed film is the hologram.

● **Figure 10.41** Energy-level transitions used by a ruby laser. The chromium atoms are excited by green pumping radiation. A transition to the metastable state follows. Stimulated emission from this level produces the red laser light.

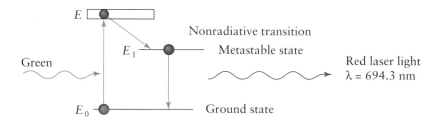

The ruby laser is an example of a pulsed solid-state laser: It produces a single, short pulse of laser light each time the flash tube fires, and the lasing material (the chromium atoms) is distributed in a solid matrix. Neodymium:yttrium-aluminum garnet (Nd:Yag) lasers are also of this type. Another common type of laser is the gas laser, of which the helium-neon (He-Ne) laser is a good example. Here, the lasing material is mixture of about 15% helium gas and 85% neon gas. In most applications, the neon gas produces

coherent radiation, in modern applications usually from a laser, is split in two, one beam traveling directly to a piece of photographic film, and the second reaching the film after being reflected off some object. At the position of the film, the two beams recombine, interfering with one another to produce a complex pattern of bright and dark areas (see "Interference" in Section 9.1.). This pattern is recorded on the film and contains information not only about the relative intensities of the object and reference beam, but also about their phase differences.

To see the image (to reconstruct the object), light of the same type as that used originally is shone on the film. Depending on the details of the original exposure process, the transmitted or reflected light is viewed obliquely to reveal the image. The resulting hologram (from the Greek word *holos*, meaning "whole") is a true 3-D image of the original object. Its aspect and/or color changes as you move your head; some parts of the object become visible while others disappear. The apparent shape, size, and brightness of the image change just as they would if you were examining the genuine article (● Figure 10.43). Gabor's research earned him the Nobel Prize in physics in 1971.

When first perfected during the 1960s, holography—the science of producing holograms—was viewed as little more than a curiosity, a clever way to generate interesting 3-D images and to illustrate simultaneously some basic physical concepts. In recent years, however, the ability of holograms to store huge quantities of information has been recognized by those in the computer industry involved in data storage and retrieval. Other applications involve component testing and analysis in which two holograms of an object taken at different times are superimposed. Any changes in the object, like deformations produced by, say, forces exerted during industrial stress tests or by tumor growth beneath the surface of the object, show up immediately as interference patterns. Recently, banks and credit unions have begun to place embossed holograms on their membership cards to reduce fraud. The presence of these highly detailed holographic images makes card duplication nearly impossible and attempts at counterfeiting readily apparent (see Explore It Yourself 10.5). Finally, researchers at the University of Michigan have developed holographic filmstrips that can display art objects, rare artifacts, and old musical instruments in full three-dimensionality for interested students, something ordinary photographs simply cannot do, 3-D glasses notwithstanding.

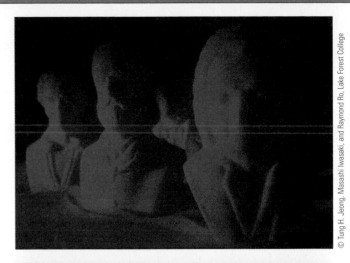

● **Figure 10.43** Two views of the image of a hologram made using a diode laser. (See "How to Make Holograms Using a Laser Pointer" by T. H. Jeong, R. J. Ro, and M. Iwasaki, *Optics and Photonics News* 11, no. 7 (July 1999).)

Explore It Yourself 10.5

Reach into your wallet or purse and take out one of your credit cards. It's a pretty good bet that, if your card isn't more than a few years old, it will show a silvery region about 2 centimeters or so in size containing a reflection hologram. Examine the hologram carefully under strong (white) light. Slowly tilt the card to change the angle of incidence of the illumination. Pay close attention to the subtle alteration in the color and appearance of the hologram image (a bird taking flight, a world map, etc.) as the card is rotated. Use a hand-magnifying lens to see the image in greater detail if necessary. Notice also the apparent depth (third dimension) of the image. The presence of such complex holographic images makes credit cards carrying them very difficult to counterfeit. Of course, the same level of complexity makes holograms suitable elements for incorporation in graphic design projects. If you are interested in exploring the full range of commercially available transmission and reflection holograms, try a web search using these terms as key words. You will be amazed at the beauty and variety of holographic art.

coherent radiation of wavelength 632.8 nanometers (red) as a result of stimulated emission from a metastable level to which it has been excited by collisions with the helium atoms. Helium-neon lasers can also be designed to yield green light with a wavelength of 543.5 nanometers. The He-Ne laser is a continuous laser in that it produces a steady beam of laser light.

Semiconductor lasers, sometimes called diode lasers, are not solid-state lasers but small, layered electronic devices that rely on a population inversion that is achieved electronically by injecting charges into the active, lasing medium (often gallium arsenide [GaAs]) from adjacent semiconducting layers ("cladding"). The cladding, whose index of refraction is lower than that of the active medium, also serves to confine the laser output from the active medium by total internal reflection (see Section 9.3). The laser light in these devices is amplified by multiple reflections between the end facets of the semiconductor crystal that have been carefully cleaved and act as cavity mirrors. Diode lasers typically operate in the near-infrared portion of the EM spectrum between about 750 and 880 nanometers. They are widely used in the telecommunications industry: Most long-distance telephone calls are carried by diode-laser-generated light signals transmitted over fiber-optic cables. The recent development of compact blue-emitting gallium nitride semiconducting lasers operating at wavelengths near 400 nanometers will likely lead to new applications of these devices in areas such as optical data storage and printing where shorter wavelengths mean higher spatial resolution.

Two other types of lasers, excimer lasers and dye lasers, are also in use. Excimer lasers use reactive gases like chlorine and fluorine mixed with inert gases (argon, krypton, or xenon) to produce laser light in the ultraviolet range. Dye lasers employ complex organic dyes in a liquid solution to produce laser light whose wavelength can be varied ("tuned") over a range of about 100 nanometers. ● Table 10.2 summarizes the characteristics of some of the most common types of lasers.

Lasers have assumed increasingly large and important roles in a variety of areas since their invention in the late 1950s. Today, lasers are used to perform surgery to correct certain vision defects (see Section 9.5) and to remove cancerous tumors of the skin, to cut metals and other materials precisely, to induce nuclear fusion reactions, to measure accurately the distances separating objects (the Earth and the Moon or two mirrors on opposite sides of a geological fault), to transmit telephonic information along optical fibers, to faithfully replay recorded music, movies, and data (CD, CD-ROM, and DVD players, see Figure 8.28), and to determine the price of goods at the supermarket checkout counter (● Figure 10.44). Beyond our own Earth, laserlike processes involving carbon dioxide and water have been discovered by astronomers in giant interstellar cloud complexes sprinkled throughout our galaxy. Regardless of the location or application, lasers all operate according to the same basic physics—physics entirely accessible and comprehensible once the structure of the atom is understood.

● **Table 10.2** Characteristics of Some Common Lasers		
Active Medium	**Wavelength (nm)**	**Type**
Argon fluoride	193	Pulsed
Nitrogen	337.1	Pulsed
Helium-cadmium	441.6	Continuous
Argon	476.5, 488.0, 514.5	Continuous
Krypton	476.2, 520.8, 568.2, 647.1	Continuous
Helium-neon	543.5, 632.8	Continuous
Rhodamine 6G dye	570–650	Pulsed
Ruby	694.3	Pulsed
Gallium arsenide	780–904* (IR)	Continuous
Neodymium:Yag	1060 (IR)	Pulsed
Carbon dioxide	10,600 (IR)	Continuous
*Depends on temperature.		

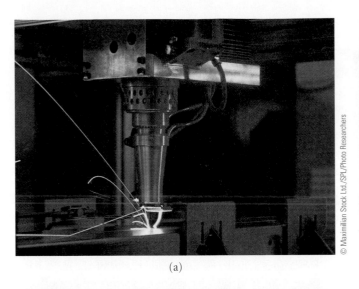

(a)

(b)

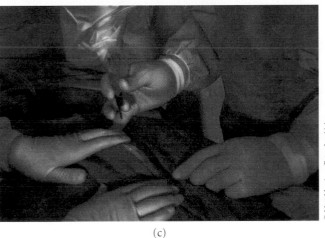

(c)

(d)

● **Figure 10.44** Applications of laser technology. (a) Laser welding. (b) Laser drilling. (c) Laser surgery. (d) Laser price scanning.

LEARNING CHECK

1. (True or False.) An x ray can be emitted only when an electron undergoes an allowed energy-level transition.
2. Rank (from lowest to highest) the frequencies of the characteristic K_α radiation emitted by the following elements:
   a) iron ($Z = 26$)
   b) zinc ($Z = 30$)
   c) titanium ($Z = 22$)
   d) copper ($Z = 29$)
3. Which of the following processes or phenomena is *not* involved in the operation of lasers?
   a) stimulated emission
   b) population inversion

   c) metastable state
   d) blackbody radiation
4. Which of the following common devices does *not* employ a laser in its operation?
   a) DVD player
   b) supermarket checkout price scanner
   c) telephonic fiber-optics communication system
   d) plasma TV
5. _____ are three-dimensional images made with the aid of lasers.

ANSWERS: 1. False **2.** (c), (a), (d), (b) **3.** (d) **4.** (d) **5.** Holograms

It is extremely difficult to describe in a few paragraphs very much of the history associated with the rapid and revolutionary developments that occurred in physics during the 40-year period from 1895 to 1935. It is equally difficult, if not impossible, in the same space to do justice to the many brilliant men and women physicists and chemists who contributed their talents to bring about these developments. The list of Nobel Prize recipients in physics and their citations for this period (Appendix A) may help to establish the significance of the work done by some of the "giants" of this era, but it does not begin to capture the excitement and spirit of adventure that existed within the scientific community at this time. Nowhere was this spirit more in evidence than at the Solvay Conferences held between 1911 and 1933.

Ernest Solvay (1838–1922) was a Belgian industrial chemist who perfected a commercial process for producing sodium carbonate and became extremely wealthy as the use of the "Solvay process" spread across the world. Solvay had long been interested in the fundamental structure of matter and by 1910 had become especially intrigued with the developing crisis between classical physics and the recently introduced quantum theories. In an effort to promote discussion and to resolve the crisis, Solvay held an international conference in October 1911 to which he invited many of the most prominent physicists of the period, including Planck, Einstein, and Marie Curie. The first conference proved to be so successful that Solvay established a foundation to sponsor similar conventions from time to time in the years that followed, right up to the present time.

The first Solvay Conference had as its topic radiation theory and quanta, while the second conference in 1913 dealt with the structure of matter. Five additional conferences were held between 1921 and 1933 on topics ranging from the electrical conductivity of metals, through magnetism, to the structure and properties of the nucleus. ● Figure 10.45 is a group portrait of the participants in the 1927 Solvay Conference. The participants in the other conferences held during this period were no less luminary than those shown in this photograph.

The progress made in the understanding and interpretation of natural phenomena during these early Solvay conferences was considerable. In what follows, we will mention some of the highlights of these sessions in an attempt to convey the importance of the work done during these meetings and the intense, spirited, and often humorous manner in which it was carried out.

During the third conference (1921), the two major subjects discussed were the nuclear model of the atom, as proposed and elaborated upon by Ernest Rutherford (more on this in Chapter 11), and Bohr's theory. During the discussion following Rutherford's report, three remarkable notions emerged—all of which were later confirmed as being substantially correct. (1) Jean Perrin suggested that the mechanism by which the Sun derives its energy is the transmutation of hydrogen into helium, whereby radiation equivalent to the mass difference between the reactants and the product is produced according to Einstein's equation $E = mc^2$ (Section 11.7). (2) Rutherford, after agreeing with Perrin that huge quantities of

● **Figure 10.45** Participants of the 1927 Solvay Conference. Front row: I. Langmuir, M. Planck, M. Curie, H. A. Lorentz, A. Einstein, P. Langevin, Ch. E. Guye, C. T. R. Wilson, O. W. Richardson. Middle row: P. Debye, M. Knudsen, W. L. Bragg, H. A. Kramers, P. A. M. Dirac, A. H. Compton, L. de Broglie, M. Born, N. Bohr. Back row: A. Piccard, E. Henriot, P. Ehrenfest, Ed. Herzen, Th. de Donder, E. Schrödinger, E. Verschaffelt, W. Pauli, W. Heisenberg, R. H. Fowler, L. Brillouin.

© Institut de Physique Solvay/AIP Niels Bohr Library

energy could be released in the combination of hydrogen nuclei to form helium, then commented:

> It has occurred to me that the hydrogen atoms of the (solar) nebula might consist of particles which one might call the "neutrons," which would be formed by a positive nucleus with an electron at a very short distance. These neutrons would hardly exercise any force in penetrating into matter. They will serve as intermediaries in the assemblage of nuclei of elements of higher atomic weights.

The neutron was later discovered in 1932 by James Chadwick. (3) Marie Curie argued that the stability of the nucleus could not be accounted for on the basis of electrostatic forces and that such stability required strong but very short-range, attractive forces of another type. These forces are now referred to as "nuclear" forces (Section 11.1).

The 1927 Solvay Conference focused on the new quantum mechanics and initiated what has been called the Einstein-Bohr dialogues concerning the implications of the probabilistic nature of this theory. Briefly, Bohr believed that at the subatomic level, the best one could do was to calculate the probabilities of finding a system in a given set of allowed states and that the very act of measurement involved interactions with the system that forced it into one of these allowed states where it is "observed." Numerous repeated measurements of the same type would reveal the system, on the average, to be in that state having the highest quantum-mechanical probability.

Einstein was not at all happy with the lack of determinism in the quantum theory and argued that the quantum mechanical description of nature may not have exhausted all the possibilities of accounting for observable phenomena. He believed that, if carried further, the analysis would reveal a means of knowing precisely, with 100% surety, the outcome of an experiment of a particular type on a given quantum system. Einstein's attitude toward quantum mechanics, as regards its being a complete description of nature on a microscopic level, may be summarized in his own words: "To believe this is logically possible without contradiction: but it is so very contrary to my scientific instinct that I cannot forgo the search for a more complete conception."

The Einstein-Bohr dialogues produced some lively debates during the 1927 Solvay meeting. The chairman of the session, H. A. Lorentz, himself an eminent physicist and master of three languages (English, German, and French), tried to maintain order but found it exceedingly difficult to do so, as one speaker after another joined in the fray, each in his own language. At one point, things became so confused that one of the participants, Paul Ehrenfest, a friend and colleague of both Einstein and Bohr, went to the blackboard and wrote: "The Lord did there confound the language of all the Earth." And, at a later time, during one of the lectures, Ehrenfest passed a note to Einstein that read: "Don't laugh! There is a special section in purgatory for professors of quantum theory, where they will be obliged to listen to lectures on classical physics ten hours every day." Present-day students of quantum theory might still find solace in Ehrenfest's remark!

The significance of the events at these meetings for later generations of physicists has been great indeed. The traditions of excellence established during these early conferences continue even to the present. As Werner Heisenberg, one of the founders of modern quantum theory and a Solvay participant in 1927, 1930, and 1933, wrote in 1974: "There can be no doubt that in those years (1911–1933) the Solvay Conferences played an essential role in the history of physics . . . the Solvay Meetings have stood as an example of how much well-planned and well-organized conferences can contribute to the progress of science. . . ."

SUMMARY

- At the turn of the last century, developments in experimental physics, particularly those related to **blackbody radiation,** the **photoelectric effect,** and **atomic spectra,** forced theoretical physicists to introduce revolutionary new concepts that defied classical physics as articulated by Newton and Maxwell.

- Key characteristics of these three processes were explained by assuming that things on the atomic scale are **quantized.** Light quanta (**photons**) have energies that are proportional to their frequencies.

- The **Bohr model** of the atom has electrons restricted to certain orbits in which the angular momentum is quantized. Photons are emitted and absorbed when electrons jump from outer orbits to inner orbits, or vice versa.

- During the 1920s and 1930s, the field of **quantum mechanics** emerged as it became clear that particles possess wavelike properties. This allowed a refinement of Bohr's model of the atom and led to a new probabilistic interpretation of physics on the atomic scale. No longer simply tiny particles, electrons are now represented as probability "clouds" surrounding nuclei.

- The atomic theory put forth by Bohr and his colleagues has been widely applied since its inception in 1913. One early use of the model was to account for the characteristic x-ray spectra from heavy elements.

- More recently, the establishment of the energy-level structure of such elements as neon and chromium, including the identification of relatively long-lived metastable states, has been instrumental in the development of laser systems.

- Today, quantum mechanics forms the foundation of our understanding of physics on the microscopic scale and stands as a monument to the combined genius of the men and women who participated in the early Solvay conferences where the foundations of this branch of physics were laid.

Equation	Comments
$E = hf$	Energy of atomic oscillator. Energy of a photon
$\Delta E = hf$	Energy of photon emitted or absorbed in energy-level transition
$\lambda = \dfrac{h}{mv}$	de Broglie wavelength of material particle
$\Delta x\, \Delta mv \geq h$	Heisenberg's uncertainty principle

MAPPING IT OUT!

1. Quantum mechanics is one of the most daunting of all areas in physics. To students encountering this subject for the first time at the introductory level, quantum mechanics can seem mysterious, even contradictory, because it is so counterintuitive to what we are used to experiencing in the macroscopic domain where Newton's laws apply. Some of the confusion can often be dispelled by organizing the material in a way that highlights some of the major connections between the principal elements of the subject. Concept maps can be of value in this regard.

   Review the reading in Sections 10.5 and 10.6. Then make a list of the main ideas or concepts introduced in these sections. Your list should include some, if not all, of the following items: uncertainty principle, exclusion principle, wave-particle duality, de Broglie wavelength, and wave function. Try to organize the concepts you have identified around the central theme of *quantum mechanics,* linking the various concepts appropriately by connecting words or short phrases to form meaningful propositions. When you have completed your concept map, compare your result with that of a fellow student. What similarities exist between the two maps? What distinc-

tions can be seen in them? If there are significant differences in the linkages between concepts held in common in the two maps, discuss them with your colleague and try to resolve them. Consult your instructor for assistance in this effort, if necessary.

2. Section 10.8 introduces the laser as a device whose development followed closely from our understanding of quantum mechanics and atomic structure. Make a list of the basic concepts related to the development, function, and application of the laser in modern society. Reexamine Section 10.8 if necessary to ensure that your list is as complete as possible. Next, prioritize the items in your list from most important to least important. If some concepts are of equal importance or significance with respect to others, be sure to identify them as such. Now exchange your list with another student in your class. Does your list contain the same elements as that of your neighbor? Are his or her items prioritized in the same manner as yours? Discuss the differences in the two lists with your colleague and attempt to reconcile them. Ask your instructor or another classmate for help if you need it.

QUESTIONS

(▶ *Indicates a review question, which means it requires only a basic understanding of the material to answer. The others involve integrating or extending the concepts presented thus far.*)

1. ▶ What does it mean when we say that the energy of something is "quantized"?
2. Of the things that a car owner has to purchase routinely—gasoline, oil, antifreeze, tires, and so on—which are normally sold in quantized units, and which are not?
3. ▶ What assumption allowed Planck to account for the observed features of blackbody radiation?
4. On an old television, there are two main control knobs, one to change the channel and another to change the loudness. Is each controlling the TV in a quantized fashion or a continuous fashion?
5. ▶ What is a photon? How is its energy related to its frequency? To its wavelength?
6. ▶ Describe the photoelectric effect. Name some devices that make use of this process.
7. If nature suddenly changed and Planck's constant became a much larger number, what effect would this have on things like solar cells, atomic emission and absorption spectra, lasers, and so on?
8. Sodium undergoes the photoelectric effect, with one electron absorbing a photon of violet light and another absorbing a photon of ultraviolet light. What is different about the two electrons afterward?
9. What aspects of the photoelectric effect can be explained without recourse to the concept of the photon? What aspects of this phenomenon require the existence of photons for their explanation?

10. Based on what you learned about image formation in Chapter 9, describe how you might design a photocopying machine that could make a copy that is enlarged or reduced compared to the size of the original.
11. If sunlight can be conceived of as a beam of photons, each of which carries a certain amount of energy and momentum, why don't we experience (or feel) any recoil as these particles collide with our bodies when, say, we're at the beach?
12. ▶ What is the difference between a continuous spectrum and an emission-line spectrum?
13. A mixture of hydrogen and neon is heated until it is luminous. Describe what is seen when this light passes through a prism and is projected onto a screen.
14. ▶ What are the basic assumptions of the Bohr model? Describe how the Bohr model accounts for the production of emission-line spectra from elements like hydrogen.
15. ▶ Discuss what is meant by the term *ionization.* Give two ways by which an atom might acquire enough energy to become ionized.
16. ▶ Compare the emission spectra of the elements hydrogen and helium (see Figure 10.12). Which element emits photons of red light that have the higher energy?
17. If an astronomer examines the emission spectrum from luminous hydrogen gas that is moving away from the Earth at a high speed and compares it to a spectrum of hydrogen seen in a laboratory on Earth, what would be different about the frequencies of spectral lines from the two sources?

18. The spectrum of light from a star that is observed by the Hubble Space Telescope is not exactly the same as that star's spectrum observed by a telescope on Earth. Explain why this is so.

19. A high-energy photon can collide with a free electron and give it some energy. (This is called the Compton effect.) How are the photon's energy, frequency, and wavelength affected by the collision?

20. ▶ What is the de Broglie wavelength? What happens to the de Broglie wavelength of an electron when its speed is increased?

21. ▶ An electron and a proton are moving with the same speed. Which has the longer de Broglie wavelength? (You may want to look ahead at some useful information in the table in the back inside cover.)

22. ▶ In an electron microscope, electrons play the role that light does in optical microscopes. What makes this possible?

23. ▶ What is the uncertainty principle?

24. Explain why the Bohr model of the atom is incompatible with the Heisenberg uncertainty principle.

25. ▶ What does it mean when a hydrogen atom is in its "ground state"?

26. ▶ Explain why a hydrogen atom with its electron in the ground state cannot absorb a photon of just any energy when making a transition to the second excited state ($n = 3$).

27. Describe the spectrum produced by ionized hydrogen, that is, a sample of hydrogen atoms all of which have lost one electron.

28. ▶ Will the energy of a photon that ionizes a hydrogen atom from the ground state be larger than, smaller than, or equal to, the energy of a photon that ionizes another hydrogen atom from the first excited state ($n = 2$)? Explain.

29. What would an energy-level diagram look like for the quantized cat in Figure 10.3?

30. ▶ In what part of the EM spectrum does the Lyman series of emission lines from hydrogen lie? The Balmer series? The Paschen series? Describe how each of these series is produced. In what final state do the electron transitions end in each case?

31. Radioactive strontium (Sr) tends to concentrate in the bones of people who ingest it. Why might one expect that strontium would behave like calcium (Ca) chemically and thus be preferentially bound in bone material, which is predominantly calcium in composition?

32. ▶ The x-ray spectrum of a typical heavy element consists of two parts. What are they? How is each produced?

33. ▶ Will the frequency of the K_α peak in the x-ray spectrum of copper (Cu) be higher or lower than the frequency of the K_α peak of tungsten (W)? Explain how you arrived at your answer.

34. ▶ Describe how the Bohr model may be used to account for characteristic x-ray spectra in heavy atoms.

35. ▶ What is the origin of the word *laser*?

36. ▶ Distinguish laser light from the light emitted by an ordinary light bulb in as many ways as you can.

37. ▶ Distinguish between a metastable state and a normally allowed energy state within an atom. Discuss the role of metastable states in the operation of laser systems.

38. ▶ Define what is meant by the term *population inversion*. Why must this condition be achieved before a system can successfully function as a laser?

39. A friend tells you that physicists have just invented a new pulsed laser that is pumped with yellow light and produces laser light in the ultraviolet portion of the EM spectrum. Why might you be a little skeptical of this claim? Explain.

40. ▶ Describe the operation of a pulsed ruby laser.

41. ▶ What are the Solvay conferences? In what ways did these meetings contribute to the development of quantum mechanics during the years 1911–1933?

PROBLEMS

1. Find the energy of a photon whose frequency is 1×10^{16} Hz.

2. If your body is emitting infrared radiation of wavelength 9.4×10^{-6} m, what is the energy of the released photons?

3. In what part of the EM spectrum would a photon of energy 9.5×10^{-25} J be found? What is its energy in electronvolts?

4. Gamma rays (γ rays) are high-energy photons. In a certain nuclear reaction, a γ ray of energy 0.511 MeV (million electronvolts) is produced. Compute the frequency of such a photon.

5. Electrons striking the back of a TV screen travel at a speed of about 8×10^7 m/s. What is their de Broglie wavelength?

6. In a typical electron microscope, the momentum of each electron is about 1.6×10^{-22} kg-m/s. What is the de Broglie wavelength of the electrons?

7. During a certain experiment, the de Broglie wavelength of an electron is 670 nm = 6.7×10^{-7} m, which is the same as the wavelength of red light. How fast is the electron moving?

8. If a proton were traveling the same speed as electrons in a TV picture tube (see Problem 5), what would its de Broglie wavelength be? The mass of a proton is 1.67×10^{-27} kg.

9. A hydrogen atom has its electron in the $n = 2$ state.
   a) How much energy would have to be absorbed by the atom for it to become ionized from this level?
   b) What is the frequency of the photon that could produce this result?

10. A hydrogen atom initially in the $n = 3$ level emits a photon and ends up in the ground state.
   a) What is the energy of the emitted photon?
   b) If this atom then absorbs a second photon and returns to the $n = 3$ state, what must the energy of this photon be?

11. The following is the energy-level diagram for a particularly simple, fictitious element, Kansasium (Ks). Indicate by the use of arrows all allowed transitions leading to the emission of photons from this atom and order the frequencies of these photons from highest (largest) to lowest (smallest).

● **Figure 10.46** Problem 11

12. An atom of neutral zinc possesses 30 electrons. In its ground configuration, how many fundamental energy levels are required to accommodate this number of electrons? That is, what is the smallest value of n needed so that all 30 of zinc's electrons occupy the lowest possible quantum energy states consistent with the Pauli exclusion principle?

13. Referring to Figure 10.35, notice that the bremsstrahlung x-ray spectrum of both Mo and W cut off (have zero intensity) at about 0.035 nm. X rays with this wavelength are produced by the target element when bombarding electrons are promptly stopped in a single collision and give up all their energy in the form of EM waves. Confirm that electrons having energies of 35,000 eV will produce x-ray photons with wavelengths near 0.035 nm by this process.

14. The characteristic K_α and K_β lines for copper have wavelengths of 0.154 nm and 0.139 nm, respectively. What is the ratio of the energy difference between the levels in copper involved in the production of these two lines?

15. Referring to Figure 10.36, we see that the atomic number Z is proportional to $f^{1/2}$ or that Z^2 is proportional to f. Since the frequency of the characteristic x-ray lines is itself proportional to the energy of the associated x-ray photon, we are led to conclude that ΔE and hence the energies of the atomic levels also scale as Z^2. Based on this analysis (and ignoring any differences due to the masses of their nuclei), how much greater is the energy associated with the ground state of helium ($Z = 2$) than that of hydrogen ($Z = 1$)? Make a similar comparison between the ground state energies for sodium ($Z = 11$) and hydrogen.

16. Characteristic x rays emitted by molybdenum have a wavelength of 0.072 nm. What is the energy of one of these x-ray photons?

17. In a helium-neon laser, find the energy difference between the two levels involved in the production of red light of wavelength 632.8 nm by this system.

18. The carbon dioxide laser is one of the most powerful lasers developed. The energy difference between the two laser levels is 0.117 eV.
    a) What is the frequency of the radiation emitted by this laser?
    b) In what part of the EM spectrum is such radiation found?

19. If you bombard hydrogen atoms in the ground state with a beam of particles, the collisions will sometimes excite the atoms into one of their upper states. What is the *minimum* kinetic energy the incoming particles must have if they are to produce such an excitation?

20. a) Which of the following elements emits a K-shell x-ray photon with the highest frequency?
    b) Which emits a K-shell photon with the lowest frequency?
    i) Silver (Ag)
    ii) Calcium (Ca)
    iii) Iridium (Ir)
    iv) Tin (Sn)

CHALLENGES

1. Can you think of a reason why metals exhibit the photoelectric effect most easily? Is there a connection between this phenomenon and properties of a good electrical conductor?

2. One serious problem with sending intense laser beams long distances through the atmosphere to receiving targets is that the beams spread out. This effect, called *thermal blooming*, is caused by the fact that the beam heats the air, thereby changing the speed that the light travels. Explain how such a change in the speed of light in the air could produce the observed effect. Will the speed increase or decrease upon heating? Why?

3. How might you explain the concept of quantization to a younger brother or sister, using money as the quantized entity?

4. The muon is a negatively charged particle that is much like an electron but with a mass about 200 times larger. A "muonic" hydrogen atom forms when a muon orbits a proton. The muonic atom's orbits and energy levels follow the basic rules of the Bohr model and the quantum mechanical model of the atom. Using the analysis in Section 10.5, explain why the size of each Bohr orbit in the muonic atom is much smaller than the corresponding orbit in an ordinary hydrogen atom. How would the energies of each of the energy levels in the muonic atom be different from those in the regular hydrogen atom? What would be different about the emission spectrum?

5. The rate at which solar wind particles enter the atmosphere is higher during the day than at night, yet the intensity of the auroral emissions remains high well after the Sun has set. Can you suggest a means by which the atmospheric molecules might be able to radiate long after the period of collisions with charged particles has ended? (*Hint:* How long does it take a typical atom to radiate from a normal allowed energy state? How could this time be lengthened?)

6. Compute a rough estimate of the number of photons emitted each second by a 100-W light bulb. You might make the simplifying assumptions that (1) all of the electrical energy is converted into radiant energy and (2) the "average" photon that is emitted has an energy corresponding to the peak of the BBR curve for an object at 3,000 K (refer to Section 8.6).

7. In the Bohr model for hydrogen, the radius of the nth orbit can be shown to be n^2 times the radius of the first Bohr orbit $r_1 = 0.05$ nm (see Example 10.3). Similarly, the energy of an electron in the nth orbit is $\frac{1}{n^2}$ times its energy when in the $n = 1$ orbit. What is the *circumference* of the $n = 100$ orbit? (This is the distance the electron has traveled after having revolved around the proton once.) For such large-n states, the orbital frequency is about equal to the frequency of the photon emitted in a transition from the nth level to an adjacent level with $n + 1$ or $n - 1$. Given this, find the frequency and corresponding period of the electron's orbit by computing the frequency associated with the transition from $n = 100$ to $n = 101$. Using your values for the electron's orbital size (distance) and travel time (period), calculate the approximate speed of the electron in the 100th orbit. How does this speed compare to the speed of light?

Akasofu, Syun-Ichi. "The Dynamic Aurora." *Scientific American* 260, no. 5 (May 1989): 90–97. A detailed look at how the northern lights and the southern lights are produced.

Beardsley, Tim. "The Dope on Holography." *Scientific American* 279, no. 3 (September 1998): 41. An update on efforts to develop holographic data-storage devices.

Berns, Michael. "Laser Scissors and Tweezers." *Scientific American* 278, no. 4 (April 1998): 62–67. Describes how lasers are used by biologists to do "surgery" on individual cells.

Cassidy, David C. "Heisenberg, Uncertainty, and the Quantum Revolution." *Scientific American* 266, no. 5 (May 1992): 106–112. A look at the life and work of Nobel laureate Werner Heisenberg.

Clark, Ronald W. *Einstein: The Life and Times.* New York: Avon, 1971. Perhaps the best biography of Einstein.

Englert, Berthold-Georg, Marlan O. Scully, and Herbert Walther. "The Duality in Matter and Light." *Scientific American* 271, no. 6 (December 1994): 86–92. A discussion of wave-particle duality, including an imagined conversation between Albert Einstein and Niels Bohr.

Gourley, Paul L., and Darryl Y. Sasaki. "Biocavity Lasers." *American Scientist* 89, no. 2 (March–April 2001): 152–159. Report on how human cells can take part in light amplification to produce a new laser that may prove useful in the detection of cancer.

Marburger, John H. III. "What Is a Photon?" *Physics Teacher* 34, no. 8 (November 1996): 482–486. A discussion of the interpretations of the concept of a photon.

McCreary, Michael D. "Digital Cameras." *Scientific American* 278, no. 6 (June 1998): 102. A short look at how digital cameras work.

Mende, Stephen B., Davis D. Sentman, and Eugene M. Wescott. "Lightning between Earth and Space." *Scientific American* 277, no. 2 (August 1997): 56–59. A report on types of recently discovered electrical discharges that go upward from thunderstorms and produce light displays.

Mort, Joseph. "Xerography: A Study in Innovation and Economic Competitiveness." *Physics Today* 47, no. 4 (April 1994): 32–38. A look at the history and current technology of photocopying.

Tegmark, Max, and John Archibald Wheeler. "100 Years of Quantum Mysteries." *Scientific American* 284, no. 2 (February 2001): 68–75. A look back (and ahead) at the field of quantum mechanics—one of the great triumphs of twentieth-century science.

For additional readings, explore InfoTrac® College Edition, your online library. Go to http://www.infotrac-college.com/wadsworth and use the passcode that came on the card with your book. Try these search terms: plasma display, photoelectric effect, Robert Bunsen, Gustav Kirchhoff, scanning tunneling microscope, aurora, diode laser, Solvay conference.

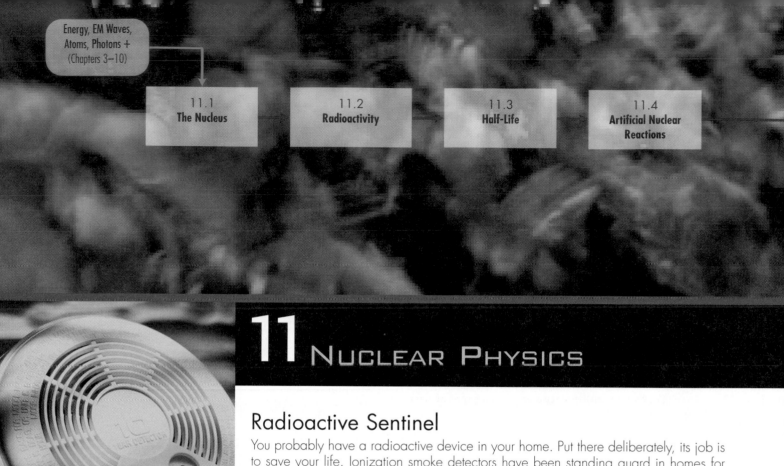

Energy, EM Waves, Atoms, Photons + (Chapters 3–10)

| 11.1 The Nucleus | 11.2 Radioactivity | 11.3 Half-Life | 11.4 Artificial Nuclear Reactions |

11 NUCLEAR PHYSICS

Ionization smoke detector.

© Photodisc Collection/Getty

Radioactive Sentinel

You probably have a radioactive device in your home. Put there deliberately, its job is to save your life. Ionization smoke detectors have been standing guard in homes for decades, ready to detect the tiny particles that comprise smoke and sound an alarm. Although an electrical device, this type of smoke detector also makes use of a small amount of radioactive material.

The heart of an ionization smoke detector is a chamber through which air can pass. Ions are continually created in the air in the chamber, and a battery causes them to carry a small electrical current through the air. This current is closely monitored by the device's electronics.

When smoke particles enter the chamber, the ions are attracted to them and become attached to them. This reduces the current, which triggers the alarm. The device is very sensitive, as anyone who has generated a small amount of smoke while cooking and was startled by the sound of the smoke alarm going off can attest.

How are these ions produced? Radiation from a tiny amount of radioactive material knocks electrons off nitrogen and oxygen atoms in the air. How is this material able to continually emit this radiation for hundreds of years with no power input? The answer lies in the phenomenon known as radioactive decay, which is one of the main subjects of this chapter.

The basic properties of nuclei, the many applications of nuclear processes, and the promises and perils of nuclear energy are the main topics of this chapter. The first two sections describe the composition of the nucleus and the process of *radioactive decay*. This leads to the concept of half-life and the use of radioisotopes as natural clocks. Most of the remainder of the chapter deals with the role of energy in nuclear physics, particularly the two principal means of tapping that energy—nuclear fission and nuclear fusion.

11.1 The Nucleus

The nucleus occupies the very center of the atom. It is tiny yet incredibly dense: More than 99.9% of the atom's mass is compressed into roughly *one-trillionth* of its total volume. If an atom could be enlarged until it were 2,000 feet across, its nucleus would only be about the size of a pea. The nucleus is impervious to the chemical and thermal pro-

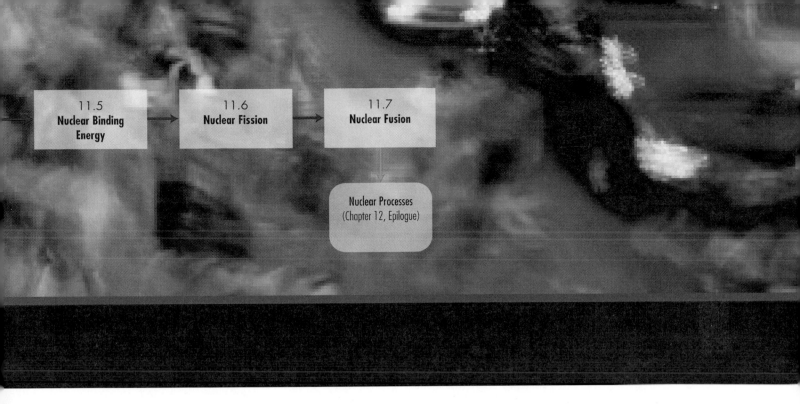

| 11.5
**Nuclear Binding
Energy** | 11.6
Nuclear Fission | 11.7
Nuclear Fusion |

Nuclear Processes
(Chapter 12, Epilogue)

cesses that affect its electrons. But it is seething with energy—energy that makes the Sun and the other stars shine.

The nucleus contains two kinds of particles, protons and neutrons. The particles have nearly the same mass, about 1,840 times that of an electron. The masses are extremely small, so it is convenient to introduce an appropriate unit of mass, the *atomic mass unit,* u.

$$1 \text{ u} = 1.66 \times 10^{-27} \text{ kg}$$

The proton carries a positive charge, and the neutron is uncharged (● Table 11.1). The number of protons in the nucleus is the **atomic number** Z of the atom. There is one proton in the nucleus of each hydrogen atom ($Z = 1$), two in that of helium ($Z = 2$), eight in oxygen ($Z = 8$), and so on. This number determines the identity of the atom.

The neutral neutrons have much less influence on the properties of the atom: their main effect is on the atom's mass. In fact, the number of neutrons in the nuclei of a particular element can vary. Most helium atoms have two neutrons in the nucleus, but some have one, three, four, or even six (● Figure 11.1). Each is still a helium atom; the only difference is the mass. This allows us to associate another number with each nucleus, the **neutron number.**

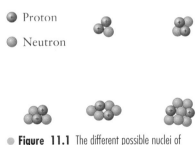

● Proton

● Neutron

● **Figure 11.1** The different possible nuclei of helium atoms. All have two protons, but the number of neutrons varies. The electrons in orbit around the nuclei are not shown. With this scale, the radius of such an orbit would be about 1,000 feet.

Neutron Number The number of neutrons contained in a nucleus.

The symbol N is used to represent the neutron number just as Z is used to represent the atomic number. For all helium atoms, $Z = 2$, but N can be 1, 2, 3, 4, or 6.

The mass of an atom is determined primarily by how many protons and neutrons there are in the nucleus. (The electrons are so light that they contribute only a negligible amount to the total mass.) Therefore, it is useful to define yet a third number for each nucleus.

● **Table 11.1** Properties of the Particles in the Atom

Particle	Mass	Charge
Electron	9.109×10^{-31} kg = 0.00055 u	-1.602×10^{-19} C
Proton	1.67262×10^{-27} kg = 1.00730 u	1.602×10^{-19} C
Neutron	1.67493×10^{-27} kg = 1.00869 u	0

Mass Number The total number of protons and neutrons in a nucleus.

The **mass number** is represented by the letter *A*. Note that

$$A = Z + N$$

The mass number indicates what the mass of a nucleus is, just as the atomic number indicates the amount of electric charge in the nucleus. Protons and neutrons are collectively referred to as **nucleons.** (The terms *nucleon number* and *atomic mass number* are sometimes used instead of mass number.) The mass number is just the total number of nucleons in the nucleus. The possible mass numbers for helium are *A* = 3, 4, 5, 6, or 8. Each different possible "type" of helium is called an **isotope.**

Isotopes Isotopes of a given element have the same number of protons in the nucleus but different numbers of neutrons.

The different isotopes of an element have essentially the same atomic properties: These are determined by the atomic number *Z*. For example, most of the carbon atoms in your body have six neutrons in their nuclei, but small numbers of the isotopes with seven and eight are also present. The three different carbon isotopes are indistinguishable as far as chemical processes (such as burning) are concerned. The different isotopes of an element often *do* have vastly different nuclear properties. For example, most nuclear power plants rely on the splitting of uranium-235 nuclei, that is, uranium atoms with mass number *A* = 235. Uranium-238 will not work.

Most of the 114 different elements have several isotopes. Some have only a few (hydrogen has 3), and others have more than 20 (iodine, silver, and mercury, to name a few). More than 2,500 different isotopes have been identified and studied. Of these, only about 300 occur naturally. The rest are produced in certain nuclear processes, such as nuclear explosions. The vast majority of isotopes are unstable; the nuclei eventually transform into different nuclei through a process known as *radioactive decay.* (More on this in Section 11.2.)

Each different isotope of an element is designated by its mass number. The most common isotope of helium has *A* = 4 (*Z* = 2 and *N* = 2) and is called helium-4. The other helium isotopes are helium-3, helium-5, helium-6, and helium-8. The three isotopes of carbon in your body are carbon-12, carbon-13, and carbon-14. Two of the isotopes of hydrogen have been given special names: hydrogen-2 is called deuterium and hydrogen-3 is called tritium. (The prefixes come from the Greek words for *second* and *third*.)

Freezing, boiling, burning, crushing, and other chemical and physical processes do not affect the nuclei of atoms. These processes are influenced by the forces between different atoms, forces that involve only the outer electrons. But nuclei are not indestructible: A number of nuclear processes do affect them. Nuclei can lose or gain neutrons and protons, absorb or emit gamma rays, split into smaller nuclei, or combine with other nuclei to form larger ones. These processes are called *nuclear reactions.* Some occur around us naturally, but most are produced artificially in laboratories. Most of the remainder of this chapter deals with the nature of nuclear reactions and their applications.

To diagram nuclear reactions, we use a special notation to represent each isotope. It consists of the atom's chemical symbol with a subscript and a superscript on the left side. The *subscript* is the atom's atomic number *Z*, and the *superscript* is the atom's mass number *A*. Some examples:

Helium-4	$^{4}_{2}\text{He}$	Carbon-14	$^{14}_{6}\text{C}$
Carbon-12	$^{12}_{6}\text{C}$	Uranium-235	$^{235}_{92}\text{U}$

The two numbers indicate the relative mass and charge that the nucleus possesses. The neutron number N can be found by subtracting the lower number (Z) from the upper number (A). For example, each uranium-235 nucleus contains 143 neutrons ($235 - 92 = 143$). Note that both the chemical symbol and the subscript indicate what the element is, so they must always agree. Regardless of what the superscript is, if the subscript is 6, the element is carbon and the symbol must be C.

This notation can be extended to represent individual particles as well. The designations for neutrons, protons, and electrons are

Neutron ^1_0n

Proton ^1_1p

Electron $^0_{-1}\text{e}$

For the electron, 0 is used for the mass number because its mass is so small. The -1 indicates that it has the same size charge as a single proton, except it is negative.

Perhaps you've wondered how protons can be bound together in a nucleus. After all, like charges do repel each other. There is indeed a large electrostatic force acting to push the protons apart, but another force many times stronger acts to hold them together inside a nucleus. This is called the *strong nuclear force,* one of the four fundamental forces in nature. (The *weak nuclear force* is involved in certain nuclear processes, but it is not important in holding the nucleus together. More on this in Chapter 12.) The strong nuclear force is an attractive force that acts between nucleons: Every proton and neutron in a nucleus exerts an attractive force on every other proton and neutron. Compared to the gravitational and electromagnetic forces, the strong nuclear force is rather strange. It is much stronger than the others, but it has an extremely short range: The attractive nuclear force between particles in the nucleus effectively disappears if the particles become more than about 3×10^{-15} meters apart. This puts an upper limit on the size that a nucleus can have and still be stable. In a large nucleus, protons on opposite sides are far enough apart that the repulsive electric force becomes important. No known stable isotope has an atomic number larger than 83.

The effectiveness of the nuclear force at holding a nucleus together also depends on the relative numbers of neutrons and protons. If there are too many or too few neutrons compared to the number of protons, the nucleus will not be stable. For example, carbon-12 and carbon-13 are stable, but carbon-11 and carbon-14 are not. The ratio of N to Z for stable nuclei is about 1 for small atomic numbers and increases to about 1.5 for large nuclei (● Figure 11.2). The stable isotope lead-208 has 126 neutrons and 82 protons in each nucleus.

In summary, the nucleus is a collection of neutrons and protons held together by the strong nuclear force. This force is limited in its ability to hold nucleons together. Nuclei that are too large or that do not have the proper ratio of neutrons to protons are unstable. They eject particles and release energy. There are several different mechanisms used by various unstable nuclei to accomplish this. The main ones are discussed in Section 11.2.

LEARNING CHECK

1. Different isotopes of an element have the same number of _____ in the nucleus but different numbers of _____ .
2. Which of the following is *not* true about a nucleus of the isotope $^{18}_8\text{O}$?
   a) It contains 8 protons.
   b) It contains 18 nucleons.
   c) It contains 8 electrons.
   d) It contains 10 neutrons.
3. (True or False.) Protons can stay close together inside a nucleus because the repulsive electric force between them disappears at short range.

ANSWERS: 1. protons, neutrons **2.** (c) **3.** False

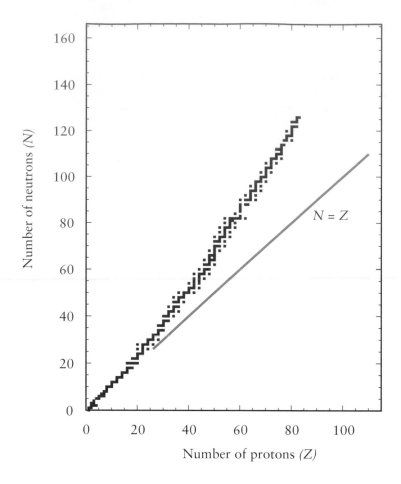

● **Figure 11.2** Plot showing the number of neutrons and the number of protons in the isotopes that are stable. Each small square represents one stable isotope. Its coordinates are the numbers N and Z. The green line indicates the nuclei that would have the same number of protons and neutrons. The points curve away from this line, indicating that larger stable nuclei have successively more neutrons than protons.

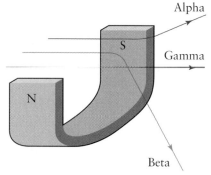

● **Figure 11.3** The three common types of nuclear radiation are affected differently by a magnetic field. Alpha and beta rays are deflected as they go through the field because they are charged particles. Gamma rays are not deflected.

11.2 Radioactivity

Radioactivity, also called *radioactive decay,* occurs when an unstable nucleus emits radiation. Isotopes with unstable nuclei are called **radioisotopes.** The majority of all isotopes are radioactive. For the moment, we will put aside the question of where radioisotopes come from and concentrate on the processes of radioactive decay.

When it was first investigated, nuclear radiation was found to be similar to x rays. For example, it exposes photographic film. Soon it was determined that there were actually three different types of nuclear radiation, named **alpha** (α), **beta** (β), and **gamma** (γ) **radiation.** Alpha rays and beta rays are actually high-speed charged particles. They can be deflected with magnetic and electric fields (● Figure 11.3). Gamma rays are extremely high-frequency electromagnetic waves (high-energy photons).*

Several different devices are used to detect radiation, the most common being the *Geiger counter.* It exploits the fact that nuclear radiation is ionizing radiation. A gas-filled cylinder is equipped with a fine wire running along its axis (● Figure 11.4). A high voltage is maintained between the outer wall and the wire, causing a strong electric field. When an alpha, beta, or gamma ray enters the cylinder and ionizes some of the gas atoms, the freed electrons are accelerated by the electric field. These in turn ionize other atoms, and an avalanche of electrons reaches the wire and causes a current pulse. Most Geiger counters emit an audible click each time a ray is detected. They also keep track of how many are detected each second. In other words, they indicate the count rate—the number of alpha, beta, or gamma rays detected each second.

The emission of each type of radiation has a different effect on the nucleus. Both alpha decay and beta decay alter the identity of the nucleus (the atomic number is changed) as well as release energy. Gamma-ray emission does not in itself change the nucleus: It sim-

---

*It is now known that there are many more ways that unstable nuclei can undergo radioactive decay. There are several types of beta decay alone. For the sake of simplicity, only the three most common forms of radioactive decay—alpha, beta (one type), and gamma decay—will be considered.

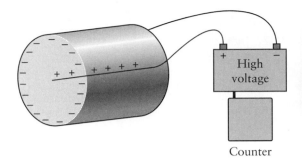

● **Figure 11.4** (Left) Simplified sketch of a Geiger counter. Nuclear radiation ionizes the gas in the cylinder. The freed electrons are accelerated to the wire and produce a current pulse. (Right) Geiger counter detecting beta rays emitted by an isotope in a lead cup (gray). The count rate is enhanced by the magnet (blue), since beta rays are deflected by the magnetic field into the detector (marked G).

ply carries away excess energy in much the same way that photon emission carries away excess energy from excited atoms. Gamma-ray emission often accompanies alpha decay and beta decay. These two processes often leave the nucleus with excess energy. ● Table 11.2 lists several radioisotopes and their modes of decay.

Alpha Decay

An alpha particle is really four particles tightly bound together: two protons and two neutrons. It is identical to a nucleus of helium-4. For this reason, an alpha particle can be represented as

$$\text{alpha particle:} \quad \alpha \quad \text{or} \quad {}^{4}_{2}\text{He}$$

A nucleus that undergoes alpha decay *does not* contain a helium-4 nucleus. This just happens to be a particularly stable combination of nuclear particles that can be ejected as a group from the nucleus.

The emission of an alpha particle reduces both the atomic number Z and the neutron number N by 2. The mass number A is reduced by 4. ● Figure 11.5 is a diagram of the

● **Table 11.2** Properties of Selected Isotopes

Element	Isotope	Decay Mode(s)	Relative Abundance (%)
Hydrogen	${}^{1}_{1}\text{H}$	. . . (stable)	99.985
	${}^{2}_{1}\text{H}$ deuterium	. . .	0.015
	${}^{3}_{1}\text{H}$ tritium	beta	. . .
Helium	${}^{3}_{2}\text{He}$	. . .	0.00014
	${}^{4}_{2}\text{He}$	. . .	99.9999
	${}^{5}_{2}\text{He}$	alpha	. . .
	${}^{6}_{2}\text{He}$	beta	. . .
	${}^{8}_{2}\text{He}$	beta, gamma	. . .
Carbon[a]	${}^{12}_{6}\text{C}$	. . .	98.90
	${}^{13}_{6}\text{C}$	. . .	1.10
	${}^{14}_{6}\text{C}$	beta	trace
	${}^{15}_{6}\text{C}$	beta	. . .
Silver[a]	${}^{107}_{47}\text{Ag}^{*}$[b]	gamma, then stable	51.84
	${}^{108}_{47}\text{Ag}$	beta, gamma	. . .
	${}^{109}_{47}\text{Ag}^{*}$[b]	gamma, then stable	48.16
	${}^{110}_{47}\text{Ag}$	beta, gamma	. . .
Uranium[a]	${}^{232}_{92}\text{U}$	alpha, gamma	. . .
	${}^{233}_{92}\text{U}$	alpha, gamma	. . .
	${}^{234}_{92}\text{U}$	alpha, gamma	0.0055
	${}^{235}_{92}\text{U}$	alpha, gamma	0.72
	${}^{236}_{92}\text{U}$	alpha, gamma	. . .
	${}^{237}_{92}\text{U}$	beta, gamma	. . .
	${}^{238}_{92}\text{U}$	alpha, gamma	99.27
	${}^{239}_{92}\text{U}$	beta, gamma	. . .

[a]Not every isotope of the element is given.

[b]Asterisks (*) indicate that the nuclei are in an excited state.

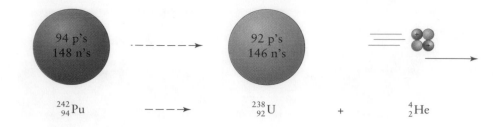

$$^{242}_{94}\text{Pu} \quad ---\rightarrow \quad ^{238}_{92}\text{U} \quad + \quad ^{4}_{2}\text{He}$$

alpha decay of a plutonium-242 nucleus. The atomic number of the nucleus decreases from 94 to 92: The nucleus is transformed from plutonium to uranium. The original plutonium-242 nucleus is called the "parent," and the resulting uranium-238 nucleus is called the "daughter."

Figure 11.5 shows why the isotopic notation introduced earlier is so convenient. When used to represent a nuclear process, like alpha decay, it clearly shows how the mass and the charge of the nucleus are affected. Both the total electric charge and the total number of protons and neutrons must be the same before and after the process. This means that the sum of the subscripts on the right side of the arrow must equal the sum of those on the left. The same is true for the superscripts. This allows one to determine what the decay product (the daughter) is when a given nucleus undergoes alpha decay.

Example 11.1

The isotope radium-226 undergoes alpha decay. Write the reaction equation, and determine the identity of the daughter nucleus.

From the periodic table of the elements, we find that the atomic number of radium is 88 and its chemical symbol is Ra. So the reaction will appear as follows:

$$^{226}_{88}\text{Ra} \rightarrow \; ? \; + \; ^{4}_{2}\text{He}$$

The mass number A of the daughter nucleus must be 222 for the superscripts to agree on both sides of the arrow. By the same reasoning, the atomic number Z must be 86. From the periodic table, we find the element radon (Rn) has $Z = 86$. The daughter nucleus is radon-222.

$$^{226}_{88}\text{Ra} \rightarrow \; ^{222}_{86}\text{Rn} \; + \; ^{4}_{2}\text{He}$$

Because alpha decay causes such a drastic change in the mass of the nucleus, it generally occurs only in radioisotopes with high atomic numbers. The alpha particle is ejected with very high speed (typically around one-twentieth the speed of light). Alpha particles are quickly absorbed when they enter matter: Even a sheet of paper can stop them.

Beta Decay

Beta decay is easily the oddest of the three kinds of radioactivity. A beta particle is simply an electron ejected from a nucleus. This means that a beta particle has the same symbol as an electron.

$$\text{Beta particle:} \quad \beta \quad \text{or} \quad ^{0}_{-1}\text{e}$$

But wait a minute: There aren't any electrons in a nucleus. During beta decay, one of the neutrons is spontaneously converted into an electron (the beta particle) and a proton. The electron is ejected with very high speed, and the proton remains in the nucleus.* We can represent this process as follows:

---

*Another particle, a *neutrino* (from the Italian for "little neutral one"), is also emitted in beta decay. Neutrinos are very strange little beasts—they have no charge, their mass is extremely small, and they rarely interact with matter. Neutrinos routinely pass through the entire Earth without being absorbed, deflected, or otherwise affected. For our purposes, the neutrino can be regarded as just part of the energy that is released during beta decay. (Chapter 12 has more to say about neutrinos.)

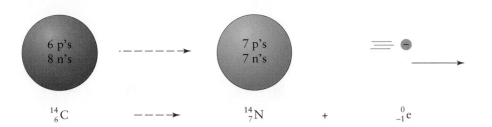

Figure 11.6 A nucleus of carbon-14 undergoing beta decay.

$$neutron \rightarrow proton + electron$$

$$_{0}^{1}n \rightarrow _{1}^{1}p + _{-1}^{0}e$$

The total electric charge remains the same—zero.

A nucleus that undergoes beta decay loses one neutron and gains one proton. • Figure 11.6 shows the beta decay of a carbon-14 nucleus. Note that the mass number A of the nucleus is unchanged.

Example 11.2

The isotope iodine-131 undergoes beta decay. Write the reaction equation, and determine the identity of the daughter nucleus.

From the periodic table, we find that iodine's chemical symbol and atomic number are I and 53. Therefore:

$$_{53}^{131}I \rightarrow ? + _{-1}^{0}e$$

The mass number stays the same, and the atomic number is increased by 1 to 54. We find that this is the element xenon (Xe). The daughter nucleus is xenon-131.

$$_{53}^{131}I \rightarrow _{54}^{131}Xe + _{-1}^{0}e$$

Often the daughter nucleus in both alpha and beta decay is itself radioactive and decays into another isotope. Plutonium-242 undergoes alpha decay to uranium-238. This is a radioisotope that undergoes alpha decay into thorium-234. This process continues until, after a total of nine alpha decays and six beta decays, the stable isotope lead-206 is reached. This is called a **decay chain.** When the Earth was formed from the debris of exploded stars, hundreds of different radioisotopes were present in varying amounts. Geological formations that were originally rich in uranium-238 now contain large amounts of lead-206 as well.

Gamma Decay

Because a gamma ray has no mass or electric charge, gamma-ray emission has no effect on the mass number A or the atomic number Z of a nucleus. Nuclei can exist in excited states in much the same way that orbiting electrons can. Gamma rays are emitted when the nucleus goes to a lower energy state. • Figure 11.7 shows the gamma decay of a strontium-87 nucleus. The identity of the nucleus is not changed during the process.

Gamma-ray photons, as indicated in Figure 10.7, have energies from about 100,000 electronvolts to more than a billion electronvolts. Most gamma-ray photons emitted in

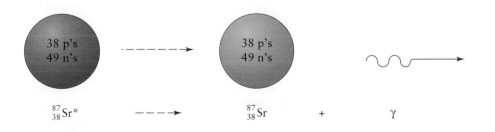

Figure 11.7 A nucleus of strontium-87 undergoing gamma decay. The asterisk (*) indicates that the nucleus is in an excited state.

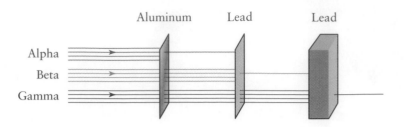

Figure 11.8 Alpha, beta, and gamma radiation differ a great deal in their ability to penetrate matter. Alpha rays are the least penetrating, while gamma rays are the most.

gamma decay are around 1 million electronvolts. (The unit of energy that is used most often in nuclear physics is the megaelectronvolt (MeV), which is 1 million electronvolts.)

One way to compare the different decay processes is to focus on three properties of the nucleus: mass, electric charge, and energy. Alpha decay alters all three: An alpha particle carries away considerable mass, two charged protons, and a great deal of kinetic energy. Beta decay has little effect on the mass of the nucleus, but it does increase the positive charge of the nucleus, and it takes away energy in the form of the beta particle's kinetic energy. Gamma decay only takes energy away from the nucleus.

The three types of nuclear radiation differ considerably in their ability to penetrate solid matter. All three are ionizing radiation. They ionize atoms as they pass through matter. Alpha particles are the least penetrating. Their positive charge causes them to interact strongly with atomic electrons and nuclei, and they are absorbed after traveling only a short distance. Beta particles are more penetrating than alpha particles. For comparison, a thin sheet of aluminum that will block essentially all alpha particles will stop only a fraction of beta particles. Gamma rays are the most penetrating of the three. Since they have no electric charge and travel at the speed of light, they interact with atoms much less frequently. It requires several centimeters of lead to block gamma rays (see ● Figure 11.8).

Radioactivity and Energy

What becomes of the energy of nuclear radiation as it is absorbed? Most of it goes to heat the material. Early experimenters with radioactivity noticed that highly radioactive samples were physically warmer than their surroundings. As long as the substance continued to emit radiation, it stayed warm. Radioactivity has been used as an energy source on spacecraft. Many interplanetary space probes, including the *Voyager* spacecraft that photographed Saturn, Jupiter, Uranus, and Neptune, were equipped with *radioisotope thermoelectric generators* (RTGs). The heat from the radioactive decay of an isotope such as plutonium-238 is converted into electricity to operate cameras, radio transmitters, and other onboard equipment.

The interior of the Earth is so hot that much of it is molten. Some of this heat reaches the surface in volcanoes, in geysers, and through other geothermal processes. Geologists are not certain why the Earth's interior is so hot, but they know what keeps it from cooling off: heat from radioactive decay. In the Earth's interior, relatively small amounts of radioisotopes are still present. The energy released as these radioisotopes decay is enough to compensate for the transfer of heat to the Earth's surface. Since much of this radioactive material is concentrated in a layer just below the Earth's outer crust, this layer is kept in a partially liquid state by heat from the radioactivity. This in turn allows the crustal plates (continental land masses) to slowly slide around over the Earth in a process called *continental drift*.

Applications

Radiation from radioactive decay is used routinely for many purposes. Nuclear medicine makes use of radioisotopes for diagnosis and treatment (see the Physics Potpourri on page 442). Dozens of industrial facilities around the world use gamma radiation from cobalt-60 and other isotopes to sterilize disposable medical products such as syringes and physician gloves. The gamma rays can penetrate deep into containers of items and kill bacteria and other microorganisms. Many of these facilities also irradiate food items,

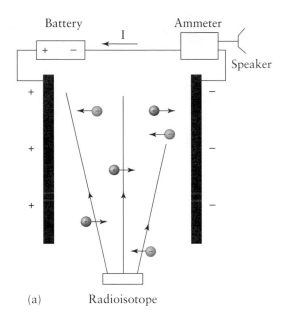

(a)   Radioisotope

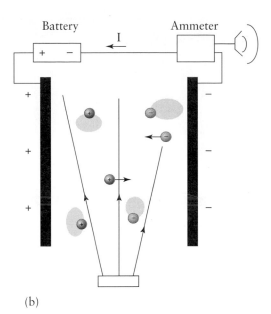

(b)

● **Figure 11.9** Simplified diagram of a smoke detector. (a) A current flows between the plates because the radiation ionizes the air. (b) Ions are attracted to the smoke particles and attach to them. This reduces the current, which triggers the alarm.

from spices and herbs to fresh meat, with gamma rays and other ionizing radiation to prevent spoilage, prolong shelf life, and kill harmful bacteria like salmonella.

Because gamma rays are absorbed in a predictable way as they pass through solid material, they can be used to measure the thickness and integrity of manufactured items. Flaws in jet engine parts and cracks in large cables can be detected because more gamma rays will pass through items with gaps in them than through solid material. The thickness of metal sheets as they are manufactured can be measured by monitoring how much gamma radiation passes through them.

Another way in which radioactivity is put to good use is in the most common type of smoke detector (see p. 428). The alpha radiation from a radioisotope (americium-241) ionizes the air between two plates connected to the terminals of a battery (● Figure 11.9). Consequently, an electric current flows through the circuit. When smoke is present, ions are attracted to the smoke particles and attach to them. The ammeter detects the resulting decrease in the current and triggers the alarm. Thousands of lives have been saved by these simple, inexpensive devices.

Analysis of radioactive decay is an important tool in nuclear physics. The type of radiation emitted by a particular radioisotope, along with the amount of energy released, provide clues to the structure of the nucleus. Since nuclei are much too small for us to examine with our eyes, we have to use indirect information, such as that carried by radiation when it leaves a nucleus, to learn about them.

Concept Map 11.1 summarizes the common types of radioactive decay.

LEARNING CHECK

1. Nuclear radiation can be detected by using a
   _____ .

2. Of the three types of nuclear decay, only
   _____ does not affect the atomic number
   of the nucleus.

3. (True or False.) Alpha decay causes the atomic number
   of the nucleus to increase.

4. Radioactive decay is involved in which of the following?
   a) Keeping the Earth's interior hot.
   b) Making smoke detectors work.
   c) Supplying energy for interplanetary space probes.
   d) All of the above.

ANSWERS: 1. Geiger counter 2. gamma decay 3. False 4. (d)

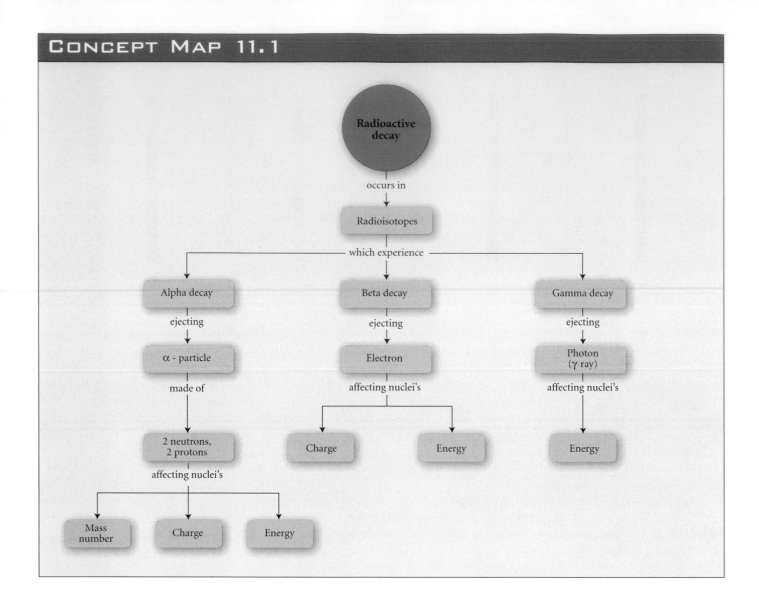

11.3 Half-Life

We now address an aspect of radioactive decay that we have avoided so far—time. If a nucleus is radioactive, when will it decay? There is no way of knowing exactly when a particular nucleus will decay: It may wait a billion years or a millionth of a second. Radioactive decay is a random process, much like throwing dice (● Figure 11.10). One can't predict exactly what will come up each throw, but one can analyze the results of hundreds of throws and indicate how likely it is that each possible value will come up. Similarly, we can predict how much time will elapse, on the average, before a nucleus of a given radioisotope will decay. There is wide variation among the thousands of different radioisotopes. Nearly all nuclei of some isotopes will decay in less than a second, while only a fraction of the nuclei of other isotopes will decay in 4.5 billion years, the age of the Earth. This leads us to the concept of **half-life.**

© Vern Ostdiek

● **Figure 11.10** The exact value of one roll of dice can't be predicted, but the approximate number of times that seven will turn up in 1,000 rolls can be predicted quite closely. The exact time at which a given nucleus will decay can't be predicted, but the fraction of 1 million nuclei that will decay during some time interval can be predicted quite closely.

DEFINITION

Half-Life The time it takes for half the nuclei in a sample of a radioisotope to decay. The time interval during which each nucleus has a 50% probability of decaying.

● Table 11.3 Half-Lives of Isotopes in Table 11.2

Element	Isotope	Half-Life
Hydrogen	^1_1H	. . .
	^2_1H deuterium	. . .
	^3_1H tritium	12.3 yr
Helium	^3_2He	. . .
	^4_2He	. . .
	^5_2He	2×10^{-21} s
	^6_2He	0.805 s
	^8_2He	0.119 s
Carbon[a]	$^{12}_6C$	. . .
	$^{13}_6C$	. . .
	$^{14}_6C$	5,730 yr
	$^{15}_6C$	24 s
Silver[a]	$^{107}_{47}Ag*$[b]	44.2 s
	$^{108}_{47}Ag$	2.42 min
	$^{109}_{47}Ag*$[b]	39.8 s
	$^{110}_{47}Ag$	24.6 s
Uranium[a]	$^{232}_{92}U$	70 yr
	$^{233}_{92}U$	159,000 yr
	$^{234}_{92}U$	245,000 yr
	$^{235}_{92}U$	704,000,000 yr
	$^{236}_{92}U$	23,400,000 yr
	$^{237}_{92}U$	6.75 d
	$^{238}_{92}U$	4,470,000,000 yr
	$^{239}_{92}U$	23.5 min

[a]Not every isotope of the element is given.

[b]Asterisks (*) indicate that the nuclei are in an excited state.

The half-lives of radioisotopes range from a tiny fraction of a second to billions of years (● Table 11.3). During the span of one half-life, approximately half of the nuclei in a sample will decay—that is, emit their radiation. Half of the remaining nuclei will decay during the span of a second half-life, leaving only one-fourth of the original nuclei. After three half-lives, one-eighth remain, and so on. After n half-lives, one-half raised to the power n of the original nuclei will remain undecayed.

For example, let's say that we start with 8 million nuclei of a radioisotope with a half-life of 5 minutes. About 4 million of the nuclei will decay in the first 5 minutes. Half of the remaining 4 million will decay during the next 5 minutes, leaving 2 million. After 15 minutes, there will be 1 million left undecayed. After 50 minutes (10 half-lives), there will be about 7,800 nuclei left (8 million $\times \left(\frac{1}{2}\right)^{10} = 7{,}800$).

Example 11.3

A pure sample of uranium-237 is prepared. As the uranium nuclei decay, the sample becomes "contaminated" with decay products. How much time will elapse before only one-fourth of the sample is uranium-237?

One-half of the uranium nuclei decay during 6.75 days, the half-life of uranium-237 (from Table 11.3). After another 6.75 days, one-half of the remaining nuclei decay, leaving one-fourth of the original amount. Therefore, after a total of *13.5 days,* only one-fourth of the sample will be uranium-237.

In reality, it isn't possible to count how many nuclei are left undecayed every 5 minutes. But with a Geiger counter, one can keep track of how the rate of decay, the number of nuclei that decay each minute, decreases. In the previous example, 4 million decay during the first 5 minutes, 2 million during the next 5 minutes, and so on. A Geiger counter placed nearby might show an initial count rate of 10,000 counts per minute. (The count rate would depend on how close the counter is to the sample.) Five minutes later, the count rate would be one-half that, about 5,000 counts per minute. In other words, the count rate also

● **Figure** **11.11** (a) Graph of the number of remaining nuclei versus time for a radioisotope with a 5-minute half-life. (N_0 is the initial number of nuclei.) (b) Graph of the count rate versus time for the same radioisotope.

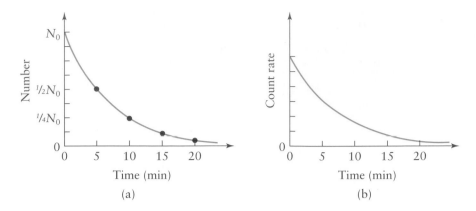

N_0

$\frac{1}{2}N_0$

$\frac{1}{4}N_0$

Number

Time (min)

(a)

Count rate

Time (min)

(b)

is halved during each half-life (● Figure 11.11). Thus, by monitoring the count rate of the nuclear radiation emitted by a radioisotope, one can determine the isotope's half-life.

The way in which the nuclei of a radioisotope decay over time, as shown in Figure 11.11a, is different from the way most manufactured items with limited life spans fail. For example, incandescent and fluorescent lamps generally function for hundreds or thousands of hours before "burning out." We can ascribe a "half-life" to them, in the sense that on average one-half of a large number of lamps will fail before that time elapses. But unlike radioactive decay in which the number of nuclei that decay per unit of time starts high and decreases steadily, the number of lamps that burn out per time usually is very low until roughly half of the "half-life" has passed. For example, a study showed that 50% of a certain type of lamp failed before 12,000 hours of use, but only 1% failed during the first 4,000 hours. Were they radioactive nuclei with a half-life of 12,000 hours, 21% would have decayed during that time.

Is it useful to know the half-lives of radioisotopes? Yes, for a number of reasons. Knowing the half-life of a radioisotope allows us to estimate over what period of time a sample will emit appreciable amounts of radiation. The isotope americium-241 is used in smoke detectors partly because its long half-life (432 years) ensures that it will continue to produce the ions needed for the detector's operation throughout the device's lifetime.

Small amounts of certain radioisotopes are sometimes used medically in the human body (see the Physics Potpourri on page 442). The flow of the isotope through the bloodstream (for example) can be monitored with a Geiger counter. The radioisotope that is used must have a long enough half-life to remain detectable during the time that it takes to move through the body. Its half-life must not be too long, however. To minimize any possible harmful effects, the body should not be subjected to the radiation any longer than necessary. One of the most commonly used radioisotopes in nuclear medicine is technetium-99, a gamma emitter with a half-life of 6 hours.

Explore It Yourself 11.1

You can simulate the decay of radioactive nuclei with a large number of pennies or similar flat objects. It's best to use 50 or more.

1. Place the pennies in a small box or bag and shake them thoroughly. Dump the pennies out on a flat surface like a desktop.
2. Treat each penny with "tails" showing as a nucleus that has decayed. Push these off to the side, and record how many pennies are left. Collect these "undecayed" pennies, and repeat the procedure several times.
3. Each penny has a 50% chance of turning up "tails" after each throw. Thus each throw represents one half-life for the pennies. On the average, half of them "decay" during each throw.
4. Notice how the number of pennies left undecayed decreases. Make a simple graph of these numbers. You should get graphs like those in Figure 11.11.
5. Think about the probabilities: After, say, the fifth throw, any surviving penny has turned up "heads" five times in a row. What is its chance of turning up "heads" on the next throw?

Neutron ●

7 p's
7 n's

6 p's
8 n's

Proton ●

Dating

The regular rate of decay of radioisotopes can be used like a clock. **Carbon-14 dating** is a good example. Carbon is a key element in all living things on Earth. Carbon enters the food chain as plants take in carbon dioxide from the air. The complex carbon-based molecules formed by plants are ingested by animals and people. About 99% of this carbon is the stable isotope carbon-12. About one out of every trillion carbon atoms is radioactive carbon-14. In the upper atmosphere, carbon-14 is constantly being produced as cosmic rays from outer space collide with atoms in air molecules. Some of the fragments are free neutrons that cause the formation of carbon-14 when they collide with nitrogen-14. The reaction goes like this:

$$\,^{1}_{0}n + \,^{14}_{7}N \rightarrow \,^{14}_{6}C + \,^{1}_{1}p$$

(See ● Figure 11.12.) The carbon-14 atoms can combine with oxygen atoms to form carbon dioxide molecules, then mix in the atmosphere and enter the food chain. A small percentage of all the carbon atoms in plants, animals, and people is carbon-14. (You are slightly radioactive! The amount of carbon-14 is so small that the radiation—beta particles—is not a hazard.) The half-life of carbon-14 is about 5,700 years, so very little of it decays during the lifetime of most organisms. An exception would be certain trees that can live thousands of years.

After an organism dies, no new carbon-14 is added, so the percentage of carbon-14 decreases as the nuclei undergo radioactive decay. This can be used to measure the age of the remains. For example, if a tree is cut down and used to build a shelter, 5,700 years later there will be one-half as much carbon-14 in the wood as in a live tree (● Figure 11.13). After 11,400 years, there would be one-fourth as much, and so on. Thus, by measuring the carbon-14 content in ancient logs, charcoal, bones, fabric, or other such artifacts, archaeologists can estimate their age. This process has become an invaluable tool to archaeologists and is quite accurate for material ranging up to about 40,000 years old.

Geologists use radioisotopes to estimate the ages of rock formations and the Earth itself. The ratio of the amount of the parent radioisotope to the amount of the daughter is used. For example, about 28% of the naturally occurring atoms of the element rubidium are the radioisotope rubidium-87. Its half-life is so long—49 billion years—that only a small percentage of it has decayed since the Earth was formed. Rubidium-87 undergoes beta decay into the stable isotope strontium-87. The age of rock formations that contain rubidium-87 can be estimated by measuring the relative amount of strontium-87 that is present. The more strontium-87 present, the older the rock.

Because of radioactive decay, most of the radioisotopes present when the Earth formed have long since decayed away. Only radioisotopes with very long half-lives, like rubidium-87 and uranium-238, have survived to this day. A few radioisotopes with short half-lives

● **Figure 11.13** (a) As soon as a tree dies, the amount of carbon-14 in it begins to decrease. (b) After 5,700 years (one half-life), it contains one-half as much. (c) After 11,400 years, it contains one-fourth as much.

(a)

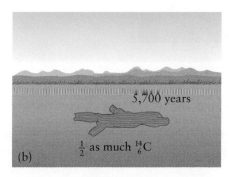

5,700 years

$\frac{1}{2}$ as much $^{14}_{6}C$

(b)

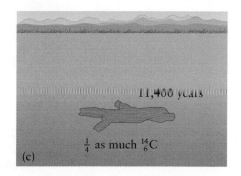

11,400 years

$\frac{1}{4}$ as much $^{14}_{6}C$

(c)

Radiation: A Killer Turned Lifesaver

As pointed out in Section 8.5, x rays, gamma rays, and other ionizing radiation can be harmful to life. They can break up the large complex molecules in living cells, causing them to die or mutate. But life on this planet thrives in an environment where there has always been small amounts of nuclear radiation present. There are cosmic rays coming from outer space, radioactive nuclei in the soil and rocks, and even radioisotopes in living tissue, like carbon-14 and potassium-40. Such radiation does kill cells, but the human body routinely replaces millions of dead cells each day, so it can repair the damage caused by low radiation levels.

In the early decades of the nuclear era, it became apparent that high levels of nuclear radiation can be harmful, even fatal. But it was also realized that there are a number of ways that radiation could be medically beneficial. Today, nuclear medicine is a major branch of the medical field. Thousands of hospitals use nuclear medicine daily. It has been estimated that one-third of all hospitalized patients benefit from the use of radioisotopes for diagnosis or treatment.

Often, diagnosing an ailment involves finding abnormally high or low activity in some part of the body or determining if substances are circulating as they should. Radioactive nuclei make excellent probes for these tasks because their location in the body can be determined by detecting the telltale radiation that they emit. A small quantity of a radioactive material can be placed in the bloodstream. Then, its movement through the circulatory system can be monitored by placing a Geiger counter near a vein or using a camera that responds to radiation instead of light (see ● Figure 11.14).

Some chemical elements naturally accumulate in specific places. For example, the body uses calcium to "build" bones and teeth. The element iodine accumulates in the thyroid gland at a rate affected by such conditions as congestive heart failure or improper thyroid function. In one standard test, a patient ingests a pill that includes a radioactive isotope of iodine, such as iodine-123 (half-life = 13.1 hours). The iodine enters the bloodstream and begins to accumulate in the thyroid after a few hours. A radiation detector placed by the patient's neck indicates if the accumulation rate is normal or not.

The fact that radiation can kill cells makes it a powerful tool for fighting cancer. The strategy is to irradiate and kill the cancer cells but at the same time limit the damage to normal tissue. External radiation therapy involves sending a beam of radiation into the body, aimed at the cancerous tissue. Exposure of healthy cells to the radiation is greatly reduced by rotating the beam so that it hits the cancer from different directions. It is like shining a flashlight at a particular spot on the floor, while at the same time walking in a circle about that spot. X rays and gamma rays are usually used for this purpose, but in some cases, a beam of high-speed neutrons or protons is more effective.

There are also therapies that place the source of radiation inside the body. In some cases, radioactive material is sealed in a small container and placed next to the cancer. This allows high doses to be administered to very specific locations. An oral dose of the radioisotope iodine-131 (half-life = 8 days) can be used to irradiate the thyroid selectively. The beta and gamma radiation it emits destroys cancerous tissue in the thyroid where the iodine accumulates. A more recent innovation is to incorporate radioactive nuclei in antibodies that naturally seek out cancer cells and attach to them. The radiation from the radioisotope can then kill the cells.

Radiation therapy is not limited to fighting cancer. It is used to treat certain blood and thyroid disorders. In Europe, it is used to treat arthritis.

Nuclear medicine relies heavily on the knowledge that has been amassed in the field of nuclear physics. For example, gamma-emitting radioisotopes are preferred for diagnostic procedures that require placing the material inside the body. Since gamma rays interact less with matter than some other forms of radiation, they do less damage to tissue and are more likely to make it out of the body where they can be detected. The half-life of an implanted radioisotope is a critical parameter for determining dosages, since it indicates over what period of time the irradiation will be appreciable.

The evolution of our attitude toward nuclear radiation during the last century is remarkable and complex. In the early 1900s nuclear radiation was a scientific curiosity, its potential for causing harm largely unknown or unappreciated. The bombing of Hiroshima and Nagasaki gave the world stark images of thousands of victims suffering from massive doses of radiation. For decades after World War II, the nuclear arms race raised the specter of most of the Earth's inhabitants being exposed to radioactive fallout if all-out war occurred. Accidents like the explosion of the Chernobyl nuclear power plant and the discovery that naturally occurring radioactive radon gas can seep into our homes showed that weapons are not the only way that radiation can reach us. But by the time the threat of global nuclear war had receded in the 1990s, nuclear medicine had quietly established itself as a true lifesaving discipline. We can hope that nuclear radiation will continue to be far more widely used as a tool for saving lives than as a weapon for taking them.

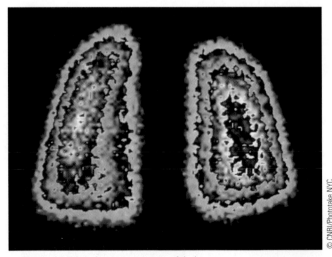

© CNRI/Photatake NYC

● **Figure 11.14** Radioisotope scan image of the lungs.

occur naturally because they are constantly being produced, carbon-14 being a good example. We humans have added a large number of radioisotopes to the environment through processes like nuclear explosions and nuclear power production. Concern over the amount of radioactive fallout in the atmosphere produced by nuclear weapons testing led to the limited Test Ban Treaty of 1963.

11.4 Artificial Nuclear Reactions

Radioactivity and the formation of carbon-14 from nitrogen-14 are examples of natural nuclear reactions. With the exception of gamma decay, each of these results in *transmutation*—the conversion of an atom of one element into an atom of another (the medieval alchemists dream come true). Many other types of nuclear reactions can be induced artificially in laboratories. A common example of this is the bombardment of nuclei with alpha particles, beta particles, neutrons, protons, or other nuclei. If a nucleus "captures" the bombarding particle, it will become a different element or isotope. (This is called *artificial transmutation*. Note that here something is added to the nucleus, whereas in radioactive decay something leaves the nucleus.)

One example of a useful artificial reaction involves bombardment of uranium-238 nuclei with neutrons. The result is uranium-239.

$$^{238}_{92}\text{U} + ^{1}_{0}\text{n} \rightarrow ^{239}_{92}\text{U}$$

This new nucleus undergoes two beta decays, resulting in plutonium-239.

$$^{239}_{92}\text{U} \rightarrow ^{239}_{93}\text{Np} + ^{0}_{-1}\text{e}$$

$$^{239}_{93}\text{Np} \rightarrow ^{239}_{94}\text{Pu} + ^{0}_{-1}\text{e}$$

The plutonium-239 can be used directly in nuclear reactors, but the original uranium-238 can't. A breeder reactor is a nuclear reactor designed to use neutrons to produce, or "breed," reactor fuel: the plutonium-239. This process also shows how elements with Z greater than 92 can be produced artificially.

Similarly, neutron bombardment can be used to produce hundreds of other isotopes, most of which are radioactive. Some of the radioisotopes used in nuclear medicine (Physics Potpourri, page 442) are produced this way. **Neutron activation analysis** is an accurate method for determining what elements are present in a substance. The material to be tested is bombarded with neutrons. Many of the nuclei in the substance are transformed into radioisotopes. By monitoring the radiation that is then emitted, it is possible to determine which elements were originally present (● Figure 11.15). For example, the only naturally occurring isotope of sodium, sodium-23, becomes radioactive sodium-24 by neutron absorption.

$$^{23}_{11}\text{Na} + ^{1}_{0}\text{n} \rightarrow + ^{24}_{11}\text{Na}$$

The sodium-24 emits both gamma and beta rays with specific energies. If sodium is present in an unknown substance, it can be detected by bombarding the substance with neutrons and then looking to see whether gamma and beta rays with the proper energies are emitted.

Neutron activation analysis played a key role in the discovery that a large asteroid or comet struck the Earth 65 million years ago, resulting in the extinction of dinosaurs. The element iridium is rare on Earth, but in the 1970s, neutron activation analysis revealed a high concentration of it in ancient sediments. Scientists concluded that a massive meteorite impact threw iridium-laden debris into the atmosphere, which eventually settled on the Earth's surface.

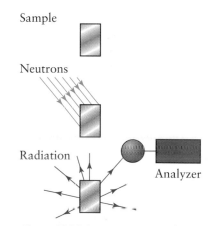

● **Figure 11.15** In neutron activation analysis, the sample is irradiated with neutrons. This produces radioisotopes in the sample. Analysis of the radiation that is emitted makes it possible to determine the original composition of the sample.

Neutron activation analysis has a number of uses related to law enforcement. It can reveal the presence of arsenic in a single hair from a victim who was poisoned. An art forgery can be detected by using neutron activation analysis to determine if the paint or other materials conform to what was known to be in use when the piece was supposedly created.

This procedure can be used to find explosives and illegal drugs—even if they are inside a locked vehicle. Most chemical explosives contain nitrogen, and some illegal drugs, including cocaine, contain chlorine. A truck-mounted neutron activation analysis system can be parked next to a suspicious vehicle and scan it for the presence of either element. Neutrons are sent into the vehicle, producing radioisotopes in the various substances present inside. Gamma rays that are emitted pass into the detecting system and are analyzed to see if their energies match those of activated nitrogen or chlorine nuclei.

LEARNING CHECK

1. (True or False.) It is possible to change a nucleus that is stable into one that is not stable.

2. _____ activation analysis is used to determine which elements are present in a sample.

ANSWERS: 1. True 2. Neutron

11.5 Nuclear Binding Energy

A very important and useful aspect of a nucleus is its binding energy. Imagine the following experiment: A nucleus is dismantled by removing each proton and neutron one at a time, and the total amount of work done in the process is measured. If these protons and neutrons are reassembled to form the original nucleus, an amount of energy equal to the work done would be released. This energy is called the **binding energy** of the nucleus. It indicates how tightly bound a nucleus is. A useful quantity for comparing different nuclei is the average binding energy for each proton and neutron, called the *binding energy per nucleon*. It is just the total binding energy divided by the total number of protons and neutrons in the nucleus—the mass number. If we measure the binding energy per nucleon for all of the elements, we find that it varies considerably, from about 1 MeV to nearly 9 MeV (see ● Figure 11.16).

Nuclei with mass numbers around 50 have the highest binding energy per nucleon: The protons and neutrons are more tightly bound to the nucleus than in larger or smaller nuclei, so these nuclei are the most stable. The high stability of the helium-4 nucleus (alpha particle) is indicated by the small peak in the graph at $A = 4$.

● **Figure 11.16** Graph of the binding energy per nucleon versus mass number. A higher binding energy means the nucleus is more tightly bound.

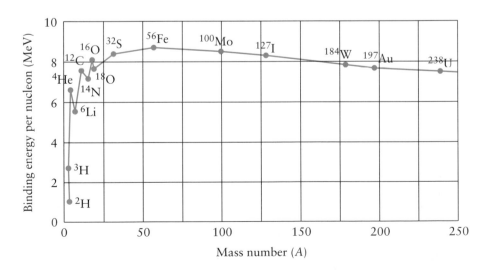

A proton or neutron bound to a nucleus is similar to a ball resting in a hole in the ground. The ball's binding energy is the amount of energy that would have to be provided (the amount of work that would have to be done) to lift it out of the hole. A deeper hole means a higher binding energy.

The graph of binding energy per nucleon suggests how one can tap nuclear energy. Imagine taking a large nucleus with A around 200 and splitting it into two smaller nuclei. The graph shows that each of the smaller nuclei, with A around 100, has a higher binding energy per nucleon than the original nucleus. All of the neutrons and protons have become more tightly bound together and have *released* energy in the process. (This is like moving the ball to a deeper part of the same hole. Its potential energy is decreased, so it has given up some energy.) The act of splitting a large nucleus, referred to as *nuclear fission*, releases energy.

Similarly, we can combine two very small nuclei into one larger nucleus and release energy. If a hydrogen-1 nucleus and a hydrogen-2 nucleus are combined to form helium-3, the binding energy of each proton and neutron is increased, and energy is again released. This process is called *nuclear fusion*.

Clearly, the graph of binding energy per nucleon versus mass number is very important in nuclear physics because it shows why nuclear fission and nuclear fusion release energy. Fission is exploited in nuclear power plants and atomic bombs, and fusion is the source of energy in the Sun, the stars, and in hydrogen bombs. But how does one actually go about measuring the binding energy of a nucleus? It is not practical to actually dismantle a nucleus and to keep track of the amount of work done. (Remember that the largest nuclei contain over 200 nucleons.) The best way to measure binding energy is to use the equivalence of mass and energy.

One of the predictions of Einstein's special theory of relativity, presented in Section 12.1, is that the mass of an object increases when it gains energy. The exact relationship between the energy E of a particle and its mass m is the famous equation

$$E = mc^2$$

When a particle is given energy, its mass increases; and when it loses energy, its mass decreases. In other words, mass can be converted into energy, and vice versa. The amount of mass "lost" (converted into energy) by things around us when we take energy from them is generally much too small to be measured. But the quantities of energy involved in nuclear processes are so great that mass-energy conversion can be measured.

For example, the mass of a hydrogen-1 atom (one proton and one electron) is 1.00785 u, and the mass of one neutron is 1.00869 u (see Table 11.1). The total mass of a hydrogen atom and a free neutron is 2.01654 u. But careful measurements of the mass of a hydrogen-2 (deuterium) atom give the value 2.01410 u. When the neutron and the proton are bound together in the hydrogen-2 nucleus, their combined mass is 0.00244 u *less* than when they are apart (Figure 11.17). This much mass is converted into energy when a proton and a neutron combine, and this energy is the binding energy of the hydrogen-2 nucleus.

In a similar way, we find that the masses of all nuclei are less than the combined masses of the individual protons and neutrons. The energy equivalent of this "mass defect" is the total binding energy of the nucleus. The binding energy divided by the mass number is the binding energy per nucleon of the nucleus. This is a good way to compute the binding energy per nucleon for the different isotopes.

Incidentally, the atomic mass unit, u, is defined to be exactly one-twelfth of the mass of one atom of carbon-12. In other words, u was defined so that the mass of an atom of carbon-12 is exactly 12 u. Other isotopes could have been used for establishing the size of u, but carbon-12 is convenient because it is very common.

Fission and fusion are processes that convert matter into energy. In both cases, the total mass of all the nucleons afterwards is less than the total mass before. Some of the original mass, typically from 0.1 to 0.3%, is converted into energy. In the next two sections, we take a closer look at fission and at fusion.

Concept Map 11.2 summarizes the concepts presented in this section.

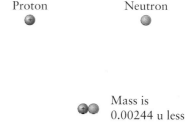

 Figure 11.17 A proton and a neutron bound together in a nucleus have *less mass* than a proton and a neutron separated. When the two combine, some of their mass is converted into energy—the binding energy of the nucleus. (The electron is not shown.)

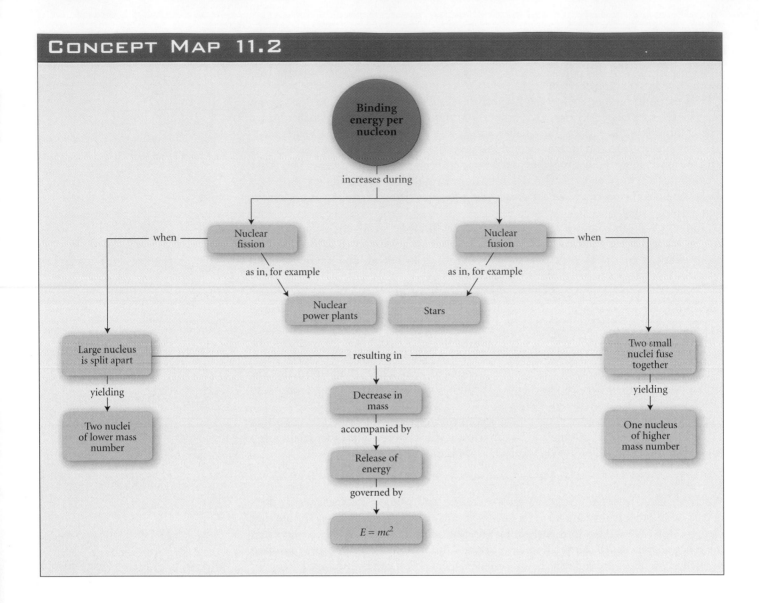

Binding energy per nucleon

increases during

when — Nuclear fission — as in, for example — Nuclear power plants

Nuclear fusion — when — as in, for example — Stars

Large nucleus is split apart — resulting in — Two small nuclei fuse together

yielding — Two nuclei of lower mass number

Decrease in mass

accompanied by

Release of energy

governed by

$E = mc^2$

yielding — One nucleus of higher mass number

LEARNING CHECK

1. The _____ of a nucleus is a measure of how tightly bound it is.

2. (True or False.) The average mass of each neutron in the nucleus of an iron atom is not the same as the average mass of each neutron in the nucleus of a gold atom.

ANSWERS: 1. binding energy 2. True

11.6 Nuclear Fission

In the 1930s, a discovery was made during neutron bombardment experiments that was one of the most fateful in human history: The nuclei of certain isotopes actually split when they absorb neutrons. Uranium-235 and plutonium-239 are the two most important nuclei that do. Energy is released in the process along with several free neutrons. This process is called **nuclear fission.**

Nuclear Fission The splitting of a large nucleus into two smaller nuclei. Free neutrons and energy are also released.

The two resulting nuclei are called *fission fragments*. The fissioning nuclei of a particular isotope will not all split the same way. There are dozens of different ways that a nucleus can split, and there are dozens of different possible fission fragments.

Fission is most commonly induced by bombarding nuclei with neutrons. Protons, alpha particles, and gamma rays also have been used. Upon absorption of a neutron, the nucleus becomes highly unstable and quickly splits (● Figure 11.18). Two of the many possible fission reactions of uranium-235 are

$$\,^1_0n + \,^{235}_{92}U \rightarrow \,^{236}_{92}U^* \rightarrow \,^{141}_{56}Ba + \,^{92}_{36}Kr + 3\,^1_0n$$

$$\,^1_0n + \,^{235}_{92}U \rightarrow \,^{236}_{92}U^* \rightarrow \,^{140}_{54}Xe + \,^{94}_{38}Sr + 2\,^1_0n$$

The asterisk (*) indicates that the uranium-236 is unstable, so much so that it splits immediately. In both cases, energy is released during the fissioning. The total mass of the fission fragments and the neutrons is less than the mass of the original uranium nucleus and the neutron. The missing mass is converted into energy. Most of this energy appears as kinetic energy of the fission fragments; they have high speeds.

The average amount of energy released by the fissioning of a uranium-235 nucleus is 215 million electronvolts (3.4×10^{-11} joules). By comparison, the amount of energy released during chemical processes like burning, metabolism in your body, and chemical explosions is typically about 10 electronvolts for each molecule involved. This is why nuclear-powered ships and submarines can go years without refueling, while ships that use diesel or oil must take on tons of fuel for each trip.

In addition to the energy released during fission, two other aspects of this process are extremely important.

1. Almost all of the possible fission fragments are radioactive. The ratio of neutrons to protons in most fission fragments is too high for the nuclei to be stable, and they undergo beta decay. These radioactive fission fragments are important components of the radioactive fallout from nuclear explosions and of the nuclear waste produced in nuclear power plants.
2. The neutrons that are released during fission can strike other nuclei and cause them to split. This process is called a *chain reaction.*

For uranium-235, 2.5 neutrons are released on the average by each fission. If just one of the neutrons from each fission induces another fission, a *stable* chain reaction results. The number of nuclei that split each second remains constant. Energy is therefore released at a steady rate. This type of reaction is used in nuclear power plants. If, on the average, more than one of the neutrons causes other fissions, an *unstable* chain reaction results—a nuclear explosion. One fission could trigger two, these two could trigger four

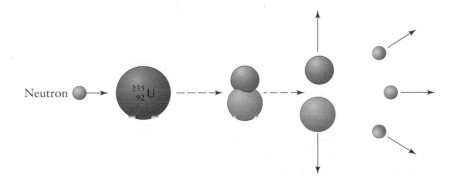

● **Figure 11.18** A nucleus of uranium undergoing fission. In this example, three neutrons are released.

Neutron

$^{235}_{92}U$

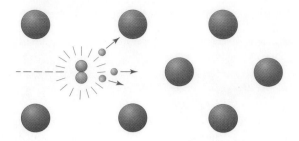

● **Figure 11.19** Fission chain reaction in pure uranium-235. Since all of the nuclei can fission, each neutron emitted by one fission is likely to strike another nucleus and cause it to fission. This leads to an explosive chain reaction.

more, then eight, sixteen, and so on. Energy is released at a rapidly increasing rate. This is the process used in atomic bombs.

We can make a comparison between a nuclear chain reaction (fission) and a chemical chain reaction (burning). A stable chain reaction is similar to the burning of natural gas in a furnace or on a cooking stove. The number of gas molecules that burn each second is kept constant, and the energy is released at a steady rate. An unstable chain reaction is like the explosion of gunpowder in a firecracker. The energy released at the start quickly causes the rest of the powder to burn rapidly until all of it is consumed.

In the remainder of this section, we look at some of the details involved in making atomic bombs and nuclear power plants. Keep in mind that there are two different types of nuclear bombs: atomic bombs, which use nuclear fission, and hydrogen bombs (also called *thermonuclear bombs*), which use both fission and nuclear fusion. The latter will be discussed in the next section.

The key raw material for both atomic bombs and nuclear power plants is uranium. Naturally occurring uranium is approximately 99.3% uranium-238 and 0.7% uranium-235. The uranium-235 fissions readily; the uranium-238 does not (unless irradiated with extremely high-speed neutrons), but it can be "bred" into fissionable plutonium-239.

Atomic Bombs

To produce an uncontrolled fission chain reaction, one must ensure that the fission of each nucleus leads to more than one additional fission. This is accomplished by using a high density of fissionable nuclei so that each neutron emitted in a fission is likely to encounter another nucleus and cause it to fission (● Figure 11.19). In other words, nearly pure uranium-235 or plutonium-239 must be used. Uranium-235 can be extracted from uranium ore through a complicated and costly series of processes referred to as *enrichment*, or *isotope separation*. Since all uranium isotopes have essentially the same chemical properties, the separation processes rely on the difference in the masses of the nuclei.

Not only is nearly pure uranium-235 or plutonium-239 necessary, there must be a sufficient amount of it and it must be put into the proper configuration—a sphere, for example. This is so that each neutron is likely to collide with a nucleus before it escapes through the surface. When these conditions are met, there is a *critical mass* of fissionable material. If there are too few fissionable nuclei present, or they are spread out too far (the density is low), too many of the neutrons will escape without causing other fissions. Also, the critical mass must be held together long enough for the chain reaction to cause an explosion. If not, the initial energy and heat from the first fissions will blow the critical mass apart and stop the chain reaction: a "fizzle." A firecracker is tightly wrapped with paper for a similar reason.

Another important consideration is timing: A premature explosion is highly undesirable, to say the least. This is prevented by keeping the fissionable material out of the critical mass configuration before the explosion. Two different techniques are used. In a gun-barrel atomic bomb, two subcritical lumps of uranium-235 are placed at opposite ends of a large tube (● Figure 11.20). The explosion is triggered by forcing the two lumps together into a critical mass. This type of bomb was dropped on Hiroshima, Japan.

The other type of bomb, the implosion bomb, is used with plutonium-239. A subcritical sphere of the plutonium is surrounded by a shell of specially shaped conventional explosives (● Figure 11.21). These explosives squeeze the plutonium-239 so much that it

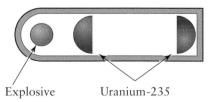

Explosive   Uranium-235

● **Figure 11.20** Gun-barrel atomic bomb. The two lumps of uranium are forced together to form a critical mass.

becomes a critical mass, and an explosion occurs. This type of bomb was dropped on Nagasaki, Japan.

A nuclear explosion releases an enormous amount of energy in the form of heat, light, other electromagnetic radiation, and nuclear radiation. The temperature at the center of the blast reaches millions of degrees. Even a mile away, the heat radiation is intense enough to instantly ignite wood and other combustible materials. The air near the explosion is heated and expands rapidly, producing a shock wave followed by hurricane-force winds. In addition to the radioactive fission fragments, large quantities of radioisotopes are produced by the neutrons and other radiation that bombard the air, dust, and debris. As this material settles back to the ground, it is collectively referred to as *fallout*.

The amount of energy released by an atomic bomb explosion is generally expressed in terms of the number of tons of TNT, a conventional high explosive, that would release the same amount of energy. One ton of TNT releases about 4.5 billion joules when it explodes. Most atomic bombs are in the 10–100 kiloton range: They are equivalent to between 10,000 and 100,000 tons of TNT. (An entire 100-unit freight train can carry only about 10,000 tons.) Some of these weapons are small enough to be carried by one person.

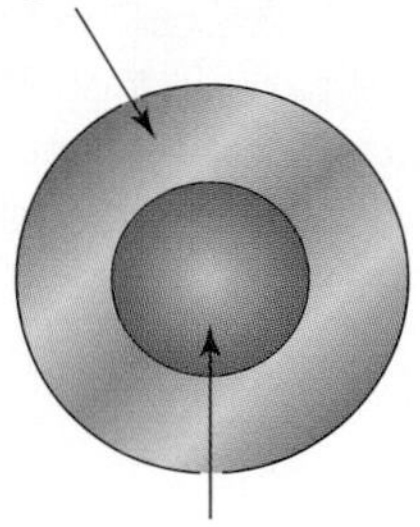

Explosive

Plutonium

● **Figure 11.21** Implosion atomic bomb. The plutonium is subcritical until it implodes under the force of the explosives. This brings it to a critical mass.

Nuclear Power Plants

About one-sixth of the world's electricity is generated by nuclear power plants. Over 400 of them are currently in operation. The reactors in these plants release the energy from nuclear fission in a controlled manner. The key to preventing a fission chain reaction from escalating to an explosion is controlling how the neutrons induce other fissions.

There are dozens of different designs for nuclear reactors. Most of them use only slightly enriched uranium fuel, typically about 3% uranium-235. This keeps the density of fissionable nuclei low so that not all neutrons from each fission are likely to induce other fissions (● Figure 11.22). This is also economical because the enrichment process is very expensive. It also makes it impossible for a critical mass configuration to arise: A nuclear reactor cannot explode like an atomic bomb.

Another important factor in producing and controlling a fission chain reaction in a nuclear reactor is the fact that slow neutrons are much better at causing uranium-235 nuclei to undergo fission than fast neutrons. Slowing down the neutrons released in the fissioning of uranium-235 makes it much more likely that they will be captured by other uranium-235 nuclei and cause them to split. This makes it possible to use a low concentration of uranium-235 in a reactor.

How are the neutrons slowed down? This is done through the use of a moderator, a substance that contains a large number of small nuclei. Small nuclei are more effective at taking kinetic energy away from the neutrons in the same way that a golf ball will lose more kinetic energy when it collides with a baseball at rest than with a bowling ball at rest. As neutrons pass through the moderator, they collide with the nuclei and lose some of their energy. They slow down. After many collisions, they have the same average kinetic energy as the surrounding nuclei. (For this reason, they are called *thermal neutrons* because their energies are determined by the temperature of the moderator.) Almost all nuclear power plant reactors in the United States use water as the moderator because of the hydrogen nuclei in the water molecules. Many reactors in England and the former Soviet Union use carbon in the form of graphite as the moderator.

Uranium-238   Uranium-235

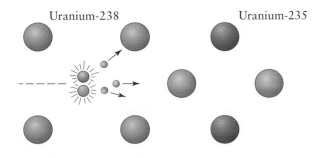

● **Figure 11.22** A controlled fission chain reaction in enriched uranium. Since only a small percentage of the nuclei are fissionable, not all of the neutrons induce other fissions. Some are absorbed by uranium-238 nuclei, which do not fission.

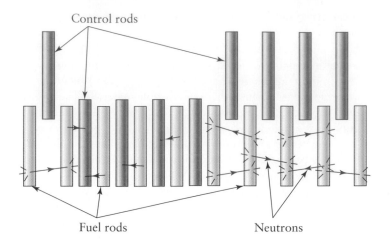

Control rods

Fuel rods   Neutrons

● Figure 11.23 Simplified diagram of the core of a nuclear reactor. Neutron-absorbing control rods can be inserted between the fuel rods to control the chain reaction. The control rods prevent neutrons released in one fuel rod from inducing fissions in another fuel rod.

The uranium fuel is shaped into long rods that are separated from each other. Large reactors have tens of thousands of individual fuel rods. Each rod relies on neutrons from neighboring rods to sustain the chain reaction. This makes it possible to control the chain reaction by lowering *control rods* between the fuel rods (● Figure 11.23). Control rods are made of materials that absorb neutrons effectively, such as the elements cadmium and boron. The rate of fissioning in a reactor and, therefore, the rate of energy release, is regulated by the number of control rods that are withdrawn.

The energy released in a nuclear reactor is mainly in the form of heat. This heat is used to boil pressurized water into high-temperature steam. The rest of a nuclear power plant is much the same as that found in coal-fired power plants (refer to Figure 5.37). The steam turns a turbine, which turns a generator, which produces electricity. In most nuclear reactors, water flows around the fuel rods and carries away the heat in much the same way that the coolant in a car's radiator carries away heat from hot engine parts. If, for some reason, the reactor loses its water (called a "loss of coolant accident"), the reactor can become dangerously overheated.

Nuclear power plants are equipped with a sophisticated array of safety features and emergency backup systems. Millions of gallons of water are poised to flood the reactor if the temperature becomes too high or if the coolant system leaks. The control rods are designed to drop into place at the slightest sign of trouble. The entire nuclear reactor and supporting systems are housed inside a huge steel-reinforced concrete containment building. This building is typically about 200 feet high, dome shaped, with walls that arc more than 1 foot thick. It is designed to seal airtight if there is a possibility that radiation has leaked from the reactor. All of this is needed because there is much more radioactivity inside a nuclear reactor than is released by a typical nuclear bomb.

Two notable accidents at nuclear power plants point out the potential for disaster that exists when human operators and designers make mistakes. In 1979, an unlikely series of errors and equipment malfunctions caused the core of a reactor at Three Mile Island, Pennsylvania, to lose most of its cooling water for 3 hours. Over half of the core melted. The containment building served its purpose, and very little radiation escaped from the plant. But the costly reactor was ruined, and the core was reduced to a pile of intensely radioactive rubble. The billion-dollar cleanup lasted for more than a decade.

A major nuclear disaster occurred in April 1986 when a graphite-moderated power plant at Chernobyl (near Kiev, Ukraine) blew itself apart. Operators performing tests on the generator pushed the reactor beyond its limits. Since the plant did not have the kind of containment building used at Three Mile Island, tons of radioactive debris were released into the atmosphere and then carried by winds literally around the world. The huge mass of graphite was ignited by the heat and burned for 10 days, adding to the severity of the accident. Thirty-one people were killed by the immediate effects of the accident, and over 300,000 were required to move out of the contaminated area and resettle elsewhere. Early predictions that thousands might eventually die from radiation exposure now appear to have been wrong. A United Nations report published in 2000 indicates that approximately

1,800 cases of thyroid cancer, mostly in individuals who were children at the time of exposure, were caused by the accident. Estimates of the total economic loss to the economies of Ukraine and neighboring countries run into hundreds of billions of dollars. Better reactor design and operator training should make a repeat of this accident unlikely.

Eventually, the fissionable isotopes in the fuel rods are consumed, and the spent rods have to be replaced. Spent fuel rods are highly radioactive because they contain large amounts of fission fragments, in the form of more than 200 different radioisotopes. So much radiation is emitted that the rods must be kept under water for months to remain cool. Even after the radioisotopes with short half-lives have decayed away and the rods have cooled, they remain dangerously radioactive for thousands of years. Finding a safe way to dispose of spent fuel rods and other nuclear waste is a major concern of government regulatory officials and nuclear power plant authorities.

LEARNING CHECK

1. Which of the following is *not* the usual outcome of nuclear fission?
   a) The binding energy per nucleon of the fission fragments is smaller than that of the original nucleus.
   b) Energy is released.
   c) Neutrons are released.
   d) The fission fragments are radioactive.
2. (True or False.) Of the thousands of different isotopes known to exist, only a couple of them can be used to make an atomic bomb.
3. In a nuclear reactor, _____ are used to limit the number of neutrons passing from one fuel rod to another.
4. (True or False.) In a worst-case accident at a nuclear power plant, the core could explode like an atomic bomb.

ANSWERS: 1. (a)   2. True   3. control rods   4. False

11.7 Nuclear Fusion

As incredible as it may seem, another source of nuclear energy exists that dwarfs nuclear fission: **nuclear fusion.** It is the source of energy for the Sun and the stars and for hydrogen bombs.

Nuclear Fusion The combining of two nuclei to form a larger nucleus.

Fusion is like the reverse of fission, although it usually involves very small nuclei. One example of a fusion reaction is shown in ● Figure 11.24. In this case, two hydrogen nuclei, one hydrogen-1 and the other hydrogen-2 (deuterium), fuse to form a nucleus of helium-3. There are many other possible fusion reactions that result in a release of energy. Some of these reactions, including the amount of energy released, are

$$\,^{2}_{1}H + \,^{2}_{1}H \rightarrow \,^{3}_{2}He + \,^{1}_{0}n + 3.3 \text{ MeV}$$

$$\,^{2}_{1}H + \,^{3}_{1}H \rightarrow \,^{4}_{2}He + \,^{1}_{0}n + 17.6 \text{ MeV}$$

$$\,^{2}_{1}H + \,^{3}_{2}He \rightarrow \,^{4}_{2}He + \,^{1}_{1}p + 18.3 \text{ MeV}$$

In each case, energy is released because the total mass of the nucleons after the fusion is less than the total mass before. As with fission, the missing mass is converted into energy. This energy appears as kinetic energy of the fused nucleus and the energy of the proton, neutron, or gamma ray.

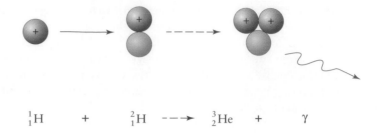

$$\ce{^1_1H} \quad + \quad \ce{^2_1H} \quad \dashrightarrow \quad \ce{^3_2He} \quad + \quad \gamma$$

Refer to Section 5.1.

Fusion can occur only when the two nuclei are close enough for the short-range nuclear force to pull them together. This turns out to be a major problem in trying to induce a fusion reaction. Nuclei are positively charged. When brought close together, they exert strong repulsive forces on each other and resist fusion.

How can this difficulty be overcome? The most common way is with extremely high temperatures. If nuclei can be given enough average kinetic energy, their inertia will carry them close enough together to fuse when they collide (● Figure 11.25). This process is called *thermonuclear fusion.*

When we say that the temperatures are high, we really mean it: The hydrogen-1 plus hydrogen-2 fusion reaction requires a temperature of about *50 million* degrees Celsius. Some fusion reactions require several hundred million degrees Celsius. These unearthly temperatures do occur in the interiors of stars and at the centers of nuclear fission explosions.

Fusion in Stars

The Sun glows white hot and emits enormous amounts of energy. This energy originates in a natural fusion reaction in the Sun's interior. The Sun is composed mostly of hydrogen in a dense, high-temperature plasma. At the Sun's core, the temperature is around 15 million degrees Celsius, and the pressure is over one billion atmospheres. Under these conditions, the hydrogen undergoes a series of fusion reactions that results in the formation of helium. Each second, hundreds of millions of tons of hydrogen fuse, and more than 4 million tons of matter are converted into energy.

Stars get their energy from nuclear fusion. Some produce energy by reactions like those in the Sun, whereas others that are larger and have hotter cores use other fusion reactions. In large stars with core temperatures in excess of 100 million degrees Celsius, the helium fuses to form larger nuclei such as carbon and oxygen, which in turn may fuse to form silicon and, eventually, iron. In this way, the elements in the periodic table are built up from the basic raw material, hydrogen. The heaviest elements are produced during *supernova explosions*—the gigantic explosions of massive stars at the ends of their life cycles (● Figure 11.26)—by successive neutron capture followed by beta decay. The elements contained in the Earth and everything on it (including yourself) were formed in the cataclysms of supernovas that occurred billions of years ago.

Life as we know it would be impossible on this planet without the Sun. Its energy supports the entire web of life. It is ironic that solar energy—that tranquil, natural energy source—comes from a violent nuclear process on a scale nearly beyond imagination.

Thermonuclear Weapons

The most destructive weapons in the world's arsenals are *thermonuclear warheads,* also called *hydrogen bombs.* These weapons get most of their explosive energy from the fusion of hydrogen. To produce the high temperatures necessary for the fusion reactions to

● **Figure** **11.25** In a high-speed collision of two nuclei, the repulsive force between them is overcome, and fusion occurs. At extremely high temperatures, nuclei possess such high speeds because of their thermal motions. This is thermonuclear fusion.

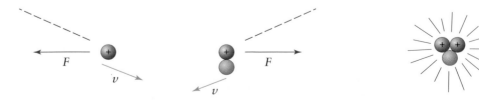

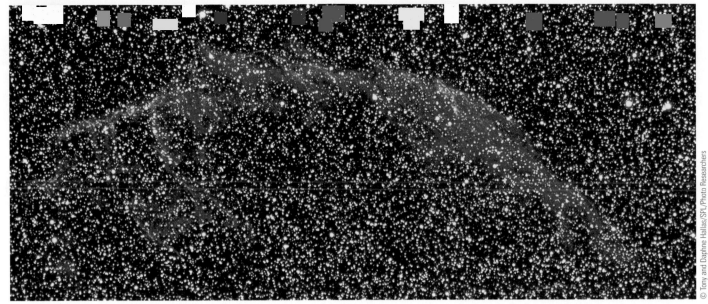

● **Figure 11.26** The Veil Nebula, part of a supernova remnant 2,500 light years away.

occur, a nuclear fission explosion is used as a trigger. The fission explosion, an incredibly huge blast in itself, is but a primer for the monstrous fusion explosion. In terms of the relative amount of energy released, a hydrogen bomb is to an atomic bomb what a stick of dynamite is to a small firecracker.

Truly enormous amounts of energy are released. Many thermonuclear devices are in the *megaton* range; their energy output is given in terms of *millions* of tons of TNT. The largest weapon ever tested was rated at more than 50 megatons. To put this in perspective, this one blast released more energy than the total of all of the explosions in all of the wars in history, including the two atomic bomb blasts in World War II. However, it wasn't the most powerful blast to have ever occurred on Earth. Some volcanic eruptions, like the one on the island of Krakatoa in 1883, released more energy.

Currently, the world's arsenals contain thousands of nuclear warheads. Indications are that if even a fraction of these were used in a short "war," much of the life on this planet, including humankind, would be threatened with extinction.

Controlled Fusion

Soon after fusion was discovered, scientists looked for ways to harness it as an energy source. The initial success of nuclear fission reactors gave them hope. But fusion presents technical challenges that have resisted solutions for decades. For a thermonuclear fusion reaction to occur, two conditions must be met:

1. The nuclei must be raised to an extremely high temperature to ignite the fusion reaction.

    It is possible to produce temperatures in the millions of degrees, but the problem is containing (referred to as "confining") a plasma at this high temperature. Containers made of conventional materials would melt long before such temperatures would be reached.

2. There must be a high enough density of nuclei for the probability of collisions and, therefore, fusions, to remain high.

    If energy is to be released at a usable rate, a large number of fusions must occur each second. This means that the density of the plasma must be kept sufficiently high for the nuclei to collide often.

Research on controlled fusion is being pursued along a number of avenues. Several of these employ *magnetic confinement* of the plasma. Specially shaped magnetic fields are used to keep the plasma confined without letting it come into contact with other matter. This is possible because the nuclei are charged particles and consequently experience a force when moving in a magnetic field. A "magnetic bottle" is formed out of magnetic

Refer to Section 8.2.

Big Bang Nucleosynthesis

In Section 11.7, we discussed how the elements of the periodic table are built up in the cores of stars from hydrogen by fusion reactions, a process called *nucleosynthesis*. But where, one might ask, did the hydrogen come from that serves as the fuel for the nuclear fires in stars? The same question may be asked about the origins of other light elements, like deuterium (hydrogen-2), helium-3, and helium-4, commonly involved in fusion reactions studied terrestrially. This is an important issue because deuterium and helium-3 are consumed, not produced, in stars during the usual sequence of nuclear reactions that occur. Even helium-4, the end result of hydrogen fusion, is largely used up in the creation of heavier elements by additional nuclear reactions. And yet we *do* find substantial amounts of these light nuclear species in the cosmos. Where did they come from if not from stars?

Because nuclear processes usually demand high temperatures and high densities for their operation, we seek another place, besides stars, where such conditions prevail *or once prevailed*. A natural source of such extreme temperatures and densities is the Big Bang (see the Physics Potpourri on page 320). Indeed, the answer to our question now seems to be that these isotopes were manufactured during the first 3 minutes after the creation of the universe. The source of these nuclei was the enormous energy present in the Big Bang. As we noted earlier in the chapter, energy and mass are alike, and one may be converted into the other. Fission and fusion reactions convert mass into energy, but during the first few seconds after the birth of the universe, energy (in the form of very short wavelength gamma rays) was converted into matter. This matter was generally in the form of particles with equal masses but opposite electric charges, so-called *particle-antiparticle pairs* (● Figure 11.27a; for more on antiparticles, see Section 12.2). Thus, for example, when the universe was only about a thousandth of a second old, with a temperature greater than 10^{12} K, it consisted of a hurly-burly of high-energy photons and fast-moving particles and their anti's, specifically protons and antiprotons, neutrons and antineutrons, and electrons and antielectrons (positrons).

As the universe expanded and cooled, a point was reached when the temperature fell below 10^{12} K and the photon energies were no longer sufficient to permit the creation of proton-antiproton and neutron-antineutron partners. The existing pairs then began to collide with and annihilate one another, transforming matter back into energy and producing more lower-frequency photons (● Figure 11.27b). If the num-

ber of particles had exactly matched the number of antiparticles, all the mass originally created would have been destroyed, leaving the universe devoid of matter. As it happened, there existed a slight excess of particles over antiparticles. The difference was small but significant: about $10^9 + 1$ protons for every 10^9 antiprotons, for example. (It turns out that this discrepancy is related to the violation of a basic conservation law involving time reversibility in particle interactions; for more see Section 12.3.) This asymmetry left a residue of protons and neutrons after the era of annihilation with about one neutron for every seven protons.

When the universe reached the age of a few seconds and its temperature fell below 10^{10} K, the average photon energy was no longer sufficient to create even the lighter electrons and positrons, and these particles began to annihilate one another like the earlier behavior of the protons and neutrons and their anti's. Again, because of the slight asymmetry of matter over antimatter, the net outcome of this era of annihilation was a residue of electrons. Some 10^5–10^6 years later when the temperature of the universe was between 3,000 and 4,000 K, these electrons combined with the protons to produce neutral hydrogen atoms.

However, when the universe was only a few seconds old, the positrons were effectively gone as were many of the original electrons. As a result, beta decay of the remaining free neutrons into protons and electrons began to dominate over the compensating production of neutrons from the merging of free protons and electrons (compare the beta decay reaction on page 434). During the same period, many of the free neutrons combined with protons to produce deuterium according to the following reaction:

$$\,^1_0n + \,^1_1p \rightarrow \,^2_1H + \gamma$$

The newly formed deuterium persisted in reasonable numbers only after the temperature in the early universe dropped to less than 10^9 K—that is, after the universe was about 100 seconds old.

Prior to this time, deuterium quickly underwent a series of nuclear reactions that converted much, but not all, of it into helium-4. A small fraction was transformed into helium-3, and some of it remained unreacted. The first and third reactions on page 451 show the types of interactions that consumed deuterium to yield helium-3 and helium-4. Thus, after about 200 seconds, all the free neutrons had disappeared, and what remained were protons, electrons, deuterium nuclei, and helium nuclei, the raw material from which future stars and galaxies

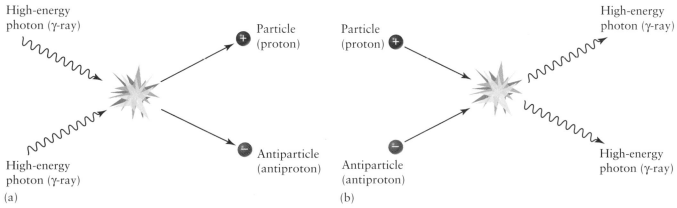

● **Figure 11.27** (a) Pair production reaction. High energy photons (gamma rays) interact to produce matter (in this case a proton) and antimatter (an antiproton). (b) Pair annihilation. A particle and an antiparticle collide and annihilate one another to produce photons (gamma rays). For each reaction to occur, the total energy of the gamma rays must at least be equal to the mass-energy of the proton and antiproton.

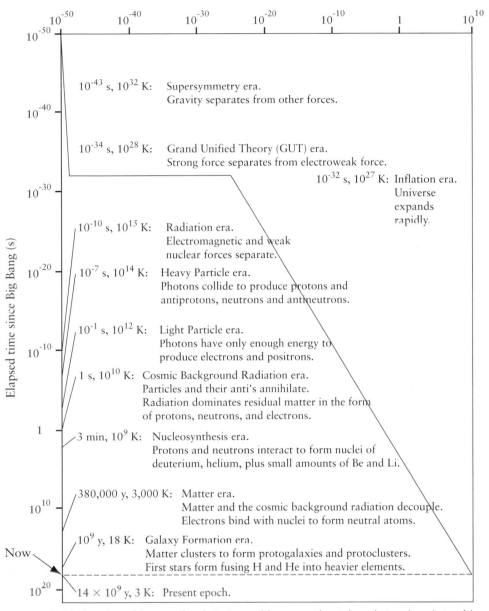

Size of universe (ly)

10^{-50} 10^{-40} 10^{-30} 10^{-20} 10^{-10} 1 10^{10}

Elapsed time since Big Bang (s)

10^{-43} s, 10^{32} K: Supersymmetry era.
Gravity separates from other forces.

10^{-34} s, 10^{28} K: Grand Unified Theory (GUT) era.
Strong force separates from electroweak force.

10^{-32} s, 10^{27} K: Inflation era.
Universe expands rapidly.

10^{-10} s, 10^{15} K: Radiation era.
Electromagnetic and weak nuclear forces separate.

10^{-7} s, 10^{14} K: Heavy Particle era.
Photons collide to produce protons and antiprotons, neutrons and antineutrons.

10^{-1} s, 10^{12} K: Light Particle era.
Photons have only enough energy to produce electrons and positrons.

1 s, 10^{10} K: Cosmic Background Radiation era.
Particles and their anti's annihilate.
Radiation dominates residual matter in the form of protons, neutrons, and electrons.

3 min, 10^{9} K: Nucleosynthesis era.
Protons and neutrons interact to form nuclei of deuterium, helium, plus small amounts of Be and Li.

380,000 y, 3,000 K: Matter era.
Matter and the cosmic background radiation decouple.
Electrons bind with nuclei to form neutral atoms.

10^{9} y, 18 K: Galaxy Formation era.
Matter clusters to form protogalaxies and protoclusters.
First stars form fusing H and He into heavier elements.

Now

14×10^{9} y, 3 K: Present epoch.

● **Figure 11.28** The evolution of the universe from the Big Bang until the present, with particular emphasis on the production of the lightest chemical elements.

would be constructed. ● Figure 11.28 summarizes the various stages of element production in the early universe.

The currently observed abundances of deuterium, helium-3, and helium-4 are all relatively small compared to the abundance of hydrogen. For example, there is only about 1 helium nucleus for every 10 hydrogen nuclei in the cosmos and a paltry 1 deuterium nucleus for every 50,000 hydrogen nuclei. The helium-3 to hydrogen number ratio is even smaller, about 1 per 500,000. The sequence of events that has been described here represents our best understanding of how the light elements were introduced into the universe. To check

whether this scenario makes sense, comparisons between the observed abundances and those calculated from models based on this picture have been made. Although the computed results for the rate of production of the light isotopes are quite sensitive to the total density of nucleons at the time of their formation, for reasonable ranges of this parameter, good agreement between the model predictions and the observations can be achieved. Such results constitute the strongest evidence for the correctness of the Big Bang origin of the universe outside that provided by the existence of the cosmic microwave background radiation (see again the Physics Potpourri on page 320).

● **Figure 11.29** Schematic of a tokamak fusion device. The plasma containing hydrogen nuclei is trapped by the magnetic field produced by the various electromagnets. [Adapted from *Scientific American* 249, no. 4 (October 1983):63.]

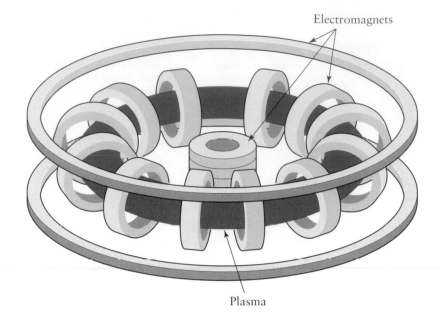

Electromagnets

Plasma

fields, and the plasma is injected into it. One of the most promising designs, called a *tokamak,* has the plasma confined inside a toroid   the shape of a doughnut (● Figure 11.29). Fusion reactions have been produced, but so far, the amount of energy released has been less than the amount used to produce the reactions.

Another approach to controlled fusion is to use extremely intense bursts of laser light to produce miniature fusion explosions. A small capsule containing deuterium and tritium is blasted from several directions simultaneously. The enormous pressure inside the capsule, over 10^{11} atmospheres, squeezes the nuclei close enough together for fusion to take place. The National Ignition Facility (NIF), under construction at Lawrence Livermore National Laboratory, is the newest and largest such facility (see ● Figure 11.30.) When completed, NIF will use short pulses from 192 laser beams to deliver 5×10^{14} watts of power to a fuel capsule.

A third technology has also achieved fusion and shows promise of one day producing a self-sustaining fusion reaction. The *pulsed power,* or "Z," *machine* makes use of an electromagnetic process similar to that shown in Figure 8.51. Hundreds of very thin wires are arranged symmetrically around, and parallel to, a central axis (the "z-axis"). A huge momentary current (thousands of amperes) is sent simultaneously through each wire.

● **Figure 11.30** The 10-meter-diameter target chamber (blue) during construction of the National Ignition Facility. One hundred ninety-two high-powered laser beams will converge on a fuel capsule at the center of the chamber.

Courtesy Lawrence Livermore National Laboratory

The wires are quickly vaporized by the ohmic heating and form a plasma. The huge magnetic field produced by the cumulative current exerts forces on the current-carrying plasma and squeezes it violently, producing an intense pulse of x rays. These x rays trigger fusion in a BB-sized deuterium pellet. It remains to be seen whether this avenue of research, or any of the others being pursued, will ever lead to commercial power from nuclear fusion.

There are good reasons why controlled fusion would make a good energy source. The main one is that the oceans contain an enormous supply of the fuel: hydrogen nuclei. Compared to fission reactors, much less radioactive waste would be generated. Chief byproducts would be helium, which has a number of uses, and tritium, which can be used as fuel. Fusion chain reactions would be easier to control than fission chain reactions. But because of the technical challenges, controlled fusion will not be a viable energy source in the near future.

Cold Fusion

There are ways to induce fusion without using extremely high temperatures. Referred to as *cold fusion,* these processes use other means to bring the fusing nuclei close together. The largest nuclei known, "superheavy" elements with atomic numbers over 100, are produced in the laboratory using one type of cold fusion. Smaller nuclei are accelerated to high speeds and collide with large nuclei. Under proper conditions, the nuclei fuse to form larger nuclei. Between 1958 and 1974, scientists in the United States and in the USSR accelerated very small nuclei to synthesize elements 102 through 106. In the 1980s, a team in Germany specialized in creating even larger nuclei by colliding not-so-small nuclei with large nuclei. For example, in 1996, they produced element 112 by bombarding lead-208 with zinc-70. The reaction equation is

$$\,^{70}_{30}\text{Zn} + \,^{208}_{82}\text{Pb} \rightarrow \,^{277}_{112}\text{XX} + \,^{1}_{0}\text{n}$$

("XX" represents the new element which, as of this writing, is unnamed.)

The kinetic energy of the incoming nucleus must be just large enough to overcome the electrostatic repulsion between the two nuclei but not so large that the resulting nucleus has enough excess energy to undergo immediate fission. Researchers are optimistic that even larger nuclei can be formed using this technique.

Another form of cold fusion seen in the laboratory is produced with the aid of an exotic elementary particle known as the *muon.* (Muons and other elementary particles are discussed in Chapter 12.) Hydrogen atoms normally combine in pairs to form H_2 molecules. If one of the two electrons is removed, the result is two hydrogen nuclei bound together by their attraction to the electron. But the nuclei are too far apart to undergo fusion. The negatively charged muon is basically an overweight electron: Its mass is about 200 times larger. Consequently, a "muonic atom" can exist that is just a muon in orbit about a hydrogen nucleus. Now, if the electron in the molecule described above is replaced by a muon, the two nuclei will be about 200 times closer together, and it is possible for them to fuse. This type of cold fusion has been observed, but it is not likely to be used as a source of energy. Muons are unstable, with a half-life of only 2.2×10^{-6} seconds. Too much energy would be needed to create a constant supply of muons.

LEARNING CHECK

1. What is the origin of solar energy?
2. A _____ weapon makes use of both nuclear fission and nuclear fusion.
3. Which of the following is *not* a method of producing nuclear fusion?
   a) Heating a magnetically confined plasma
   b) Forcing nuclei together using lasers

   c) Cooling nuclei to very cold temperatures
   d) Forcing nuclei together with a nuclear fission explosion
4. (True or False) Fusion is used to produce the largest nuclei known to exist.

ANSWERS: 1. fusion of hydrogen   **2.** thermonuclear   **3.** (c)   **4.** True

The history of nuclear physics began over 100 years ago, at a time when the very existence of atoms was still disputed. It is a remarkable history for many reasons. The pace of discovery, compared to the days of Galileo and Newton, was swift. The way in which physics research was done changed during the period. Instead of lone scientists working in small laboratories, most of the work was done by research groups using increasingly sophisticated equipment. The classification of physicists into experimentalists and theoreticians, starting in the nineteenth century with Michael Faraday and James Clerk Maxwell, shaped the growing physics community.

● **Figure 11.31** French physicist Henri Becquerel, the discoverer of radioactivity.

The first discoveries in nuclear physics were made before the structure, or even the existence, of the nucleus was known. Henri Becquerel (1852–1908) was the third in a line of four generations of prominent French physicists (see ● Figure 11.31). Like his father before him, Becquerel studied fluorescence—the emission of visible light from a substance when it is irradiated with ultraviolet light. (This process is exploited in fluorescent lights.)

Becquerel heard of the new x rays, discovered by Wilhelm Roentgen in 1895, and in early 1896, he began to test fluorescent substances to see if they too emitted x rays. In one experiment, he placed uranium on a photographic film plate that was wrapped to keep out visible light. After exposing the uranium to sunlight, he found that the film had been irradiated with what he thought to be x rays emitted by the uranium because of the sunlight. He tried to repeat the experiment, but the late winter weather in Paris turned cloudy for several days. Becquerel checked the film and expected very little exposure because of the weak sunlight. To his surprise, he found that the radiation from the uranium had been just as strong as before. During the following weeks, he determined that the uranium constantly emitted the radiation and needed no external stimulation. Clearly these were not x rays. Nuclear radiation, dubbed "Becquerel rays," had been discovered.

At this point, another famous family of physicists arrived on the scene, the Curies (see ● Figure 11.32). Marie Curie (1867–1934), originally Maria Sklodowska, grew up in Russian-occupied Warsaw. Her scientific career began when she joined her sister in Paris in 1891. While working toward her doctorate, Marie met Pierre Curie (1859–1906), a French chemist. They were married in July 1895. Two years later, a daughter, Irène, was born. She would follow in her mother's footsteps and become a renowned scientist.

At about this time, Marie, at the suggestion of Pierre, undertook an investigation of "Becquerel rays" as a thesis project. She tested all of the known elements and found that only uranium and thorium emitted the radiation. Upon testing ore samples from a museum, she found that some minerals were more radioactive than could be accounted for by uranium and thorium. She suspected that some new substance was emitting the radiation. Her husband joined the investigation, and in July of 1898, they announced the discovery of a new radioactive element, which Marie patriotically named polonium, for her native Poland.

Similarly, the Curies discovered another radioactive element, radium, in September of the same year. The task of actually isolating pure samples of the new elements was enormous. To get a tiny sample of radium for analysis, the Curies spent years extracting it from a ton of ore. The conditions in their crude laboratory were primitive, and no one realized the danger of radiation. Marie eventually died from a condition probably caused by radiation poisoning. Even her research notes were found to be radioactive.

● **Figure 11.32** Pierre and Marie Curie shortly after their marriage.

● **Figure 11.33** Ernest Rutherford.

Marie and Pierre Curie shared the 1903 Nobel Prize in physics with Henri Becquerel. Three years later, Pierre was killed by a runaway carriage. Marie carried on her research and received a second Nobel Prize in 1911 in chemistry.

The most prominent figure in the early years of nuclear physics was Ernest Rutherford (1871–1937; ● Figure 11.33). Born into a Scottish family in New Zealand, Rutherford developed an interest in physics at an early age. He attended college in New Zealand and won a scholarship to Cambridge. There his brilliance was soon recognized. Rutherford was an excellent experimenter with a reputation for simplicity and elegance. In 1897, he began to study the newly discovered nuclear radiation. The next year he made his first discovery: Two different types of radiation are emitted by uranium. These he named alpha and beta. The third type of nuclear radiation, gamma, was discovered later in France.

In 1898, Rutherford took a position at McGill University in Montreal and continued his work. In 1900, he began collaborating with a chemist, Frederick Soddy, and together they established that radioactivity causes transmutation of an element. They began to discover other radioactive substances that were later identified as different isotopes of radium. During this time, Rutherford reached the controversial conclusion that alpha particles are ionized helium atoms.

In 1907, Rutherford returned to England and took a position at the University of Manchester. The following year he won the Nobel Prize in chemistry. (The fields of chemistry and physics overlap a great deal, particularly in the study of the nucleus.) At this time, he experimentally verified the identity of alpha particles. His laboratory flourished, and much of his experimentation was carried out with the help of students and assistants. One such experiment led Rutherford to his most noted discovery, the nuclear model of the atom.

At this time, the structure of the atom was a topic of speculation. The accepted model held that the electrons were embedded in some kind of positively charged sphere, somewhat like raisins in a cake (refer to Figure 6.24). In an experiment performed by a student in Rutherford's laboratory, it was observed that some alpha particles were deflected as they traveled through a thin gold foil. The amount of deflection was much larger than would occur if the positive charge of an atom were spread evenly throughout its volume. Rutherford concluded, and then verified with experiments, that the positive charge is confined to a small region at the center of the atom, which he called the nucleus. This was in 1911, 15 years after the discovery of nuclear radiation.

The work in nuclear physics continued, with an understandable lull during World War I. Rutherford found that alpha particles could be used to disintegrate nuclei of nitrogen. He gave the proton its name and speculated that a similar particle that had no charge might exist in the nucleus. The eventual identification of the neutron was made in 1932 by James Chadwick (1891–1974), a former student of Rutherford.

Two of the most prominent experimenters in the 1930s were Irène Curie (1897–1956) and her husband Frédéric Joliot (1900–1958; ● Figure 11.34). They met when they both worked for Irène's mother, Marie. Early in 1932, Joliot and Curie reported results from experiments with a newly discovered type of radiation

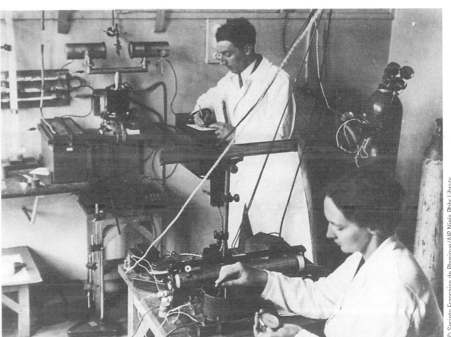

● **Figure 11.34** Husband-and-wife team Frédéric Joliot and Irène Curie, daughter of Marie and Pierre.

that they could not explain. They missed their chance: The radiation was high-speed neutrons. Chadwick quickly performed the necessary experiments and verified their identity. The discovery of the neutron completed Rutherford's nuclear model of the atom and solved some important mysteries in nuclear physics. Chadwick received the 1935 Nobel Prize in physics for his work.

The greatest discovery of Joliot and Curie, which they did interpret correctly, was artificially induced radioactivity. In a one-page paper dated 10 February 1934, they announced that bombarding certain kinds of nuclei with alpha particles resulted in their transmutation into radioactive nuclei. This earned them a Nobel Prize.

● **Figure 11.35** Nuclear physicist Enrico Fermi.

One of the first to exploit the discovery of artificial radioactivity was the Italian physicist Enrico Fermi (1901–1954; ● Figure 11.35). Fermi was a brilliant student and earned his doctorate at the age of 20. He became the key figure in a push to restore Italy's greatness in physics. Already famous at the time of Joliot and Curie's discovery, Fermi decided to use neutrons to induce artificial radioactivity. Alpha particles are positively charged and are therefore repelled by nuclei. The uncharged neutrons are much more likely to enter a nucleus. Fermi and his group in Rome started irradiating all of the known elements and soon produced about 40 new radioisotopes (see Section 11.4). This brought Fermi the 1938 Nobel Prize in physics.

Fermi and others missed an opportunity to make a critical discovery when the element uranium was irradiated with neutrons. Although they reported strange results, it was Otto Hahn (1879–1968; see ● Figure 11.36), once a student of Rutherford in Canada, and Fritz Strassman (1902–1980) who discovered barium and other midsized elements in an irradiated uranium sample. The correct interpretation of this finding, that some of the uranium had undergone fission, was made by Lise Meitner (1878–1968; see Figure 11.36) and her nephew, Otto Frisch (1904–1979). (Meitner was a colleague of Hahn but had fled to Sweden because of the rise of the Nazis in Germany.)

The news spread to physicists around the world, and many groups quickly confirmed the result. Soon uranium fission was recognized as a potential source of energy, as bombs or in some controlled process. Coming as it did, just as the world was being plunged into the worst war in history, one has to marvel at the timing of this discovery. The incentive to develop nuclear weapons was great.

Volumes have been written about the great atomic bomb project, called the *Manhattan Project,* during World War II. Some of the greatest physicists of the time, many of them refugees from Hitler's madness, were gathered in top-secret laboratories around the United States. The technical challenges were enormous, but the thought of a German atomic bomb was a powerful motivator. Three bombs were exploded: One was tested near Alamogordo, New Mexico, and the other two were used to destroy the Japanese cities of Hiroshima and Nagasaki. The hundreds of thousands of casualties caused by these explosions closed the final chapter on a war that saw civilians killed by the millions.

The impact of the Manhattan Project on physics was enormous. On the one hand, its result was a triumph in nuclear physics and engineering, but it left a cruel legacy for all involved. More clearly than ever before, it showed how the work of scientists can be used to destroy life. This vivid demonstration of the destructive aspects of applied physics spurred many of the principal scientists to seek careful control of nuclear weapons after the war. But Pandora's box had been opened, and the nuclear arms race was on.

● **Figure 11.36** Otto Hahn and Lise Meitner, two of the discoverers of nuclear fission.

SUMMARY

- The nucleus is the tiny, extremely dense core of an atom. It is composed of protons and neutrons held together by the short-range **strong nuclear force.**

- The different **isotopes** of a given element have the same number of protons in their nuclei but different numbers of neutrons.

- Isotopes are designated by the name of the element and the **mass number,** A, the total number of protons and neutrons in the nucleus, as in uranium-235.

- The nuclear force cannot hold a nucleus together if the ratio of protons to neutrons is too high or too low or if the nucleus is too large. In such cases, the nuclei are **radioactive** and emit nuclear radiation.

- There are three common types of radioactive decay—**alpha, beta,** and **gamma**—and each has a different effect on a nucleus.

- The **half-life** of a **radioisotope** is the time it takes for one-half of the nuclei in a given sample to decay. There are over 2,000 different radioisotopes with half-lives that range from tiny fractions of a second to billions of years.

- Some radioisotopes, most notably carbon-14, are quite useful for determining the ages of organic matter and of geological formations.

- Measurements of the **binding energy** per nucleon for different elements indicate that large and small nuclei are not as tightly bound as those with A around 50 to 60. This explains why splitting large nuclei and fusing small nuclei release energy.

- Energy is so concentrated in nuclear processes that Einstein's equivalence of mass and energy can be observed.

- Bombarding nuclei with neutrons and other particles can induce a variety of nuclear reactions.

- **Nuclear fission,** the splitting of a nucleus into two smaller nuclei, occurs when certain large nuclei are struck by neutrons. The energy released is exploited in atomic bombs and in nuclear power plants.

- **Nuclear fusion** is the combining of two small nuclei to form a larger nucleus. The Sun, the stars, and thermonuclear warheads use energy released by nuclear fusion.

- Efforts to harness fusion as a source of energy are hampered by the extreme conditions that are required to induce fusion.

MAPPING IT OUT!

1. Reread the material presented in Section 11.3 on the concept of *half-life,* and, following the principles for concept mapping that we have applied previously throughout the book, extend Concept Map 11.1 on p. 438 to include the notion of the half-life of a radioactive decay process. Be sure to incorporate the application of nuclear half-life to the dating of rocks and artifacts of early civilizations. Try to link the half-life concept and its ancillary properties not only to the main concept of radioactive decay but to the particular types of decay processes already expressed in Concept Map 11.1.

QUESTIONS

(▶ Indicates a review question, which means it requires only a basic understanding of the material to answer. The others involve integrating or extending the concepts presented thus far.)

1. ▶ Why do different isotopes of an element have the same chemical properties?

2. The atomic number of one particular isotope is equal to its mass number. Which isotope is it?

3. A mixture of two common isotopes of oxygen, oxygen-16 and oxygen-18, is put in a chamber that is then spun around at a very high speed. It is found that one isotope is more concentrated near the axis of rotation of the chamber and the other is more concentrated near the outer part of the chamber. Why is that, and which isotope is where?

4. ▶ What is the name of the force that holds protons and neutrons together in the nucleus?

5. ▶ What aspects of the composition of a nucleus can cause it to be unstable?

6. ▶ Describe the common types of radioactive decay. What effect does each have on a nucleus?

7. A nuclear explosion far out in space releases a large amount of alpha, beta, and gamma radiation. Which of these would be detected first by a radiation detector on Earth?

8. A standard treatment for some cancers inside the body is to use nuclear radiation to kill cancer cells. If the radiation has to pass through normal tissue before reaching the cancer, why would alpha radiation not be a good choice?

9. A concrete wall in a building is found to contain a radioactive isotope that emits alpha radiation. What could be done to protect people from the radiation (short of razing the building)? What if it were gamma radiation that was being emitted?

10. ▶ Explain the concept of half-life.

11. ▶ One-half of the nuclei of a given radioisotope decays during one half-life. Why doesn't the remaining half decay during the next half-life?

12. Suppose you have a large number of regular six-sided dice. How could you use them to simulate the decay of a "radioisotope" with a half-life longer than one-half of a throw? (Hint: See Explore It Yourself 11.1.) How could you simulate one with a half-life shorter than one-half of a throw?

13. ▶ How is carbon-14 used to determine the ages of wood, bones, and other artifacts?

14. One cause of uncertainty in carbon-14 dating is that the relative abundance of carbon-14 in atmospheric carbon dioxide is not always constant. If it is discovered that during some era in the past carbon-14 was more abundant than it is now, what effect would this have on the estimated ages of artifacts dated from that period?

15. The half-life of plutonium-238, the isotope used to generate electricity on the *Voyager* spacecraft, is about 88 years. What effect might this have on the spacecraft's anticipated useful lifetime?

16. ▶ The half-lives of most radioisotopes used in nuclear medicine range between a few hours and a few weeks. Why?

17. ▶ What are the principal steps in neutron activation analysis?

18. During the normal operation of nuclear power plants and nuclear processing facilities, machinery, building materials, and other things can become radioactive even if they never come into physical contact with radioactive material. What causes this?

19. If the binding energy per nucleon (see the graph in Figure 11.16) increased steadily with mass number instead of peaking around $A = 56$, would nuclear fission and nuclear fusion reactions work the same way they do now? Explain.

20. ▶ How can a nucleus of uranium-235 be induced to fission? Describe what happens to the nucleus.

21. ▶ What aspect of nuclear fission makes it possible for a chain reaction to occur? What is the difference between a chain reaction in a bomb and one in a nuclear power plant?

22. ▶ Explain how materials that absorb neutrons are used to control nuclear fission chain reactions.

23. ▶ What are fission fragments, and why are they so dangerous?

24. ▶ There is much more uranium-235 in a typical nuclear power plant than there was in the bomb that destroyed the city of Hiroshima. Why can't the reactor explode like an atomic bomb?

25. ▶ Why is a nuclear fusion reaction so difficult to induce?

26. If the strong nuclear force had a longer range than it does, what effect (if any) would that have on efforts to harness controlled fusion as an energy source?

27. ▶ Why are extremely high temperatures effective at causing fusion? What is used to produce such temperatures in a thermonuclear warhead?

28. ▶ Why is magnetic confinement being used in fusion research?

29. ▶ What is meant by the term "cold fusion"?

30. ▶ What was the source of the lightest nuclei, like deuterium, in the universe? Where were the rest of the elements in the periodic table through uranium produced?

PROBLEMS

(Note: In some of these, you may need to use the Periodic Table of the Elements in the back inside cover.)

1. Determine the composition of the nuclei of the following isotopes.
   a) carbon-14
   b) silver-108
   c) plutonium-242

2. The isotope helium-6 undergoes beta decay. Write the reaction equation, and determine the identity of the daughter nucleus.

3. The isotope silver-110 undergoes beta decay. Write the reaction equation, and determine the identity of the daughter nucleus.

4. The isotope polonium-210 undergoes alpha decay. Write the reaction equation, and determine the identity of the daughter nucleus.

5. The isotope plutonium-239 undergoes alpha decay. Write the reaction equation, and determine the identity of the daughter nucleus.

6. The isotope silver-107* undergoes gamma decay. Write the reaction equation, and determine the identity of the daughter nucleus.

7. The following is a possible fission reaction. Determine the identity of the missing nucleus.

$$\,_{0}^{1}n + \,_{92}^{235}U \to \,_{92}^{236}U^{*} \to \,_{39}^{95}Y + \,? + 2\,_{0}^{1}n$$

8. The following is a possible fission reaction. Determine the identity of the missing nucleus.

$$\,_{0}^{1}n + \,_{92}^{235}U \to \,_{92}^{236}U^{*} \to \,_{57}^{143}La + \,? + 3\,_{0}^{1}n$$

9. Two deuterium nuclei can undergo two different fusion reactions. One of them is given at the beginning of Section 11.7. In the second possible reaction, two deuterium nuclei fuse to form a new nucleus plus a lone proton. Write the reaction equation, and determine the identity of the resulting nucleus.

10. Iron-58 and lead-208 fuse into a large nucleus plus a neutron. Write the reaction equation, and determine the identity of the resulting nucleus.

11. A Geiger counter registers a count rate of 4,000 counts per minute from a sample of a radioisotope. Twelve minutes later, the count rate is 1,000 counts per minute. What is the half-life of the radioisotope?

12. Iodine-131, a beta emitter, has a half-life of 8 days. A 2-gram sample of initially pure iodine-131 is stored for 32 days. How much iodine-131 remains in the sample afterwards?

13. An accident in a laboratory results in a room being contaminated by a radioisotope with a half-life of 3 days. If the radiation is measured to be eight times the maximum permissible level, how much time must elapse before the room is safe to enter?

14. The amount of carbon-14 in an ancient wooden bowl is found to be one-half that in a new piece of wood. How old is the bowl?

15. A nucleus of element 112 is formed using the reaction equation given near the end of section 11.7. It then undergoes 6 successive alpha decays. Give the identity of the isotope that results after each step of this process.

16. A nucleus of element 114 is produced by fusing calcium-48 with plutonium-244. Write the reaction equation assuming three neutrons are also released.

CHALLENGES

1. Geiger counters are not very accurate when the count rates are very high; they indicate a count rate lower than the actual value. Explain why this is so.

2. The deflection of an alpha particle as it passes through a magnetic field is much less than the deflection of a beta particle (see Figure 11.3). Why?

3. One of the types of radioactive decay not discussed in this chapter is *electron capture*. One of the electrons orbiting the nucleus actually enters the nucleus and a "reverse beta decay" takes place. What effect does electron capture have on the parent nucleus? A nucleus of oxygen-15 undergoes electron capture. Write out the reaction equation, and determine the identity of the daughter nucleus.

4. As a general rule, the radioactivity from a particular radioisotope is considered to be reduced to a safe level after 10 half-lives have elapsed. (Obviously, the initial quantity of the isotope is also important.) By how much is the rate of emission of radiation reduced after 10 half-lives? Plutonium-239 is considered to be one of the most dangerous radioisotopes. Its half-life is about 25,000 years. How long would a sample of plutonium-239 have to be kept isolated before it could be considered safe?

5. The naturally occurring radioisotopes uranium-238 and uranium-235 have decay chains that end with the stable isotopes lead-206 and lead-207, respectively. Natural minerals such as zircons contain these uranium and lead isotopes. Careful measurements of the relative amounts of the isotopes can be used to estimate the ages of such minerals.
   a) How many separate alpha decays are there in each decay chain?
   b) For a mineral sample that is 4.5 billion years old, what would be the approximate ratio of the number of uranium-238 nuclei to the number of lead-206 nuclei? (Assume the initial decay of uranium-238 is the slowest in the decay chain.)

c) Repeat part (b) for uranium-235 and lead-207.

d) If you could only measure the ratio of the number of lead-206 nuclei to the number of lead-207 nuclei, what would the approximate value be, assuming the normal relative abundance of uranium-238 and uranium-235?

6. After a fuel rod reaches the end of its life cycle (typically 3 years), most of the energy that it produces comes from the fissioning of plutonium-239. How can this be?

7. When the plutonium bomb was tested in New Mexico in 1945, approximately 1 gram of matter was converted into energy. How many joules of energy were released by the explosion?

8. Using the information given in Section 11.5 and the mass-energy conversion equation, compute the binding energy (in MeV) and the binding energy per nucleon for hydrogen-2 (deuterium).

SUGGESTED READINGS

Armbruster, Peter, and Fritz Peter Hessberger. "Making New Elements." *Scientific American* 279, no. 3 (September 1998): 72–77. Describes how new, "superheavy" elements are produced using cold fusion.

Badash, Lawrence. "The Discovery of Radioactivity." *Physics Today* 49, no. 2 (February 1996): 21–26. A close look at Becquerel's discovery.

Crawford, Elisabeth, Ruth Lewin Sime, and Mark Walker. "A Nobel Tale of Postwar Injustice." *Physics Today* 50, no. 9 (September 1997): 26–32. An account of Lise Meitner's flight from Hitler's Germany, the discovery of nuclear fission, and her being passed over for the Nobel Prize.

Horgan, John. "But Is It Art?" *Scientific American* 257, no. 4 (October 1987): 48–52. Describes several techniques used to test the authenticity of artworks.

Jaworowski, Zbigniew. "Radiation Risk and Ethics." *Physics Today* 52, no. 9 (September 1999): 24–29. A critical look at the current standards for protecting people from radiation. Are they too strict?

Kirshner, Robert P. "The Earth's Elements." *Scientific American* 271, no. 4 (October 1994): 58–65. Describes how various elements are created via fusion processes in stars and supernovae, with stunning images from the *Hubble Space Telescope*.

Lake, James A., Ralph G. Bennett, and John F. Kotek. "Next-Generation Nuclear Power." *Scientific American* 286, no. 1 (January 2002): 72–81. Examines old and new designs for nuclear fission power plants.

Oganessian, Yuri Ts., Vladimir K. Utyonkov, and Kenton J. Moody. "Voyage to SUPERHEAVY Island." *Scientific American* 282, no. 1 (January 2000): 63–67. Describes how element 114 was first synthesized, and the prospects for creating larger, more stable nuclei.

Rhodes, Richard. *Dark Sun: The Making of the Hydrogen Bomb.* New York: Simon & Schuster, 1995. A best-selling history of the development of thermonuclear weapons.

Rhodes, Richard. *The Making of the Atomic Bomb.* New York: Simon and Schuster, 1986. Award-winning best-seller about the World War II bomb project.

"Special Issue: Radioactive Waste." *Physics Today* 50, no. 6 (June 1997): 22–62. Five articles on various aspects of radioactive waste and how to deal with it.

Sullivan, Jeremiah D. "The Comprehensive Test Ban Treaty." *Physics Today* 51, no. 3 (March 1998): 24–29. Includes descriptions of ways to detect nuclear explosions at great distances.

York, Derek. "The Earliest History of the Earth." *Scientific American* 268, no. 1 (January 1993): 90–96. Describes how radioactive decay, neutron activation analysis, and lasers are used to date geological samples.

For additional readings, explore InfoTrac® College Edition, your online library. Go to http://www.infotrac-college.com/wadsworth and use the passcode that came on the card with your book. Try these search terms: radioactivity, radioisotope thermoelectric generator, carbon-14 dating, nuclear medicine, neutron activation analysis, nuclear energy, Chernobyl accident, fusion energy, Marie Curie, Ernest Rutherford, Lise Meitner.

Newton's Laws, Momentum and Energy, Nuclear Physics (Chapters 1–3, 11)

Physics of Matter, EM Waves, Quantum Mechanics (Chapters 4, 8, 10)

12.1 Special Relativity: The Physics of High Velocity

12.2 Forces and Particles

control   on cocaine

12 SPECIAL RELATIVITY AND ELEMENTARY PARTICLES

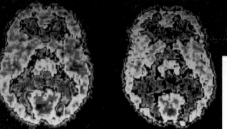

© Lynne Sanders, permission of NIDA

Positron emission tomography (PET) scans of a normal human brain (left) and that of a cocaine abuser (right) showing where the cocaine interferes with glucose metabolism. Red shows the highest level of glucose utilization and blue the least. Comparison of the two scans reveals that the cocaine-addicted brain does not use glucose nearly as effectively (note the reduction of red areas on the right), a circumstance that can lead to the disruption of brain function.

Antimatter: Available at a Medical Facility Near You

Antimatter. The word conjures up propulsion systems in science fiction tales of deep-space exploration or exotic particles produced by giant atom smashers. For most of us, antimatter remains the stuff of imagination, the purview of scientists who study it in isolated laboratories. Yet antimatter is finding its way into our daily experience because of its potential to save human life through *positron emission tomography* or PET.

PET is a medical imaging technique that uses positrons—antielectrons—to monitor biochemical processes in the body. Small amounts of tracer compounds containing positron-emitting isotopes with short half-lives, like carbon-11, oxygen-15, and fluorine-18, are ingested or inhaled by the patient. As the radioactive nuclei decay, they release positrons that quickly annihilate with electrons in the surrounding tissue to produce two high-energy gamma rays. The gamma rays are observed with specialized detectors, and an image is created that reveals the location and concentration of the radioisotope in the body part scanned.

PET has revolutionized medical research and treatment in several areas, especially neurology, where it has dramatically enhanced our understanding of brain chemistry and function. Because they expose patients to very low doses of radiation for short spans of time, PET scans can be performed rapidly and repeatedly, and offer the potential for monitoring the efficacy of new drug treatments. And all this takes place while the patient remains comfortable and alert.

For patients, PET provides a means for diagnosing and treating their diseases. For students, PET offers a chance to study the application of particle physics in everyday life and to investigate such questions as:

What distinguishes a positron from an electron, and more generally, antimatter from matter? Does every particle have an antiparticle? How are positrons and other antimatter particles produced, and how do they interact with normal matter?

These questions and others related to the properties of elementary particles and their interactions will be considered in this chapter. Our goal is to develop an appreciation for what particle physicists call the Standard Model and to forge links between submicroscopic physics and cosmology. As we shall see, the holy grail of high-energy physicists and cosmologists is the unification of physics, including all the known forces of nature, into one grand theory of everything. We start with one of the seminal theories in this unification process developed by Albert Einstein—the special theory of relativity.

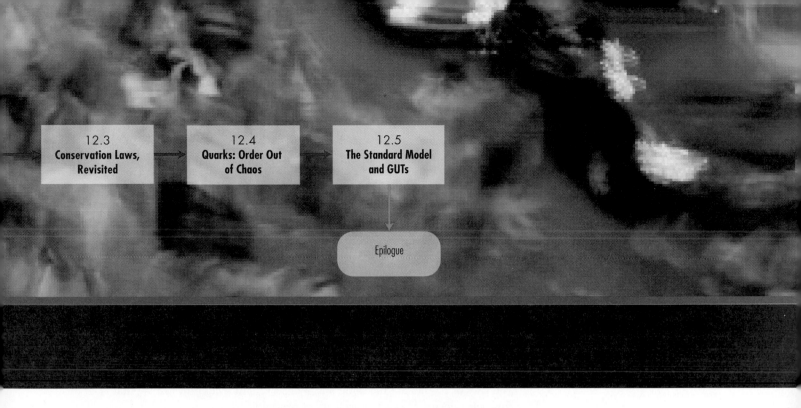

12.1 Special Relativity: The Physics of High Velocity

Imagine the following experiment: You are seated in the cargo area of a small pickup truck moving directly away from a companion at a constant velocity of 20 km/h. Your friend tosses a baseball to you with a horizontal velocity of 50 km/h (● Figure 12.1). From your point of view, what is the ball's velocity? If you answered 30 km/h, you're right. Clearly, the velocity of the ball with respect to you depends upon *your own velocity*—that is, *the velocity of the observer*. This is common sense: Galileo and Newton both would have agreed with you completely.

Now consider a second hypothetical experiment: You enter a spacecraft and leave the Earth, traveling uniformly at a velocity of 200,000 km/s. After a time, a friend sends out a light ray, which moves at the speed of light—300,000 km/s—in your direction (● Figure 12.2). When the light ray reaches you, what would you measure its speed to be? If you answered 100,000 km/s, you are wrong. Strange as it may seem, you would find the speed of the light to be 300,000 km/s, just what it is for your friend back on Earth. Evidently, *the speed of light does not depend upon the motion of the observer.* Light (and all other forms of electromagnetic radiation), behaving as Maxwell predicted, does *not* act as we would expect, based on the physics of Newton and Galileo. The theory of Newtonian mechanics and the theory of electromagnetism appear to be in conflict.

Postulates of Special Relativity

Albert Einstein recognized the contradiction between the predictions of classical mechanics and those of electromagnetism as regards the propagation of light. He set about reconciling the two by adopting these two postulates.

1. The speed of light, $c = 300,000$ km/s, is the same for all observers, regardless of their motion.

   Like the gravitational constant G in Newton's law of universal gravitation, c is a fundamental constant of nature. The fact that the speed of light is constant was just

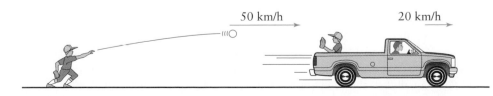

50 km/h   20 km/h

● **Figure 12.1** If you are traveling at 20 km/h, and a friend throws a ball toward you at 50 km/h, you see the ball approaching at 30 km/h.

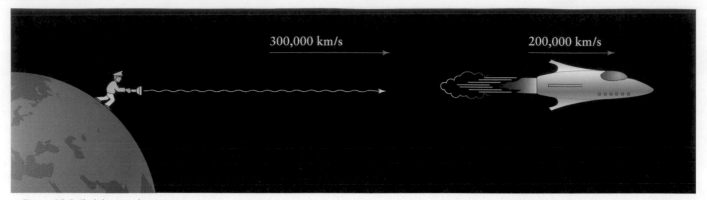

● **Figure 12.2** The light approaches you at 300,000 km/s, *even if you are moving away* from the source at 200,000 km/s. (Drawing not to scale.)

starting to be accepted when Einstein began his studies in the early 1900s. Over the past one hundred years, precise experiments have demonstrated beyond doubt that the speed of light is invariant under all circumstances. For example, measurements of the speed of photons emitted in the decay of subatomic particles called *pions* (π mesons, see page 477) originally traveling at 0.9998c, instead of producing ~2c as expected from Newtonian kinematics, yielded c to within 0.02%, in excellent agreement with predictions based on Einstein's first postulate.

2. The laws of physics are the same for all observers moving uniformly—that is, at a constant velocity. This is the *principle of relativity.*

This means that if two observers traveling toward one another at a constant speed perform identical experiments, they will get identical results. Moreover, *no* experiment can be performed by either observer that will indicate whether they are moving or what their speed is. Two people playing air hockey in the lounge of a 747 jet plane cannot tell from the motion of the puck on the table whether they are aloft and traveling uniformly at 800 km/h or sitting at rest on the airport tarmac. You can play billiards or shuffleboard on a cruise ship and not be able to tell the ship is in motion so long as its velocity is constant. In each case, the results of the "experiments" (playing air hockey or shuffleboard) cannot be used to demonstrate uniform motion relative to the Earth because the laws of mechanics (and of physics more generally) are the same for you and an Earth-bound observer. But, you say, you can look out the window of the jet or the ship and see that you are in motion. True, but how can you *prove* that you are not really at rest and that, by some magic, the clouds, trees, mountains, and so forth are not moving uniformly past you in the opposite direction? In point of fact, you can't! The principle of relativity has been confirmed numerous times in particle-collision experiments where energy and momentum measurements show excellent agreement with predictions based on this postulate.

Based on these two postulates, Einstein developed his **special theory of relativity,** which was published in 1905. It describes how two observers, in uniform relative motion, perceive space and time differently. One of the interesting aspects of this theory is that once you have accepted the experimentally verified postulates on which it is based, the fundamental predictions can be understood with only elementary algebra. The equations of special relativity are only a little more difficult than Newton's law of universal gravitation.

Predictions of Special Relativity

Let's consider one of the predictions of special relativity, something called **time dilation.** Imagine that we construct two identical "clocks" consisting of a flashbulb, a mirror, and a light-sensitive detector (● Figure 12.3). The flashbulb emits flashes of light at some pre-set rate. Each light pulse is reflected by the mirror into the detector. Each time the detector receives a pulse, an audible click is emitted like that of a standard wall clock. Let's now synchronize our two clocks and give one to some friends who are to travel in a spaceship with velocity *v* relative to you on Earth. The question is, Will the two clocks keep the same time? That is, will they continue to tick at the same rate? The answer seems obvious: Yes!

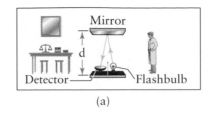

(a)

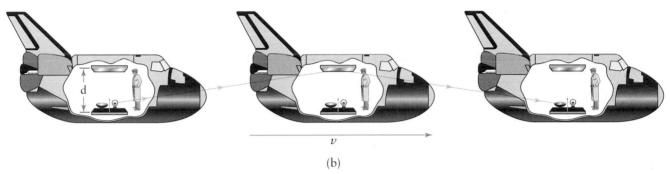

v

(b)

● **Figure 12.3** (a) A "light clock" at rest in a laboratory on Earth. (b) The light clock on board a spacecraft traveling uniformly with velocity *v*, as it would be seen by an observer on Earth.

This is the answer Newton would have provided. Unfortunately, given the postulates of special relativity, it is the wrong answer. Let's see why.

Consider the clock in the spaceship. When your friends took it aboard, everyone agreed that it was a properly working "standard" clock. Consequently, they note nothing peculiar in its performance as they travel along. Indeed, they *cannot* identify anything different about the clock, because if they did, they could know they were moving. Such a circumstance would violate the principle of relativity, which says that physics is the same for all uniformly moving observers. The clock on the spacecraft, *as seen by your friends aboard the craft,* ticks along at the same rate it did when they first received it.

But what about the clock on board the spaceship as seen by you, an external observer? If you track the motion of the clock, say, with a powerful telescope, you see that the light, in going from flashbulb to detector, follows a zigzag path, since the pulse (moving with the spaceship) has a sideways component to its velocity in addition to a vertical component (see Figure 1.12). Evidently, the path that the light travels in the moving clock is longer than the path it follows in your laboratory clock. Consequently, since the speed of light is the same in both cases, you conclude that the time it takes the light to reflect back to the detector is longer for the moving clock than for your clock. In other words, the moving clock is running slow: The rate at which it ticks is smaller than that for your clock. Or, put yet another way, the time *intervals* between ticks on the moving clock have been dilated or expanded.

Of course, if your friends read your clock from their spaceship, it is *your* clock that appears to be running slow, because from their vantage point, your laboratory appears to be moving with uniform velocity *v* in the *opposite direction.* The symmetry between the observations made on Earth and in the spaceship is guaranteed by the principle of relativity. But who is *really* right, you ask? Whose clock is *really* running slow? Both observers are right, and each clock is really running slow when compared to the other. The observers perceive the rate of flow of time differently because of their relative motion. But because no experiment can determine which observer is really in uniform motion, each observer's perception is as good or as true as the other's. *Time, then, is not an absolute:* It depends on the observer and their state of *relative* motion.

By how much will the interval between successive ticks differ for the two clocks discussed above? Not very much, it turns out, unless *v* is very close to the speed of light. If we let Δ*t* be the time between ticks on the clock at rest with an observer and let Δ*t'* be the observed time between ticks on the clock moving with velocity *v* relative to the observer, then Δ*t* and Δ*t'* are related by the following equation:

$$\Delta t' = \frac{\Delta t}{\sqrt{1 - v^2/c^2}}$$

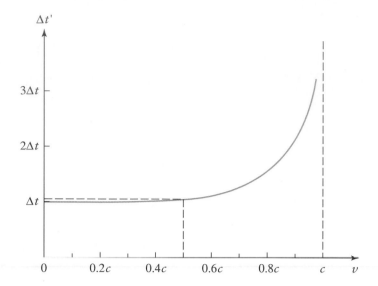

● **Figure 12.4** Graph of the time between clicks ($\Delta t'$) on a clock moving relative to an observer versus the clock's velocity v. At speeds even up to $0.5c$, the time $\Delta t'$ is nearly the same as Δt, the time between ticks when the clock is at rest with respect to the observer.

● Figure 12.4 is the graph of $\Delta t'$ for different velocities v. Even for velocities as high as one-half the speed of light, 150,000 km/s, $\Delta t'$ is not much larger than Δt; the clocks tick at very nearly the same rate. Effects of time dilation are virtually unknown in our daily lives, because we do not experience extremely high speeds. However, these effects are as real as any other phenomena in physics, and they have been observed in a very interesting manner.

A subatomic particle called the *muon* decays spontaneously into an electron plus some other particles in an average time of 0.000002 seconds. These muons are produced in large numbers by collisions between high-energy particles from space (*cosmic rays*) and atmospheric molecules some 10 or more kilometers above the Earth. Given their short lifetimes, we should find very few muons reaching the ground, even though they cover the distance between their place of production and the Earth's surface at nearly the speed of light. To the contrary, however, we detect great numbers of muons at ground level. How can this be?

A resolution to this paradox comes by applying special relativity theory. Because the muons are traveling so rapidly, their internal clocks, which regulate their rate of decay, appear to us to be running some ten times too slow. Consequently, from our perspective, there is ample time for them to reach the ground and to be detected—which is what happens. Of course, from the point of view of the muons, they do decay in 0.000002 seconds according to their own clocks, and, again from their perspective, it is our clocks that are running ten times too slow.

Example 12.1   What is the mean lifetime of a muon as measured in the laboratory if it is traveling at $0.90c$ with respect to the laboratory? The mean lifetime of a muon at rest is 2.2×10^{-6} seconds.

If an observer were moving along with the muon, then the muon would appear to be at rest to such an observer. The muon would decay in an average time $\Delta t = 2.2 \times 10^{-6}$ seconds as seen by this observer. For an observer in the laboratory, the muon lives longer because of time dilation. Applying our equation, we find that the average muon lifetime, $\Delta t'$, as determined in the laboratory is

$$\Delta t' = \frac{\Delta t}{\sqrt{1 - v^2/c^2}} = \frac{2.2 \times 10^{-6}\ \text{s}}{\sqrt{1 - (0.90c)^2/c^2}} = \frac{2.2 \times 10^{-6}\ \text{s}}{\sqrt{0.19}} = 5.0 \times 10^{-6}\ \text{s}$$

This is about 2.3 times the average lifetime of a muon at rest. To the laboratory observer, the muon's clock appears to be running more than two times too slow. How fast would the muon have to be traveling for its clock to appear to be running ten times too slow, as mentioned in the text?

Time dilation is one verified prediction of special relativity. There are two other important ones. The first is **length contraction,** in which moving rulers are shortened in the direction of motion. A convenient way to measure a distance is to time how long it takes light to traverse it. But if moving clocks run slow, so that the elapsed light-travel time is smaller, then moving rulers must be too short in the direction of motion. (Remember, distance = velocity × time.) The length of a meter stick moving relative to an observer is decreased by the same factor as the time between ticks in the two light clocks. Thus, moving observers disagree on issues involving *both* length *and* time.

Like time dilation, length contraction isn't ordinarily observed because the speeds at which you and I normally travel are so very small compared to the speed of light. But it is a real effect. When high-speed electrons move through the Stanford Linear Accelerator in California, their electric field lines are compressed in the direction of motion by length contraction. As the relativistic electrons pass through coils of wire arrayed along the accelerator, they produce a brief signal that is demonstrably different from that of more slowly moving electrons. The observed difference is precisely accounted for in terms of the special relativistic predictions of length contraction.

Another consequence of special relativity, the most important in particle physics, is the equivalence of energy and mass that was introduced in Section 11.5. If moving observers disagree on matters involving length and time, then they will also disagree on the velocities of material particles. For example, if a collision between two electrons occurs in one laboratory setting, the initial and final velocities of the particles as determined by an observer in that laboratory will not agree in general with those determined by another observer moving uniformly relative to the first. Yet both observers *must* agree that the physics of the collision is the same. In particular, both must agree that momentum and energy are conserved during the collision.

It turns out that for the laws of conservation of momentum and energy to be preserved in such cases, the observers must each include in their calculations the **rest energy** E_0 of the particles, given by Einstein's famous equation

$$E_0 = mc^2$$

where m is the ordinary mass of each particle (sometimes called the *rest mass*) and c is the speed of light. Thus, as Einstein wrote in 1921, "Mass and energy are therefore essentially alike; they are only different expressions for the same thing."

In special relativity theory, then, the total relativistic energy of a particle as measured by an observer is comprised of two parts: the rest energy, mc^2, of the particle plus whatever additional energy the particle has due to its motion—that is, its kinetic energy. The rest energy of a particle is clearly the same for all observers, but the kinetic energy (and hence the total energy) of the particle is not. It depends upon the frame of reference of the observer. Specifically, the energy of a particle of mass m moving with velocity v relative to a particular observer is equal to

$$E_{rel} = KE_{rel} + mc^2 = \frac{mc^2}{\sqrt{1 - v^2/c^2}}$$

Solving for the relativistic kinetic energy, we find

$$KE_{rel} = \frac{mc^2}{\sqrt{1 - v^2/c^2}} - mc^2$$

For very low particle speeds, this equation reduces to the familiar Newtonian form $\frac{1}{2}mv^2$ (see Challenge 1 at the end of this chapter). However, for velocities approaching the speed of light, the energy increases without limit. Thus, to accelerate a particle to the speed of light would require an infinite amount of energy. This is the reason why no material particle can ever travel at the speed of light. The energy and power demands for doing so simply cannot be met! The speed of light is not only a constant for all observers, it is also an absolute speed barrier that no object can cross (● Figure 12.5).

The fact that the (rest) mass of a particle is the same (that is, *invariant*) for all observers, does *not* mean that the mass cannot change. Quite the contrary! In an inelastic collision, mass frequently changes and is transformed into energy. Conversely, in the course of such encounters, energy may be converted into mass. This state of affairs is the direct result of Einstein's recognition of the equivalence of these two things. Thus, when

"It's not just a good idea, it's the law."

SPEED LIMIT
c = 300,000 km/s

● **Figure 12.5**

two balls of clay collide and stick together, some of the initial energy is converted to internal energy. Within the framework of special relativity, the energy that has gone into internal energy (and any other forms of internal excitation present in the final system) is measured exactly by the increase in the rest mass of the final system over that of the initial. Now in most practical situations in everyday life, the changes in mass that accompany interactions of this type are far too small to be detected. However, we have discussed some examples in which this type of conversion does lead to very dramatic effects in connection with nuclear reactions in Section 11.5. As we shall see, similar conversions taking place in inelastic collisions involving high-speed particles are the bread and butter of experimental high-energy physics and lead to the creation of exotic new species seldom seen in nature.

Example 12.2   In an x-ray tube (Figure 8.37), an electron with mass $m = 9.1 \times 10^{-31}$ kilograms is accelerated to a speed of 1.8×10^8 m/s. How much energy does the electron possess? Give the answer in joules and in MeVs (million electronvolts).

The total relativistic energy of the electron, E_{rel}, is its relativistic kinetic energy plus its rest energy. From our equation, we see that

$$E_{rel} = KE_{rel} + mc^2 = \frac{mc^2}{\sqrt{1 - v^2/c^2}}$$

Let's first determine at what fraction of the speed of light the electron is moving.

$$\frac{v}{c} = \frac{1.8 \times 10^8 \text{ m/s}}{3.0 \times 10^8 \text{ m/s}} = 0.60$$

$$v = 0.60c$$

The electron travels at 60% of the speed of light.

Evaluating the square root in the equation for E_{rel} gives

$$\sqrt{1 - v^2/c^2} = \sqrt{1 - (0.60)^2} = 0.80$$

The energy of the electron is then given by

$$E_{rel} = \frac{(9.1 \times 10^{-31} \text{ kg})(3.0 \times 10^8 \text{ m/s})^2}{0.80} = 1.02 \times 10^{-13} \text{ J}$$

But 1 joule $= 6.25 \times 10^{18}$ electron volts (see the conversion table on inside back cover), so

$$E_{rel} = (1.02 \times 10^{-13})(1 \text{ J}) = (1.02 \times 10^{-13})(6.25 \times 10^{18} \text{ eV})$$

or

$$E_{rel} = 637{,}500 \text{ eV}$$

Since 1 MeV $= 1 \times 10^6$ eV, the energy of the electron is approximately 0.638 MeV.

Notice, with E in MeV, the *equivalent* mass of the electron could be given as 0.638 MeV/c^2. This is a frequently used and very convenient way of representing subatomic particle masses because it eliminates the small numbers that necessitate the cumbersome exponential notation.

Let's compare the relativistic kinetic energy of the electron to that given by classical physics. The rest energy of the electron, mc^2, may be easily shown to be 0.511 MeV following the model above. Then

$$KE_{rel} = E_{rel} - mc^2 = 0.638 \text{ MeV} - 0.511 \text{ MeV} = 0.127 \text{ MeV}$$

According to Newtonian mechanics,

$$KE_{classical} = \frac{1}{2}mv^2 = \frac{1}{2}(9.1 \times 10^{-31} \text{ kg})(1.8 \times 10^8 \text{ m/s})^2$$

$$= 1.47 \times 10^{-14} \text{ J} = 92{,}100 \text{ eV} = 0.092 \text{ MeV}$$

The classical result *underestimates* the electron's kinetic energy by almost 30%.

If we reflect on the special theory of relativity, we see that it accomplishes a profound unification in physics: It reconciles the physics of low speeds with that of high speeds. It is a better, more comprehensive system of mechanics than that formulated by Newton because it works for all particles, regardless of their relative velocities. In the limit of small velocities, we recover the laws of classical mechanics as we specified them in Chapters 1–3; for high velocities, we find that Einstein's theory predicts new effects not contained in Newton's physics that are confirmed experimentally. We will return to this theme of unification in Section 12.5 after we consider elementary particles and the forces they mediate, since it has been, and remains today, one of the overriding goals of physical science.

Concept Map 12.1 summarizes the basic postulates and implications of Einstein's special theory of relativity.

CONCEPT MAP 12.1

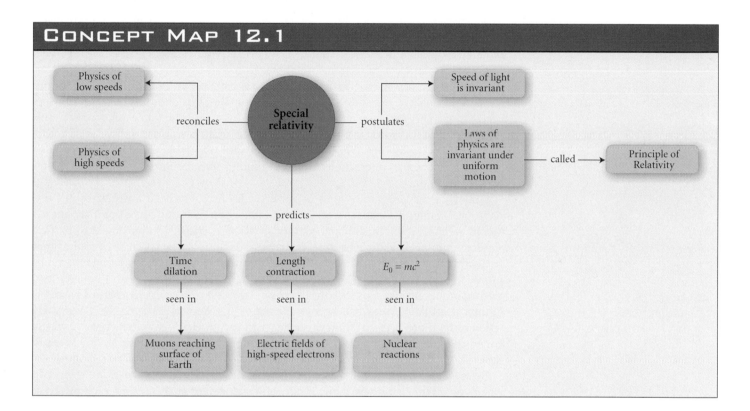

LEARNING CHECK

1. According to Einstein's special theory of relativity, when compared to an identical clock at rest, a clock moving with constant velocity will run
   a) slow.
   b) fast.
   c) at the same rate.
   d) alternately fast and slow.

2. The _____ states that the laws of physics are the same for all observers moving uniformly.

3. An observer traveling with speed 0.5c moves directly toward a beacon emitting photons with speed c in her direction. The observer measures the speed of the approaching photons to be
   a) 0.25c
   b) 0.5c
   c) c
   d) 1.5c

4. (True or False.) To someone moving horizontally at a speed of 0.99c past a vertical meter stick, the stick will appear to be 100 cm long.

5. A horizontal stretched spring weighs _____ (more than, less than, the same as) an identical spring that is unstretched.

ANSWERS: 1. (a) **2.** principle of relativity **3.** (c) **4.** True **5.** more than

12.2 Forces and Particles

The Four Forces: Natural Interactions among Particles

At the end of Section 2.8, we introduced the four fundamental forces of nature. ● Table 12.1 lists these forces and includes some properties of each. It is important to acknowledge that *all* the interactions that occur in our environment are due to these forces. They produce the beauty, variety, and change that we witness daily in the world around us.

In Chapter 2, we defined a force as a push or pull acting on a body that usually causes a distortion or a change in velocity (or both). This is a perfectly good description of what we mean by a force in classical physics, but to investigate the realm of particle physics, we must broaden our definition to include every change, reaction, creation, annihilation, disintegration, and so on, that particles can undergo. Thus, when a radioactive nucleus spontaneously decays (see Section 11.2), we will describe this decay in terms of a force that acts between the parent nucleus and its decay products. Similarly, when two particles collide and undergo a nuclear reaction to create new particles (see Section 11.4), we say that there is a force responsible for this transformation.

Because the roles played by forces in particle physics are somewhat different from those traditionally assigned to them in classical physics, it is often the case that they are referred to as the *four basic interactions* of nature instead of the four basic forces. In this context, we use the word "interaction" to mean the mutual action or influence of one or more particles on another. With this in mind, we return to Table 12.1 and discuss each of the four fundamental interactions briefly, beginning with the most familiar, the gravitational interaction.

Gravity, a very important force in our everyday lives, has been investigated at some length in Chapter 2. Several aspects of this interaction as it pertains to particle physics should be reviewed. First, although gravity affects *all* particles, its importance in particle physics is entirely negligible because its strength is so feeble when compared to the other interactions that can occur. To get a feel for just how inconsequential gravity is on a subatomic level, we can compare the strength of the gravitational attraction between two protons bound in a nucleus with the electrical repulsion between these charged particles. A simple calculation (see Challenge 2 at the end of this chapter) shows the electrical interaction to be more than 10^{36} times stronger than the gravitational interaction. Comparisons between the strength of gravity and the other interactions of nature are given in Table 12.1. In each case, the effects of gravity are too small to be considered seriously.

Before moving to a discussion of the other forces, it is worth remarking upon two aspects of gravity that do have important consequences for *large-scale* interactions: First, gravitational interactions may dominate in circumstances where charge neutrality prevails. If many particles interact together at once and the number of positive charges balances the number of negative ones, electrical forces may cancel out, leaving gravity the dominant interaction. Unlike the electrical interaction, gravity cannot be shielded out or

---

● Table 12.1 The Four Fundamental Forces or Interactions

Name (example)	Relative Strength	Range (m)	Carrier Particle	Carrier Mass (MeV/c^2)	Spin
Strong (binds the nucleus)	1	$\approx 10^{-15}$	Meson[a]	$> 10^2$	1
Electromagnetic (binds atoms and molecules)	10^{-2}	Infinite	Photon	0	1
Weak[b] (produces beta decay)	10^{-7}	$\leq 10^{-18}$	Z^0, $W^\pm$	$\leq 10^5$	1
Gravitational (binds planets to the Sun)	10^{-38}	Infinite	Graviton	0	2

[a]At the level of the nucleons, we may regard the messengers of the strong force to be the mesons, although, as discussed in Section 12.5, the true carriers of this interaction are the massless, chargeless *gluons*.

[b]All the messengers of the basic forces are uncharged, except for two of those of the weak force. The W^+ particle carries one unit of positive charge, while the W^- holds one unit of negative charge. The other carrier of the weak force, the Z^0, is electrically neutral.

eliminated because there is only one kind of mass. And second, gravity is a long-range interaction. The gravitational force varies inversely as the distance squared, and although it grows ever weaker with separation, it never completely disappears. Thus, gravitational effects may reach over vast regions of space, accumulating in such a fashion as to affect the structure and evolution of the entire universe.*

Aside from gravity, the next most familiar force or interaction is the **electromagnetic interaction,** discussed in Chapter 8. Unlike the gravitational force that is always attractive, the electromagnetic force can be either attractive or repulsive, depending on the signs of the interacting charges. But, like gravity, the electromagnetic force is a long-range force, becoming smaller as the distance separating the charges increases. The electromagnetic force also manifests itself in the magnetic forces associated with moving charges. This interaction is responsible for all the various kinds of electromagnetic radiation, from gamma rays to radio waves, investigated in Chapter 8.

It is important to note that although the electric and magnetic forces act only between charged particles, the electromagnetic interaction can have an influence on uncharged particles as well. For example, a photon is not a charged particle, but the absorption or emission of a photon by an atom is an electromagnetic process.

Next among nature's interactions is the **weak nuclear force,** which is responsible for beta decay (the conversion of a neutron to a proton within the nucleus; see Section 11.2). The term *weak* may be interpreted in a variety of ways. For example, this interaction is weak in the sense that it is effective only over very short distances: at least 100 times smaller than the range of the strong nuclear force and infinitesimal compared to the ranges of gravity and electromagnetism. The weak force is also "weak" because the probability is quite small that interactions involving this force occur. Indeed, particle interactions involving the weak force generally happen only as a last resort when all other interaction mechanisms are blocked.

Although it is not apparent from what we have said so far, a very close relationship exists between the electromagnetic and weak interactions. The similarity between these two was first noted in the late 1950s, and further studies of the connections between the weak force and electromagnetism have led to a unification of these two interactions into one, the **electroweak interaction.** This is much like the unification between electricity and magnetism in Maxwell's theory. It is now recognized that the source of the electromagnetic and weak forces is the same but that their practical manifestations differ considerably, leading to a separate classification for each. More will be said about the issue of unification of forces in Section 12.5.

The last of the forces of nature is the **strong nuclear force.** The strong force is responsible for holding the nuclei of atoms together and is involved in nuclear fusion reactions (see Section 11.7). It is a short-range, attractive interaction that does not depend upon electric charge. Considering the probability that two colliding particles will interact by the strong force as opposed to any one of the other three basic interactions, the strong force is indeed quite "strong," some 100 times as effective in bringing about a reaction as the electromagnetic force. Comparisons like these between the relative probability that a reaction will occur via a particular interaction have been used to establish the measures of the relative strengths of the four forces given in Table 12.1.

Having compared the properties of the four forces, let's now consider *how* we have acquired our knowledge about their characteristics. Einstein's theory of special relativity has as one of its postulates the finiteness of the speed of light: 300,000 km/s is the maximum speed attainable by particles in the universe. This is also the maximum speed at which information may be propagated through the universe. The fact that a star 150 million kilometers away suddenly explodes *cannot* be known to us until at least 500 seconds (about 8 minutes) later because the particles ejected from the event require that long to make their way to us. Information about this explosion thus comes to us through the

---

* To properly account for the interaction between massive objects like stars and galaxies or to develop models for the global structure of the universe that are consistent with observation, use must be made of Einstein's *general theory of relativity.* This is a more comprehensive and accurate theory of gravity than that originally proposed by Newton, and it permits scientists to interpret correctly such phenomena as the deviation in the path of starlight passing near the Sun and the wavelength shifts in the spectral lines of certain very dense stars called *white dwarfs.* Additional discussion of general relativity is given in the Epilogue.

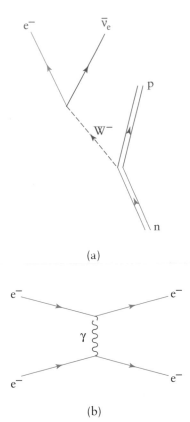

(a)

(b)

● **Figure 12.6** (a) Modern representation of the beta decay of a neutron. The neutron transmutes into a proton after emitting a W⁻ particle, and the W⁻ subsequently decays into an electron (e⁻) and an antineutrino ($\overline{\nu}_e$). (b) In particle physics, the electromagnetic repulsion between two electrons is viewed as being caused by the exchange of (virtual) photons (γ). This process is shown in what is called a "Feynman diagram," after theoretical physicist Richard P. Feynman (1912–1988).

intermediary of particles that race out from the interaction site carrying data about the nature of the event to our location.

In the same way that we come to understand the details of a stellar explosion by the particles emitted during the event, particle physicists come to know the characteristics of the four forces of nature by the particles produced in these interactions. In fact, current theories associate with each force a **carrier,** or mediator, of the interaction. These carriers are exchanged between the particles experiencing the forces, and they communicate the interaction between the reactants and the products. For example, if we wiggle an electron, the change in its electric field will propagate outward as a wave traveling at the speed of light. The disturbance in the field produces forces on other charged particles in the neighborhood of the electron and communicates to them information about the electron's motion. The role of the wave is that of a messenger, and this has led physicists to conceive of the influence of the electric field as being conveyed or carried by a messenger particle—the photon (recall Section 10.5 and the discussion of wave-particle duality). In this view, all the effects of an electromagnetic field may be explained by the exchange of photons.*

Table 12.1 includes the carriers of the four basic interactions as well as their masses (measured in equivalent energy units; see Example 12.2). In the next section, we will explore the characteristics of these and other elementary particles in greater detail. Before doing so, we show ● Figure 12.6, which depicts how a particle physicist might represent the weak interaction that converts a neutron inside a nucleus into a proton (a beta decay) and how modern physics views the repulsion between two electrons.

Classification Schemes for Particles

In Chapter 4, we classified matter into solid, liquid, gas, and plasma phases. We also classified matter according to the number and kinds of atoms that are present: elements, compounds, mixtures, and so forth. We were further able to categorize the properties of matter on the basis of the forces that acted between the constituents: large forces in solids, smaller forces in liquids, and so on.

Just as we could classify bulk matter in several different ways, it is possible to classify elementary particles using different schemes. In this section, we will consider three ways of doing so: on the basis of spin, on the basis of interaction, and on the basis of mass. Before going any further in this discussion, however, it is worth defining what we mean by an **elementary particle** and by an **antiparticle.**

> **DEFINITION**
>
> **Elementary Particles** The basic, indivisible building blocks of the universe. The fundamental constituents from which all matter, antimatter, and their interactions derive. They are believed to be true "point" particles, devoid of internal structure or measurable size.

> **DEFINITION**
>
> **Antiparticle** A charge-reversed version of an ordinary particle. A particle of the same mass (and spin) but of opposite electric charge (and certain other quantum mechanical "charges").

Every known particle has a corresponding antiparticle. There are antielectrons (**positrons**), antiprotons, antineutrons, and so on. Collections of antiparticles form antimatter, just as collections of ordinary particles form (ordinary) matter. The first antiparti-

---

* The photons that mediate the electromagnetic interaction are not *real* photons like those produced in the emission of light by excited atoms but what are called *virtual photons.* These particles are undetectable in the traditional sense but are responsible for the transmission of forces between other ordinary, observable particles. Thus, while the virtual photons themselves cannot be directly observed, the *effects* of the virtual photons as carriers of the electromagnetic interaction can be directly seen.

cle, the positron, was discovered in 1932 by Carl Anderson in cosmic rays. The antiproton was discovered in 1955, and the antineutron the year following. Particle physicists now routinely create and even store small quantities of antimatter in high-energy accelerators.

When matter and antimatter meet, mutual annihilation results, accompanied by a burst of energy in the form of gamma rays (see Figure 11.27). The positrons used in PET scans are produced by the decay of radioactive isotopes, like fluorine-18, which are first generated by bombarding other nuclei with light particles. For example, fluorine-18 may be created by colliding oxygen-18 nuclei with high-speed protons. In 1995, using similar high-energy physics techniques, scientists at the Conseil Européen pour la Recherche Nucléaire (CERN; see chapter 3 opening photo) were successful in producing antihydrogen, an atom of antimatter composed of an antiproton and a positron.

In what follows, we shall have several occasions to examine the creation and annihilation of particles and antiparticles. In those reactions, we will distinguish antiparticles using the same symbol as that for the corresponding particle but with a "bar" over it. Thus, for example, if n designates a neutron, then $\bar{n}$ (pronounced "en-bar") represents an antineutron.*

The kinds of particles that have been termed "elementary" have gradually changed with time. Before 1890 and the discovery of the electron, atoms were regarded as the smallest units of matter, and they were believed to possess no internal structure of their own. In the 1930s, it was believed that the basic building blocks of nature consisted of the proton, the neutron, the electron, the positron, the photon, and the neutrino (low-mass particles, typically produced in beta decay reactions, that have no charge and only very weakly interact with ordinary matter). Circa 1934, these were the "atoms" (indivisible particles) sought by the ancient Greeks. In the last 70 years, particle physicists have discovered that not only are there more than six "elementary" particles in nature but that some of the original six are not really elementary at all! They themselves are composed of still more basic and elusive particles. At the time of this writing, there exist well over 100 different subatomic particles (● Figure 12.7), and it is well accepted that the proton and the neutron (among others) are not the ultimate constituents of matter. The majority of the remainder of this chapter will be spent developing the story of our changing perspective on what constitutes a truly "elementary" particle. But, first, a bit more about the properties of subatomic particles.

Refer to the discussion of Dalton's atomic theory in the Physics Family Album in Chapter 4.

"PARTICLES, PARTICLES, PARTICLES."

● **Figure 12.7**

Cartoon by Sidney Harris. Used by permission.

Spin

Spin, like mass and charge, is an intrinsic property of all elementary particles and measures the angular momentum carried by the particle. If we treat an elementary particle as a simple hard sphere, then we can picture spin as due to the rotation of the particle about an axis through itself. Unlike, say, a basketball whirling at the end of one's finger, which can have any amount of spin, the spin of elementary particles is quantized (see Section 10.5) in units of $h/2\pi$, where h is Planck's constant. The spins of all known particles are either integral or half-integral multiples of this basic unit. In other words, spin can take on values of $0, \frac{1}{2}, 1, \frac{3}{2}$, and so on, in units of $h/2\pi$. Experiments have shown, for example, that the spin of the electron and the proton is $\frac{1}{2}$, but that of the photon is 1.

Stephen Hawking has provided another view of spin that may be helpful in thinking about this property of elementary particles. It has the advantage of avoiding any conflicts with quantum mechanics that arise from thinking of particles as little marbles or ball bearings. In Hawking's approach, the spin of a particle tells us what the particle looks like from different directions. Thus, a particle of spin zero is like the period at the end of this sentence. It looks the same from all directions (● Figure 12.8). A spin 1 particle is like an arrow. It looks different when seen from different directions, but a rotation through a complete circle (360°) restores the original view. A spin 2 particle is similar to a two-headed arrow; rotation by 180° gives the same appearance. In this picture, spin $\frac{1}{2}$ particles,

---

* There are several noteworthy exceptions to this rule, however. For example, an antielectron (or positron) is symbolized e^+, not $\bar{e}^-$. Likewise, an antimuon is denoted μ^+, not $\bar{\mu}^-$. See Table 12.3 for some other cases in which the "bar" rule is not adhered to strictly.

● **Figure 12.8** Illustrations of objects exhibiting symmetry properties analogous to those of elementary particles, as described by Stephen Hawking. (a) Dot (spin = 0); (b) ace of clubs (spin = 1); (c) jack of hearts (spin = 2).

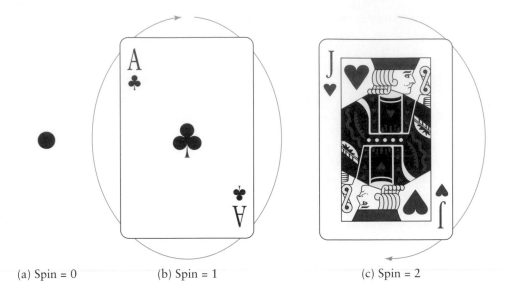

(a) Spin = 0   (b) Spin = 1   (c) Spin = 2

like electrons, have the remarkable property that *two* complete rotations through 360° are required for the particle to "look" the same. Can you think of any common object or figure that exhibits this type of symmetry?

Particles possessing half-integral spins are called **fermions,** after Enrico Fermi (see page 460), who carefully investigated the behavior of collections of such particles. Particles with integer spins are called **bosons,** after Satyendra Bose, who, with Einstein, developed the laws describing their collective behavior. The main difference between these two types of particles is that the former obey the Pauli exclusion principle (page 413) while the latter do not. This law, for which Wolfgang Pauli won the Nobel Prize in 1945, states that no two interacting fermions of the same type can be in exactly the same state. They must be distinguishable in some manner. Thus, in a normal helium atom, when the two electrons are in the lowest atomic energy state (the ground state), the exclusion principle demands that these spin $\frac{1}{2}$ particles differ in some way. How can this be achieved? Isn't one electron just like any other? Same mass, same charge, same spin? Yes. But let's return to our rotating sphere model for a moment. Relative to the axis of rotation, the marble may spin either clockwise or counterclockwise. For a given total angular momentum, two distinct spin states associated with the directions of rotation exist (● Figure 12.9). In the same way, one can associate with the electron two different spin configurations, called *spin-up* and *spin-down,* each with the same total amount of spin, $h/4\pi$. With this addition, it is now possible to satisfy the Pauli principle for helium by requiring one of the electrons to be in a spin-up state while the other is in a spin-down state. The state of every electron in every atom can be accounted for by this principle.

Bosons, by contrast, do not obey the Pauli principle. They are truly indistinguishable. An unlimited number of bosons can be concentrated in any given volume of space without violating any physical laws. This accounts for the fact that there is no restriction on

● **Figure 12.9** Spin-up, (a), and spin-down, (b), configurations for a rotating marble.

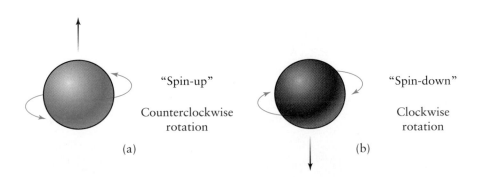

"Spin-up"

Counterclockwise rotation

(a)

"Spin-down"

Clockwise rotation

(b)

the number of photons (spin 1 particles) that can be packed into a beam of light and hence no (theoretical) limit to the intensity of the beam. There is also no limit to the number of force-carrying particles that can be exchanged in a given reaction, because *all* the mediators of the fundamental interactions have integral spins (Table 12.1). It is worth mentioning that the phenomena of superfluidity and superconductivity both result from the collective behavior of large numbers of bosons occupying the same state in what is called a *Bose–Einstein condensate*; see the Physics Potpourri on superfluids on page 156, and Section 7.3.)

Elementary Particle Lexicon

Separating particles according to spin divides them into two groups. Establishing additional selection criteria further subdivides these two groups. A very useful way of doing so is to identify those particles that participate in strong force interactions and those that do not. We thus distinguish the four groups of particles given in ● Table 12.2: the baryons, the leptons, the mesons, and the **intermediate (or *gauge*) bosons.** Examples of several particles of each type are included in the table. An electron is a lepton with spin $\frac{1}{2}$ that is unaffected by the strong force; a proton is a baryon with spin $\frac{1}{2}$ that does interact via the strong force. Photons are bosons that do not participate in strong interactions.

The names for these groups derive largely from earlier classification schemes based on experimentally determined masses for these particles. The word **baryon** comes from the Greek *barys* meaning heavy, while the word **lepton** means "light one" in Greek. **Mesons** refers to the "middle ones" with intermediate mass. At the time these groups were named, the heaviest particles known were found among the baryons and the lightest were included among the leptons. Recent discoveries, however, have revealed mesons and at least one lepton with masses larger than those of protons and neutrons. Thus, it is no longer possible to specify the correct class of an elementary particle by its mass alone, although for most species this is still a useful guide.

The force-carrying particles, the intermediate bosons, exhibit a wide variety of mass. Because they are bosons, there is no limit to the number that can be exchanged in any interaction, but there is a close correlation between the range of the force they mediate and their mass. If the carrier particles have a high mass, it will generally be difficult to produce them and to exchange them over long distances. Thus, a force carried by massive particles will have only a short range. This is the case for the W and Z particles that mediate the weak interaction. However, if the carrier particles have no mass of their own (like the photon and the graviton), the forces associated with them (in these cases, the electromagnetic and gravitational forces, respectively) will be of long range.

Before leaving this section, we introduce one additional bit of nomenclature that is used commonly in connection with elementary particles, the word *hadron.* This word is also of Greek origin, coming from *hadros,* meaning "thick" or "strong." Baryons and mesons are collectively referred to as hadrons (● Table 12.3) because they interact by the strong force, a distinction not shared by the leptons or the gauge particles.

Concept Map 12.2 shows ways elementary particles may be classified, as well as the connections among them.

● **Table 12.2** Classification of Particles		
Spin Group	**Particles Interacting Via the Strong Force**	**Particles Not Affected by the Strong Force**
Fermions (Half integer spin)	*Baryons* (Protons, neutrons, lambdas, sigmas, . . .)	*Leptons* (Electrons, muons, neutrinos, . . .)
Bosons (Integer spin)	*Mesons* (Pions, kaons, etas, . . .)	*Intermediate (or gauge) bosons* (Photons, Z^0, $W^{\pm}$, gravitons)

● Table 12.3 Properties of Long-Lived Hadrons

Class	Particle Name	Symbol[a]	Antiparticle	Mass (MeV/c^2)	B[b]	S[c]	Lifetime (s)
Baryons	Proton	p	$\bar{p}$	938.3	+1	0	Stable(?)
	Neutron	n	$\bar{n}$	939.6	+1	0	886
	Lambda	Λ^0	$\bar{\Lambda}^0$	1,115.7	+1	−1	2.6×10^{-10}
Spin = $\frac{1}{2}$	Sigma	Σ^+	$\bar{\Sigma}^-$	1,189.4	+1	−1	0.8×10^{-10}
		Σ^0	$\bar{\Sigma}^0$	1,192.6	+1	−1	7.4×10^{-20}
		Σ^-	$\bar{\Sigma}^+$	1,197.5	+1	−1	1.5×10^{-10}
	Xi	Ξ^0	$\bar{\Xi}^0$	1,315	+1	−2	2.9×10^{-10}
		Ξ^-	$\bar{\Xi}^+$	1,321	+1	−2	1.6×10^{-10}
Spin = $\frac{3}{2}$	Omega	Ω^-	$\bar{\Omega}^+$	1,672	+1	−3	0.8×10^{-10}
Mesons	Pion	π^+	π^-	139.6	0	0	2.6×10^{-8}
		π^0	Self[e]	135.0	0	0	8.4×10^{-17}
	Kaon[d]	K^+	K^-	493.7	0	+1	1.2×10^{-8}
		K^0	$\bar{K}^0$	497.7	0	+1	0.9×10^{-10}
Spin = 0							5.2×10^{-8}
	Eta	η^0	Self	547.8	0	0	5.5×10^{-19}
	"Dee-plus"	D^+	D^-	1,869	0	0	1.0×10^{-12}
	"Bee-plus"	B^+	B^-	5,279	0	0	1.7×10^{-12}
Spin = 1	Psi	ψ^0	Self	3,097	0	0	7.6×10^{-21}
	Upsilon	Υ^0	Self	9,460	0	0	1.2×10^{-20}

[a]Superscripts to the right of the particle symbols indicate the charge carried by the particle in units of the proton charge (see Section 12.3).

[b]Baryon number (see Section 12.3).

[c]Strangeness (see Section 12.3).

[d]There are actually two different types of K^0 particles, one a short-lived particle, K^0_S, and another of somewhat longer lifetime, K^0_L. For this reason, two values for the particle lifetime are given in the final column of the table.

[e]Some neutral particles are their own antiparticles. Thus, when two π^0's meet, they annihilate one another to form γ rays.

LEARNING CHECK

1. Which of the following is/are *not* fundamental forces of nature?
   a) friction
   b) gravity
   c) tension (as in a spring)
   d) strong nuclear
   e) electromagnetism

2. Particles with integer spins are called _____, while those with half-integer spins are called _____.

3. Match each item in column A with its description from column B. Each entry in column A has only one correct match from column B.

A	B
(i) baryons	(a) strongly interacting, spin 0 or 1 particles
(ii) mesons	(b) carrier particles for the fundamental forces
(iii) leptons	(c) strongly interacting, spin $\frac{1}{2}$ or $\frac{3}{2}$ particles
(iv) intermediate bosons	(d) spin $\frac{1}{2}$ particles that do not interact by the strong force

4. (True or False.) A positron has the same mass as an electron, but its charge is $+e$ instead of $-e$.

5. Interactions that have a range of only 10^{-18} m or less are most likely ones that involve
   a) the gravitational force.
   b) the electromagnetic force.
   c) the weak force.
   d) the strong force.
   e) None of the above.

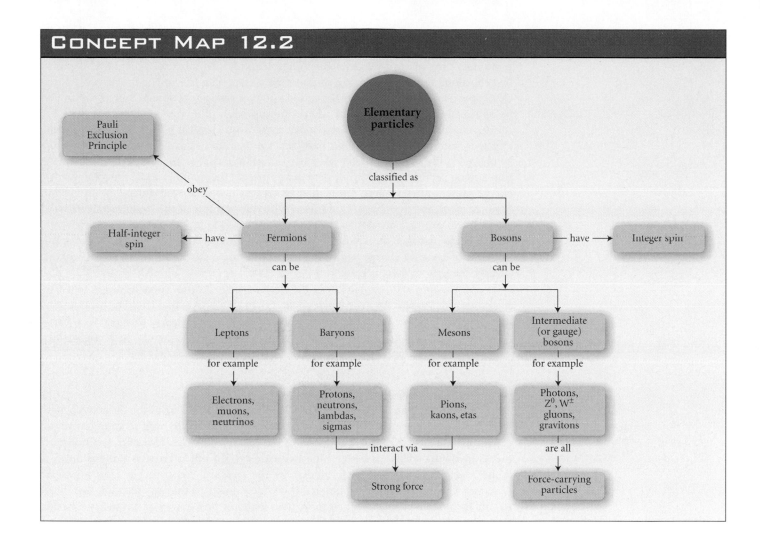

12.3 Conservation Laws, Revisited

In Chapter 3, we emphasized the importance of conservation laws in physics. We used the laws of conservation of mass, energy, linear momentum, and angular momentum to derive information about physical systems "before-and-after." Such an approach permitted us to draw some general conclusions about the systems without knowing all the details of the physics of the interaction taking place. A good example of this is the application of the principle of conservation of linear momentum during a collision. Here, information about the initial speeds of the colliding objects may be extracted from that concerning the final speeds without knowing the details of the very complicated interactions occurring in the crash itself.

In elementary particle physics, the same technique may be applied because the same physical laws are at work. When two particles collide, linear momentum must be conserved: The net momentum present before the collision must equal that after the collision. Likewise, mass-energy must be conserved: The total energy present at the start of an interaction between particles, including that stored as mass, must be the same as that present after the interaction is completed. These two laws mandate that no single isolated particle can spontaneously decay into particles whose mass exceeds its own, and that a particle cannot decay into a *single* particle lighter than itself. A high-energy photon (γ ray) may spontaneously break up into an electron and a positron, each of which has more (rest) mass than the photon, but this can happen only in the vicinity of another particle, usually an atomic nucleus, that absorbs the excess momentum to keep the total constant.

In particle physics, the law of conservation of angular momentum that we used to analyze spinning ice skaters and orbiting spacecraft in Chapter 3 sees application largely in the preservation of total spin in particle reactions. For example, when a photon turns into an electron-positron pair, spin must be conserved. This means that since the spin of the photon is 1, the spins of the departing electron and positron (each of magnitude $\frac{1}{2}$) must both lie along the same direction so that their sum equals 1 ($\frac{1}{2} + \frac{1}{2}$). Similarly, if a particle of spin zero is spontaneously converted into two particles of spin 1/2, their spins must be oppositely aligned so that they add up appropriately $\left[\frac{1}{2} + \left(-\frac{1}{2}\right) = 0\right]$.

The conservation laws discussed above come from classical physics but work equally effectively at the level of subatomic particles. Another classical conservation law also must be obeyed at the submicroscopic level: conservation of charge. In any reaction or interaction among particles, the total (or net) charge present before must equal the total (or net) charge present after. Charge can neither be created nor destroyed during an interaction between elementary particles. Continuing with our example of the spontaneous decay of a photon to produce an electron and a positron, we see that the net charge before the decay is zero and that the net charge afterward is also zero, since the electron possesses one unit of negative charge while an antielectron possesses one unit of positive charge. Clearly, the sum of these is zero.

It is interesting that the stability of the electron against decay owes its origin to a classical conservation law: All the particles lighter than an electron that could be produced by its decay are uncharged. If the electron were to undergo a change to produce such particles, a violation of conservation of charge would occur. Consequently, such conversions are forbidden in nature, and the electron is believed to be absolutely stable.

New Conservation Laws

The four conservation laws of classical physics were developed from the observed behavior of macroscopic systems, but they are found to apply equally well to submicroscopic systems, in particular to the interaction of elementary particles. However, when experiments involving collisions between particles are carried out, a curious thing is noticed: Some reactions that are *not* forbidden by the classical conservation laws are *never* observed. Currently, in particle physics, it is widely assumed that any reaction that is not strictly forbidden *will* occur, albeit sometimes with low probability or frequency. The fact that, regardless of the number of times a particular collision happens, a certain outcome is never seen has led physicists to suspect that there are *additional* conservation laws that operate *only* at the submicroscopic scale. These new conservation laws may be *ad hoc*, approximate rules, or they may be fundamental statements about nature like that of charge conservation. In either case, the recognition, statement, and use of these additional conservation laws have led to remarkable progress in particle physics, and we take time now to consider several of them. To describe these new regulations on particle reactions, new "charges" or quantum numbers (see p. 408) have had to be invented.

Conservation of Baryon and Lepton Numbers

Consider a collision between two protons. Of the many particles that can be produced in such an interaction, mesons are never seen as the *sole* products, even in cases like the one shown below, which does not violate any of the four classical conservation laws.*

$$p + p \not\rightarrow \pi^+ + \pi^+ + \pi^0$$

(You'll have to take our word for it that this reaction doesn't violate momentum or mass-energy conservation, but you can verify for yourself that it conserves charge and spin using the information in Table 12.3.) Alternatively, mesons *may* be the sole products of the collision between a proton and an antiproton. A possible reaction of this type is the following:

$$p + \bar{p} \rightarrow \pi^+ + \pi^- + \pi^0$$

Evidently, there seems to be a hidden conservation law that prevents the first reaction from happening yet allows the second.

---

* In this chapter, we dispense with the superscripts and subscripts giving the mass and atomic number, respectively, for the elementary particles as was done in Chapter 11. Thus, the proton will be designated p instead of $^1_1$p, and the neutron will be symbolized n instead of $^1_0$n. We will denote the electron as e$^-$.

Investigations of instances similar to that discussed above have led particle physicists to formulate the **law of conservation of baryon number.**

> **LAWS**
>
> **Law of Conservation of Baryon Number** In a particle interaction, the baryon number (B) must remain constant. The number of baryons going into the reaction must equal the number emerging.

Like the law of conservation of charge, the law of conservation of baryon number is a simple counting rule. Each baryon is assigned baryon number $B = +1$; each antibaryon is assigned baryon number $B = -1$. Leptons and mesons have baryon number zero. To apply the conservation of baryon number, simply add up the baryon numbers of the reactants, and compare that value to the total baryon number of the products. If the two agree, then baryon number is conserved in the reaction, and the reaction may occur (unless forbidden by other conservation laws). If the values of B before and after the reaction are not equal, baryon conservation is violated, and the reaction cannot occur.

Looking back at the proton-proton reaction given above, we see immediately why it is not observed: The total baryon number at the start is 2, while that at the end is zero—mesons have no baryon number. However, in the interaction between a proton and an antiproton, the initial baryon number is zero. This fact permits the reaction to yield only mesons under certain circumstances.

Conservation of baryon number is obeyed in all interactions—strong, weak, or electromagnetic. It has never been observed to be violated, and it dictates that baryons are created and destroyed in pairs. When a baryon is born in a reaction, an antibaryon must also be created to conserve baryon number. For this reason, the number of baryons in the universe minus the number of antibaryons is constant.

Just as conservation of charge is responsible for the stability of the electron, conservation of baryon number is the cause of the stability of the proton. The proton is the lightest baryon known. It cannot decay to any other lighter baryons, because none exist. But conservation of baryon number prevents a proton from decaying to any other type of particle, like mesons or leptons, too. Thus it appears that the proton is absolutely stable *if* baryon conservation is a fundamental law and not an approximate one. Some recent theories that attempt to unify the strong force with the electroweak force require that the proton decay spontaneously to lighter particles, in violation of baryon conservation, in an average time somewhere between 10^{30} and 10^{32} years. This result, if true, may not seem to have much relevance for you and me here and now—we are in no immediate danger of spontaneously disappearing in a burst of radioactivity due to the decay of protons in our bodies.* But the stability of the proton does have considerable impact on the fate of the universe, whose age is currently estimated to be around 10^{10} years.

A conservation law similar to that adhered to by baryons exists for leptons. Electrons, muons, and neutrinos all are created and annihilated in lepton-antilepton pairs. For example, a photon, a nonlepton with lepton number zero, decays into an electron and a positron to conserve lepton number. In beta decay, when a neutron decays into a proton and an electron, an antilepton—in this case, an antineutrino—*must* accompany the process to conserve lepton number:

$$n \rightarrow p + e^- + \bar{\nu}_e$$

There is one additional complication to the application of this lepton counting rule that does not appear when using conservation of baryon number, however. Specifically, leptons belong to *families,* and within each family, lepton number is conserved separately. Thus the six known leptons (and their anti's) divide into three families: the electron fam-

---

* Physicist Maurice Goldhaber once said that "we know in our bones that the proton's lifetime is very long if not infinite. If the lifetime were shorter than 10^{16} years, the radioactivity stemming from the decay of protons in our body would imperil our lives."

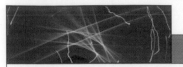
Symmetry and Conservation Laws

In quantum physics, governed as it is by probabilities, it is sometimes more important to know what absolutely cannot happen than to know anything else. In this context, conservation laws play an important role in elementary particle physics. Knowing that no process that contradicts a conservation law can occur, we infer that any process that does *not* contradict a conservation law has some probability of happening. Thus, in the world of particle physics, any reaction that is not forbidden may be assumed to take place. Having said this, let's explore an interesting connection between conservation laws and symmetry.

We generally use the word *symmetry* to express an arrangement characterized by some sort of geometrical regularity. For example, the relative positions of atoms in a crystalline solid may repeat over and over in three dimensions so that no matter where you are in the material, things always look the same. Or the locations of objects with respect to a plane may be such that those on one side of the plane appear as mirror images of those on the other (● Figure 12.10).

Mathematicians discuss symmetry in terms of the operations under which the form of the object remains unchanged. A crystal may exhibit symmetry under the operation of translation, a simple shift in position from one point to another within the crystal. The human body shows symmetry upon reflection about a vertical plane that bisects perpendicularly the shoulder line. Systems like these, which remain unchanged

upon translation or reflection (or even rotation), are said to be *invariant* under the application of these operations. Almost every conservation law in physics stems from some fundamental symmetry in nature or, alternatively, from some basic principle of invariance.

The connections between conservation laws, symmetries, and properties of invariance were formulated by one of the twentieth century's foremost mathematicians, Emmy Noether (● Figure 12.11). A resident of Germany until 1933 when she emigrated to the United States, Noether expressed the relationship between these quantities in a theorem stating that, because the laws of physics are invariant under certain symmetry operations, physical quantities related to those laws must be conserved—that is, remain constant. Let's now illustrate by example the underlying symmetries associated with some of the better-known conservation laws from classical physics.

One of the most important conservation laws is that of mass-energy. In any physical interaction, the total energy, including that bound up as mass, must stay constant: The amount of energy present before the interaction must equal the amount available after the interaction. What fundamental symmetry of nature provides the basis for this conservation law?

To answer this question, consider an analogy proposed by Heinz Pagels. Suppose the law of gravity were time dependent in such a way that on every Monday the gravitational force was a little bit weaker than it was the rest of the week. Then on Monday you could

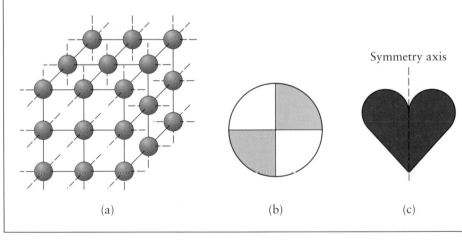

(a)   (b)   (c)

Symmetry axis

● **Figure 12.10** Examples of objects that show symmetry that is (a) translational, (b) rotational, and (c) reflective. From inside the (infinite) crystal lattice, the structure looks the same regardless of where the observer is located. The segmented circle is symmetric with respect to a 180° rotation about its center. The idealized heart shape is symmetric with respect to reflection about the dashed line, called the *symmetry axis.*

ily," consisting of the electron and its neutrino; the "tau family," containing the tau and its neutrino;* and the "muon family," with the muon and the muon neutrino (● Table 12.4). Within each family, the particles are assigned lepton number +1, and the anti's have lepton number −1. In elementary particle physics reactions, the electron, muon, and tau lepton numbers before must balance those after.

> **LAWS**
>
> **Law of Conservation of Lepton Number** In particle interactions, lepton number (L) is conserved; within each lepton family (electron, muon, and tau) the value of the lepton number at the start of a reaction must equal its value at the end.

---

* Although its existence was never in doubt, the first direct evidence of the tau neutrino was reported only in July 2000 by an international team of scientists in experiments conducted at Fermilab outside Chicago. High-energy protons were driven into a tungsten target to produce (among numerous other particles) tau neutrinos that were detected through their rare interactions with nuclei in a specially prepared photographic emulsion. Of the roughly 200 candidate interactions captured in the emulsion after 5 months of experimentation, 4 had all the characteristics to mark them as involving tau neutrinos.

● **Figure 12.11** Emmy (Amalie) Noether (1882–1935).

pump water up to a reservoir high on a hill and on Tuesday let the water run back down to, say, turn a turbine. In the process, you would get out more energy than you put in. Energy would be created! This is because the amount of work you would have had to do to pump the water upward against a weakened gravitational force on Monday would be smaller than the work done the next day by the strengthened gravitational force in driving the water downhill. This, of course, cannot happen in practice because the law of gravity is time invariant. It doesn't change from day to day or year to year. Conservation of energy, then, derives from the fact that physical laws are invariant with respect to time. Any deviation from this symmetry principle can be shown to lead to the creation of mass or energy where there was none before, in contradiction to observation.

Another of the classical conservation laws that has seen wide use throughout our discussions is the conservation of angular momentum. To appreciate the symmetry principle on which this conservation law is based, imagine a basketball spinning in empty space. We describe this motion relative to some axis initially fixed in space. But does it matter how we pick this axis? Is one axis to be preferred over another? If you think about it, the answers to these questions should be "No!" In empty space there is no "up" or "down" to provide any guidance when picking an axis about which to characterize the ball's rotation. Any axis will do just as well as another. Indeed, if we choose one axis and then rotate it by some amount, we'd expect to find the equation describing the basketball's motion to remain unchanged. Conservation of angular momentum or spin arises because empty space is isotropic: It looks the same in all directions. It is only when forces are present that certain "preferred directions" develop and angular momentum fails to remain constant. This was why we emphasized the notion of an "isolated system" when applying this conservation law to the motion of spacecraft (see Figure 3.41).

The other classical conservation laws of charge and linear momentum may be similarly interpreted in terms of symmetry principles inherent in nature. For example, conservation of linear momentum occurs because empty space is homogeneous. The invariance principles underlying many of the conservation laws used in elementary particle physics involve "internal" symmetries associated with the components of the fields used to describe the particles. Recognition of these new symmetries has permitted particle physicists to discover the existence of new conserved quantities and new conservation laws that have allowed them to better understand why some things happen in nature and others never do.

Emmy Noether was one of many women who made important contributions to nuclear and particle physics during the twentieth century. Marie Curie, Lise Meitner, and C. S. Wu (see the Physics Potpourri on page 486) also played significant roles in the development of this branch of physics. Those interested in learning more about the achievements of women physicists should check out the University of California–Los Angeles site (http://www.physics.ucla.edu/~cwp/) devoted to "Contributions of Twentieth Century Women to Physics." It offers profiles, photos, and interviews of 83 women physicists, as well as an extensive set of references for those seeking more detailed information on any of the individuals highlighted on the Web site.

Conservation of lepton number in this form explains why the following reaction is not observed, even though it does not appear to violate any other conservation laws:

$$\mu^- \not\to e^- + \gamma$$

If all leptons had the same kind of lepton number, then this reaction should be possible, since the lepton number at the start is $+1$, and that at the finish is also $+1$. However, experiments have clearly demonstrated that the properties of leptons differ according to family and that because of these differences, a reaction of this type is precluded: It violates lepton number conservation within each lepton family, since the muon lepton number on the left is $+1$, while that on the right is zero. Thus the muon cannot disintegrate as shown but may decay to an electron by the following route, which *does* satisfy all the relevant conservation laws:

$$\mu^- \to e^- + \bar{\nu}_e + \nu_\mu$$

Family	Particle Name	Symbol[b]	Antiparticle	Mass (MeV/c^2)	Lepton Number L_e	L_μ	L_τ
Electron	Electron	e^-	e^+ (positron)	0.511	+1	0	0
	Neutrino	ν_e	$\bar{\nu}_e$	$\sim 0 (\lesssim 3 \times 10^{-6})$	+1	0	0
Muon	Muon	μ^-	μ^+	105.7	0	+1	0
	Neutrino	ν_μ	$\bar{\nu}_\mu$	$\sim 0 (<0.19)$	0	+1	0
Tau	Tau	τ^-	τ^+	1,777	0	0	+1
	Neutrino	ν_τ	$\bar{\nu}_\tau$	$\sim 0 (<18.2)$	0	0	+1

[a]The spins of all the leptons, regardless of family, are $\frac{1}{2}$. For this reason, we have not separately listed this property for each particle.

[b]Superscripts to the right of the particle symbols indicate the charge carried by the particle in units of the proton charge. Thus, the electron possesses one unit of negative charge, while the antimuon carries one unit of positive charge. The neutrinos are chargeless.

Conservation of Strangeness

Beginning in the early 1950s, particle physicists began to detect new particles, which they labeled kaons (K), lambdas (Λ), and sigmas (Σ). These new members of the elementary particle zoo exhibited some very strange properties. First, they were always observed to be formed in pairs. The following two reactions are typical of the way these *strange particles*, as they came to be called, are produced:

$$\pi^- + p \rightarrow \Lambda^0 + K^0$$

$$\bar{p} + p \rightarrow K^- + K^0 + \pi^+ + \pi^0$$

Conversely, the following reaction, which did not seem to violate any then-known conservation laws, was found not to occur:

$$p + p \nrightarrow p + \Lambda^0 + \pi^+$$

Reactions like this that produced single, strange hadrons were never observed.

The second strange aspect of these particles involved their decay rates. The production of these strange species occurred with very high frequency or probability, provided enough energy was available in the collisions. This seemed to indicate that the strong interaction was their source. However, once produced, the lifetimes of these strange particles were much too long for strongly interacting entities: Unstable, strongly interacting particles typically decay to other strongly interacting particles on time scales of $\sim 10^{-23}$ seconds. The strange particles decayed to other hadrons only after enormously longer times on the order of 10^{-10} to 10^{-8} seconds. These times turn out to be more characteristic of particles that decay not by the strong interaction but by the weak one!

In 1953, Murray Gell-Mann and Kazuhiko Nishijima independently proposed that certain particles possess another type of "charge" or another quantum number termed **strangeness** (S). This quantity is conserved in strong and electromagnetic interactions, but is not conserved in weak interactions. Strangeness is a partially conserved quantity. Table 12.3 lists the strange charges for some of the better-known strange particles. Their anti's possess strangeness in equal magnitude but of opposite sign. Nonstrange particles, like protons and neutrons, have zero strangeness.

DEFINITION

Strangeness In strong and electromagnetic interactions, strangeness (S) is conserved; in weak interactions, strangeness may change by ± 1 unit. Strangeness is a partially conserved quantity.

The impact of the Gell-Mann–Nishijima theory was immediate and revolutionary. It completely solved all the puzzles presented by the strange particles. When strange particles are produced in strong interactions, the total strangeness of the products must be

● Table 12.5 Some Important Conservation Laws in Particle Physics	
Physical Quantity or Operation	**Interactions in Which Quantity Is Conserved**
Mass-energy	Strong, electromagnetic, and weak
Linear momentum	Strong, electromagnetic, and weak
Angular momentum	Strong, electromagnetic, and weak
Electric charge	Strong, electromagnetic, and weak
Baryon number	Strong, electromagnetic, and weak
Electron lepton number	Electromagnetic and weak
Muon lepton number	Electromagnetic and weak
Strangness	Strong and electromagnetic only
Parity (see Physics Potpourri on p. 486)	Strong and electromagnetic only

zero to conserve strangeness. This guarantees that strange particles will be formed in strange-antistrange pairs. (Recall this same behavior is found in strong interactions that produce baryons to conserve baryon number.) Thus in the reaction

$$\pi^- + \mathrm{p} \to \Lambda^0 + \mathrm{K}^0$$

$S = 0$ initially, because the pion and proton are nonstrange, but $S = 0$ afterward also, because the strangeness of Λ^0 is -1 while that of K^0 is $+1$. Once created, a strange particle cannot decay via the strong or electromagnetic force to other particles with no strangeness or even to particles of lower strangeness than its own, since this would constitute a violation of the conservation of strangeness. Consequently, the only decay channel left to a strange particle is the weak interaction, which, because of its lower probability of occurrence, simply takes longer to happen. In weak processes, strangeness is not conserved, and there can be a net change in the strangeness in such reactions. An example of this is shown below:

$$\begin{array}{cccccc} \Lambda^0 & \longrightarrow & \mathrm{p} & + & \pi^- & \\ S = & -1 & \text{(weak interaction)} & 0 & & 0 & (\Delta S = +1) \end{array}$$

● Table 12.5 summarizes our current knowledge of some of the conservation laws. Those listed were adequate to explain all the reactions between particles studied up to about 1970. After this date, as we shall discover in the next section, several new conservation laws and "charges" had to be invented by particle physicists to explain the observation of new species of elementary particles. The following examples demonstrate how conservation laws may be used to determine whether a given reaction can occur and to predict the identity of "missing" particles in simple reactions.

Example 12.3

Identify the conservation law(s) that would be violated in each of the following reactions: (a) $\pi^+ \to \mathrm{e}^+ + \gamma$; (b) $\pi^- + \mathrm{p} \to \mathrm{K}^0 + \mathrm{p} + \pi^0$; (c) $\Lambda^0 \to \pi^- + \pi^+$.

The conservation laws that can be checked easily are those involving charge, baryon number, lepton number, and strangeness. Let's examine each reaction in the light of these regulations.

(a) Since the photon (γ) is neutral, the decay does not violate charge conservation. None of the reactants is a strange particle or a baryon, so conservation laws involving these quantum numbers are irrelevant. But the reaction does violate conservation of lepton number.

Because mesons have lepton number 0, as do photons, lepton number is not conserved: $L_e = 0$ initially and $L_e = -1$ finally. (Remember, an antielectron has a lepton number opposite that of an electron.)

(b) This interaction clearly violates charge conservation because the total charge at the start of the reaction is zero, but the net charge at the finish is $+1$, the K^0 and π^0 being neutral.

Because none of the particles in this reaction are leptons, lepton number conservation need not be considered. Although baryon number is conserved in this interaction (mesons have $B = 0$ while protons have $B = +1$), strangeness is not. Protons are strongly

Does Nature Distinguish Left from Right?

In an earlier Physics Potpourri (page 482), it was argued that, underlying every conservation law in physics, a symmetry principle exists. One important, but not yet mentioned, conservation law in elementary particle physics, first carefully studied by Eugene Wigner in 1927, is the *conservation of parity*. Expressed in a manner that emphasizes the controlling symmetry, the law states that for every reaction or interaction observed to occur, the spatially inverted reaction or interaction is also equally likely to occur. In this context, *spatial inversion* is a mathematical operation that changes (inverts) the direction of the coordinates used to describe the reaction/interaction: $x \rightarrow -x$, $y \rightarrow -y$, and $z \rightarrow -z$. To see the effects of this operation as applied to particle interaction, you must imagine that you are standing on your head and watching the reaction going on in a mirror (● Figure 12.12). The physical quantity that is conserved (remains constant) in such interactions is the *chirality* or handedness of the system. According to this principle, nature is ambidextrous. Anything that can be accomplished in a left-handed way can be achieved just as easily and effectively in a right-handed way. To deny the conservation of parity would be to allege that nature somehow differentiates between left and right.

● **Figure 12.13** Chien-Shiung Wu (1912–1997) in her laboratory at Columbia University in 1963.

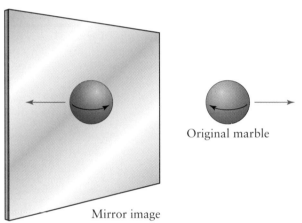

Original marble

Mirror image

● **Figure 12.12** Illustration of the parity (or spatial inversion) operation. The marble moves to the right and spins clockwise (has spin-down). Its spatially inverted partner (or mirror twin) travels to the left and rotates counterclockwise (spin-up).

By the 1950s, parity conservation had been confirmed in many experiments involving the strong and electromagnetic interactions. But by the middle of the decade, evidence began to accumulate that parity was not conserved in all reactions. In 1956, two young theorists, Tsung Lee and Chen Yang, carefully analyzed all the available data and concluded that parity might not be conserved in weak interactions. They even went so far as to suggest some experiments that could be done to test their hypotheses. One such investigation, carried out by Chien-Shiung Wu (● Figure 12.13), involved beta decay in radioactive cobalt-60.

A sample of cobalt-60 was cooled to a very low temperature and placed in a strong magnetic field. In the absence of thermal agitation, the external magnetic field aligned the magnetic fields associated with the spinning nucleons, producing an effect much like the alignment found in ferromagnetic materials used to make bar magnets (see Section 8.1). With this "preferred" direction established in the sample, Wu and her collaborators checked the directions in which the electrons were emitted during the beta decay of the cobalt nuclei. What

interacting particles, so this represents a strong interaction that must conserve strangeness. The π's and the p's are nonstrange particles (see Table 12.3); the K^0 has one unit of strangeness. Before the reaction $S = 0$; afterward, $S = +1$. Strangeness is not constant in this *strong* interaction as it must be. (Put another way, since strange particles must be produced in pairs, this reaction obviously violates conservation of strangeness as there is only one strange particle (the K^0) created in this interaction.)

(c) In this decay, a baryon ($B = +1$ for Λ^0) is converted into two mesons ($B = 0$ for $\pi^{\pm}$). This is forbidden by conservation of baryon number. Again, because this reaction involves no leptons, conservation of lepton number is irrelevant. However, charge obviously is conserved: $Q_{initial} = 0$ (the Λ^0 is uncharged); $Q_{final} = -1 + 1 = 0$.

This reaction also *would not* violate conservation of strangeness because the decay of the strange particle Λ^0 occurs by the *weak* interaction for which S may change by one unit, as it does here: $S_{initial} = -1$; $S_{after} = 0$ (since the mesons are nonstrange); and $\Delta S = +1$, which is permitted in weak interactions (see Table 12.5).

In this chapter, we use Q to represent charge, reserving the symbol q for generic reference to quarks beginning in Section 12.4. This notation differs from what we adopted in Chapter 7 and should not be confused with the symbol for *heat* used in Chapter 5.

they found astonished everyone. Most of the electrons were emitted in a direction *opposite* to that of the aligning magnetic field. This finding was a clear violation of spatial-inversion invariance because the "mirror image" of this reaction required that most of the particles be emitted *along* the direction of the applied magnetic field in contradiction to what was actually observed (see ● Figure 12.14.) Thus, parity is *not* conserved in the weak interaction, and the weak force *does* distinguish left from right: That is, it distinguishes a reaction from its spatially inverted twin. For their pioneering work in this area, Lee and Yang received the 1957 Nobel Prize in physics.

Since that time, other elementary particle experiments have provided additional evidence that parity is not conserved in the weak interaction and that nature truly does make a distinction between left and right. (For example, all neutrinos are left-handed, their spins always directed opposite to their momenta; no right-handed neutrinos have been observed.) The effect has now been seen in systems with atomic and molecular dimensions. For example, atoms are "right-handed" due to the weak force's effect on the electrons orbiting the nucleus, and otherwise identical mirror-twin amino acids can now be distinguished on the basis of small energy differences brought about by the weak interaction. It is now evident that the world is chirally asymmetric on all scales, from elementary particles on up.

What if nature didn't distinguish left from right? What if parity *were* strictly conserved? Then there'd be no fundamental physical difference between left and right, only one based on human convention. If this were true, ask yourself how you would be able to communicate the distinction between left and right to inhabitants of some alien (non-human) civilization light-years away. The question turns out to be harder to answer than you might think, and if you're curious to learn more, check out Martin Gardner's book *The New Ambidextrous Universe* (3rd revised edition, W. H. Freeman and Co., New York, 1990). In it, Gardner provides an excellent discussion of symmetry principles and their connection to conservation laws. He also addresses the right-left problem in connection with his consideration of Project OZMA, one of the early attempts to make contact with extraterrestrials by radio means. Enjoy!

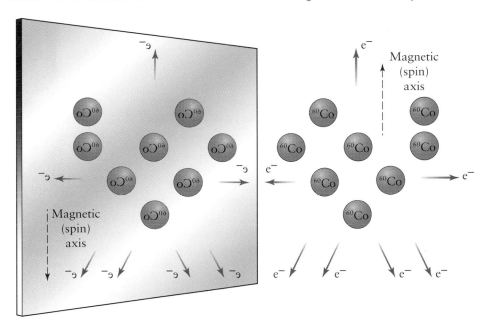

● **Figure 12.14** Illustration of the violation of conservation of parity in the decay of cobalt-60. Magnetically aligned cobalt nuclei emit electrons predominantly in a direction opposite to that associated with their magnetic field. In the mirror-image experiment, most of the electrons are ejected in the direction of the (reversed) magnetic field.

Example 12.4

Each of the reactions below is missing a *single* particle. Figure out what it must be if these interactions are permitted.

(a) $p + \bar{p} \rightarrow n + $ _____

(b) $\bar{\nu}_\mu + p \rightarrow n + $ _____

(c) $p + p \rightarrow p + \Lambda^0 + $ _____

(a) Here we have a proton encountering an antiproton and annihilating. The annihilation energy is used to create two new particles. Since the initial charge is zero (the antiproton has one unit of *negative* charge), the missing particle must be neutral like the neutron. But the starting baryon number is also zero ($B = +1$ for p and $B = -1$ for $\bar{p}$). Thus, the missing particle must be a baryon with $B = -1$ to cancel the baryon number ($+1$) of the neutron.

As this is a strong interaction, strangeness must be conserved. Since $S = 0$ at the beginning and n itself is nonstrange (see Table 12.3), the absent particle must also have strangeness zero. The only particle satisfying *all* these criteria is $\bar{n}$, the antineutron.

(b) The key conservation law to use to identify the missing particle is conservation of muon lepton number. The $\bar{\nu}_\mu$ is a muon antineutrino with $L_\mu = -1$. The p and the n are clearly baryons with $L = 0$. The absent species must belong to the muon family and have $L_\mu = -1$.

To conserve charge, the missing lepton must also possess one unit of positive charge to balance the initial charge of the proton. (Remember, neutrinos have no charge.) The only particle that meets these requirements is the μ^+, an antimuon.

(c) To fill in the missing entity in this reaction, we rely upon conservation of strangeness. This is a strong interaction, so we know strangeness must remain constant: $S_{initial} = 0$; therefore S_{final} must be zero as well. As it stands, the sum of the strangeness of the products, *excluding* the missing particle is $0 + (-1) = -1$. The particle that is lacking *must* have $S = +1$.

Conservation of charge demands that the missing particle have one unit of positive charge because $Q_{initial} = +2$, while $Q_{final} = +1$ *without* the additional contribution of the missing particle.

The missing particle *cannot* be a baryon, however, because adding another baryon to the products destroys the equality that already exists in baryon number. We are obviously searching for a singly charged, positive meson having one unit of positive strangeness. Examining Table 12.3 shows that we need a K^+, a kaon.

LEARNING CHECK

1. In interactions taking place by the strong force, which of the following quantities are conserved? Which are conserved in weak interactions?
   a) electric charge
   b) baryon number
   c) mass-energy
   d) strangeness
   e) linear momentum
   f) angular momentum (spin)

2. (True or False.) When strange particles are produced in weak interactions, they always come in pairs with equal and opposite strangeness, S.

3. The reaction $p \rightarrow e^+ + \gamma$ is
   a) possible because it conserves everything.
   b) impossible because it does not conserve strangeness.

   c) impossible because it does not conserve lepton number.
   d) impossible because it does not conserve charge.
   e) None of the above.

4. The six leptons are organized into three _____: electron, muon, and tau.

5. (True or False.) The electron is believed to be absolutely stable because all the known lighter particles to which it might decay are uncharged, and the conversion of an electron to such particles would violate conservation of charge.

ANSWERS: 1. strong force: (a), (b), (c), (d), (e), (f); weak force: (a), (b), (c), (e), (f) **2.** False **3.** (c) **4.** families **5.** True

12.4 Quarks: Order out of Chaos*

The rapid proliferation of subatomic particles during the 1960s and early 1970s caused many physicists to wonder if "elementary particles" were really so "elementary" after all. Maybe they themselves were composites of still smaller entities, the "*really* elementary" particles. But even before this, in 1956, a Japanese physicist named Shoichi Sakata had proposed a model in which all hadrons were made out of just six: the proton, the neutron, the lambda, and their anti's. Although there were some peculiar aspects to this

---

* As used here, the word *chaos* is synonymous with confusion: It does not refer to that branch of dynamics wherein experimental outcomes become highly unpredictable due to their extreme sensitivity to the precision of the starting parameters.

model concerning its predictions for the masses and binding energies of some hadrons, it did a pretty good job of accounting for the properties of the then-known baryons and mesons, and it violated no then-recognized laws of physics.

In 1961, Murray Gell-Mann and Yuval Ne'eman independently gave a new way to classify hadrons. They introduced an additional property of subatomic particles called *unitary spin*. This quantum characteristic is conserved in strong interactions and has eight components, each of which is a combination of the quantum numbers we have seen before (plus a few others that are too esoteric for us to consider at this level). Because of the eight-part structure of the basic "charge" in this theory and the theory's potential for leading to a deeper understanding of nature, Gell-Mann christened this the "eightfold way," in analogy with "the noble eightfold way" of Buddhism that leads to nirvana.

Many of the particles listed in Table 12.3 had not yet been discovered at the time these two competing theories for the composition of hadrons were proposed. In the best tradition of the scientific method, a test was soon devised to distinguish which of the two, if either, gave the better description of nature. Before 1963, the spin of the sigma particles was not known. The Sakata model predicted that these particles should have spin $\frac{3}{2}$. The eightfold way predicted a spin of only $\frac{1}{2}$. When the spins of the Σ^0 and $\Sigma^\pm$ were finally measured, they were found to be $\frac{1}{2}$, in agreement with the eightfold way.

Additional support for the Gell-Mann–Ne'eman model came a little later, with the discovery of the omega minus, Ω^-. Based on the eightfold way, the existence of a single, massive particle with charge -1, strangeness 3, and spin $\frac{3}{2}$, which had never been seen before, was predicted. Particle physicists began to search for this species in new, higher-energy collision experiments, and in February 1964, the successful detection of a particle having all the predicted characteristics was announced by scientists at the Brookhaven National Laboratory. ● Figures 12.15 and ● 12.16 show the original photograph of the particle tracks in the discovery experiment and the analyses that led to their interpretation in terms of the Ω^-.

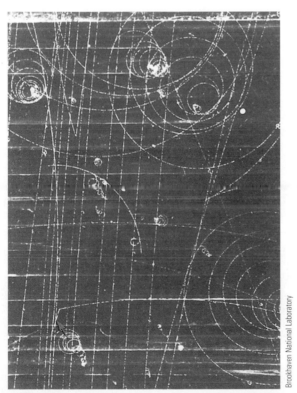

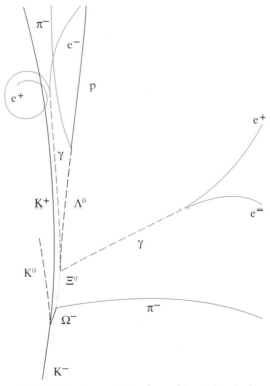

● **Figure 12.15** The first photograph of the Ω^- particle, taken at the Brookhaven National Laboratory in 1964. The path of the Ω^- is marked with an arrow at the lower left of the picture.

Brookhaven National Laboratory

● **Figure 12.16** A schematic reconstruction of some of the particle tracks shown in Figure 12.15. Solid lines indicate the trajectories of charged particles, while dashed lines give the paths of neutral particles (which do not show on the original photograph). The formation and decay of the Ω^- involve the following reactions: (1) $K^- + p \rightarrow \Omega^- + K^+ + K^0$; (2) $\Omega^- \rightarrow \Xi^0 + \pi^-$.

• **Table 12.6** Some Properties of the Originally Proposed Quarks[a]

Quark	Electric Charge (Q)[b]	Baryon Number (B)	Strangeness (S)
u	$+\frac{2}{3}$	$+\frac{1}{3}$	0
d	$-\frac{1}{3}$	$+\frac{1}{3}$	0
s	$-\frac{1}{3}$	$+\frac{1}{3}$	-1
$\bar{u}$	$-\frac{2}{3}$	$-\frac{1}{3}$	0
$\bar{d}$	$+\frac{1}{3}$	$-\frac{1}{3}$	0
$\bar{s}$	$+\frac{1}{3}$	$-\frac{1}{3}$	$+1$

[a]The quarks are all spin $\frac{1}{2}$ particles.

[b]These values are in units of the proton charge. Thus the charge of an up quark in the international system of units would be $(\frac{2}{3})(1.6 \times 10^{-19}\,\text{C}) = 1.07 \times 10^{-19}\,\text{C}$.

Quarks

Further investigation of the implications of the eightfold way led to a refinement of the theory in 1964. At that time Gell-Mann and George Zweig postulated that all hadrons were formed from three fundamental particles, which Gell-Mann called **quarks,** and their anti's. The quarks were designated u (for "up"), d (for "down"), and s (for "strange"). • Table 12.6 gives the properties of these three particles and their anti's. Notice that all the quarks have spin $\frac{1}{2}$, baryon number $\pm\frac{1}{3}$, and charge $\pm\frac{1}{3}$ or $\pm\frac{2}{3}$ (in units of the proton charge). These are fractionally charged particles, unlike anything we have dealt with before!

Surprising as this may appear, the introduction of these noninteger charged particles enabled physicists to describe perfectly all the hadrons discovered prior to about 1970 and, with some extensions, all the heavy particles found since. The two rules governing the formation of hadrons from quarks are simple:*

1. Mesons are composed of quark-antiquark pairs, like $u\bar{d}$ and $d\bar{s}$.
2. Baryons are constructed out of three-quark combinations; antibaryons are made up of three antiquarks.

For example, a π^+ meson is equivalent to a $u\bar{d}$ pair with their spins oppositely directed. This combination gives a particle of spin 0, baryon number 0 ($= \frac{1}{3} - \frac{1}{3}$), and charge $+1$ ($= \frac{2}{3} + \frac{1}{3}$; see Table 12.6) as required. A K^0 meson may be shown to consist of a combination of a d quark and an $\bar{s}$ quark.

Similarly, a proton is a collection of uud quarks, while the neutron is a udd quark combination. (You should take a few minutes to assure yourself that the addition of quarks as indicated gives the usual properties of the proton and the neutron that you are familiar with. The data in Tables 12.3 and 12.6 will be helpful in this regard.) Other three-quark arrangements give other baryons: $\Sigma^+ = $ uus, $\Lambda^0 = $ uds, and $\Omega^- = $ sss.

Example 12.5

To what hadron does the combination of a d and a $\bar{u}$ quark correspond? Assume the spins of the quarks are antiparallel.

A quark-antiquark pair gives a meson, so that is the type of hadron we're looking for. Now a d quark has charge $-\frac{1}{3}$ and a $\bar{u}$ quark has charge $-\frac{2}{3}$. The total charge of our mystery particle is thus -1. Since neither the d nor the $\bar{u}$ quark is strange, their combination must yield a nonstrange particle. If the spins of the d and $\bar{u}$ quarks are oppositely aligned, the net spin will be zero for the meson. Evidently, we seek a spin zero meson with strangeness 0 and charge -1. Consulting Table 12.3, we find that the particle in question is the π^- meson.

---

* In 2003, evidence of a new kind of elementary particle—an *exotic baryon*—consisting of five quarks was reported by four experimental groups in Japan, the United States, Germany, and Russia. The Θ^+, as the new particle is now generally called, has a mass of 1540 MeV/c^2, and is believed to be composed of two ud quark pairs plus an $\bar{s}$ quark. Theorists, extending the original model of Gell-Mann and Zweig, had predicted the existence of such exotic hadrons more than five years ago, and experimental searches are now under way to discover other "pentaquark" combinations to shed more light on baryon structure.

Example 12.6

Give the quark combination associated with the xi minus, Ξ^-, baryon.

Baryons are composed of three quarks. The Ξ^- has charge -1, strangeness -2, and spin $\frac{1}{2}$. Because the quarks each are spin $\frac{1}{2}$ particles, we see immediately that the spins of two of the quarks making up the Ξ^- must be paired off (spin-up plus spin-down) so that the net spin is only $\frac{1}{2}$. Since the Ξ^- is a strange baryon, it must contain s quarks; they are the only quarks that carry this property. Given that an s quark possesses -1 unit of strangeness, the Ξ^- must contain two of these quarks to have net strangeness -2. If the Ξ^- has two s's, then together they contribute a total charge of $-\frac{2}{3}$ ($= -\frac{1}{3} - \frac{1}{3}$). To make up the additional $-\frac{1}{3}$ charge needed to obtain the total charge of -1 for the Ξ^- requires a nonstrange quark of charge $-\frac{1}{3}$. The only candidate is the d quark. Thus a Ξ^- is a dss quark combination.

It is important to emphasize at this point that the quark model *applies only* to hadrons. Leptons, like the electron, the muon, and the neutrinos, are *not* made of quarks. Indeed, the status of the leptons and the quarks is much the same within the realm of particle physics: Both are groups consisting of fundamental, irreducible spin $\frac{1}{2}$ fermions that together make up the matter in the universe.

The quark model has several distinct advantages over any competing theories of subatomic particles. First, it explains why mesons all have integral spins—that is, why mesons are bosons. Because mesons are two-quark combinations, the mesons, as a class, can only have total spin 0 (when the spins of the two constituent quarks point in opposite directions) or 1 (when the two spins are aligned). Second, the model also accounts for the fact that baryons all are fermions, half-integral spin particles: Any arrangement of three quarks will always yield either a spin $\frac{1}{2}$ or a spin $\frac{3}{2}$ particle, depending on whether the spins of two of the three are paired (spin $\frac{1}{2}$) or whether all three spins are parallel (spin $\frac{3}{2}$). By the same token, the quark model permits a new interpretation of strangeness and its conservation. Strangeness is just the difference between the number of strange antiquarks and the number of strange quarks making up the particle. Conservation of strangeness in strong interactions may now be seen as the prohibition of the conversion of an s quark to a d or a u quark during a reaction.

Given the abundant success of the quark model, it was not too long before particle physicists began looking for evidence of the existence of quarks. Despite many careful searches among many different types of particle interactions, no free quarks have ever been found.* According to prevailing theories of quark confinement, quarks are inescapably bound within hadrons by what is called the *color force*. Within each hadron, the color force is mediated by exchange particles called *gluons,* and the "strings" that bind the individual quarks comprising the hadrons are called *gluon tubes.* (See Section 12.5 for more on gluons and the color force.) Thus, the strong force that exists between hadrons is now revealed to be just a shadow of the "*really* strong force" (the color force) that binds the quarks. But, if all this is true, we are still left with our original question: What evidence is there for the existence of quarks?

The experimental evidence for quarks comes from two sources: first, the scattering of high-energy electrons off protons and, second, the observation of jets of hadrons coming from collisions between electrons and positrons or protons and antiprotons. In the first instance, a case for protons being composed of smaller point particles has been made on much the same basis as that used by Rutherford to argue for the existence of the nucleus (see Physics Family Album for Chapter 11). If the proton were a uniform spherical distribution of positive charge, then the deflection suffered by a high-speed electron penetrating the interior of a proton would be dependent in large part on how much charge the electron "saw" as it passed through the proton. On the other hand, if the proton were

---

* Tantalizing evidence for the existence of free quarks in what is called a *quark-gluon plasma* (QGP) is beginning to emerge, however, from the Brookhaven National Laboratory's Relativistic Heavy Ion Collider (RHIC) experiments conducted in 2003. By colliding opposing beams of high-speed gold ions, scientists have succeeded in creating, for a brief instant, conditions like those that existed in the universe less than a second after the Big Bang, when the temperature was about 2×10^{12} K. In such circumstances, the protons and neutrons in the gold nuclei "melt," releasing the quarks and gluons that compose them to produce a tiny sample of "soup"—the QGP. Like bound quarks, evidence for a QGP comes from studying the particle jets that emerge from the head-on collisions of the gold ions, and further experimentation will be necessary to confirm these early results.

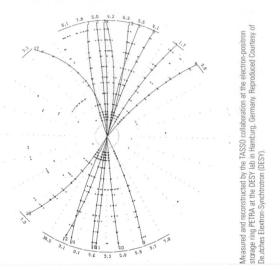

Figure 12.17 Two jets of particles (mostly pions) emerge from the collision and annihilation of electrons and positrons in this computer reconstruction of an experiment done in Hamburg, Germany, with an e⁻-e⁺ collider. The focused nature of these jets and the fact that they contain mostly pions suggest that each one developed from a single precursor, a quark or an antiquark, instead of directly from the annihilation energy of the e⁻-e⁺ pairs.

Measured and reconstructed by the TASSO collaboration at the electron-positron storage ring PETRA at the DESY lab in Hamburg, Germany. Reproduced Courtesy of Deutsches Elektron-Synchrotron (DESY).

made up of three small, fractionally charged quarks, the deflection of the incoming electron would be slight, unless it happened to hit one of the quarks "head on." Then the electrical force between the electron and the quark would be substantial, and the electron might be scattered through a very large angle. This would seldom happen, of course, since the proton in the quark model is mostly empty space. And what is observed? Just what is expected from the quark picture! In particular, the number of electrons deflected by large amounts is in good agreement with the predictions of the quark model.

A second type of evidence for quarks involves the products of high-energy collisions of beams of electrons and positrons. In these reactions, the energy derived from the e⁻-e⁺ annihilation goes into the creation of q-q̄ pairs. As these particles move off in opposite directions (as required to conserve linear momentum), they produce a stream of new q-q̄ pairs. The quarks in this bidirectional jet quickly combine to give various hadrons, so that at the level of the experiment, what is seen is two trains of heavy particles, oriented at 180°, traveling away from one another (● Figure 12.17). Similar jets of hadrons are observed in high-energy collisions between protons and antiprotons. Coupled with the data from the electron-scattering experiments described in the previous paragraph, these observations provide persuasive evidence for the overall correctness of the quark model and the existence of these elusive particles.

The quark model also provides a handy way of understanding and analyzing decays and reactions in particle physics. Only two rules need to be remembered:

1. Quark-antiquark pairs can be created from energy in the form of gamma rays or collisional kinetic energy.
2. The weak interaction alone can change one type of quark into another. The strong and electromagnetic interactions cannot cause such changes. Thus we see quark types are conserved by the strong and electromagnetic forces but not by the weak force.

Example 12.7

Analyze the decay n → p + e⁻ + ν̄$_e$ in terms of the quark model.

The beta decay of a neutron may be rewritten in terms of the quarks making up the various hadrons present in the reaction as follows. (*Remember:* The electron and its antineutrino, being leptons, are not composed of quarks.)

$$udd \rightarrow uud$$

Canceling a u and a d quark on both sides of the arrow yields:

$$d \rightarrow u$$

The fundamental decay is the conversion of a d quark to a u quark by the weak force. A W⁻ particle is created in the process from the mass deficit between the d and u quarks; this carrier particle subsequently decays into an electron and an antineutrino (see ● Figure 12.18).

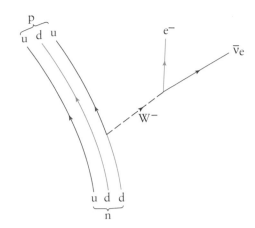

● **Figure 12.18** An alternative representation of the beta decay of a neutron emphasizing the quark content of the neutron and the proton. The weak interaction transforms a d quark into a u quark, thereby converting the original neutron (udd) into a proton (udu). Compare this description of the reaction to that depicted in Figure 12.6a.

Explore It Yourself 12.1

Want to build your own baryons? Play particle pinball? Find out what life might be like without some of the fundamental forces? All this and more awaits the curious student in the Fermilabyrinth at the online Lederman Science Center (http://www-ed.fnal.gov/ed_lsc.html). Pay a visit to this site and play some of the many educational games provided to learn more about elementary particle physics and especially about the detection and analysis techniques used to probe the details of particle-collision reactions.

Explore It Yourself 12.2

Probing the hidden structure of a system by bombarding it with small but highly energetic particles is nothing new. After all, this is how we learn about the condition of teeth and bones from a medical or dental "x ray" (see *X Rays* in Section 8.5). Rutherford used this technique to explore the structure of the atom and to estimate the size of the nucleus (see Physics Family Album on page 459). Particle physicists have used similar means to discern the existence of pointlike quarks within hardrons. Let's try a scattering experiment of our own to illustrate the principles employed in such investigations and to "discover" the size of two-dimensional circular "atoms" in a rectangular target.

Using a photocopier, make an enlargement at least twice the size of ● Figure 12.19, including the rectangular boundary enclosing the circles (the atoms). Attach a piece of carbon paper, carbon side down, to the pattern so that it is completely covered, and place the two sheets on a smooth, flat, hard surface. A tabletop or a tiled floor will do. Drop a marble (or other small hard sphere) onto the paper stack from a height of 30 to 60 cm, *making sure to catch the marble on the rebound so that it only strikes the sheets once.* Repeat this process at least 100 times, covering the entire area of the pattern as completely as possible.

Remove the carbon paper from the target sheet and count the total number of marble "hits" (indicated by black carbon dots) that lie within the rectangular boundary defining the target. Next, count only the "hits" that lie wholly within the circles in the target. Using a millimeter ruler, measure the length and width of the target and compute the total target area. Finally, count the number of circles present in the target.

If the circles/atoms are all uniform and the "hits" randomly distributed, then we expect the ratio of the area of all the circles to the total area of the target to be equal to the ratio of the number of hits within the circles to the total number of hits on the target. Compute the ratio of hits from your data. Find the area of all the circles by multiplying this result by the total target area you determined from your millimeter measurements. Dividing by the number of circles yields the area of a single circle/atom in the target. Because the area of a circle is π times the radius of the circle squared, we are now in a position to compute the radius, r, of one of the circles/atoms in the target.

What do your measurements give for this value? How does this indirectly determined value of r compare to what you get from a direct measurement of the radius of one of the circles using a ruler?

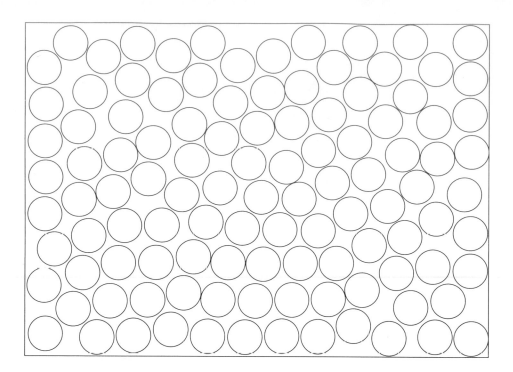

LEARNING CHECK

1. Three quarks combine to form
   a) a baryon.
   b) an electron.
   c) a meson.
   d) a photon.
2. The _____ interaction alone can change one type of quark into another.
3. (True or False.) Quarks are elementary particles possessing electric charge equal to $\frac{1}{3}$ or $\frac{2}{3}$ the magnitude of the charge on a proton.

4. Mesons are
   a) always composed of either two u quarks or two d quarks.
   b) always composed of two different types of quarks.
   c) always composed of a quark and an anti-quark.
   d) always composed of a quark and a lepton.
   e) None of the above.
5. _____ are the carriers of the color force that binds the quarks inside hadrons.

ANSWERS: 1. (a)   2. weak   3. True   4. (c)   5. Gluons

12.5 The Standard Model and GUTs

Quarks are fermions. They have half-integral spins, and they should therefore obey the Pauli exclusion principle. But if this is true, how do we explain the existence of the Ω^-? This particle is composed of three strange quarks, all of which share the same mass, spin, charge, and so on—that is, the same quantum numbers. It would appear that the Ω^- is made up of three identical, interacting s quarks. Doesn't this violate the exclusion principle? How can we reconcile these circumstances?

One way would be to argue that quarks are somehow different from other fermions and, consequently, don't have to conform to the Pauli principle. When particles like the Ω^- were discovered, some theoretical physicists did suggest this explanation to resolve the dilemma. However, other scientists were reluctant to make exceptions to the exclusion principle and instead proposed that quarks carry an additional property that makes them distinguishable within hadrons like the Ω^-. For this scheme to work, this new characteristic had to come in three varieties to permit, for example, the three s quarks in the Ω^- to be different from each other. The name that particle theorists gave to this new quantum property or number was *color*, although it has absolutely nothing to do with what we commonly refer to as color—that is, our subjective perception of certain frequencies in the electromagnetic spectrum. The three quark colors were labeled after the three primary colors of the artist's palette: red, blue, and green. Antiquarks are colored antired, antiblue, and antigreen (● Figure 12.20).

The fact that color is not an observed property of hadrons indicates that these particles are "colorless." If they are composed of colored quarks, then the way the colors come together within the hadron must be such as to produce something with no net color, something that is color neutral. For this to be true, the three quarks that make up baryons must each possess a color different from their companions, one red, one blue, and one green. The addition of the primary colors produces the result we call "white," so baryons containing three different colored quarks are considered to be white or neutral as concerns the color charge. In an analogous manner, for mesons to be color-neutral requires them to be made up of a quark of one color and an antiquark possessing the corresponding anticolor. Returning to our previous example of the π^+ meson, we see in the light of this new color physics that if the u quark is red, the $\bar{\text{d}}$ quark must be antired.

You have probably noticed that the language of particle physics is rather whimsical in comparison with the other areas of physics that we have studied. To carry this whimsy one step further, we note that often the different types of quarks, u, d, s (and others that we will shortly introduce), are designated quark *flavors*. Quarks come in six flavors (not counting the antiflavors), and each flavor comes in three colors. Complicated as this may seem to you now, its still less troublesome than making a choice from among 31 varieties at a Baskin-Robbins® ice cream store.

Color is another example of an internal quantum number. It is not directly observed in hadrons, and it cannot be used to classify them. Moreover, it does not influence the interactions among hadrons. So what is the significance of the color charge? What possible role does it play in particle physics?

Current theories now suggest that the color charge is the source of the interquark force, much as ordinary electric charge is the source of the electrical force. And just as the electrical force between charged particles is carried by the photon, the color force between the bound quarks is carried by what are called *gluons*. Gluons are electrically neutral, zero mass, spin 1 particles that are exchanged between the bound quarks in hadrons. There are eight gluons in all, and the emission or absorption of a gluon by a quark can lead to a change in the color of the quark. Such changes occur very rapidly and continuously among the quarks forming a given hadron, but always in such a way that the overall color of the hadron remains neutral. The decay of a single meson ($q\bar{q}$) into two may also be interpreted in terms of the emission of a gluon by one quark bound in the original meson. The radiated gluon is here quickly converted into matter and antimatter—namely, a new quark and antiquark, which subsequently bind to the existing pair to form two mesons. Unlike virtual photons, which mediate the electrical interaction but carry no electric charge themselves, gluons, the carriers of the color force, are themselves colored, that is, they possess color charge. This fact permits the gluons to interact among themselves by the attractive color force and form what are known as *glueballs*.

The color force is sometimes referred to as the *chromodynamic force*, from the Greek word *chroma*, meaning "color." The theory of color and its connection to gluons is called *quantum chromodynamics*. As is the case for the quarks, evidence for the existence of gluons may be found in high-energy particle-collision experiments, particularly ones in which three or four hadron jets are seen (● Figure 12.21). Data to support the existence of glueballs are much less compelling, although some physicists believe that the iota (ι) particle, produced in the decay of the ψ (see the following subsection), may represent a candidate for a glueball. Other glueball candidates include the short-lived f_0 (1500) and f_J (1710) particles (the numbers in parentheses refer to the mass of the particle in MeV/c^2) produced at CERN, although again the evidence is inconclusive. At the GSI in Darmstadt, Germany (page 137), new experiments are being planned involving proton-antiproton beam collisions that hold great promise for providing more definitive data on the existence and properties of glueballs.

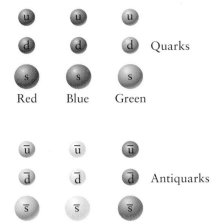

u	u	u	
d	d	d	Quarks
s	s	s	
Red	Blue	Green	

$\bar{\text{u}}$	$\bar{\text{u}}$	$\bar{\text{u}}$	
$\bar{\text{d}}$	$\bar{\text{d}}$	$\bar{\text{d}}$	Antiquarks
$\bar{\text{s}}$	$\bar{\text{s}}$	$\bar{\text{s}}$	
Antired (Cyan)	Antiblue (Yellow)	Antigreen (Magenta)	

● **Figure 12.20** Each quark (and antiquark) comes in three colors (or anticolors).

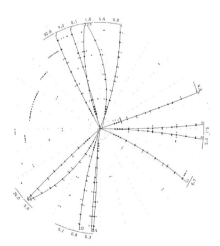

● **Figure 12.21** Electron-positron annihilation sometimes gives rise to three jet events. Such events are thought to be due to the fragmentation into ordinary hadrons of a quark and an antiquark, together with a gluon. This image was produced with the same collider used in Figure 12.17.

Measured and reconstructed by the TASSO collaboration at the electron-positron storage ring PETRA at the DESY lab in Hamburg, Germany. Reproduced Courtesy of Deutches Elextron-Synchrotron (DESY).

Charm, Truth, and Beauty

Recall the symmetries mentioned earlier as regards the leptons and the quarks (page 491). We have repeatedly seen how symmetry principles have guided particle physicists through the maze of possibilities associated with high-energy collision reactions. They also provided guidance to theoreticians attempting to answer the question: How many quarks are there in nature? In the early 1960s, three quarks were enough to account for the known hadrons, but at the time, four different leptons were known: the electron and its neutrino and the muon and its neutrino. (The tau was not discovered until 1975.) Because of this asymmetry in the numbers of quarks and leptons, it was suggested that there should exist a fourth quark to balance out the leptons. Other, "harder" evidence for the existence of a fourth quark came in the form of certain rare, weak interactions involving hadrons in which their charges remained constant but their strangeness changed. Together, these bits of data convinced particle physicists that another quark, the c quark, was present in nature and that it carried a new quantum "charge" called *charm*. Theory indicated that the charmed quark should have electric charge $+\frac{2}{3}$, strangeness 0, and a mass greater than any of the previously identified quarks.

The search for particles containing the c quark—so-called charmed particles—began in earnest in 1974 at Brookhaven and at the Stanford Linear Accelerator Center (SLAC) in California. In November 1974, Samuel Ting, leader of the Brookhaven group, and Burton Richter, head of the SLAC team, jointly reported the discovery of a short-lived particle produced in e^--e^+ annihilation events that seemed to "fit the bill." The particle, called J by Ting's group and ψ (psi) by Richter's, is now believed to be a $c\bar{c}$ combination—one state of several belonging to what has been referred to as "charmonium"—having a mass of 3.1 GeV/c^2. In 1976, Ting and Richter shared the Nobel Prize in physics for their work.

The existence of charmonium clearly indicated the presence of a new quark and its anti, but to prove beyond doubt that this special type of quark existed required the identification of a particle, say, a meson, composed of one charmed quark (or its anti) plus another, different quark, like a u or a $\bar{d}$. This was necessary so that the property of charm could be explicitly manifested and not cancel out as it does in the ψ. After two years of searching, in 1976, a charmed meson, labeled D^0, was found; it appears to be made up of a $c\bar{u}$ quark combination. By the end of the 1970s, other mesons possessing both charm and strangeness were discovered, and even a charmed baryon—the Λ_c = udc! The existence of charm in the universe was secure.

But even more surprises were in store for particle physicists. In 1977, in experiments conducted at Fermilab outside Chicago, a very heavy meson, called the Υ (upsilon), was discovered. It could not be accounted for in terms of the four previously postulated quarks. So to explain the properties of this hadron, researchers proposed that a fifth quark, the b or "bottom" quark (early on referred to as *beauty*), existed. The Υ was taken to be a $b\bar{b}$ pair.

The b quark then carries yet another special quantum charge. The discovery of the Υ meson spurred the search for additional mesons composed of a b (or a $\bar{b}$) quark and a lighter u or d quark, much as the identification of ψ led to the search for charmed mesons. Because the masses of these mesons were expected to be very large and their lifetimes extremely short, many, many ultrahigh-energy collisions had to be carried out and carefully studied before enough evidence to corroborate the existence of such particles could be gathered. Finally, in 1982, after examining over 140,000 e^--e^+ collisions, investigators at Cornell University identified 18 events that they concluded corresponded to decays involving B mesons, hadrons with masses near 5.3 GeV/c^2 composed of b quarks: $B^+ = \bar{b}u$; $B^- = b\bar{u}$; and $B^0 = \bar{b}d$. Here was strong confirmation of the reality of the fifth quark.

In 1975, the announcement of the discovery of the tau and its (presumed) neutrino (recall this particle was not observed in collision reactions until 2003) brought the number of leptons to six. With the identification of the b quark in 1982, the quarks once again lagged the leptons by one. To maintain the symmetry between the two groups, the need to have a sixth quark was pointed out. The presence of this newest quark, the t or "top" quark (sometimes called *truth*), permits the quarks to be paired off into three families

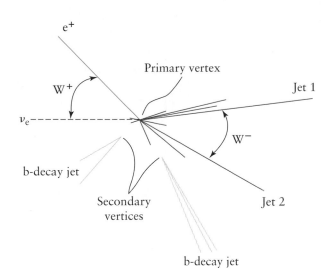

● **Figure 12.22** Candidate event for top-antitop production. Each top quark decays at the p p̄ primary collision vertex into a W boson plus a bottom quark. The W⁺ decays into a positron e⁺ plus an invisible neutrino (ν_e), and the W⁻ decays into a quark and an antiquark, which appear as two jets of hadrons (jets 1 and 2). Each bottom (b) quark becomes a neutral B meson that travels a few millimeters from the production vertex before its decay creates a hadron jet. (From "Top-ology" by Chris Quigg, *Physics Today* 50, no. 5 (May 1997), p. 24.)

much as the leptons are: (u,d); (s,c); and (b,t). In 1985, in experiments done at CERN, some preliminary evidence for a meson containing the top quark was found.

In March 1995, based on data from proton-antiproton collision experiments carried out using the Fermilab Tevatron accelerator, representatives of a consortium of over 400 scientists and engineers announced the discovery of more than 100 events out of several trillion associated with the production of the top quark with a probable error of less than 1 in 500,000. ● Figure 12.22 shows a schematic representation of a typical "top event" resulting from a p-p̄ collision and the subsequent decay of the top and antitop quarks into other particles, including other quarks. Based on the analyses of these events, the current value of the top quark mass is 174.3 GeV/c^2, a value greater than the mass of the nucleus of an atom of gold!

If current plans are followed, by 2007, the $3.0 billion Large Hadron Collider (LHC) at CERN will begin operation with proton beams having energies up to 14 TeV (*tera* = 10^{12}), and will produce about 8×10^6 t t̄ pairs per experiment. Within a very short time, physicists will be able to use the top quark to answer many questions that still remain about matter and the forces that govern the physical universe. ● Table 12.7 summarizes the properties of the six quarks.

Are there more than six quarks in nature? More than six leptons? Should we expect to see an expansion in the numbers of families of these particles beyond three in the future? And if so, does this provide evidence that these "elementary" particles may themselves be composites of still more fundamental species? The answers to all of these questions cannot be given at this time, but the results of experiments done within the last few years seem to indicate that there cannot be more than three distinct families of quarks and leptons. With the confirmation of the existence of the top quark, it appears that we have found *all* of the fundamental building blocks of nature (● Figure 12.23). ● Figure 12.24

● **Figure 12.23**

● **Table 12.7** Summary of Basic Characteristics of u, d, s, c, b, and t Quarks

Quark	Electric Charge	Strangeness	Charm	Bottomness	Topness	Mass[b] (MeV/c^2)
		Quantum Number or Charge[a]				
u	$+\frac{2}{3}$	0	0	0	0	5
d	$-\frac{1}{3}$	0	0	0	0	8
c	$+\frac{2}{3}$	0	1	0	0	1,500
s	$-\frac{1}{3}$	−1	0	0	0	160
t	$+\frac{2}{3}$	0	0	0	1	174,000
b	$-\frac{1}{3}$	0	0	1	0	4,250

[a]All the quarks have baryon number $\frac{1}{3}$ and spin $\frac{1}{2}$.

[b]Values given are approximate upper limits to quark mass ranges.

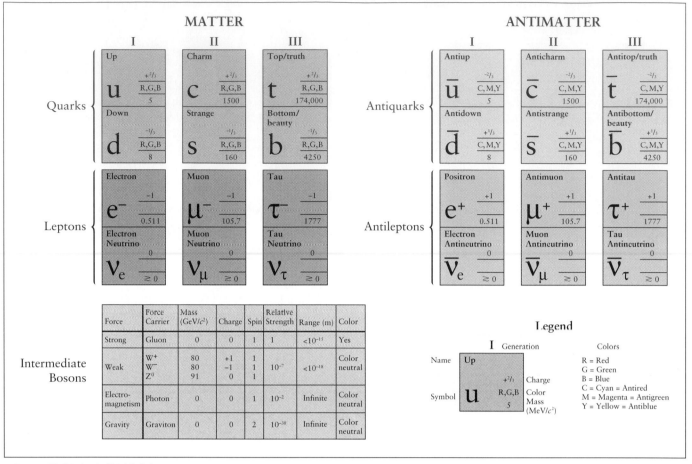

● **Figure 12.24** Standard Model of elementary particles.

summarizes the properties of the three quark and lepton families, as well as those of their anti's, in what is now referred to as the *Standard Model* of elementary particle physics.

Although highly successful at accounting for much of what is known about elementary particles, the Standard Model has limitations. For example, it predicts that the masses of the neutrinos should all be precisely zero. Laboratory experiments have placed upper limits on these masses (see Table 12.4), but the most compelling evidence that neutrinos must have mass comes from recent observations at the Sudbury Neutrino Observatory (SNO). Operated by a consortium of scientists from 15 institutions from the United States, Great Britain, and Canada, the heart of this facility consists of 1000 tons of "heavy" water (D_2O) in a 12-m-diameter containment vessel buried 2 km underground. Nearly 9500 light-sensitive detectors placed inside the vessel look for faint bursts of light that accompany the interaction of neutrinos (mainly emitted by nuclear reactions in the core of the Sun) with the heavy water. Unlike other neutrino detectors, SNO has the capacity to distinguish the type of neutrino (electron, muon, or tau) involved in the interaction, and recent results indicate that only about one-third of the total number of neutrinos reaching the Earth are of the electron variety.*

Almost *all* the neutrinos produced by the Sun are electron neutrinos, so two-thirds of them must have changed flavor during their approximately eight-minute flight to Earth. Since massless neutrinos cannot spontaneously change type, the implication of this discovery is that neutrinos must have non-zero rest mass. Theoretical models to explain these *neutrino oscillations,* as they are called, do not provide explicit values for the indi-

---

* This confirms results reported by Raymond Davis Jr. and collaborators more than forty years ago showing that the number of electron neutrinos detected from the Sun is smaller than what is expected on the basis of our best models for the solar interior. For his pioneering work in what was called the "solar neutrino problem," Davis shared the 2002 Nobel Prize in physics.

vidual neutrino masses, only their mass differences. Current values for the mass differences, although highly model dependent, are generally small, ranging from about 10^{-2} to as little as 10^{-5} eV/c^2. If the scale of the neutrino masses is set by the electron neutrino (~ 1 eV/c^2), then all the neutrino masses are very small. But the fact that these elusive particles have any mass at all remains something that the Standard Model cannot explain. At present, much of the effort in theoretical particle physics is devoted to revising the Standard Model to remedy this and other deficiencies.

The Electroweak Interaction and GUTs

In Section 12.2, we mentioned that similarities between the electromagnetic and the weak nuclear forces first noted in the 1950s have led scientists to the conclusion that these two interactions are really just two different manifestations of a more basic interaction called the **electroweak interaction.** The theory describing this interaction was developed independently by Steven Weinberg (1967) and Abdus Salam (1968), and is often referred to as the Weinberg-Salam theory. Weinberg and Salam, together with Sheldon Glashow, who had also made contributions to the unification of these two forces, shared the Nobel Prize in physics for 1979.

According to the Weinberg-Salam model, at moderate to high energies (of the order of 100 billion electronvolts—100 GeV), the electroweak interaction is carried by four massless bosons and is described by equations that are symmetric (i.e., they remain unchanged) when we perform mathematically well-defined "rotations" on some specific characteristics of these particles. As the energy of the system is lowered, however, the symmetry is broken spontaneously, with the result that the original family of four massless bosons separates into two subfamilies, the first consisting of the massless photon, which mediates what we call the electromagnetic interaction, and the second including the very massive intermediate (or gauge) bosons ($W^{\pm}$, Z^0), which carry the weak interaction.

The reasons why some of the originally massless carriers of the electroweak interaction acquire mass as the symmetry of the system is broken involves what are called *gauge fields* and *Higgs mechanisms* and are too complicated for us to explore in any detail, but an example involving broken symmetry in another context might help remove some of the mystery associated with this concept.*

This example has been given by Stephen Hawking and connects the idea of symmetry within a system to its energy. Consider a roulette ball on a roulette wheel (● Figure 12.25). At high energies when the wheel is spun rapidly, the ball behaves in basically one way: it rolls around and around in one direction in the groove of the wheel. We might say that at high energies there is only one state in which the ball can be found, and all rapidly spin-

---

* In the simplest model, the initial symmetry of the system is broken by a new particle, the *Higgs*, a massive spin 0 boson, which interacts with the intermediate gauge bosons, as well as the leptons and the quarks. These latter particles acquire mass in the process, the values being determined by how strongly each "couples" to or interacts with the Higgs. For the recently confirmed top quark whose mass is very large, the coupling with the Higgs is relatively strong. The LHC's ability to produce large numbers of top quarks makes it ideal for exploring possible connections between the t and the eagerly sought-after Higgs.

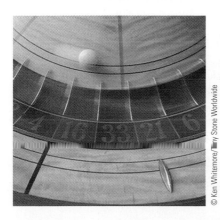

● **Figure 12.25** High (left) and low (right) energy states at the roulette table. When the wheel spins rapidly, there exists only one state of the system—with the ball whirling around in the groove. When the wheel slows down, there appear to be 38 possible states for the system corresponding to the 38 slots into which the ball can lodge.

ning roulette wheels display the same state: They're all symmetrical with respect to one another. However, as the wheel slows and the ball's energy decreases, it eventually drops into one of the 38 slots molded into the wheel; at low energies, it appears that there are 38 different states in which the ball can exist. If we surveyed roulette wheels solely at low energies, we might be led to the conclusion that these 38 possibilities were the only ones allowed to the ball. We would miss the fact that at high energies these different states merge into one. This is analogous to what happens to the carrier family for the electroweak interaction. At high energy, these carriers are all massless and behave like one another. But at low energy when the symmetry is broken, this group appears as two families of completely different particles.

At the time Weinberg and Salam first published their theory, particle accelerators did not exist that could produce the energies required to test their predictions, particularly those concerning the existence of the massive $W^\pm$ and Z^0 particles that carried the weak nuclear force. By the early 1980s, however, machines were beginning to come on line that could achieve the needed energies, and, in January 1983, Carlo Rubbia of CERN reported the detection of the W particles in nine events out of over a million recorded p-$\bar{\text{p}}$ collisions. In July 1984, the CERN group also announced the discovery of the Z^0 particle together with additional evidence to support their earlier detection of the W particles. The 1984 Nobel Prize in physics was awarded to Rubbia and his collaborator Simon van der Meer for their search and discovery of the carriers of the weak interaction. These results completely vindicated the Weinberg-Salam model and left little doubt about its correctness.

The success of the unification of the electromagnetic and the weak nuclear forces caused many physicists to wonder whether it might not also be possible to unite the strong nuclear force with the electroweak force to produce what has come to be called a **grand unified theory,** or **GUT.** As Hawking has pointed out, these theories (there are actually several competing ones) are neither all that grand nor fully unified. For example, they do not include gravity. Nevertheless, they represent the first steps toward a true unification of the interactions of nature, and they do present some interesting implications for the long-term future of the universe.

The basic idea of GUTs is the following: Just as the electromagnetic and weak forces represent different aspects of the same basic force that coalesce at high energies, the electroweak and strong nuclear forces are thought to be different manifestations of yet another even more fundamental force that emerges at still higher energies. The energy at which all three of these forces become fused into a single force is called the *grand unification energy*. Its value is not well known but would probably have to be at least 10^{15} billion electronvolts, far above the energies ever likely to be attainable with terrestrial laboratory particle accelerators. The GUTs also predict that at this energy the different spin $\frac{1}{2}$ particles, such as quarks and leptons, would all behave in the same manner. This is similar to the formation of a single family of electroweak carrier particles from the photons and the intermediate bosons that mediate the separate electromagnetic and weak forces.

Because of the impossibility of reaching the energies needed to test GUTs directly in the laboratory, scientists have begun searching for indirect, low-energy consequences of these theories. One search pertains to the possible decay of the proton. In Section 12.3, we argued that conservation of baryon number in strong interactions demands that the proton be absolutely stable, since there exist no lighter baryons to which it can decay. But at grand unification energies, the quarks that make up the proton are indistinguishable from leptons, so that the spontaneous decay of a proton (through quark conversion) to lighter particles such as positrons becomes possible. The three quarks bound inside a proton do not normally have sufficient energy to make this transition. We live in an environment in which the available energies are typically far below the threshold energy for grand unification. However, like all interactions governed by quantum mechanics, a very, very small, but non-zero, probability exists that such a transition can occur, in which case, the proton would decay. The chances of this happening, as you might guess, are very low. If you could watch a single proton, hoping to catch it in the act of decaying as described, you'd have to wait and watch for something like a million million million million million (or 10^{30}) years! Clearly, this is not practical. The entire universe has been in existence for only 14 thousand million years or so.

To increase your chances of witnessing the decay of the proton and, hence, of validating one version of GUTs, instead of watching only one proton, you could watch lots of them. You could monitor a large quantity of matter containing an enormous number of protons—maybe 10^{31} of them. If this were done for a year, you would expect, on average, to see as many as 10 proton decays. Increasing the number of protons in the sample obviously improves your chances of seeing a decay.

This is precisely the approach taken by researchers running the Super-Kamiokande experiment in Japan. Using an underground tank filled with 50 kilotons of water (each water molecule contributes 10 protons) and equipped with thousands of light-sensitive devices, experimenters look for the flashes of radiation emitted by the high-speed positrons believed to accompany the decay of the protons. The results of this experiment have been discouraging: No clearly defined proton-decay events have been recorded since data began being collected in 1996. This makes the average lifetime of the proton greater than 10^{33} years. This value is greater than that predicted by the simplest GUTs, but there are more elaborate versions of these theories that give longer proton lifetimes.

One thing is clear. If the proton is found to be unstable, even on such long time frames, then eventually all the matter in the universe will evaporate. This is so because the positrons, as products of the proton decay, will annihilate with electrons to produce gamma rays. Assuming the universe survives for the requisite amount of time, it will ultimately be stripped of all matter and reduced to a cold, dark space filled with photons (which steadily lose energy) and a few isolated electrons and neutrinos (which escaped destruction)—a grim fate indeed, but one that may well await our now-glorious universe. More so than most, this example demonstrates how the physics of the subatomic domain can influence that of the largest scale known. But the reverse is also true.

We have seen that it is not feasible to expect that terrestrial laboratory experiments will ever achieve the energies required to verify directly the predictions of GUTs. Indeed, one might well ask, "Have such energies ever been available in the entire history of the universe?" And the answer is, "Yes, in the very first few moments after the birth of the universe in the Big Bang!" (See the Physics Potpourri on page 454 in Chapter 11.)

Thus we begin to see that the ultimate tests of GUTs may rest in the field of cosmology, the study of the structure and evolution of the universe on the largest scale possible (see the Epilogue). The justification of theories involving the smallest of physical entities, quarks and electrons, may depend on our understanding of the largest of physical structures, galaxies and clusters of galaxies.

LEARNING CHECK

1. A quark can change color by
   a) absorbing a W^+ boson.
   b) emitting a photon.
   c) absorbing a gluon.
   d) emitting a neutrino.
   e) None of the above.

2. Gluons can attract each other to form glueballs because
   a) gluons are sticky.
   b) gluons exchange (virtual) photons.
   c) gluons possess color charge.
   d) gluons emit Z^0 bosons.
   e) None of the above.

3. Grand unified theories (GUTs) have as their goal the merging of which of the following fundamental forces of nature:
   a) the electric and magnetic forces.
   b) the strong and weak nuclear forces.
   c) the strong, weak, and electromagnetic forces.
   d) the strong, weak, electromagnetic, and gravitational forces.

4. (True or False.) Recent experiments suggest that the lifetime of the proton is greater than 10^{33} years, a factor of more than 10^{20} longer than the current age of the universe.

5. The ultimate tests of grand unified theories may come in the field of _____, the study of the structure and evolution of the universe as a whole.

ANSWERS: 1. (c) 2. (c) 3. (c) 4. True 5. cosmology

Physics, like almost any other discipline, is full of colorful individuals who, by their powerful personalities and intellects, have exerted far-reaching influence on the progress of the science. In elementary particle physics, few people have done more to shape the development of the field or the language in which it is expressed than Murray Gell-Mann (● Figure 12.26).

Murray Gell-Mann was born in New York City and entered Yale University at the age of 15. After graduating in physics, he attended the Massachusetts Institute of Technology (MIT), where he received his doctorate in theoretical physics when he was only 21. After a brief stay at the University of Chicago, Gell-Mann moved to Caltech in 1957, where he has remained. A man of great intellectual prowess and wide-ranging interests, Gell-Mann is well acquainted with Eastern and Western religions and literature, knows several languages, including Swahili, and, of course, is highly skilled in applying abstract mathematical theory to problems in physics. He is a refined, urbane individual not given to expansive gestures but to calm, orderly discourse peppered with literary references and clever witticisms. Logical progression and mathematical rigor are the hallmarks of his work.

● **Figure 12.26** Theoretical physicist Murray Gell-Mann, one of the architects of the quark model.

© American Institute of Physics

A brief summary of the major contributions to theoretical particle physics made by Murray Gell-Mann is appropriate here: 1953—introduced the idea of "strangeness" to explain the unusual properties of kaons and the lambda; 1961—developed the eightfold way to organize the proliferating "zoo" of elementary particles into a few families having simple, regular symmetries; 1964—introduced quarks as a more fundamental way of explaining the symmetries exhibited in the eightfold way; 1971—introduced "colored" quarks to differentiate three possible varieties of each basic quark type; 1973—recognized "color" as the source of the strong force and coined the name *quantum chromodynamics* to describe the theory of the strong interaction. For his accomplishments up to that time, Gell-Mann received the 1969 Nobel Prize in physics, and he justly deserves to be called "one of the heroes of our story," as John Polkinghorne refers to him in his chronicle of the evolution of particle physics in the years since the Second World War entitled *Rochester Roundabout*.

As an illustration of the deeply perceptive yet whimsical way in which Gell-Mann approaches physics, the tale of the quarks, as told by Michael Riordan in his book *The Hunting of the Quark*, is perhaps the best. In March 1963, Gell-Mann visited Columbia University to give a series of three talks. At a luncheon during his stay in New York, Gell-Mann was asked why a fundamental group of three particles—a triplet—did not appear in nature, while larger groupings, like those containing eight particles (an octet), did. Gell-Mann's reply was "that would be a funny quirk," and he proceeded to show why: Such a triplet, if it existed, would have to be composed of particles with fractional charges equal to $\frac{2}{3}$, $-\frac{1}{3}$, and $-\frac{1}{3}$ the magnitude of the electron charge, respectively. No such fractionally charged entities had ever been observed.

Later that day and into the following morning, Gell-Mann reflected on this situation and recognized that such fractionally charged species were not completely out of the question *if* they remained permanently trapped inside hadrons and never appeared as free particles in nature. In his second lecture at Columbia, Gell-Mann referred to the members of this triplet as "quarks," a nonsense word he had used before and by which he meant "those funny little things." Although relatively little time was devoted to "quarks" in the talk, they were very much the subject of the postlecture coffee hour.

In the fall of that year, Gell-Mann worked out the details of how to construct baryons and mesons out of this fractionally charged triplet of particles and wrote a short, two-page article that was published in a new European journal, *Physics Letters*. (Gell-Mann, when asked why he had not submitted the paper to more prestigious American journals, replied that it "probably would have been rejected.") By this time, Gell-Mann had discovered a passage from James Joyce's novel *Finnegans Wake*, which lent legitimacy to his use of the term *quarks* to designate the fictional triplet:

> Three quarks for Muster Mark!
> Sure he hasn't got much of a bark,
> And sure any he has it's all beside the mark.
> But O, Wreneagle Almighty, wouldn't un be a sky of a lark
> To see that old buzzard whooping about for uns shirt in the dark
> And he hunting round for uns speckled trousers around by Palmerston Park?

The poem relates the dream of one of Joyce's protagonists as he lies passed out on the floor of a Dublin tavern. The last three lines had, no doubt, a particularly appealing irony in them for Gell-Mann in view of the parallel circumstances of the hapless Muster Mark vainly searching for his clothes in the dark and of the frantic experimentalists equally vainly searching for free quarks. Indeed, Gell-Mann was one of the last to believe that quarks were anything but mathematical entities. At a conference in Berkeley in 1966, he spoke of "those hypothetical and probably fictitious spin $\frac{1}{2}$ quarks."

Murray Gell-Mann's achievements have been compared to those of Dmitri Mendeleev (refer to Physics Potpourri in Section 4.1). Both men brought order out of confusion, Mendeleev by organizing the elements into the periodic table based on their chemical properties and Gell-Mann by establishing a kind of periodic table of mesons and baryons based on their group symmetry properties.

In each case, gaps in the patterns led to predictions of missing members that were subsequently discovered experimentally. The direction and impetus provided to chemists and atomic physicists by Mendeleev's work is strongly paralleled by that given to elementary particle physicists by Gell-Mann's.

SUMMARY

- The **special theory of relativity** extends classical mechanics to systems moving with speeds near that of light. It postulates that the speed of light is a constant for all observers and that the laws of physics are the same for all observers moving uniformly.

- **Time dilation,** in which moving clocks run slow, and **length contraction,** in which moving rulers appear shortened in the direction of motion, are experimentally confirmed predictions of this theory.

- Special relativity also establishes the equivalence of mass and energy, according to which the **rest energy** of a stationary object of mass m is given by $E_0 = mc^2$.

- There are **four fundamental forces** or interactions in nature: **gravity, electromagnetism,** the **weak nuclear force,** and the **strong nuclear force.** Only the last three of these are important in elementary particle physics. These forces are mediated through the exchange of **carrier particles** like photons and gravitons.

- All subatomic particles may be classified as either **bosons** or **fermions,** depending on whether their spins are integral or half-integral multiples of the basic unit of intrinsic angular momentum, $h/2\pi$. Particles may also be segregated by mass and by the types of fundamental interactions in which they participate. For every particle, there exists an **antiparticle,** having the same mass, but opposite electric charge.

- Underlying each conservation law in physics is a fundamental symmetry property of the universe. By studying what does and doesn't occur in particle interactions, physicists can infer the existence of new conservation laws. Conservation of **baryon number** and **strangeness** in the strong interaction and conservation of **lepton number** in the weak interaction are examples of such laws.

- **Quarks** are fractionally charged particles that come in six **flavors:** up, down, strange, charmed, bottom, and top. **Baryons** are composed of three quarks, while **mesons** are made up of a quark-antiquark pair. The six quarks and their anti's, together with the electron, the **positron,** the tau, and their associated neutrinos, comprise all the matter and antimatter in the universe according to the **Standard Model** of particle physics.

- An important goal in modern physics is the unification of the fundamental forces of nature. In the 1960s, the electromagnetic force was successfully joined with the weak force to produce the **electroweak interaction.** Scientists are now seeking ways to unify the strong force with the electroweak force in what are called **grand unified theories (GUTs).** Although far from complete, current versions of GUTs make predictions that may ultimately be tested only by cosmological observations of the universe that connect to its origin in the Big Bang.

IMPORTANT EQUATIONS

Equation	Comments
$\Delta t' = \dfrac{\Delta t}{\sqrt{1 - v^2/c^2}}$	Time dilation
$E_0 = mc^2$	Definition of rest energy; equivalence of mass and energy
$E_{rel} = \dfrac{mc^2}{\sqrt{1 - v^2/c^2}}$	Relativistic total energy
$KE_{rel} = \dfrac{mc^2}{\sqrt{1 - v^2/c^2}} - mc^2$	Relativistic kinetic energy

The development of the *standard model* of elementary particles is one of the most significant achievements in physics in the twentieth century. While refinements of this description of the subatomic world continue to be made, an understanding of the basic components and structure of this model serves as a foundation for appreciating the new discoveries in particle physics that regularly occur.

One way to organize the elements of the standard model is in a display like that shown in Figure 12.24. Another way is to use a concept map. Examine Figure 12.24 closely and, if necessary, reread the material leading up to it. Then, using the principles of concept mapping that we have employed so often in past chapters, try to construct a map that captures the essential features and relationships embodied in the standard model of particle physics. When you have completed your map, exchange it for one that a classmate has prepared. Critically examine your friend's map. Does it include the most important concepts contained in the standard model? Does it properly exhibit/explain the relationships between the various elements of this model in a manner consistent with Figure 12.24 and the material introduced in the chapter? What are the strengths and weaknesses of your colleague's map? Discuss these issues with the person with whom you exchanged maps and together try to come to a better mutual understanding of the standard model.

(▶ *Indicates a review question, which means it requires only a basic understanding of the material to answer. The others involve integrating or extending the concepts presented thus far.*)

1. ▶ What does the acronym PET stand for? Why is PET a good example with which to begin a discussion of elementary particle physics?
2. ▶ Describe the two fundamental postulates underlying Einstein's special theory of relativity.
3. Suppose you were traveling toward the Sun at a constant velocity of $0.25c$. With what speed does the light streaming out from the Sun go past you? Explain your reasoning.
4. Light travels in water at a speed of 2.25×10^8 m/s. Is it possible for particles to travel through water at a speed $v > 2.25 \times 10^8$ m/s? Why or why not? Explain.
5. ▶ In your own words, define what is meant by *time dilation* in special relativity theory. Provide a similar definition for *length contraction*. Give an example in which the effects of time dilation are actually observed.
6. Galileo used his pulse like a clock to measure time intervals by counting the number of heartbeats. If Galileo were traveling in a spaceship, moving uniformly at a speed near that of light, would he notice any change in his heart rate, assuming the circumstances of his travel produced no significant physiological stress on him? If someone on Earth were observing Galileo with a powerful telescope, would he or she detect any change in Galileo's heart rate relative to the resting rate on Earth? Explain your answers.
7. Newton wrote: "Absolute, true, and mathematical time, of itself, and from its own nature, flows equally without relation to anything external." Comment on the significance of this statement for two timekeepers in relative motion. In the light of special relativity, is Newton's statement valid? Explain.
8. Why don't we generally notice the effects of special relativity in our daily lives? Be specific.
9. Does $E_0 = mc^2$ apply only to objects traveling at the speed of light? Why or why not?
10. If a horseshoe is heated in a blacksmith's furnace until it glows red hot, does the mass of the horseshoe change? If a spring is stretched to twice its equilibrium length, has its mass been altered in the process? If so, explain how and why in each case.
11. ▶ List the four fundamental interactions of nature, and discuss their relative strengths and effective ranges.
12. ▶ What common feature of the electromagnetic and gravitational interactions requires that their carrier or exchange particles be massless?
13. ▶ What is an antiparticle? What may happen when a particle and its anti collide?
14. What is the antiparticle of a photon? How does the charge of the antiphoton compare to that of the photon?
15. According to Table 12.4, the rest mass of an electron is 0.511 MeV/c^2. What is the rest mass of a positron?
16. ▶ Distinguish between fermions and bosons in as many different ways as you can.
17. ▶ Give some ways by which physicists classify elementary particles.
18. ▶ In which of the four basic interactions does an electron participate? A neutrino? A proton? A photon?
19. ▶ Baryon conservation and lepton conservation are laws used frequently by particle physicists to decide whether a reaction involving elementary particles is possible or not. Explain how these laws are applied to make such determinations.
20. ▶ In your own words, describe what a physicist means when using the term *strangeness*.
21. ▶ What is a quark? How many different types of quarks are now known? What are some of the basic properties that distinguish these quarks?
22. ▶ Describe the kinds of evidence that have led scientists to conclude that quarks exist.
23. ▶ How many quarks form a baryon? A meson? What is the relationship (if any) between a quark and a lepton (e.g., an electron)?
24. In the quark model, is it possible to have a baryon with strangeness -1 and electric charge $+2$? Explain.
25. What kind of a particle (baryon, meson, or lepton) corresponds to a $t\bar{t}$, that is, to a top-antitop quark combination? Describe some of the properties such a particle would have.
26. ▶ Quarks are said to possess "color." What does this mean? Are physicists really suggesting that quarks look red like ripe strawberries or green like healthy grass? Explain.
27. ▶ Describe the *Standard Model* of elementary particle physics.
28. A bumper "snicker" on a car belonging to the chairperson of a physics department reads: "Particle physicists have GUTs!" Explain in your own words the meaning of this little joke or "play on words."
29. ▶ Unification of its basic laws and theories has long been a goal in physics. Describe some ways in which physicists have been successful in unifying certain forces and theories. In what area(s) of physics is the process of unification still ongoing?
30. If a proton can decay, then its lifetime is of the order of 10^{33} years, far longer than the current age of the universe. Does this necessarily imply that a proton decay has not yet occurred in the entire history of the universe? Explain.

[For many of the exercises, referring to Table 12.3 and Figure 12.24 will be very helpful.]

1. The lifetime of a certain type of elementary particle is 2.6×10^{-8} s. If this particle were traveling at 95% the speed of light relative to a laboratory observer, what would this observer measure the particle's lifetime to be?

2. How fast would a muon have to be traveling relative to an observer for its lifetime as measured by this observer to be 10 times longer than its lifetime when at rest relative to the observer?

3. The lifetime of a free neutron is 886 s. If a neutron moves with a speed of 2.9×10^{8} m/s relative to an observer in the lab, what does the observer measure the neutron's lifetime to be?

4. A computer in a laboratory requires 2.50 μs to make a certain calculation, as measured by a scientist in the lab. To someone moving past the lab at a relative speed of $0.995c$, how long will the same calculation take?

5. The formula for length contraction gives the length of an interval on a ruler moving with velocity v relative to an observer as $\sqrt{1 - v^{2}/c^{2}}$ times the length of the same interval on a ruler at rest with respect to the observer. By what fraction is the length of a meter stick reduced if its velocity relative to you is measured to be 95% the speed of light?

6. If an electron is speeding down the two-mile-long Stanford Linear Accelerator at 99.98% the speed of light, how many meters long is the trip as seen from the perspective of the electron? (See Problem 5.)

7. Calculate the rest energy of a proton in joules and MeVs. What is the mass of a proton in MeV/c^{2}?

8. The tau is the heaviest of all the known leptons, having a mass of 1777 MeV/c^{2}. Find the rest energy of a tau in MeVs and joules. What is the mass of the tau in kilograms? Compare your result with the mass in kilograms of an electron.

9. If a 1.0-kg mass is completely converted into energy, how much energy, in joules, would be released? Compare this value to the amount of energy released when 1.0 kg of liquid water at 0°C freezes.

10. In a particular beam of protons, each particle moves with an average speed of $0.8c$. Determine the total relativistic energy of each proton in joules and MeVs.

11. A particle of rest energy 140 MeV moves at a sufficiently high speed that its total relativistic energy is 280 MeV. How fast is it traveling?

12. If the relativistic kinetic energy of a particle is 9 times its rest energy, at what fraction of the speed of light must the particle be traveling?

13. If a proton and an antiproton, both at rest, were to completely annihilate each other, how much energy would be liberated?

14. Indicate which of the following decays are possible. For any that are forbidden, give the conservation law(s) that is/are violated.
    a) $\Sigma^{+} \rightarrow n + \pi^{0}$
    b) $n \rightarrow p + \pi^{-}$
    c) $\pi^{0} \rightarrow e^{-} + e^{-} + \nu_{e}$
    d) $\pi^{-} \rightarrow \mu^{+} + \nu_{\mu}$

15. Indicate which of the following reactions may occur, assuming sufficient energy is available. For any that are forbidden, give the conservation law(s) that is/are violated.
    a) $p + p \rightarrow p + n + K^{+}$
    b) $\gamma + n \rightarrow \pi^{+} + p$
    c) $K^{-} + p \rightarrow n + \Lambda^{0}$
    d) $p + p \rightarrow p + \pi^{+} + \Lambda^{0} + K^{0}$

16. Supply the missing particle in each of the following *strong* interactions, assuming sufficient energy is present to cause the reaction:
    a) $n + p \rightarrow p + p + \underline{\hspace{1cm}}$
    b) $p + \pi^{+} \rightarrow \Sigma^{+} + \underline{\hspace{1cm}}$
    c) $K^{-} + p \rightarrow \Lambda^{0} + \underline{\hspace{1cm}}$
    d) $\pi^{-} + p \rightarrow \Xi^{0} + K^{0} + \underline{\hspace{1cm}}$

17. Identify the missing particle(s) in each of the following *weak* interactions:
    a) $\Omega^{-} \rightarrow \Xi^{0} + \underline{\hspace{1cm}}$
    b) $K^{0} \rightarrow \pi^{+} + \underline{\hspace{1cm}}$
    c) $\pi^{+} \rightarrow \mu^{+} + \underline{\hspace{1cm}}$
    d) $\tau^{+} \rightarrow \mu^{+} + \underline{\hspace{1cm}} + \underline{\hspace{1cm}}$

18. A 0.1-kg ball connected to a fixed point by a taut string whirls around a circular path of radius 0.5 m at a speed of 3 m/s. Find the angular momentum of the orbiting ball. Compare this with the intrinsic angular momentum of an electron. How many times larger is the former than the latter?

19. What is the quark combination corresponding to an antineutron? An antiproton?

20. Give three (3) combinations of quarks (not antiquarks!) that will give baryons with charge: (a) $+1$; (b) -1; and (c) 0.

21. Give all the quark-antiquark pairs that result in mesons that have no charge, no strangeness, no charm, and no bottomness. How do you think such particles might be distinguished from one another?

22. Distinguish a particle composed of a combination of $\bar{d}\,\bar{u}\,\bar{s}$ quarks from one composed of dus quarks in as many ways as you can. Do the same for a dss combination and a $\bar{d}\,\bar{s}\,\bar{s}$ combination.

23. The Δ^{-} (delta minus) is a short-lived baryon with charge -1, strangeness 0, and spin $\frac{3}{2}$. To what quark combination does this subatomic particle correspond?

24. The Ξ^{0} (xi zero) is a baryon having charge 0, strangeness -2, and spin $\frac{1}{2}$. What quarks combine to form this subatomic particle?

25. The F^{-} meson possesses -1 unit of charge, -1 unit of strangeness, and -1 unit of charm. Identify the quark combination making up this rare subatomic particle.

26. The D^{+} meson is a charmed particle with charge $+1$, charm $+1$, and strangeness 0. Work out the quark combination for this species of elementary particle.

27. What is the quark composition of the Δ^{++}? It has no strangeness, no charm, and no topness or bottomness. Its spin is $\frac{3}{2}$.

28. Analyze the following reactions in terms of their constituent quarks:
    a) $\pi^{+} + p \rightarrow \Sigma^{+} + K^{+}$
    b) $\gamma + n \rightarrow \pi^{-} + p$

29. Analyze the following decays in terms of the quark contents of the particles:
    a) $\Sigma^{-} \rightarrow n + \pi^{-}$
    b) $\Lambda^{0} \rightarrow p + \pi^{-}$

30. If the average lifetime of a proton was 10^{33} years, about how many protons would you have to assemble together and observe simultaneously to witness a total of 100 proton decays in one year? Explain the reasoning that led to your conclusion.

1. Show that in the limit of low speeds the expression for the relativistic kinetic energy (see page 469 or Important Equations) reduces to the familiar one from classical mechanics, $KE = 1/2mv^2$. [*Hint:* For speeds v much smaller than c, the quantity $(1 - v^2/c^2)^{-1/2}$ is approximately equal to $1 + v^2/2c^2$. You should check this for yourself using typical speeds given in the exercises in Chapter 3.]

2. Carry out the calculation suggested in Section 12.2 on page 472. In particular, find the ratio of the Coulomb force of repulsion between two protons separated by 10^{-15} m to their gravitational force of attraction for the same distance. The mass and charge of the proton are given in Table 11.1.

3. When a proton-antiproton pair at rest annihilates, two photons are created. Find the wavelengths of these two light quanta, given that they share equally in the annihilation energy. In what portion of the electromagnetic spectrum do such wavelengths appear?

4. If the average lifetime of the proton were 10^{16} years, estimate the number of protons per kilogram of human body mass that would decay radioactively each year. Assume that a human being is made entirely of water molecules and that each molecule contributes 10 protons. (*Hint:* The molar weight of water is 18 g, and each mole of any substance contains 6.02×10^{23} particles.)

   If each proton decay released 1 MeV of energy, how much energy per kilogram would be produced *in total* by the decaying protons in your body in 1 year? Give the result in joules per kilogram. If the energy released is mostly in the form of kinetic energy of the positrons produced in the decay, then 1 J/kg of this type of ionizing radiation is rated at about 1 Sv (the *sievert* (Sv) is the SI unit of radiation dose). Compare the total exposure in sieverts of a person to the radiation produced by his/her own body to the maximum safe dose of 1 mSv/yr for the general public from all artificial sources.

5. Although doubly charged baryons have been found (baryons with net charge +2), no doubly charged mesons have yet been identified. What effect on the quark model would there be if a meson of charge +2 were discovered? How could such a meson be interpreted within the quark model?

6. In the electroweak theory, symmetry breaking occurs on a length scale of about 10^{-17} m. Compute the *frequency* of particles whose de Broglie wavelength (see Section 10.5) equals this value, and, using this result, determine their energies by the Planck formula (page 392). Show that the mass of particles having such energies is on the order of the mass of the $W^{\pm}$ particles (see Figure 12.24).

7. One interesting (unbelievable?) implication of time dilation is contained in what is called the *twin paradox*. Imagine identical twins who decide to perform an experiment to test the accuracy of Einstein's predictions about time dilation. After synchronizing their (identical) clocks, one of the twins enters a spaceship and travels to a distant star at a speed 95% that of light. Upon arrival, the traveling twin immediately turns around and heads home at the same speed. According to the twin on Earth, the clock aboard the spaceship runs slower than the one on Earth, so that the traveling twin ages less than the Earthbound sibling. But according to the space-faring twin, it is the clock on Earth that is running too slowly, making the stay-at-home twin the younger of the two. This is the paradox: Each twin argues that the other will be younger at the end of the trip. How can this paradox be resolved? Which of the two twins is correct in their analysis, and why? If the trip requires a total time of 5 years to make as measured by the clock on Earth, what will be the time recorded for the trip by the clock on board the spaceship?

8. Suppose you perform a scattering experiment in which you fire BBs into a slab of jello in which a layer of randomly distributed but identical marbles are suspended. Of the BBs entering the jello, 90% pass cleanly through without scattering off any marbles. If the total target area presented to the incoming projectiles is 200 cm^2 and there are 25 marbles hidden in the jello, what is the cross-sectional area for scattering for *each* marble? What is the radius of a typical marble? Describe how such scattering experiments could be used to probe the structure of subatomic particles. (See Explore It Yourself 12.2.)

Close, F. E., and P. R. Page. "Glueballs." *Scientific American* 279, no. 5 (November 1998): 80–85. A description of theory and experiment bearing on the characteristics of glueballs.

Franklin, A. "The Road to the Neutrino." *Physics Today* 53, no. 2 (February 2000): 22–28. This article describes the early history of β-decay experimentation and the introduction of the neutrino.

Hecht, E. "On Morphing Neutrinos and Why They Must Have Mass." *The Physics Teacher* 41, no. 3 (March 2003): 164–168. This paper explores the recently confirmed hypothesis that neutrinos have mass and spontaneously transform from one type to another.

Kane, G. "The Dawn of Physics Beyond the Standard Model." *Scientific American* 288, no. 6 (June 2003): 68–75. A review of the Standard Model of particle physics, including its strengths and weaknesses.

Kane, G, and E. Witten. *Unveiling the Ultimate Laws of Physics.* Cambridge, MA: Perseus Publishing, 2001. Clearly written description of recent advances beyond the Standard Model, including an excellent discussion of experimental methods of particle physics.

McDonald, A. B., J. R. Klein, and D. L. Wark. "Solving the Solar Neutrino Problem." *Scientific American* 288, no. 4 (April 2003): 40–49. A description of the Sudbury Neutrino Observatory and of how experiments there have solved the 30-year-old mystery of the "missing" solar neutrinos.

Quinn, H. R. "The Asymmetry between Matter and Antimatter." *Physics Today* 56, no. 2 (February 2003): 30–35. A look at how phase transitions and the decay of massive neutrinos could lead to a universe where matter dominates antimatter.

Smith, C. L. "The Large Hadron Collider." *Scientific American* 283, no. 1 (July 2000): 70–77. A discussion of the design and promise of the LHC, particularly as relates to the discovery of the Higgs boson.

Smith, T. P. *Hidden Worlds: Hunting for Quarks in Ordinary Matter.* Princeton: Princeton University Press, 2003. A nonmathematical tour of the quark world presenting what we know about quarks and how we know it, with special emphasis on everyday particles like protons and neutrons.

Veltman, M. *Fact and Mysteries in Elementary Particle Physics.* New York: World Scientific Press, 2002. A comprehensive overview of modern particle physics by a Nobel laureate in the field. The book includes biographical sketches of many of the prominent particle-physics personalities.

Wilczek, F. "QCD Made Simple." *Physics Today* 53, no. 8 (August 2000): 22–28. This article explores the fundamentals of QCD and circumstances under which this theory is needed to explain physical reality.

Web Sites of Interest

In addition to the Lederman Science Center site maintained by Fermi Lab (see Explore It Yourself 12.1), here are a few additional locations on the World Wide Web that provide authoritative and interesting information about elementary particle physics:

quarknet.fnal.gov/—Another website maintained by Fermi Lab devoted to involving teachers and students in cutting-edge research on quarks and other elementary particles.

particleadventure.org/—Award-winning site developed by the Particle Data Group at the Lawrence Berkeley National Laboratory that contains many entertaining and educational activities focusing on the fundamentals of matter and forces.

www.cpepweb.org/—Official site of the Contemporary Physics Education Project, which provides instructional material to help teachers and students learn the basics of elementary particle physics.

EPILOGUE

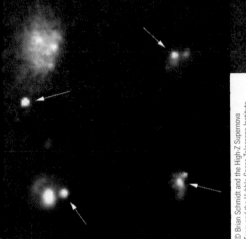

Mosaic of four type Ia supernovae in distant galaxies discovered with the Hubble Space Telescope (HST).

© Brian Schmidt and the High-Z Supernova Team and the Hubble Space Telescope Institute

Albert Einstein called it "the greatest blunder of my life." But was it? The *cosmological constant*, Λ, has a long and checkered history in physics. Introduced by Einstein in 1917 as part of his *general theory of relativity*, it provided a repulsive force between galaxies, keeping them from collapsing together under their mutual gravitational attraction and thereby stabilizing the universe on the largest possible scale. Ten years later, when Edwin Hubble discovered that the galaxies were all rushing apart and that the universe was expanding (see the Physics Potpourri on p. 228), Einstein discarded his cosmological constant as being inconsistent with observation. It was reintroduced in 1948 by Fred Hoyle (who coined the term "Big Bang") and his collaborators, only to be rejected again. Now, this cosmological parameter has once more taken center stage as a result of work by two independent research groups seeking to understand the nature of the expansion of the universe by studying supernova explosions.

The groups, the High-Z Supernova Search Team and the Supernova Cosmology Project (SCP), use complementary but independent analysis techniques to determine the distances and redshifts,* z, to particular types of supernovae. Based on the work of these scientists, it is now possible to extend the Hubble plot (see Figure 6.18) out to distances and look-back times (remember, since the speed of light is constant, the farther out in space we look, the further back in time we see) where departures from the observed straight-line relationship between speed and distance would be expected. The question is, how do we interpret such departures?

If gravity were the only influence on the expansion of the universe, then we'd expect to see the galaxies gradually slowing down as time goes on due to the attractive forces between them. The greater the mass-energy density in the universe, the greater the attraction and the more rapid the slowing. Because looking farther out in space is equivalent to looking further back in time to earlier epochs in the universe's history, we should see the most remote galaxies moving faster—that is, having higher redshifts—than the nearby ones. A Hubble plot of velocity (y-axis) versus distance (x-axis) should begin to curve upward. The greater the mass-energy density, the greater the upward curvature.

But look at ● Figure E.1! It shows a Hubble plot that includes data for the so-called type Ia supernovae (see opening photograph), investigated by the High-Z and SCP teams, and

---

* Since the galaxies are all receding from us, the frequencies of their spectral lines are all reduced from what they would be if the galaxies were stationary. The wavelengths of the lines are thus increased and shifted to the red end of the spectrum. The amount of the redshift is given by $z = (\lambda_{observed} - \lambda_{emitted})/\lambda_{emitted}$. A redshift of $z = 1$ implies that the observed wavelength is twice as large as the emitted wavelength.

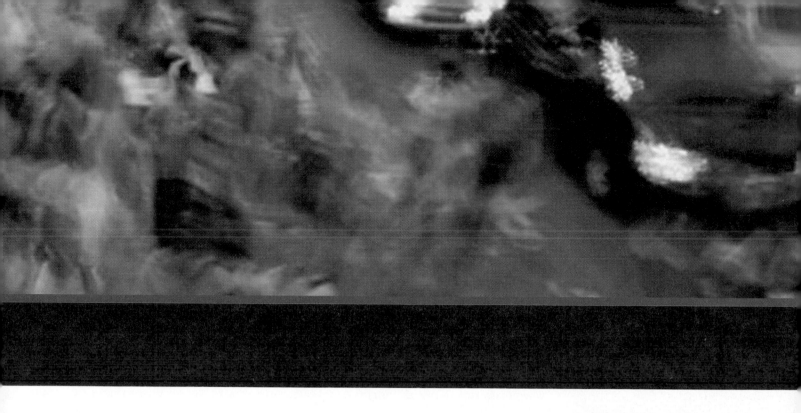

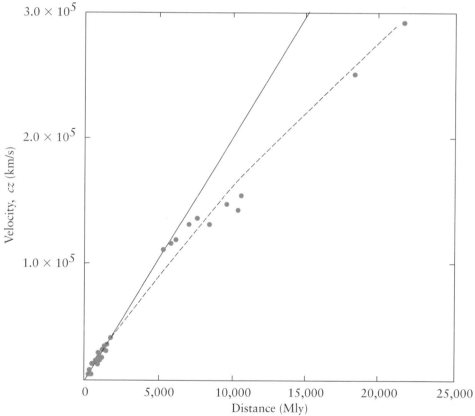

extends out to redshifts near $z = 1.0$. (For reference, $z = 0.5$ corresponds to a look-back time extending to about one-third the present age of the universe, or ~5 billion years.) Clearly, the data appear to be turning unmistakably *downward*, not upward, suggesting that the cosmic expansion has been *accelerating*, not decelerating, at least in the epoch since $z \leq 1$. Even for unreasonably small values of the universe's mass-energy density, the data seem to demand that a *repulsive* force be included in our cosmological models. This is precisely the kind of term Einstein envisioned when he introduced the cosmological constant.

Although the negative cosmic pressure seemingly required by the type Ia supernova observations is probably not constant over all space and time like Einstein's Λ, the fact of its existence, now confirmed by supernova observations out to $z > 1.5$, will continue to fuel what Washington University physicist Clifford Will has called a "renaissance" in general relativity. Among the most interesting questions spawned by this work concerns the source of this cosmic repulsion. And, while there are even now many speculative answers to this question, at this juncture in our inquiry into physics, the bigger question is what exactly is this general theory of relativity that forms the basis for any discussion of cosmology?

In what follows, we will consider several important features of general relativity, as well as some implications of this theory for the structure and evolution of the universe. In so doing, we will discover how the various branches of physics (mechanics, thermodynamics, electromagnetism, quantum mechanics, etc.) must be drawn together in order to even begin to address the most fundamental and interesting questions about the cosmos.

General Relativity

The **general theory of relativity,** developed by Einstein beginning in 1915, is basically a theory of gravity. It incorporates special relativity theory (see Section 12.1), and permits us to understand the motion of material particles and photons traveling in strong gravitational fields where Newton's law of universal gravitation (compare Section 2.8) gives only approximately correct answers. There exist several well-documented examples wherein the superiority of Einstein's theory of gravity to that of Newton's has been clearly demonstrated. We will consider some of these in due course. Let's first consider some ways in which Einstein's conception of gravity differs fundamentally from Newton's.

A basic postulate of general relativity is the *principle of equivalence.* It asserts that in a uniform gravitational field, all objects, regardless of their size, shape, or composition, accelerate at the same rate. Einstein recognized that this condition would make it impossible to physically distinguish the motion of a freely falling object near the Earth's surface from the motion of the same object released in a laboratory that was accelerating upward at a rate of 9.8 m/s² (1g) *in the absence of a local gravitational field.* As far as the laws of free fall are concerned, an accelerated laboratory is equivalent to one that is unaccelerated but possesses gravity. Einstein even went one step further and argued that not only are the laws of freely falling bodies the same for two such laboratories, but *all* the laws of physics are the same in such circumstances. The adoption of this so-called *strong* principle of equivalence leads to a very striking consequence: If the path of a light ray is observed to deviate from a straight line in an accelerating "laboratory" like the elevator in ● Figure E.2, then the path of a light ray will be similarly deflected in a gravitational field produced, say, by a massive object like the Sun.

● **Figure E.2** Two views of the same experiment involving the propagation of a pulse of light. Light, which travels in a straight line according to an observer at rest outside an elevator rising rapidly with an acceleration of 1g, will follow a curved trajectory according to an observer in the elevator. Both observers agree that the light enters the elevator near the top ledge of the left-hand window (1), and exits near the bottom ledge of the right-hand window (2'). The external observer (left) associates this fact with the upward motion of the elevator during the time required for the light pulse to cross the elevator car. The observer in the elevator (right), believing himself to be in an Earthlike gravitational field, interprets the observation to mean that gravity has caused the light path to bend downward. Both descriptions are equally valid. (This image is not drawn to scale and greatly exaggerates the amount of bending an actual light ray would suffer under these circumstances.)

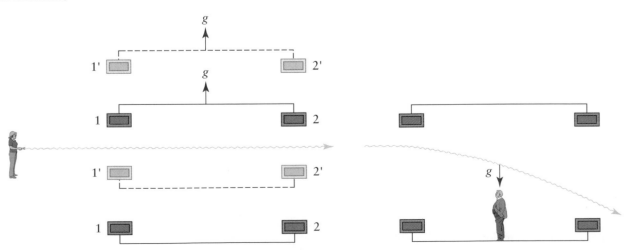

Einstein's interpretation of the cause of the deflection of a light ray in a gravitational field was quite different from that which Newton might give. Newton would likely speak of the attractive force acting between the mass of the gravitating body and the effective mass of the photons making up the light (compare Section 10.2). (Recall that photons have no rest mass. Their mass derives from the energy they possess according to Einstein's equation $E_0 = mc^2$; see Section 12.1.) This force produces a centripetal acceleration of the photon and a subsequent deviation in its motion from the straight line predicted by Newton's first law (see Section 2.2). Indeed, one can calculate, using Newtonian methods, the expected deflection for such a light ray, but the result, sadly, turns out to be equal to only one-half the observed deflection.

The remaining amount of deviation, and the correct explanation for the effect, belong to Einstein, who viewed the situation in a revolutionary way. In particular, he argued that there was no need to describe the deflection in terms of forces but noted instead that the paths followed by the photons are merely their natural or *geodesic* trajectories in a curved space (or, more properly, *spacetime*, because in relativity space and time are inextricably linked) that has been distorted by the presence of mass (and/or energy). Princeton physicist John A. Wheeler has contrasted the views of Einstein and Newton on this subject as follows.

According to Newton: *Force tells mass how to accelerate.*
                          Mass tells gravity how to exert force.

According to Einstein: *Curved spacetime tells mass-energy how to move.*
                          Mass-energy tells spacetime how to curve.

Perhaps a fable will help distinguish these two viewpoints more clearly. Imagine you are covering the final hole of the Master's Golf Championship from a helicopter above the eighteenth green. Sharing the reporting responsibilities with you are none other than Isaac Newton and Albert Einstein. As the three of you watch, Tiger Woods prepares to putt out for a birdie. After lining up his shot, he holes the putt and waves triumphantly to the cheering crowd. The trajectory followed by the ball to the cup, as seen from your aerial vantage point, is shown in ● Figure E.3a.

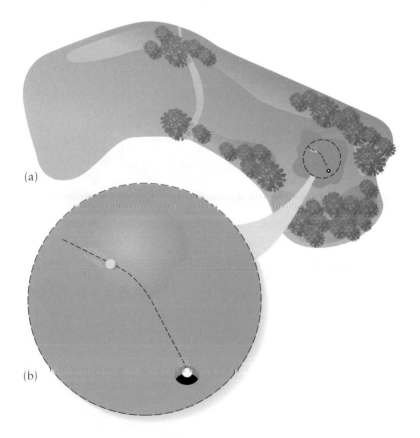

(a)

(b)

● **Figure E.3** (a) Path of a golf ball as seen from high above the surface of the green. The deviation from straight-line motion appears to require the action of some (unknown) force. (b) Close-up view of the surface of the green along which the ball moved. In rolling into the cup, the ball followed the natural contours of the green, including a small mound that deflected it toward the hole.

Noting that the path deviated from a straight line, Newton remarks that there must have been a force acting on the ball to accelerate it along the curved trajectory it followed to the cup. He speculates that perhaps there was a strong cross wind, which diverted the ball; or that Woods was using a steel-core ball and there was an iron ore deposit beneath the surface of the green that provided a magnetic force on the ball; or . . . At this juncture, we can imagine Einstein interrupting to point out that all this talk of forces is unnecessary because the ball was just following the natural contours of the curved surface of the eighteenth green on which it moved. To prove this, Einstein suggests that the helicopter land and the green be inspected close up. The result is depicted in Figure E.3b.

Einstein knew what all good golfers know: The surfaces of most greens are not flat planes but are instead rolling contours designed to challenge the skill of the players at "reading the green" and putting the ball so as to take advantage of the dips and curves. In making his final shot, Tiger Woods relied on no mysterious forces to move the ball but simply recognized that the ball would be traveling along a warped surface. He accurately assessed the natural path the ball should take in this space to reach the hole. This is how Einstein would have us view the effects of what we call "gravity."

As in all parables, there are limitations to the fictional story just told, but it serves to emphasize the fundamental differences in the world views of Einstein and Newton. Thus enlightened, we may now return to the issue of the deflection of light and address the source of the curvature: mass-energy. In Einstein's theory of general relativity, matter and energy warp space (and alter time). The greater the density of matter and energy present, the greater the warping—that is, the stronger the curvature. Again, an analogy in two dimensions may be helpful. Consider a large rubber sheet, stretched taut between supporting posts. If we place a marble near the center of the sheet, the weight of the marble will deform the surface only very slightly in its immediate vicinity (● Figure E.4a). The space surrounding the marble will remain flat, and the path of a small ball bearing rolled past the marble will be deviated by just the tiniest amount from a straight line due to the bending of the surface of the sheet. However, if we replace the marble with a bowling ball, the rubber will be stretched to a much greater degree, and the deformation of the space surrounding the bowling ball from a flat plane will be far larger and extend more widely around the ball. A ball bearing now rolled past the bowling ball will experience a space

● **Figure E.4** The warping of space by massive objects causes deviations in the paths of particles moving nearby. The greater the mass, the larger the deformation of the space surrounding it. (a) The marble produces only a small warp in the two-dimensional surface of a rubber sheet, and consequently a negligible alteration in the ball bearing's motion from a straight line. (b) The bowling ball creates a considerable depression in the surface of the sheet so that the path of the ball bearing deviates significantly from that of a straight line. In each case, the ball bearing is following a *geodesic* path along the surface of the sheet, but in (b), the surface has been warped from a flat plane by the action of a massive object.

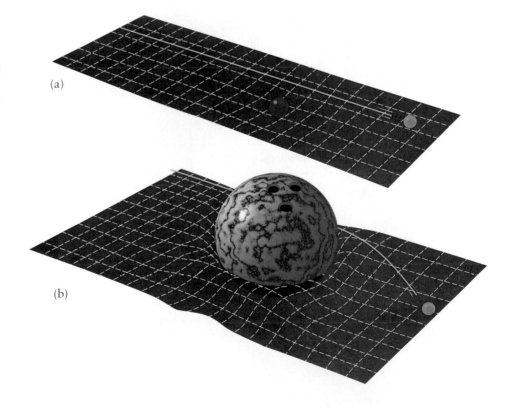

(a)

(b)

(a)

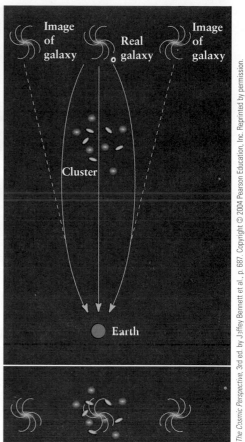

View as seen from Earth: Multiple images of same galaxy.

(b)

● **Figure E.5** (a) HST picture of a cluster of yellow, elliptical galaxies acting like a gravitational lens. The blue ovals surrounding the cluster are distorted images of a more distant galaxy lying almost directly behind the cluster. (b) The large mass of the cluster warps the space in its vicinity and bends the paths of light rays from a background galaxy passing nearby. Light arriving at Earth from different directions produces multiple, distorted images of the more distant galaxy. (From Bennett, *The Cosmic Perspective.*)

exhibiting considerable curvature, and its path will deviate significantly from a "straight" line (Figure E.4b).

A similar deflection effect happens in three dimensions for the light from distant stars that passes near the Sun or the light from remote galaxies that passes near a foreground cluster of galaxies. The first detection of this phenomenon was made in 1919 when the positions of stars close to the Sun in the sky were measured during the relative darkness provided by a solar eclipse. The observed differences in the positions of the stars during and outside the eclipse (when the Sun was far removed from them in the sky) agreed extremely well with the predictions of general relativity but only poorly with those of Newtonian gravitation.

More recently, astronomers have turned the deflection of light by massive objects from a test of the validity of general relativity to a tool to assess mass. ● Figure E.5a shows an HST image of a yellow cluster of galaxies acting as what is called a *gravitational lens.* The mass of this cluster has distorted the "fabric" of space in its vicinity and, like a lens, bends the light from a more distant blue galaxy to produce multiple images of it arrayed along circular arcs surrounding the cluster (Figure E.5b). Careful analyses of the images of a lensed galaxy in pictures like this using general relativity permit astronomers to establish the mass of the foreground cluster. In cases where comparisons can be made between this method and others used to determine cluster masses, the agreement has been found to be highly satisfactory. This lends credence to the view that we are correctly interpreting the nature of the observations, and it gives us confidence in the masses established via gravitational lensing alone in circumstances where other methods cannot be employed.

Another prediction of Einstein's general theory is that, because of the warping of space by mass, the orbital path of one spherical object about another will not be a closed ellipse as predicted by the inverse square law of Newton. Instead, it will be an open curve for which the point of closest approach between the two bodies slowly advances in space (an

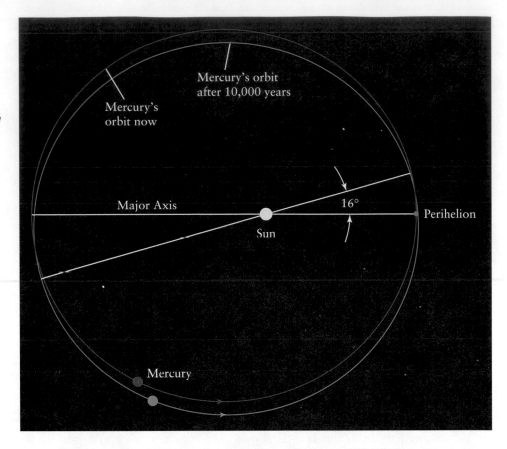

● **Figure E.6** Display of the precession of the perihelion of Mercury. This effect arises because the gravitational attraction of the Sun for Mercury deviates from a strict $1/r^2$ force law due, in part, to general relativistic effects. The actual precession of Mercury is small, accumulating to only about 16° every 10,000 years from all sources (including contributions caused by gravitational perturbations associated with nearby planets).

Peri- means "close to," and *helion* refers to the Sun.

Again, peri- means "close to"; *astro* means "star."

effect called precession). This is observed in the solar system in the motion of Mercury about the Sun (● Figure E.6): The *perihelion* point of Mercury's orbit precesses at a rate of 43 arcseconds (0.012 degrees) per century due to general relativity. The agreement between the predictions of general relativity and the observations for Mercury's orbit was one of the first successful tests of Einstein's theory.*

A more dramatic example of this is seen in the motion of two pulsars (see Physics Potpourri in Section 3.8) discovered to be orbiting one another by Russell Hulse and Joseph Taylor in 1974. The objects in this system (catalogued as PSR 1913 + 16) experience gravitational fields 10,000 times stronger than the Sun's field at Mercury, rendering the general relativistic effects far more significant than they are in the solar system. Indeed, the precession rate of the *periastron* point in PSR 1913 + 16 is a bit more than 4° *per year!*

By associating the observed precession in this binary with general relativistic effects, Hulse and Taylor were able to deduce the masses of the two stars and then conduct a very sensitive test of another aspect of general relativity theory: The prediction of a gradual decrease in the orbital period of the system due to energy losses in the form of *gravitational radiation.* This is something that has no analog in Newtonian theory.

Recall from Section 8.5 that when an electric charge is accelerated, it radiates electromagnetic waves, which propagate outward at the speed of light. Similarly, according to general relativity, if a massive object is accelerated, it, too, radiates energy in the form of *gravitational waves* that also travel at the speed of light. Compared to more common electromagnetic waves, gravitational waves are much less intense and very difficult to detect directly. Some very sophisticated experiments have been conducted in laboratories around the world to detect gravitational waves from celestial sources, but none has succeeded unambiguously.

---

* Some controversy surrounds the interpretation of the precession of Mercury's perihelion wholly in terms of general relativity because deviations from perfect sphericity in the shape of the Sun can produce similar effects. Measurements of the Sun's oblateness have yielded small enough values that no significant disagreements with general relativity have arisen, but some uncertainty remains.

For a system like the binary pulsar, general relativity predicts that, because of the gravitational radiation emitted from the centripetally accelerated masses, the two stars should gradually spiral inward toward one another, their orbital period getting shorter by a minuscule 75 *micro*seconds per year. Astonishingly, by 1983, data on this highly unusual double star had accumulated to yield a measured orbital decay rate of 76 ± 2 microseconds per year in near perfect agreement with general relativity. Thus, although the direct detection of gravitational waves remains an elusive goal,* their existence, based on the precise agreement between the predictions of general relativity and the observations of PSR 1913 + 16, can scarcely be in doubt. For these achievements, Taylor and Hulse were selected as the 1993 recipients of the Nobel Prize in physics.

Once we acknowledge the validity of general relativity and that mass and energy deform space(time), it becomes possible to inquire about the effect that *all* the matter and energy contained in the universe have on its global structure, evolution, and geometry. After all, the largest possible concentration of matter and energy is that present in the *entire* universe! Is the total amount of "stuff" in the universe large enough to significantly alter the "shape" or geometry of the universe from the flat space of Euclid, much as the presence of a heavy bowling ball deforms a flat rubber sheet? Questions like this one bring us into the realm of cosmology: The search for an understanding of what our universe is like on the grandest of scales, how it evolved to this state from its beginnings in the Big Bang (see Physics Potpourri on page 320), and what the future holds for us some 5 or 10 or 100 billion years from now. The answers to all these questions are not known with certainty, and it is not possible in the space that remains here for us to provide more than a glimpse of some of the current thinking about these issues. We shall thus end with a beginning, and focus on what is called the *flatness problem*.

Explore It Yourself

You can reproduce the effects of space warping in a manner similar to that shown in Figure E.4 as follows: Remove the bed linens from a mattress with a smooth surface. (A baby's crib mattress or the mattresses in many hide-a-beds work well.) Place a heavy object whose dimensions are small compared to those of the mattress at the center of the surface. A bowling ball serves the purpose very nicely, but it need not be a spherical object; a large brick or even a heavy book will do, depending on the firmness of the mattress. What is important is that the mattress surface be noticeably depressed near the object. Take a small, lightweight, round object, like a marble or a ping-pong ball, and roll it across the mattress surface. Begin by rolling the ball at a moderate speed parallel and close to one edge of the mattress. Describe the path followed by the object as it moves across the surface. Gradually set the starting point for the ball's motion nearer and nearer to the central axis of the mattress. What happens to the trajectory of the object? Describe its motion now. Vary the speed with which the object rolls. How does this affect the object's path along the surface? Can you explain what is happening using the principles of general relativity?

Cosmology

Before the voyages of exploration made by Columbus and others more than 500 years ago, many otherwise educated and urbane people believed that the world was flat. They shunned traveling too far from home for fear that it would lead them to fall off the edge of the world. The view of the global geometry of the world that they adhered to was strictly a bounded Euclidean one. (You may recall from your high school math courses that the geometry of Euclid is based on five postulates, one of which states that through a point not on a line, one and only one line may be drawn that is parallel to the original

---

* In 2002, the U.S.-based Laser Interferometer Gravitational Wave Observatory (LIGO) began operation. Its purpose is to directly detect gravitational radiation from space. Using two installations 2,000 miles apart, each with 1.2-meter-diameter vacuum pipes arranged in an L with 4-kilometer-long arms, small test masses fitted with mirrors that reflect ultrastable laser beams are monitored for motion attributable to passing gravitational waves. The use of widely separated stations is necessary to rule out test-mass motion due to local effects (vibrations and the like) that mimic the response to gravitational radiation. In its first year or so of operation, LIGO has not recorded a positive detection of gravity waves, but scientists are optimistic that this situation will change when LIGO is joined with other similar units in operation in Europe and Japan to improve the sensitivity of the experiments.

● **Figure E.7** Unlike the individual shown in this woodcut, it is impossible to encounter, much less poke through, a bounding surface to the universe. Modern cosmology teaches that any model universe, regardless of its specific geometrical properties, is unbounded in the sense of having no edges. The universe encompasses the entire spacetime volume that exists. There is nothing outside the universe to be experienced as suggested by this image.

line.) Although our appreciation for the vastness and complexity of the cosmos has increased during the last 500 years, it is remarkable that humankind is still struggling to understand the geometry of the universe, and that the early savants may have been correct after all: On the largest of scales, the universe may well be Euclidean although it is clearly without boundaries or edges (● Figure E.7).

According to general relativity, the curvature of the universe is dependent on its total mass-energy density, Ω_T, including the energy contained in the field associated with the cosmic repulsion, now commonly called *dark energy.* If, in the dimensionless units employed in general relativity, this value equals unity, space is flat on a cosmological scale. Any deviation from $\Omega_T = 1$ means that the geometry of the universe will not be that of a three-dimensional Euclidean space(time). Besides the geometry of Euclid, two other general types of geometry exist that could describe the global structure of the universe. They differ from Euclidean geometry only in terms of the parallel-line postulate. One is called *Riemannian geometry,* after mathematician Georg F. B. Riemann, and it adopts the postulate that through any point not on a line, *no* parallels may be drawn. This leads to geometrical characteristics in three dimensions analogous to those in two dimensions on the surface of a sphere. The other type of geometry demands that through any point not on a line, an *infinite* number of parallel lines may be constructed. The characteristics of this geometry were developed by Nikolai Lobachevski and, in three dimensions, are similar to the properties of two-dimensional surfaces shaped like the bells of trombones.

A variety of observations in the last few years suggests that the contribution to Ω_T from matter (particles), Ω_M, is about 30% of the critical value, that is, $\Omega_M \sim 0.3$. Of this, most (perhaps 99%) of the matter is "dark" (that is, nonluminous), and a significant fraction (more than about 85%) is likely to be nonbaryonic (compare with the discussion in Section 12.2). Particle physics offers three candidates for such material: (1) *axions,* hypothetical particles thought to account for nature's distinction between left and right (see the Physics Potpourri on page 486), whose masses range from ~ 1 eV/c^2 to as low as 10^{-16} eV/c^2 depending on the model; (2) *neutralinos,* hypothetical so-called *weakly interacting massive particles* (WIMPs), with masses between about 10 and 500 GeV/c^2; and (3) ordinary neutrinos of mass ≤ 1 eV/c^2. Current estimates place the neutrino contribution to the mass-energy density of the universe at between 0.1 and 5%, far too small to account for the bulk of the nonbaryonic contribution to Ω_T. Neutralinos are electrically neutral particles that interact with normal matter only through gravitation. Such particles, if they exist, would likely attach themselves to galaxies and form halos around them. All searches for WIMPs, including neutralinos, have failed to produce any conclusive results to date.

The CERN Axion Solar Telescope is currently looking for these elusive particles that may provide the explanation for the dark-matter component of the universe.

Matter provides only about a third of the critical mass-energy density required for a flat universe. What about the dark-energy contribution, Ω_Λ? Based on observations of the type Ia supernovae, the High-Z and SCP groups independently arrive at a value for Ω_Λ of about 70%. Remarkably, the sum $(\Omega_M + \Omega_\Lambda)$ is very close to unity (● Figure E.8). Consequently, the universe is very close to being Euclidean in structure. If the present mass-energy density of the universe is even within a factor of 2 of the critical density after ~ 14 billion years of expansion, then, extrapolating back to the time of the Big Bang, the original mass-energy density must have agreed with the critical value to better than a few parts in 10^{61}! Thus, it would seem that the universe appears flat now because it was flat at the beginning.

The precise nature of the dark-energy field that appears to be responsible for the observed acceleration in the expansion of the universe is not yet known. Clues to the nature of the dark-energy source will likely be found in the fine details of the history of the cosmos, because the different dark-energy models currently under investigation provide different expansion rates. The differences, however, are extremely small and will require measurements of supernovae brightnesses that are more accurate by up to a factor of ten, as well as the extension of the observations to larger redshifts. This is a big challenge to observational cosmologists but one that may well be met in the next decade with the advent of new large telescopes with adaptive optics capabilities (see Section 9.2) and the launch in 2010 of NASA's replacement for HST, the 6-meter James E. Webb Space Telescope.

But why should the universe have started out with just exactly the right amount of matter and energy to render it flat instead of some other value that would have caused it to be curved in a different manner? How are we to interpret the fact that the universe seems to have begun with this carefully selected initial condition? The original Big Bang scenarios had no good answers to these questions. And physicists are always suspicious of specially arranged circumstances that do not themselves emerge naturally from the physics of the situation but have to be imposed from outside the theory. This is the essence of the flatness problem.

The currently proposed solution to this problem avoids the necessity of prescribing special initial conditions by introducing some new physics. It provides the missing link that led to the deficiency in the explanatory power of the initial Big Bang models of the universe. Elementary particle physicists have come to the rescue of cosmologists and have developed a new approach that has solved the flatness problem (and several other difficulties inherent in the original Big Bang scenarios). The model for the universe that has emerged from the marriage of the ultrasmall and the ultralarge is known as the *inflationary universe*.

As discussed in Section 12.5, grand unified theories (GUTs) seek to amalgamate the fundamental interactions of nature into a unified whole. Such a unification, in which the four forces become indistinguishable, can occur only at very high energies, about 10^{16} to 10^{19} GeV. Such energies are unattainable in terrestrial laboratories but were amply present at the creation of the universe in the Big Bang. The era in which these enormous energies prevailed, some 10^{-43} s after the Big Bang, is known as the *TOE* (or Theory of Everything) *epoch*. Initial attempts to develop a TOE focused on *superstring theory*, in which the fundamental particles correspond to different vibrational frequencies of 10-dimensional strings. Since space and time together provide only four dimensions, the remaining six dimensions in these theories (there are five variants) remain curled up into knots with sizes of the order of 10^{-33} meters, much too tiny to see.

More recent efforts have succeeded in merging the five competing theories into a grander, 11-dimensional theory called *M-Theory* (for membrane, matrix, magic, or mystery, depending on whom you read). According to this model, our universe is "glued" to a three-dimensional "brane" (short for membrane) embedded in a higher-dimensional hyperspace. Many other parallel universes may exist on other branes, but they remain invisible to us because photons, in this conception, cannot cross brane boundaries. Only gravitons can do so, indicating that we may be able to "feel" the influence of these other

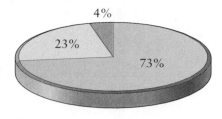

Dark energy ☐ Baryonic matter

☐ Nonbaryonic dark matter

● **Figure E.8** Pie chart showing the composition of the universe. Most of the mass-energy density of the cosmos is tied up in dark energy, the source of the negative pressure responsible for the acceleration of the universe. The remainder is mostly composed of nonbaryonic dark matter. Only about 4% of the mass-energy of the universe is made up of ordinary matter, and most of it also is "dark," consisting of gas distributed around and between galaxies. The luminous material in the universe (stars, nebulae, and galaxies) constitutes only about 1% of its total mass-energy density.

universes (and the matter in them) through the gravitational force. This has been a suggested solution for the dark-matter problem.

Fantastic as these predictions are, M-Theory does have one distinguishing characteristic lacking in the earlier string theories: Some of its predictions appear to be testable. For example, if strong sources of gravitational radiation are ever detected with LIGO that cannot be identified optically, these sources could be candidates for concentrations of gravitons transmitted to our brane from normal matter on another adjacent brane. As further refinements in both theory and observation are made, we can expect to learn more about the degree to which M-Theory fulfills its promise to unify the realm of microphysics (quantum mechanics) with that of macrophysics (general relativity).

At the very beginning of time, then, we believe there was complete symmetry among the forces of nature. As the universe evolved, the energy density declined, and a spontaneous break in the symmetry occurred, as one after the other of the forces was "frozen out" and assumed its own identity. These episodes of spontaneously broken symmetry corresponded to phase changes in the universe. Several things happened to the structure and evolution of the early universe assuming that such symmetries existed but then were broken. One of them was that the universe suffered a period of rapid and enormous inflation in size, which effectively rendered it flat, regardless of its initial curvature. Let's see how this developed.

As the universe expanded and cooled, the first phase transition that occurred was the one that split off the gravitational force. The next one, and the one that is important for our discussion, was that which pared the strong nuclear force from the remaining electroweak force. This phase change may be compared to the solidification of water (refer to Sections 4.1 and 5.6). If a vessel of water is slowly cooled and the container is not disturbed during the process, it is possible to *supercool* the water below 0°C without it freezing. The system gets stuck in a metastable liquid phase. Eventually, some perturbation will occur, and nucleation will begin. Very quickly, ice crystals will form and spread through the volume, converting the symmetric liquid phase (in which the molecules show no preferred orientation with respect to any particular direction) to a distinctly asymmetric solid phase (which exhibits characteristic anisotropies along different crystalline directions). The same kind of thing happened in the universe when the strong nuclear force was "frozen out."

Like the water in the example above, we believe the universe got stuck in a metastable state, called a *false vacuum state,* as it expanded and cooled. If this phase was prolonged, the energy density (there was no mass in the universe at this very early period) consisted principally of that associated with the vacuum state. Only a small part of the energy density was tied up in radiation. The nature of a vacuum energy density may be likened to that of the latent heat of fusion in a liquid (see page 199). The latent heat of fusion is the amount of energy that must be present at the melting point to maintain the molecules of a substance in the liquid phase. That is, it is the amount of internal energy required to prevent the molecules from binding together to form a solid. Similarly, the vacuum energy density was the energy needed to maintain the universe in a symmetric phase wherein the strong and electroweak forces were joined. It was this vacuum energy that ultimately drove the inflation.

When the universe eventually underwent the phase transition that severed the strong interaction from its electroweak sibling, enormous amounts of energy were released, exactly as happens when water freezes and heat is removed to the surroundings. The energy thus provided produced a rapid and extraordinary expansion of the scale of the universe. In fact, the equations of general relativity show that the inflation occurred in an exponential fashion (refer to Physics Potpourri on page 112) causing the universe to grow by a factor of at least 10^{25} (and perhaps as much as 10^{50}) in the period when the universe was about 10^{-32} seconds old. The era of inflation was very short (only about 10^{-34} seconds or so long) and ended when the transition to the asymmetric state was fully achieved and the vacuum energy was completely depleted. After this, the universe resumed its normal expansion characteristics, but the effects of the inflationary period were profound.

In particular, if regions of spacetime in the very early universe were highly curved before the phase change, the inflation reduced their curvature, rendering them flat. A

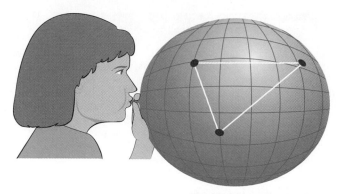

● **Figure E.9** By inflating a balloon quickly to very large dimensions, the curvature of the surface can be reduced to a point where it is difficult, by local measurements, to distinguish it from a flat, Euclidean space. The same effect may have occurred in the early universe if the inflationary hypothesis is correct. If so, then the flatness problem that plagued early versions of the Big Bang can be solved.

vivid two-dimensional analogue that helps explain how this could happen involves the surface of a balloon that is being inflated (● Figure E.9). At the beginning of the process, the balloon's surface is quite small and highly curved. As air gradually enters the balloon, the surface expands and, as perceived by a local observer who samples only a small portion of the space, becomes less highly curved. Imagine now blowing into the balloon very hard so that it inflates rapidly to extremely large dimensions. After the period of inflation is over, the surface of the balloon is so distended that *all* local observers *anywhere* in the two-dimensional space of the surface of the fabric see a flat space. Regardless of what the curvature of the balloon's surface was at the start, once the inflation is complete, it looks flat. A similar circumstance is believed to have occurred in the early universe, and it is for this reason that in the present epoch we see the mass-energy density so close to the critical density for a flat, Euclidean world. A nice aspect about this explanation for the flatness problem is that it requires no special initial conditions for its success. *Whatever* the initial state of the universe, after the inflation, it always arrives at a final state that is flat in terms of its spacetime curvature. Put another way, Ω_T naturally approaches unity due to the inflation with no need for *ad hoc* assumptions about what conditions in the Big Bang were like prior to an age of 10^{-34} seconds.

The inflationary cosmology, developed in the 1980s by Alan Guth, Paul Steinhardt, and Andre Linde, has also provided solutions to several other nagging problems associated with the Big Bang (such as the relative smoothness in the distribution of cosmic microwave background radiation; page 321). The success of this model derives partly from its reliance on physical theories designed to explain the fundamental structure of matter on the tiniest of dimensions. The physics of the smallest scale has thus led to the elucidation of the physics of the largest scale. Elementary particle physics and cosmology have been inextricably linked, and advances in one area have fostered advances in the other. This type of synergy is at the heart of modern scientific exploration, whether in physics, chemistry, biology, medicine, environmental science, or any other field.

The word *epilogue* literally means "to say in addition." It frequently refers to the concluding section of a piece of literature or a play that summarizes or comments on the design or action of the work. In this epilogue, we have focused on certain elements of general relativity theory that have opened up new avenues for exploring and understanding the cosmos on both local and global scales. If you carefully review the discussion, particularly noting the many references to earlier sections and topics in the text, you can see just how much "other" physics is required to fully appreciate the perspectives on the universe that general relativity offers us. It is not easy to acquire such an appreciation by considering the various physical concepts introduced in this book in isolation from one another. To succeed in this endeavor demands an integration and a synthesis of principles and concepts from the entire body of theory and experiment that constitutes what we call "physics." And so it is, or should be, in all scientific endeavors. This, then, is the summary statement that we wish to leave you with. We hope we have piqued your curiosity about the natural universe so that, confident in the knowledge you have gained from reading this text, you will continue your own personal inquiry into physics.

Adams, F., and G. Laughlin. *The Five Stages of the Universe: Inside the Physics of Eternity.* New York: The Free Press, 1999. A thorough and entertaining discussion of the long-term future of the universe in which the nature of complexity at each of the five stages in its development from creation to the final "Dark Age" is investigated.

Arkani-Hamed, N., S. Dimopoulos, and G. Dvali. "The Universe's Unseen Dimensions." *Scientific American* 283, no. 2 (August 2000): 62–69. A description of brane theory that postulates the existence of additional dimensions for the universe that can be sensed only through gravity and may offer the chance to unify the four fundamental forces of nature.

Cline, D. B. "The Search for Dark Matter." *Scientific American* 288, no. 3 (March 2003): 50–59. The dynamics of clusters of galaxies suggests that the universe is filled with dark matter. This article explores the possible candidates for this material and the ways in which it might be detected.

Duff, M. J. "The Theory Formerly Known as Strings." *Scientific American* 278, no. 2 (February 1998): 64–69. This article discusses how attempts to develop a Theory of Everything have advanced beyond strings to include membranes: p-branes, black-branes, five-branes, and so on.

Freedman, W. L., and M. L. Turner. "Cosmology in the New Millenium." *Sky & Telescope* 106, no. 4 (October 2003): 30–41. Two of the country's leading astrophysicists discuss the age, make-up, and eventual fate of the universe, emphasizing the precision with which recent observations have allowed us to determine the fundamental parameters that quantify the cosmos.

Gibbs, W. W. "Ripples in Spacetime." *Scientific American* 286, no. 4 (April 2002): 62–71. Review of ongoing attempts to discover gravity waves using interferometric techniques. This piece provides a nice description of the major international initiatives already in operation or soon to be completed, including LIGO.

Greene, B. *The Elegant Universe: Superstrings, Hidden Dimensions, and the Quest for the Ultimate Theory.* New York: W. W. Norton & Co., 1999. Written by a professor of physics and mathematics at Columbia University, this is a lively and accessible discussion exploring recent advances in string theory with clarity and depth.

Guth, A. H., and A. P. Lightman. *The Inflationary Universe: The Quest for a New Theory of Cosmic Origins.* New York: Perseus Publishing, 1998. An in-depth discussion of the inflationary universe by one of the originators of the theory.

Kirshner, R. P. *The Extravagant Universe: Exploding Stars, Dark Energy, and the Accelerating Cosmos.* Princeton, NJ: Princeton University Press, 2002. As one of the leaders of the High-Z Supernovae Search team, the author provides an insider's view of the discovery of type Ia supernovae and their role in providing evidence for dark energy and the acceleration of the universe.

Krauss, L. M. "The History and Fate of the Universe: Guide to Accompany the Contemporary Physics Education Project Cosmology Chart." *The Physics Teacher* 41, no. 3 (March 2003): 146–155. The author provides a set of explanatory notes to assist students and teachers in interpreting the CPEP's wall chart describing the history and fate of the universe. The chart is available through the CPEP's web site: *http://CPEPweb.org.*

Milgrom, M. "Does Dark Matter Really Exist?" *Scientific American* 287, no. 2 (August 2002): 42–52. A discussion of a controversial alternative explanation for the acceleration of the universe based on a modification of the law of gravity instead of the putative existence of dark matter and dark energy.

Naeye, R. "Delving into Extra Dimensions." *Sky & Telescope* 105 no. 6 (June 2003): 39–44. A highly readable article covering elements of string theory and M-Theory. It emphasizes new developments like finite-sized extra dimensions and the cyclic model universe that may have observational consequences that can be tested in the foreseeable future.

Perlmutter, S. "Supernovae, Dark Energy, and the Accelerating Universe." *Physics Today* 54, no. 4 (April 2003): 53–60. The leader of the Supernova Cosmology Project reviews the evidence supporting the current view that the expansion of the universe is accelerating and discusses some aspects of the dark-energy field that may be driving it.

Tegmark, M. "Parallel Universes." *Scientific American* 288, no. 5 (May 2003): 40–51. The author describes how current cosmological observations may imply the existence of other, "parallel" universes and discusses the properties of four such possibilities.

Winners of the Nobel Prize in Physics

Since 1901, the Nobel Prizes have been awarded nearly annually (some prizes were not presented during the two world wars or when suitable candidates were not identified) in the fields of physics, chemistry, medicine and physiology, and literature, as well as for peace, to individuals whose work has

● **Figure A.1** Alfred Nobel

been of major benefit to humankind. (A Nobel Prize in economics was added in 1969.) The prizes commemorate the life and work of the Swedish chemist and industrialist Alfred B. Nobel (1833–1896; ● Figure A.1). Nobel, who had little more than an elementary school education, was a highly imaginative and restless person. At the time of his death, he had never held permanent residency in any country, had become a noted linguist (reading, writing, and speaking five languages) who wrote poetry in English, and had accumulated

more than 355 patents in different countries around the globe. His best-known inventions were the "Nobel lighter," a device developed in 1862 that made it possible to safely and reliably harness the explosive power of nitroglycerine,* and later a product he called *dynamite,* or nitroglycerine in the form of a paste wrapped in a paper cylinder. However, he also invented other types of high explosives (including blasting gelatin and *ballistite,* a smokeless power and precursor to today's cordite), designed some prototypes for rockets and aerial torpedoes, and worked on producing artificial rubber and leather, as well as artificial silk (which was later perfected by others as rayon). During his life, Nobel amassed great wealth from the royalties on his many inventions, and, in his will dated 27 November 1895, he stipulated that the income from his estate should be used each year to support the awarding of what have come to be the most prestigious civic honors in the world.

To date, more than 600 Nobel Prizes have been awarded to recipients representing over 40 countries. (Nobel expressly stated that in awarding the various prizes, "no consideration whatever be paid to the nationality of the candidate, so that the most worthy receives it.") Each prize winner receives a specially designed diploma, a gold medal, and a cash award. In physics and chemistry, the selection of Nobel Prize recipients is made by the Royal Swedish Academy of Science, based on nominations presented by former prize winners and members of recognized institutions of higher learning and official national academies of science of any country. The following list gives the Nobel laureates in physics, the year of their award, and the discovery or achievement that led to their selection.

1901 *Wilhelm Roentgen* Discovery of x rays.

1902 *Hendrik A. Lorentz and Pieter Zeeman* Research on the relationship between magnetism and radiation.

1903 *Antoine Henri Becquerel* Discovery of radioactivity.
Pierre Curie and Marie Curie Research in radioactivity.

1904 *John William Strutt (Lord Rayleigh)* Discovery of argon.

1905 *Philipp E. A. von Lenard* Research on cathode rays.

1906 *Joseph J. Thomson* Studies on conduction of electricity in gases.

1907 *Albert A. Michelson* Optical instruments and investigations using them.

1908 *Gabriel Lippmann* Interference method in color photography.

1909 *Guglielmo Marconi and Carl Ferdinand Braun* Contributions to wireless communications.

1910 *Johannes D. van der Waals* Equation of state for gases and liquids.

1911 *Wilhelm Wien* Studies of blackbody radiation.

1912 *Nils G. Dalén* Automatic regulators for use in coastal lighting.

1913 *Heike Kamerlingh Onnes* Research in low-temperature physics.

1914 *Max von Laue* Diffraction of x rays by crystals.

1915 *William H. Bragg and William L. Bragg* Studies of crystal structure using x rays.

1916 No prize

1917 *Charles G. Barkla* Discovery of characteristic x-ray spectra of elements.

1918 *Max Planck* Discovery of quantization of energy.

1919 *Johannes Stark* Effect of electric fields on spectral lines.

1920 *Charles-Édouard Guillaume* Discovery of anomalies in nickel-steel alloys.

1921 *Albert Einstein* Explanation of the photoelectric effect.

1922 *Niels Bohr* Quantum model of the atom.

1923 *Robert A. Millikan* Work on the electron's charge and on the photoelectric effect.

1924 *Karl M. G. Siegbahn* Research in x-ray spectroscopy.

1925 *James Franck and Gustav Hertz* Studies of collisions between electrons and atoms.

1926 *Jean B. Perrin* Studies on the discontinuous structure of matter.

1927 *Arthur H. Compton* Scattering of electrons and photons.
Charles T. R. Wilson Invention of the cloud chamber to study the paths of charged particles.

1928 *Owen W. Richardson* Work on the emission of electrons from hot objects.

1929 *Louis-Victor de Broglie* Discovery of the wave nature of electrons.

1930 *Chandrasekhara Venkata Raman* Studies of the scattering of light by atoms and molecules.

1931 No prize

1932 *Werner Heisenberg* Creation of quantum mechanics.

1933 *Erwin Schroedinger and Paul A. M. Dirac* Contributions to quantum mechanics.

1934 No prize

1935 *James Chadwick* Discovery of the neutron.

1936 *Victor F. Hess* Discovery of cosmic rays.
Carl D. Anderson Discovery of the positron.

1937 *Clinton J. Davisson and George P. Thomson* Discovery of diffraction of electrons by crystals.

1938 *Enrico Fermi* New radioisotopes produced by neutron irradiation.

1939 *Ernest O. Lawrence* Invention of the cyclotron.

1940–
1942 No prize

1943 *Otto Stern* Discovery of the magnetic moment of the proton.

1944 *Isador I. Rabi* Nuclear magnetic resonance.

1945 *Wolfgang Pauli* Discovery of the exclusion principle.

1946 *Percy W. Bridgman* Contributions to high-pressure physics.

1947 *Edward V. Appleton* Reflection of radio waves by the ionosphere.

1948 *Patrick M. S. Blackett* Discoveries in nuclear physics.

1949 *Hideki Yukawa* Prediction of mesons.

---

*This development followed an explosion at one of Nobel's manufacturing plants that killed five people, including his youngest brother, Emil.

1950 *Cecil F. Powell*   Photographic method of studying nuclear processes.

1951 *John D. Cockroft and Ernest T. S. Walton*   Transmutation of atomic nuclei by accelerated particles.

1952 *Felix Bloch and Edward M. Purcell*   Research on the magnetic fields of nuclei.

1953 *Frits Zernike*   Invention of the phase-contrast microscope.

1954 *Max Born*   Research in quantum mechanics.
Walther Bothe   Development of the coincidence method.

1955 *Willis E. Lamb*   Discoveries concerning hydrogen spectrum.
Polykarp Kusch   Determination of the magnetic moment of the electron.

1956 *William Shockley, John Bardeen, and Walter H. Brattain*   Discovery of the transistor.

1957 *C. N. Yang and T. D. Lee*   Prediction of the violation of conservation of parity.

1958 *Pavel A. Čerenkov, Ilya M. Frank, and Igor Tamm*   Discovery and interpretation of the Čerenkov effect.

1959 *Emilio G. Segrè and Owen Chamberlain*   Discovery of the antiproton.

1960 *Donald A. Glaser*   Invention of the bubble chamber.

1961 *Robert Hofstadter*   Studies of electron scattering by nuclei.
Rudolf L. Mössbauer   Studies of emission and absorption of gamma rays in crystals.

1962 *Lev D. Landau*   Theoretical work on condensed matter, especially liquid helium.

1963 *Eugene P. Wigner*   Laws governing interactions between protons and neutrons in the nucleus.
Maria Goeppert-Mayer and J. Hans D. Jensen   Discoveries related to nuclear shell structure.

1964 *C. H. Townes, Nikolai G. Basov, and Alexander M. Prochorov*   Discovery of the principle behind lasers and masers.

1965 *Sin-Itiro Tomonaga, Julian Schwinger, and Richard P. Feynman*   Development of quantum electrodynamics.

1966 *Alfred Kastler*   Optical methods for studying atomic structure.

1967 *Hans A. Bethe*   Theory of nuclear reactions in stars.

1968 *Luis W. Alvarez*   Contributions to elementary particle physics.

1969 *Murray Gell-Mann*   Discoveries concerning elementary particles.

1970 *Hannes Alfvén*   Work in magnetohydrodynamics.
Louis Néel   Discoveries concerning ferrimagnetism and antiferromagnetism.

1971 *Dennis Gabor*   Holography.

1972 *John Bardeen, Leon N. Cooper, and J. Robert Schrieffer*   Explanation of superconductivity.

1973 *Leo Esaki*   Discovery of tunneling in semiconductors.
Ivar Giaever   Discovery of tunneling in superconductors.
Brian D. Josephson   Prediction of properties of supercurrent through a tunnel-barrier.

1974 *Antony Hewish*   Discovery of pulsars.
Martin Ryle   Pioneering work in radio astronomy.

1975 *Aage Bohr, Ben Mottelson, and James Rainwater*   Discovery of the relationship between collective and particle motion in nuclei.

1976 *Burton Richter and Samuel C. C. Ting*   Discovery of the J/ψ particle.

1977 *John N. Van Vleck, Nevill F. Mott, and Philip W. Anderson*   Studies of electrons in magnetic solids.

1978 *Arno A. Penzias and Robert W. Wilson*   Discovery of the cosmic background radiation.
Peter L. Kapitza   Work in low-temperature physics.

1979 *Sheldon L. Glashow, Abdus Salam, and Steven Weinberg*   Work on the unification of the weak and electromagnetic interactions.

1980 *Val L. Fitch and James W. Cronin*   Experimental proof of charge-parity (CP) violation, which has implications for cosmology.

1981 *Nicolaas Bloembergen and Arthur L. Schawlow*   Contributions to the development of laser spectroscopy.
Kai M. Siegbahn   Contributions to the development of high-resolution electron spectroscopy.

1982 *Kenneth G. Wilson*   Theory of critical phenomena in connection with phase transitions.

1983 *Subrahmanyan Chandrasekhar and William A. Fowler*   Work relating to the evolution of stars.

1984 *Carlo Rubbia and Simon van der Meer*   Discovery of the W and Z particles.

1985 *Klaus von Klitzing*   Discovery of the quantized Hall effect.

1986 *Ernst Ruska*   Design of the first electron microscope.
Gird Binning and Heinrich Rohrer   Design of the scanning tunneling microscope.

1987 *J. George Bednorz and K. Alex Müller*   Discovery of superconductivity in a new class of materials.

1988 *Leon Lederman, Melvin Schwartz, and Jack Steinberger*   Experiment that established the existence of a second kind of neutrino and that employed the first beam of neutrinos produced in a laboratory.

1989 *Norman Ramsey, Hans Dehmelt, and Wolgang Paul*   Contributions to the development of precision atomic spectroscopy, which laid the basis for cesium atomic clocks and other devices.

1990 *Jerome Friedman, Henry Kendall, and Richard Taylor*   Experimental verification of the existence of quarks.

1991 *Pierre-Gilles de Gennes*   Development of methods for studying complex systems at the atomic level.

1992 *Georges Charpak*   Invention and development of particle detectors.

1993 *Russell Hulse and Joseph Taylor*   Discovery of a binary pulsar.

1994 *Bertram N. Brockhouse and Clifford Shull*   Development of neutron-scattering techniques.

1995 *Frederick Reines and Martin L. Perl*   Pioneering experimental contributions to lepton physics.

1996 *David M. Lee, Douglas D. Osheroff, and Robert C. Richardson*   Discovery of superfluidity in helium-3.

1997 *Steven Chu, Claude Cohen-Tannoudji, and William Phillips*   Development of methods to cool and trap atoms with laser light.

1998 *Robert Laughlin, Horst Störmer, and Daniel Tsui*   Discovery and explanation of the fractional Hall effect.

1999 *Gerardus 't Hooft and Martinus J. G. Veltman*   Contributions to mathematical aspects of the standard model of particle physics.

2000 *Zhores I. Alferov, Herbert Kroemer, and Jack S. Kilby*   Invention and development of devices at the heart of modern information technology.

2001 *Eric A. Cornell, Wolfgang Ketterle, and Carl E. Wieman*   Creation of Bose-Einstein condensates.

2002 *Raymond Davis Jr., Masatoshi Koshiba, and Riccardo Giacconi*   Detection of cosmic neutrinos and contributions to x-ray astronomy.

2003 *Alexei A. Abrikosov, Vitaly L. Ginzburg, and Anthony J. Leggett*   Contributions to the theoretical understanding of superconductivity and superfluidity.

Math Review

The following is a review of the math used in this book that goes beyond the simple arithmetic you use nearly every day. Although most of it should be familiar to you, you may have forgotten some of it if you have not used it in some time. Keep in mind that mathematics plays a relatively small role in this book. It is used to express concisely and to apply interrelationships that must first be understood conceptually.

Basic Algebra

Algebra is arithmetic with letters used to represent numbers. Let's say, for example, that you form an equal partnership with three other people. Your share of any income or expense is then one-fourth of each amount. When an item is purchased for $60, your share of its cost is $15. This situation can be represented by an algebraic relationship—an equation. If we let the letter S stand for your share, and the letter A stand for the amount of the income or expense, then S is equal to one-fourth of A, or 0.25 multiplied by A. The equation is

$$S = 0.25 \times A$$

or

$$S = 0.25A$$

The multiplication sign, $\times$, is usually omitted in algebra. Whenever a number and a letter, or two letters, appear side by side, they are multiplied by one another.

S and A are called *variables* because they can take on different values (that is, they can be different numbers). An algebraic equation expresses the mathematical relationship that exists between variables.

To find your share, S, of a $200 payment, you replace A with $200 and perform the multiplication:

$$S = 0.25A$$
$$S = 0.25 \times \$200$$
$$S = \$50$$

The use of algebra in this situation really isn't necessary. All you have to do is to remember to multiply the total payment by 0.25 to compute your share. The algebraic equation is just a shorthand way of expressing the same thing.

This approach can be broadened for use with larger or smaller partnerships. We can let F represent the fraction for the partnership. In the above example, F is 0.25. For a partnership of two people, F is one-half, or 0.5. For 10 people, it is one-tenth or 0.1. Then the share, S, of an amount, A, for each partner is given by

$$S = FA$$

For a partnership of 10 people, the share of a $200 income is

$$S = 0.1 \times \$200$$
$$S = \$20$$

You might be able to compute this mentally, but the equation is a formal statement of the rule that you use to compute the answer.

This type of equation, in which one quantity equals the product of two others, is the one that appears most often in this book. A large number of important relationships in physics take the form of this equation.

The above equation can be used in additional ways. For example, if the share S is known, the equation can be used to find the original amount A. This can be done because any equation can be manipulated in certain ways without altering the precise relationship it expresses. The basic rule is:

A mathematical operation can be performed on an equation without altering its validity as long as *the exact same thing is done to both sides.*

The following examples illustrate the kinds of operations that can be useful. Numerical values are also used in the examples to show that the relationship expressed by each equation is still correct.

Addition

A number or a variable can be added to both sides of an equation. Let's say we have the equation

$$A = B - 6$$

This means that when B is known to be 8, A is 2:

$$2 = 8 - 6$$

If B is 11, A is 5; if $B = 83$, $A = 77$; and so on.

We can add any quantity to both sides of the equation without affecting its validity. For example, we can add 4 to both sides of the equation $A = B - 6$:

$$A + 4 = B - 6 + 4$$
$$A + 4 = B - 2$$

The relationship has not been changed. This means that any pair of values for A and B that satisfied the original equation will also satisfy the "new" equation. The values $B = 8$ and $A = 2$, which worked in the first equation, also work in the second:

$$2 + 4 = 8 - 2$$
$$6 = 6$$

Why would we want to add something to both sides of an equation? We can do this to change the focus of the equation. The original equation is used to find A when B is the known, or given, quantity. When B is known (or chosen) to be 8, the equation indicates that A must then be 2. If, instead, the value of A is known and we wish to find B, we can change the equation so that B equals something in terms of A. In this case we simply add 6 to both sides:

$$A = B - 6$$
$$A + 6 = B - 6 + 6$$
$$A + 6 = B$$

or

$$B = A + 6$$

The original relationship is preserved because we get 8 for B when A is known to be 2:

$$B = 2 + 6$$
$$B = 8$$

What we have done is *solve the equation for* B, or, equivalently, *express B in terms of* A. This sort of thing is generally the goal when equations are manipulated.

Here's another example, except in this case a variable is added:

$$C = D - E$$

For example, if $D = 13$ and $E = 7$, then C is 6:

$$C = 13 - 7$$
$$C = 6$$

If, instead, we know C and E, then we might want to find the value of D. To do so, we add E to both sides:

$$C + E = D - E + E$$
$$C + E = D$$

or

$$D = C + E$$

With C equal to 6 and E equal to 7, we find that D is 13, as before:

$$D = 6 + 7$$
$$D = 13$$

Subtraction

The same variable or number can be subtracted from both sides of an equation.

Subtraction is very similar to addition. We can use it, for example, to reverse the first addition example.

$$B = A + 6$$

We can subtract 6 from both sides.

$$B - 6 = A + 6 - 6$$
$$B - 6 = A$$

or

$$A = B - 6$$

Multiplication

Both sides of an equation can be multiplied by the same variable or number. For example,

$$G = \frac{H}{5}$$

The right side is the standard way to represent division in algebra. In this case it means "H divided by 5."

If H is known to be 45, then

$$G = \frac{45}{5}$$
$$G = 9$$

We can multiply both sides of the original equation by 5 if we wish to solve it for H:

$$G \times 5 = \frac{H}{5} \times 5$$
$$5G = \frac{H5}{5}$$

The 5's can be canceled, leaving H (5 divided by 5 equals 1).

$$5G = \frac{H5}{5} = \frac{H\cancel{5}}{\cancel{5}}$$
$$5G = H$$

or

$$H = 5G$$

For $G = 9$, H has to be 45, as before.

$$H = 5 \times 9$$
$$H = 45$$

Choosing to multiply both sides by 5 allows us to solve the equation for H.

This process also works for multiplication by a variable.

$$I = \frac{J}{B}$$

This equation gives a value for I when the values for J and B are known. If $J = 42$ and $B = 7$, then the equation tells us that I must be 6. If we want to express J in terms of I and B, we multiply both sides by B:

$$I \times B = \frac{J}{B} \times B$$

As before, the Bs can be canceled.

$$IB = \frac{JB}{B} = \frac{J\cancel{B}}{\cancel{B}}$$
$$IB = J$$
$$J = IB$$

If $I = 6$ and $B = 7$, then J will be 42, as required.

Division

We can divide both sides of an equation by the same variable or number.

Let's go back to the partnership. Suppose one of your partners tells you that your share of a bill is \$75. If you want to know what the amount of the bill is, you can solve the "share equation" for A by using division. (Remember that A = amount, F = fraction, and S = your share.)

$$S = FA$$

We can divide both sides by F.

$$\frac{S}{F} = \frac{FA}{F}$$
$$\frac{S}{F} = \frac{\cancel{F}A}{\cancel{F}} = A$$
$$A = \frac{S}{F}$$

So, the original amount A when your share is \$75 is

$$A = \frac{\$75}{0.25}$$
$$A = \$300$$

Sometimes it is necessary to use different operations in sequence. For example,

$$T = \frac{U}{C}$$

This tells us that T must be 50 if $U = 400$ and $C = 8$. But what if we are given the values of T and U so that C is unknown? We solve the equation for C by first multiplying both sides by C and then dividing both sides by T:

$$C \times T = \frac{U}{C} \times C$$
$$CT = U$$
$$\frac{CT}{T} = \frac{U}{T}$$
$$C = \frac{U}{T}$$

If we knew that $U = 400$ and $T = 50$, we could find the correct value for C:

$$C = \frac{U}{T}$$
$$C = \frac{400}{50}$$
$$C = 8$$

There is a nice shorthand way of looking at what we've done. The two forms of the equation (before and after) are

$$T = \frac{U}{C}$$
$$C = \frac{U}{T}$$

Note that in the first equation, C is below on the right and T is above on the left. In the second equation, the C and T have been switched: C is above on the left, and T is below on the right. This is a general result for this kind of equation: To move a variable (or number) to the other side of an equation, you must change it from above to below, or from below to above.

Here is another example of using operations in sequence. We will solve the following equation for M. To do this, we use the previous rules to move everything to the other side of the equation except M.

$$P = 3M + 7$$

First, we subtract 7 from both sides:

$$P - 7 = 3M + 7 - 7$$
$$P - 7 = 3M$$

Then we divide both sides by 3:

$$\frac{P-7}{3} = \frac{3M}{3} = M$$
$$M = \frac{P-7}{3}$$

Powers

Both sides of an equation can be raised to the same power. That is, we can square or cube both sides, we can take the square root of both sides, and so on.

If we have

$$V = WD$$

Then

$$V^2 = (WD)^2$$

Using numbers, if $W = 3$ and $D = 2$,

$$V = 3 \times 2$$
$$V = 6$$

After the equation is squared:

$$V^2 = (3 \times 2)^2$$
$$V^2 = 6^2$$
$$V^2 = 36$$

So $V = 6$ also fits this equation.

If we start with the equation

$$X^2 = YZ$$

we can take the square root of both sides to find X.

$$\sqrt{X^2} = \sqrt{YZ}$$
$$X = \sqrt{YZ}$$

Or X could be

$$X = -\sqrt{YZ}$$

In most situations in physics, the positive square root is used.

Exercises

Solve the following equations for X.

1. $A = X - 7$
2. $B = X - 25.2$
3. $C = 12 + X$
4. $C = X + 87$
5. $E = X - F$
6. $G = X + 2H$
7. $I = \dfrac{X}{12}$
8. $J = -6X$
9. $K = \dfrac{X}{L}$
10. $M = NX$
11. $O = \dfrac{4}{X}$
12. $P = \dfrac{20}{X}$
13. $R = 3X - 5$
14. $S = 2X + T$
15. $U = \sqrt{X}$
16. $V = 4X^2$

Proportionalities

Much of physics involves studying the relationships between different variables such as speed and distance. It is very useful to know, for example, how the value of one quantity is affected if the value of another is changed. To illustrate this, imagine that you go into a candy store to buy several chocolate bars of the same kind. Since we can choose from different sized bars with different prices, the number of bars that you buy will depend on the price of each as well as on the amount that you spend. The following equation can be used to tell you the number N of bars costing C dollars each that you will get if you spend D dollars:

$$N = \frac{D}{C}$$

If you spend \$2 on chocolate bars that cost \$0.50 (50 cents) each, the number that you buy is

$$N = \frac{\$2}{\$0.50}$$
$$N = 4$$

Let's look at this relationship in terms of how N changes when one of the other quantities takes on different values. For example, how is N affected when you increase the amount you spend on a particular type of chocolate? Clearly, the more you spend, the more you get. At 50 cents each, \$2 will let you buy four bars, \$3 will let you buy six bars, \$4 will let you buy eight bars, and so on. This relationship is called a *proportionality:* we say that "N is proportional to D," (or "N is directly proportional to D"). *If D is doubled*—as from \$2 to \$4—*then N is also doubled*, from 4 to 8 in this case. If D is tripled, N becomes three times as large also.

In a similar way, if the amount you spend, D, is decreased, the number N that you get is also decreased. One dollar will let you buy only two bars at 50 cents each. If D is halved, then N is halved.

$D(\$)$	N
1	2
2	4
3	6
4	8

Now let's say you decide to spend only \$2. How does the cost of the bar you decide to buy affect the number of bars that you get? We know that you get four bars when they cost 50 cents each. If you buy bars that cost \$1 each instead, the number you get is two.

$$N = \frac{\$2}{\$1}$$
$$N = 2$$

If you choose bars that cost \$2 each, you can get only one bar. Thus, when C is increased, N is decreased. In particular, *if C is doubled, N becomes one-half as large.* This is called an *inverse proportionality.* We say that "N is inversely proportional to C." If C is decreased, N is increased. At 25 cents each, you can buy eight chocolate bars for \$2.

$C(\$)$	N
0.25	8
0.50	4
1	2
2	1

Scientific Notation

Scientific notation is a convenient shorthand way to represent very large and very small numbers. It is based on using powers of 10. Since 100 equals 10×10, and 1,000 equals $10 \times 10 \times 10$, we can abbreviate them as follows:

$$100 = 10^2 \quad (= 10 \times 10)$$
$$1,000 = 10^3 \quad (= 10 \times 10 \times 10)$$

The numbers 100, 1,000, and so on, are powers of 10 because they are equal to 10 multiplied by itself a certain number of times. The number 2 in 10^2 is called the *exponent*. To represent any such power of 10 in this fashion, the exponent equals the number of zeros. So:

$$10 = 10^1$$
$$10,000 = 10^4$$
$$1,000,000 = 10^6$$

In this same fashion, since 1 has no zeros it is represented as

$$1 = 10^0$$

Decimals like 0.1 and 0.001 can also be represented this way by using negative exponents: 10 with a negative exponent equals 1 divided by 10 to the corresponding positive number.

$$0.01 = \frac{1}{100} = \frac{1}{10^2} = 10^{-2}$$

Similarly,

$$0.001 = \frac{1}{1,000} = \frac{1}{10^3} = 10^{-3}$$

To represent decimals by 10 to a negative exponent, the number in the exponent equals the number of places that the decimal point is to the left of the 1. For 0.01, the decimal point is two places to the left of 1, so $0.01 = 10^{-2}$. Other examples are as follows:

$$0.1 = 10^{-1}$$
$$0.0001 = 10^{-4}$$
$$0.000000001 = 10^{-9}$$

To review the process, 10 with a positive number exponent equals 1 followed by that number of zeros. Ten with a negative-number exponent equals a decimal in which the number of zeros between the decimal point and the 1 is equal to the number in the exponent *minus one*. So, 10^5 equals 1 followed by five zeros—100,000. The number 10^{-5} equals the decimal with four zeros (5 − 1) between the decimal point and 1—0.00001.

$10^4 = 10,000$	$10^{-1} = 0.1$
$10^3 = 1,000$	$10^{-2} = 0.01$
$10^2 = 100$	$10^{-3} = 0.001$
$10^1 = 10$	$10^{-4} = 0.0001$
$10^0 = 1$	

When two powers of 10 are multiplied, their exponents are added:

$$100 \times 1,000 = 100,000$$
$$10^2 \times 10^3 = 10^5$$

Similarly,

$$10^4 \times 10^8 = 10^{12}$$
$$10^{-3} \times 10^7 = 10^4$$
$$10^{-2} \times 10^{-5} = 10^{-7}$$

When a power of 10 is divided by another power of 10, the second exponent is subtracted from the first:

$$10^6 \div 10^2 = \frac{10^6}{10^2} = 10^{6-2} = 10^4$$
$$10^4 \div 10^9 = 10^{4-9} = 10^{-5}$$
$$10^3 \div 10^{-5} = 10^{3-(-5)} = 10^8$$
$$10^{-7} \div 10^{-1} = 10^{-7-(-1)} = 10^{-6}$$

Scientific notation utilizes powers of 10 to represent any number. A number expressed in scientific notation consists of a number (usually between 1 and 10) multiplied by 10 raised to some exponent. For example

$$23,400 = 2.34 \times 10,000$$
$$23,400 = 2.34 \times 10^4$$

When expressing a number in scientific notation, the exponent equals the number of times the decimal point is moved. For numbers like 23,400 the decimal point is understood to be to the right of the last digit—as in 23,400.—so it was moved four places to the left to get it between the 2 and the 3. *If the decimal point is moved to the left, the exponent is positive*, as in the above example. Similarly,

$$641 = 6.41 \times 10^2$$
$$497,500,000 = 4.975 \times 10^8$$

If the decimal point is moved to the right, the exponent is negative.

$$0.0053 = 5.3 \times 0.001 = 5.3 \times 10^{-3}$$
$$0.07134 = 7.134 \times 10^{-2}$$
$$0.000000964 = 9.64 \times 10^{-7}$$

When two numbers expressed in scientific notation are multiplied or divided, the "regular" parts are placed together, and the powers of 10 are placed together.

For multiplication

$$(4.2 \times 10^3) \times (2.3 \times 10^5) = (4.2 \times 2.3) \times (10^3 \times 10^5)$$
$$= 9.66 \times 10^8$$
$$(1.36 \times 10^4) \times (2 \times 10^{-6}) = (1.36 \times 2) \times (10^4 \times 10^{-6})$$
$$= 2.72 \times 10^{-2}$$

For division

$$(6.8 \times 10^7) \div (3.4 \times 10^2) = (6.8 \div 3.4) \times (10^7 \div 10^2)$$
$$= 2.0 \times 10^5$$
$$(7.3 \times 10^{-5}) \div (2 \times 10^{-8}) = (7.3 \div 2) \times (10^{-5} \div 10^{-8})$$
$$= 3.65 \times 10^3$$

Exercises

Express the following numbers in scientific notation.

17. 253,000
18. 93,000,000
19. 5632.5
20. 0.0072
21. 0.00000000338

Use scientific notation to find answers to the following:

22. $(4 \times 10^2) \times (2.1 \times 10^7)$
23. $(6.1 \times 10^{11}) \times (1.4 \times 10^{-5})$
24. $(3.2 \times 10^4) \times (5 \times 10^{-6})$
25. $(7 \times 10^{-3}) \times (8.3 \times 10^{-8})$
26. $(6 \times 10^9) \div (2 \times 10^2)$
27. $(9.3 \times 10^7) \div (4.1 \times 10^{-4})$
28. $(5.23 \times 10^{-8}) \div (3 \times 10^{-6})$
29. $(4.5 \times 10^{-3}) \div (2 \times 10^{-5})$
30. $(2.4 \times 10^4) \div (6 \times 10^9)$

Significant Figures

The numbers that are used in physics and other sciences represent measurements, not simply abstract values. A measured distance of, say, 96.4 miles is really some value between 96.35 miles and 96.45 miles. This imprecision must be kept in mind when such numbers are multiplied, divided, and so on. For example, if a car travels 96.4 miles in 3 hours, the speed is found by dividing the distance by the time:

$$\text{speed} = \frac{\text{distance}}{\text{time}} = \frac{96.4 \text{ miles}}{3 \text{ hours}}$$

The strictly mathematical answer is

$$\text{Speed} = 32.1333333333 \text{ mph}$$

But the uncertainty in the distance measurement means that the exact speed can't be known precisely. Most likely it is some value between 32.1 mph and 32.2 mph. The value for the time is also imprecise. For this reason, the answer should not indicate more precision than the input numbers warrant.

In this case, the first three digits (3, 2, and 1) are said to be *significant figures,* while the remaining digits (all 3s) are *insignificant*. The significant figures represent the limit to the accuracy or precision in the answer. Our proper answer is one that shows only the significant figures—the number 32.1333333333 rounded off to 32.1:

$$\text{Speed} = 32.1 \text{ mph}$$

Mathematical rules indicate how many significant figures there are in an answer, based on the accuracy of the input data. But for the sake of simplicity, you may assume there are three or four significant figures in the answers to the Problems and Challenges. This means that you should round off your answers to three or four digits. Some examples:

$$6.4768 \rightarrow 6.48$$
$$25,934 \rightarrow 25,900$$
$$0.4575 \rightarrow 0.458$$

Answers to Odd-Numbered Exercises

1. $X = A + 7$
3. $X = C - 12$
5. $X = E + F$
7. $X = 12I$
9. $X = KL$
11. $X = \frac{4}{O}$
13. $X = \frac{R + 5}{3}$
15. $X = U^2$
17. 2.53×10^5
19. 5.6325×10^3
21. 3.38×10^{-9}
23. 8.54×10^6
25. 58.1×10^{-11} or 5.81×10^{-10}
27. 2.27×10^{11}
29. 2.25×10^2

Answers

This appendix contains answers to the Explore It Yourself activities (beginning on page C-9), to the odd-numbered Problems, and to the odd-numbered Challenges that are problems. Worked-out solutions are also given for the odd-numbered Problems, except those that are nearly identical to Examples or other Problems.

Answers to Odd-Numbered Problems and Challenges

CHAPTER 1 Problems

1. $d = 20$ m

   From conversion table: 1 m = 3.28 ft

   $d = 20$ m $= 20 \times 1$ m $= 20 \times 3.28$ ft

   $d = 65.6$ ft

3. milli- means $\frac{1}{1000}$ or 0.001; so: 1 second = 1000 milliseconds

   0.0452 s $= 0.0452 \times 1$ s $= 0.0452 \times 1000$ ms

   0.0452 s $= 45.2$ ms

5. $T = 0.8$ s (period)

   $f = \dfrac{1}{T} = \dfrac{1}{0.8 \text{ s}}$

   $f = 1.25$ Hz

7. average speed $= \dfrac{\text{distance}}{\text{time}}$

   $v = \dfrac{d}{t} = \dfrac{1{,}200 \text{ mi}}{2.5 \text{ h}}$

   $v = 480$ mph

9. In Figure 1.13c, the length of the resultant velocity arrow is 0.66 times the length of the arrow representing 8 m/s (v_1). Therefore the resultant velocity is 0.66×8 m/s $= 5.3$ m/s.

   In Figure 1.13d, the resultant velocity is 12.5 m/s.

11. $d = vt$

    $d = 25$ m/s $\times 5$ s

    $d = 125$ m

    for $v = 250$ m/s: $d = vt$

    $d = 250$ m/s $\times 5$ s

    $d = 1{,}250$ m

13.

(graph: d (meters) vs t (seconds), straight line from origin to about (5, 125))

   slope = speed = 25 m/s (See Figure 1.22).

15. (a) $a = 4.06$ m/s^2 (See Example 1.3)

    (b) $a = -8$ m/s^2 (See Example 1.4)

17. $a = 200$ m/s^2 (See Example 1.5)

19. $a = 2.86$ m/s^2 (See Example 1.5)

21. (a) $v = at \qquad a = 60$ m/s$^2 \qquad t = 40$ s

    $v = 60$ m/s$^2 \times 40$ s

    $v = 2{,}400$ m/s

    (b) $v = at$

    $7{,}500$ m/s $= 60$ m/s^2 t

    $\dfrac{7{,}500 \text{ m/s}}{60 \text{ m/s}^2} = t \qquad t = 125$ s

23. (a)

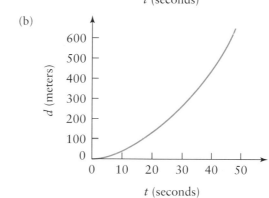

(b)

(graph: d (meters) vs t (seconds), curved line rising)

25. (a) $v = at \qquad a = g \qquad t = 3$ s

    $v = gt = 9.8$ m/s$^2 \times 3$ s

    $v = 29.4$ m/s

    (b) $d = \dfrac{1}{2}at^2 \qquad d = \dfrac{1}{2}gt^2$

    $d = \dfrac{1}{2}(9.8 \text{ m/s}^2) \times (3 \text{ s})^2$

    $d = 4.9$ m/s$^2 \times 9$ s^2

    $d = 44.1$ m

27. $a = 4.9$ m/s^2

    (a) $v = at \qquad t = 3$ s

    $v = 4.9$ m/s$^2 \times 3$ s

    $v = 14.7$ m/s

    (b) $d = \dfrac{1}{2}at^2$

    $d = \dfrac{1}{2}(4.9 \text{ m/s}^2) \times (3 \text{ s})^2$

    $d = 2.45$ m/s$^2 \times 9$ s^2

    $d = 22.05$ m

29. at "a": $a = \text{slope} = \dfrac{500 \text{ m/s}}{0.001 \text{ s}}$

$\qquad\quad a = 500,000 \text{ m/s}^2$

$\quad$ at "b": $a = 0 \text{ m/s}^2$

$\quad$ at "c": $a = -2,500,000 \text{ m/s}^2$

31. $a = \dfrac{\Delta v}{\Delta t} = \dfrac{300 \text{ mph}}{5 \text{ s}} = 60 \text{ mph/s} = 2.7\,g$

CHAPTER 1   Challenges

1. $t = 0.237 \text{ s}$

3. (a) $v = 1,021 \text{ m/s}$

$\quad$ (b) $a = 0.00271 \text{ m/s}^2$

5. $a = 7.9 \text{ m/s}^2$

7. $v = \sqrt{2ad}$

CHAPTER 2   Problems

1. One example: $W = 150 \text{ lb} \qquad 1 \text{ lb} = 4.45 \text{ N}$

$\qquad\qquad\qquad W = 150 \times 1 \text{ lb} = 150 \times 4.45 \text{ N}$

$\qquad\qquad\qquad W = 667.5 \text{ N}$

$\qquad\qquad\qquad W = mg$

$\qquad\qquad\qquad 667.5 \text{ N} = m\,9.8 \text{ m/s}$

$\qquad\qquad\qquad \dfrac{667.5 \text{ N}}{9.8 \text{ m/s}^2} = m$

$\qquad\qquad\qquad m = 68.1 \text{ kg}$

3. (a) $W = mg = 30 \text{ kg} \times 9.8 \text{ m/s}^2$

$\qquad W = 294 \text{ N}$

$\quad$ (b) $W = 294 \text{ N} \qquad 1 \text{ N} = 0.225 \text{ lb}$

$\qquad W = 294 \times 1 \text{ N} = 294 \times 0.225 \text{ lb}$

$\qquad W = 66.15 \text{ lb}$

5. $F = 600 \text{ N} \qquad$ (See Example 2.1)

7. $F = ma$

$\qquad 10 \text{ N} = 2 \text{ kg} \times a$

$\qquad 10 \text{ N}/2 \text{ kg} = a$

$\qquad a = 5 \text{ m/s}^2$

9. $F = ma$

$\qquad 20,000,000 \text{ N} = m \times 0.1 \text{ m/s}^2$

$\qquad \dfrac{20,000,000 \text{ N}}{0.1 \text{ m/s}^2} = m$

$\qquad\qquad m = 200,000,000 \text{ kg}$

11. $m = 4,500 \text{ kg} \qquad F = 60,000 \text{ N}$

$\quad$ (a) $F = ma$

$\qquad 60,000 \text{ N} = 4,500 \text{ kg} \times a$

$\qquad \dfrac{60,000 \text{ N}}{4,500 \text{ kg}} = a$

$\qquad\qquad a = 13.3 \text{ m/s}^2$

$\quad$ (b) $v = at$

$\qquad v = 13.3 \text{ m/s}^2 \times 8 \text{ s}$

$\qquad v = 106.4 \text{ m/s}$

$\quad$ (c) $d = \dfrac{1}{2}at^2$

$\qquad d = \dfrac{1}{2}(13.3 \text{ m/s}^2) \times (8 \text{ s})^2$

$\qquad d = 6.65 \text{ m/s}^2 \times 64 \text{ s}^2$

$\qquad d = 425.6 \text{ m}$

13. (a) $a = 3 \text{ m/s}^2 \qquad$ (See Example 1.3)

$\quad$ (b) $F = 240 \text{ N} \qquad$ (See Example 2.1)

$\quad$ (c) $d = \dfrac{1}{2}at^2 = \dfrac{1}{2}(3 \text{m/s}^2) \times (3 \text{ s})^2$

$\qquad d = 1.5 \text{ m/s}^2 \times 9 \text{ s}^2$

$\qquad d = 13.5 \text{ m}$

15. (a) $a = 28 \text{ m/s}^2 = 2.86\,g \qquad$ (See Example 1.3)

$\quad$ (b) $d = 87.5 \text{ m} \qquad$ (See solution of 13 (c))

$\quad$ (c) $F = 504,000 \text{ N} \qquad$ (See Example 2.1)

17. $a = 6\,g \quad g = 9.8 \text{ m/s}^2$

$\qquad = 6 \times 9.8 \text{ m/s}^2$

$\quad a = 58.8 \text{ m/s}^2$

$\quad F = ma \quad m = 1,200 \text{ kg}$

$\qquad = 1,200 \text{ kg} \times 58.8 \text{ m/s}^2$

$\quad F = 70,560 \text{ N}$

19. $v = 60 \text{ m/s} \quad r = 400 \text{ m} \quad m = 600 \text{ kg}$

$\quad$ (a) $a = \dfrac{v^2}{r} = \dfrac{(60 \text{ m/s})^2}{400 \text{ m}}$

$\qquad a = \dfrac{3,600 \text{ m}^2/\text{s}^2}{400 \text{ m}}$

$\qquad a = 9 \text{ m/s}^2$

$\qquad g = 9.8 \text{ m/s}^2 \qquad 1 \text{ m/s}^2 = \dfrac{g}{9.8}$

$\qquad a = 9 \times 1 \text{ m/s}^2 = 9 \times \dfrac{g}{9.8}$

$\qquad a = \dfrac{9}{9.8}g$

$\qquad a = 0.918\,g$

$\quad$ (b) $F = ma$

$\qquad = 600 \text{ kg} \times 9 \text{ m/s}^2$

$\qquad F = 5,400 \text{ N}$

21. $F = m\dfrac{v^2}{r}$

$\quad 60 \text{ N} = 0.1 \text{ kg} \times \dfrac{v^2}{1 \text{ m}}$

$\quad v^2 = \dfrac{60 \text{ N-m}}{0.1 \text{ kg}} = 600 \text{ m}^2/\text{s}^2$

$\quad v = 24.5 \text{ m/s}$

23. $F = m\dfrac{v^2}{r}$

$\quad 200 \text{ N} = 1,000 \text{ kg} \times \dfrac{(5,000 \text{ m/s})^2}{r} = \dfrac{2.5 \times 10^{10}}{r}$

$\quad r = \dfrac{2.5 \times 10^{10}}{200} = 1.25 \times 10^8 \text{ m}$

25. (a) At *twice* as far away the force is *one-fourth* as large.

$\qquad F = 150 \text{ lb.}$

$\quad$ (b) $F = 66.7 \text{ lb}$

$\quad$ (c) $F = 6 \text{ lb}$

CHAPTER 2   Challenges

7. (a) $F = 44.7 \text{ N}$

$\quad$ (b) $v = 3,075 \text{ m/s}$

$\quad$ (c) $t = 86,400 \text{ s}$

CHAPTER 3   Problems

1. $mv = 650 \text{ kg m/s} \qquad$ (See Section 3.2)

3. $F = \dfrac{\Delta mv}{\Delta t} = \dfrac{(mv)_f - (mv)_i}{\Delta t}$

$\quad mv_i = 1,000 \text{ kg} \times 0 \text{ m/s}$

$\quad mv_i = 0 \text{ kg m/s}$

$\quad mv_f = 1000 \text{ kg} \times 27 \text{ m/s}$

$\quad mv_f = 27,000 \text{ kg m/s}$

$\quad \Delta t = 10 \text{ s}$

$\quad F = \dfrac{(mv)_f - (mv)_i}{\Delta t} = \dfrac{27,000 \text{ kg m/s} - 0 \text{ kg m/s}}{10 \text{ s}}$

$\quad F = \dfrac{27,000 \text{ kg m/s}}{10 \text{ s}}$

$\quad F = 2,700 \text{ N}$

5. $v = 39 \text{ m/s} \qquad$ (See Example 3.2)

7. $mv_{\text{before}} = 50 \text{ kg} \times 5 \text{ m/s}$

$\quad mv_{\text{before}} = 250 \text{ kg m/s}$

$\quad mv_{\text{after}} = mv = (40 \text{ kg} + 50 \text{ kg}) \times v$

$$mv_{after} = 90 \text{ kg} \times v$$
$$mv_{before} = mv_{after}$$
$$250 \text{ kg m/s} = 90 \text{ kg} \times v$$
$$\frac{250 \text{ kg m/s}}{90 \text{ kg}} = v$$
$$v = 2.78 \text{ m/s}$$

9. $$mv_{before} = 0 \text{ kg m/s}$$
$$mv_{after} = (mv)\text{gun} + (mv)\text{bullet}$$
$$mv_{after} = 1.2 \text{ kg} \times v + 0.02 \text{ kg} \times 300 \text{ m/s}$$
$$mv_{after} = 1.2 \text{ kg} \times v + 6 \text{ kg m/s}$$
$$mv_{before} = mv_{after}$$
$$0 \text{ kg m/s} = 1.2 \text{ kg} \times v + 6 \text{ kg m/s}$$
$$-6 \text{ kg m/s} = 1.2 \text{ kg} \times v$$
$$\frac{-6 \text{ kg m/s}}{1.2 \text{ kg}} = v$$
$$v = -5 \text{ m/s} \quad \text{(opposite direction of bullet)}$$
or use: $\dfrac{v \text{ gun}}{v \text{ bullet}} = \dfrac{-m \text{ bullet}}{m \text{ gun}}$

11. Work = 30,000 J   (See Example 3.3)
13. PE = 2,156 J   (See Example 3.7)
15. KE = 4,500 J   (See Example 3.6)
17. $KE = \dfrac{1}{2}mv^2$

$$60,000 \text{ J} = \frac{1}{2} \times 300 \text{ kg} \times v^2$$
$$\frac{60,000 \text{ J}}{150 \text{ kg}} = v^2$$
$$400 \text{ J/kg} = v^2$$
$$v = 20 \text{ m/s}$$

19. v = 111 m/s   (See Example 3.8)
21. $$v^2 = 2gd$$
$$(7.7 \text{ m/s})^2 = 2 \times 9.8 \text{ m/s}^2 \times d$$
$$59.3 \text{ m}^2/\text{s}^2 = 19.6 \text{ m/s}^2 \times d$$
$$\frac{59.3 \text{ m}^2/\text{s}^2}{19.6 \text{ m/s}^2} = d$$
$$d = 3 \text{ m} \quad \text{(or use Table 3.2)}$$

23. (a) PE = 323,000 J (See Example 3.7)
    (b) v = 46.4 m/s (104 mph; See Example 3.8)
25. (a) KE = 4,000 J   (See Example 3.6)
    (b)

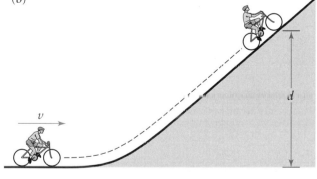

KE at bottom = PE when stopped on hill
$$4,000 \text{ J} = mgd = 80 \text{ kg} \times 9.8 \text{ m/s}^2 \times d$$
$$\frac{4,000 \text{ J}}{784 \text{ kg m/s}^2} = d$$
$$d = 5.1 \text{ m} \quad \text{or use } d = \frac{v^2}{2g}$$

27. It would have to be thrown vertically, with just enough speed that it would slow to a stop just as it reached the ceiling: $v = \sqrt{2gd}$
$= \sqrt{2 \times 9.8 \text{ m/s}^2 \times 20 \text{ m}} = 19.8 \text{ m/s } (=44 \text{ mph})$

29. $KE_{before} = \dfrac{1}{2}mv^2 = \dfrac{1}{2} \times 50 \text{ kg} \times (5 \text{ m/s})^2$

$KE_{before} = 25 \text{ kg} \times 25 \text{ m}^2/\text{s}^2$

$KE_{before} = 625 \text{ J}$
$KE_{after} = \dfrac{1}{2}mv^2 = \dfrac{1}{2} \times 90 \text{ kg} \times (2.78 \text{ m/s})^2$
$KE_{after} = 45 \text{ kg} \times 7.73 \text{ m}^2/\text{s}^2$
$KE_{after} = 348 \text{ J}$
$KE_{lost} = 625 \text{ J} - 348 \text{ J}$
$KE_{lost} = 277 \text{ J}$

31. $$P = \frac{work}{t}$$
$$200 \text{ W} = \frac{10,000 \text{ J}}{t}$$
$$t = \frac{10,000 \text{ J}}{200 \text{ W}}$$
$$t = 50 \text{ s}$$

33. $$P = \frac{work}{t} = \frac{PE}{t}$$
$$PE = mgd = 1,000 \text{ kg} \times 9.8 \text{ m/s}^2 \times 30 \text{ m}$$
$$PE = 294,000 \text{ J}$$
$$P = \frac{PE}{t} = \frac{294,000 \text{ J}}{10 \text{ s}}$$
$$P = 29,400 \text{ W}$$

35. P = 217 W   (See Problem 33)

CHAPTER 3   Challenges

3. v_b = 281.4 m/s

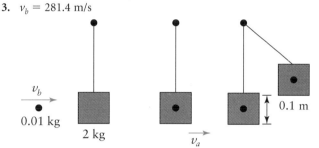

7. (a) KE = 715.4 J
    (b) F = 238.5 N
    (c) P = 1,431 W
         P = 1.92 hp

CHAPTER 4   Problems

1. $$p = \frac{F}{A} = \frac{2,000,000 \text{ lb}}{400 \text{ ft}^2}$$
$$p = 5,000 \frac{\text{lb}}{\text{ft}^2}$$
$$1 \text{ ft}^2 = 144 \text{ in.}^2 \qquad p = 5,000 \frac{\text{lb}}{\text{ft}^2} = 5,000 \frac{\text{lb}}{144 \text{ in.}^2}$$
$$p = \frac{5,000 \text{ lb}}{144 \text{ in.}^2}$$
$$p = 34.7 \text{ psi}$$

3. $F = pA$
$F = 80 \text{ psi} \times 1,200 \text{ in.}^2$
$F = 96,000 \text{ lb}$

5. F = 1,176 lb   (See Example 4.2)

7. (a) $D = \dfrac{m}{V} = \dfrac{393 \text{ kg}}{0.05 \text{ m}^3}$

$D = 7,860 \text{ kg/m}^3$
$D_w = D \times g = 7,860 \text{ kg/m}^3 \times 9.8 \text{ m/s}^2$
$D_w = 77,028 \text{ N/m}^3$
    (b) Iron   (From Table 4.4)

9. (a) $D_w = \dfrac{W}{V}$

water: $D_w = 62.4 \text{ lb/ft}^3$ (See Table 4.4)

$$62.4 \text{ lb/ft}^3 = \frac{40,000 \text{ lb}}{V}$$

$$V = \frac{40,000 \text{ lb}}{62.4 \text{ lb/ft}^3}$$

$$V = 641 \text{ ft}^3$$

(b) gasoline: $D_w = 42 \text{ lb/ft}^3$

$$D_w = 42 \text{ lb/ft}^3 = \frac{40,000 \text{ lb}}{V}$$

$$V = \frac{40,000 \text{ lb}}{42 \text{ lb/ft}^3}$$

$$V = 952 \text{ ft}^3$$

11. $m = 162 \text{ kg}$ (See Example 4.4 and Table 4.4)

13. (a) $m = 2,190 \text{ kg}$ (See Example 4.4 and Table 4.4)

(b) $W = mg = 2,190 \text{ kg} \times 9.8 \text{ m/s}^2$
     $W = 21,500 \text{ N} = 4,830 \text{ lb}$

15. $p = 0.433 \ d = 0.433 \times 12 \text{ ft}$
     $p = 5.196 \text{ psi}$

17. $p = 3,030,000 \text{ Pa}$ (See Example 4.6)

19. About 7 psi

21. $F_b = W_{\text{water displaced}}$ $W = (D_w)_{\text{water}} \times V$
     $F_b = 62.4 \text{ lb/ft}^3 \times 12 \text{ ft}^3$
     $F_b = 749 \text{ lb}$

23. (a) $W_{He} = D_W \times V = 0.011 \text{ lb/ft}^3 \times 200,000 \text{ ft}^3 = 2,200 \text{ lb}$
     (D_W from Table 4.4)

(b) $F_b = W_{\text{air}} = D_W \times V = 0.08 \text{ lb/ft}^3 \times 200,000 \text{ ft}^3 = 16,000 \text{ lb}$

(c) $F_{\text{net}} = F_b - W_{He} = 16,000 \text{ lb} - 2,200 \text{ lb} = 13,800 \text{ lb}$

25. (a) $W = 468 \text{ lb}$ (See Example 4.5 and Table 4.4)

(b) $F_b = W_{\text{water}} = (D_w)_{\text{water}} \times V = 62.4 \text{ lb/ft}^3 \times 3 \text{ ft}^3$
     $F_b = 187.2 \text{ lb}$

(c)

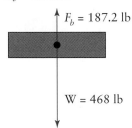

$F_b = 187.2 \text{ lb}$

$W = 468 \text{ lb}$

$$F_{\text{net}} = 468 \text{ lb} - 187.2 \text{ lb}$$
$$F_{\text{net}} = 280.8 \text{ lb}   \text{(downward)}$$

27. (a) $W = 5,720,000 \text{ lb}$ (See Example 4.5 and Table 4.4)

(b) $W_{\text{seawater}} = F_b = W_{\text{ice}}$ (since it floats)
     $W_{\text{seawater}} = D_w V$ $D_w(\text{seawater}) = 64.3 \text{ lb/ft}^3$
     $5,720,000 \text{ lb} = 64.3 \text{ lb/ft}^3 \times V$

$$\frac{5,720,000 \text{ lb}}{64.3 \text{ lb/ft}^3} = V$$

$$V_{\text{seawater}} = 88,960 \text{ ft}^3$$

(c) $V_{\text{out}} = V_{\text{total}} - V_{\text{underwater}}$
     $= 100,000 \text{ ft}^3 - 88,960 \text{ ft}^3$

$$V_{\text{out}} = 11,040 \text{ ft}^3 = \frac{1}{9} V_{\text{total}}$$

29. Scale reading $= 100 \text{ N} - F_b$
     $F_b = W_{\text{water}} = D_{\text{water}} \times g \times V_{\text{al}}$
     $W_{\text{al}} = D_{\text{al}} \times g \times V_{\text{al}}$ $V_{\text{al}} = 0.00378 \text{ m}^3$
     $F_b = 37.0 \text{ N}$
     Scale reading $= 63.0 \text{ N}$

CHAPTER 4   Challenges

3. $D = 2,700 \text{ kg/m}^3$
    specific gravity $= 2.7$
    $(D_w)_{\text{moon}} = 27.6 \text{ lb/ft}^3$

5. About $0.014 \text{ lb} = 0.062 \text{ N}$ lighter.

CHAPTER 5   Problems

1. $30°C = 86°F$ (no jacket needed)

3. $\Delta l = -0.336 \text{ ft}$ means it is shorter   (See Example 5.1 and Table 5.2)
    $l = 699.664 \text{ ft}$

5. Need to shorten diameter from 5 mm to 4.997 mm
    $\Delta l = -0.003 \text{ mm}$
    $\Delta l = \alpha l \Delta T$ $\alpha = 12 \times 10^{-6}/°C$
      $- 0.003 \text{ mm} = 12 \times 10^{-6}/°C \times 5 \text{ mm} \times \Delta T$

$$\frac{-0.003 \text{ mm}}{60 \times 10^{-6} \text{ mm/°C}} = \Delta T$$

$$\frac{-0.003}{60} \times 10^6 °C = \Delta T$$

$$-0.00005 \times 10^6 °C = \Delta T$$
$$\Delta T = -50°C$$

7. $\Delta U = \text{work} + Q$ $Q = -2 \text{ J}$ (heat lost)
       Work $= Fd = 50 \text{ N} \times 0.1 \text{ m}$
       Work $= 5 \text{ J}$ (Work done *on* gas)
          $\Delta U = \text{Work} + Q = 5 \text{ J} + -2 \text{ J}.$
          $\Delta U = 3 \text{ J}$

9. $Q = 1,081,000 \text{ J}$ (See Example 5.2 and Table 5.3)

11. (a) $Q = Cm\Delta T$
          $C = 2,000 \text{ J/kg-°C}$ for ice
          $\Delta T = 25°C$ from Figure 5.34
          $Q = Cm\Delta T = 2,000 \text{ J/kg-°C} \times 1 \text{ kg} \times 25°C$
          $Q = 50,000 \text{ J}$

(b) $Q = Cm\Delta T$
     $C = 4,180 \text{ J/kg-°C}$ for water
     $\Delta T = 100°C$ from Figure 5.34
     $Q = Cm\Delta T = 4,180 \text{ J/kg-°C} \times 1 \text{ kg} \times 100°C$
     $Q = 418,000 \text{ J}$

13. (a) $KE = 375,000 \text{ J}$ (See Example 3.6)

(b) $Q = Cm\Delta T$ $Q = KE$
     $KE = Cm\Delta T$ $C = 460 \text{ J/kg-°C}$ for iron
     $375,000 \text{ J} = 460 \text{ J/kg-°C} \times 20 \text{ kg} \times \Delta T$

$$\frac{375,000 \text{ J}}{9,200 \text{ J/°C}} = \Delta T$$

$$\Delta T = 40.8°C$$

15. $\Delta T = 47.4°C$ (See Example 5.3 and Table 5.3)

17. (a) Rel. Hum. $= 62.5\%$ (See Example 5.6 and Table 5.5)

(b) Rel. Hum. $= 5.78\%$ (See Example 5.6 and Table 5.5)

19. The amount of water vapor in the air is given by the water vapor density, which is the humidity.

$$\text{Rel. Hum.} = \frac{\text{Humidity}}{\text{Sat. Den.}} \times 100\%$$

at 20°C, Sat. Den. $= 0.0173 \text{ kg/m}^3$

$$40\% = \frac{\text{Humidity}}{0.0173 \text{ kg/m}^3} \times 100\%$$

$$\frac{40\% \times 0.0173 \text{ kg/m}^3}{100\%} = \text{Humidity}$$

Humidity $= 0.00692 \text{ kg/m}^3$
There is 0.00692 kg of water vapor in each m^3 of air.

21. $m = DV$ $V = 150 \text{ m}^3$

$$\text{relative humidity} = \frac{\text{humidity}}{\text{saturation density}} \times 100\%$$

$$60\% = \frac{D}{0.0228} \times 100\%$$ $D = 0.0137 \text{ kg/m}^3$

$m = 0.0137 \text{ kg/m}^3 \times 150 \text{ m}^3$
$m = 2.1 \text{ kg}$

23. Humidity $= 0.0094 \text{ kg/m}^3$ (See solution of 19)
     From Table 5.5, air with this humidity will be saturated when cooled to 10°C. Therefore the air must be cooled 20°C, which means it must rise 2,000 m.

25. Carnot eff. = 34.9%   (See Example 5.7)

27. (a) Eff. = $\dfrac{\text{work output}}{\text{energy input}} \times 100\%$

Eff. = $\dfrac{3{,}000 \text{ J}}{15{,}000 \text{ J}} \times 100\%$

Eff. = 20%

(b) Carnot Eff. = 88%   (See Example 5.7)

CHAPTER 6   Problems

1. (a) $\rho = 0.167$ kg/m   (See Example 6.1)

(b) $v = 15.5$ m/s   (See Example 6.1)

3. $v = 388$ m/s   (See Example 6.2)

5. $v = f\lambda$

$v = 4$ Hz $\times 0.5$ m

$v = 2$ m/s

7. $v = f\lambda$

80 m/s $= f \times 3.2$ m

$\dfrac{80 \text{ m/s}}{3.2 \text{ m}} = f$

$f = 25$ Hz

9. for $f = 20$ Hz:

$\lambda = 17.2$ m

$\lambda = 56.4$ ft   (See Example 6.3)

for $f = 20{,}000$ Hz

$\lambda = 0.0172$ m

$\lambda = 0.677$ in.   (See Example 6.3)

11. (a) $\lambda = 1.315$ m   (See Example 6.3)

(b) $v = f\lambda$ $v = 1{,}440$ m/s   (water, from Table 6.1)

1,440 m/s $= 261.6$ Hz $\times \lambda$

$\dfrac{1{,}440 \text{ m/s}}{261.6 \text{ Hz}} = \lambda$

$\lambda = 5.505$ m

13. $v = 347$ m/s   (See Problem 5)

$v = 20.1 \times \sqrt{T} = 347$ m/s

$(20.1)^2 \times T = (347 \text{ m/s})^2$

$T = 298$ K $= 25°$C

15. This is destructive interference, so the difference in the two distances equals one half the wavelength of the sound.

7.2 m $-$ 7 m $= 0.2$ m $= \dfrac{1}{2}\lambda$

$\lambda = 0.4$ m

$f = 860$ Hz   (See Problem 7)

17. (a) Total distance sound travels: $d = vt = 320$ m/s $\times 0.03$ s $= 9.6$ m (v from Table 6.1)

Distance to snow $= \dfrac{1}{2}d = 4.8$ m

(b) Depth of snow $= 5$ m $- 4.8$ m $= 0.2$ m

19. $d = vt$ $t = \dfrac{d}{v} = \dfrac{4{,}782{,}000 \text{ m}}{344 \text{ m/s}}$

$t = 13{,}900$ s $= 3.86$ h

21. (a) $d = vt$ $v = 344$ m/s   (air)

8,000 m $= 344$ m/s $\times t$

$\dfrac{8{,}000 \text{ m}}{344 \text{ m/s}} = t$

$t = 23.2$ s

(b) $d = vt$ $v = 4{,}000$ m/s   (granite, Table 6.1)

8,000 m $= 4{,}000$ m/s $\times t$

$\dfrac{8{,}000 \text{ m}}{4{,}000 \text{ m/s}} = t$

$t = 2$ s

23. Distance sound travels underwater in 0.01 second:

$d = vt$ $v = 1{,}440$ m/s

$d = 1{,}440$ m/s $\times 0.01$ s

$d = 14.4$ m

Sound travels to fish, reflects, and then returns. Distance to fish is ½ the distance sound traveled.

$d = 7.2$ m

25. For each 10 dB, sound is about 2 times louder.

40 dB $= 4 \times 10$ dB

$2 \times 2 \times 2 \times 2 = 16$

Sound is 16 times louder.

27. (a) Harmonics of note have frequencies equal to 2, 3, 4, . . . times frequency of note. Harmonics of 4,186 Hz note have frequencies: 8,372 Hz, 12,558 Hz, 16,744 Hz, 20,930 Hz, etc. We can't hear frequencies over 20,000 Hz. So, highest harmonic we can hear is 16,744 Hz. Including the note itself (4,186 Hz), we can hear 4 harmonics.

(b) The note one octave below has 1/2 the frequency $f = 2{,}093$ Hz Harmonics: 4,186 Hz, 6,279 Hz, 8,372 Hz, 10,465 Hz, 12,558 Hz, 14,651 Hz, 16,744 Hz, 18,837 Hz, 20,930 Hz. The highest harmonic we can hear is 18,837 Hz. Including the note itself (2,093 Hz), we can hear 9 harmonics.

CHAPTER 6   Challenges

1. Amplitude is decreased.

3. In air: $t = 0.581$ s

In steel: $t = 0.038$ s

Jack hears the sound in the rail 0.543 seconds before he hears the sound in the air.

7. 1,000,000 times

CHAPTER 7   Problems

1. $I = \dfrac{q}{t} = \dfrac{5 \text{ C}}{20 \text{ s}}$

$I = 0.25$ A

3. $I = \dfrac{q}{t}$ $t = 1$ min $= 60$ s

0.7 A $= \dfrac{q}{60 \text{ s}}$

0.7 A $\times 60$ s $= q$

$q = 42$ C

5. $V = IR$

120 V $= 12$ A $\times R$

$\dfrac{120 \text{ V}}{12 \text{ A}} = R$

$R = 10 \ \Omega$

7. $I = 1.8$ A   (See Example 7.1)

9. $V = 20$ V   (See Example 7.2)

11. $P = 1{,}440$ W   (See Example 7.4)

13. $P = 1{,}500{,}000$ W   (See Example 7.4)

15. (a) $P = 40$ W $I = 3.33$ A   (See Example 7.5)

$P = 50$ W $I = 4.17$ A   (See Example 7.5)

(b) $P = 40$ W: $R = 3.6 \ \Omega$ (See solution of 5)

$P = 50$ W: $R = 2.88 \ \Omega$ (See solution of 5)

17. $E = 9{,}600{,}000$ J   (See Example 7.6)

or: $P = 4$ kW $t = 40$ min $= \dfrac{2}{3}$ h

$E = 4$ kW $\times \dfrac{2}{3}$ h

$E = 2.67$ kWh

19. Hair dryer: $E = 360{,}000$ J   (See Example 7.6)

Lamp: $E = 2{,}160{,}000$ J   (See Example 7.6)

21. (a) $P = 1{,}080$ W   (See Example 7.4)

(b) $E = 64{,}800$ J   (See Example 7.6)

or: $P = 1.08$ kW $t = 1$ min $= \dfrac{1}{60}$ h

$E = Pt = 1.08 \text{ kW} \times \frac{1}{60} \text{ h}$

$E = 0.018 \text{ kWh}$

23. (a) $P = IV$

$1,000,000,000 \text{ W} = I \times 24,000 \text{ V}$

$\dfrac{1,000,000,000 \text{ W}}{24,000 \text{ V}} = I$

$I = 41,670 \text{ A}$

(b) $E = Pt$ $\quad t = 24 \text{ h} = 24 \times 3600 \text{ s}$

$t = 86,400 \text{ s}$

$E = 1,000,000,000 \text{ W} \times 86,400 \text{ s}$

$E = 8.64 \times 10^{13} \text{ J}$

in kWh: $\quad E = 1,000,000 \text{ kW} \times 24 \text{ h}$

$E = 24,000,000 \text{ kWh}$

(c) $10\cancel{c} = \$0.1$ $\quad$ Revenue $= \$2,400,000$

25. (a) $E = 28,800,000 \text{ J}$ $\quad$ (See Example 7.6) or:

$E = Pt = 4 \text{ kW} \times 2\text{h} = 8 \text{ kWh}$

(b) $I = 133 \text{ A}$ $\quad$ (See Example 7.5)

(c) $E = Pt$ $\quad P = \dfrac{E}{t} = \dfrac{8 \text{ kWh}}{1 \text{ h}}$

$P = 8 \text{ kW} = 8,000 \text{ W}$

CHAPTER 7 Challenges

1. $F = -8.2 \times 10^{-8} \text{ N}$

3. Number of electrons $= 1.25 \times 10^{18}$

7. $V = 12 \text{ V}:$ $\quad I = 100 \text{ A}$

$V = 30 \text{ V}:$ $\quad I = 40 \text{ A}$

$V = 60 \text{ V}:$ $\quad I = 20 \text{ A}$

$V = 120 \text{ V}:$ $\quad I = 10 \text{ A}$

The lower voltage dryers would have to have successively larger wires to handle the larger currents.

9. $P = I^2 R$

4 times as much (40 kWh per day)

CHAPTER 8 Problems

1. $\dfrac{V_o}{V_i} = \dfrac{N_o}{N_i}$

120 V input, 9 V output

$\dfrac{9 \text{ V}}{120 \text{ V}} = \dfrac{N_o}{N_i}$

ratio $= \dfrac{N_o}{N_i} = 0.075$

3. Example: $\quad f = 92.5 \text{ MHz} = 92,500,000 \text{ Hz}$

$c = f\lambda$

$3 \times 10^8 \text{ m/s} = 9.25 \times 10^7 \text{ Hz} \times \lambda$

$\dfrac{3 \times 10^8 \text{ m/s}}{9.25 \times 10^7 \text{ Hz}} = \lambda$

$\lambda = 3.24 \text{ m}$

5. $c = f\lambda$

$3 \times 10^8 \text{ m/s} = f \times 0.0254 \text{ m}$

$\dfrac{3 \times 10^8 \text{ m/s}}{0.0254 \text{ m}} = f$

$f = 1.18 \times 10^{10} \text{ Hz} = 11,800 \text{ MHz}$

7. UV band: $\quad f = 7.5 \times 10^{14} \text{ Hz to } f = 10^{18} \text{ Hz}$

λ's:

$c = f\lambda$ $\qquad\qquad\qquad\qquad c = f\lambda$

$3 \times 10^8 \text{ m/s} = 7.5 \times 10^{14} \text{ Hz} \times \lambda$ $\quad$ $3 \times 10^8 \text{ m/s} = 10^{18} \text{ Hz} \times \lambda$

$\dfrac{3 \times 10^8 \text{ m/s}}{7.5 \times 10^{14} \text{ Hz}} = \lambda$ $\qquad\qquad$ $\dfrac{3 \times 10^8 \text{ m/s}}{1 \times 10^{18} \text{ Hz}} = \lambda$

$\lambda = 4 \times 10^{-7} \text{ m}$ $\qquad\qquad$ $\lambda = 3 \times 10^{-10} \text{ m}$

9. Energy emitted $\propto T^4$. From 300 K to 3000 K, T increases 10 times.

Energy output increases $10^4 = 10,000$ times

11. $\lambda_{max} = \dfrac{0.0029 \text{ m-K}}{T}$

$\lambda_{max} = \dfrac{0.0029 \text{ m-K}}{10,000,000 \text{ K}}$

$\lambda_{max} = 2.9 \times 10^{-10} \text{ m}$ $\qquad$ X ray (See Figure 8.30)

13. $\lambda = \dfrac{0.0029}{T}$

$T \times \lambda = T \times 0.0000012 \text{ m} = 0.0029$

$T = 2,420 \text{ K}$

CHAPTER 9 Problems

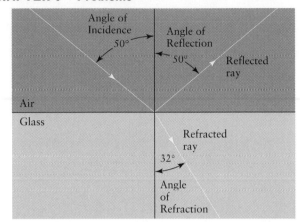

1. By the Law of Reflection, the angle of reflection equals the angle of incidence, 50°. Therefore the reflected ray makes an angle of 50° with respect to the normal.

For the refracted ray, the light slows upon entering the glass, so the ray is bent toward the normal—the angle of refraction is smaller than the angle of incidence. The actual angle is found using Figure 9.34. The refracted ray makes an angle of about 32° with respect to the normal.

3.

Angle of Incidence (°)	Angle of Refraction (°)*
5	3
10	7
20	14

*From Figure 9.34

Doubling the angle of incidence approximately doubles the angle of refraction. Doubling the angle of incidence from 20° to 40° should result in an angle of refraction of about $2 \times 14°$ or 28°. Figure 9.34 gives an angle of refraction of about 27° for this result, in good agreement with this estimate.

5. Rays skimming the water's surface are refracted to the fish's eye at the critical angle for an air-water interface, about 49° (see Example 9.2). The opening angle of the cone of light thus seen by the fish is twice this angle or about 98°. (See the accompanying diagram.)

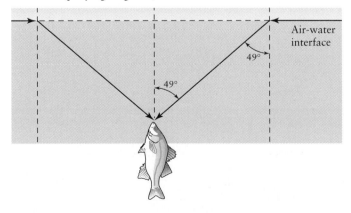

7. $M = \dfrac{-\text{image distance}}{\text{object height}} = \dfrac{-p}{s}$

The virtual image formed by the lens is upright, so $M = +2.5$. For a virtual image, the image is on the same side of the lens as the object: $p = -15$ cm. (See Example 9.4)

$2.5 = \dfrac{-(-15 \text{ cm})}{s} = \dfrac{15 \text{ cm}}{s}$

$s = \dfrac{15 \text{ cm}}{2.5} = 6.0$ cm

9. $M = \dfrac{\text{image height}}{\text{object height}} = \dfrac{-0.005 \text{ m}}{2 \text{ m}}$

$M = -0.0025$

11. $p = 15$ cm (to the right of the lens)   (See Example 9.3)
Image is real and inverted.

13. $p = \dfrac{sf}{s-f} = \dfrac{50 \text{ cm} \times (-20 \text{ cm})}{50 \text{ cm} - (-20 \text{ cm})} = \dfrac{-1000 \text{ cm}^2}{70 \text{ cm}}$

$= -14.3$ cm (to the left of the lens)

$M = \dfrac{-p}{s} = \dfrac{-(-14.3 \text{ cm})}{50 \text{ cm}} = 0.29$ The image is virtual, upright, and reduced in size.

15. $p = \dfrac{sf}{s-f} = \dfrac{16 \text{ cm} \times (8 \text{ cm})}{16 \text{ cm} - (8 \text{ cm})} = \dfrac{128 \text{ cm}^2}{8 \text{ cm}}$

$= 16$ cm (to the left of the lens)

$M = \dfrac{-p}{s} = \dfrac{-16 \text{ cm}}{16 \text{ cm}} = -1$ The image is inverted but the same size as the object.

17. Light scattered per second $\propto \lambda^{-4}$. (See p. 378.) If the wavelength, λ, is doubled, the amount of scattered light is *reduced* by a factor of $(2)^4 = 16$.

CHAPTER 9   Challenges

1. The point in an interference pattern where constructive interference occurs depends on the wavelength of the light. Since different colors have different wavelengths, interference with white light results in each color having its interference maxima occur at slightly different locations on the observing screen. The colors no longer overlap completely at all points (as in white light), but are separated on the observing screen: Dispersion has taken place. The shorter wavelengths yield constructive interference at points closer to the middle of the pattern, so the spectrum of colors starts with violet closest to the middle of the pattern and red farthest away.

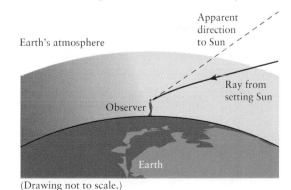

(Drawing not to scale.)

3. The ray bends slightly downward as it passes through the atmosphere (see the accompanying figure, which greatly exaggerates the effect). By tracing back along the direction from which the rays reaching the eye of the observer appear to have come, we find the apparent position of the Sun (or any other astronomical object) is slightly higher than its true

position. Thus, the Sun can be seen in the sky for a brief time after it has actually set below the horizon.

5. Apply Snell's law (see the footnote on p. 352):
$n_1 \sin A_1 = n_2 \sin A_2$
$(1.0) \sin 45° = (2.42) \sin A_2$
$0.707 = (2.42) \sin A_2$
$\sin A_2 = \dfrac{0.707}{2.42} = 0.293$
$A_2 = \sin^{-1}(0.293) = 17°$

7. (a) $\dfrac{1}{f} = \dfrac{1}{s} + \dfrac{1}{p}$

$\dfrac{1}{f} - \dfrac{1}{p} = \dfrac{1}{s}$

$\left(\dfrac{1}{f} \times \dfrac{p}{p}\right) - \left(\dfrac{1}{p} \times \dfrac{f}{f}\right) = \dfrac{1}{s}$

$\dfrac{1}{s} = \dfrac{p-f}{fp}$

$s = \dfrac{fp}{p-f}$ (Compare with the formula on p. 362)

(b) $s = \dfrac{50 \text{ mm} \times 60 \text{ mm}}{60 \text{ mm} - 50 \text{ mm}} = \dfrac{3,000 \text{ mm}^2}{10 \text{ mm}} = 300$ mm

$= 0.30$ m

$M = -p/s = \dfrac{-60 \text{ mm}}{300 \text{ mm}} = -1/5 = -0.20$

(c) $s = 100$ mm $= 0.10$ m

$M = -1$

9. (a) $p = 58.3$ mm

(b) $M = -0.167$ image height $= -16.7$ mm

(c) $p = 50.7$ mm $M = -0.0145$ image height $= -1.45$ mm

11. A rainbow is not fixed in space: It always has the same location and appearance relative to you no matter where you are (within broad limits). When you move, the rainbow moves with you, insofar as the colors you see are coming from different droplets. Thus, no matter how earnestly you try, you can never reach the end of the rainbow.

CHAPTER 10   Problems

1. $E = 41.4$ eV   (See Example 10.1)

3. First find the frequency; then use Figure 10.7.
$E = hf$ $h = 6.63 \times 10^{-34}$ J/Hz
9.5×10^{-25} J $= 6.63 \times 10^{-34}$ J/Hz $\times f$

$\dfrac{9.5 \times 10^{-25} \text{ J}}{6.63 \times 10^{-34} \text{ J/Hz}} = f$

$\dfrac{9.5}{6.63} \times \dfrac{10^{-25}}{10^{-34}}$ Hz $= f$

$f = 1.43 \times 10^9$ Hz   low-frequency microwave

$1 \text{ eV} = 1.6 \times 10^{-19}$ J $1 \text{ J} = \dfrac{1 \text{ eV}}{1.6 \times 10^{-19}}$

$E = 9.5 \times 10^{-25}$ J

$E = 9.5 \times 10^{-25} \times \left(\dfrac{1 \text{ eV}}{1.6 \times 10^{-19}}\right)$

$E = \dfrac{9.5 \times 10^{-25}}{1.6 \times 10^{-19}} = \dfrac{9.5}{1.6} \times \dfrac{10^{-25}}{10^{-19}}$

$E = 5.94 \times 10^{-6}$ eV

5. $\lambda = 9.1 \times 10^{-12}$ m   (See Example 10.2)
$\lambda = 0.0091$ nm

7. $\lambda = \dfrac{h}{mv}$

$v = \dfrac{h}{m\lambda} = \dfrac{6.63 \times 10^{-34} \text{ J/Hz}}{(9.11 \times 10^{-31} \text{ kg}) \times (6.7 \times 10^{-7} \text{ m})}$

$v = 0.109 \times 10^{(-34 + 31 + 7)}$ m/s

$v = 1,090$ m/s

9. (a) To be ionized, the atom's energy must be $E_\infty = 0$ since the electron goes to the $n = \infty$ level. If it starts in the $n = 2$ level, the energy it must gain, ΔE, is:
$\Delta E = E_\infty - E_2$
From Figure 10.30:
$E_2 = -3.40$ eV
$\Delta E = 0 - (-3.40$ eV$)$
$\Delta E = 3.40$ eV

(b) The energy of a photon that would ionize the atom equals ΔE.
$E_{photon} = \Delta E = 3.4$ eV
To find the frequency:
$E_{photon} = hf$
3.40 eV $= 4.136 \times 10^{-15}$ eV/Hz $\times f$
$\dfrac{3.40 \text{ eV}}{4.136 \times 10^{-15} \text{ eV/Hz}} = f$
$f = 8.22 \times 10^{14}$ Hz

11. Using Figure 10.32 as a model, the possible energy level transitions are shown in the figure.

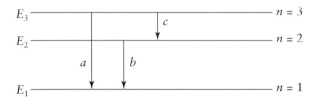

Only *downward* transitions lead to *emission* of photons.
   The frequency of each photon is proportional to the change in energy of the electron during the transition. On the energy level diagram, this energy is proportional to the length of the arrow showing the transition. Since a's arrow is longest and c's is shortest,
$E_a > E_b > E_c$
$f_a > f_b > f_c$
   The photon emitted in the $n = 3$ to $n = 1$ level transition has the highest frequency, followed by the $n = 2$ to $n = 1$ photon, and the $n = 3$ to $n = 2$ photon.

13. See Example 10.4

15. $E \propto Z^2 \Rightarrow$ Energy of He ground state $= \left(\dfrac{2}{1}\right)^2 = 4$ times energy of H ground state
Energy of Na ground state $= \left(\dfrac{11}{1}\right)^2 = 121$ times energy of H ground state

17. First find the frequency:
$c = f\lambda$ $\qquad \lambda = 632.8$ nm $= 6.328 \times 10^{-7}$ m
3×10^8 m/s $= f \times 6.328 \times 10^{-7}$ m
$\dfrac{3 \times 10^8 \text{ m/s}}{6.328 \times 10^{-7} \text{ m}} = f$
$f = 4.74 \times 10^{14}$ Hz
   The energy of such a photon, which equals the energy level difference in the atom, is:
$E = hf = 4.136 \times 10^{-15}$ eV/Hz $\times 4.74 \times 10^{14}$ Hz
$E = 1.96$ eV

19. The minimum energy that will excite a hydrogen atom in the ground state is the energy that will put it in the $n = 2$ state.
$\Delta E = E_2 - E_1$
$= -3.40$ eV $- (-13.6$ eV$)$
$\Delta E = 10.2$ eV
   Incoming particles with this much kinetic energy could cause a $n = 1$ to $n = 2$ transition.

CHAPTER 10   Challenges

7. $Cir. = 3{,}140$ nm $= 3.14 \times 10^{-6}$ m
$f = 6.5 \times 10^9$ Hz, $T = 1.5 \times 10^{-10}$ s
$v = 21{,}000$ m/s   much smaller than c.

CHAPTER 11   Problems

1. (a) carbon-14
$A = 14 \qquad Z = 6 \qquad N = 8$
6 protons and 8 neutrons
(b) silver-108
$A = 108 \qquad Z = 47 \qquad N = 61$
47 protons and 61 neutrons
(c) plutonium-242
$A = 242 \qquad Z = 94 \qquad N = 148$
94 protons and 148 neutrons

3. $^{110}_{47}\text{Ag} \rightarrow {}^{110}_{48}\text{Cd} + {}^{0}_{-1}e$
$\qquad\qquad\qquad \uparrow$
$\qquad\qquad\quad$ daughter

5. $^{239}_{94}\text{Pu} \rightarrow {}^{235}_{92}\text{U} + {}^{4}_{2}\text{He}$
$\qquad\qquad\qquad \uparrow$

7. $^{1}_{0}\text{n} + {}^{235}_{92}\text{U} \rightarrow {}^{236}_{92}\text{U}^* \rightarrow {}^{95}_{39}\text{Y} + {}^{A}_{Z} + 2{}^{1}_{0}\text{n}$
$A: \qquad 1 + 235 - 236 - 95 + A + 2$
$\qquad\qquad 236 - 97 = A$
$\qquad\qquad\qquad A = 139$
$Z: \qquad 0 + 92 = 92 = 39 + Z + 0$
$\qquad\qquad 92 - 39 = Z$
$\qquad\qquad\qquad Z = 53$
$^{139}_{53}\text{I}$ Iodine

9. $^{2}_{1}\text{H} + {}^{2}_{1}\text{H} \rightarrow {}^{3}_{1}\text{H} + {}^{1}_{1}\text{p}$

11. 4,000 cts/min to 1,000 cts/min in 12 min
$\frac{1}{4}$ as large $\quad$ 2 half lives $\quad$ 12 min $= 2t_{1/2}$
$t_{1/2} = 6$ min.

13. After 1 half-life, count rate is reduced to $\frac{1}{2}$.
After 2 half-lives, count rate is reduced to $\frac{1}{4}$.
After 3 half-lives, count rate is reduced to $\frac{1}{8}$.
Three half-lives must elapse for count rate to drop to a safe level.
$t_{1/2} = 3$ days
$\qquad$ 9 days

15. $^{273}_{110}\text{Ds}, {}^{269}_{108}\text{Hs}, {}^{265}_{106}\text{Sg}, {}^{261}_{104}\text{Rf}, {}^{257}_{102}\text{No}, {}^{253}_{100}\text{Fm}$ (See Example 11.1)

CHAPTER 11   Challenges

3. $^{15}_{8}\text{O} + {}^{0}_{-1}e \rightarrow {}^{15}_{7}\text{N}$

5. (a) uranium-238: 8 alpha decays
uranium-235: 7 alpha decays
(b) 1:1
(c) approximately 1:80
($80 {}^{207}_{82}\text{Pb}$ nuclei for each $^{235}_{92}\text{U}$ nucleus)
(d) approximately 1:1.7
($138 {}^{207}_{82}\text{Pb}$ nuclei for every $80 {}^{206}_{82}\text{Pb}$ nuclei)

7. $E = 9 \times 10^{13}$ J

CHAPTER 12   Problems

1. $\Delta t' = 8.3 \times 10^{-8}$ s $\quad$ (See Example 12.1)

3. $\Delta t' = 3460$ s $= 57.7$ min. $\quad$ (See Example 12.1)

5. $L' = L\sqrt{(1 - v^2/c^2)} = 1.0$ m $\sqrt{(1 - (0.95c)^2/c^2)}$
$= 1.0$ m $\sqrt{(1 - 0.9025)} = 0.312$ m
The meter stick's length is reduced to 31.2 cm, or shortened by a factor of about 3.2.

7. $E_0 = mc^2 = (1.673 \times 10^{-27}$ kg$) \times (3 \times 10^8$ m/s$)^2$
$\qquad\qquad = 1.506 \times 10^{-10}$ J

$E_0 = 1.506 \times 10^{-10}$ J $\times (6.25 \times 10^{18}$ eV/J$) = 9.41 \times 10^8$ eV

$\qquad = 9.41 \times 10^8$ eV $\times 10^{-6}$ MeV/eV

$\qquad = 9.41 \times 10^2$ MeV $= 941$ MeV

$\quad m = 941$ MeV/c^2 (See Table 12.3)

9. $E_0 = mc^2 = 1.0$ kg $\times (3 \times 10^8$ m/s$)^2$

$\qquad = 9.0 \times 10^{16}$ J

$\quad E = $ (mass of water) $\times$ (latent heat of fusion)

$\qquad = 1.0$ kg $\times (334{,}000$ J/kg$) = 3.34 \times 10^5$ J

$\quad \dfrac{E_0}{E} = \dfrac{9.0 \times 10^{16} \text{ J}}{3.34 \times 10^5 \text{ J}} = 2.69 \times 10^{11}$ (269 billion times larger)

11. $E_{\text{rel}} = \dfrac{mc^2}{\sqrt{(1 - v^2/c^2)}} = \dfrac{140 \text{ MeV}}{\sqrt{(1 - v^2/c^2)}} = 280$ MeV

$\quad \sqrt{(1 - v^2/c^2)} = 0.5$

$\quad 1 - v^2/c^2 = 0.25$

$\quad v^2/c^2 = 0.75$

$\quad v = 0.87\ c$

13. $E_0 = mc^2 = (2m_p c^2) = 2(1.673 \times 10^{-27}$ kg$) \times$

$\quad (3 \times 10^8$ m/s$)^2 = 3.01 \times 10^{-10}$ J

$\quad = 3.01 \times 10^{-10}$ J $\times (6.25 \times 10^{18}$ eV/J$) = 1.88 \times 10^9$ eV

$\quad = 1.88 \times 10^9$ eV $\times 10^{-6}$ MeV/eV $= 1.88 \times 10^3$ MeV

$\quad = 1{,}880$ MeV

15. (a) Not possible; violates strangeness conservation

(b) Not possible; violates charge conservation

(c) Not possible; violates baryon number conservation and spin conservation

(d) Possible

17. (a) π^-

(b) π^-

(c) ν_μ

(d) $\nu_\mu; \bar{\nu}_\tau$

19. $\bar{n} = \bar{d}\,\bar{d}\,\bar{u}; \bar{p} = \bar{u}\,\bar{u}\,\bar{d}$

21. $u\bar{u}, d\bar{d}, t\bar{t}, c\bar{c}, s\bar{s}, b\bar{b}, u\bar{t}$, and $\bar{u}t$

23. $d\,d\,d$; all three quarks must have their spins aligned to yield a net spin of 3/2

25. $\bar{c}s$

27. $u\,u\,u$; all three quarks must have their spins aligned to yield a net spin of 3/2

29. (a) $d\,d\,s \rightarrow u\,d\,d + d\,\bar{u}$ Or,

$\quad s \rightarrow (u + \bar{u}) + d$ (A strange quark is converted into a u-$\bar{u}$ pair plus a d quark by the weak interaction.)

(b) $u\,d\,s \rightarrow u\,u\,d + d\,\bar{u}$ Or,

$\quad s \rightarrow (u + \bar{u}) + d$ (See part (a) above.)

CHAPTER 12   Challenges

1. $KE_{\text{rel}} = \dfrac{mc^2}{\sqrt{(1 - v^2/c^2)}} - mc^2$

For $v \ll c$, $(1 - v^2/c^2)^{-1/2} = 1 + \frac{1}{2}v^2/c^2$

Thus, $KE_{\text{rel}} = mc^2 \times (1 + \frac{1}{2}v^2/c^2) - mc^2$

$\qquad = mc^2 + \frac{1}{2}mv^2 - mc^2 = \frac{1}{2}mv^2$

3. $E_p + E_{\bar{p}} = 2E_\gamma = 2hf$

$m_p c^2 + m_{\bar{p}} c^2 = 2hf$

Since $m_{\bar{p}} = m_p$, $2m_p c^2 = 2hf$ Or,

$m_p c^2 = hf$

1.67×10^{-27} kg $\times (3 \times 10^8$ m/s$)^2 = 6.63 \times 10^{-34}$ J/Hz $\times f$

$\dfrac{1.67 \times 10^{-27} \text{ kg} \times (3 \times 10^8 \text{ m/s})^2}{6.63 \times 10^{-34} \text{ J/Hz}} = f$

2.27×10^{23} Hz $= f$

Since $\lambda f = c$, $\lambda = c/f = \dfrac{3 \times 10^8 \text{ m/s}}{2.27 \times 10^{23} \text{ Hz}}$

$\lambda = 1.32 \times 10^{-15}$ m

These photons have wavelengths in the gamma-ray region of the spectrum.

5. Mesons are now believed to be composed of a quark and an antiquark. The largest possible magnitude of total charge (in units of the basic charge $e = 1.60 \times 10^{-19}$ C) for such a combination is $(\frac{2}{3} + \frac{1}{3}) = 1$. If mesons with charge $+2$ were discovered, the composition of this class of subatomic particles within the quark model would demand 4 quarks, combined in $q\bar{q}$ pairs in ways that preserved the properties of the already known mesons having charge 0 and ±1. Such a discovery would also permit the existence of mesons with spin $= 2$, formed by the 4 quarks, all with spin ½, aligned together in spin-up or spin-down configurations.

7. If the clock at rest on Earth measures an interval of time equal to 5 years, the shipboard clock of the traveling (younger) twin will record a time of only 5 years $\times$ $\sqrt{1 - (0.95)^2}$ or 1.56 years.

Answers to Explore It Yourself Activities (Some do not have answers.)

Chapter 1

Explore It Yourself 1.1

1. The number is the period of the motion (see p. 15).

2. The number is the frequency of the motion (see p. 17).

3. The period $= \dfrac{1}{\text{frequency}}$, and the frequency $= \dfrac{1}{\text{period}}$ (see p. 17).

4. A longer string makes the period larger and the frequency smaller.

Explore It Yourself 1.2

Runners can keep up their fastest pace for only about 20 seconds; then fatigue sets in. Friction makes swimming speeds much lower than running speeds. (More on these matters in Ch. 2 and 3.)

Explore It Yourself 1.3

Light travels almost a million times faster than sound. The lightning is approximately 1 mile away for every 5 seconds it takes the sound to get there. (Or 1 kilometer away for every 3 seconds.)

Explore It Yourself 1.4

The acceleration in g's $= 1 \div$ number of seconds it takes the speed to change by 22 mph.

Chapter 2'

Explore It Yourself 2.1

1. Yes. Things with more friction (less "slick") require a steeper book.

2. Usually much less steep. Force of rolling friction is usually much smaller.

Explore It Yourself 2.2

1. Kinetic (see Figure 2.5)

2. Static (see Figure 2.5)

3. Force of kinetic friction is usually less than the maximum force of static friction.

Explore It Yourself 2.3

To the left or right, depending on which direction you whirled the sock. (See Figure 2.12.)

Explore It Yourself 2.4

Faster requires larger force on string. Centripetal force increases if speed increases.

Explore It Yourself 2.5

45°, if the nozzle is near the ground.

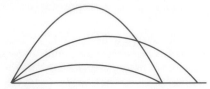

Explore It Yourself 2.6

  1. No. The tighter one gets there first. No. Size/area also matters.

  2. No. Larger object can fall faster than smaller one if it is much heavier.

Explore It Yourself 2.7

Forward. Backward. No. Your acceleration was caused by the wall's force on you—backward. (See Figure 2.30.)

Explore It Yourself 2.9

  2. Place the pins right next to each other.

Chapter 3

Explore It Yourself 3.1

The speed of the two skaters afterward is lower than that of the one before. The speed of the two skaters afterward is higher if the heavier person is moving initially.

Explore It Yourself 3.2

The two skaters have the same speed only if they have the same mass (and weight). If not, the lighter person will be going faster. (See Figure 3.7.)

Explore It Yourself 3.3

As you are moving horizontally, your legs do no work against gravity. As you are going up stairs, your legs do work lifting your body, so you start to tire—depending on what shape you are in and how high you climb.

Explore It Yourself 3.4

  1. The farther up the ramp the battery is when released, the faster it is going when it reaches the bottom. It starts with more potential energy, so it gains more kinetic energy.

  2. It is a tie. The lower battery ends up going one-half as fast as the upper one because its vertical drop is one-fourth as much ($v = \sqrt{2gh}$). But the total distance it has to travel is considerably less. The total time it takes each battery to get to the backstop is the same (ignoring kinetic friction). The lower battery wins (barely) if it is started farther up the ramp, and the upper one wins if the lower one is started below one-fourth of the way up.

Explore It Yourself 3.5

The power output depends on how much you weigh and how fast you go up the stairs. It is typically between 250 W (one-third hp) and 1,500 W (2 hp).

Explore It Yourself 3.6

You speed up—spin faster—when you pull your arms in. Each part of your arms has a certain amount of angular momentum as it moves in a circle with some radius. When you pull your arms in, the radius decreases, which makes the angular momentum decrease. Consequently, the rest of your body and the chair spin faster so that the total angular momentum remains constant. The reverse happens when you move your arms out and slow down.

Chapter 4

Explore It Yourself 4.1

Depending on how hard you squeeze, the point will cause a stronger sensation or even pain. The area of the point is much smaller than that of the other end, so the pressure is much larger. Nerves in your fingers respond to this higher pressure.

Explore It Yourself 4.2

The can is crushed by the atmosphere. The boiling causes steam in the can to drive out much of the air. When the can is inverted into the water, much of the steam quickly condenses back to liquid water, causing the pressure inside the can to drop. The air pressure on the outside of the can is no longer balanced by pressure on the inside, so the can is crushed.

Explore It Yourself 4.3

One simple way is to choose packages with the same weight (or mass), like one pound of flour and one pound of cereal. Then the material in the package with the smaller volume has the higher density. Glass containers should be avoided because glass adds a lot more to the mass and weight of the package than does cardboard or plastic. Substances with nearly the same density would require measurements (weight, volume) to determine which has higher density.

Explore It Yourself 4.4

The water shoots out farthest, and therefore is going fastest, from the bottom hole. It is going slowest from the top hole. This is because the pressure is highest at the lowest hole.

Explore It Yourself 4.5

The weight of the water causes the cardboard or paper to bulge downward slightly. This increase in volume lowers the pressure inside the glass, so the (higher) atmospheric pressure outside causes a large enough upward force on the cardboard or paper to support the water.

Explore It Yourself 4.6

The rubber band is stretched less with the cup in the water, because the water causes an upward buoyant force on the cup. Therefore, the rubber band has to supply a smaller upward force to support the cup.

Explore It Yourself 4.7

The ice cube floats at the boundary of the gasoline and the water. Since ice's density is greater than that of gasoline but less than that of water (see Table 4.4), it sinks in gasoline and floats in water.

Explore It Yourself 4.8

The moving air above the paper is at a lower pressure, so the paper is lifted up by the higher pressure below it.

Chapter 5

Explore It Yourself 5.1

Unless some of them are heated and others not (for example, some in sunlight, or some near a heating or cooling vent), they should show the same temperature. Typically, the readings will vary by a degree or two because the thermometers are not designed to be precision instruments.

Explore It Yourself 5.2

See Figure 5.8. The two pieces of roadway pull away from each other a bit as their temperature drops.

Explore It Yourself 5.3

The water level is lower after the balloon has been removed from the hot water because the air in the balloon occupies a larger volume (so it displaces more water) when its temperature is higher. The air expands as it is warmed. (See Question 11 at the end of this chapter.)

Explore It Yourself 5.4

Parts of the pump should be warmer after you pump up the tire. The work you do in compressing the air heats it. The heated air, in turn, warms the metal parts of the pump.

Explore It Yourself 5.5

Your hand should warm up faster in #2. In #1, your hand warms up because of heat transferred by radiation. In #2, both radiation and convection of the heated air warm your hand.

Explore It Yourself 5.6

The time in #4 should be roughly seven times that in #3. To boil a given amount of water to steam, it must gain much more internal energy than it takes to raise its temperature from room temperature to the boiling point. Note: While the water is boiling, the can and water stay at about 100°C, so the paint doesn't get hot enough to smoke. Once the water is boiled away, the can's temperature rises rapidly.

Chapter 6

Explore It Yourself 6.1

1. They travel with the same speed.
2. Depending on how much friction there is between the Slinky and the surface, the pulse can reflect and return—perhaps only partway—to your end. The pulse is "inverted" by the reflection: It switches from right to left or from left to right (look ahead to Figure 6.13).

Explore It Yourself 6.2

The tone or pitch of the bottle with CO_2 in it is lower, because sound travels slower in CO_2 than it does in air.

Explore It Yourself 6.3

The echoes arrive sooner after the clap as you get closer to the building. (When 15 m away the echo arrives about 0.09 s after the clap; at 10 m, the time is about 0.06 s.)

The ear can distinguish two successive sound pulses only if the second comes about 0.02 s or more after the first. That corresponds to a distance of about 3.5 m from the building.

Explore It Yourself 6.4

1. The sound is loudest when you are directly in front of the opening.
2. The diffracted sound—what you hear when you are off to one side—has less treble: the higher frequencies are not diffracted as much.

Explore It Yourself 6.5

2. Your voice should sound muffled, with much less treble in it than in #1.
3. Your voice should have much more treble in it than in #1.

Some of the sound that you hear from your voice travels through the air to your ears, and some travels through the bones and other tissues in your head. In the latter case, much of the higher-frequency sound does not make it to your inner ear. When you cover your ears, you block the sound reaching your ears through the air.

Explore It Yourself 6.6

2. Each vowel sound requires a unique positioning of the jaw, tongue, and lips. When you try to make the sound of one vowel using the positioning for another, the result usually does not sound like what you intended.

Explore It Yourself 6.7

2. As you move farther apart, most of the sound each of you hears is reverberant sound. Each successive spoken syllable is drowned out by the reverberating sound from the preceding one.
3. When you speak slowly, the syllables don't "overlap" as much and it is easier to be understood.
4. The reverberation time is shorter. You can understand each other when you are farther apart.

Explore It Yourself 6.8

1. The tone quality should be different when your ears are blocked because the higher frequencies present in the complex tone are attenuated.
2. Since a steady whistle is a pure tone consisting of just one frequency, the tone quality should be the same when your ears are blocked.

Chapter 7

Explore It Yourself 7.1

The pieces of paper and thread should be attracted to the cup, perhaps strongly enough to be pulled up to the cup and stick to it. If the latter happens, the electric force was stronger than the force of gravity.

Explore It Yourself 7.2

The foil wads should repeatedly jump up to the cup, and then back to the metal. This should last a few seconds. The speed depends on how heavy the wads are and how well charged the cup is. The wads are attracted to the cup, but as soon as they touch it some of the extra electrons on it move onto the wads. This makes the wads and the cup both negatively charged, so they repel each other and the wads go back down (see Section 7.2). When they touch the metal, the extra electrons flow into it so the wads are again attracted to the cup. The fact that electrons can readily move in metals is the key to this exercise.

Explore It Yourself 7.3

The cups repel each other because they are both negatively charged. The hanging cup should swing away from the other cup a bit. The closer you place the cups, the larger the force and the farther the string will be deflected from the vertical.

Explore It Yourself 7.4

The stream of water should be deflected toward the cup.

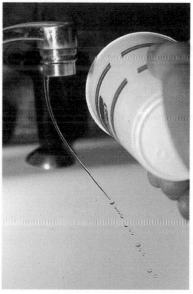
© Vern Ostdiek

Explore It Yourself 7.5

The rising smoke should be attracted to the Styrofoam cup. This is the principle behind electrostatic precipitators. (See the paragraph following Explore It Yourself 7.5.)

Explore It Yourself 7.6

Fluorescent lights emit light only when a current is flowing through them. When AC is used, the current flows for a short time, is nearly zero momentarily, then flows in the other direction. Because of this, a fluorescent light actually emits "bursts" of light 120 times each second (two bursts during each cycle, 60 cycles each second—in the U.S.A.). This causes the "stroboscope" effect you see. Incandescent light bulbs do not show this effect nearly as well because they rely on a glowing filament to emit the light. In the short time that the alternating current is nearly zero, the filament does not cool off enough to reduce its light output appreciably.

Chapter 8

Explore It Yourself 8.1

Depending on how good your compass is, it should agree within 5 degrees or so.

Explore It Yourself 8.2

The needle should turn quickly to a direction perpendicular to the cable. As an automobile is being started, a huge current flows through the battery and starter motor—typically around 100 amperes. This produces strong magnetic fields around the cables that carry the current, which will deflect a compass needle.

Explore It Yourself 8.4

The paper should appear yellowish when the lights are very dim. The dimmer reduces the current to the lights, which makes the filaments cooler. This shifts the peak wavelength of their equivalent blackbody radiation to larger values, so there is relatively more red light present.

Explore It Yourself 8.5

If you examine Aldebaran and Sirius, you should find that the former appears orange in color, while the latter looks blue-white. Aldebaran is about 1000 K hotter than Betelgeuse, and Sirius has about the same temperature as Rigel. The majority of the brightest-appearing stars in the sky will look blue-white in color. Although relatively rare, because of their high surface temperatures, these stars are very luminous and can be seen over great distances. By contrast, the more numerous cool stars are not intrinsically very bright and so cannot be seen very far out in space. Our visual perception of the dominant type of stars in the sky is biased by how strongly the power emitted by a star depends on its temperature (P is proportional to T^4).

Chapter 9

Explore It Yourself 9.1

Align the images with your head on one side of the normal (perpendicular) direction to the mirror. Use a straightedge to draw a line on the cardboard base along your line of sight. Move your head to the opposite side of the normal and realign the images. Again, use a straightedge to draw a line on the cardboard along your new sight line. Extend the two lines until they meet at the mirror's surface. Now draw in the normal to the mirror at the point of intersection. Use a protractor to measure the angle of incidence and the angle of reflection. You should find they are equal to within small errors of measurement.

Explore It Yourself 9.2

The green and red beams combine to give the color yellow (additive mixing). Looking through a red filter at a brightly lit scene produces a red image of the scene. If a green filter is now placed behind the red one, the image will go dark (black): No light gets through the second filter.

The red filter passes only light with wavelengths near 650 nm. The green filter passes only light with wavelengths near 520 nm. The light emerging from the red filter does not have the proper wavelengths to be transmitted by the green filter, and little or no light passes through it.

Explore It Yourself 9.3

The intensity of the overhead lights should not change as you rotate the Polaroid lens because the light emitted by fluorescent tubes or incandescent bulbs is unpolarized. The intensity of the reflected light ("glare") should vary as you turn the Polaroid lens, with the relative amount of change depending on the angle of incidence of the light on the reflecting surface. The greatest extinction of the reflected light should occur when the sunglasses are oriented in their "normal" fashion for wearing. The extinction should be the least when the glasses are at 90° to this orientation. Using two pairs of polarizing lenses, you should see a similar effect (see the accompanying photograph). This suggests that the reflected light is polarized, although generally not to the same extent as the light leaving the first lens of the crossed sunglasses.

Explore It Yourself 9.4

The light intensity from the LCD should vary much like that of the reflected light in EIY 9.3, indicating that the LCD emission is polarized too. Performing the same experiment using sky light as the source, you should once again see the same variation in intensity as the Polaroid lens is rotated. This shows that sky light is also partially polarized.

Explore It Yourself 9.5

The image of your right hand in a mirror is that of a left hand. As you move your finger toward the mirror's surface, the image of your finger also moves toward the surface, but from the opposite direction. Just as your real finger is about to touch the mirror, the image of your finger is also poised to touch the surface. Images formed by plane mirrors are as far behind the mirror's surface as their corresponding objects are in front of the mirror.

Explore It Yourself 9.6

With an angle of 45°, you should see 8 images, 7 reflected ones and the real one of the original object. At 60°, you should see 6 images, 5 reflections and the original. For a mirror angle of 30°, a total of 12 images should be seen. The total number of images (including the original) equals 360° divided by the angle between the mirrors. The accompanying figure shows a simple design for a kaleidoscope using two mirrors at 60° cemented in a cylindrical tube. For convenience of viewing, one end of the tube is closed off, save for a small peephole that permits visual access to the region between the mirrored surfaces.

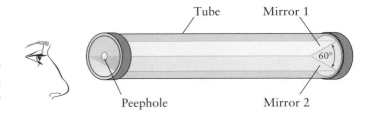

Explore It Yourself 9.7

If your finger is close enough to the spoon's concave surface, you should see an upright, magnified image. With the spoon at a comfortable distance from your face, the image of your face should be inverted. Using the reverse (convex) side of the spoon, you will see an upright image of your face (and surroundings) that is reduced in size because the image is further behind the spoon's surface than your face is in front of it.

Explore It Yourself 9.8

You should see a magnified image of your hand or the lamp (see the accompanying figure). As you move your hand/lamp up and down, the size of the image (its magnification) will change accordingly. At higher revolution speeds, the curvature of the oil's surface is greater, yielding a smaller focal length for the concave mirror. For a given object distance, the image distance will then be smaller and so will the resultant magnification.

© Vern Ostdiek

Explore It Yourself 9.9

At the water-air interface, the pencil appears to bend or kink so that its image appears below what would be the line of extension of the length of the pencil in air. See the accompanying figure.

© Eugene Hecht

Explore It Yourself 9.10

Looking obliquely through the side of the beaker at the light emerging from the water's surface, you will not be able to see the entire circle; a portion of it along the perimeter closest to you will be unobservable. As you move your head up and down, the fraction of the circle that "goes missing" decreases and increases, respectively. Light from the unobservable part of the circle strikes the water's surface at an angle greater than the critical angle for total internal reflection at the air-water interface. This light is reflected back into the water and does not cross the boundary between the two media. Thus, no rays from this portion of the circle reach your eyes.

Explore It Yourself 9.11

The image of the source should be inverted and reduced in size.

Explore It Yourself 9.12

Employing the technique shown in Figure 9.50, the following table summarizes what you will see as the distance between the source and the magnifying (converging) lens is changed.

Object	Image			
Location	Type	Location	Orientation	Relative Size
$\infty > s > 2f$	Real	$f < p < 2f$	Inverted	Reduced
$s = 2f$	Real	$p = 2f$	Inverted	Same size
$f < s < 2f$	Real	$\infty > p > 2f$	Inverted	Magnified
$s = f$	——	∞	——	——
$s < f$	Virtual	$-p > s$	Upright	Magnified

When the lens is closer than one focal length to the source, the image is virtual and not able to be projected on the screen. Looking back through the lens at the source yields an upright, magnified image of it.

Explore It Yourself 9.13

The image of the distant, brightly lit scene as observed with the camera obscura should be inverted and reduced in size (see Figure 9.49). If the pinhole is too small, the image may be too faint to be seen easily. If the pinhole is too large, the image may not be terribly sharp because light reaches the screen from different parts of the object and overlaps to produced a blurry picture. Pinholes of a few millimeters in diameter should give pretty good results.

Explore It Yourself 9.14

With the screen inside the position of best focus, the outside edges of the blurred image should appear reddish-orange. Moving the screen beyond the point of best focus results in a blurred image whose edges are bluish in color. Compare with Figure 9.56.

Explore It Yourself 9.15

To create your own personal rainbow, the Sun must be behind you and the water spray generally in front of you (in the direction of your shadow). The rainbow is most easily seen using a mist of small droplets that tend to remain suspended in air and produce a more stable display. The outer portion of the rainbow arc will be colored red, while the inside will have a blue-violet hue. The angular radius of the bow should be about 40° or about two spread-hands in size (see Figure 9.69) providing that the drops are more than a few arm-lengths distant.

Chapter 10

Explore It Yourself 10.1

You should see a continuous spectrum of all the colors (as in Figure 9.67), from blue (closest to the image of the bulb) through red.

Explore It Yourself 10.2

You should see separate images of the compact fluorescent lamp (with some overlap), each a different color: blue, two shades of green, yellow, and red. The light from compact fluorescent lamps mainly consists of several narrow bands of color, rather than just lines (as in an emission-line spectrum) or all colors (as in a continuous spectrum).

Chapter 11

Explore It Yourself 11.1

5. Still 50%. It's the same for each throw, regardless of how many times in a row a coin has turned up "heads."

This glossary includes highlighted definitions and laws from the text, along with definitions of selected terms that are found in more than one place in the text. Terms that are narrow in scope or are used in only one section or chapter (such as optic fiber and reverberation) are not included.

Acceleration (*a*)   Rate of change of velocity. The change in velocity divided by the time elapsed.

Alternating current (AC)   An electric current consisting of charges flowing back and forth.

Amplitude   Maximum displacement of points on a wave, measured from the equilibrium position.

Angular momentum   The mass of an object times its speed, times the radius of its path.

Antiparticle   A charge-reversed version of an ordinary particle. A particle of the same mass (and spin) but of opposite electric charge.

Archimedes' principle   The buoyant force acting on a substance in a fluid at rest is equal to the weight of the fluid displaced by the substance.

Atom   The smallest unit or building block of a chemical element.

Baryon   A strongly interacting particle that is composed of 3 quarks. Protons and neutrons are examples. Baryons are fermions.

Bernoulli's principle   When the speed of a moving fluid *increases,* the pressure *decreases,* and vice versa.

Big Bang   The explosive event at the beginning of the universe nearly 14 billion years ago. The expansion produced by this event continues today and is reflected in the movement of distant galaxies away from one another.

Blackbody   A hypothetical object that absorbs all light and other electromagnetic waves that strike it. The radiation it emits is called blackbody radiation (BBR).

Boson   A particle with integral spin. Bosons do not obey the Pauli Exclusion Principle.

Buoyant force   The upward force exerted by a fluid on a substance partly or completely immersed in it.

Centripetal acceleration   Acceleration of an object moving along a circular path. It is directed toward the center of the circle.

Centripetal force   Name applied to the force acting to keep an object moving along a circular path. It is directed toward the center of the circle.

Compound   A pure substance consisting of two or more elements combined chemically.

Conduction   The transfer of heat by direct contact between atoms and molecules.

Convection   The transfer of heat by buoyant mixing in a fluid.

Cosmology   The study of the structure and evolution of the universe as a whole.

Coulomb's law   The force acting on each of two charged objects is directly proportional to the net charges on the objects and inversely proportional to the square of the distance between them.

Current (*I*)   Rate of flow of electric charge. The amount of charge that flows past a given point per second. The unit of measure is the ampere.

Diffraction   The bending of a wave as it passes around the edge of a barrier. Diffraction causes a wave passing through a gap or a slit to spread out into the shadow regions.

Direct current (DC)   An electric current that is a steady flow of charge in one direction.

Dispersion   The variation in the speed of propagation of a periodic wave with the frequency of the wave. Dispersion is responsible for the separation of white light into a spectrum of colors by a prism.

Doppler effect   The apparent change in frequency of a wave due to motion of the source of the wave, the receiver, or both.

Echolocation   Process of using the reflection of a wave to locate objects.

Elastic collision   Collision in which the total kinetic energy of the colliding objects is the same before and after the collision.

Electric charge (*q*)   An inherent physical property of certain subatomic particles that is responsible for electrical and magnetic phenomena. The unit of measure is the coulomb.

Electric field   The agent of the electric force. It exists in the space around charged objects and causes a force to act on any other charged object.

Electromagnetic force (or interaction)   The force that acts between charged particles or objects. It can be either attractive or repulsive, and it is responsible for most of the common forces in our everyday life like contact forces, friction, and elastic forces.

Electromagnetic wave   A transverse wave consisting of a traveling combination of oscillating electric and magnetic fields.

Electron   Negatively charged particle, usually found orbiting the nucleus of an atom.

Element   One of the 114 different fundamental substances that are the simplest and purest forms of matter.

Elementary particles   The basic, indivisible building blocks of the universe. The fundamental constituents from which all matter, antimatter, and their interactions derive. They are believed to be true "point" particles, devoid of internal structure or measurable size.

Energy (*E*)   The measure of a system's capacity to do work. That which is transferred when work is done.

Fermion   A particle with half-integral spin. Fermions obey the Pauli Exclusion Principle.

First law of thermodynamics   The change in internal energy of a substance equals the work done on it plus the heat transferred to it.

Force (*F*)   A push or a pull acting on a body. Usually causes some distortion of the body, a change in its velocity, or both.

Frequency (*f*)   The number of cycles of a periodic process that occurs per unit time. The unit of measure is the hertz (Hz).

Frequency (*f*) **(of a wave)**   The number of cycles of a wave passing a point per unit time. It equals the number of oscillations per second of the wave.

Friction   A force of resistance to relative motion between two bodies or substances.

Gamma rays   The highest-frequency electromagnetic waves. One type of nuclear radiation.

Gravitational force   The attractive force that acts between all pairs of objects.

Half-life   The time it takes for one-half of the nuclei of a radioisotope to decay. The time interval during which each nucleus has a 50% probability of decaying.

Heat (*Q*)   The form of energy that is transferred between two substances at different temperatures.

Heat engine   A device that converts heat into mechanical energy or work. It absorbs heat from a hot source such as burning fuel, converts some of this energy into usable mechanical energy or work, and outputs the remaining energy as heat at some lower temperature.

Hubble relation (or law)   A mathematical expression showing that the farther a galaxy is from us, the faster it is moving away. One implication of this relation is that the universe is expanding.

Humidity   The mass of water vapor in the air per unit volume. The density of water vapor in the air.

Inelastic collision   Collision in which the total kinetic energy of the colliding bodies after the collision is not equal to the total kinetic energy before.

Infrared (IR)   Electromagnetic waves with frequencies just below those of visible light. Infrared is the main component of heat radiation.

Interference   The consequence of two waves arriving at the same place and combining. Constructive interference occurs wherever the two waves meet in phase (peak matches peak); the waves add together. Destructive interference occurs wherever the two waves meet out of phase (peak matches valley); the waves cancel each other.

Intermediate boson   A force-carrying elementary particle that mediates the fundamental interactions of nature. The photon is an example.

Internal energy (*U*)   The sum of the kinetic energies and potential energies of all the atoms and molecules in a substance.

Isotopes   Isotopes of a given element have the same number of protons in the nucleus but different numbers of neutrons.

Kinetic energy (*KE*)   Energy due to motion. Energy that an object has when it is moving.

Kinetic friction   Friction between two substances that are in contact and moving relative to each other.

Laser   A beam of coherent monochromatic light. Acronym for "light amplification by stimulated emission of radiation."

Law of conservation of angular momentum The total angular momentum of an isolated system is constant.

Law of conservation of energy Energy cannot be created or destroyed, only converted from one form to another. The total energy in an isolated system is constant.

Law of conservation of linear momentum The total linear momentum of an isolated system is constant.

Law of fluid pressure The (gauge) pressure at any depth in a fluid at rest equals the weight of the fluid in a column extending from that depth to the "top" of the fluid divided by the cross sectional area of the column.

Law of reflection The angle of incidence equals the angle of reflection.

Law of refraction A light ray is bent toward the normal when it enters a transparent medium in which light travels more slowly. It is bent away from the normal when it enters a medium in which light travels faster.

Lepton An elementary particle that does not interact via the strong nuclear force. Electrons and neutrinos are examples. Leptons are fermions.

Linear momentum The mass of an object times its velocity.

Longitudinal wave Wave in which the oscillations are along the direction the wave travels.

Magnetic field The agent of the magnetic force. It exists in the space around magnets and causes a force to act on each pole of any other magnet.

Mass (m) A measure of an object's resistance to acceleration. A measure of the quantity of matter in an object.

Mass density (D) The mass per unit volume of a substance. The mass of a quantity of a substance divided by the volume it occupies.

Mass number (A) The total number of protons and neutrons in a nucleus.

Meson A particle that is composed of a quark and an antiquark. Mesons are bosons.

Microwaves Electromagnetic waves with frequencies between those of radio waves and infrared light.

Molecule Two or more atoms combined to form the smallest unit of a chemical compound.

Neutrino An elementary particle with no charge and little mass. Many important processes in astronomy release neutrinos, including supernova explosions and fusion reactions in stars. Neutrinos are fermions.

Neutron Electrically neutral particle residing in the nucleus of an atom.

Neutron number (N) The number of neutrons contained in a nucleus.

Newton's first law of motion An object will remain at rest or in motion with constant velocity unless acted on by a net external force.

Newton's law of universal gravitation Every object exerts a gravitational pull on every other object. The force is proportional to the masses of both objects and inversely proportional to the square of the distance between them.

Newton's second law of motion An object is accelerated whenever a net external force acts on it. The net force equals the object's mass times its acceleration.

Newton's third law of motion Forces always come in pairs: when one object exerts a force on a second object, the second exerts an equal and opposite force on the first.

Nuclear fission The splitting of a large nucleus into two smaller nuclei. Free neutrons and energy are also released.

Nuclear fusion The combining of two small nuclei to form a larger nucleus.

Ohm's law The current in a conductor is equal to the voltage applied to it divided by its resistance.

Orbit The path of a body as it moves under the influence of a second body. An example is the path of a planet or a comet as it moves around the Sun.

Pascal's principle Pressure applied to an enclosed fluid is transmitted undiminished to all parts of the fluid and to the walls of the container.

Pauli Exclusion Principle Two interacting fermions of the same type, such as electrons in an atom, cannot be in exactly the same state.

Period (T) The time for one complete cycle of a periodic process.

Photon The quantum of electromagnetic radiation. The energy of a photon equals Planck's constant times the frequency of the radiation.

Plasma An ionized gas, often referred to as the fourth state of matter. It consists of positive ions and electrons.

Polarization A term used to describe the orientation of the plane of vibration of a transverse wave. Light can be polarized by passing it through a Polaroid filter that absorbs all waves except those vibrating along a direction parallel to its transmission axis.

Potential energy (PE) Energy due to an object's position. Energy that a system has because of its configuration.

Power (P) The rate of doing work. The rate at which energy is transferred or transformed. Work done divided by the time. Energy transferred divided by the time.

Pressure (p) The force per unit area for a force acting perpendicular to a surface. The force acting on a surface divided by its area.

Proton Positively charged particle residing in the nucleus of an atom.

Quantum mechanics A mathematical model that describes the behavior of particles on an atomic and subatomic scale. This theory demonstrates that matter and energy are quantized (come in small discrete bundles) on smallest scales imaginable.

Quark A fractionally charged elementary particle found only in combination with other quarks in mesons and baryons. Quarks are fermions.

Refraction The bending of a wave as it enters a different medium.

Relative humidity Humidity expressed as a percentage of the saturation density.

Relativity Theory developed by Einstein connecting measurements of space, time, and motion made by one observer to those made by another observer in a different environment. The special theory of relativity relates experimental results for two observers moving at high speed with respect to one another; the general theory of relativity relates observations made in strong gravitational fields to those found in weak fields.

Resistance (R) A measure of the opposition that a substance has to current flow. The unit of measure is the ohm.

Scanning tunneling microscope (STM) Device that uses the tunneling of electrons to form images of individual atoms at the surface of a solid.

Second law of thermodynamics No device can be built that will repeatedly extract heat from a heat source and deliver mechanical work or energy without ejecting some heat to a lower-temperature reservoir.

SI An internally consistent system of units within the metric system.

Speed (v) Rate of movement. Rate of change of distance from a reference point. The distance that something travels divided by the time elapsed.

Spin Intrinsic angular momentum possessed by subatomic particles. It is quantized in integral or half-integral multiples of $h/2\pi$.

Strong nuclear force One of the four fundamental forces in nature. It is the force that holds neutrons and protons together in the nucleus.

Superconductor A substance that has zero electrical resistance. Electrical current can flow in it with no loss of energy to heat.

Temperature (T) The measure of hotness or coldness. The temperature of matter depends on the average kinetic energy of atoms and molecules.

Transverse wave Wave in which the oscillations are perpendicular (transverse) to the direction the wave travels.

Ultraviolet radiation (UV) Electromagnetic waves with frequencies just above those of visible light.

Vacuum Completely empty space devoid of all matter, including air.

Velocity Speed with direction.

Voltage (V) The work that a charged particle can do divided by the size of the charge. The energy per unit charge given to charged particles by a power supply. The unit of measure is the volt.

Wave A traveling disturbance consisting of coordinated vibrations that carry energy with no net movement of matter.

Wavelength (λ) The distance between two successive "like" points on a wave. An example is the distance between two adjacent peaks or two adjacent valleys.

Weak nuclear force One of the four fundamental forces in nature. It is responsible for beta decay.

Weight (W) The force of gravity acting on a substance.

Weight density (D_w) The weight per unit volume of a substance. The weight of a quantity of a substance divided by the volume it occupies.

Work The force that acts times the distance moved in the direction of the force.

X rays Electromagnetic waves with frequencies in the range from higher-frequency ultraviolet to low-frequency gamma rays.

CREDITS

This page constitutes an extension of the copyright page. We have made every effort to trace the ownership of all copyrighted material and to secure permission from copyright holders. In the event of any question arising as to the use of any material, we will be pleased to make the necessary corrections in future printings. Thanks are due to the following authors, publishers, and agents for permission to use the material indicated. Photos not credited were taken by the authors, Vern Ostdiek and Don Bord.

Note: Tables are indicated by *t* and footnotes by *n* following the page number.

TABLE OF CONVERSION FACTORS AND OTHER INFORMATION

DISTANCE:

1 m = 3.28 ft	1 ft = 0.305 m
1 cm = 0.394 in.	1 in. = 2.54 cm
1 km = 0.621 mi	1 mi = 1.61 km
1 light-year = 5.9×10^{12} mi	1 light-year = 9.5×10^{15} m

AREA:

1 m^2 = 10.8 ft^2	1 ft^2 = 0.093 m^2
1 cm^2 = 0.155 in.^{2}	1 in.^{2} = 6.45 cm^2
1 km^2 = 0.386 mi^2	1 mi^2 = 2.59 km^2

VOLUME:

1 m^3 = 35.3 ft^3	1 ft^3 = 0.028 m^3
	= 7.48 gallons
1 cm^3 = 0.061 in.^{3}	1 in.^{3} = 16.4 cm^3
1 m^3 = 1,000 liters	
1 liter = 1.06 qt	1 qt = 0.95 liter

TIME:

1 min = 60 s
1 h = 60 min = 3,600 s
1 day = 24 h = 86,400 s
1 year = 3.16×10^7 s

SPEED:

1 m/s = 2.24 mph	1 mph = 0.447 m/s
	= 1.47 ft/s
1 m/s = 3.28 ft/s	1 ft/s = 0.305 m/s
= 3.60 km/h	= 0.682 mph
1 km/h = 0.621 mph	1 mph = 1.61 km/h

ACCELERATION:

1 m/s^2 = 3.28 ft/s^2	1 ft/s^2 = 0.305 m/s^2
1 m/s^2 = 2.24 mph/s	1 mph/s = 0.447 m/s^2
1 g = 9.8 m/s^2	1 g = 32 ft/s^2 = 22 mph/s
1 m/s^2 = 0.102 g	

MASS:

1 kg = 0.069 slug	1 slug = 14.6 kg
1 kg = 6.02×10^{26} u	1 u = 1.66×10^{-27} kg

FORCE:

1 N = 0.225 lb	1 lb = 4.45 N
	1 ton = 8,896 N
	= 2,000 lb

ENERGY, WORK, AND HEAT:

1 J = 0.738 ft-lb	1 ft-lb = 1.36 J
= 0.239 cal	= 0.0013 Btu
1 cal = 4.184 J	1 Btu = 778 ft-lb
1 cal = 0.004 Btu	1 Btu = 252 cal
1 J = 0.00095 Btu	1 Btu = 1,055 J
1 kWh = 3,600,000 J	
1 kWh = 3,413 Btu	1 Btu = 0.00029 kWh
1 food calorie = 1,000 cal	
1 eV = 1.6×10^{-19} J	1 quad (Q) = 10^{15} Btu
1 J = 6.25×10^{18} eV	

POWER:

1 W = 0.738 ft-lb/s	1 ft-lb/s = 1.36 W
1 W = 0.00134 hp	1 hp = 550 ft-lb/s
	1 hp = 746 W
1 W = 3.41 Btu/h	1 Btu/h = 0.293 W

PRESSURE:

1 Pa = 1.45×10^{-4} psi	1 psi = 6,900 Pa
1 Pa = 9.9×10^{-6} atm	1 psi = 0.068 atm
1 atm = 1.01×10^5 Pa	1 atm = 14.7 psi
= 76 cm Hg	= 29.92 in. Hg
= 10.3 m H$_2$O	= 33.9 ft H$_2$O

MASS DENSITY:

1 kg/m^3 = 0.0019 slug/ft^3	1 slug/ft^3 = 515 kg/m^3

WEIGHT DENSITY:

1 N/m^3 = 0.0064 lb/ft^3	1 lb/ft^3 = 157 N/m^3

METRIC PREFIXES:

Power of 10	Prefix	Abbreviation
10^{15}	peta	P
10^{12}	tera	T
10^{9}	giga	G
10^{6}	mega	M
10^{3}	kilo	k
10^{-2}	centi	c
10^{-3}	milli	m
10^{-6}	micro	μ
10^{-9}	nano	n
10^{-12}	pico	p
10^{-15}	femto	f

IMPORTANT CONSTANTS AND OTHER DATA

Symbol	Description	Value
c	Speed of light in vacuum	3×10^8 m/s
g	Acceleration of gravity (Earth)	9.8 m/s^2
G	Gravitational constant	6.67×10^{-11} N-m^2/kg^2
h	Planck's constant	6.63×10^{-34} J-s
q_e	Charge on electron	-1.60×10^{-19} C
m_e	Mass of electron	9.11×10^{-31} kg
m_p	Mass of proton	1.673×10^{-27} kg
m_n	Mass of neutron	1.675×10^{-27} kg

SOLAR SYSTEM DATA

Sun:	mass	1.99×10^{30} kg
	radius (average)	6.96×10^8 m
Earth:	mass	5.979×10^{24} kg
	radius (average)	6.376×10^6 m
Moon:	mass	7.35×10^{22} kg
	radius (average)	1.74×10^6 m
Earth-Sun distance		1.496×10^{11} m
Earth-Moon distance		3.84×10^8 m
Earth orbital period		3.16×10^7 s
Moon orbital period (average)		2.36×10^6 s